SURVEYING
and
LEVELLING

SECOND EDITION

R. Subramanian

Formerly, Professor and Head
Department of Civil Engineering
NITTTR, Chandigarh

OXFORD
UNIVERSITY PRESS

OXFORD
UNIVERSITY PRESS

Oxford University Press is a department of the University of Oxford.
It furthers the University's objective of excellence in research, scholarship,
and education by publishing worldwide. Oxford is a registered trademark of
Oxford University Press in the UK and in certain other countries.

Published in India by
Oxford University Press
Ground Floor, 2/11, Ansari Road, Daryaganj, New Delhi, 110 002, India

ISBN-13: 978-0-19-808542-3
ISBN-10: 0-19-808542-7

Typeset in Times New Roman
by Anvi Composers, New Delhi 110063
Printed in India by Gopsons Papers Ltd, Noida 201301

To my parents

Preface to the Second Edition

The fundamental techniques of surveying have remained unaltered, while the equipment employed in various methods has seen constant improvements. Total stations, GPS survey equipment, laser distomats, and remote sensing digital equipment are becoming less bulky and more accurate in order to facilitate better handling. Miniaturization of equipment, coupled with automated functions, better precision, and improved on-board software are the most noticeable developments in this field. Based on such developments and the requirements spelt out by many users, the book has been revised to include expanded material on several topics.

The first edition of *Surveying and Levelling*, published in 2007, received a warm welcome from faculty and students alike. Responding to this enthusiastic acceptance, the second edition has been designed based on the feedback received from different categories of users to make the book more useful.

New to the Second Edition

- Revised and expanded sections on digital levels, digital theodolites, total stations, aerial surveying methods, image processing, remote sensing, GPS, and GIS
- End-chapter exercises now include MCQs along with their answers
- Numerous additional solved examples
- Colour plates depicting modern surveying instruments

Extended Chapter Material

Presented in a single volume, the chapters gradually move from basics to advanced surveying techniques, covering chain surveying, compass surveying, plane tabling, theodolite surveying using conventional instruments, curve surveying, tacheometry, trigonometrical levelling, triangulation, photogrammetry, hydrographic surveying, and astronomical surveying. The text discusses the calculation of areas and volumes using survey data and the working of instruments such as automatic levels, digital instruments, electronic theodolites, and total stations.

Chapter 1 has expanded coverage on map and project systems.

Chapter 4 has a new topic on digital theodolites.

Chapter 7 describes the operating functions and salient features of a digital level with an example.

Chapter 17 provides a brief preview of how the Great Trignometrical Survey of India (GTS) and the survey of Great Britain were conducted in an age of limited equipment and transportation facility.

Chapter 19 provides data pertaining to India as listed in the global data bank of Permanent Service for Mean Sea Level (PSMSL).

Chapter 20 covers underground surveys in more detail. There are new topics on different methods of control such as mechanical method, the Weisback triangle, and optical transfer of control.

Chapter 22 has new topics such as aerial cameras, scale of vertical and tilted photographs, ground control, ground coordinates, and photomaps and mosaics.

Chapter 23 elaborates on total stations including their major functions, how to set up the instrument, the role of various soft keys, and their salient features.

Chapter 24 provides a detailed discussion on various modern methods of surveying such as remote sensing, image processing, and GPS including their application to surveying and GIS.

I sincerely hope this enlarged and revised edition of *Surveying and Levelling* will appeal more to students and faculties alike and it will receive the same enthusiastic response the first edition enjoyed.

Acknowledgements

I would like to gratefully acknowledge the feedback and suggestions given by various faculty members for the improvement of the book.

I am obliged to the editorial team of Oxford University Press India for bringing out the second edition in quick time and in a very elegant format.

Suggestions for improving the presentation and contents can be sent to the publishers through their website www.oup.com or to the author at rsmani2k@gmail.com.

R. Subramanian

Preface to the First Edition

Surveying and levelling is the art and science of locating points on a surface as well as above and below it. It essentially involves locating a point or finding its coordinates. The two terms 'surveying' and 'levelling' are conventionally used separately; however, the term surveying technically includes levelling, which, in simple terms, means determining the heights of points.

Surveying is fundamental to all civil engineering activities, including the construction of buildings, bridges, roads and railways, pipelines, dams, ports, and harbours as also important in the social context-for ownership and/or delimitation of land, property, etc. Engineering projects cannot commence without survey data, which is an integral part of the design and execution of engineering projects. Surveying techniques are used to collect accurate data for planning, designing, and the precise realization of a project on ground.

During the last two decades, developments in electronics and optics have led to significant advancements in surveying instruments and techniques. The old compass, dumpy level, and theodolite have given way to versatile digital instruments such as digital levels, theodolites, electromagnetic distance measuring devices (EDMs), and total stations. The techniques of surveying have also undergone changes with the introduction of remote sensing techniques and satellite-based positioning systems. Error-prone activities such as bisecting an object and reading and recording data have given way to automatic targeting and digital display, storage, and retrieval of data.

About the Book

Presented in a single volume, this book has been specially written keeping in mind the requirements of undergraduate students of engineering. Most of the subject matter prescribed in undergraduate curricula in civil engineering has been covered in an easy-to-comprehend manner. The book will also be useful to students preparing for AMIE examinations and to professionals.

Keeping in mind the requirements of students and the curricula followed in several universities, the book deals with both old and new instruments and techniques. While the syllabi followed in different universities do vary to some extent, a common thread and the minimum requirements of a survey professional were worked out to prepare the contents.

Written in a student-friendly manner, the book incorporates a large number of solved examples after every key concept as well as review questions and exercise problems at the end of each chapter.

Content and Structure

The book has been organized into 24 chapters, beginning with conventional topics and leading to more advanced instruments and techniques. The structure of the book is as follows.

Chapter 1 presents the basic principles of surveying and certain general topics such as scales and verniers.

Chapter 2 discusses the measurement of distances using conventional methods such as chains and tapes. The chapter deals in detail with the instruments and methods used in chain surveys.

Chapter 3 deals with direction measurement by measuring magnetic bearings. The prismatic compass, methods using this compass, and the precautions to be taken are dealt with in this chapter.

Chapter 4 details a second method of direction measurement, that of measuring angles. The chapter discusses conventional and modern (optical and digital) theodolites along with different techniques for accurately measuring angles.

Chapter 5 presents plane tabling, an old form of surveying that combines fieldwork-measurements and plotting-and paperwork.

Chapter 6 is devoted to traverse computations. It presents methods for adjusting traverses and determining missing sides and angles.

Chapter 7 covers the important aspect of levelling, which is determining elevations of points relative to each other or with respect to a chosen datum. It discusses levelling instruments, methods, and the computational work involved in detail.

Chapter 8 deals with the computation of areas. It describes different methods for computing areas from survey data.

Chapter 9 focuses on an extension of levelling methods known as contouring. It covers methods of contouring, computations related to plotting contours, and applications.

Chapter 10 deals with the computation of volumes from survey data for earthwork, reservoir capacity, etc.

Chapter 11 describes minor instruments that have found extensive use in earlier times for rapid and rough surveys, such as clinometers and sextants. The chapter also discusses some office equipment such as the eidograph and the pantograph.

Chapter 12 provides an account of the permanent adjustments of levels and theodolites required to keep the instruments in good working condition.

Chapter 13 explains the important aspect of setting out works, which is actually a part of construction survey.

Chapter 14 describes tacheometric surveying, in which horizontal and vertical distances are measured without using chains or tapes.

Chapter 15 elaborates on surveys required for the design and layout of curves. It describes simple circular curves, transition curves, and vertical curves in detail.

Chapter 16 covers trigonometric levelling, which is concerned with determining the heights and distances of points by observing vertical angles and using trigonometric principles.

Chapter 17 details geodetic surveying, especially using triangulation. It presents the principles and methods of triangulation and precise levelling.

Chapter 18 focuses on the important aspects of adjusting distances and angles measured for triangulation. It explains the theory of errors and the different methods of adjustment.

Chapter 19 deals with hydrographic surveying, which is the survey of water bodies. It discusses the methods of sounding and determining the profile of the surface underneath water bodies.

Chapter 20 covers engineering surveys such as profile levelling, cross-sectioning, and topographical surveying.

Chapter 21 is devoted to astronomical surveying. It discusses the salient aspects of spherical trigonometry and the determination of meridian, azimuth, latitude, longitude, and time.

Chapter 22 presents the elementary aspects of aerial surveying, which has become quite viable with the availability of aerial platforms for taking photographic and digital images of the earth's surface.

Chapter 23 details some modern instruments of surveying such as electromagnetic distance measuring equipment and total stations, which are being extensively used in the field for surveying.

Chapter 24 introduces the modern methods of surveying based essentially upon satellite technology, such as remote sensing and global positioning systems. It briefly introduces geographical information systems as well.

Acknowledgements

I am extremely grateful to the editorial team at Oxford University Press India for the excellent effort put in to bring out the book. I am deeply indebted to my family members for bearing with me during the writing of this book.

I am grateful to Toshni-Tek International, Chennai, and Tirupati Drawing and Survey Stores, Pune, for providing catalogues of survey instruments, which were used in the preparation of the manuscript.

Despite our best efforts, it is possible that some errors and misprints may have gone unnoticed. I shall be grateful if these are brought to our notice.

R. Subramanian

Features of the Book

Every chapter begins with a list of learning objectives. These act as signposts that indicate the topics you will learn in the chapter.

The book is replete with solved examples to encourage self-study.

Example 16.1   A theodolite was set up at station A and the staff reading on a benchmark of elevation 380.355 was 2.785 m. The staff was then placed at station P and the vertical angle reading at the 3.0 mark of the staff was 8° 28′ 40″. Find the reduced level of station P if the distance between the instrument and P was 185 m.

Solution   The RL of station P is determined as follows:

RL of benchmark = 380.355 m

Staff reading = 2.785 m

RL of line of sight (horizontal) = 380.355 + 2.785 = 383.14 m

Height of staff mark above line of sight = $D \tan \alpha$ = 185 tan(8° 28′ 40″)

$= 27.575$ m

RL of station P = 383.14 + 27.575 − 3 = 407.715 m

There are more than 600 illustrations to explain the concepts discussed.

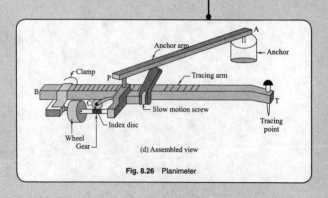

(d) Assembled view

Fig. 8.26   Planimeter

End-chapter exercises include objective questions with answers to provide an easy and quick way for self-assessment.

Review questions are designed to test your understanding of the chapter.

There are numerous unsolved problems in each chapter, the answers of which have been provided in the Appendix.

Brief Contents

Colour Plates

Plate 1
- Optic Theodolite (Chapter 4, p. 191)
- Digital Theodolite (Chapter 4, P.194)
- Digital Level (Chapter 7, p. 272)
- Digital Aerial Camera (Chapter 22, p. 860)

Plate 2
- Sokkia Total Station (Chapter 23, p. 904)
- Distomat (Chapter 23, p.900)
- GPS Receiver(Chapter 24, p.931)

Detailed Contents

CHAPTER 1

Introduction and Overview

Learning Objectives

After going through this chapter, the reader will be able to

- appreciate the importance of surveying
- state the objectives of surveying
- explain the basic principles of surveying
- classify surveys according to instruments, methods, purpose, and place
- explain the design and construction of different types of scales
- explain the purpose, parts, and working of telescopes and bubble tubes
- list the types of errors encountered in survey work

Introduction

Surveying is the art of determining the relative positions of points on the surface of the earth. Relative position means location with respect to a reference point obtained by measuring distances—both horizontal and vertical—and angles or directions. Distances may be measured directly or indirectly. Direct measurements were the most common means in the earlier times, whereas indirect measurements, particularly suitable in difficult terrains, have developed considerably over the years. Angles between the lines can be measured either directly or by measuring the directions of the lines with respect to some fixed line.

Levelling is the operation done to determine elevations or vertical distances of points. However, in practice, levelling alone is not very useful unless it is combined with horizontal measurements to locate points. The term levelling is sometimes used as different from surveying. In general, the term surveying includes levelling. Survey work is performed with different types of instruments. These include instruments used to measure distances directly or indirectly and angle-measuring instruments.

The instruments used in surveying have undergone unprecedented developments in the past two decades. This has been mainly possible with the advent of electronics and other technological developments in this field. Considerable improvement has also been achieved in the quality of optical parts, which are vital to most types of surveying equipment. The range of surveying equipment in use today is very wide. Starting with the old prismatic compasses, dumpy levels, and vernier theodolites, developments in various fields have led to automatic and digital levels, micro-optic and electronic theodolites, electromagnetic distance-measuring

(EDM) equipment, and total stations (a combination of electronic theodolites and EDMs). Satellite-based technologies such as remote sensing and global positioning systems have also entered the field of surveying. Modern instruments have built-in software that enables display and storage of data as well as transfer of data to other systems for further processing. These have helped eliminate errors due to wrong reading and recording of data or missing data.

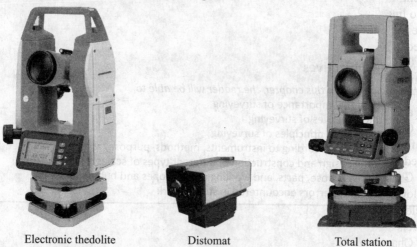

Electronic thedolite   Distomat   Total station

In India, both old instruments and modern equipment find use today. This book, therefore, deals with both the old and modern equipment used in surveying. These technologies and methods are continuously changing and some of the material in the book may become completely obsolete in the times to come. Since many students will train with old equipment, these are dealt with for the sake of immediate use and continuity.

Surveying is a core subject for civil engineers. It is the starting point for many projects. A good surveyor is an asset to any company. Transportation systems such as roads and railways, buildings, bridges, and dams, and many other projects have to start with surveying as the basic operation. Knowledge of surveying is thus fundamental and very useful to all engineering professionals.

1.1 Organization of the Book

The book is organized in such a way that it more or less traces the history of developments in surveying. It comprises 24 chapters covering a wide variety of methods. While the earlier chapters cover the conventional methods of surveying, the last two chapters outline methods of very recent origin and those that are suitable for future use. Each chapter has a large number of worked out examples, review questions, and problems.

Chapter 1 is an introduction to surveying. It begins with the basic objectives and principles of surveying. Some basic definitions are followed by the classification of surveys. Next some common elements such as scales and symbols are discussed. Telescopes and spirit levels are common parts of many surveying instruments. The surveying telescope and spirit levels are discussed in this chapter.

Chapter 2 deals with the measurement of distances. Chain and tape surveys without using an angle-measuring instrument is the basic theme of this chapter.

Chapter 3 deals with the measurement of directions. This essentially means measurement of angles. Measuring bearings using a magnetic compass, giving the relative directions of lines, is discussed in detail in this chapter. The chapter also deals with plotting traverses using bearings.

Chapter 4 is again a chapter on measuring directions. A different type of instrument, the theodolite, is discussed here. A theodolite measures the angle between two lines directly. The vernier theodolite is a versatile instrument that has been in use for many years. The modern forms of this basic instrument such as the optical theodolite and the electronic theodolite are also discussed in this chapter.

Chapter 5 deals with a method of surveying known as plane table surveying. This is a graphical method of surveying wherein measurement and plotting are done simultaneously. Plane table surveying is generally used for filling details into an already plotted traverse.

Chapter 6 deals with the paperwork after a field survey has been completed. Different methods of plotting survey work are dealt with here. In addition, methods used to calculate the missing parameters of a traverse, if any, are discussed.

Chapter 7 deals with levelling, which is a method of finding elevations of points. A level is used for this purpose. Methods used to find the elevations of points from fieldwork are dealt with in detail. Modern levels such as automatic levels and digital levels are also discussed here.

Chapter 8 deals with the computation of areas. Partitioning and allocation of land of a given area are common dealings in public life. The methods used to calculate the area of a traverse are dealt with here. Methods used to calculated the areas of pieces of land with irregular boundaries are also discussed. The instrumental method of measuring areas using a planimeter is discussed. The modern digital planimeter is also discussed.

Chapter 9 deals with a form of levelling known as contouring. While determining the elevations of points with a level, as discussed in Chapter 7, it is necessary to locate these points on the ground with horizontal distances for a complete survey. A survey in which points are located both horizontally and vertically is known as contouring. Many methods of contouring and use of such data are discussed in this chapter.

Chapter 10 deals with computation of volumes. Computation of earth work is important in many engineering applications. The data from a contouring survey is primarily used for computing volumes.

Chapter 11 deals with some minor instruments used in surveying. These instruments are used as accessories to main survey equipment for various purposes.

Chapter 12 deals with permanent adjustments of levels and theodolites. Instruments in continuous use lose their calibration over a period of time. Thus, correcting adjustments are done periodically to keep the instruments in good working condition and ensure accurate results.

Chapter 13 focuses on setting out works, which means transferring points from a plan to the ground in their correct positions. Laying out buildings and other engineering works are discussed.

Chapter 14 discusses tacheometric surveying, which is one of the indirect methods of linear measurement. Stadia tacheometry using a stadia diaphragm and other methods of tacheometry are discussed.

Chapter 15 deals with a very important aspect of roads and railways—curves. Curves are important components of such engineering works. The basic elements of their design and the survey work required to lay out curves are discussed in this chapter.

Chapter 16 deals with trigonometrical levelling, a method whereby the elevations and distances of points are determined by measuring the vertical angles to staves held over stations. This is required in cases where the difference in elevation is large and one cannot use the normal method of levelling using a horizontal line of sight.

Chapter 17 deals with geodetic surveying. The method of triangulation is dealt with in detail here. Both the instruments and methods used for geodetic surveying are precise, as these provide the control points for all other surveys.

Chapter 18 deals with errors. A brief mention of the possible errors and mistakes are briefly discussed in many chapters. However, this chapter discusses in detail the errors encountered in geodetic surveying, and the methods to correct the random errors that creep into values.

Chapter 19 discuses hydrographic surveying, which is the survey conducted under water to determine the profile of the seabed or ocean floor. Such surveys are important for designing ports, harbours, waterways for navigation, etc.

Chapter 20 deals with some important engineering applications of surveys such as tunnel, mine, and route surveys.

Chapter 21 deals with astronomical surveying. The method used is essentially based on observations of celestial bodies to determine time and various other factors of importance in surveying.

Chapter 22 deals with terrestrial and aerial surveys using photographic techniques. Terrestrial photogrammetry uses photo-theodolites whereas aerial surveying uses aircrafts to photograph the earth's surface. Such photographs can be used to determine the complete topography of large tracts of land.

Vernier theodolites and levels have more or less given way to modern instruments. Electronic theodolites and electromagnetic distance-measuring instruments have facilitated the fieldwork in surveying in a big way. Electromagnetic distance-measuring instruments use electromagnetic waves to find distances. The combination of an electromagnetic distance-measuring instrument and an electronic theodolite is called a total station and finds wide application in geodetic surveying. Chapter 23 discusses these two instruments.

Chapter 24 deals with the new entrants in the field of surveying—satellite-based technologies such as remote sensing and global positioning systems.

1.2 Objectives of Surveying

As mentioned earlier, surveying enables us to acquire data on the relative positions, horizontal distances, and elevations of points. This data is then used for operations such as preparing plans, maps, and land records, partitioning land, laying out engineering works, and calculating areas and volumes. The objectives of surveying can thus be stated as follows.

(a) Collect and record data on the relative positions of points on the surface of the earth.

(b) Compute areas and volumes using this data, required for various purposes.

(c) Prepare the plans and maps required for various activities.

(d) Lay out, using survey data, the various engineering works in correct positions.

(e) Check the accuracy of laid-out lines, built-up structures, etc.

1.3 Basic Principles of Surveying

Two basic principles of surveying need to be followed for accurately locating points on the surface of the earth. These are explained in this section.

1.3.1 Working from Whole to Part

A fundamental principle of surveying is working from whole to part. All survey work should invariably follow this principle unless a situation arises for which there is no alternative except to violate this. The basic objective of this principle is to prevent accumulation of errors. In working from part to whole, any error in a part gets magnified and in the end the errors become large and uncontrollable.

In large survey works conducted by state agencies or Survey of India, control points at large distances are first established very accurately by triangulation. From these control points, smaller areas are surveyed by setting out a series of triangles. Such subdivision can go on till all the details are surveyed, each with reference to the previously established systems of points.

This basic principle can also be explained with reference to the method of measuring a line with a tape, the line being longer than the tape (refer to Fig. 1.1). As shown in Fig. 1.1(a), we establish a number of points R, S, T, etc. between P and Q. This is done so that PR, RQ, QS, etc. are shorter than the length of a chain or tape. The method of establishing intermediate points should follow the principle of working from whole to part.

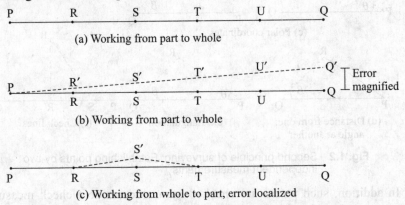

(a) Working from part to whole

(b) Working from part to whole

(c) Working from whole to part, error localized

Fig. 1.1 First principle of surveying: Working from whole to part

If we establish R in line with P and Q, S with reference to P and R, T with reference to Q and R, etc., any error in setting the point R will lead to a further magnified error as we set further points [Fig. 1.1(b)]. The line obtained finally will not be a true measure of the length of the line.

On the other hand, if we follow the principle of working from whole to part, we need to establish points R, S, T, etc. independently with reference to P and Q. Any error in establishing a point is localized and will not affect the setting of other points.

Thus, if point S is established incorrectly as S′, only the lengths RS′ and S′T will be measured incorrectly. PR, TU, etc. will be measured correctly. The error thus gets localized and leads to better accuracy in measurement [Fig. 1.1(c)].

1.3.2 Establishing Any Point by At Least Two Independent Measurements

Points in surveying are located by linear or angular measurement or by both. If two control points are established first, then a new station can be located by two linear or two angular measurements, or by one linear and one angular measurement. In Fig. 1.2, P and Q are control points. A new station R can be established by any of the following means. In Fig. 1.2(a), R is established using the distances PR′ and R′R (rectangular coordinates from P). In Fig. 1.2(b), R is established using the distances PR and QR. This method is known as *trilateration*. In Fig. 1.2(c), R is established by the angle θ and the distance PR. These are the polar coordinates of R from P. Polar coordinates from Q may also be used. In Fig. 1.2(d), R is established using the distance PR and the angle PQR. In Fig. 1.2(e), R is established using the angles PQR and QPR. This is the method of *triangulation*.

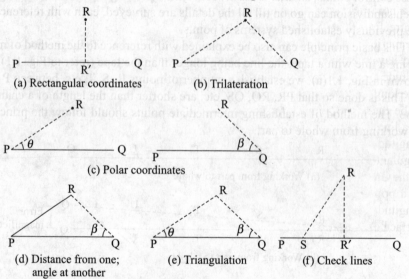

(a) Rectangular coordinates   (b) Trilateration

(c) Polar coordinates

(d) Distance from one;   (e) Triangulation   (f) Check lines
angle at another

Fig. 1.2   Second principle of surveying: Establishing points by two independent measurements

In addition, such points are checked by an independent check measure to ensure that there has been no error in establishing the required point. As an example, a point established at right angles to a line is checked by a linear measurement from another point on the line. This is particularly important if the perpendicular line RS is long. In Fig. 1.2(f), point R is established by rectangular coordinates. The distance from a known point S on the line PQ to R is measured as a check.

1.4 Basic Definitions

The following basic definitions are fundamental to the various surveying methods.

Earth   The earth is one of the planets of the solar system, third planet from the sun. The details of the planets of the solar system are given in Appendix 5. The earth is considered to be an oblate spheroid. This is so because the equatorial diameter is more than the polar diameter (Fig. 1.3). Any plane parallel to the equatorial plane cuts a circle on the earth's surface. A plane passing through the poles will give an ellipse. Very accurate measurements have shown that the equatorial plane also does not give a circle, but an ellipse. The earth can thus be considered to be an ellipsoid.

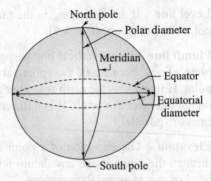

Fig. 1.3   Earth

Meridian   A meridian is a line of intersection of a plane passing through the poles with the surface of the earth.

Latitude and longitude   The latitude of a point is the angle subtended at the centre from the equatorial plane along the meridian through that point. A point may have a northern latitude or a southern latitude (Fig. 1.4). One of the longitudes (the longitude at Greenwich) is assumed to be the zero longitude. The longitude of a line is measured with respect to this longitude. The longitude of a point is the angular distance of that point east or west of the Greenwich meridian. The longitude of a point can thus be a west or a east

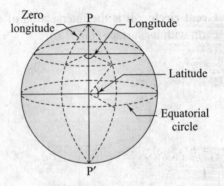

Fig. 1.4   Latitude and longitude

longitude. The latitude and longitude of a point fix the location of that point on the surface of the earth.

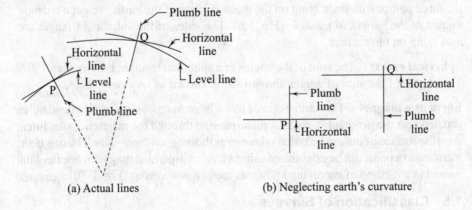

(a) Actual lines   (b) Neglecting earth's curvature

Fig. 1.5   Level line, plumb line, and horizontal line

Level line It is a line lying in the mean spheroidal surface of the earth. This is not a line but an arc (Fig. 1.5).

Plumb line It is a vertical line perpendicular to the level line at a point. All the plumb lines, because of the spherical surface of the earth, should intersect at a point. However, as the earth is neither a true spheroid nor an oblate spheroid, they do not intersect at a point. When the curvature of the earth is neglected, all plumb lines are parallel.

Elevation The elevation of a point is the vertical height along the plumb line through the point above any datum level plane. The datum can be the mean sea level or any assumed datum plane.

Horizontal plane It is a plane tangential to a point on the earth's surface.

Horizontal line A horizontal line through a point is the line lying in a horizontal plane through the point. Strictly speaking, this line is tangential to the curved surface of the earth (Fig. 1.5).

Great circle It is the intersection of a plane passing through the centre of the earth with its surface (Fig. 1.6).

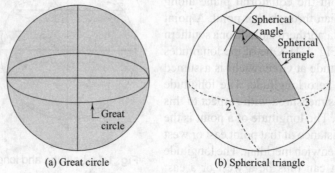

(a) Great circle (b) Spherical triangle

Fig. 1.6 Great circle and spherical triangle

Spherical triangle Consider any three points on the surface of the earth. If we join the three points with lines lying on the mean surface of the earth, we get a triangle known as the *spherical triangle* (Fig. 1.6). The sides of the spherical triangle are arcs lying on the surface.

Spherical excess The sum of the angles of a spherical triangle is more than 180°. The excess of the sum of angles above 180° is known as *spherical excess*.

Earth as a magnet The earth behaves like a huge magnet with its magnetic poles situated near the geographic poles. A plane passing through the magnetic poles intersects the surface of the earth in what is known as the *magnetic meridian*. The magnetic meridian at a point can be established using a freely suspended magnetic needle. This is used in a method of surveying known as *compass surveying* (Fig. 1.7).

1.5 Classification of Surveys

Surveys can be classified into two classes depending upon whether the spherical shape of the earth is taken into account or not. In survey work requiring very high precision, as in triangulation for setting control stations, the curved shape of the earth's surface is taken into account. This is known as *geodetic surveying*. Survey conducted neglecting the curved shape of the earth's surface is known as *plane surveying*.

1.5.1 Plane Survey

In plane surveying, the curved nature of the earth's surface is neglected. A line on the earth's surface is taken to be a straight line. When three points lying on the surface of the earth are joined, they form a plane triangle. Consequently, all plumb lines are assumed to be parallel. The methods of plane surveying are used when the extent of the survey is small and high precision is not required. This assumption is justified in the case of surveys of small areas. This is so because the difference between an arc of length 25 km and the corresponding chord is only 10 mm. Further, the sum of the angles of a spherical

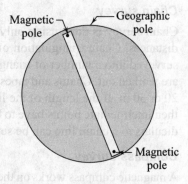

Fig. 1.7 Earth's magnetism

triangle about 195 km^2 in area exceeds 180° (spherical excess) by just 1″. Surveys conducted for engineering projects fall under this category.

1.5.2 Geodetic Survey

Geodetic survey is survey done for works requiring high precision, as in triangulation surveys to establish control points and in surveys of large areas. The equipment used for geodetic surveying is of very high precision and the methods used also ensure a high degree of accuracy in the measurements. The curvature of the earth is accounted for in the measurements taken during such surveys.

Surveys can be further classified into categories depending on the purpose, instruments, techniques used, etc. These classifications are shown in Table 1.1.

Table 1.1 Classification of surveys

Instrument-based	Method-based	Purpose-based	Place-based
Chain or tape	Triangulation	Reconnaissance	Land
Compass	Traversing	Preliminary	Water
Plane table	Levelling	Engineering	Aerial
Theodolite	Tacheometry	Geographical	Underground
Level	Trigonometrical levelling	Mine survey	
Tacheometer		Route survey	
EDM		Location	
Total station		Geological	
Satellite-based		Defence Archaeological	

1.5.3   Instrument-based Surveys

In earlier times, few surveying instruments were in use, and surveys were classified based upon the instruments used. Chains, compasses, plane tables, and theodolites were the only instruments used, either alone or in conjunction with one another. Today, many types of instruments are available.

Chain survey

Chain survey is done using only a chain or tape. A chain or tape measures linear distances. Chain triangulation of a small area is done by dividing the area to be surveyed into a number of triangles. Using the sides of the triangles, other details are worked out. Chains and tapes meant for survey work are available in lengths of 20 or 30 m. If the length of the line to be measured is more than the chain length, then intermediate points have to be set up by ranging. For locating details, perpendiculars to a chain line can be set by many methods.

Compass survey

A magnetic compass works on the principle that a freely suspended magnetic needle points in the magnetic north–south direction. This gives a reference direction, which remains parallel at all stations if the area is free from magnetic influences. A compass, together with a chain or tape, can be used to survey a given area by many methods such as triangulation or traversing.

Plane table survey

Plane table survey essentially combines fieldwork and plotting work. A plane table is a drawing board with modifications for attaching it to a tripod. With an alidade providing the direction, and with a chain or tape, points can be plotted on a sheet. A plan can be prepared or, with stations already marked on the table, further details can be filled in.

Theodolite survey

A theodolite measures horizontal and vertical angles. It has a telescope which provides the line of sight and graduated horizontal and vertical circles, both with verniers. Vernier theodolites with a least count of 20″ are commonly used for ordinary work. Precision theodolites are made of larger diameter circles. Optic and electronic theodolites have greater precision and are easy to operate. The theodolite is a very versatile instrument and is used for all types of survey work, from triangulation to traversing and filling in details.

Levelling

A level has a telescope for providing the line of sight and a supporting system for ensuring that the line of sight remains horizontal in all directions. A graduated rod, known as the *levelling staff*, is used with a level to find the difference in elevation between points.

Tacheometry

A tacheometer is very similar to a theodolite but has a stadia diaphragm having three cross hairs. The reading taken by a stadia rod or levelling staff against all the three cross hairs enables the horizontal and vertical distances to be calculated.

EDM survey

Electromagnetic distance measurement involves the generation, transmission, and reception from a reflector at a station of light, radio, or microwave signals. The phase difference between the transmitted and received signals enables the distance between the instrument station and the reflector station to be calculated and displayed or stored.

Total station survey

A total station is a combination of an electronic theodolite and an EDM. Horizontal distances and horizontal and vertical angles are determined using a total station. Total stations are used in triangulation surveys and other forms of surveys needing a very high level of precision. A total station can display and store the values required as well as transfer the data to a computer for further processing.

Satellite-based instrument survey

The instruments and methods of remote sensing can be used for surveys conducted for various purposes. Global positioning systems, on the other hand, use an array of satellites strategically placed around the globe. Handheld transreceivers or total stations with GPS capability receive signals from a minimum of four satellites and can determine the position of the receiver very accurately. Depending upon the atmospheric conditions and the terrain, such instruments can be used for surveys.

1.5.4 Method-based Surveys

There are different methods of surveying depending upon the instruments available, the terrain, and the purpose of the survey. The following are common methods.

Triangulation

Triangulation is a basic method of surveying. A triangle is a stable figure, and measuring one of the sides, known as the *base line* and the three angles establishes a triangle on the ground. The other sides can be calculated using several formulae. If the three sides are measured, the method is known as trilateration. The triangle can be extended by adding two more sides each time to form another triangle, thus increasing the area covered.

Traversing

A traverse can be open or closed. The sides and their directions are measured with a compass and chain or a theodolite and tape. A traverse can enclose a large area and the details along the sides or within it can be measured from the sides. An open traverse is run along narrow stretches of land such as river banks, roads, or railway lines. A closed traverse is a closed polygon-shaped figure and is run to enclose a large tract of land.

Levelling

Levelling is a method of surveying used for determining the elevations of points. A level is used for this type of survey. When, along with the elevations, the horizontal locations of points are also determined, the method is known as *contouring*.

Tacheometry

Tacheometry is a method of surveying wherein horizontal distances and differences in elevation between survey stations are determined without directly measuring the distances. Tacheometry uses the principle of *stadia surveying*. The stadia diaphragm has three cross hairs and the readings taken on a staff against all the three hairs enable the calculation of distances and elevations.

Trigonometrical levelling

In trigonometrical levelling, distances and elevations are determined by measuring vertical angles to graduations on staves using the principles of trigonometry. Elevations in many cases cannot be determined by ordinary levelling because of the large differences in elevation. Trigonometrical levelling provides a solution in such cases.

1.5.5 Purpose-based Surveys

Based upon their purpose, surveys can be classified into the following types.

Reconnaissance

Reconnaissance operations are conducted to get an idea about the terrain and any special or difficult features that may be encountered during a regular survey. Generally, no instruments are used. Distances may be roughly estimated by pacing. Sufficient notes and sketches are drawn for future use during such surveys.

Preliminary surveys

A preliminary survey is more detailed in its scope. It locates all the prominent features of a terrain as well as any particular features to be represented on a map.

Engineering surveys

Engineering surveys are very detailed surveys required to locate engineering projects such as roads, railways, factories, and dams. More precise instruments and methods are used for such surveys.

Geographical surveys

Geographical surveys are conducted to collect data for the preparation of geographical maps. These maps may be prepared to serve different purposes such as specifying national boundaries, land use, contours, and resources.

Mine surveys

Mine surveys include both surface and underground surveys. Special techniques are required to transfer surface data to underground points. The surface maps show the general layout of the mine. Subsequently, underground maps are prepared for the design of tunnels and shafts in the mine, underground plans, geological maps, etc.

Route surveys

Route surveys are surveys conducted for locating road or railway networks. The most convenient alignment of roads is decided based on such surveys. Other engineering aspects such as the road profile, earth work in cutting and banking, and road curves and superelevation are taken care of subsequently.

Location surveys

Location surveys are conducted to locate points on the ground based on the plans prepared.

Geological surveys

Geological surveys are of economic importance, as both surface and sub-surface surveys are conducted to locate ores and mineral deposits. In addition, geological features of the terrain such as folds and faults are located.

Defence surveys

Defence surveys are conducted by the military establishment to locate strategic positions in the enemy area. Aerial surveys are conducted for this purpose. Knowledge of the terrain features from ground surveys are also important to prepare strategic plans for defence and attack.

Archaeological surveys

Archaeological surveys are conducted to locate relics of antiquity, civilizations, etc.

1.5.6 Place-based Surveys

Based upon the place of survey, the following types of surveys can be identified.

Land survey

Land surveys are done on land to prepare plans and maps of a given area. Such surveys are conducted for the purpose of partitioning land, determining their areas, locating boundaries of properties, etc. Topographical surveys are surveys conducted to prepare plans or maps indicating the location of important features such as buildings, rivers, and woods, including the elevation of points from some datum. Cadastral surveys or public land surveys are conducted for the purpose of locating land features such as agricultural fields, buildings, houses, and other property lines. Both urban and rural areas are surveyed extensively to obtain such data. City surveys are similar surveys conducted in cities for similar purposes but with great refinement, as the cost of the land is very excessive and the exact location and demarcation of features become important.

Hydrographic survey

A hydrographic survey deals with water bodies such as lakes, rivers, streams, and coastal areas. The objective is to obtain data to design water navigation systems, determine shorelines, help in the design of structures built in water or along shore lines, and obtain data about the ground surface underneath water.

Aerial survey

An aerial survey is done from aircraft, which take photographs of the surface of the earth in overlapping strips of land. Extensive areas can be covered by such surveys. This form of survey is also known as a photogrammetric survey. The method is very expensive. It is, however, recommended in large projects in difficult terrain where ground surveys may be difficult or impossible.

Underground survey

Underground surveys are done in the case of mines and tunnels. These are done by transferring ground points to the underground level and conducting surveys of tunnels, caves, mines, etc.

1.6   Horizontal Control

Horizontal control is the establishment of points that can be used for referencing further work. Control points are established with great precision using highly precise equipment and methods. Once these points are established, they are marked permanently by suitable means. They are generally triangulation points established by state agencies. Control points can be later used by other agencies wanting to conduct more detailed surveys.

1.7   Vertical Control

Vertical control is the establishment of points of known elevation above a national datum such as the mean sea level at a point. Like the points established for horizontal control, such points are established with highly precise equipment and methods. These points of known elevation, known as *benchmarks*, can be used by other agencies wanting to carry out levelling or contouring.

1.8   Surveying Practice

Surveying practice involves planning the survey work, selecting appropriate instruments depending upon the nature of survey, organizing field survey operations, collecting and recording data from the field by linear and angular measurements, paperwork, and taking care of instruments through periodic maintenance.

1.8.1   Fieldwork

Fieldwork in surveying involves selecting instruments for a particular job, taking linear and angular measurements in the field, and recording data in proper forms along with field notes and sketches. After a reconnaissance of the field to be surveyed, the surveyor approximately locates stations for setting up the instrument, ensuring that they command a clear view of the area and their intervisibility. Linear and angular measurements are done as per requirement and the data is recorded in a field book. Check measurements are also taken to check the accuracy of measurements.

Field records are a very important part of the fieldwork. They are the only source of survey data available after the fieldwork is over. Different formats are used for recording field data obtained using different types of instruments and different methods of survey. Enough notes and sketches should be made for relating any future work to the surveyed field stations.

1.8.2   Care in Using Instruments

Most surveying instruments are delicate and sensitive to environmental factors. The following points should be noted while using these instruments.

(a) The manufacturer's instructions must be carefully studied before using the equipment.

(b) The procedure for keeping the instrument within the box and taking it out should be carefully followed.

(c) The instrument should be protected from sun and rain by shading it with a field umbrella.

(d) Many instruments have optical parts such as lenses which are damaged by moisture. Proper care should be taken to keep moisture away from such parts.

(e) The instrument should be protected in heavy windy conditions to prevent it from swaying and falling down.

(f) The procedure for using the instrument should be carefully studied and understood.

(g) While operating the screws, no excessive force should be used. They must be delicately handled.

1.8.3 Maintenance of Instruments

Instruments should be periodically checked to ensure that they function properly. They should be stored properly when not in use. The optical parts of the instruments such as lenses and mirrors should be kept clean and free from dust and moisture. The manufacturer's instructions should be followed in using and storing the instruments. The instruments must be properly checked periodically for permanent adjustments. The procedure for permanent adjustments should be followed carefully.

1.8.4 Paperwork

Once the survey fieldwork is over, the data collected in terms of linear and angular measurements along with field notes and sketches are used for various purposes. The paperwork involves the following.

(a) Analysing the data, making various calculations to check the data, and correcting the data as per standard practice. Sometimes, it may be necessary to redo the fieldwork for checking or for collecting missing data.

(b) Making plans or maps to standard scale using the collected data.

(c) Calculating the areas and volumes for various uses.

(d) Locating various engineering structures and projects in the plans and maps, drawing contours, locating gradient lines, etc.

1.8.5 Qualities of a Good Surveyor

A competent surveyor is an asset to any enterprise. The qualities of a good surveyor are the following.

(a) Sound knowledge of the theory of surveying

(b) Thorough knowledge of the instruments he/she is handling, limits of precision, possible errors, and the methods to maintain the instruments in good condition

(c) Sufficient field practice, as this alone can give him/her the confidence, judgement, and proficiency essential for accurate survey work

(d) In addition to technical knowledge, leadership qualities to extract quality work from his subordinates

(e) Ability to maintain a good professional relationship with his subordinates and superiors

1.9 Plans and Maps

One of the basic objectives of surveying is to prepare plans and maps. Plans and maps are graphical representations of the actual site to some scale. A graphical representation is called a *plan* when the scale is large and a *map* when the scale is small.

A plan is a representation of points on the surface of the earth on a horizontal plane by projecting them onto the plane. As the earth's surface is curved, the representation is distorted. The distortion will be imperceptible if the area covered is small. Generally, the scale will be such that a small length in the plan represents a larger length on the ground. A certain length on the plan representing a smaller length is used in mechanical drawings to represent small parts.

A plan generally shows horizontal distances. A topographical map represents both horizontal and vertical distances. This is achieved by hachures or colour schemes. The scales normally used in survey work are given in Table 1.2.

Table 1.2 Common scales for plans and maps

Type of map	Representative fraction	Numerical scale
Preliminary survey map for route surveys	1/1000 to 1/6000	1 cm = 10 m to 60 m
Mine survey maps	1/1000 to 1/25,000	1 cm = 10 m to 25 m
Buildings	1/1000	1 cm = 10 m
Town planning	1/5000 to 1/10,000	1 cm = 50 m to 100 m
Cadastral maps	1/1000 to 1/5000	1 cm = 10 m to 50 m
Forest map	1/25,000	1 cm = 250 m
Location plans	1/500 to 1/2500	1 cm = 5 m to 25 m
Topographical maps	1/250,000	1 cm = 2.5 km
Geographical maps	1/16,000,000	1 cm = 160 km

1.9.1 Map Projections

Map projections are attempts to project a set of spherical coordinates on to a plane to make it two dimensional. There are many projection methods used to represent the features on the surface of the earth on to a plane. The different coordinate systems are shown in Fig. 1.8.

The rectangular coordinate system is the most common method of representing points, lines, and planes. For points and lines in a plane, the coordinates are (x,y). For points in three-dimensional space, a point will have coordinates (x,y,z,).

Polar coordinate system is another method of representation. Polar coordinates of a point are (r,θ) for points and lines in a plane.

Spherical polar coordinates are used to represent the surface features of the earth. These coordinates are the latitude and longitude of a point. These along with the elevation of the point above a chosen datum completely defines the point.

The following terminology is important for representing features on the surface of the earth assuming it to be a sphere (Fig. 1.9).

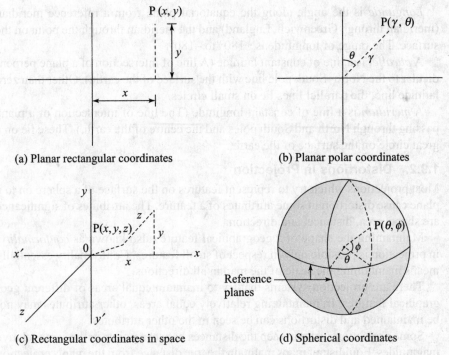

(a) Planar rectangular coordinates   (b) Planar polar coordinates

(c) Rectangular coordinates in space   (d) Spherical coordinates

Fig 1.8   Coordinate systems

The reference plane for latitude is the equatorial plane. When a point lies above this plane, the point has North latitude. When the point lies below the equatorial reference plane, the point has South latitude.

The reference meridian is the meridian that passes through Greenwich, England. The points lying to the east of reference meridian have East longitude. Similarly, points lying to the west of reference meridian have West longitude.

Latitude of a point on the surface of the earth is the angle along the meridian (the line of intersection of plane passing through the North–South poles and the point) from the point to the equatorial plane. The range of latitude is $+90°$ (North pole) to $-90°$ (South pole).

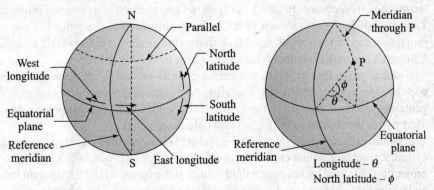

Fig 1.9   Latitude and longitude

Longitude is the angle along the equatorial plane from a reference meridian (meridian through Greenwich, England) and the meridian through the point on the surface. The range of longitude is +180° to –180°.

A *parallel* is a line of constant latitude (A line of intersection of a plane perpendicular to the North–South pole line with the surface of the earth.). Other than zero latitude line, the parallel lines lie on small circles.

A *meridian* is a line of constant longitude (The line of intersection of a plane passing through North and South poles and the centre of the earth.). These lie on a great circle on the surface of the earth.

1.9.2.   Distortions in Projection

Most projections which try to represent features on the surface of a sphere on to a plane cause distortion in some attributes of a feature. The attributes of significance are shape, area, distance, and direction.

Maintaining the shape of a geographical feature (also known as *conformality*) in projection is possible only in respect of small features. Conformality essentially means maintaining the scale of the map in all directions.

There are projection systems that try to maintain equal areas of different geographical features. In maintaining relatively equal areas, other attributes may not be maintained and distortions can be seen in the other attributes.

Some projection systems keep the distances between points in the same relative magnitudes. Equidistant maps maintain the true distance from the map's projection centre to all points.

Another important attribute is the directions between points. It is important in many cases particularly for navigation. Projections that maintain direction gives the true bearings of lines between any two points.

None of the projection systems maintains all the attributes in a single map but make compromises with one or more attributes to maintain one.

Various projection methods are used to represent features on the surface of the earth. *Orthographic projection* is the commonly used method. In this method the curved surface is projected onto a plane and there is distortion if the distances are large. Other projection systems include azimuthal, conical, and cylindrical systems (Fig. 1.10).

Azimuthal projection   Figure 1.10(a) shows one method of azimuthal projection. The plane of projection shown in the figure is tangential to the pole. The plane of projection can be tangential, normal or transverse as shown in Fig. 1.10(a). Johann Lambert's azimuthal projection maintains area as well as direction.

Conical projection systems, wherein a conical surface is used, have also been popular. The conical projection can also use a cone in three modes—normal, tangential or transverse as shown in Fig. 1.10(b). Equidistant conical projections are used where the distances between points are maintained.

Cylindrical projection shown in Fig. 1.10(c) is one method of projection using a cylindrical surface which encloses the entire sphere. The projection is similar to the projection obtained by keeping a light source at the centre of the transparent sphere and projection obtained on the curved surface of the sphere. The curved surface can then be opened as a planar surface. The meridians are equally spaced but the distance between parallels seems to be distorted (are more) as they approach the poles.

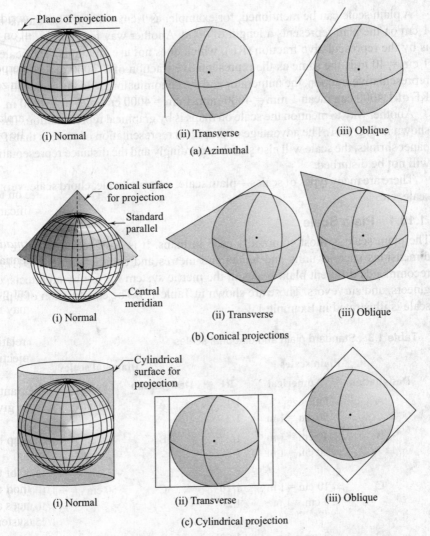

Fig. 1.10 Projection systems

The Mercator projection system which is commonly used in atlases is a cylindrical projection system. The cylinder used in cylindrical projection can also be normal, transverse or tangential as shown in Fig. 1.10(c).

Survey of India is the government agency in India that prepares maps. The maps are prepared in a series. Each map covers a small area of the country.

1.10 Scales

Scales are the basic requirement for the preparation of plans and maps. Scales are used to represent large distances on paper. All plans and maps should mention the scale to which they are drawn.

Scales can be mentioned directly or represented graphically on the plan or map.

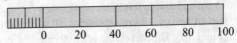

Fig 1.11 Graphical representation of scale (1:2000 or 1 cm = 20 m)

A plain scale can be mentioned, for example, as 1 cm = 40 m. This means that 1 cm on the plan represents a length of 40 m. Another way to represent the scale is by the representative fraction (RF), which does not use any units. The scale of 1 cm = 40 m is the same as the representative fraction of 1:4000 or 1/4000. In a representative fraction, the numerator and the denominator are in the same unit. An RF of 1/4000 can mean 1 mm = 4000 mm, 1 cm = 4000 cm, or 1 m = 4000 m.

Another way to mention the scale on maps is by graphical representation. This is shown in Fig. 1.11. The advantage of graphical representation of scale is that if the paper shrinks, the scale will also shrink accordingly and the distance representation will not be disturbed.

There are many types of scales—plain scale, diagonal scale, chord scale, vernier scale, etc.

1.10.1 Plain Scale

The plain scale is most commonly used in maps. It is designed to measure two dimensions such as units and tenths and metres and decimetres. IS 1491:1959 recommends different plain scales in the metric system for use by architects, engineers, and surveyors. These are shown in Table 1.3. The construction of a plain scale is illustrated in Example 1.1.

Table 1.3 Standard plain and diagonal scales

Plain scales			Diagonal scales		
Designation	Numerical	RF	Designation	Length	RF
A	Full size	1/1	A	150 cm	1/1
	50 cm = 1 m	1/2			
B	40 cm = 1 m	1/2.5	B	100 cm	1/100,000
	20 cm = 1 m	1/5			1/50,000
					1/25,000
C	10 cm = 1 m	1/10	C	50 cm	1/100,000
	5 cm = 1 m	1/20			1/50,000
					1/25,000
D	2 cm = 1m	1/50	D	150 cm	1/100,000
	1 cm = 1 m	1/100			1/8000
					1/4000
E	5 mm = 1m	1/200			
	2 mm = 1 m	1/500			
F	1 mm = 1 m	1/1000			
	0.5 mm = 1 m	1/2000			

Example 1.1 Construct a plain scale of RF 1/400 to measure upto a metre and represent 36 m on the scale.

Solution RF = 1/400 is the same as 1 cm = 4 m or 2.5 cm = 10 m. Take a line of length 15 cm and divide it into six parts. Each part of length 2.5 cm represents 10 m (Fig. 1.12). The part on the left extreme is divided into 10 equal parts and each division represents 1 m. The zero of the

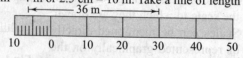

Fig. 1.12 Plain scale (Example 1.1)

scale starts from the end of this part and the numbering at each part is marked as shown in the figure. To show or measure 36 m, we start from the 30-m mark and take six divisions towards the left from the zero. This gives 36 as shown.

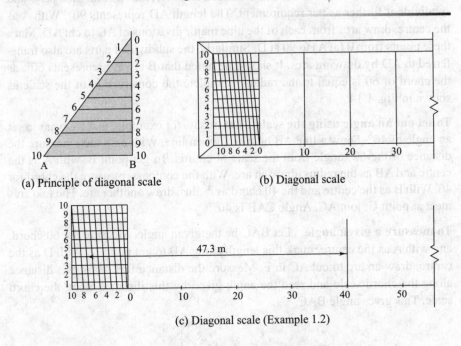

(a) Principle of diagonal scale

(b) Diagonal scale

(c) Diagonal scale (Example 1.2)

Fig. 1.13 Diagonal scale

1.10.2 Diagonal Scale

Using a diagonal scale, one can measure three dimensions such as units, tenths, and hundredths. The diagonal scale is made on the principle of similar triangles as shown in Fig. 1.13(a). A short length AB can be divided into 10 parts as shown in the figure. From similar triangles, line 1-1 is equal to one-tenth of AB, line 2-2 is two-tenths of AB, and so on. Diagonal scales recommended by IS 1562:1962 are given in Table 1.3. The construction of a diagonal scale is illustrated with Example 1.2.

Example 1.2 Construct a diagonal scale 1 cm = 4 m to read metres and decimetres. Represent 47.3 m on the scale.

Solution 1 cm = 4 m means 2.5 cm = 10 m. We construct a small scale 25 cm long as shown in Fig. 1.11(c). The length is divided into 10 parts. Each part represents 10 m. The leftmost segment is divided into 10 parts. The vertical width of the scale may be kept at 5 cm and divided into 10 parts. The divisions on the horizontal legs are joined diagonally as shown. The length of 47.3 m can be measured as shown.

1.10.3 Chord Scale

A scale of chords is prepared either to measure a given angle or to set off a given angle. This is done essentially by measuring or setting the chord forming the angle.

The basic principle is illustrated in Fig. 1.13. Take a convenient length AB. At B draw a perpendicular and make AB = BC. Join AC. With A as the centre and AC as the radius draw an arc to cut AB extended at D. Divide AC into nine parts and subdivide it further as per requirement. The length AD represents 90°. With A as the centre, draw arcs from each of the nine major divisions of AC to cut AD. Mark these points from 0 (at A) to 90 at D. Similarly, the subdivided parts are also transferred to AD by drawing arcs. It should be noted that B always represents 60°, as the chord of 60 is equal to the radius. Complete the construction of the scale as shown in Fig. 1.14.

To set out an angle using the scale   Let us say, for example, that we want to set an angle of 40°. Draw a line AB or take a given line. With a divider, measure the distance of the 60° angle from the scale of chords. To get point B, with A as the centre and AB as the radius, draw an arc. With the compass, measure the chord for 40. With B as the centre and the 40 chord as radius, draw another arc. The two arcs meet at point C; join AC. Angle CAB is 40°.

To measure a given angle   Let BAC be the given angle. Measure the 60 chord, and with A as the centre, mark this length along AB to get point D. With D as the centre, draw an arc to cut AC in E. Measure the distance DE. Mark this distance along the chord scale and read the angle given by this distance along the chord scale. This gives angle BAC.

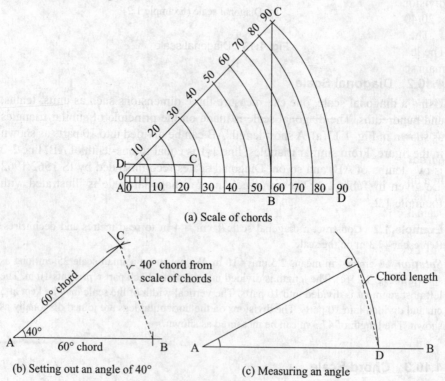

(a) Scale of chords

(b) Setting out an angle of 40°

(c) Measuring an angle

Fig. 1.14   Scale of chords

1.10.4 Vernier Scale

Vernier scales are sliding scales along a main scale and are used to measure parts of a main scale division. Vernier scales are designed on the assumption that the human eye can perceive accurately the coincidence of two lines, a marking on the vernier scale and one on the main scale (Fig. 1.15). The vernier scale is commonly employed in theodolites. A theodolite has a circle graduated in degrees and in 10 or 20 minutes, and an angle is read with the help of a vernier to measure a part of the main scale division. The following terminology must be understood with respect to a vernier scale.

The *main scale* is the scale to which a vernier is attached. If the main scale is divided into tenths of a unit, a vernier is attached to the main scale to read finer than is possible with the main scale.

The *vernier scale* is a sliding scale that slides along the main scale. A certain number of divisions are taken from the main scale and divided into a greater (or smaller) number of parts, so that the vernier scale division will have less (or more) value than that of the main scale division.

The *value of a main scale division* is the value in some unit of one division of the main scale. If the main scale is divided into 20′ divisions, then 20′ is the value of a main scale division.

The *value of a vernier scale division* is the value in some unit of one division on the vernier scale. For example, if the main scale division is 20′ and 39 such divisions are taken and divided into 40 parts, then the vernier scale division = 39 × 20/40 = 39′/2 = 19.5′.

The *least count of a vernier* is the least value that can be read using the vernier. The least count is given by the difference between the value of a division on the main scale and the value of a division on the vernier scale. In the case just described above, least count = 20′ − 19.5′ = 0.5′ = 30″. There are essentially two types of vernier scales—direct reading vernier scale and retrograde vernier scale. The direct reading vernier scale is more common. Other types of verniers are double, extended, and circular.

Direct reading vernier

In a direct reading vernier, the graduations in the main scale and vernier scale run in the same direction. Figure 1.15 shows such a vernier. A direct reading vernier scale is designed with the knowledge of the main scale division and the least count required. The procedure for designing such a vernier is as follows.

1. Let *n* be the number of divisions on the vernier.
2. Take a length equal to (*n* − 1) divisions of the main scale and divide it into *n* parts.

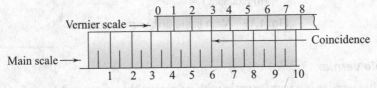

Fig. 1.15 Main scale and vernier scale—coincidence of graduations

3. Let the value of one main scale division be m. Value of $(n-1)$ main scale divisions $= (n-1)m$.

4. Value of one vernier scale division, $v = (n-1)m/n$ (as the number of vernier scale divisions is n).

5. The two lengths being equal,

$$(n-1)m = nv \quad \text{or} \quad v = (n-1)m/n$$

6. Least count = value of one main scale division – value of one vernier scale division $= m - v = m - (n-1)m/n = m/n$.

The least count is thus equal to the value of the main scale division divided by the number of divisions on the vernier scale. Vernier scales are designed on this principle. This is illustrated with the following examples.

Example 1.3   A metre scale is divided into divisions such that each division is equal to 1/100th of a metre. Design a vernier scale having a least count of one millimetre.

Solution   Each main scale division is equal to one centimetre. The least count is equal to the main scale division divided by the number of divisions on the vernier scale. The main scale division being 1 cm, the number of divisions on the vernier scale = 10, giving a least count of 1 mm. Thus, nine divisions on the main scale are taken and divided into 10 parts. The vernier is shown in Fig. 1.16(a).

Example 1.4   Design a vernier scale for a theodolite in which the main scale division is equal to $10'$ (each degree is divided into six parts). The theodolite should have a least count of $10''$.

Solution   Value of main scale division = $10'$. Least count required = $10''$. Least count $= 10'' = m/n = 10'/n$. It is clear that n should be 60 to get the least count required. 59 main scale divisions are taken and divided into 60 parts. The main and vernier scales are shown in Fig. 1.16(b).

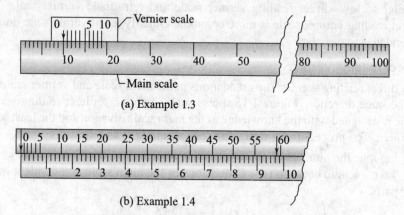

(a) Example 1.3

(b) Example 1.4

Fig. 1.16   Main scale and vernier scale

Double vernier

A double vernier is a vernier scale with its zero at the centre and extending in both directions from the centre (Fig. 1.17). It is used when there are two main scales running in opposite directions. The vernier to be used will run in the same direction as the main scale. Thus in Fig. 1.17, the left vernier is used with the top main scale. The right vernier will be used with the bottom main scale.

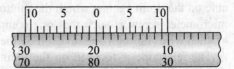

Fig. 1.17 Double vernier

Retrograde vernier

In a retrograde vernier, the graduations increase in the direction opposite to that of the main scale. In such a vernier, the value of the vernier scale division is more than that of the main scale. n divisions of the vernier scale are equivalent in length to $(n + 1)$ divisions of the main scale. The least count can be determined as follows. Using usual notations,

$$nv = (n + 1)m, \quad v = (n + 1)m/n$$

The least count in this case is $v - m = (n + 1)m/n - m = m/n$, same as in a direct vernier.

Extended vernier

Extended verniers are used when the main scale divisions are very close and it is difficult to judge the coincidence. In this case $(2n - 1)$ divisions of the main scale are divided into n divisions of the vernier scale (Fig. 1.18). The least count in this case is also m/n.

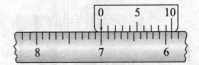

Fig. 1.18 Retrograde vernier

Vernier division $v = (2n - 1)m/n$

Least count = difference between two main scale divisions and one vernier division

$$= 2m - (2n - 1)m/n = m/n$$

Circular scale vernier

In most survey instruments, the main scale is of angles and is engraved or etched on a circular plate. The main scale being a circular scale, the vernier is also a circular scale. The vernier scale of a normal vernier theodolite

Fig. 1.19 Circular scale vernier

reading 20″ is shown in Fig. 1.19. The main scale is divided into 20′ divisions. In this case, 59 main scale divisions are taken and divided into 60 divisions on the vernier, giving a least count of 20″.

1.10.5 Micrometer Microscope

Vernier scales cannot be used for very fine reading as the divisions become too close. Angles up to 10″ can be read from a vernier scale. If greater precision is required, a micrometer microscope is used (Fig. 1.20). A micrometer microscope uses a small low-power telescope having an objective and an eyepiece. A frame carrying a single wire or two vertical closely spaced wires is kept within the telescope. This frame can be moved at right angles to the telescope axis with the micrometer drum, which is graduated along the perimeter. An index is available on the drum against which the micrometer reading is taken. The pitch of the micrometer screw is such that one complete revolution of the drum moves the vertical wires through one division on the main scale.

The telescope is focused on the main scale. The view from the telescope is shown in Fig. 1.20. When an angle is read, the two vertical wires may fall within two graduations of the main scale. The micrometer drum is then rotated to bring the nearest vertical scale graduation exactly within the two wires. The movement of the drum gives the fractional part of the angle to be added to the main scale reading.

1.11 Telescope

The surveyor's telescope is a component common to many instruments such as levels, alidades, and theodolites. A telescope is used to view distant objects clearly. The basic principle of a telescope is shown in Fig. 1.21.

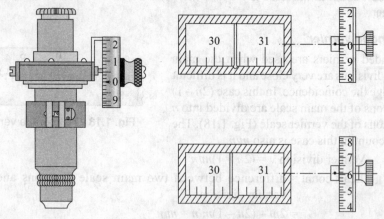

Fig. 1.20 Micrometer microscope

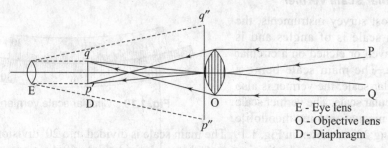

E - Eye lens
O - Objective lens
D - Diaphragm

Fig. 1.21 Surveyor's telescope—Optical diagram

The main parts of a telescope are the following.

Objective lens This is a concave lens placed in front of the telescope. The objective lens forms an image at the diaphragm, which has cross hairs at right angles. To place the image at the diaphragm, the telescope has to be focused.

Diaphragm This is a circular glass piece in which the cross hairs are etched. The diaphragm is a module with adjustment screws placed along the horizontal and vertical diameters. The diaphragm is placed at the focal distance of the objective. From the optical ray diagram, it is clear that the image is inverted and real.

Eyepiece The eyepiece at the rear end of the telescope is again a lens that focuses on the diaphragm. When focused, the cross hairs and the image are clearly visible to the observer. The telescope tube is a cylindrical metal tube which houses these parts.

Focusing Focusing the telescope means bringing the image within the plane of the cross hairs. There are two types of telescopes depending upon how the focusing is done. These are discussed in the followings sections.

1.11.1 External Focusing Telescope

The external focusing telescope has two separate tubes—one holding the objective lens and the other holding the diaphragm and the eyepiece. Focusing is done by increasing or decreasing the distance between the objective and the diaphragm [Fig. 1.22(a)]. This is achieved by a focusing screw which activates a rack and pinion arrangement. The length of the telescope changes with focusing.

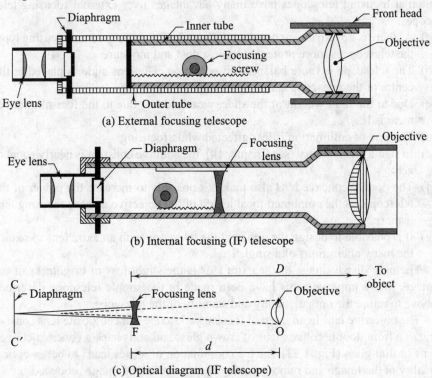

(a) External focusing telescope

(b) Internal focusing (IF) telescope

(c) Optical diagram (IF telescope)

Fig. 1.22 External and internal focusing telescopes

1.11.2 Internal Focusing Telescope

The internal focusing telescope uses a different mechanism for focusing. An additional double concave lens is used between the objective and the diaphragm [Fig. 1.22(b)]. The movement of this lens with the focusing screw brings the image in the plane of the diaphragm. The additional lens means that the length of the telescope does not change while focusing. At the same time the brightness of the image is slightly reduced due to absorption in the additional lens. The optical diagram of the internal focusing telescope is shown in Fig. 1.22(c). O is the objective lens, F is the double concave focusing lens, and CC′ is the plane of the diaphragm.

The conjugate focal lengths of the objective lens are f_1 and f_2. Conjugate focal lengths are the distances of the object and its image from the optical centre of the lens. By moving the focusing lens F suitably, the image is brought within the plane of the diaphragm. If d is the distance between the two lenses, $(D - d)$ and $(f_2 - d)$ are the conjugal focal lengths of the double concave lens. From the lens formula, if f and f' are the focal lengths of the objective and focusing lenses, we have $1/f = 1/f_1 + 1/f_2$ for the objective lens and $1/f' = 1/(D - d) + 1/(f_2 - d)$ for the focusing lens.

The design of the focusing lens and its distance from the objective can be determined using this formula.

Advantages of internal focusing telescopes

Internal focusing telescopes have many advantages over external focusing telescopes.

(a) As there is no movement between sliding tubes, as in an external focusing type, the telescope is more protected against dust and moisture.

(b) The telescope is more balanced in operation. The lens slide is placed at the centre of the telescope and is lightweight.

(c) Due to the light weight of the slide, wear and tear due to the focusing movement is less.

(d) The line of collimation is less affected while focusing.

(e) In stadia tacheometry (see Chapter 14), the additive constant is nearly equal to zero.

(f) The double concave lens also makes it possible to increase the power of the telescope, as the combined focal length of the objective and the focusing lens is more.

(g) It is possible to design the optics of the telescope with an extra lens to reduce the many aberrations of a single lens.

The only disadvantage of the extra lens is the slight loss of brightness of the image. Many improvements have been made in the simple telescope described above to refine the optical quality and stability of the telescope.

The objective lens in an internal focusing telescope is a composite lens consisting of a front double convex lens of crown glass and at its back a concavo-convex lens of flint glass (Fig. 1.23). Such a combination of lenses leads to better optical quality of the image and removes many aberrations in the image produced.

1.11.3   Eyepieces

There are many types of eyepieces used. The following is a discussion on some of the common types of eyepieces.

Ramsden's eyepiece

Ramsden's eyepiece is the most commonly used eyepiece. It consists of two equal plano-convex lenses as shown in Fig. 1.24(a). The

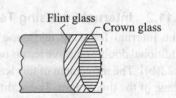

Flint glass
Crown glass

Fig. 1.23   Compound objective lens

two lenses are placed two-thirds of their focal length apart. This eyepiece is a non-erecting eyepiece as the image formed by the objective is seen as it is formed by the objective.

Huygen's eyepiece

Huygen's eyepiece consists of two plano-convex lenses but the two lenses are not of equal size as shown in Fig. 1.24(b). One is larger than the other. The lenses are so adjusted that the distance between them is equal to two-thirds of the focal length of the larger convex lens or twice the focal length of the smaller lens. This is also a non-erecting eyepiece but is not very common in survey equipment.

Erecting eyepiece

An erecting eyepiece consists of four lenses as shown in Fig. 1.24(c). The four lenses provide a magnified and erect image of the object formed by the objective lens. The four lenses of the eyepiece themselves invert the image formed by the objective. There is a loss of brilliance due to the additional lenses.

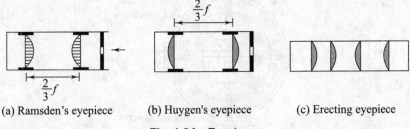

(a) Ramsden's eyepiece   (b) Huygen's eyepiece   (c) Erecting eyepiece

Fig. 1.24   Eyepieces

Diagonal eyepiece

The diagonal eyepiece (Fig. 1.25) is useful when the line of sight becomes very steeply inclined as in astronomical observations or even during ordinary survey work when the difference in elevation between points is very large. Using a reflecting prism the image formed at the cross hairs is seen at an angle of 45° to the telescope axis.

Ramsden's eyepiece is preferred because it forms a brighter image and the inversion of the image is not a problem. The person looking through the telescope gets used to the inverted image very soon.

1.11.4   Diaphragm

The diaphragm in modern telescopes consists of a reticule held in a circular frame with four adjusting screws diametrically placed along the horizontal and vertical directions (Fig. 1.26). In levels, the diaphragm may have screws only along the vertical diameter, as the adjustment of the horizontal hair is only important in levelling. The four capstan headed screws are used to adjust the position of the cross hairs during permanent adjustment. One screw is loosened and the diagonally

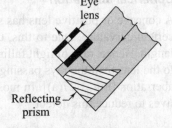

Fig. 1.25   Diagonal eyepiece

opposite screw is tightened. The cross hair moves in the direction of screw being tightened.

The cross hairs may be of many different patterns as shown in Fig. 1.26(b). Generally, the central cross hair is used in levelling operations. The two outer horizontal cross hairs are required for tachometry. Often, two vertical cross hairs are provided so that the plumbness of a vertical staff held at a station can be checked. The cross hairs can be platinum wires or lines etched in glass. When the eyepiece is focused on the diaphragm, the cross hairs are seen magnified. The focusing of the objective brings the image in the plane of the diaphragm so that the image can be viewed along with the cross hairs.

The line joining the optical centre of the objective lens to the optical centre of the eyepiece forms the axis of the telescope. The line joining the centre of the cross hairs to the objective forms the line of sight. These two axes coincide in the telescope if it is poorly adjusted.

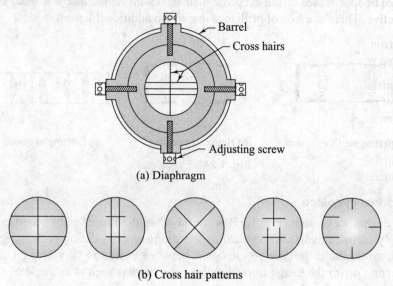

(a) Diaphragm

(b) Cross hair patterns

Fig. 1.26 Diaphragm and cross hairs

1.11.5 Optical Defects

The following is a discussion on two important optical defects of a single lens that affect the quality of image.

Spherical aberration

A concave or positive lens has surfaces that are ground to spherical surfaces of a certain curvature. Due to this, the thickness of the lens is not uniform across its length. Hence, rays of light falling near the edges are refracted more and fall closer to the lens than the rays passing through lower layers. This is known as spherical aberration [Fig. 1.27(a)]. In most telescopes, compound lenses are used as objectives to reduce this defect.

Chromatic aberration

The refraction by a lens depends upon the frequency of the light. As white light is composed of many colour components, from red to violet, a beam of white light does not converge at a point but is dispersed at many focal points. The violet ray is refracted the most and converges closer to the lens. The red ray is refracted the least and converges farther than the violet ray. This defect is known as chromatic aberration. Positive and negative lenses have opposite effects of chromatic aberration. A compound lens consisting of a concave lens and a convex lens makes it possible to reduce this defect.

1.11.6 Qualities of a Good Telescope

The following qualities are desirable in a good surveyor's telescope.

Absence of spherical aberration   The objective lens must be a compound lens to be free from spherical aberration. Invariably surveyor's telescopes have a compound objective lens. The compound lens is known as an *aplanatic combination*.

Achromatism   It refers to the absence of chromatic aberration. A compound lens is used in most telescopes to obtain aberration-free images.

Definition   It is the power of the telescope to produce a sharp and clear image. The line of collimation or axis of the telescope should contain the optical centres of all the lenses. The grinding and polishing of the lens also affect the definition.

Brightness   Brightness or *illumination* is the power of the telescope to produce a bright image. It varies inversely with the magnifying power. A higher magnification produces a faint and dull image.

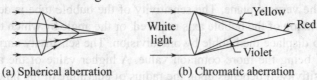

(a) Spherical aberration   (b) Chromatic aberration

Fig. 1.27 Lens aberrations

Size of field   Size of field or *field of view* is the whole circular area seen through the telescope. This is indicated by the angle formed at the optical centre of the objective lens by the diameter of the field of view. This angle varies from 1° to 2°. A large size of field is desirable as it helps to see distant objects in the field of view before bisecting them finely. The size of field is inversely proportional to the magnifying power. It does not depend upon the size of the objective lens and increases with the size of the eyepiece.

Magnification   The magnifying power of a telescope is taken as the ratio of the focal length of the objective to that of the eyepiece. The magnifying power should match the aperture of the instrument. A high value of the magnifying power for the aperture makes an image too faint. On the other hand, if the magnifying power is too low for the aperture, the objects will be too small to be sighted clearly. The magnifying power varies from 20 to 30 diameters but telescopes with magnifying power of 80 diameters are also available.

1.12 Spirit Level

A spirit level, also known as a *level tube* or *bubble tube*, is a part of many surveying instruments. A level must have a level tube for the purpose of levelling. A theodolite has

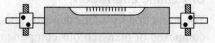

Fig. 1.28 Level tube

a plate bubble tube and an altitude bubble tube for the same purpose.

A level tube consists of a sealed glass tube fixed to a brass tube with plaster of Paris or any other adhesive (Fig. 1.28). The glass tube is nearly filled with ether, alcohol, or a mixture of similar spirit-like liquids, the remaining part being an air bubble. The air bubble occupies the highest point of the glass tube, the centre of the bubble always coming to rest at the highest point. The upper surface of the glass tube is finely ground to a curve such that the section through its centre is an arc of a circle. The glass tube is graduated, generally in 2-mm divisions from the centre in both directions. From the graduations, one can ascertain whether or not the bubble is in the centre, and if off-centre, by how many divisions. The level tube is fixed to the instrument using capstan-headed nuts which can be adjusted.

The level tube gives a horizontal line because a liquid surface is a level surface and is always at right angles to the direction of gravity. When the bubble is in the centre of its run, the line tangential to the bubble at the centre is a horizontal line. This line, tangential to the circular curve of the glass tube, is known as the *bubble line* or *axis of the bubble tube*. The length of the bubble is affected by temperature. The bubble is reduced in length by a rise in temperature, as the liquid expands.

Sensitivity of the bubble

Sensitivity is the ability of the bubble tube to show a small angular deviation of its axis in the vertical plane. The sensitivity of the bubble tube is defined as the angular deviation of the bubble axis required, or the angle by which the telescope is turned, to displace the bubble by one division. The sensitivity varies from $10''$ to $50''$, $20''$ being the more common value. A higher value of the angle means greater sensitivity. It depends upon the radius of curvature of the inner surface of the longitudinal section of the bubble tube. Greater the radius of curvature, greater the sensitivity. In addition, longer the bubble, greater the sensitivity. The sensitivity and radius of curvature (R) can be determined from a simple test as follows (Fig. 1.29).

1. Set up the instrument on fairly level ground and level the bubble to bring it to the centre. This can be done using the foot screws of the instrument (see Chapters 4 and 7). Keep a staff at about 50 m from the instrument and take a reading.

2. Using the foot screws shift the bubble by a number of divisions. Let n be the number of divisions by which the bubble is off-centre.

3. Take the reading on the staff in this position. Let the difference between the two readings be s.

4. If D is the distance between the instrument and the staff, then $\tan\alpha = s/D$ gives the value of the angle by which the line of sight has been turned. Since α is a very small angle, $\tan\alpha = \alpha$. If n is the number of divisions and l is the value of

one division, then $nl = R\alpha$. From the two equations, $R = nl/\alpha = nlD/s$. Sensitivity is the angular value of one division. Therefore,

$$\text{Sensitivity} = l/R = s/Dn \quad \text{(in radians)}$$
$$= (206{,}265)s/Dn \quad (1 \text{ rad} = 206{,}265\,\text{s})$$

Alternative method An alternative procedure (Fig. 1.30) for determining the sensitivity is as follows.

1. Set up the instrument on fairly level ground and level it. Hold a staff at 50 m from the instrument and take a reading.
2. Using the foot screws, shift the bubble to the left extreme end and take a reading on the staff held at the same point. Let the reading on the staff be s_1. Read the ends of the bubble. Let l_1 and r_1 be the readings at the left and right ends of the bubble, respectively.

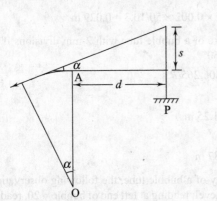

Fig. 1.29 Determining the sensitivity of the bubble tube

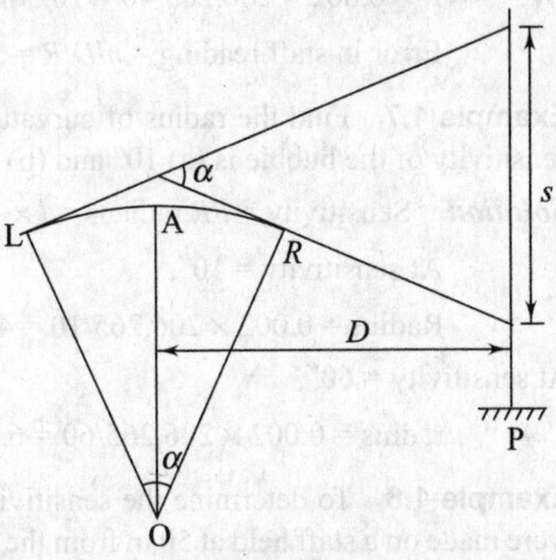

Fig. 1.30 Determining the sensitivity of the bubble tube—Alternative method

3. Shift the bubble to the extreme right end with the foot screws and take a reading on the staff. Let s_2 be the reading on the staff. Read the ends of the bubble. Let l_2 and r_2, respectively, be the readings at the left and right ends of the bubble in this case.
4. $(s_1 - s_2)$ is the change in the staff reading. The shift of the bubble is $(l_1 - r_1)/2$ in the first case and $(l_2 - r_2)/2$ in the second case. The total shift of the bubble is the arithmetic sum of the two.
5. If α is the angle by which the line of sight is rotated, then from Fig. 1.28, $\tan \alpha = \alpha$ (in radians) $= s/D = nl/R$. Here, $s = (s_1 - s_2)$ and $nl = (l_1 - r_1)/2 + (l_2 - r_2)/2$.

Calculations of sensitivity and radius of curvature are illustrated with the following examples.

Example 1.5 To find the sensitivity and radius of curvature of a bubble tube, the following observations were made: bubble at the centre, reading on staff held at 80 m from instrument = 1.675 m; bubble shifted 5 divisions off-centre, reading on staff = 1.705 m; length of a division of the bubble = 2 mm. Find the radius of curvature and sensitivity of the bubble.

Solution   Radius of curvature $R = nlD/s$. $n = 5$ divisions, $l = 2$ mm $= 0.002$ m, $D = 80$ m, and $s = 1.705 - 1.675 = 0.03$ m.

$R = 5 \times 0.002 \times 80/0.03 = 26.67$ m

Sensitivity (angular deviation per division) $= 206{,}265s/nD$

$= 206{,}265 \times 0.03/(5 \times 80)$

$= 15.5$ s

Example 1.6   The sensitivity of the bubble tube in a level is 40″ per 2 mm division. Calculate the error in the staff reading taken at a distance of 50 m if the bubble was off-centre by three divisions at the time of reading. Also find the radius of curvature of the bubble.

Solution $n = 3, l = 2$ mm $= 0.002$ m, $D = 50$ m.

Sensitivity $= l/R$ radians $= l \times 206{,}265/R$ s $= 40″$

$R = 0.002 \times 206{,}265/40 = 10.3$ m

Error in staff reading $= nlD/R = 3 \times 0.002 \times 50/10.3 = 0.029$ m

Example 1.7   Find the radius of curvature of a bubble tube with 2-mm divisions if the sensitivity of the bubble is (a) 10″ and (b) 60″.

Solution   Sensitivity $= l/R$ radians $= l \times 206{,}265/R$ s.

At sensitivity $= 10″$,

Radius $= 0.002 \times 206{,}265/10 = 41.25$ m

At sensitivity $= 60″$,

radius $= 0.002 \times 206{,}265/60 = 6.87$ m

Example 1.8   To determine the sensitivity of a bubble tube, the following observations were made on a staff held at 50 m from the level: reading at left end of bubble $= 20$, reading at right end $= 10$, staff reading $= 1.865$ m; reading at left end $= 10$, reading at right end $= 20$, staff reading $= 1.785$ m. Find the sensitivity and radius of curvature of the bubble tube.

Solution   The bubble is shifted by $(20 - 10)/2 = 5$ divisions in each case. Total shift of the bubble is 10 divisions.

Change in staff reading, $s = 1.865 - 1.785 = 0.08$ m

Radius of curvature $= nlD/s$ m, $n = 10$, $l = 0.002$ m, $D = 50$ m

Radius $= 10 \times 0.002 \times 50/0.08 = 12.5$ m

Sensitivity $= l/R$ radians $= l \times 206{,}265/R = 0.002 \times 206{,}265/12.5 = 33″$

1.13   Mistakes and Errors

Errors and mistakes do happen in measurement. Mistakes happen due to poor judgement or carelessness. These can be corrected if detected. To avoid mistakes, care should be taken while using the instruments and taking and recording readings. Errors, on the other hand, occur due to many reasons. Errors can be minimized but not avoided. These can be classified as follows.

Instrumental errors   These are errors which occur due to the malfunctioning of the instrument. Incorrect graduations on a tape is an example. These can be brought within the desired limits of accuracy by applying proper corrections.

Personal errors   These errors are due to limits on human perception. Reading of a vernier scale by judging the coincidence of graduations can be a source of error. Similarly, sighting and bisecting an object accurately can be a source of error.

Errors due to natural causes   These occur due to temperature variations, magnetic substances, refraction, etc. Such errors can be controlled by using proper methods, corrections to measurements, and precautionary measures.

In any measurement taken in survey work, the true value of the quantity is never known. Therefore, the true error is also unknown. By employing proper instruments, methods, and precautions, one can minimize the errors and obtain values very close to the true values. The accuracy required to be obtained in different types of survey work is different. The surveyor should be familiar with such requirements, the sources of errors, and the appropriate precautions to be taken. These are detailed out in most of the following chapters while dealing with different instruments and the methods to use them.

Precision and accuracy   The terms precision and accuracy are commonly used with surveying instruments and the measurements made using them. Precision is the degree of fineness with which instruments are manufactured, while accuracy is the precision actually obtained by using the instruments.

Random errors   These are the errors that remain after all other types of errors are eliminated. These errors can be positive or negative. These are analysed using the theory of probability (see Chapter 18).

1.14   Units of Measurement

In survey work lengths, angles, and time are the main quantities measured. SI units are used and are mandatory in maps. For a description of SI units, refer to Appendix 1.

Lengths are expressed in metre and its multiples such as kilometre. Angles are measured in two units—the hexagesimal system and the centesimal system. In the hexagesimal system, the units used are degrees (°), minutes ('), and seconds ("). $60' = 1°$ and $60'' = 1'$. 1 circumference = 360°. The centesimal system is becoming more popular now, as the degree can be expressed easily in decimals. In this system, 1 circumference = 400 grads represented as 400^g. One grad = 100 centigrads (100^{cg}) and 1 centigrad = 100 centi-centigrads (100^{ccg}). The centesimal system is more popular in European countries whereas the hexagesimal system is widely used in USA, UK, India, and many other countries. Another unit of angle is the radian. One radian is the angle subtended by an arc of length equal to the radius of a circle at the centre of the circle.

$$1 \text{ rad} = 180/\pi = 57.2957° \text{ in the hexagesimal system}$$
$$= 200/\pi = 63.662^g \text{ in the centesimal system}$$

1.15   Plotting

Survey data can be transferred onto paper by plotting. The plan or map is drawn on a tracing paper to some standard scale. The drawing is first done with pencil. It is then inked to get darker lines for the multiplication process. The drawing can be multiplied by ammonia printing or other multiplying processes to get a number of hard copies. Standard symbols, are used in plotting.

With the advent of computers, colour printers, and user-friendly graphic software, the drawing work has become greatly simplified. Plans are now drawn with computer software, edited electronically, and modified and stored very easily. Hard copies can be obtained with a printer at the press of a button. Such processes have not only simplified the drawing work, but also enabled corrections to be applied easily. Storing the drawings in any auxiliary storage device such as a CD also takes less space and makes them more secure from loss. Manual drawing and inking will be completely replaced by such techniques in the future.

Symbols These are used to represent various features in a map or plan such as roads, railway lines, buildings, and fences. The symbols used have become more or less standardized by custom. Some of the symbols used are given in Appendix 2. It is important to ensure that the size of the symbol matches the scale of the drawing.

Numerical values In surveying we deal with numbers such as values of distances and angles. For very accurate computations, seven-figure logarithm tables were used in earlier times. With the easy availability of calculators now, manipulations using numbers have become very simple. Twelve-digit calculators are quite inexpensive and easily available. However, one should realize the significance of figures in the computed value. In field measurements, when a quantity is measured, say, the length of a line, the rightmost digit is not very reliable and is subjected to error. For example, if the length of the line is 321.45 m, the digit 5 is prone to error and is not very reliable. The maximum error in the value is 0.05, while the probable error is 0.025.

The number of significant figures in values computed by the mathematical processes of addition, subtraction, multiplication, division, squaring, or raising to a power cannot be more than the lowest number of significant digits used in arriving at the answer. Similarly, the values of angles and distances measured must be consistent. Thus, if angles are measured to a certain accuracy, the distances should also be measured to an accuracy level consistent with it. For example, if angles are measured to an accuracy of $1''$, distances need to be measured to a greater accuracy than when the angles are measured to an accuracy of $1'$. The time and effort spent in measuring can be saved if this fact is kept in mind.

Summary

Surveying is the art of locating points on the surface of the earth in terms of their horizontal and vertical distances. Any part of the surface of the earth is a spherical surface. In ordinary surveying, the spherical nature, or curvature, of the earth's surface is neglected. In precise surveying, called geodetic surveying, this curvature of the surface is taken into account. The objectives of surveying are to collect data about points on the surface of the earth, prepare plans and maps, calculate areas and volumes, and transfer points from plans to the ground.

The art of surveying is based on two cardinal principles—working from whole to part and locating points by at least two measurements. Surveying can be classified into four categories—(i) based on the instruments, (ii) based on the methods, (iii) based on the purpose, and (iv) based on the place of survey.

A basic objective of surveying is to prepare plans and maps of areas. These are prepared to the standard scales prescribed. Scales are used for representing horizontal distances in plan, either as numerical scales or as representative fractions. They are also represented graphically on maps. Scales are of many types—plain scale, diagonal scale, chord scale, and vernier scale. Vernier scales find use in many surveying instruments, to read the fractional part of a main scale division. When greater accuracy than what can be provided by a vernier is needed, micrometer microscopes are used.

Telescopes are a common component in many surveying instruments. Telescopes basically consist of an objective lens, forming an image in the plane of the diaphragm, and an eyepiece that helps to see the magnified image. Modern telescopes are internal focusing telescopes that use a moving double concave lens for focusing, that is, converging the images of objects at different distances to the plane of the diaphragm with cross hairs. The objective lens is made using a compound lens in order to obtain a sharp and clear image. The image is seen magnified and inverted. The qualities of a good telescope are achromatism, absence of spherical aberration, good definition, magnification, and brightness of image.

A bubble tube is also a necessary part of many surveying instruments. The bubble tube contains a bubble which when centred indicates that the instrument is level. The sensitivity of a bubble tube is the angular value of one division of the bubble tube.

Plans and maps are plotted to standard scale. These should use standard and suitable symbols. Plans and maps are now being prepared using graphic software, thus eliminating pencil or inked drawings.

Errors in surveying can be classified into instrumental errors, personal errors, and errors due to natural causes. While mistakes can be eliminated by taking care, errors can only be reduced but not eliminated. Random errors are analysed using statistical methods.

While expressing numerical values from computations, one should determine the number of significant digits that can be relied upon in the answer. The figures should be rounded off to the lowest number of significant digits used in arriving at the answer.

Exercises

Multiple-Choice Questions

1. A scale of 1 cm = 3 km is represented as a representative fraction as
   (a) 1: 3000   (b) 1: 30000   (c) 1: 300000   (d) 1: 3000000
2. The scale for a map is given as 1/100000. The scale can be represented as 1 cm =
   (a) 1 m   (b) 10 m   (c) 100 m   (d) 1000 m
3. A diagonal scale is used for measuring
   (a) units and one-tenths of units   (c) diagonals of closed polygons
   (b) units, tenths and hundredths of units (d) angles between lines in plan
4. A chord scale is used for measuring
   (a) angles or setting out angles   (c) curved lines
   (b) straight lines   (d) areas
5. The least count of a vernier is the
   (a) value of the division on the main scale
   (b) value of division on the vernier scale
   (c) sum of values of one vernier scale and one main scale divisions.
   (d) difference between the values of one main scale division and one vernier scale division
6. A vernier scale is made using a main scale of one metre to read mm. If the vernier scale is divided into cm divisions, the vernier will have

     (a) 10 divisions for 9 main scale divisions
     (b) 11 divisions for 10 main scale divisions
     (c) 20 divisions for 19 main scale divisions
     (d) 21 divisions for 20 main scale divisions

7. In a theodolite, the circular main scale is divided into degress and each degree is divided into three parts. If the theodolite is to have a least count of 20″, then _____ main scale divisions are divided into _____ vernier scale divisions.
     (a) 19, 20   (b) 29, 30   (c) 49, 50   (d) 59, 60

8. In a theodolite, the main scale is divided into 10′ divisions. If the least count has to be 10″, then _____ main scale divisions are divided into _____ vernier scale divisions.
     (a) 59, 60   (b) 49, 50   (c) 39, 40   (d) 29, 30

9. Which of the following statements is true?
     (a) An external focusing telescope has an internal lens for focusing.
     (b) An internal focusing telescope is focused by the sliding of two tubes of the telescope.
     (c) An internal focusing telescope has an internal lens for focusing.
     (d) Both external and internal focusing telescopes use the same focusing mechanism.

10. Chromatic aberration in a telescope is reduced by using
     (a) a convex lens   (c) a compound lens of convex and concave lenses
     (b) a concave lens   (d) two convex lenses

11. For a telescope, the size of field is defined as the
     (a) diameter of the objective lens
     (b) diameter of the eye piece
     (c) whole circular area seen through the telescope
     (d) length of the telescope

12. Sensitivity of a spirit level is defined as the
     (a) error in the reading on a staff held at 100 m per one division
     (b) size of the bubble in the tube
     (c) angle by which the telescope is turned when the bubble becomes off-centre by one division
     (d) number of divisions by which the telescope turns when telescope is turned by one second

13. Spherical aberration is the
     (a) variation in the shape of the earth from a true sphere
     (b) error in survey data due to curved surface of the earth
     (c) change in the image of a sphere seen through the telescope
     (d) variation in the refraction of rays on the different parts of the lens

14. In a spirit level having 2 mm divisions, the readings on a staff held at 100 m were (*i*) 1.875 m when the bubble is central and (*ii*) 1.905 when the staff is 4 divisions off centre. The sensitivity of the spirit level is
     (a) 15.5″   (b) 19.35″   (c) 16.5″   (d) 17.2″

15. A spirit level has 2-mm divisions. The radius of curvature of the tube with sensitivity of 20″ is
     (a) 18.65 m   (b) 19.35 m   (c) 20.62 m   (d) 21.58 m

16. The sensitivity of a bubble tube is 30″ per 2-mm divisions. If the bubble was off-centre by 4 divisions, the error in the reading taken at a distance of 80 m is
     (a) 0.038 m   (b) 0.046 m   (c) 0.051 m   (d) 0.058 m

Review Questions

1. Write a short note on the objectives of surveying.
2. Explain the two basic principles of surveying.
3. Give a classification of surveys and explain briefly the various types of surveys.

4. What is the difference between a plan and a map?
5. Explain the terms scale and representative fraction.
6. Explain the construction of a diagonal scale.
7. Explain the construction of a chord scale. Add a note on its use.
8. Write a brief note on surveying practice.
9. Describe the qualities of a good surveyor.
10. Draw a neat sketch and explain the function of a telescope.
11. Write briefly about the objective lens, diaphragm, and eyepiece of a telescope.
12. Explain the term sensitivity of the bubble tube.
13. Write a brief note on errors in surveying.
14. Distinguish between precision and accuracy.
15. Enlist the steps to be taken to minimize errors in surveying.

Problems

1. Construct a plain scale 1 cm = 5 m to read upto a metre and represent 34 m on the sketch.
2. Construct a diagonal scale 1:200,000 and represent 36.8 km on it.
3. If the RF of a plain scale is 1/800, construct the scale and show 156 m on the scale.
4. A metre scale is divided into 10 parts and each part is further divided into 10 divisions. Design a vernier to read up to 1/1000th of a metre.
5. If the main scale is divided into centimetres and millimetres, design a vernier to read up to 0.1 mm.
6. The horizontal plate of a theodolite is divided into degrees and each degree is further divided into 3 parts. Construct and show in a neat sketch a vernier to read up to (a) 30″ and (b) 20″.
7. If the main scale of a theodolite is divided into degrees and each degree is divided into 6 parts, design and sketch a vernier to read upto 10″.
8. A bubble tube has 2 mm divisions. When tilted through an angle of 0.5′, the bubble moves through 3 divisions. The change in the staff reading at a distance of 50 m is 0.03 m. Find the sensitivity and radius of curvature of the tube.
9. When the bubble is central, the reading on a staff held at 100 m is 1.685 m. When the bubble is off-centre by 4 divisions, the reading on the staff held at the same point is 1.715 m. If the length of a division on the tube is 2 mm, find the sensitivity and radius of curvature of the tube.
10. The sensitivity of the bubble tube of a level is 25″ per 2-mm division. Determine the error in the staff reading when the staff is held at 100 m and the bubble is off-centre by 4 divisions. Also find the radius of curvature of the tube.
11. Find the angular value of one 2-mm division of a bubble tube if its radius of curvature is 20 m.
12. Find the radius of curvature of a bubble tube for which the sensitivity of a 2-mm division is (a) 10″ and (b) 40″.
13. Find the radius of curvature and error in the staff reading of a bubble tube due to bubble being out of centre by 6 divisions, if the sensitivity of the bubble tube is (a) 30″ and (b) 1′. The staff is held at a distance of 50 m and a division of the bubble tube is 2-mm long.
14. When a telescope was tilted in one direction, the left and right end readings of the bubble were 18 and 6. The staff reading was 2.815 at 100 m. When tilted in the opposite direction, the left and right end bubble readings were 6 and 18 and the staff reading was 2.735. Find the sensitivity and radius of curvature of the bubble tube.

CHAPTER 2

Measurement of Distance—Chain Surveying

Learning Objectives

After going through this chapter, the reader will be able to

- explain the methods of measurement of distance
- explain the construction of a standard chain and tape
- explain the method for ranging and measuring the length of a survey line
- enlist the different types of errors that can occur during the measurement of distance using a chain or tape
- explain the corrections in measurement with a chain or tape
- explain the different types of equipment and their functions for chain triangulation or traversing
- describe the methods of laying out chain angles to determine the directions of chain lines
- describe the methods of conducting a chain survey, booking field notes, and taking offsets
- explain the methods of overcoming obstacles to chaining
- explain the methods of plotting chain survey data
- explain the limits of precision in chain surveying

Introduction

The measurement of distances and angles is the key element of fieldwork in surveying. Distances can be measured by many methods. Using a chain or tape is the most common method, which will be detailed in this chapter. Angles are measured with respect to some reference line or even another line (any other selected direction, an arbitrary meridian). These two attributes of a line., i.e., its distance and angle, fix a line on the ground. In this chapter we will study the measurement of distance.

The SI unit of distance is the metre, which is used to measure small distances. Large distances can be measured in multiples of this unit, such as the kilometre. Very small distances can be measured in millimetres or centimetres. In regular survey work, the unit metre is commonly used. If the distances or lengths are specified in other systems of units, they can be converted into SI units using the conversion factors outlined in Appendix 1.

The term *chaining* in survey work is used to refer to measurement of distances with a chain or tape. The term *taping* can also be used for measurements with a tape. There are also techniques which measure distances with electronic instruments using waves. Tapes are more accurate and easier to carry and use than a chain. However, chains are ideally suited for rough use in difficult terrain.

2.1 Methods of Measuring Distance

There are basically two methods of measuring distances—direct measurement and indirect measurement. Direct measurement of distance uses instruments which directly measure distances along the ground, such as chains, tapes, and odometers. Indirect measurement of distance is done by instruments using radio or light waves such as electromagnetic distance measurement instruments (EDMs), geodimeters, and tellurometers and techniques using optical methods such as tacheometry.

2.1.1 Direct Measurement

Direct measurement of distance means actually measuring the distance along the ground by some means. The methods include counting of paces and using a wheel for measuring.

Counting of paces

One method is to directly count the number of paces while walking along the length to be measured. The distance covered by one pace by an individual more or less remains constant and can be used for a rough estimation of the distance.

Passometer   It is an instrument used to help in counting the number of paces made by an individual. It can be kept in the shirt or trouser pocket while pacing. The instrument records the number of paces made by the bearer.

Pedometer   It is similar to the passometer but has the facility of setting the pace distance of the individual. It records the number of paces and directly indicates the distance based on the pace distance input by the user.

Use of wheel-based instruments

A wheel can be rolled along the distance to be measured and the number of revolutions noted. Knowing the circumference of the wheel, the distance can be calculated as the product of the number of revolutions and the circumference. The perambulator and odometer, are wheel based.

Perambulator   It is an instrument which measures distance based on the number of revolutions of a wheel. It has a wheel connected to a handle (Fig. 2.1), as in a cycle wheel, for rolling the wheel along the distance. A counter is fitted to count and record the number of revolutions made by the wheel as it is rolled along. The counter can be set to zero before starting the measurement.

Odometer   It is an instrument fitted to most automobiles to measure the distance travelled by the vehicle. It records the number of revolutions made by the wheel of the vehicle and computes the distance as the product of the circumference of the wheel and the number of revolutions.

Speedometer   The speedometer is an instrument attached to most automobiles. As distance is a product of speed and time, if a constant speed can be maintained, then by observing the time of travel over the length, the distance can be calculated.

This of course needs a smooth path for maintaining the speed.

A chain or tape can be used to measure the distance more accurately than any of the above methods. The types of chains and tapes used are discussed in detail in the following sections.

2.1.2 Indirect Measurement

Fig. 2.1 Wheel-based distance measurement

Indirect measurement of distance involves methods such as tacheometry (Chapter 14), electromagnetic distance measurement (Chapter 23), and trigonometric levelling (Chapter 16). Tacheometry uses a theodolite fitted with a stadia diaphragm or a tacheometer to compute distances from intercepts of cross hairs on a staff or stadia rod. Electromagnetic distance measurement (EDM) uses light waves, radio waves, or microwaves to measure distances. Waves are transmitted and made to reflect from reflectors at a station and the phase difference between the transmitted and received waves is measured to compute the distance. Distomats and total stations are used for this. Trigonometric levelling involves measuring the vertical angle to a target and calculating the horizontal distance using trigonometry.

2.2 Metric Chain and Other Equipment

Decades ago, chains of 100 ft length were in common use. With the changeover to the metric system, metric chains have completely replaced the 100-ft chain. Metric chains can come in lengths of 5 m, 10 m, 20 m, and 30 m. The Bureau of Indian Standards (BIS) has published a code, IS 1492:1954, which provides the standards for the metric chain.

The metric chain consists of a 4-mm-diameter galvanized mild steel wire bent into rings at its ends. Two such links are connected by three circular or oval rings to provide flexibility in handling. The links, except the end links, are 200 mm long between the central rings at the ends [see Fig. 2.2(e)]. The end links are shorter in length due to the provision of handles. The two end links [Fig. 2.2(e)] are provided with brass handles connected by swivel joints to the links. This provides enough flexibility to hold, straighten, and turn the chain. The handle has a groove at its outer end to facilitate holding arrows and fixing them to the ground at the end or the beginning of the chain. To facilitate reading the fractions of a metre, appropriate markings are provided depending on the length of the chain.

5-m chain Tallies are provided at every metre length. Each link is 20 cm long. The shape of the tally will be different at different metre lengths as shown Fig. 2.2(f).

10-m chain Tallies are provided at every metre length. Each link is 20 cm long. The shape of the tallies is similar to that of a 5-m chain.

20-m/30-m chains These chains have 20-cm-long links and are provided with tallies at every 5 m length. To facilitate reading the fractions, brass rings are provided at every metre length (except where the 5-m tally is provided).

Flexibility in handling and operating with the chain is a prime factor to be considered for its manufacture. The brass handles with swivel joints and links with open rings at the ends provide for flexibility of operation. This also reduces the risk of links getting kinks easily. In some chains, the rings are welded. Though this reduces flexibility, it reduces the risk of stretching of the chain.

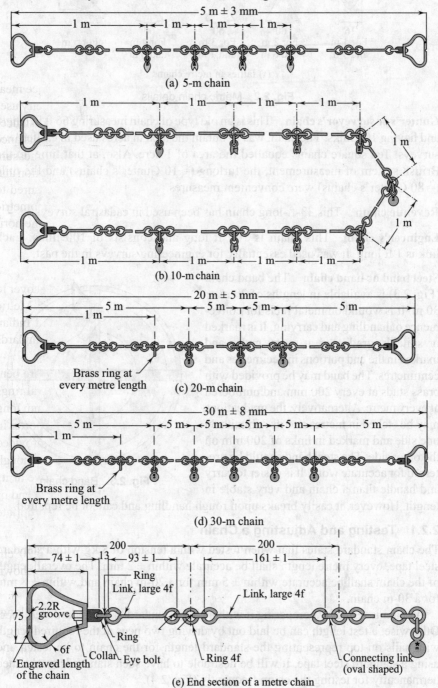

(a) 5-m chain

(b) 10-m chain

(c) 20-m chain

Brass ring at
every metre length

Brass ring at
every metre length

(d) 30-m chain

(e) End section of a metre chain

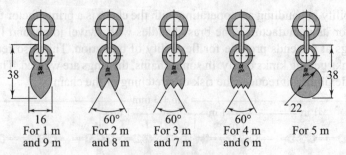

38				38
				22
16	60°	60°	60°	
For 1 m	For 2 m	For 3 m	For 4 m	For 5 m
and 9 m	and 8 m	and 7 m	and 6 m	

(f) Tallies in metre chains

Fig. 2.2 Metric chain details

Gunter's or surveyor's chain This is an old type of chain measuring 66 ft in length and having 100 links. The chain was to obtain the area in acres used in earlier land surveys. Ten square chains equalled the area of 1 acre. Also, at that time, in the British system of measurement, the furlong (= 10 Gunter's chains) and the mile (= 80 Gunter's chains) were convenient measures.

Revenue chain This 33-ft-long chain has been used in cadastral surveys.

Engineer's chain This chain is 100 ft long and consists of 100 links. Each link is 1 ft long. It was used essentially for engineering surveys in the past.

Steel band or band chain The band chain (Fig. 2.3) is available in lengths of 20 m or 30 m. It is wound on metal reels for convenience of handling and carrying. It is marked or suitably graduated at every metre and marked in the end portions in decimetres and centimetres. The band may be provided with brass studs at every 200 mm and numbered at every metre. Alternatively, the graduations may be etched in metres and its subunits on one side and marked in links of 200 mm on the other side. The steel band should be preferred for accurate work. It is easier to carry and handle than a chain and very stable in length. However, it easily breaks upon rough handling and cannot be repaired.

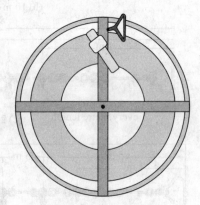

Fig. 2.3 Band chain

2.2.1 Testing and Adjusting a Chain

The chain standard states that when tested with a tension of 8 kg with a standard steel tape, every metre length shall be accurate within ± 2 mm. The overall length of the chain shall be accurate within ± 5 mm for a 20-m chain and within ± 8 mm for a 30-m chain.

A chain is tested by comparing it with a chain standard or a standard steel tape. Otherwise, a test length can be laid out by driving two pegs at the required length with nails on top representing the standard length for the chain to be tested and using a standard steel tape. It will be desirable to have such standards established permanently for testing the chain frequently (Fig. 2.4).

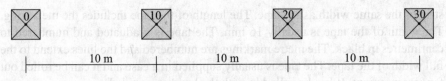

Fig. 2.4 Test bed for chains

A chain can get shortened by the bending of links and clogging of the connecting links, thus preventing them from stretching fully. It can get elongated by the wear and tear of the many wearing surfaces, by the stretching of the links, and the opening out of the connecting rings. The chain thus needs to be tested frequently for its reliabilty.

An elongated chain can be adjusted by
(a) closing up the connecting rings if they have opened out;
(b) hammering to bring back into shape,
(c) adjusting the end links,
(d) replacing the existing rings by new rings,
(e) removing some small rings.

If the chain has become shorter, it can be adjusted by
(a) straightening the links that have become bent,
(b) cleaning the joints between the links if they are clogged with mud,
(c) stretching some rings,
(d) replacing some rings,
(e) using the adjustable links at the ends.

The adjustments should be done symmetrically with respect to the centre so that the centre and the tags in the rings retain their correct positions.

2.2.2 Tapes

Tapes are made of many materials and have the advantage of being lightweight and providing greater precision. They cannot stand very rough handling as can chains. They can be divided into four classes as detailed below (see Fig. 2.5).

Cloth or linen tape

Tapes made of cloth or linen can be used for taking offset measurements. However, they find little use in surveying due to their tendency to stretch easily, twist, and shrink when wet.

Metallic tape

Metallic tape is made of yarn and metallic wire, the yarn and metallic wire in the warp and only the yarn in the weft. It comes in lengths of 2, 5, 10, 20, 30, and 50 m. Longer tapes have a metal ring attached at the outer end of the tape, which is fastened to the tape by a metal strip of the same width as the tape. The outer end is also reinforced for about 10 cm by a leather or plastic

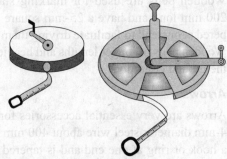

Fig. 2.5 Steel tape

strip of the same width as the tape. The length of the tape includes the metal ring. The width of the tape is usually 16 mm. The tape is graduated and numbered to centimetres in black. The metre markings are numbered and the lines extend to the full width of the tape. The tape is usually supplied in a case and it can be rolled out for measurement. It can be rolled in with the handle provided.

Steel tape

Metric steel tapes can come in denominations of 1 to 50 m. The tape is made of steel or stainless steel and can have a protective vinyl coating to prevent rust. The tape has a metallic ring at the outer end, which is fastened with a metal strip of the same width as the tape. Steel tape is costly compared to the former types but is a very accurate distance measuring instrument. The length of the tape includes the metal ring.

The tape is suitably numbered and graduated to metres, decimetres, and centimetres. The outer end of the tape may also be marked to millimetres. The tape comes in a suitable case and can be pulled out to the required length with the metallic ring. The tape winds back automatically when a button on the case is pressed. The steel tape should be preferred for precision work in distance measurement.

Invar tape

Invar tape is made of an alloy of steel and nickel (about 36 per cent) and is used for very high precision work, as in base line measurement for triangulation. Invar tape, though costly, is characterized by a very low thermal coefficient of expansion, about one-tenth of that of steel. The coefficient of thermal expansion can be as low as 0.6×10^{-6} per degree centigrade. It comes in lengths of 30 m, 50 m, and 100 m and is about 6 mm wide. It is graduated throughout and the end lengths are graduated to millimetres. It is very delicate requiring careful handling and is hence used for high precision work only.

2.2.3 Other Accessories for Chaining

While measuring distances with a chain or tape some accessories are essential. Some of these are described below (Fig. 2.6).

Wooden pegs

Wooden pegs are used for marking stations. They are generally 150 mm to 200 mm long and have a 25-mm square cross section [Fig. 2.6(a)]. They are tapered at one end to facilitate driving them into the ground with a wooden hammer. Wooden pegs of longer lengths and heavier sections are also used depending upon the ground conditions.

Arrows

Arrows are very essential accessories for chaining a distance. They are made of 4-mm diameter steel wire about 400 mm long [Fig. 2.6(b)]. The wire is bent into a hook or ring at one end and is tapered to a point at the other. The ring assists in carrying the arrows while the pointed end helps in driving it into the ground.

Usually, a set of 10 arrows is carried with a chain and inserted at the end of a chain length. Chains have a groove on the handle to receive the arrows.

Ranging rods

Ranging rods, as the name itself indicates, are used for ranging. They are made of well-seasoned wood of deodhar, pine, sissoo, or other similar varieties of wood. They come in lengths of 2 m and 3 m and generally have circular cross sections 30 mm in diameter [Fig. 2.6(c)]. Consistent with their function of being distinctive and conspicuous, they are painted in alternate bands of white and red or white and black. The bands are 200 mm wide. The rod may also have a flag of a distinctive colour attached at the top for visibility from long distances. The lower end of the ranging rods are shod with a steel shoe for a length of 150 mm to protect the rod from wear and tear. Ranging rods are used for ranging and for marking stations for the purpose of intervisibility and ranging.

Ranging poles

Ranging poles serve the same purpose as ranging rods, except that they are longer, 4 m to 6 m in length, and have a larger cross section, with the diameter ranging from 60 mm to 100 mm. For distinctiveness and visibility, they too are painted in alternate bands of white and red or black and usually have a flag of distinctive colour attached to the top.

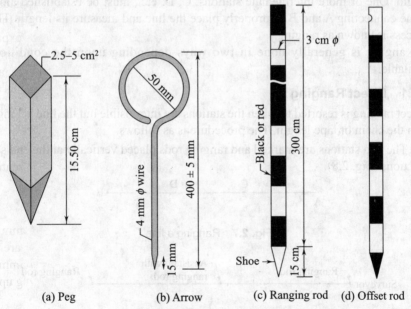

(a) Peg   (b) Arrow   (c) Ranging rod   (d) Offset rod

Fig. 2.6 Accessories for chaining

Offset rods

Offset rods are used to measure short offsets. They are similar in construction to ranging rods but are generally longer and have a larger cross section [Fig. 2.6 (d)]. Offset rods are generally 3 m in length and divided into 200-mm sections and painted in alternate bands of black and white. They have iron shoes at the bottom and may be provided with a hook or notch on one side to pull or push the chain. They also may have rectangular holes running through at right angles to help aligning offsets at right angles.

Laths and whites

Laths are short sticks, about 1 m long, made of soft wood used for marking the ends of chain lengths during ranging in place of a ranging rod. Their ends can be easily sharpened with a knife for driving into the ground. Whites are similarly rough sticks made out of wood, with sharp ends, and are used for the same purpose as laths. A groove is made on the top cross section for inserting a paper or card, making them more visible from a distance.

2.3   Ranging a Survey Line

Ranging a survey line means placing a line on the ground along the shortest distance between two points. When the line is shorter than a chain or tape and the two stations are intervisible, ranging is not a problem. However, if the line is longer than a chain or tape length or when the stations are not intervisible, ranging must ensure that the intermediate points established lie along the line joining the two stations. In Fig. 2.7, the two stations A and B are survey stations, are not clearly intervisible, and the distance between them is more than a chain or tape length. One or more intermediate stations, C, D, etc., must be established along a line connecting A and B to properly place the line and measure its length. This process is known as ranging.

Ranging is generally done in two ways depending upon the conditions available.

2.3.1   Direct Ranging

Direct ranging is resorted to when the stations are intervisible but the line is longer than the chain or tape length. The procedure is as follows.

1. The two stations are marked and ranging rods placed vertically at the end stations (Fig. 2.8).

C   D

A   • B

Fig. 2.7   Ranging a line

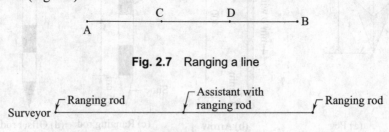

Fig. 2.8   Direct ranging

2. With the chain connected to the ranging rod at one end, an assistant holds a third ranging rod at the loose end of the chain vertically and moves it as per the instructions given by the surveyor at the other end. The instructions for the movement of the assistant to be in line with the ranging rods at the two

ends are given as per mutually agreed signs. Some suggested signals and the corresponding actions are listed in Table 2.1.

3. The surveyor should stand about 60 cm behind the end survey station (to which the chain is connected) in line with the ranging rod at the other end and direct the assistant. The assistant should hold the ranging rod vertically at arm's length.

4. With a few trials, the third ranging rod can be placed between the end ranging rods, in line with them.

5. The surveyor finally checks the alignment of the three ranging rods before chaining the line.

Table 2.1 Suggested signals and their meanings

Signal	Action
Rapid sweeps with the right hand	Move right fast
Slow sweeps with the right hand	Move right slowly
Right arm extended	Continue moving to the right
Right arm up and moved to the right	Plumb the ranging rod by moving to the right
Rapid sweeps with the left hand	Move left fast
Slow sweeps with the left hand	Move left slowly
Left arm extended	Continue moving to the left
Left arm up and moved to the left	Plumb the ranging rod by moving to the left
Both hands above head and moved down	Position correct
Both hands forward and brought down	Fix the point

Using a line ranger

To facilitate ranging a line which is longer than the chain or tape length with intervisible stations, a simple instrument called a line ranger can be used. A line ranger (Fig. 2.9) has two right-angled, isosceles prisms, one placed over the other. To establish an intermediate point P between and in line with two stations A and B, the observer has to approximately stand in line with the two stations, holding the line ranger at eye level. The two sides of the prisms are at right angles to the two stations. As the observer looks through the prisms, he sees two images of the ranging rods. The ranging rod at A is seen after reflection from the hypotenuse of one of the prisms. The ranging rod at B is similarly seen after reflection from the hypotenuse of the other prism. Two images will appear when the observer is not aligned with A and B. By moving with the line ranger, the observer can make the two images coincide and the point below the centre of the prism is a point in line with A and B.

One of the prisms is adjustable and can be used to test the accuracy of the instrument. To test the line ranger, establish three points along a line. Set up ranging rods at the end stations. Holding the line ranger over the middle point, check whether the ranging rods at the end stations are seen in one line. If not, adjust the prism to make their images coincide.

The line ranger is thus useful for establishing intermediate points without going to the end of the line and sighting from there.

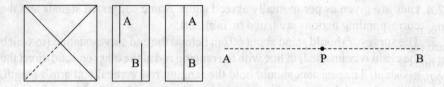

Fig. 2.9 Line ranger

2.3.2 Indirect Ranging

Indirect ranging is a method used when the stations are far apart and not intervisible. The method is also known as *reciprocal ranging*. The stations may not be intervisible due to intervening high ground or small hillocks. Consider the situation shown in Fig. 2.10. A and B are two stations with a large distance between them. A and B are not intervisible. The method is to select two intervening stations C and D such that from C both D and B are intervisible and from D both A and C are visible. The procedure is as follows.

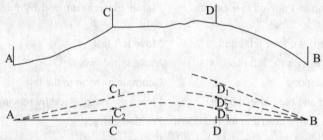

Fig. 2.10 Indirect ranging

1. With ranging rods placed vertically at A and B, two persons take up positions with ranging rods at C and D as closely aligned as possible with the line joining A and B.

2. The person holding the ranging rod at C then directs the person holding the ranging rod at D to come in line with A and B. The new position of the ranging rod at D is D_1.

3. Now the person with the ranging rod at D_1 directs the person holding the ranging rod at C to come in line with A and B. C takes up the new position C_1.

4. This process is repeated alternately by the persons holding the ranging rods at C and D. With a few trials, the ranging rods at C and D will be placed in the line of A and B.

5. Each person can finally check that the ranging rods are placed in line with AB by seeing that the two stations visible to them are in line.

2.4 Measuring Distances

In this section, we will study the procedures involved in measuring distances with a chain and a tape.

2.4.1 Measuring Distances with Chains

For measuring the distance between two points with the chain or tape, two persons are required. They are called *chainmen*. One of them is the *leader* or the person who is placed in the forward direction of the line. The other chainman is called the *follower* and he follows from behind. The follower has a much crucial role to play and his responsibilities are (i) to direct the leader to be in line with the ranging rod at the other end of the line, (ii) to give instructions to the leader, (iii) to hold the rear handle of the chain as it is being moved, and (iv) to pick up arrows or other marking accessories left by the leader on the ground. The duties of the leader are (i) to carry a ranging rod, a set of arrows, and the forward handle of the chain, (ii) to drag the chain forward, (iii) to attend to instructions from the follower, and (iv) to mark the length of a chain by inserting arrows.

The procedure to measure a distance with a chain is the following.

1. The two terminal stations are selected and marked with ranging rods placed vertically over the station points.
2. Both the leader and the follower begin from the starting point of the line. The first step is to unfold the chain. This is done by removing the leather strap or any other tying material used in folding the chain, holding both the handles, and throwing the chain in the forward direction of the line. If the chain has been folded correctly, it will unfold without entangling. The leader takes one end of the chain and walks forward, stretching the chain to its full length.
3. The chain must be inspected for any kinks or other faults that may change its length.
4. The follower keeps the chain at the station point, holds the chain firmly with both hands and directs the leader to stretch the chain fully and approximately in line with the ranging rod at the other end.
5. The leader will have a set of arrows, a ranging rod, and the forward handle of the chain. Bringing the chain approximately in line with the terminal stations, the leader holds a ranging rod at the end of the chain, standing to one side and facing the follower for instructions for ranging.
6. The follower, based upon the agreed signals, directs the leader to come in line with the terminal ranging rods.
7. Putting the chain in that direction, the leader inserts an arrow at the end of the chain length. If the ground is hard and the arrow cannot be instructed, the point is marked by cross lines made with the arrow or by using chalk.
8. The follower, holding the rear handle of the chain, and the leader, holding the front handle of the chain, both move forward in the direction of the line. The leader has to move slightly to one side of the line so that the arrow placed at the chain end is not disturbed by the moving chain. The follower now fixes the rear end of the chain at the arrow left by the leader.
9. The process of ranging, bringing the chain in line, and inserting the arrow is repeated, if required (when the line is more than two chain lengths long).
10. The last part of the line, less than one chain length long, is measured by laying out the chain till the end station and taking the reading along the chain.
11. The length of the line should be noted in the field book.

The line may be very long and the leader should start chaining with ten arrows in hand. The arrows are removed by the follower and the number of arrows in his hand indicates the number of full chain lengths measured. If all the ten arrows are used by the leader and collected by the follower (the line is more than ten chain lengths long), this should be noted down. The follower then hands over the ten arrows to the leader and the work is continued.

2.4.2 Measuring Distances with Tapes

The procedure for measuring lines with tape is exactly the same as that of chaining. Tape measurements are preferred nowadays due to the higher accuracy and precision in linear measurements achieved and the ease of handling due to the light weight of tape. The procedure is the following.

1. Two persons are required for measurement as in the case of chaining. One is the follower and the other is the leader. The two end stations of the line are carefully marked and ranging rods placed vertically at these points.
2. The follower holds the outer end of the tape and the leader moves forward holding the tape, 10 arrows, and one ranging rod. When the tape is wound out fully, the follower holds the outer end of the tape or metal ring if one is attached and sets it at the beginning of the line marked by a peg or arrow or any other marking.
3. The leader holds a ranging rod at arm's length at the end of the tape and stands facing the follower for directions for ranging the line.
4. The follower then directs the leader to be in line with the ranging rod at the other end. The leader fixes an arrow at the zero end of the tape.
5. The leader and the follower now move forward and repeat the procedure. The arrows left by the leader are collected by the follower. The number of arrows in the hand of the follower is an indication of the number of chain lengths measured.
6. If the leader uses up all the 10 arrows, this point is noted in the field book. The follower hands over all the arrows to the leader and the end of 10 tape lengths is marked by a ranging rod.
7. The fractional length (less than one tape length) at the end of the line is directly measured by the leader using the tape and noted in the field book.

2.5 Error Caused by Wrong Chain Length

The lengths 20 m or 30 m of chains are known as the designated lengths of the chains. Over a period of time the length of a chain changes due to the various reasons mentioned earlier. The length of the chain can increase or decrease. Consider the cases shown in Fig. 2.11.

In Fig. 2.11(a), the chain has a designated length of L. The actual length of the chain is L'. $L' < L$, which means that the chain has become shorter. *When a chain is shorter, a shorter distance is measured and incorrectly recorded as longer (the designated length).* Let L be 20 m and L' be 19.9 m. When the chain is laid out straight, we are actually measuring a length of 19.9 m only, which is the actual length of the chain. However, it is recorded as 20 m. Thus, *with a shorter chain the error has to be deducted or the error is positive and the correction is negative.*

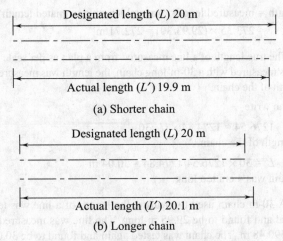

Fig. 2.11 Errors due to incorrect length

In Fig. 2.11(b), the chain has a designated length of L. The actual chain is longer, i.e., $L' > L$. *When a chain is longer, a longer distance is measured and incorrectly recorded as the designated length.* Thus, if $L = 20$ m and $L' = 20.1$ m, when we measure one chain length, we measure 20.1 m and record it as 20 m. Thus, *if the chain is longer, the error has to be added to the measured length or the error is negative and the correction is positive.*

If a line is measured with an incorrect chain, shorter or longer,

$$\text{Correct length} = (L'/L) \times \text{measured length}$$

where L' is the actual length of the chain and L is the designated length.

Correct area when lengths are measured with an incorrect chain Areas are measured in square units. If an area is measured with an incorrect chain, shorter or longer, the correct area can be determined as follows:

$$\text{Correct area} = (L'/L)^2 \times \text{measured (or calculated) area}$$

where L' is the actual length of the chain and L is the designated length of the chain.

Correct volume when lengths are measured with an incorrect chain When distances (length, breadth, and width) are measured with an incorrect chain, the volume calculated in cubic units will be incorrect. The correct volume can be calculated as

$$\text{Correct volume} = (L'/L)^3 \times \text{measured (or calculated) volume}$$

where L' is the actual length of the chain and L is the designated length of the chain.

The following examples illustrate these principles.

Example 2.1 A 30-m chain was tested before a survey and found to have a length of 29.93 m. If the length of a line measured with this chain was 273.35 m, find the true length of the line.

Solution As explained earlier, using a short length chain means that 29.93 m is being recorded as 30 m. Therefore, the true length of the line is less than the measured length.

True length = measured length × (actual length/designated length)

= 273.35 × (29.93/30) = 272.71 m

Example 2.2 The true length of a line measured from a plan as per scale was 1276.54 m. When the line was measured with a 30-m-long chain, the length was measured as 1274.84 m. Find the true length of the chain.

Solution We can write

1276.54 = 1274.84 × (L'/30)

where L' is the length of the chain.

L' = 30 × 1276.54/1274.84 = 30.04 m

Therefore, the chain was 4 cm too long.

Example 2.3 A 30-m chain used to measure the length of a line was tested before the line was measured and found to be 29.95 m long. The line was measured and the length was recorded as 590.48 m. The chain was tested again and found to be 30.08 m long. Find the true length of the line.

Solution The length of the chain would have changed gradually during measurement and hence the average length of the chain must be used to find the true length.

Average length of the chain = (29.95 + 30.08)/2

= 30.015 m

True length = measured length × (actual length/designated length)

= 590.48 × 30.015/30 = 590.77 m

Example 2.4 A rectangular plot was measured with a 20-m chain 12 cm too long. The lengths of the sides were recorded as 280 m and 420 m. Find the true area of the plot.

Solution The actual length of the chain = 20.12 m. The true lengths of the sides of the plot can be determined as follows:

Width recorded = 280 m

True width = 280 × 20.12/20 = 281.68 m

Measured length = 420 m

True length = 420 × 20.12/20 = 422.52 m

True area = 281.68 × 422.52 = 119,015.43 sq. m

Alternatively,

Area of the plot as measured = 280 × 420 = 117,600 sq. m

True area = measured area × (actual length/designated length)^2

= 117,600 × (20.12/20)^2 = 119,015.43 sq. m

Example 2.5 A 30-m chain was tested before starting the day's work and found to be 20 cm too short. After measuring a length of 1200 m, the chain was tested again and was found to be 10 cm too long. At the end of the day's work the chain was tested again and was found to be 30 cm too long. Find the true length of the line if the total length measured was 2648 m.

Solution The chain length being different at the beginning and the end of the measurement, the average length will be used as the actual length. Also, since the measurement has been done in two stages, the average length in the two stages will be used separately to calculate the true length.

(a) For the first 1200 m:

      Length before measurement = 30 – 0.2 = 29.8 m

      Length after measuring 1200 m = 30 + 0.1 = 30.1 m

      Average length = (29.8 + 30.1)/2 = 29.95 m

      True length of the first portion = 1200 × 29.95/30 = 1198 m

(b) For the next 1448 (i.e., 2648 – 1200) m:

      Length of chain at the beginning = 30.1 m

      Length of chain at the end = 30.3 m

      Average length = (30.1 + 30.3)/2 = 30.2 m

      True length of the second portion = 1448 × (30.2/30) = 1457.65 m

      True length of the line = 1198 + 1457.65 = 2655.65 m

Example 2.6 All the dimensions of an embankment were measured with a 20-m chain and the volume was calculated as 486.95 cu. m. It was then found that the chain was 10 cm too long. Find the true volume of the embankment.

Solution Measurement of volume involves three dimensions—length, breadth, and width. The linear error thus propagates in the cubical dimension. Thus,

$$\text{True volume} = \text{measured volume} \times (\text{actual length/designated length})^3$$
$$= 486.95 \times (20.1/20)^3 = 494.3 \text{ m}^3$$

Example 2.7 A plan drawn to a scale of 1:4000 was measured by a scale of 1:5000. Find the % error in the length and area measured.

Solution Let L be the length of the line in the plan.

      Actual length = $4000L$ m

      Measured length = $5000L$ m

      Difference = $5000L - 4000L = 1000L$ m

      % error = $(1000L/4000L) \times 100 = 25\%$

Since area is measured in terms of the square of length,

      Actual length = $4000L$ m

      Actual area ∝ $(4000L)^2$

      Measured area ∝ $(5000L)^2$

      Let actual area be $K \times (4000L)^2$

      Measured area = $K \times (5000L)^2$

      Difference = $K \times [(5000L)^2 - (4000L)^2]$

      % error in area = $\{K \times [(5000L)^2 - (4000L)^2]/[K \times (4000L)^2\} \times 100$

                 = $[(25 - 16)/16] \times 100 = 56.25\%$

Example 2.8 A plan was plotted to a scale of 1:2500. The paper has shrunk over a period of time so that a line 15 cm long originally now measures only 14.76 cm. It is also mentioned that the data used in plotting was measured with a 30-m chain 15 cm too long. If the area of the plotted plan now measures 98.68 sq. cm, find the true area of the land represented by the plot.

Solution Area has dimensions of square metres and thus corrections will be in terms of the square of the linear error. The correction due to the shrinking of the paper is calculated as follows:

True area= measured area × (true length of line/measured length of line)^2

True area of plan = 98.68 × (15/14.76)^2 = 101.915 cm^2

Correction to area due to incorrect length of chain is calculated as follows:

Actual length of chain = 30.15 m, designated length = 30 m

True area = measured area × (actual length/designated length)^2

$\qquad$ = 101.915 × (30.15/30)^2 = 102.936 cm^2

True area of land = 102.936 × (25)^2 = 64,335 m^2

2.6 Measurements Along Slopes

The terrain that has to be surveyed in general will not be a horizontal plane, whereas the distance required for the preparation of a plan or map is the horizontal distance. When distance is measured along a slope, it has to be converted into horizontal distance for the purpose of plotting. In general, there are two methods of chaining along a slope—one is to measure distances horizontally in steps and transfer the points to the ground and the second is to measure along the slope and convert the distances into horizontal distances.

2.6.1 Stepping or Measuring Horizontally

Suppose one has to measure the distance between the stations A and B shown in Fig. 2.12. The procedure is as follows.

1. At least two persons are required for the chaining. One end of the chain is held at A and a convenient length is selected and the chain is held horizontally. The chain tends to sag due to its own weight and has to be counteracted by applying sufficient pull to it so that it remains horizontal.

2. The follower holds the end of the chain at A. The leader goes along the line with a selected length of the chain and a ranging rod and faces the follower. The follower directs the leader to be in line (ranging) and both pull the chain to eliminate the sag of the chain.

3. The length of the chain is selected such that it can be held truly horizontal with the pull applied by hand. Once the chain has been stretched to be horizontal, the point C′ of the end of the chain is transferred to the ground using a plumb bob or a drop arrow or by dropping a stone. A *drop arrow* is an arrow weighted at the bottom so that it falls vertically when dropped. The point C is thus obtained. The distance between A and C is the length of the chain held horizontally.

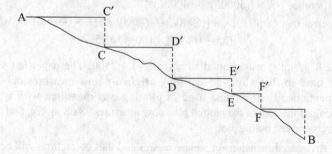

Fig. 2.12 Slope measurement by stepping

4. The process can be repeated starting at C to get points D, E, F, etc. till the end B is reached.
5. The total horizontal distance is the sum of the lengths of the number of steps taken to reach B from A.
6. In general, measurement using this method should be done downhill. If the distance has to be measured uphill, the second method is preferable, as the follower, while going uphill, has to direct the leader for ranging, hold the chain, and transfer the point to the ground, which is very difficult.
7. The horizontal distance (selected as the step distance) also depends upon the slope. For steep slopes, the distance has to be small.

2.6.2 Measuring Along the Slope

Length can be measured along the slope and the slope distance converted into horizontal distance. There are two ways to achieve this—one is to measure the slope either as an angle or a gradient and calculate the horizontal distance; the other is to measure a larger distance along the slope for a given horizontal distance.

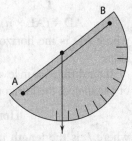

Fig. 2.13 Clinometer

The slope of the terrain may be determined as an angle θ with an angle-measuring instrument. The simplest instrument is a clinometer, (shown in Fig. 2.13), which consists of a graduated arc or a semicircle with the line of sight provided by two pins. The vertical line is indicated by a thread attached to the centre of the semicircle, which is weighted to remain vertical. When the line of sight is horizontal the thread reads zero. When the line of sight is inclined, the semicircle is rotated to get the line of sight along the slope, but the thread remains vertical. After bisecting the object along the line of sight, the thread is held pressed to the graduated arc and the reading taken, which gives the slope of the ground.

By measuring the slope

When the slope of the ground has been measured, it is easy to find the horizontal distance, as shown in Fig. 2.14.

$$\text{Horizontal distance} = l\cos\theta$$

where l is the length along the slope and θ is the angle made by the ground with the horizontal. The slope has to be measured whenever the slope angle changes, which is judged by the eye. If the slope is gentle and uniform for a long distance, this method can be used. For very undulating ground, this method is not suitable.

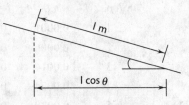

Fig. 2.14 Ground slope and horizontal length

Another way to express the slope of the ground is to express it as a gradient of $1:n$, which means that the ground rises or falls 1 unit for every n units measured along the horizontal (see Fig. 2.15). When the horizontal distance AB is n, BC = 1, and AC = $(1^2 + n^2)^{1/2}$. For any length measured along the slope this relationship can be similarly used. Thus triangles ABC and ADE are similar.

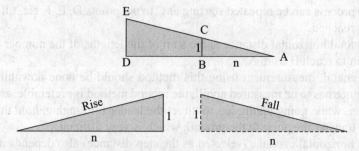

Fig. 2.15 Gradient and length

From the similarity of triangles,

$$AD = (AE \times n)/[\sqrt{(1^2 + n^2)}]$$

where AD is the horizontal distance and AE is the distance measured along the slope.

If the difference in level between the end points of a line along the slope is measured or calculated, the horizontal distance can be calculated as follows (see Fig. 2.16):

$$\text{Horizontal distance} = \sqrt{(L^2 - h^2)}$$

where L is the length measured along the slope and h is the difference in level between the end points A and B.

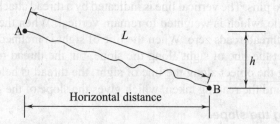

Fig. 2.16 Slope length and height difference

Table 2.2 Hypotenusal allowance

Slope (°)	secθ – 1	l = 20 m	l = 30 m	100 links	150 links
1	0.0001	0.002	0.003	0.01	0.015
2	0.0006	0.0012	0.018	0.06	0.09
3	0.0014	0.028	0.042	0.14	0.21
4	0.0024	0.048	0.072	0.24	0.36
5	0.0038	0.076	0.114	0.38	0.57
6	0.0055	0.11	0.165	0.55	0.825
7	0.0075	0.15	0.225	0.75	1.125
8	0.0098	0.196	0.294	0.98	1.47
9	0.0124	0.248	0.372	1.24	1.86
10	0.0154	0.308	0.462	1.54	2.31

By providing for hypotenusal allowance

The length along a slope is always more than the corresponding horizontal distance. Thus, it is possible to measure a larger distance along a slope such that the equivalent horizontal distance is known.

In Fig. 2.17, let the slope be equal to θ. AC is the sloping terrain, AB is the horizontal distance, and θ is the angle made by AC with the horizontal. With A as the centre and AB as the radius, draw an arc to cut AC in B′. AB = AB′. The distance B′C by which the slope distance exceeds the horizontal distance is known as the *hypotenusal allowance*. Let AB = l be the horizontal length required. Then AC = $l/\cos\theta = l\sec\theta$. The hypotenusal allowance B′C = $l\sec\theta - l = l(\sec\theta - 1)$. The hypotenusal allowance can be calculated as shown in Table 2.2.

If one uses a 20-m chain, the procedure is to measure $20 + 0.076 = 20.076$ m along a 5° slope. The corresponding horizontal distance will be 20 m. By providing for hypotenusal allowance, the slope correction is done simultaneously with chaining. If θ is measured in radians, $\sec\theta$ can be expanded as

$$\sec\theta = 1 + \theta^2/2 + 5\theta^4/24 + \cdots$$

Neglecting higher powers of θ,

$$\sec\theta = 1 + \theta^2/2 \text{ and } \sec\theta - 1 = \theta^2/2$$

As an example, for a 5° slope, $\theta = 0.0873$ rad, hypotenusal allowance = $0.0873^2/2 = 0.0038$ m per metre, and $20 \times 0.0038 = 0.076$ m for a 20-m chain length, as obtained in the table.

If instead of the slope angle, the difference in height for a certain length is known, then the hypotenusal allowance is calculated as follows. If h is the difference in level, l is the horizontal distance, and L is the distance along the slope, then $L = (l^2 + h^2)^{1/2}$; hypotenusal allowance = $L - l = (l^2 + h^2)^{1/2} - l$. This is the exact value of the hyotenusal allowance.

$$\sqrt{(l^2 + h^2)} = \sqrt{l[1 + (h/l)^2]} = l(1 - h^2/2l^2 - h^4/8l^4 - \cdots)$$

Neglecting higher powers of h,

$$\sqrt{(l^2 + h^2)} = l(1 - h^2/2l^2)$$

Hypotenusal allowance = $l(1 - h^2/2l^2) - l = h^2/2l$ (approximately)

For example, a 5° slope can be written as 1:11.43 or as 1.7498:20 (tan 5° = 0.0875 and 1/0.0875 = 11.43).

Hypotenusal allowance = $1.7498^2/2 \times 20 = 0.076$ m

This is same as the earlier obtained value.

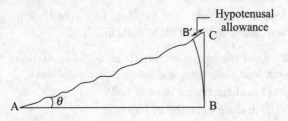

Fig. 2.17 Hypotenusal allowance

If the distance is measured along the slope, 20.076 m can be measured to get 20 m horizontal length. Alternatively, the correction $h^2/2l$ can be subtracted from the length measured along the slope.

2.6.3 Advantages and Disadvantages of Measuring Along the Slope

The ranging of the line has to be done before measuring the line. Measuring uphill thus becomes difficult. With highly undulating ground and frequent changes in slope, it is possible to measure by stepping as the slope changes do not affect the measurement. In gentle and uniform slopes, other methods can be used. The stepping method requires careful transfer of points to the ground and one has to ensure that the sag in the chain or tape is minimal, by applying sufficient pull. Slopes less than about 3° can be neglected, as the error in the measurement is small. The following examples illustrate chaining along a slope.

Example 2.9 The length of a chain line measured along a slope is 327.5 m. Find the horizontal distance if (a) the slope of the ground is 7.5°, (b) the ground rises by 30 cm for every 10 m along the slope, and (c) the gradient is 1:4.

Solution (a) When the slope is 7.5°, the horizontal distance is given by $l \cos\theta$, where l is the distance along the slope and θ is the slope.

Horizontal distance = 327.5 × cos (7.5°) = 324.7 m

(b) Horizontal distance for 10 m along the slope = $[(10)^2 + (0.3)^2]^{1/2}$ = 9.9955 m. Horizontal distance for 327.5 m along the slope = 327.5 × 9.9955/10 = 327.35 m.

(c) The gradient of 1:4 means a rise of 1 m for every 4 m horizontally. For 1 m rise, the distance along the slope = $[(4)^2 + (1)^2]^{1/2}$ = $(17)^{1/2}$. Horizontal distance for 327.5 m along the slope = 327.5 × 4/(17)$^{1/2}$ = 317.7 m.

Example 2.10 If the correction for slope is (a) 0.06 m and (b) 0.04 m for 10 m length, find the slope and the hypotenusal allowance for a 30-m chain.

Solution (a) If θ is the slope, $10(1 - \cos\theta) = 0.06$. This gives $\theta = 6.28°$. Hypotenusal allowance = 30 (sec $\theta - 1$) = 0.181 m.

(b) $10(1 - \cos\theta) = 0.04$. This gives $\theta = 5.126°$. Hypotenusal allowance for a 30-m chain = 30(sec$\theta - 1$) = 0.12 m.

Example 2.11 Find the maximum slope of a terrain if the error due to slope has to be limited to (a) 1 in 1000 and (b) 1 in 500 for a 30-m chain.

Solution (a) Error is limited to 30/1000 = 0.03 m; $30(1 - \cos\theta) = 0.03$; $\theta = 2.56°$. The slope should not be more than 2.56°.

(b) Error is limited to 30/500 = 0.06 m; $30(1 - \cos\theta) = 0.06$; $\theta = 3.62°$. Therefore, the slope should not be more than 3.62°.

Example 2.12 Find the horizontal distance between stations A and D if the measurements were done along rising and falling slopes as follows:

AB: 500.65 m along a rising slope of 5.65°
BC: 700.35 m along a gradient of 1:18
CD: 400 m along a falling slope of 2.56 m in 20 m

Solution   The horizontal distance of the different sections can be determined as follows:
(a) AB: Slope of 5.65°

      Horizontal distance = 500.65 × cos (5.65°) = 498.22 m

(b) BC: Gradient of 1:18 for a gradient of 1:18, hypotenuse = $(1^2 + 18^2)^{1/2}$ = 18.0277 m.

      Horizontal distance = 700.35 × 18/18.0277 = 699.27 m

Alternatively, the slope is given by

      $\theta = \tan^{-1}(1/18) = 3.18°$

      Horizontal distance = 700.35 cos (3.18°) = 699.27 m

(c) CD: Slope of 2.56 m in 20 m

      Horizontal distance = $400 \times [20^2 - 2.56^2]/20$ = 396.71 m

Alternatively, the slope is given by

      $\theta = \sin^{-1}(2.56/20) = 7.354°$

      Horizontal distance = 400 × cos(7.354°) = 396.71 m

      Total horizontal distance = 498.22 + 699.27 + 396.71 = 1594.2 m

2.7 Errors in Chaining

While measuring distances with a chain, one has to avoid the many errors likely to creep into the measurement. Errors can be cumulative or compensating. Cumulative errors tend to accumulate as the measurements continue. An example is the incorrect length of a chain. As more measurements are taken the error due to incorrect length increases. Compensating errors tend to become positive sometimes and negative at other times. An example is the careless marking of stations which can go either way and will not always be positive or negative. The following errors in chaining should be avoided.

Incorrect length of chain

The actual length of a chain used for measurement may be different from the designated length. The chain may be shorter or longer than the designated length. A shorter chain will cause the recorded measurements to be longer than what actual lengths are. A longer chain will make the recorded lengths shorter than the actual length of the line. The error can thus be positive or negative. It tends to accumulate as more and more lengths are measured; it will be in one direction only. The length of the chain should be checked before the start of the survey and appropriate corrections applied to the measured lengths. The chain should be checked once or twice during the course of the day to see whether the length of the chain has changed during measurement. Appropriate corrections can be made as indicated earlier if the actual length of the chain is known.

Incorrect ranging

If the chain is put out of line while measuring a line, the measured length will always be more. Such error accumulates over the number of chain lengths measured. Serious distortion will occur if offsets are also taken with such defectively ranged chains. The remedy is to check the ranging accurately before taking measurements.

Loose chain

If the chain is not pulled correctly and straightened before measurement, the measured length will always be more. The error is cumulative and can be serious if repeated a number of times.

Temperature change

If a chain or tape is standardized at some temperature and the temperature during measurement is very much different from the standardizing temperature, there will be errors due to the change in length of the chain or tape owing to temperature change. With increase in temperature, the length of the chain or tape increases and a larger length is measured and recorded as a smaller (designated) length. If the temperature is less, the length decreases and a smaller length is measured and recorded as a larger (designated) length. The error can thus be positive or negative but is cumulative. This type of error is particularly serious with steel tape measurement and should be compensated for by correcting the length measured.

Variation in pull

A chain or tape generally has a designated length under a standard pull. If the pull applied during measurement is very different from the standard pull, the length of the tape changes. The error due to this is similar to that due to temperature and should be taken care of particularly when measuring with a tape. This error is generally cumulative, either positive or negative . The error can be compensating if the pull applied varies, resulting in the sagging of the tape sometimes and the stretching of the tape at other times.

Errors in slope measurements

Two serious errors can take place while measuring along a slope using the stepping method. These are sag of the tape or chain and the chain not being horizontal during stepping. If the chain or tape sags, the error is positive and cumulative. Sufficient pull must be applied to keep the chain or tape stretched. If the chain or tape is not horizontal, it results in a similar error. The error is positive and cumulative.

Incorrect marking

While measuring a long line with a chain or tape, the starting and intermediate points have to be marked. If the points are not marked accurately, errors can occur. These errors may not be too large and are compensating.

Personal mistakes

Personal mistakes can be avoided with care during chaining, reading, and recording. Some of these mistakes are reading from the wrong end of the chain, failing to correctly observe the zero position on the tape, omitting a full chain length, reading incorrectly the numbers on the tape, and wrong booking of the length. In ranging and measuring a long line, the number of arrows with the follower is taken as the number of chain lengths measured. If one arrow is misplaced or lost, a serious mistake can occur in measuring the length of the line. The reading should be checked with the tallies and the length should be spoken out loud and clean for recording.

It can be seen that cumulative errors are serious and should be carefully avoided. The following is a summary of likely errors in chaining.

Cumulative errors   The following errors are positive, making the measured length more than what it should be, and hence the correction is negative—the chain is shorter than the designated length, sag of chain or tape, measuring along incorrect alignment, slope correction not applied, and temperature lower than the calibration temperature. The following errors are negative, making the measured length smaller than what it should be, and hence the correction is positive—actual chain length is more than the designated length, and higher temperature than calibration temperature during measurement.

Compensating errors   Incorrect setting of the chain or tape and variation in pull during measurement.

2.8   Corrections to Length Measured with a Tape

A tape is a very accurate length measuring instrument and needs careful handling. The length measured with a tape has to be corrected appropriately for many factors. These are temperature variation, variation in pull, sag of tape when suspended in air, correction for incorrect length, correction for slope, and correction for incorrect alignment. The corrections can be computed as follows.

2.8.1   Correction for Incorrect Length

The tape has a designated length, which will be recorded when one tape length has been measured. The actual length of the tape may be different from the designated length. The correction to be applied is similar to that in the case of a chain. If the actual length is more, the recorded distance will be less, and if the actual length is less, the recorded distance will be more. In either case, the correction can be applied as follows: if the designated length is L and the actual length is L', the measured length ML will be corrected as

$$\text{Corrected length} = (L'/L) \times \text{ML}$$

2.8.2   Correction for Incorrect Alignment

Correction for incorrect alignment has to be applied when the line is not aligned correctly. In Fig. 2.18, AB is the correct alignment of the line while the tape is laid out along AC. The distance BC is the perpendicular deviation from the chain line. ABC is a right-angled triangle with a right angle at B. Then, $BC^2 = AC^2 - AB^2$. If $AB = l$, $AC = L$, and $BC = d$, then $L^2 - l^2 = d^2$ or $(L + l)(L - l) = d^2$. $L + l$ is approximately equal to $2L$ and $L - l$ is the correction. Thus, the correction for incorrect alignment is $d^2/2L$. Alternatively,

$$\text{Correction} = AC - AB = L - \sqrt{(L^2 - d^2)}$$

$$\sqrt{(L^2 - d^2)} = (L^2 - d^2)^{1/2} = L[1 - (d/L)^2]^{1/2}$$
$$= L(1 - d^2/2L - d^4/8L - \cdots) = L(1 - d^2/2L)$$

neglecting the higher powers of d. Therefore, the correction for wrong alignment is $d^2/2L$. This correction is always negative.

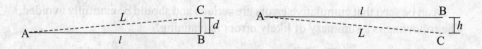

Fig. 2.18 Incorrect alignment of tape   **Fig. 2.19** Slope correction

2.8.3 Correction for Slope

When the two ends of the tape are not at the same level, the tape is not aligned vertically. In Fig. 2.19, let h be the difference in the vertical direction between the ends of the tape. AC is the tape, AB is the horizontal, and BC = h is the deviation from the horizontal. This case is very similar to the case of wrong alignment. It can be easily proved on similar lines that the correction is $h^2/2L$, where L = AC. This correction is always negative.

2.8.4 Correction for Temperature

Tapes are calibrated or standardized at a specific temperature when they are of the designated or nominal length. At other temperatures, the length of the tape will be different from the designated length. The correction is obtained from the coefficient of thermal expansion of the tape material. Steel tapes normally used for precision work have a coefficient of thermal expansion varying from 10.6×10^{-6} to 12.2×10^{-6} per degree centigrade. The average value of 12×10^{-6} may be used. If α is the coefficient of thermal expansion, l is the length, T_s is the calibration temperature, and T is the temperature during measurement, then increase or decrease in the length of the tape = $l\alpha(T - T_s)$. If T is less than T_s, the tape has contracted, and if T is more than T_s, the tape has expanded. The correction to the length measured is positive (correction has to be added) when $(T - T_s)$ is positive. This is so because the tape length has increased and thus the measured length is more than the nominal length but recorded as the nominal length. On similar lines, the correction is negative (to be subtracted) when $T - T_s$ is negative. Therefore, the correction = $\pm l\alpha T$, where l is the length, α is the coefficient of thermal expansion, and T is the difference in temperature.

2.8.5 Correction for Pull

Tapes are standardized (have the designated length) at a pull known as *standardizing pull*. If the pull applied during measurement is more or less than this pull, correction will have to be applied. If P_0 is the pull during calibration and P is the pull applied in the field, then the correction for pull can be derived as

$$\text{Correction for pull} = (P - P_0)L/AE$$

(same as the formula PL/AE for elongation of a bar), where L is the length of the tape, A is the area of cross section of the tape, and E is the Young's modulus of elasticity of the tape material. For steel tapes, the standard value of E = 200 GN/ m^2 can be used. The correction for pull can be positive or negative. If P is more than P_0, the correction is positive and should be added to the length measured. If $P - P_0$ is negative, the correction is negative and should be subtracted from the measured length.

2.8.6 Correction for Sag

When the tape or any part of it is supported between end points only, it sags due to self-weight. The sag can be eliminated by applying sufficient pull. Sag due to self-weight takes a form known as a catenary. However, it is taken as a parabola since the difference between the two shapes is small. In Fig. 2.20, the chord length AB is less than the arc length AB. The arc length AB is the length of the tape and the actual horizontal length measured is always less. The correction due to sag is thus always negative.

$$\text{Sag correction} = l(wl)^2/24P^2$$

where l is the horizontal length of the tape between the supports, w is the weight per metre length of the tape, and P is the pull applied.

In Fig. 2.20, the tape is supported at A and B and the pull applied horizontally is P. w is the weight per metre length of the tape. The length of the curve is L and the chord length is l. The curve actually will not be a parabola but a catenary but is assumed to be a parabola. This is equivalent to assuming that the tape weight acts as a uniformly distributed load on the length l. y_c is the sag at the centre. The loads acting on half the length of the tape are shown in Fig. 2.20(c). The reaction at A is $R_a = wl/2$. Taking the moment about C and equating it to zero,

$$wl/2 \times l/2 - wl/2 \times l/4 - Py_c = 0, \quad y_c = wl^2/8P$$

The equation for the curve ACB, a parabola, can be derived as

$$y = 4y_c x \, (l - x)/l^2$$

The length of the curve can be obtained by integration:

$$dS = \sqrt{[1+(dy/dx)^2]} \cong 1+(1/2)[4y_c x(l-2x)/l^2]^2$$

Length ACB $= \int ds = \int [(1+(1/2)(16y_c^2/l^4)(l^2 - 4lx + 4x^2)]$, the integration being done from 0 to l. This reduces to $L = l + 8y_c^2/3l$. Putting the value of $y_c = wl^2/8P$, $L = l + l(wl)^2/24P^2$. Therefore, the sag correction is $L - l = l(wl)^2/24P^2$. In this equation, l is the length between supports, w is the weight per unit length, and P is the pull applied.

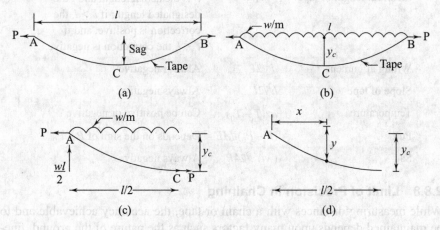

Fig. 2.20 Sag correction

When the two ends of the tape are not at the same level, the sag correction can be modified as follows: If the ends of the tape are held such that the line joining them makes an angle θ with the horizontal, the sag correction can be reduced by multiplying the above value for supports at the same level with $\cos^2\theta$.

If a long tape of length l is supported at $n + 1$ points giving n bays of equal length l_0, then sag correction for the total length can be worked out as follows:

$$\text{Sag correction} = nl_0\,(wl_0)^2/24P^2 = l(wl_0)^2/24P^2 \text{ (as } nl_0 = l)$$
$$= l(wl)^2/24n^2P^2 \text{ (as } l_0 = l/n)$$
$$= lW^2/24n^2P^2$$

where $W = wl$ is the total weight of the tape.

2.8.7 Normal Tension

When a tape is pulled, it tends to elongate and straighten. The weight of the tape on the other hand tends to slacken and sag the tape. These two effects are of opposite nature. *When the pull applied is such that it equalizes the effect due to sag, it is known as normal tension.* The normal tension can be worked out as follows:

$$\text{Correction due to pull} = (P - P_0)l/AE$$
$$\text{Correction due to sag} = l(wl)^2/24P^2$$

Assuming $P > P_0$, the correction for pull is positive and that for sag is negative. The two can be equated to obtain the normal pull. Equating the two,

$$(P - P_0)l/AE = l(wl)^2/24P^2$$

from which

$$P = 0.204\,(wl)\,\sqrt{(AE)}/\sqrt{(P - P_0)}$$

As P appears on both sides of the equation, it can be calculated by trial and error only. Table 2.3 gives a summary of tape correction formulae.

Table 2.3 Summary of tape corrections

Type	Correction	Remarks
Incorrect length of tape	$l' - l$	l' is the actual length and l the designated length; if $l' > l$, the correction is positive, and if $l' < l$, the correction is negative.
Wrong alignment	$d^2/2l$	Always negative.
Slope of tape	$h^2/2l$	Always negative.
Temperature	$l\alpha(T - T_0)$	Can be positive or negative.
Pull	$(P - P_0)l/AE$	Depends on the sign of $P - P_0$.
Sag	$(wl)^2l/24P^2$	Always negative.

2.8.8 Limit of Precision in Chaining

While measuring distances with a chain or tape, the accuracy achievable and to be maintained depends upon many factors such as the nature of the ground, fineness of graduations, equipment and accessories available, importance of the job and consequently the time and resources available, and weather conditions. The limits of precision are thus expressed in terms of the quantity of error per length, generally expressed as $1:n$. Thus an error of 1:4000 is more accurate than an error of 1:2000. If the error allowed is 0.1 m in 200 m, this can be expressed as 1:2000. The following is a list of achievable accuracy under conducive conditions:

Chain on rough terrain:	1 in 250
Chain under good conditions:	1 in 500
Tested chain under excellent conditions:	1 in 1000
Steel tape with accessories under good conditions:	1 in 2000
Invar tape with accessories:	1 in 10,000

The following examples illustrate the corrections to lengths measured with a tape.

Example 2.13   The length of a line measured with a 30-m-long tape was 368.64 m. The tape was standardized at a temperature of 20°C. If the temperature during measurement was 42°C, find the correct length of the line. The coefficient of thermal expansion of the tape material is $12 \times 10^{-6}/°C$.

Solution   Temperature correction is additive if the temperature is more than the standardizing temperature.

Temperature correction = $l\alpha(T - T_0)$

where l is the length, α is the coefficient of thermal expansion, T is the temperature during measurement, and T_0 is the standardizing temperature. Thus,

Correction for temperature = $368.64 \times 12 \times 10^{-6} \times (42 - 20) = 0.097$ m

Corrected length of the line = $368.64 + 0.097 = 368.737$ m

Example 2.14   A 30-m-long tape is held between supports under a tension of 120 N. If the tape weighs 10 N, find the horizontal distance between the supports.

Solution   The tape has a sag when supported between the end points only. The correction due to sag is given by $l(wl)^2/24P^2$, where l is the length, w is the weight per unit length of the tape, and P is the pull during measurement. Therefore,

Correction due to sag = $30(10)^2/24(120)^2 = 0.009$ m

Horizontal distance between supports = $30 - 0.009 = 29.991$ m

Example 2.15   A 30-m tape is used to measure a line along a gradient of 1:12. The tape was also kept 0.6 m out of alignment. Find the horizontal distance for one tape length along the gradient.

Solution   When the tape is kept out of alignment, the correction is given by $d^2/2l$, where d is the distance by which the tape was out of alignment and l is the length.

Correction for wrong alignment = $(0.6)^2/2 \times 30 = 0.006$ m

Corrected length along the slope = $30 - 0.006 = 29.994$ m

This is the corrected distance measured along the slope of gradient 1:12. The gradient 1:12 means 1 vertical to 12 horizontal. For this gradient, the length along the gradient = $(12^2 + 1^2)^{1/2} = 12.0416$ units. Therefore,

Horizontal distance = $29.994 \times 12/12.0416 = 29.89$ m

Example 2.16   A 30-m-long tape was standardized at 20°C and under a pull of 100 N. The tape was used to measure a distance AB when the temperature was 45°C and the pull was 150 N. The tape was supported at the ends only. Find the corrections per tape length if the cross section of the tape was 4 mm^2, the unit weight of the tape material is 0.0786 N/mm^3, and the coefficient of thermal expansion of the tape material is $11.5 \times 10^{-6}/°C$. $E = 2,000,000$ kN/m^2.

Solution Corrections have to be worked out for temperature, pull, and sag of the tape.

Correction for temperature $= l\alpha T = 30 \times 11.5 \times 10^{-6} \times (45 - 20)$

$$= 0.0086 \text{ m}$$

Correction for pull $= Pl/AE = (150 - 100) \times 30/(4 \times 10^{-6} \times 200 \times 10^9)$

$$= 0.0018 \text{ m}$$

Correction for sag $= l(wl)^2/24P^2$

Weight of tape $wl = 30 \times 4 \times 10^{-6} \times 78,600 = 9.432$ N

(unit weight $= 0.0786$ N/mm$^3 = 0.0786 \times 10^6$ N/m$^3 = 78,600$ N/m^3)

Correction for sag $= 30 \times (9.432)^2/(24 \times 150^2) = 0.005$ m

Total correction per tape length $= 0.0086 + 0.0018 - 0.005 = 0.0054$ m

Example 2.17 A 30-m-long steel tape is supported at the ends. Find the normal tension for the tape with the following data: Cross section of the tape $= 4$ mm^2, weight of tape material $= 0.0786$ N/mm^3, $E = 200$ GN/m^2. The pull at which the tape is standardized is 100 N.

Solution The normal tension is given by

Normal tension $P = 0.204W \sqrt{AE}/\sqrt{(P - P_0)}$

W is the weight of the tape and is given by

$$W = 4 \times 10^{-6} \times 78,600 \times 30 = 9.432 \text{ N}$$

$$A = 4 \text{ mm}^2 = 4 \times 10^{-6} \text{m}^2, E = 200 \text{ GN/m}^2 = 2 \times 10^{11} \text{ N/m}^2$$

$$0.204W\sqrt{AE} = 0.204 \times 9.432 \times \sqrt{(4 \times 10^{-6} \times 2 \times 10^{11})} = 1721$$

Normal tension $P = 1721/\sqrt{(P - P_0)}$

(where $P_0 = 100$ N)

$$P = 1721/\sqrt{(P - 100)}$$

This equation can be solved by trial and error.

$$P = 131.5 \text{ N}$$

Normal tension $= 131.5$ N

Example 2.18 A 50-m-long steel tape is standardized at a pull of 150 N. The tape is held such that it is supported at the ends and at its midpoint. The cross section of the tape is 5 mm^2 and its unit weight can be taken as 78,600 N/m^3. Find the normal tension of the tape if $E = 200$ GN/m^2.

Solution In this case, $P_0 = 150$ N, $l = 25$ m, $W = 5 \times 10^{-6} \times 25 \times 78,600 = 9.825$ N, $A = 5 \times 10^{-6}$ m^2, $E = 2 \times 10^{11}$ N/m^2. Both the spans are equal.

Normal tension $P = 0.204W \sqrt{AE}/\sqrt{(P - P_0)}$

$$= 0.204 \times 9.825 \sqrt{(5 \times 10^{-6} \times 2 \times 10^{11}}/\sqrt{(P - 150)}$$

$$= 2004.3/\sqrt{(P - 150)}$$

Solving by trial and error, $P = 227.55$ N.

2.9 Chain Surveying

Chain or tape surveying is done without the use of an angle-measuring instrument. Though chain surveying by itself is not common without a compass or theodolite, a chain or tape still forms the basis of distance measurement in many forms of surveys. Here we discuss chain or tape surveying when no angle-measuring instrument is at hand. The basis of chain surveying is *triangulation*. The term triangulation is generally used in connection with a form of surveying where the base line is very accurately measured with tape and the angles of the triangle are measured for the purpose of plotting the triangle. The form of triangulation used in chain surveying is different because no angle-measuring instrument is available. *Chain triangulation* is the term used for this form of surveying. In this form, the three sides of a triangle are measured with a chain or tape. The survey will thus consist of a network of triangles, and only the sides of the triangles are measured. *Chain triangulation* is suitable when the ground is reasonably open without too much detail around, the area is small and plotting is done on a large scale.

The form of the triangle is also important. The triangles should be well conditioned. A well-conditioned triangle is an equilateral triangle or very nearly so. An ill-conditioned triangle has very obtuse angles and one side is much longer than the other sides. Figure 2.21 shows both types of triangles. Better accuracy in measuring and plotting the sides of a triangle is possible if they are well conditioned. The angles of the triangles should not be less than 30° or more than 120°. However, field conditions may not always allow a selection of stations such that well-proportioned triangles are possible. If ill-conditioned triangles are unavoidable, greater care must be taken in measuring the sides of the triangles.

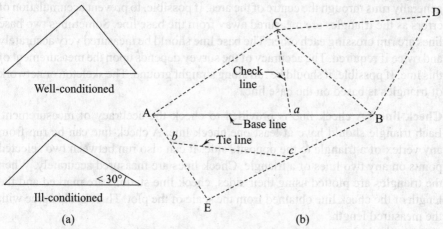

Fig. 2.21 Chain triangulation

Main stations

The vertices of the triangles are the *main survey stations* in chain triangulation. The sides of the triangles are the main chain lines and connect the main survey stations. There can be *subsidiary* or *secondary stations* in the triangulation. Secondary stations are known as *tie stations* and are generally located on the main survey lines. These tie stations are used to run tie lines mainly for the purpose of locating details which are far from the main stations. They can also be used as a check on the measurements of the main lines and offsets.

The main survey stations (vertices of triangles) are denoted by capital letters. The subsidiary stations can be denoted by small letters or numbers. All these numbers or letters should be carefully noted down in a rough sketch prepared in the field book. The factors to be considered while selecting the main and subsidiary stations are the following.

(a) Stations should be intervisible.
(b) One or two main and long lines should be run through the area, and the stations should be selected to facilitate this.
(c) The triangles should be built upon the main base lines, and stations should be selected to facilitate this.
(d) Each triangle should have tie lines, and subsidiary stations should be established for this.
(e) Stations should be so selected that well-proportioned triangles are obtained.
(f) Stations should be so selected that there is least difficulty in ranging and chaining.
(g) Stations should be so selected that they run as close to the details to be established as possible and only short offsets are needed.

Survey lines

Three types of survey lines are normally used in chain triangulation. These are base line(s), check lines, and tie lines (see Fig. 2.21).

Base line The base line is generally the longest and the most important line. It generally runs through the centre of the area, if possible, to prevent accumulation of errors as the triangles are measured away from the base line. Sometimes two base lines are run crossing each other. The base line should be measured very accurately and twice if required. The accuracy of the survey depends upon the measurement of this line. If possible, it should be run along straight ground. The skeleton or network of triangles is based on the base line.

Check line A check line is provided to check the accuracy of measurement. Each triangle should have at least one check line. A check line can be run from any vertex of a triangle to the opposite side. It can also run between two selected points on any two lines of a triangle. Check lines are measured accurately. When the triangles are plotted using their sides, check line stations are marked and the length of the check line obtained from the scale of the plot. This should agree with the measured length.

Tie lines These are run for the purpose of locating details which are far from the main lines. They also serve the purpose of check lines. A tie line is run between fixed points on the main lines.

2.10 Locating Details with Offsets

Offsets, or short lines to locate points such as building corners, hedges, and boundaries, are run either from the main survey lines or from tie lines. Offsets are lateral lines to main lines or tie lines. They are generally run perpendicular to the line from which they are taken, known as *rectangular offsets*, but can also run obliquely, in which case they are called *oblique offsets*. In case there is some difficulty in taking perpendicular offsets due to some obstacles, etc., oblique offsets are taken as shown in Fig. 2.22(b).

2.10.1 Taking offsets

Offsets, if very short, can be taken with offset rods. Longer offsets are taken with a metallic or steel tape. Offsets are recorded with two quantities—length along the chain line and length of the offset. They are generally less than 15 m long when the chain lines and tie lines are planned well in advance. However, sometimes they have to be taken longer than 15 m and are then called *long offsets*. Depending upon the accuracy required, the perpendicularity of the offset to the chain or tie line is judged visually or measured using right-angle measuring instrument. One way is to swing the tape along the chain line and note down the least measurement that gives the perpendicular distance. The perpendicularity of the offset becomes significant when the offset is long. Invariably, an instrument should be used to ensure the perpendicularity of long offsets.

Figure 2.22 shows examples of how offsets are taken. Figure 2.22(a) shows a simple perpendicular offset taken and the minimum reading on the tape taken by swinging the tape. Two people are required to take offsets. The point on the chain line is located either by judging it visually or by measuring a right angle. To take the offset, the leader holds the zero end of the tape at the point to be located. The follower moves to the chain line unwinding the tape. At the chain line, he swings the tape to find the least value of the offset. An arrow is inserted at this point on the chain line. Both the measurements—chainage along the chain line and the length of the offset—should be recorded.

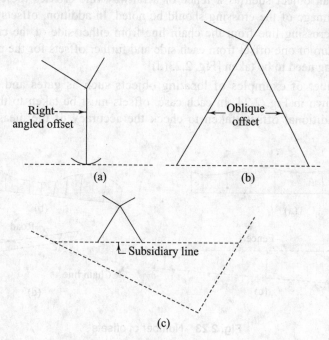

Fig. 2.22 Offsets

Figure 2.22(b) shows an example of an oblique offset. Oblique offsets are taken for important points and to check the accuracy of the long offsets taken. In the case of oblique offsets, four quantities are to be recorded—lengths along the chain line at which the offsets are taken and the length of the offsets. The points along the line are so chosen that the two offset lengths are nearly equal. Long, oblique offsets should be measured very carefully to ensure accuracy.

Very long offsets should be avoided as they are prone to error. This can be done by suitably running a subsidiary line and taking offsets from the sides of these lines as shown in Fig. 2.22(c).

2.10.2 Number of Offsets

In locating a boundary, hedge, or fence line, a number of offsets are to be taken as shown in Fig. 2.23(a). The survey data may later on be used to calculate areas, etc. For this purpose, offsets should be taken at equal intervals and the distance between offsets should be as small as possible. The following factors should be considered while fixing the number of offsets.

(a) Offsets should be short and taken at regular intervals to fix the long lines of an irregular boundary [Fig. 2.23(b)].

(b) Offsets should be taken at closer intervals for a curved boundary. However, the distances between the offsets should be such that they are adequately representable in the plan as per scale.

(c) For a straight boundary, only two offsets are required to its end points. However, it is advisable to take at least one more offset to the centre as a check [Fig. 2.23(c)].

(d) When an object such as a fence or a railway line crosses the chain line, the chainage of the crossing should be noted. In addition, offsets are taken to the crossing line from the chain line from either side of the crossing. A minimum of one offset from each side and further offsets for the purpose of checking need to be taken [Fig. 2.23(d)].

(e) A number of examples of locating objects such as gates and buildings are shown in Fig. 2.24. In each case, offsets must be taken to the corners and additional offsets taken to check the accuracy of the measurements made.

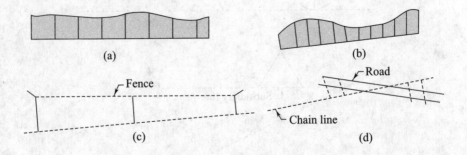

Fig. 2.23   Number of offsets

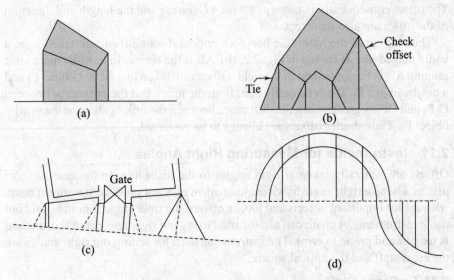

Fig. 2.24 Number of offsets

2.10.3 Accuracy in Taking Offsets

The accuracy with which offsets are taken depends upon the scale of the plan to be plotted, the importance of the object, and the lengths of the offsets. Long offsets should be measured very accurately as the error can distort the detail in the plan.

All important objects should be located by the required number of offsets and additional check offsets. The scale of the plan is generally known in advance. The accuracy with which the offset needs to be measured can be determined from the scale.

The minimum distance that can be distinguished in a plan is generally taken to be 0.25 mm. Thus, if the scale is 1:1000 or 1 cm = 10 m, the minimum length that can be conveniently plotted is 1000×0.025 cm = 25 cm on the ground. Offsets should be measured to an accuracy of one link of the chain.

Effect of incorrect ranging of the chain line on the offsets

As shown in Fig. 2.25(a), consider a main line or tie line AB from which an offset is taken at C. The line is incorrectly ranged to take the direction AD. The point C gets shifted to C′ and the offset taken is C′P while the offset should have been CP.

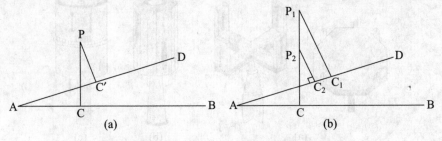

Fig. 2.25 Offset errors in long and short offsets

The offset consequently is taken at a wrong chainage and the length and direction of the offset are also incorrect.

The ranging of the main line becomes critical if long offsets are taken from a badly ranged line as shown in Fig. 2.25(b). AB is the correct line, while due to bad ranging AD is the line taken in the field. Offsets are taken to a distant object P_1 and a nearby object P_2. It can be easily seen from the figure that the difference between C_1P_1 and CP_1 is greater than the difference between the offsets taken to the nearby object P_2. Thus shorter offsets are always to be preferred.

2.11 Instruments for Measuring Right Angles

Offsets are generally taken at right angles to the main line. In the case of short offsets, laying a right angle by visual judgement may not produce significant error, whereas for important objects and longer offsets, the right angle should be laid out using instruments. A chain can also be used to lay out right angles but this method is tedious and prone to error. The instruments used for setting out right angles are the cross staff and the optical square.

2.11.1 Cross Staff

A cross staff is a simple instrument with lines of sight at right angles so that if one line of sight is placed along the chain line, the other line of sight is at right angles to the chain line. Three forms of cross staff are commonly used.

Open cross staff

The open cross staff is shown in Fig. 2.26(a). The cross staff is used to set out offsets at right angles or to set out a line at right angles to the chain line. The open cross staff is the simplest form of the cross staff. It consists of a round or an octagonal wooden head on which two lines of sight are fitted at right angles. It may also be made of a steel plate with four projecting arms, each having a fine slit. It is attached to a vertical rod with a pointed end at the bottom.

To set a right angle, the cross staff is held at the point on the line where the right angle is to be set. One set of slits is directed to the ranging rod at the end of the line, and holding it firmly in position, the observer looks through the other line of sight. He then directs another person with a ranging rod to come in line. The point is then marked. The line from the foot of the cross staff to the ranging rod is a perpendicular line to the chain line.

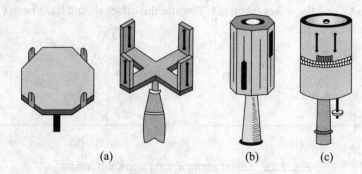

(a)   (b)   (c)

Fig. 2.26 Cross staff

To set out a rectangular offset to an object, one chainman roughly estimates the position of the perpendicular from the object to the chain line. The cross staff is then placed at that point. The staff is rotated to bisect the ranging rod at the end of the line or some point on the line. Holding the cross staff firmly, without turning it, the person moves to the other line of sight to check whether the object is sighted. If not, the cross staff should be moved along the line and the first line of sight ranged along the line. The other line of sight is then checked to see whether the object is bisected. The process is repeated till the position of the cross staff is correct. The measurement along the chain line (chainage) and distance to the object can then be measured.

French cross staff

A French cross staff is shown in Fig. 2.26(b). The French cross staff is an octagonal brass tube with a vertical sight slit on one side and an open rectangular window with a fine hair hung vertically on the opposite side. A pair of these are set at right angles and can be used for setting right angles. On the other four sides are vertical slits which are used to set angles of 45°. The brass tube has a socket at its base which can receive a long rod with a pointed end to place it on the ground while it is being used. The French cross staff is a better instrument than the open cross staff and is used in the same way.

Adjustable cross staff

Figure 2.26(c) shows a cross staff which is adjustable. It is a cylindrical brass tube 80 mm in diameter and about 100 mm long. The cylinder is divided into two parts at its centre. The upper part can be rotated with reference to the lower part using the screw provided. The lower part is divided into degrees and half degrees and the upper part has a vernier. The cross staff can thus be used to set any angle. The upper and lower parts have sighting slits. The accuracy is, however, limited by the fact that the two slits are very close to each other. It is also provided with a magnetic compass at the top face to facilitate taking the bearing of a line.

The open cross staff is a simple and easy-to-use instrument. It is not a very accurate instrument. If greater accuracy is required, an optical square is used.

2.11.2 Optical Square

An optical square is a better and more accurate instrument than a cross staff. It is a very handy and lightweight instrument. It consists of a metal round box about 50 mm in diameter and about 12.5 mm deep. It has a metal cover on its side which can slide and cover the opening of the box. The cover is used when the instrument is not in use and prevents dust accumulation on the mirrors inside.

A diagrammatic internal view is given in Fig. 2.27. Inside the circular box are fitted two mirrors marked H and I. H is known as the horizon glass and is half-silvered, the other half being unsilvered or plain. This glass is fitted to a frame which in turn is fitted to the bottom of the box. The index glass I is fully silvered and again fitted to a frame which is fixed to the base of the box. The index glass may be adjusted (can be turned) with a screw provided at the bottom. The glasses are set at an angle of 45°.

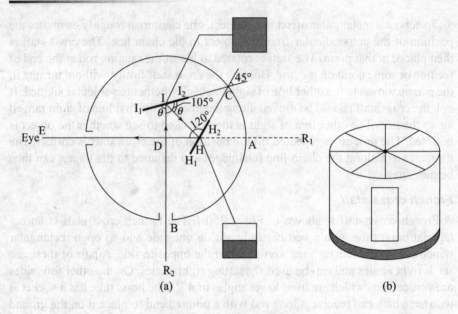

Fig. 2.27   Optical square

On the circular side of the box are three openings cut alike. The opening E is for the eye, through which the observer using the instrument sights the objects. A is a rectangular slit or aperture placed diametrically opposite to E and is used to sight objects through the horizon glass H. B is a wide rectangular opening placed at right angles to the line joining E and A. The horizon glass is set at an angle of 120° to the line joining E and A. The index glass is set at an angle of 105° to the line joining B and I. The angle between the two glasses is thus 45°. The line joining E and A is known as the *horizon sight* and the line through B at right angles to AE is known as the *index sight*.

The principle on which the optical square functions is the following: *If there are two reflecting surfaces set at an angle to each other and a ray of light in a plane perpendicular to the mirror surfaces gets reflected successively from both the mirrors, it undergoes a deviation equal to twice the angle between the mirrors.* This is illustrated in Fig. 2.27(a). A ray of light from R_2 enters through B, gets reflected from I, then gets reflected from H, and reaches the eye of the observer through E. Angle ICH = 45°. The ranging rod R_1 is on the chain line along EA and the ranging rod R_2 is along IB. Let $\angle$HIC be θ; then $\angle$IHC = $(180° - 45° - \theta)$ = 135° − θ. From the law of reflection, $\angle$EHH$_1$ = 135° − θ.

$$\angle IHE = 180° - \angle IHC - \angle EHH_1 = 180° - 2(135° - \theta) = 2\theta - 90°$$

From the law of reflection,

$$\angle I_1IB = \theta, \angle BIH = 180° - 2\theta$$

From triangle IDH,

$$\angle IDH = 180° - \angle DIH - \angle IHD$$
$$\angle IDH = 180° - (180° - 2\theta) - (2\theta - 90°) = 90°$$

Thus, when the images of ranging rods R_1 and R_2 are coincident while being observed from E, lines R_1E and R_2I are at right angles. The two mirrors I and H should have an angle of 45° between them. For this, I is inclined at 105° to the line BI and H is inclined at 120° to EH.

The optical square can be used to set out a right angle from a particular point on the chain line or to set an offset at right angles from the chain line to an object. To set a right angle to a chain line, the observer holds the optical square in hand in such a way that its centre is on the chainage at which a right angle is to be erected. A second person holds a ranging rod at a distance approximately at right angles to the chain line at the chainage. The observer will see the ranging rod at the end of the chain line directly through the unsilvered portion of the mirror H. He/she will also see the ranging rod held by the chainman by reflection through the two mirrors I and H. He/she will direct the chainman to move back and forth parallel to the chain line till the two images coincide. The foot of the ranging rod can be then marked.

To set an offset at right angles to an object, the chainman holds the optical square in hand and moves along the chain line. Either a ranging rod denoting the location of the object or the object itself is seen by reflection and the ranging rod at the end of the chain line is seen directly. When these images coincide, the point along the chain line directly below the centre of the instrument is marked. The offset is then measured from this point to the object.

Testing and adjusting the optical square

The optical square has one adjustable mirror, mirror I. The instrument should be tested and adjusted frequently in case the mirrors are not in alignment. To test and adjust the optical square, do the following.

1. Make a chain line AB on the ground (Fig. 2.28). Select a convenient point C on the chain line. Fix ranging rods at points A and B. A third person is to hold a ranging rod and move at a distance from the chain line, parallel to it.
2. Hold the optical square at C and while observing the image of the ranging rod at A, direct a chainman with a ranging rod to move and come at right angles to the chain line. When both the images coincide, fix the ranging rod at D.
3. Turn around and sight the ranging rods at B and D. The instrument should be held at the same point C. If these two ranging rods coincide, the instrument has been adjusted. If not, direct the chainman to hold another ranging rod and bring him in line at right angles to E and fix the ranging rod.

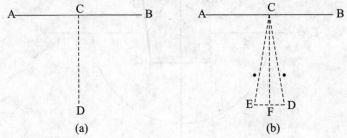

Fig. 2.28 Testing the optical square

4. The distance DE is twice the error in alignment. Take the midpoint of DE at F and fix a ranging rod. Hold the instrument at C and view either A or B and the ranging rod at F. The images should coincide now. If not repeat the test and adjustment.

Indian optical square

The Indian optical square shown in Fig. 2.29 is a simple version of the optical square. It is a brass box in the form of a wedge, with the two sloping sides forming an angle of 45°. Two mirrors are fixed to these sides which have openings above

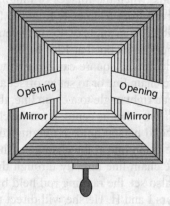

Fig. 2.29 Indian optical square

them. The larger end is fully open and faces one of the objects or ranging rods. This instrument works on the same principle and can be held in hand with the handle provided at the base of the instrument. It is used for setting out right angles or for taking perpendicular offsets.

Prism square

The prism square is a modern version of the optical square and is more accurate. It works on the same principle (Fig. 2.30) but instead of mirrors, a prism is used with its reflecting surfaces at an angle of 45°. It need not be adjusted as the angle is set at the factory and is not variable. The surfaces I_1I_2 and H_1H_2 are the reflecting surfaces and are placed at 45°. The prism square can be used to set a right angle at a point on a chain line or to run an offset at right angles from the chain line to any object.

To run a line at right angles to a chain line, the observer stands with the centre of the instrument at the chainage where the right angle is to be set. He observes the ranging rod at the end of the chain line directly over the prism. A person with a ranging rod stands at some distance from the chain line at a point approximately at

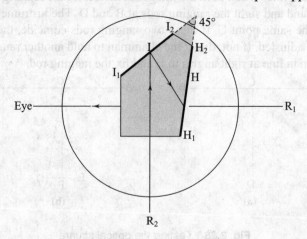

Fig. 2.30 Prism square

right angles. The observer can see this ranging rod after reflection through the prism surfaces I and H. He then directs the person to move back and forth so that the two images of the rods are coincident. The person is then directed to fix the ranging rod at that point. A line joining the point where the prism square is kept and the ranging rod fixed by the second person will be at right angles to the chain line.

To run an offset at right angles to the chain line and bisecting any object, the observer holding the prism square moves along the chain line facing the ranging rod at the end of the line. He can see the ranging rod directly over the prism and the object or a ranging rod kept at that point after reflection through the two prisms. He moves along and reaches a point where the two images coincide. The point directly below the prism square is then marked as the foot of the perpendicular from the object. The offset is then measured.

2.12 Angular and Linear Errors and Limiting Length of Offset

When an offset is laid out there are two errors that can creep in—an angular error in laying the right angle and an error in measuring the length of the offset itself. The seriousness of the effect of these errors depends upon the scale of the plan to be plotted and the length of the offset.

Let α be the angular error, $1/r$ the accuracy of linear measurement, s the scale (as 1:1000 or 1 cm = 10 m), and l the length of the offset. In Fig. 2.31(a), an offset is laid out to an angular error of α. AB is the chain line and the offset is taken at C. The offset as measured in the field is CD while the true perpendicular is CD_1. Draw an arc with C as the centre and CD as the radius. This cuts the true perpendicular at D_1. $CD = CD_1 = l$, the length of the offset. The deviation of the offset parallel to the chain line $DD_2 = l \sin\alpha$. The deviation of the offset parallel to the perpendicular is $CD_1 - CD_2 = l - l \cos\alpha = l(1 - \cos\alpha)$.

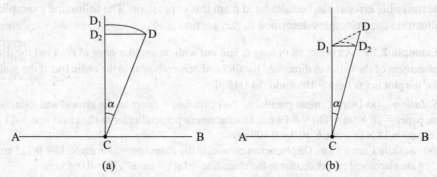

Fig. 2.31 Linear and angular error

Deviation perpendicular to the chain line = $l(1 - \cos\alpha)/s$ on paper; deviation parallel to the chain line = $l(\sin\alpha)/s$ on paper. If 0.25 mm or 0.025 cm is taken as the smallest distance that can be distinguished in plotting, then for the error to be insignificant, $l (\sin\alpha)/s = 0.025$. Therefore, $l = 0.025s/(\sin\alpha) = 0.025s \times \mathrm{cosec}\,\alpha$ is the limiting length of the offset.

2.12.1 Angular and Linear Errors Combined

The case is shown in Fig. 2.31(b). AB is the chain line and C is the point at which the offset is taken. The offset laid out in the field is CD. CD_1 is the true perpendicular. The angular error is α equal to $\angle DCD_1$. The linear accuracy of measuring the offset is 1 in r and the true length of the offset is CD_2. If l is the length of the offset, then the linear error is l/r.

CD_2 is the correct length of the offset. With C as the centre, draw an arc to cut the true perpendicular at D_1. DD_1 is the total displacement. Then, the angular error is $D_1D_2 = l \sin\alpha$ (very nearly). Taking the angular error and linear error in measurement equal, $l \sin\alpha = l/r$. This gives $r = \mathrm{cosec}\,\alpha$.

We can take D_1D_2D as a right-angled triangle with a right angle at D_2, as the angular error α is generally small. Thus,

$$DD_1 = \sqrt{[D_1D_2)^2 + (DD_2)^2]} = \sqrt{2}\,(l\sin\alpha) = \sqrt{2}\,(l/r)$$

as $D_1D_2 = DD_2 = l \sin\alpha = l/r$. If the smallest distance distinguishable = 0.025 cm and the scale = 1 cm = s m,

$$DD_1 = \sqrt{2}\,(l\sin\alpha)/s = \sqrt{2}\,(l/r)/s \quad \text{(on paper)}$$

These must be equal to 0.025 cm.

$$0.025 = \sqrt{2}\,(l/r)/s = \sqrt{2}\,(l/r)/s$$

This gives the limiting length of the offset as

$$l = 0.025rs/\sqrt{2} \text{ or } l = rs/40\sqrt{2}$$

If e is the maximum error in the length of the offset, then the total displacement $DD_1 = [e^2 + (l \sin\alpha)^2]^{1/2}$ on the ground and $[e^2 + (l \sin\alpha)^2]^{1/2}/s$ on paper. If for a given value of e and length l, the permissible error in angle is to be found, $[e^2 + (l \sin\alpha)^2]^{1/2}/s = 0.025$, from which $\sin^2\alpha = [6.25 \times 10^{-4}s^2 - e^2]/l^2$. The permissible error α can be calculated from this expression. The following examples illustrate the principles described in this section.

Example 2.19 An offset 18 m long is laid out with an angular error of 6°. Find the displacement of the offset in directions parallel and perpendicular to the chain line if the scale for the plot is (a) 1 cm = 10 m and (b) 1:1500.

Solution (a) Displacement parallel to the chain line = $l\sin\alpha$ on the ground and $l(\sin\alpha)/s$ on paper = $18 \times \sin 6°/10 = 0.19$ cm. Displacement perpendicular to the chain line = $l(1 - \cos\alpha)/s = 18 \times (1 - \cos 6°)/10 = 0.0098$ cm.

(b) Scale is 1 cm = 15 m. Displacement parallel to the chain line = $18 \times \sin 6°/15 = 0.125$ cm. Displacement perpendicular to the chain line = $18(1 - \cos 6°)/15 = 0.0065$ cm.

Example 2.20 An offset is laid with an angular error of 4° from its true direction. Find the maximum length of the offset if the error is to be negligible on the plan with a scale of (a) 1 cm = 10 m and (b) 1 cm = 30 m.

Solution The plotting accuracy is taken as 0.025 cm.

(a) Displacement on paper due to angular error = $l(\sin\alpha)/s = l(\sin 4°)/10$ cm. This must be equal to 0.025 cm. Therefore, $l(\sin 4°)/10 = 0.025$ or $l = 7.17$ m, which is the maximum offset length.

(b) As before, $l(\sin 4°)/30 = 0.025$ or $l = 10.75$ m, which is the maximum offset length.

Example 2.21 Determine the accuracy to which an offset needs to be measured so that the errors from both the sources are equal. Angular error = 3°.

Solution If l is the length of the offset, the linear error is l/r. Displacement due to angular error = $l \sin\alpha$. These two errors are to be equated. Therefore,

$$l \sin\alpha = l/r \text{ or } r = \text{cosec}\,\alpha = \text{cosec}\,3° = 19.1$$

The accuracy in linear measurement required can be taken as 1 in 19.

Example 2.22 Find the maximum length of the offset so that the combined error due to angular and linear measurements is negligible. The minimum length that can be plotted on paper is 0.25 mm. (a) The linear accuracy is 1 in 30 and the scale is 1 cm = 20 m. (b) The linear accuracy is 1 in 50 and the scale factor is 1 in 3000.

Solution Displacement due to angular error = $l \sin\alpha$. Displacement due to linear error = l/r. If the errors from both the sources are equal, $l \sin\alpha = l/r$. Combined error = $[(l\sin\alpha)^2 + (l/r)^2]^{1/2} = \sqrt{2}\,l/r$ on the round and $\sqrt{2}\,(l/rs)$ on paper.

(a) In this case $r = 30$, $s = 20$. Therefore, $\sqrt{2}\,(l/rs) = 0.025$ or $l = 0.025 \times 30 \times 20/\sqrt{2} = 10.6$ m. Therefore, the maximum length of the offset = 10.6 m.

(b) In this case $r = 5$, $s = 30$. $\sqrt{2}\,l/(50 \times 30) = 0.025$. Therefore $l = 0.025 \times 30 \times 50/\sqrt{2}$ = 26.5 m. Therefore, the maximum length of the offset = 26.5 m.

Example 2.23 An offset is 20 m long and the scale of plotting is 1:4000. If the accuracy in linear measurement is 1 in 40, find the maximum permissible error in the angular measurement.

Solution Displacement due to angular error = $l \sin\alpha = 20 \sin\alpha$. Displacement due to linear error = $l/r = 20/40 = 0.5$. The combined error is equal to $[(20\sin\alpha)^2 + 0.5^2]^{1/2}$ on the ground and $1/40\,[(20\sin\alpha)^2 + 0.5^2]^{1/2}$ on paper. The combined error on paper should be equal to 0.025 which is the plotting accuracy. There fore, $1/40\,[(20\sin\alpha)^2 + 0.5^2]^{1/2} = 0.025$, from which $\sin\alpha = 0.0577$ and $\alpha = 3°\,18'$.

Therefore maximum error in angular measurement = 3° 18′.

2.13 Some Basic Problems in Chaining

In chain surveying, no angle-measuring instrument is used. Only a chain or tape is used to measure the sides of the triangle and the offsets. In the absence of angle-measuring instruments including those for setting right angles, work should proceed with a chain or tape only. Thus, there are basically two sets of problems in chain surveying—one set of problems deals with setting perpendiculars, running parallel lines, etc. and the second set deals with obstacles to chaining in locations where either chaining is not possible (e.g., continuing the chain line across a lake) or stations are not intervisible (as in the case of an intervening hillock) or in situations where both chaining and visibility are obstructed (as in the case of a building). Most of these problems are solved using well-known geometrical principles, as described below.

Erecting a perpendicular at a point to a chain line

There are many methods of doing this with the chain or tape alone. Some of these are described here.

1. *Method of 3/4/5* This method is based on the fact that a triangle with sides in the ratio 3:4:5 is a right-angled triangle with the 5-m side as the hypotenuse.

For this purpose, 30, 40, and 50 links or 15, 20, and 25 links of the chain may be used. Three people are required to set a right angle. A length of 30 links may be taken along the chain line. In Fig. 2.32(a), AB is the chain line and C is the point where a right angle is to be set. CD = 30 links. Leaving 10 links free, hold one end of the chain at C and the other end at D. Holding the 50-link mark in hand, pull the chain taut at E. The chain lengths ED and EC must be straight and pulled taut. Mark the point E; EC is a perpendicular to the chain line at C. It is assumed that a 20-m chain is being used. Appropriate modification can be done when working with a 30-m chain. In working with a steel tape, as the tape cannot be bent sharply, a different technique is used. Set out CD = 4 m. Hold the 20-m end of the tape at E and the zero end of the tape at C. Holding the 6-m and 10-m marks together gives a loop of 4 m. If the tape is now held taut, CE = 4 m, CD = 6 m, and ED = 10 m. The angle at C is a right angle.

2. Another method is shown in Fig. 2.32(b). C is the point on the chain line AB where a perpendicular is to be erected. D and E are two points at equal distances from C. The distances CD and CE can be conveniently chosen. Holding the ends of a chain or tape at D and E, take the middle point of the tape and pull it straight so that there are equal lengths of tape on each side and the chain or tape is taut. Mark the point corresponding to the point F, where the middle length of the tape rests. CF is the perpendicular to the chain line at C. This can be proved easily. CD = CE and FD = FE by construction. In triangles FCD and FCE, CD = CE, FD = FE and FC is a common side. The two triangles are congruent. Therefore, ∠FCD = ∠FCE = 90°.

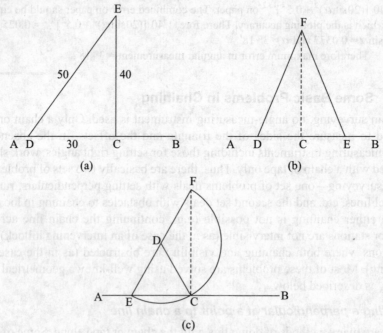

Fig. 2.32 Erecting a perpendicular at a point

3. Another method is shown in Fig. 2.32(c). Select a point D at a convenient distance from C, the point on the chain line AB where the perpendicular is to be erected. Set the zero end of the tape at D and measure DC. Keeping the zero end of the tape at D and holding the tape at length DC, draw an arc with the tape to cut the chain line at E. Mark the point E with an arrow. Range the line ED and extend it to F such that ED = DF. Then FC is the perpendicular to the chain line AB at C. This can be proved easily. If the arc EC is extended, it will pass through F because DC = EC = DF. D is the centre of the arc. Arc ECF is a semicircle and hence ∠ECF is a right angle.

Erecting a perpendicular to a chain line from an outside point

This is a problem related to setting offsets at right angles. When the point C, from which a perpendicular is to be drawn, is accessible, the following methods can be used.

1. AB is the chain line [Fig. 2.33(a)] and a perpendicular is to be drawn to it from a point C outside it. Keep the zero end of the tape at C and taking a convenient length (longer than the length of the perpendicular) swing the tape to cut the chain line at two points D and E. Measure and bisect the length DE at F. CF is the perpendicular from C to the chain line AB. This can be proved easily. CD = CE as both are radii of the same arc. DF = FE as F is obtained by bisecting the length DE. CF is a common side for the triangles CFD and CFE. The two triangles are congruent. Therefore, the angle at F is a right angle.

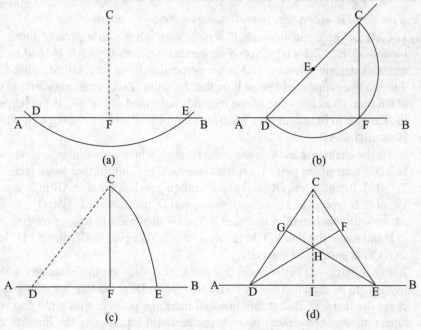

Fig. 2.33 Erecting a perpendicular at a point

2. Another method is shown in Fig. 2.33(b). AB is the chain line and C is a point outside the chain line from which a perpendicular is to be drawn to it. Measure a line CD, D lying on the chain line, and bisect CD at E. Holding one end of

the tape at E and taking a length equal to ED or EC, swing the tape to cut the chain line at a second point F. F is the foot of the perpendicular from C to AB. The proof is the same as given earlier. It is clear that the angle at F is an angle in a semicircle and hence is a right angle.

3. Yet another method is illustrated in Fig. 2.33(c). Select a convenient point on the chain line (point D). Holding the end of the tape at D, measure the length DC. With D as the centre, swing the tape with the length DC as the radius to cut the chain line at E. Measure EC. Mark a point F along the chain line such that $EF = CE^2/2CD$. F is the foot of the perpendicular from C to AB.

When the point is inaccessible, the following method can be adopted.

4. In Fig. 2.33(d), AB is the chain line and C is an inaccessible point from which a perpendicular is to be drawn to AB. Select two convenient points on the chain line, points D and E, and range the lines DC and EC. With the lines DC and EC marked on the ground, draw perpendiculars from D to EC and E to DC. DF and EG are the perpendiculars. Range the lines DF and EG to find their intersection point H. If C is visible from H, range the line CH and continue the line to intersect AB at I. I is the foot of the perpendicular from C to AB. If C is not visible from H, use the methods described earlier to drop a perpendicular HI from H to AB. I is also the foot of the perpendicular from C to AB.

Drawing a parallel line to a given chain line through a point

When the point is accessible, the following methods are possible.

1. Let AB be the chain line and C a point through which a parallel line is to be drawn [Fig. 2.34(a)]. Drop a perpendicular from C to AB by one of the methods described above. CD is the perpendicular to line AB. Measure CD. Take another convenient point E on the chain line. Erect a perpendicular to the chain line AB at E by one of the methods described above. EG is the perpendicular at E to the chain line. Measure EH = CD. Run a line along CH, which is parallel to AB.

2. AB is the chain line and C is the point through which a parallel is to be drawn. Select a convenient point D on the same side as C but further away [see Fig. 2.34(b)]. Range a line DC and place a ranging rod on the line AB in line with DC at E. Keeping the zero end of the tape at D, measure the length DE. Swing an arc with this length to intersect AB at another point F. Keep a ranging rod at F and range the line FD. Measure a length FG equal to CE along FD. Join CG. CG is parallel to line AB.

3. In Fig. 2.34(c), AB is the chain line and C is a point through which a parallel to AB is to be drawn. Select a convenient point D on the line AB. Range and chain the line CD. Bisect this line and mark the point E with a ranging rod. Select another convenient point F on the chain line. Range the line FE and extend the line to G such that FE = EG. Join CG, which is parallel to AB.

When the point is not accessible, the following method can be employed [(Fig. 2.34(d)].

4. AB is the chain line and C is an inaccessible point. A line parallel to AB is to be drawn through C. It is assumed that chaining is not possible between C

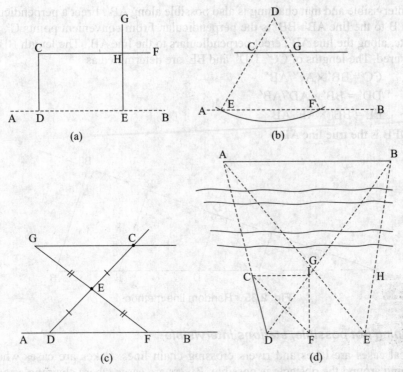

Fig. 2.34 Setting out a parallel line

and A or B but A and B can be seen from C. Range the line AC and fix a point D behind C in line with AC. Select another point E conveniently such that ED is approximately parallel to AB. Range the line EA and mark a point G on it . Now draw a line parallel to EG through C to intersect the line DE at F. Similarly, range the line EB and fix a point H on it. Now draw a line parallel to EB through F. Mark the point I where this line intersects the line DB. Join CI, which is parallel to AB.

2.13.1 Obstacles to Chaining

Despite best efforts during the reconnaissance and marking of stations, one does come across situations in which stations are not intervisible or chaining is not possible between stations or situations in which both visibility and chaining are not possible. Such constraints are known as obstacles to chaining. There are many methods to overcome such obstacles.

Chaining possible, no intervisibility

Such situations can exist due to intervening hillocks or similar obstacles. If the stations are visible from intermediate stations, reciprocal ranging, as explained earlier, is a solution to the problem. But if no intermediate points can be found, the following method is available.

Random line method   AB is the true direction of the line (Fig. 2.35). A and B are not intervisible and also no suitable intermediate points can be found due to wooded area, trees, etc. For ranging the line, a random chain line is run from A, along direction AB′, keeping B′ as close to B as possible but ensuring that A and B′

are intervisible and that chaining is also possible along AB′. Erect a perpendicular from B to the line AB′. BB′ is the perpendicular. From convenient points C′, D′, E′, etc. along the line AB′, erect perpendiculars to the line AB′. The length BB′ is measured. The lengths of CC′, DD′, and EE′ are determined as

$$CC' = BB' \times AC'/AB'$$
$$DD' = BB' \times AD'/AB'$$
$$EE' = BB' \times AE'/AB'$$

ACDEB is the true line AB.

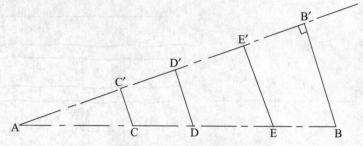

Fig. 2.35   Random line method

Chaining not possible, stations intervisible

Typical cases are lakes and rivers crossing chain lines. Lakes are cases where chaining around the obstacle is possible. Rivers are cases where chaining around the obstacle is not possible.

When chaining around the obstacle is possible, several methods are possible based on geometrical principles.

1. Figure 2.36(a) describes one such method. A and B are two stations on either side of the obstacle. The obstacle is a lake or pond, which does not obstruct the ranging of the line. The length AB cannot be measured. The simplest technique is to set up perpendiculars of equal length at A and B. This can be done with a cross staff or optical square. AD and BC are the perpendiculars and AD = BC. The length AD is so chosen that CD is beyond the obstacle. CD can then be measured, which is equal to AB, as ABCD is a rectangle.

2. Figure 2.36(b) illustrates another method. Set up a right angle at A using a cross staff or optical square. Measure a line AC of sufficient length along this perpendicular so that CB can be ranged and chained. Measure the length CB. AB can then be calculated as $(BC^2 - AC^2)^{1/2}$.

3. Another method is illustrated in Fig. 2.36(c). This method is more difficult than the other two. With a cross staff or optical square in hand, go to a point on one side of AB such that A and B are visible. Keep a ranging rod each at A and B. By moving along a line approximately parallel to AB, locate a point C such that the angle formed at C by the lines AC and BC is a right angle. This happens when the ranging rods at A and B are seen and are coincident. Mark a point below the instrument centre. Measure AC and BC. The length AB = $(AC^2 + BC^2)^{1/2}$.

4. Another simple method is described in Fig. 2.36(d). Select a point C on either side of AB such that the points A and B are visible from it and AC and BC can be ranged. Range the line AC and extend it to D such that AC = CD. Similarly, range the line BC and extend it to E such that BC = CE. ED can then be measured and is equal in length to AB. This is so because triangles ABC and EDC are congruent.

5. Yet another method is depicted in Fig. 2.36(e). This method is used when a cross staff or other similar equipment is not available. Range a line CAD across the chain line such that B is visible from C and D and both the lines CB and DB can be ranged and chained. Measure the lengths CA and AD. Measure both the lengths BC and BD. Two equations can be framed for any two triangles and can be solved for the length AB and one of the angles. For example, if ∠BCD is α, then from triangle BCA, applying the cosine formula for a triangle,

$$AB^2 = BC^2 + CA^2 - 2 \times BC \times CA \cos\alpha$$

Similarly, from triangle BCD, applying the cosine formula,

$$BD^2 = BC^2 + CD^2 - 2 \times BC \times CD \times \cos\alpha$$
$$\cos\alpha = (BC^2 + CD^2 - BD^2)/2 \times BC \times CD$$

cos α can be calculated from this formula since all the quantities on the right-hand side are known. Knowing cos α, the length AB can be calculated from the first equation.

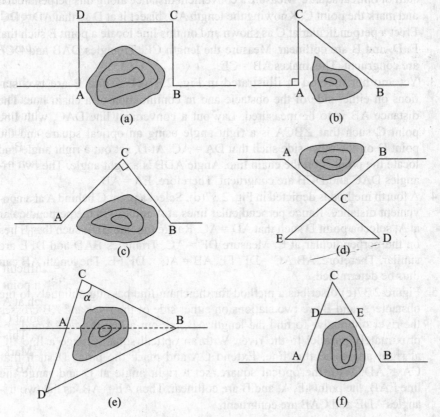

(a)   (b)   (c)   (d)   (e)   (f)

Fig. 2.36   Obstacles to chaining

6. Figure 2.36(f) depicts yet another method of determining the length AB. A and B are points on either side of the obstacle. Select a point C such that lines CA and CB can be ranged and chained. Measure the lengths CA and CB. Get a point D on CA such that CD = $k \times$ CA. Mark a point E on CB such that CE = $k \times$ CB. Triangles ACB and DCE are similar, as CD and CE are in the same proportion to the sides on which they lie. Therefore, AB/DE = AC/DC, from which AB = (AC/DC) × DE = DE/k, as AC/DC = 1/k.

When chaining around the obstacle is not possible, as in the case of a river, the following methods may be adopted to continue the chain line and measure the length across the obstacle.

1. The first method is illustrated in Fig. 2.37(a). A and B are on either side of a river and the length AB has to be measured. Erect a perpendicular at A using any of the methods described earlier or with a cross staff or prism square. Take a convenient distance AC along this perpendicular. At C set out a line CD such that CD is perpendicular to BC and D lies on the chain line of which AB is a part. Measure the length DA. This can be done with a cross staff or prism square. ABC and ADC are similar triangles. Therefore, AB/AC = AC/AD and AB = AC²/AD.

2. A second method is illustrated in Fig. 2.37(b). A and B are on either side of a river and the length AB to be measured. Erect a perpendicular at A using a cross staff or optical square. Measure a convenient distance along this perpendicular and mark the point C. Knowing the length AC, bisect it at D so that AD = DC. Erect a perpendicular at C as shown and on this line locate a point E such that E, D, and B are collinear. Measure the length CE. Triangles DAB and DCE are congruent. This makes AB = CE.

3. Yet another method is illustrated in Fig. 2.37(c). A and B are two stations on either side of the obstacle and in continuation of a chain line. The distance AB is to be measured. Lay out a convenient line DAC, with line point C such that ∠BCA is a right angle using an optical square and the point D on the other side such that DA = AC. At D, set out a right angle and locate the point E on the chain line. Angle ADE is a right angle. The two triangles DAC and DCB are congruent. Therefore, EA = AB.

4. A fourth method is depicted in Fig. 2.37(d). Select a point C behind A at a convenient distance. Range perpendicular lines at A and C. On the perpendicular at A, select a point D such that AD = AC. Range the line BDE such that E lies on the perpendicular at C. Measure DF = AC. Triangles BAD and DFE are similar. Therefore, AB/AC = DF/ FE; AB = AC × DF/FE. The length AB can thus be determined.

5. Figure 2.37(e) discribes a method for the chain line passing obliquely to the obstacle. A and B are two stations on either side of the river and AB crosses the river obliquely. To find the length AB, lay out a line through A and approximately parallel to the river. With an optical square, range a line BC at right angles to this line. Extend CA and mark the point D such that CA = AD. With the optical square, set a right angle at D and range the line EAB, the points E, A, and B are collinear. Then AE = AB, as the two triangles ADE and CAB are congruent.

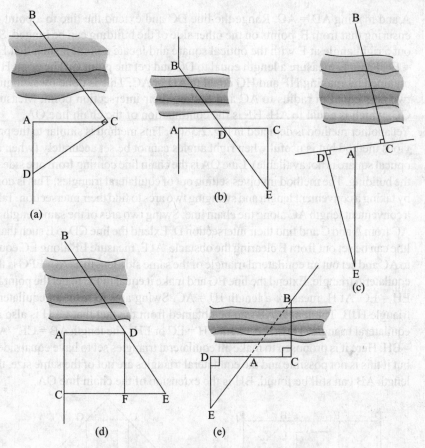

Fig. 2.37 Obstacles to chaining

When chaining and ranging are obstructed

An example of this is a building which obstructs a chain line. As both ranging and chaining are not possible, one has to use methods using geometrical principles. Many methods are available and some of these are described here.

1. Figure 2.38(a) illustrates one method. Chain line OA has been ranged up to the building. Mark a point C at a convenient distance behind A. With the optical square, set out right angles at A and C. CD and AE are made equal to any convenient length, ensuring that the line DE clears the building. Extend the line DE to B to clear the building. Make BF equal to any convenient length. Set out right angles at B and F and obtain the points G and H, making BG = FH = AE. Then AG = EB. In setting out right angles and measuring equal lengths to form rectangles, the diagonals should be measured and checked to see that they are equal. Thus AD = CE and BH = FG. It should also be verified DE = AC and BF = GH.

2. Figure 2.38(b) illustrates another method. Line OA comes from one side of the building and the objective is to continue the line across the building and measure the length along the chain between two points on either side. At A erect a perpendicular with the help of an optical square or otherwise. Measure a convenient length, locating point C. Measure a length AD, D being behind

A and making AD = AC. Range the line DC and extend the line to a point E, ensuring that from E, points on the other side of the building can be ranged. Set out a right angle at E with the optical square and locate a point F such that ED = EF. From F, measure a length equal to DC and get the point G. The point H is obtained by making HF and HG equal to AD or AC. This is done by swinging two arcs equal in radius to AC and finding their intersection point. Measure GC, which is equal to AH. HF is the continuation of the chain line OA.

3. Yet another method is depicted in Fig. 2.38(c). This method is similar to the previous method but is useful when right angles cannot be set accurately (when an optical square is not available). Line OA is the chain line coming from one side of the building. The method involves setting out of equilateral triangles. This is done by taking a convenient length and swinging two arcs to find their intersection. Take a convenient length AC along the chain line. Swing two arcs of the same length as AC from A and C and find their intersection D. Extend the line CD to E such that a line can be set out from E clearing the obstacle. At E, measure EF along EC equal to AC and set out an equilateral triangle of the same side length as AC. EFG is the equilateral triangle. Extend the line EG and make it equal to EC to get the point H. EH = EC. At H, measure a length HJ = AC. Swing arcs to form an equilateral triangle HJB. The length AB can be obtained from the fact that CEH is also an equilateral triangle. Thus, the length CH = EC or EH. The length AB = CE – AC – BH. Here it is proposed to make all equilateral triangles set to have equal sides, but if this is not possible and the equilateral triangles are not of the same size, the length AB can still be found. BH is the extension of the chain line OA.

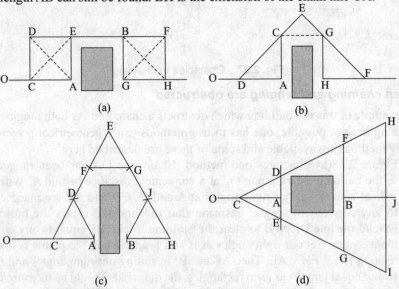

Fig. 2.38 Obstacles to chaining and ranging

4. A fourth method of chaining past a building and finding the length and direction on the other side is illustrated in Fig. 2.38(d). The chain line is on one side of the building. At A, set out a line DAE at any angle. Make the lengths AD and AE such that it is possible to range the lines on either

side of the obstacle. Take a point C behind the chain line at a convenient distance, range the lines CD and CE, and extend them past the obstacle. On these lines, make CF = $k \times$ CD and CG = $k \times$ CE to get points F and G. The line FG, on the other side of the obstacle is divided to get point B such that FB = k $\times$ AD and BG = $k \times$ AE. B is a point on the continuation of the chain line CA. To get another point on the chain line so that the chain line can be continued, continue the lines CF and CG to H and I, making CH = $n \times$ CD and CI = $n \times$ CE. Join HI and divide to get a point J such that HJ = $n \times$ AD and JI = $n \times$ AE. BJ is the continuation of the chain line CA. To get the length AB, from the similarity of triangles CDA and CFB, CB = $k \times$ CA and AB = $(k - 1) \times$ CA.

Distance between two points past an obstacle

A fourth kind of problem is to find the distance between two points which are past an obstacle such as a river. One method is to go across and measure the distance. If the distance has to be measured while on the other side of the obstacle, such as when there is an *obstruction* in the form of a river as shown in Fig. 2.39, we proceed as follows. AB is the distance to be measured and the survey team is on the other side of the river. Select a point C approximately midway between A and B. Measure the distances CA and CB by the methods described above. Range the line CA and measure a convenient length CD along CA. Range the line CB and measure a proportional length CE

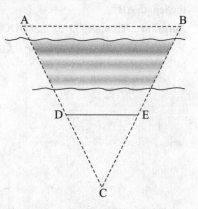

Fig. 2.39 Distance between inaccessible points

along CB. This means that CE = CD × CB/CA. Measure the length DE. AB can be calculated using the fact that CAB and CDE are similar triangles, therefore AB = DE × CA/CD. The distance AB can thus be calculated.

The following examples illustrate these above principles.

Example 2.24 During a chain survey, a pond came in the way of a chain line, making direct length measurement impossible. Two lines, AC and AD, were set out on either side of the chain line and measured 430 m on the left and 320 m on the right (Fig. 2.40). The triangle was completed by a straight line CBD, wherein CB measured 292 m and BD measured 348 m, B being on the continuation of the chain line. Find the length AB.

Solution In triangle ACD, either of the angles ACD or ADC can be calculated using the cosine formula, $a^2 = b^2 + c^2 - 2bc \cos A$, where A is the angle opposite the side a (or the angle between b and c). Let us calculate $\angle$ADC. In triangle ADC,

$$AC^2 = AD^2 + DC^2 - 2AD \times DC \times \cos\alpha \quad \text{(where } \alpha \text{ is } \angle ADC\text{)}$$

$$\cos\alpha = (AD^2 + DC^2 - AC^2)/ 2AD \times DC$$

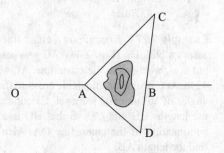

Fig. 2.40 Example 2.24

Substituting values,

$$\cos\alpha = (320^2 + 640^2 - 430^2)/2 \times 320 \times 640 = 0.7986$$

From triangle ADB, applying the cosine formula,

$$AB^2 = AD^2 + DB^2 - 2AD$$
$$\times DB \times \cos\alpha$$
$$= 320^2 + 348^2 - 2 \times 320 \times 348 \times 0.7986$$
$$= 213.61 \text{ m}$$

Example 2.25 To determine the length of a chain line crossing a river, a line AD was set at right angles to the chain line at A. At D, a right angle was set (Fig. 2.41) to obtain a point C with the angle BDC being a right angle. If the length AD = 65 m and AC = 22.5 m, find the length AB.

Solution Triangles DAC and BDC are similar. Therefore,

$$AC/AD = AD/AB \quad \text{or} \quad AB = AD \times AD/AC = 65 \times 65/22.5 = 187.8 \text{ m}$$

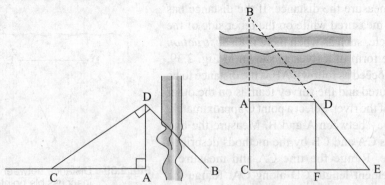

Fig. 2.41 Example 2.25   **Fig. 2.42** Example 2.26

Example 2.26 In order to determine the length across a river of a continuing chain line, the following measurements were made (Fig. 2.42): CA = AD = 40 m, ∠CAD = 90°, CE = 76.8 m, ∠ACE = 90° = ∠DFE. Find the length AB.

Solution In this case, triangles ABD and DFE are similar. Therefore,

$$AB/AD = DF/FE \quad \text{or} \quad AB$$
$$= AD \times DF/FE$$

AD = 40 m, DF = CA = 40 m, and FE = 75 − 40 = 35 m. Therefore,

$$\text{Length } AB = 40 \times 40/35$$
$$= 45.7 \text{ m}$$

Example 2.27 To continue a chain line across a building (Fig. 2.43), AC was set out at right angles to the chain line. AC = 42 m. At C, using an adjustable cross staff, angles of 30° and 60° were set. Calculare the lengths CB and CD so that BD is a continuation of the chain line OA. Also find the length AB.

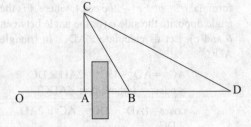

Fig. 2.43 Example 2.27

Solution

Length CB = AC/cosα,   AC = 42 m,   and α = 30°

Length CB = 42/cos 30° = 48.5 m

Length CD = AC/cosβ,   AC = 42 m,   and β = 60°

Length CD = 42/cos 60° = 84 m

Length AB = 42 × tan 30°   (AB/AC = tanα and AB = AC tanα)

    = 24.25 m

Example 2.28 To find the distance between two points on the other side of a river, a surveyor made the measurements (see Fig. 2.44) DE = 32 m, CE = 40 m, FG = 36 m, and CG = 44 m to calculate the distances CA and CB. Angles CDA, DEC, CFG, and BGC are right angles. 25% of the distances CA and CB were measured along the respective lines to get points H and J. If HJ = 23.6 m, find the lengths CA, CB, and AB.

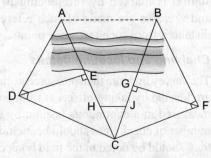

Fig. 2.44   Example 2.28

Solution First the distances CA and CB should be calculated. To find CA, note that triangles AED and CED are similar. Therefore, AE/DE = DE/EC and AE = DE^2/EC. Substituting the given values, AE = 32 × 32/40 = 25.6 m. CA = CE + EA = 32 + 25.6 = 57.6 m.

Similarly, triangles BGF and CGF are similar. Therefore, BG/GF = GF/CG and BG = GF × GF/CG = 36 × 36/44 = 29.45 m. CB = CG + GB = 44 + 29.45 = 73.45 m. Triangles CAB and CHJ are similar as CH/CA = CJ/CB = 0.25; therefore, HJ/AB = 0.25. From this, AB = 23.6/0.25 = 94.4 m.

2.14 Fieldwork

Fieldwork for surveying an area includes the following.

Reconnaissance

Reconnaissance of the area to be surveyed has to be undertaken first. This identifies key features of the area where the survey stations are to be located and determines the kind of equipment needed to be carried to complete the survey. Appropriate sketches are made during this preliminary survey to plan the actual survey work. These sketches should clearly indicate the key features, important details to be located, any obstacles such as buildings, lakes, or rivers coming in the way of chaining and ranging, and the approximate lengths of the lines obtained by judging or by pacing. The entire area should be seen to get familiar with the area. During this phase, ranging rods are carried to check the intervisibility of stations. The stations are numbered or named in the sketches prepared during the reconnaissance. The sketches are to be made in the field book.

Equipment

Generally the following equipment will be required. These include a chain with at least 10 arrows, a metallic or steel tape, a dozen ranging rods, an offset rod, pegs, a plumb bob, a cross staff or optical square, and a survey field book and other stationery. Generally, the surveying party should take along other items such as an axe, hammer, string, and chalk for various purposes.

Marking stations

Survey stations should be marked on the ground as per the plan prepared. Ranging rod or poles can be fixed if possible, otherwise pegs can be driven into the ground using a hammer. On hard ground, spikes can also be driven. The idea must be to make the stations sufficiently stable for future identification if required. Each station is marked by referencing it to permanent features such as buildings and sketches are made of such reference features for each station, marking the distances from such reference points.

Chaining and locating details

The survey lines are then measured accurately, starting from the base line. The details are located by taking offsets at right angles or oblique offsets. The lines should be measured in a continuous fashion by completing triangles one by one. Adequate number of check lines should be included for each triangle. When tie lines are run, they should be noted in the field book of such lines in terms of the chainages along the chain lines. If the work cannot be completed in a day, adequate references must be left for continuing the work on the following day.

2.15 Booking Field Notes

Booking field notes is a very important part of survey work. After the survey is over, one has to refer to the field notes to plot the data. Field notes start with a rough sketch of the area with all the salient features and main stations marked on it. This is done during the reconnaissance phase. The data of the main chain lines and offsets taken from it must be recorded in the field book with appropriate symbols and sketches as may be required. In the case of large areas many such sketches and details need to be recorded.

In recording chain line data, each line should start on a new page. Double-line booking of data is more common as shown in Fig. 2.45. The start and end points of the line are marked appropriately with the main station names. Chainages are marked between the two lines drawn in the centre of the page. Chainages along the chain lines are marked between the lines and the offset length to any point is marked on the side (to the right or left) where the object lies. Sketches can be drawn if required with appropriate symbols on the side.

2.16 Cross Staff Survey

Cross staff survey is a form of survey done with a chain or tape and only a cross staff. As shown in Fig. 2.46(a), a long line is laid out approximately through the centre of the area. Lines at right angles are laid out from this line, forming triangles and trapeziums. The area of these figures can be determined using simple formulae.

To conduct the survey of an area by this method, the area is inspected first and a location and direction are determind for the main chain line. Once the two end stations are marked and located, chaining is started along this line. Offsets are to be laid to each of the vertices from the chain line. The offsets are measured at right angles to the chain line and it is desirable to use an optical square for this purpose. The fieldwork is recorded as shown in Fig. 2.46(b). A plan of the area can be drawn to any desired scale from this data. The area can be computed as the sum of the areas of triangles and trapeziums using the following simple formulae:

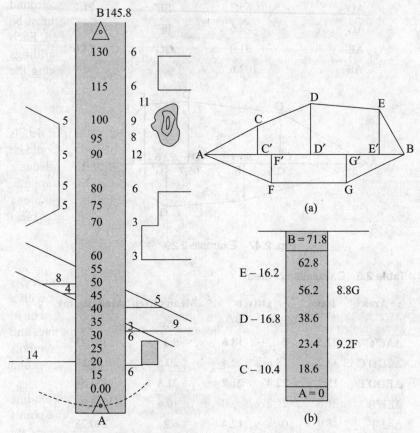

Fig. 2.45 Field book entries   **Fig. 2.46** Cross staff survey

Area of a triangle = base × altitude/2 = base × average altitude

Area of trapezium = base × (sum of parallel altitudes)/2
= base × average altitude

The following examples illustrate this principle.

Example 2.29 Figure 2.47 shows a cross staff survey. Find the area of the survey from the data given in Table 2.4.

Solution As shown in Fig. 2.47, the area is divided into triangles and trapeziums by the offsets. The calculations can be conveniently done in tabular form, as shown in Table 2.5.

Table 2.4   Data for Example 2.29

Length along the chain line AB (m)		Offset (m)	
Line	Length	Line	Length
AC′	32.5	CC′	18.6
AF′	46.9	DD′	22.4
AD′	65.7	EE′	21.2
AG′	72.6	FF′	12.4
AE′	81.0	GG′	14.6
AB	102.6		

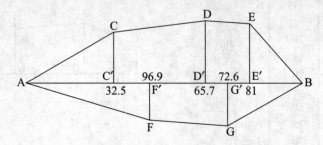

Fig. 2.47   Example 2.29

Table 2.5   Calculation of area

Area	Base	Offsets		Mean offset	Area (sq. m)
		Left	*Right*		
ΔAC′C	32.5	0	18.6	9.3	302.25
ΔCDD′C	33.2	18.6	22.4	20.5	680.6
ΔEDD′E	15.3	22.4	21.2	21.8	333.54
ΔEE′B	21.6	21.2	0	10.6	228.96
ΔAF′F	46.9	0	12.4	6.2	290.78
ΔF′FGG′	25.7	12.4	14.6	13.5	346.95
ΔG′GB	30.0	14.6	0	7.3	219.00

In Table 2.5, the base length for each area is calculated from the chainage along AB at which offsets are taken. For example, for the triangle AC′C, the base length = AC′ = 32.5 m. For the trapezium CDD′C, the base length AD′ – AC′ = 65.7 – 32.5 = 33.2 m, and so on for the other figures.

The offsets are written as the left offset and right offset. For the triangle AC′C, the left offset = 0 and the right offset = CC′ = 18.6 m. For the trapezium CDD′C, the offsets are CC′ = 18.6 m and DD′ = 22.4 m, and so on for other figures. The entries of the last column, area, are obtained by multiplying the base length by the mean offset. The total area is the sum of the figures in the last column = 2402.1 sq. m.

Example 2.30   The data from a cross staff survey is given in Table 2.6. Compute the area of the field (see Fig. 2.48).

Table 2.6 Data for Example 2.30

Length along the chain line AB (m)		Offset (m)	
Line	Length	Line	Length
AG′	28.4	CC′	10.8
AD′	59.6	DD′	23.6
AH′	78.8	EE′	26.4
AE′	106.6	FF′	29.6
AI′	121.8	GG′	8.4
AF′	134.9	HH′	16.6
AB	162.2	II′	27.4

Solution   The solution is shown in Table 2.7. In Fig. 2.48, we notice that the chain line is crossed by the line CG. Let the point of crossing be 'O'. The lengths C′O and G′O need to be calculated. Triangles CC′O and GG′O are similar. From the property of similar triangles, C′O/OG′ = CC′/GG′. Therefore, (C′O/OG′) + 1 = (CC′/GG′) + 1, from which

$$C′G′/OG′ = (CC′ + GG′)/GG′ \quad \text{and} \quad OG′ = C′G′ \times GG′/(CC′ + GG′)$$
$$OG′ = 28.4 \times 8.4/(10.8 + 8.4) = 12.425 \text{ m} \quad \text{and} \quad C′O = 15.975 \text{ m}$$

Table 2.7 Calculation of area

Area	Base	Offset		Mean offset	Area (sq. m)
		Left	Right		
ΔCC′O	12.425	10.8	0	5.4	(−) 67.095
ΔOGG′	15.975	0	8.4	4.2	67.095
ΔC′CDD′	59.6	10.8	23.6	17.2	1025.12
ΔD′DEE′	47	23.6	26.4	25	1175
ΔE′EFF′	28.3	26.4	29.6	28	792.4
ΔFF′B	27.3	29.6	0	14.8	404.04
ΔG′GHH′	50.4	8.4	16.6	12.5	630
ΔHH′II′	43	16.6	27.4	22	946
ΔII′B	40.4	27.4	0	13.7	553.48

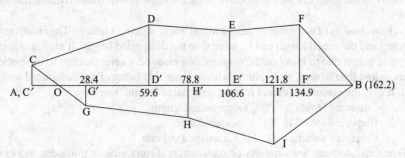

Fig. 2.48 Example 2.30

The area of triangle CC′O is taken to be negative and subtracted. This is done to compensate for that area included in the area of the trapezium C′CDD′. The rest of the procedure is the same as in the previous case.

The total of the last column is 5526, which is the area of the survey, that is, 5526 sq. m.

2.17 Plotting Chain Survey Data

From the field book, survey data is taken and a plan or map is prepared. This is one of the main objectives of the survey. Depending upon the nature of the survey and the purpose for which it is being prepared, the data is plotted to an appropriate scale.

Standard symbols are used in plotting the survey data. The sheet sizes should be as per BIS specifications. The layout of the sheet, title box, marking of the north direction, etc. should be as per standard practice. The drawings are generally done on tracing paper or cloth for the purpose of duplication, as a number of copies of the plot may be required. Duplication is done by ammonia printing or by photocopying techniques. Drawings are generally done in pencil with dark lines so that inking is not required. Many drawings are now prepared using computer software.

The main framework of triangles is plotted first to an appropriate scale depending upon the sheet size. Offsets are plotted using a set square in the case of right-angle offsets or an offset scale. An offset scale has two scales at right angles, so that both the chainages along the chain line and the perpendicular offsets are measured simultaneously. The plotting should be done with a dark pencil with a sharp point. Lettering in the drawings should be as per the standard practice followed in preparing drawings.

Summary

Chaining or taping is the most accurate method of directly measuring distance. The chains and tapes are made as per the relevant Indian standard. Chains are available in lengths of 5 to 30 m. Tapes are available in varying lengths of 2 m to 50 m and even more. Tapes are made of cloth, metal, woven linen, and steel. Invar tape is a very accurate tape made of an alloy of steel and nickel and has a very low coefficient of thermal expansion.

When tapes are used to measure distances along sloping ground, the distances are reduced to horizontal distances for plotting. There is also a method of making hypotenusal allowance for distances measured along the slope. *Hypotenusal allowance* is the extra distance measured along the slope so that the horizontal distance is known.

A chain consists of a number of links and has many wearing surfaces. The chain may get stretched and the actual length can be more than the designated length. It may also become shorter in length due to kinks or the rings getting clogged. Correction has to be applied to the measured length, area, and volume when a chain that is not of the designated length is used. If L' is the actual length and L is the designated length, then

Corrected length $= (L'/L) \times$ measured length
Corrected area $= (L'/L)^2 \times$ measured area
Corrected volume $= (L'/L)^3 \times$ measured volume

Tape measurements are similarly prone to errors. Errors include those due to incorrect length of the tape, slope in keeping the tape, wrong alignment, temperature, and pull and sag when it is not supported over the entire length. Corrections can be applied to find the correct length measured. Pull stretches the tape and sag reduces the measured length and these two opposite effects can be neutralized by applying normal tension.

In surveying an area, main stations are fixed after considering the visibility and facilities for chaining. *Main survey lines* join the main stations of a survey. Check lines are measured to check the accuracy of the work and generally connect a main station to a point on another line. *Tie lines* are lines set out for taking offsets; these also act as check lines.

Offsets are taken to objects or details by measuring their distances from some point of a survey line. Offsets are generally taken at right angles from the main lines or tie lines and the length is generally restricted to less than 15 m. Sometimes, offsets longer than this length are taken and can be oblique to the chain lines. Right angles in the case of short offsets are set by visual judgement. For accurate work and in the case of long offsets, right angles are set out with an instrument.

To set right angles, many instruments are used. The cross staff is the simplest of such instruments. Other instruments include the optical square, which works on the principle of reflection from two mirrors inclined at 45°, and the prism square, which has a prism with its sides inclined at 45°.

While measuring offsets, both linear and angular errors can creep in. Corrections can be applied to the measured lengths to correct these errors. Depending upon the errors, the limiting length of the offset can be calculated to ensure that the error is not significant when plotted in the plan.

Right angles at a point of a chain line or to a chain line from a point outside can be set out with a chain or tape using many geometrical principles. Similarly, many obstacles such as lakes, rivers, or buildings which obstruct a chain line can be crossed with similar geometrical constructions.

Fieldwork for chain surveying includes reconnaissance. In order to get familiar with the area, fix the main stations, and collect the equipment required for the survey. The equipment required includes two chains or one chain and one tape for measuring offsets, a cross staff or optical square, pegs, arrows, and ranging and offset rods. One may also require some equipment for clearing bushes and vegetation for ease of measurement. A surveyor with two to three assistants forms the surveying party. Depending upon the nature and importance of the survey, lines should be measured accurately and twice to ensure accuracy. Enough number of check lines and tie lines should be planned to ensure adequate checks and accuracy of measurement. Long oblique offsets should be taken wherever required.

Field notes should include sketches of the area indicating important details such as buildings, names of stations, and sketches of other details. Capital letters are used for main stations and small letters are used for check line stations and tie stations. Each survey line should begin on a separate page. Double-line booking of field notes is very convenient.

A cross staff survey is one where a long survey line is laid across the area, approximately dividing it into two equal parts. The survey essentially consists of taking right-angled offsets to the vertices, dividing the area into triangles and trapeziums. The area can easily be calculated with such data.

Plotting of the survey should be done on good quality paper with appropriate symbols for each detail as per standard practice. Good quality pencils should be used for plotting. Inking of drawings is not common now. Pencil drawings can be duplicated by ammonia printing or photocopying. The use of computer aided drafting software is also becoming common.

Exercises

Multiple-Choice Questions

1. A 20-m Indian standard chain has
   (a) tallies at every metre length
   (b) rings at every metre length
   (c) rings at every metre length and tallies at every 5 metres
   (d) tallies at every metre length and rings at 5 metres
2. A metallic tape is made of
   (a) steel
   (b) Invar
   (c) a composite material of steel and brass
   (d) of cloth interwoven with metallic fibres
3. If the actual length of a 20-m chain is found to be 19.8 metres, then the actual length of a line measured as 100 m with that chain will be

(a) 99 m   (b) 100 m   (c) 101.01 m   (d) 102 m

4. A 30-m chain was used to measure a line AB which was found to be 205 metres long. If the chain was found to be 2 cm too long, then the actual length of the line AB is

(a) 203.86 m   (b) 204.86 m   (c) 205.13 m   (d) 206 m

5. A 20-m chain was found to be 1.5 cm too short. If the measured length of a line with this chain was 108 metres, the actual length of the line will be

(a) 107.985 m   (b) 107.92 m   (c) 108.08 m   (d) 110 m

6. The actual length of a line measured with a chain was known to be 75 metres. When measured with a 20-m chain, the length was measured as 75.4 metres, The actual length of the chain is

(a) 19.7 m   (b) 19.8 m   (c) 20.1 m   (d) 20.4 m

7. A chain was used to mark a distance of 150 metres. The designated length of the chain was 30 metres. On testing the chain was found to be 30.01 m long. The actual length measured is

(a) 149.95 m   (b) 150 m   (c) 150.05 m   (d) 151 m

8. The area of a plot of land was found to be 204.5 m^2 according to the scale mentioned in plan. It was also mentioned that the 20-m chain used for measuring the plot was 0.015 m too long. The actual area of the plot of land is

(a) 204.2 m^2 (b) 204.34 m^2 (c) 204.65 m^2 (d) 204.8 m^2

9. An area in plan was found to be 104 m^2 as per scale. But the actual area of this land was known to be 104.8 m^2. The actual length of the 20-m chain used for measurement for plotting was

(a) 2 cm too long   (b) 7 cm too long   (c) 15 cm too long   (d) 15 cm too short

10. The sides of an embankment were measured with a 20 m chain. The volume calculated was 20.5 m^3. If the chain used was found to be 2 cm too short, the actual volume of the embankment is

(a) 20.54   (b) 20.52   (c) 20.48   (d) 20.44

11. The sides of a cube were measured with a 5-m tape and the volume calculated was 185 m^3. If the actual side of the cube was known to be 5.75 m, the actual length of the tape is

(a) 5.14 m   (b) 5.07 m   (c) 5.04 m   (d) 5.02 m

12. A rectangular block of wood was measured with a 5-m tape. The tape was found to be 0.01 m too long. If the volume calculated by measurement was 2.8 m^3, the actual volume will be

(a) 2.805 m^3 (b) 2.81 m^3 (c) 2.816 m^3 (d) 2.85 m^3

13. For a 20-m chain of 100 links, if the hypotenusal allowance was 1 link, then the slope of the ground in degrees is

(a) 6.4°   (b) 6.9°   (c) 7.2°   (d) 8.07°

14. The length measured along a slope is 128 m and the horizontal distance is 126.8 m, then the slope in gradient is

(a) 1 in 7.25   (b) 1 in 9.8   (c) 1 in 12.3   (d) 1 in 13.5

15. If a 30 m tape is 0.3% too short, then the correction per tape length is

(a) 0.03 m   (b) 0.06 m   (c) 0.09 m   (d) 0.1 m

16. If the slope of a ground is 3°, the gradient can be represented as

(a) 1:3   (b) 1:9   (c) 1:12   (d) 1:19

17. If the gradient of a sloping ground is 1:25, then the slope of the ground in degrees is

(a) 25°   (b) 12.5°   (c) 2.29°   (d) 1.25°

18. If the measured distance along the slope is 18.5 m and the gradient is 1:16, then the horizontal distance is

(a) 18.5 m   (b) 18.46 m   (c) 17.9 m   (d) 16 m

19. For a chain length of 20 m, the hypotenusal allowance on a 5° slope is
    - (a) 0.07 m
    - (b) 0.7 m
    - (c) 1.4 m
    - (d) 1.75 m
20. The hypotenusal allowance for a 30-m chain on a gradient of 1:10 is
    - (a) 0.15 m
    - (b) 0.2 m
    - (c) 0.3 m
    - (d) 0.4 m
21. If a ground slopes at an angle of 6°, the length to be measured along the slope to get a horizontal distance of 20 m is
    - (a) 20.01 m
    - (b) 20.05 m
    - (c) 20.11 m
    - (d) 20.22 m
22. If the hypotenusal allowance is 0.08 m for a chain length of 20 m, then the slope of the ground in degrees is
    - (a) 4.8°
    - (b) 5.11°
    - (c) 5.8°
    - (d) 6.2°
23. If the hypotenusal allowance is 0.1 m for a chain length of 30 m, the gradient of the ground can be expressed as
    - (a) 1 in 12.24
    - (b) 1 in 11.11
    - (c) 1 in 10
    - (d) 1 in 9.8
24. The hypotenusal allowance for ground A for a 20-m chain and for ground B for a 30-m chain is the same. The ratio of the gradient of ground A to that of ground B is
    - (a) 1.0
    - (b) 1.1
    - (c) 1.22
    - (d) 1.4
25. The maximum slope of a ground for which the error due to slope is limited to 1 in 600 for a 30-m chain is
    - (a) 2.05°
    - (b) 3.3°
    - (c) 4.05°
    - (d) 5°
26. The maximum slope of a ground for which the error due to slope is limited to 1 in 800 for a 20-m chain can be expressed in gradient as
    - (a) 1 in 41
    - (b) 1 in 20
    - (c) 1 in 18
    - (d) 1 in 12.7
27. If the distance between station A and station B measured along a slope of 5° is 4 chain lengths of 20 m, then the horizontal distance is
    - (a) 79.69 m
    - (b) 80.30 m
    - (c) 81.58 m
    - (d) 91.31 m
28. The distance measured along the slope of gradient 1:20 between stations P and Q is 280 m. The horizontal distance PQ is
    - (a) 275 m
    - (b) 279.65 m
    - (c) 280.5 m
    - (d) 281.8 m
29. Normal tension for a tape is defined as the pull
    - (a) applied by an average adult
    - (c) which equalizes the effect due to sag
    - (b) applied for standardizing the tape
    - (d) that equalizes the effect due to slope
30. While placing along a line AB. A 30-m tape was out of alignment by 10 cm, then the actual length along AB is
    - (a) 29.9998 m
    - (b) 30.0002 m
    - (c) 29.9666 m
    - (d) 30.0444 m
31. A 20-m steel tape has a cross section of 4 mm^2 and E for steel is 200 GN/m^2. The tape was standardized under a pull of 120 N and the pull applied during measurement was 80 N. The correct length of 20 m measured is
    - (a) 19.999 m
    - (b) 20.001 m
    - (c) 20.01 m
    - (d) 19.99 m
32. A 30-m steel tape was standardized at 20° C. The tape was used when the ambient temperature was 45°C. A 30-m length measured with the tape will actually be (take coefficient of expansion of tape material as 12×10^{-6})
    - (a) 30.009 m
    - (b) 29.991 m
    - (c) 30.09 m
    - (d) 29.91 m
33. Correction due sag of a tape is
    - (a) always positive
    - (b) always negative
    - (c) sometimes negative and sometimes positive
    - (d) dependent on the temperature conditions

34. Correction due to wrong alignment of the tape
    (a) is always positive
    (b) is always negative
    (c) canbe positive or negative
    (d) is depends upon whether the alignment is wrong to the right or left of the line.
35. Correction due to the tape being laid along a slope
    (a) is always positive
    (b) is always negative
    (c) can be positive or negative
    (d) depends upon whether the measurement is down the slope or upwards the slope
36. A cross staff is used for
    (a) alignment of a survey line
    (b) setting perpendicular lines to survey lines
    (c) marking of survey stations
    (d) setting a line at an angle to a survey line at a point
37. In an optical square, the two mirrors are placed at an angle of
    (a) 30°   (b) 45°   (c) 60°   (d) 90°
38. A prism square is used to
    (a) check the alignment of survey lines
    (b) get an enlarged view of station marks
    (c) mark survey stations in between survey lines
    (d) set a line at right angles to a survey line
39. In measuring an offset, if the angular error is 4°, the accuracy of linear measurement to get the linear and angular errors equal is
    (a) 1 in 10   (b) 1 in 12   (c) 1 in 14   (d) 1 in 20
40. If in laying an offset, the likely angular error is 3°, then the limiting length of the offset for it to be negligible in a plan drawn to a scale of 1 cm = 20m is (take the plotting accuracy as 0.025 cm)
    (a) 6.75 m   (b) 8.46 m   (c) 9.55 m   (d) 12 m

Review Questions

1. Draw a neat sketch of a 20-m chain to show its salient features.
2. Give a list of signals and their meanings to direct the chainman to come in line and fix a station.
3. What is hypotenusal allowance? Derive a formula for calculating it when the slope is given in degrees and in terms of the gradient.
4. Correction for slope is negative (subtracted from the measured length along the slope). Will it have the same sign while measuring uphill and downhill?
5. What is the hypotenusal allowance for 10 m if the ground has a gradient of 1:14?
6. Give a list of corrections, their values, and signs for lengths measured with a tape.
7. Explain the term normal tension and the method used to calculate it.
8. Explain the term offset. Give a sketch to show right-angled and oblique offsets and explain their use.
9. Give a list of instruments used for setting right angles. Explain the construction and function of a cross staff.
10. With a neat sketch explain the working principle of a prism square.
11. Give a brief description of the optical square and the method of using it.
12. What is meant by limiting length of offset?

13. Describe briefly the method of conducting the survey of an area, explaining the steps involved.
14. Give a sample page of a field book to explain how the entries are made.
15. Explain the meaning of cross staff survey. Where is it used?
16. Explain the likely errors in chain surveying and the precautions that should be taken to eliminate them.
17. Describe the different methods of setting out a right angle at a point on a chain line using a chain/tape only.
18. Explain the different methods used to run a line parallel to a chain line through a given point.
19. Explain the different obstacles encountered in chain surveying.
20. Explain at least one method each to continue and measure the distance between points on either side of the obstacle in the case of (a) a pond (b) a river, and (c) a building.

Problems

1. The length of a survey line was measured with a 30-m chain and found to be 128.65 m. Later it was observed that the chain was 0.02 m short. Find the true length of the line.
2. A 20-m tape was tested before starting the day's work and found to be 0.02 m too short. At the end of the day it was tested again and found to be 0.06 m too long. If the total length measured during the day was 1243.5 m, find the true length.
3. Before starting the day's work, a 30-m chain was tested and found to be 0.035 m too short. After measuring 880 m, the chain was tested again and found to be 0.015 m too long. After measuring another 750 m, the chain was tested at the end of the day and found to be 0.025 m too long. Find the true length measured during the day.
4. A plan drawn to a scale of 1 cm = 20 m has shrunk such that a line originally 10 cm long has shrunk to a length of 9.78 cm. A line AB which measures 18.7 cm on paper now has to be set out on the ground. To what length should it be set, if the chain available for measurement is 0.015 m too long?
5. A plan drawn to a scale of 1:3000 shows a rectangular tank 4 cm × 6 cm on paper. The plan has shrunk such that the lines have decreased in length by 5%. To what dimensions should the tank be set up in the field now if the 20-m chain used for setting up is 0.02 m too short?
6. The area of a plot of land was measured from a plan drawn to a scale of 1 cm = 40 m. The area was found to be 125 sq. cm. If the 30-m chain used for the survey was 0.01 m too short, find the true area of the land.
7. A plan plotted to a scale of 1:4000 has shrunk such that a line 10 cm long now measures only 9.8 cm. The 20-m chain used for conducting the survey tested to be 20.02 m too long. Find (a) the length of a line measuring 8.5 cm on plan and (b) the area of a field measuring 10 cm × 8 cm on plan.
8. If the length, breadth, and depth of a water tank were measured with a 20-m chain that is 0.1% too short and were recorded as 54 m, 36 m, and 6.5 m, respectively, find the true capacity of the tank.
9. Calculate the hypotenusal allowance for a 30-m chain (a) along a 5° slope, (b) along a gradient of 1 in 12, and (c) if the ground rises 1.8 m per chain length.
10. For a 30-m chain of 150 links, the hypotenusal allowance was 1 link for a sloping terrain. Find the slope of the ground in degrees.
11. A ground was rising 1:15 for half the length and 1:10 for the remaining half of a chain line. What is the slope correction for a 30-m chain in both cases? If the total length measured was 268.8 m, find the corresponding horizontal length.

12. If a line measured along a slope had a length of 148.8 m and the horizontal distance was 143.4 m, what is the slope in degrees?

13. The length of a line AB was measured in parts due to different slopes. The lengths measured and slopes are given below. Find the horizontal distance between A and B.

Line	AC	CD	DE	EB
Length (m)	87.2	43.4	106.8	74.2
Slope	2°	3° 30′	4° 15′	6°30′

14. Determine the horizontal distance AD with the following data:

Line	AB	BC	CD
Length (m)	80	76	106
Gradient	1:18	1:14	1:12

15. A 30-m tape was used to measure a line AB and the measured length was 86.76 m. If the tape was 0.02 m too long, what is the correct length of the line?

16. A 20-m tape was 0.5% too short. What is the correction per tape length? If the length and breadth of a rectangular plot measured with the tape were 80 m and 35 m, find the correct area of the plot.

17. A 30-m tape was standardized at 20°C. What is the correction per tape length if the temperature during measurement was 45°C. Coefficient of thermal expansion = 12×10^{-6} per degree centigrade.

18. Find the sag correction for a tape 30 m long if it is supported at the ends under a pull of 140 N. The weight per metre length of the tape is 3 N.

19. A 50-m tape was supported at the ends and at the middle. Find the sag correction if the pull during measurement was 150 N and the tape had a cross section of 3 mm^2. The unit weight of the tape is 78.6 kN/m^3.

20. In measuring distance along a sloping ground, a 20-m tape was held 0.2 m out of level and 0.3 m out of line. What is the resulting error per tape length?

21. A steel tape was used to measure distances along slopes as given below:

Slope	2° 30′	4° 15′	5° 30′
Length (m)	320	254	186

The tape was standardized at a temperature of 20°C and the temperature during measurement was 45°C. Find the correct horizontal distance.

22. Find the normal tension for a tape 30 m long if the calibrating pull was 100 N. The weight of the tape was 15 N, its area of cross section was 4 mm^2, and E = 200 GN/m^2.

23. A 30-m-long steel tape weighed 10 N and was standardized at a temperature of 20°C and with a pull of 120 N. If the temperature during measurement was 45°C and the pull applied was 180 N, find the correction per tape length if it was supported at the end points only. The coefficient of thermal expansion is $11.5 \times 10^{-6}/°C$.

24. An offset 18 m long is laid out 4° from its true direction. Find the displacement of the point parallel and perpendicular to the chain line on paper and on ground if the scale of the plan is 1:1000.

25. Find the maximum length of an offset so that the displacement of a point on paper is not more than 0.25 mm. The offset is laid 5° from its true direction and the scale of the plan is 1 cm = 20 m.

26. Find the accuracy in linear measurement of an offset such that the error in the linear and angular measurements is the same if the angular error is 5°.

27. Find the maximum length of an offset so that the error from both the sources is the same if the scale is 1 cm = 25 m and the linear accuracy is 1 in 30. Take the plotting accuracy as 0.25 mm.

28. Find the maximum permissible angular error in laying an offset of 18 m if the linear accuracy is 1 in 30 and the scale is 1:4000. 0.025 cm is the plotting accuracy.

29. The length of an offset is 20 m. The maximum error in measuring the length is 0.4 m. If the displacement on paper is not to exceed 0.25 mm, find the maximum permissible error in laying the direction of the offset. Scale: 1 cm = 30 m.

30. To continue a chain line OA beyond a pond, two lines AC and AD were set out on either side of the chain line such that CBD is a straight line. B is a point on the other side in continuation of OA. Find the length AB if AC = 321.8 m, AD = 228.7 m, BC = 124.5 m, and BD = 108.8 m.

31. To determine the distance across a river, a point A on the near side was taken and a perpendicular was set to one side of the chain line. A point C on the chain line behind A was located and a point D was chosen on the perpendicular such that ∠CDB is a right angle, B being a point in continuation of the chain line OA on the far side of the river. If AC = 61.85 m and AD = 93.8 m, find the length AB.

32. To continue a chain line past a building, from a point A, a right angle was set and a distance of 80 m taken on the perpendicular. With an adjustable cross staff, CB was set at 50° and CD was set at 75°, B and D being points on the other side of the obstacle. Find the lengths CB, CD, and AB such that B and D are on the continuation of the chain line from the other side.

33. To continue a chain line OA across a lake and find the intervening distance, two lines AC and AD were set at convenient angles so that C and D are past the obstacle and in line with B, a point on the other side and on the line OA. The distance CB also could not be measured due to an intervening pond. Lines BE and EC were laid with right angles at E. The following are the lengths measured: BE = 108 m, EC = 78 m, AC = 318.6 m, AD = 228.8 m, BD = 142.6 m. Find the distance AB.

34. In order to find the width of a river, two points A and B were taken on one bank so that AB is approximately parallel to the river. A well-defined point C on the other side was marked. With an adjustable cross staff, ∠BAC = 40 and ∠ABC = 55 were measured. Find the width of the river if AB = 200 m.

35. Data from a cross staff survey is given below. Calculate the area. L and R indicate that the offset is taken to the left and right of the chain line as we go from A to B.

Chainage	Point	Offset
0.00	A	0.0
26.1	C′	14.5 (L)
44.2	G′	18.6 (R)
56.4	D′	21.3 (L)
82.5	H′	23.6 (R)
101.3	E′	22.8 (L)
122.4	I′	20.6 (R)
136.8	F′	20.4 (L)
148.4	B	0

36. A cross staff survey was conducted and the data obtained is recorded as shown below. Find the area of the survey.

Chainage	Point	Offset
0.0	A	11.6 (L)
31.4	G′	12.4 (R)
47.6	D′	18.6 (L)
61.4	H′	25.6 (R)
88.6	E′	23.6 (L)
106.8	I′	30.2 (R)
138.4	J′	24.6 (R)
152.5	F′	20.6 (L)
174.8	B	8.8 (R)

CHAPTER

3

Measurement of Directions— Compass Surveying

Learning Objectives

After going through this chapter, the reader will be able to
- define and explain the term magnetic bearing
- explain the construction of a prismatic compass and its use
- explain the difference between a prismatic compass and a surveyor's compass
- explain the method of traversing with a compass and chain
- convert whole circle bearings to reduced bearings and vice versa and find internal angles from bearings
- explain the methods used to plot and make adjustments to a traverse
- explain the terms local attraction, magnetic declination, and dip and the precautions to be taken to eliminate errors
- explain the upkeep and maintenance of a compass

Introduction

In the previous chapter, we have studied the measurement of distances with a chain or tape and the methods of laying out angles (chain angles) by measuring distances only. However, these methods are crude, prone to errors, and not generally used in traversing. It is desirable that angles between survey lines be measured directly using angle-measuring instruments.

There are many angle-measuring instruments, some measuring the angles directly and some measuring the relative angles between a reference direction and the lines. While a theodolite, to be described in Chapter 4, measures the angle between lines directly, a compass measures the angle between a survey line and a reference line, which is usually the magnetic north–south direction at the place. The angle measured is known as *magnetic bearing*, and it is possible to calculate the angle between lines using their bearings. In this chapter, we will study the prismatic compass in detail, methods to use this instrument for traversing, and methods to calculate internal angles from such measurements for plotting and further calculations.

3.1 Magnetic Bearing

The bearing of a line as mentioned earlier is the angle that a line makes with a reference direction. When the reference direction is the magnetic north–south direction at a place, it is known as the magnetic bearing. The magnetic north–south direction at a place can be determined from the direction in which a freely suspended magnetic needle comes to rest. This is shown in Fig. 3.1(a).

The *whole circle bearing* (WCB) of a survey line is the angle made by line with the *north* direction as shown in Fig. 3.1(b). The whole circle bearing can have a value between 0° and 360°. Since for calculation purposes, especially for trigonometric ratios using log tables, we need acute angles, the whole circle bearing is reduced to an acute angle and such a bearing is known as the *reduced bearing* (RB) or *quadrantal bearing* (QB).

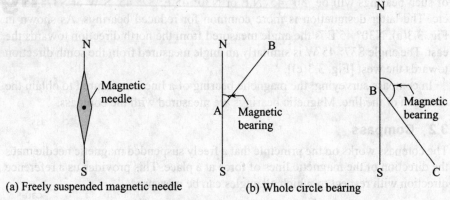

(a) Freely suspended magnetic needle   (b) Whole circle bearing

Fig. 3.1   Magnetic bearing

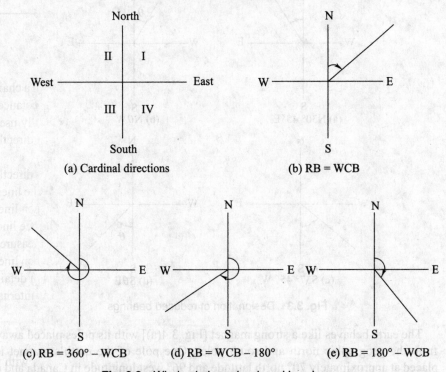

(a) Cardinal directions   (b) RB = WCB

(c) RB = 360° − WCB   (d) RB = WCB − 180°   (e) RB = 180° − WCB

Fig. 3.2   Whole circle and reduced bearings

The reduced bearing of a line is the angle made by the line with the *north* or *south* direction. The north–south and east–west directions are at right angles as shown in Fig. 3.2(a). Figures 3.2(b)–(e) show the relation between the whole circle bearings of lines in the four quadrants (made by the north–south and east–west directions) and their reduced bearings. The calculation of reduced bearings from whole circle bearings will be explained later.

Reduced bearings, being acute angles made with the north or south direction, need to be designated with the letters north or south and the direction in which the angle is measured (towards the west or east). For example, the designations of such bearings will be 30° 45′ N,E or N30° 45′E; 37° 45′ S,W or S37° 45′W, etc. The latter designation is more common for reduced bearings. As shown in Fig. 3.3(a), N30° 45′E is the angle measured from the north direction towards the east. The angle S37° 45′W is similarly an angle measured from the south direction towards the west [Fig. 3.3(c)].

In compass surveying, the magnetic bearing of a line is measured to obtain the direction of the line. Magnetic bearings are measured with the compass.

3.2 Compass

The compass works on the principle that a freely suspended magnetic needle takes the direction of the magnetic lines of force at a place. This provides us a reference direction with respect to which all angles can be measured.

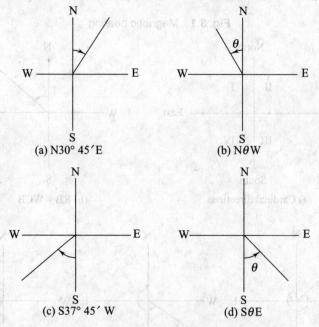

(a) N30° 45′E

(b) NθW

(c) S37° 45′ W

(d) SθE

Fig. 3.3 Designation of reduced bearings

The earth behaves like a strong magnet [Fig. 3.4(a)] with its poles placed away from the geographic north and south poles. One pole of the earth's magnet is placed at approximately 70° north latitude and 96° west longitude in Canada and a similar pole exists at a diametrically opposite location in the Southern hemisphere. A magnetic needle supported in such a way that it can rotate in a vertical plane will take up a vertical position at such a place. Since one end of a magnetic needle points to the north direction and is designated as the north pole of the needle, it is clear that the imaginary magnet inside the earth has its south pole there. This is because unlike poles attract each other. The north pole of a magnet is strictly the north-seeking pole.

The magnetic lines of force due to earth's magnetism generally go from near the South pole to the North pole. Such lines of force are parallel to the surface (horizontal) only near the equator. At other places, as these lines converge to the poles, they are inclined to the horizontal. A freely suspended magnetic needle will also take the same direction as the lines of force; it will dip (from the horizontal) by a small angle. This is known as the *dip angle*. The dip angle increases as we go from the equator to the poles.

A magnetic needle is generally made perfectly symmetrical and supported on a hard, pointed pivot. To make it take up a horizontal position, it is generally weighted with an adjustable weight. As the north pole or the north-seeking pole of the needle dips down in the Northern hemisphere, the needle is weighted in the southern segment of the needle there. Similarly, the northern segment of the needle is weighted in the Southern hemisphere. Figure 3.4(b) shows the plan and section of two forms of magnetic needles commonly used—one with pointed ends (edge-bar needle) and the other of uniform width (broad form needle). The north pole, or the north-seeking pole, of a magnetic needle when freely suspended gives us the direction of the magnetic lines of force and this is used as a reference direction in compass surveying.

There are two types of compasses—the prismatic compass and the surveyor's compass. The surveyor's compass is rarely used and is discussed here only for comparison purposes. The principle of operation of both the compasses is the same, but they are made differently and used differently in the field.

We start with the discussion of the prismatic compass.

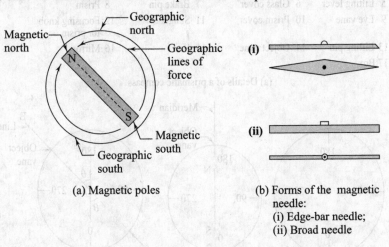

(a) Magnetic poles

(b) Forms of the magnetic needle:
(i) Edge-bar needle;
(ii) Broad needle

Fig. 3.4 Earth's magnetism

3.2.1 Prismatic Compass

A prismatic compass is shown in Fig. 3.5(a). It is made up of the following parts.

Magnetic needle The magnetic needle is the most important component of the instrument. The needle, generally of the broad form, is supported on a hard, steel pivot with an agate tip. When not in use, the needle can be lifted off the pivot, by a lifting needle, actuated by the folding of the objective vane. This is done to ensure that the pivot tip is not subjected to undue wear. The magnetic needle should be perfectly symmetrical and balanced at its midpoint on the hard pointed pivot. It should be weighted with an adjustable weight to compensate for the dip angle. The needle should be sensitive and take up the north–south direction speedily. The needle should lie in the same horizontal plane as the pivot point, and a vertical plane should pass through the pivot and the ends of the needle. For stability the needle should be made in such a way that the centre of gravity of the needle lies as much below the pivot point as possible.

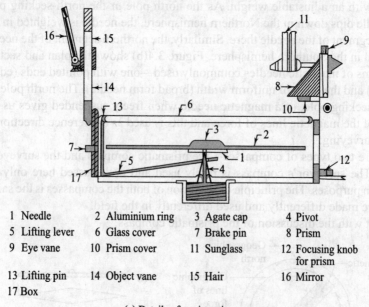

1 Needle	2 Aluminium ring	3 Agate cap	4 Pivot
5 Lifting lever	6 Glass cover	7 Brake pin	8 Prism
9 Eye vane	10 Prism cover	11 Sunglass	12 Focusing knob for prism
13 Lifting pin	14 Object vane	15 Hair	16 Mirror
17 Box			

(a) Details of a prismatic compass

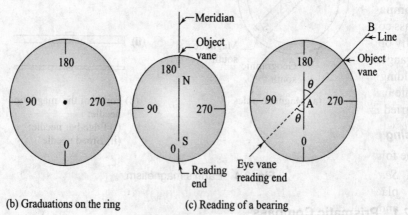

(b) Graduations on the ring   (c) Reading of a bearing

Fig. 3.5 Prismatic compass

Graduated ring An aluminium graduated ring 85 to 110 mm diameter is attached to the needle on its top to a diametrical arm of the ring. Aluminium, being a non-magnetic substance, is used to ensure that the ring does not influence the behaviour of the needle. The graduations of the ring are from 0 to 360°. 0°/360° is marked on the south end of the needle and the graduations go in a clockwise direction, with 90° marked on the west, 180° on the north, and 270° on the east directions. The graduations are marked to half degrees, but it is possible to read the angle to one-fourth of a degree by judgement. The reason for marking the graduations this way will be explained later. The graduations on the ring are inverted as they are to be read by a prism.

Eye vane and prism To measure an angle we have to have a line of sight from one station to the other. The point on the prismatic compass from where the sight-ing is done is known as the eye vane, which is made up of a rectangular frame to which a prism is hinged. A narrow slit in the prism holder enables the reading of the graduated ring when it is folded over the glass plate cover of the compass. The prism has convex surfaces, which magnify the graduations on the ring. A metal cover is used to cover the reading face of the prism when it is not in use. The prism can be raised or lowered on the metal frame for adjusting to the eye of the observer. Dark glasses may be provided on the frame, which can be brought in view while sighting bright objects to reduce glare.

Object vane Diametrically opposite the eye vane is the object vane, which is a metal frame hinged at the bottom for folding over the glass cover when it is not in use. A fine silk thread or hair is fitted on the frame vertically, which can be used to bisect a ranging rod or other objects. When the frame is folded over the glass cover, it presses against a pin, which actuates the lifting lever of the needle and lifts the needle off the pivot. Also fitted below this frame on the box is a brake pin, which, when, gently pressed, stops the oscillations of the needle by pressing against the graduated aluminium ring. The object vane may be provided with mirrors, which can be moved over the frame for sighting objects at a height or far below.

Compass box The needle and other fittings are enclosed in a metal box with a glass cover to prevent dust. The two vanes are also attached to the box at diametri-cally opposite ends. The box is attached to a metal plate through a ball and socket arrangement for levelling the compass. While the compass may also be used by holding it in the hand, it is preferable to use it with a tripod, for which the metal plate has a screwed end that can be attached to a tripod. The compass box can be carried in a leather pouch when not in use.

Using prismatic compass

The following steps are required in using the prismatic compass.

1. *Setting up and centring:* Screw the prismatic compass onto the tripod and place the tripod over the station. There is no centring device in the compass and it is centred over the station either with a plumb bob or by dropping a stone from the centre of the tripod. Centring is done by adjusting the tripod legs.
2. *Levelling:* Level the compass using the ball and socket arrangement. Gen-erally, there is no centring head provided for the compass. Levelling is done

approximately so that the needle can move freely in a horizontal plane, after opening the objective and eye vanes.

3. *Sighting the object:*   Open the object vane and eye vane and see that the needle moves freely. Direct the object vane towards the ranging rod or any other object at the next station. Sighting is done by bisecting the object with the cross hair on the object vane while looking through the eye vane. The prism of the eye vane has to be adjusted for a clear view of the graduations by moving it up or down. It is clear that the graduated ring along with the attached needle always points to the north direction while the box is rotated with the vanes. The line of sight between the stations is through the eye vane and the cross hair of the object vane and should pass through the centre of the pivot.

4. *Taking readings:*   Once the object has been clearly sighted, damp the oscillations of the needle with the braking pin if required. Once the needle comes to rest, looking through the prism, record the reading at the point on the ring corresponding to the vertical hair seen directly through the slit in the prism holder.

Graduations on ring

Suppose the line of sight points towards the north direction. The reading is taken through the prism at the eye vane. It is clear from the graduations that the prismatic compass gives the whole circle bearings of the lines. The reading taken through the prism has to be zero when the line of sight is pointing to the north. The reading end is the south end of the needle. Therefore, the zero graduation is marked at the south end as shown in Fig. 3.5(b).

Figure 3.5(c) illustrates how the bearing of a line is taken. The instrument is at station A and the line of sight has bisected the ranging rod at station B. The needle points to the north always and the reading is taken from the south end. The graduations made in the clockwise direction from the south end give the whole circle bearing of line AB, as can be seen from Fig. 3.5(c).

Temporary adjustments

At every station where the prismatic compass is placed, the following adjustments, as described above, have to be made: centring, levelling, and focusing the prism. The prism has to be focused only once if the same person has to take the readings. Centring is done by adjusting only the legs to bring the compass exactly over the station. Levelling is done to ensure that as the compass is rotated it moves very nearly in a horizontal plane and the needle moves freely.

3.2.2 Surveyor's Compass

The surveyor's compass is an old type of instrument finding rare use today. A brief description of the instrument is given below. The surveyor's compass has the following components (refer to Fig. 3.6).

Magnetic needle   The edge bar magnetic needle rests on a pivot of hard metal and floats freely.

Graduated ring   The graduated ring is not attached to the needle but to the cover box of the compass and inside it. The graduations are in the quadrantal system. The letters N, W, S, and E are marked on the ring along with graduations from 0° to 90° in each quadrant. The graduations are marked to half-degrees but can be read to one-fourth of a degree by judgement. The E and W directions are reversed on the ring. The ring moves with the compass as the box is rotated for sighting, the needle pointing to the north always.

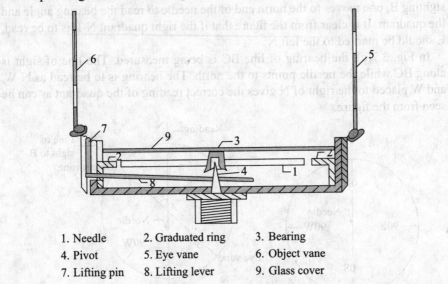

1. Needle	2. Graduated ring	3. Bearing
4. Pivot	5. Eye vane	6. Object vane
7. Lifting pin	8. Lifting lever	9. Glass cover

Fig. 3.6 Surveyor's compass

Object vane and eye vane   The object vane consists of a fine thread or hair fitted onto a metal frame for sighting objects. The eye vane is a similar frame with a fine slit but has no prism to read the graduations.

Base and tripod   The surveyor's compass cannot be used without a tripod. A base with a ball and socket arrangement and a screwing end for the tripod is used.

An arrangement for lifting the needle off the pivot is provided. This is actuated when the object vane is folded onto the cover glass.

Using surveyor's compass

The following steps are required.

1. Attach the compass box to the tripod. Place the tripod over the station and centre and level the instrument.
2. Rotate the instrument to bring the object vane in line with the ranging rod at the adjacent station. Looking through the eye vane, finely bisect the ranging rod.
3. Note the reading, by going around to the objective vane side, at the north end of the needle by looking through the glass. Take the reading along with the quadrant by noting down the letters on either side of the reading.

Graduations on ring

Figures 3.7(a)–(c) explain the graduations on the ring. N and S are marked along the north–south direction. E and W are marked along the east–west direction but their positions are interchanged, with E marked to the left of N and W to the right of N. This is done to ensure that the correct quadrant is noted when the reading is taken at the north end of the needle.

Figure 3.7(b), shows the bearing of line AB being measured. The compass is at A and the line of sight is towards B. The needle points to the north direction. After sighting B, one moves to the north end of the needle to read the bearing angle and the quadrant. It is clear from the figure that if the right quadrant N-E is to be read, E should be marked to the left N .

In Fig. 3.7(c), the bearing of line BC is being measured. The line of sight is along BC while the needle points to the north. The bearing is to be read as N-W, and W placed to the right of N gives the correct reading of the quadrant as can be seen from the figure.

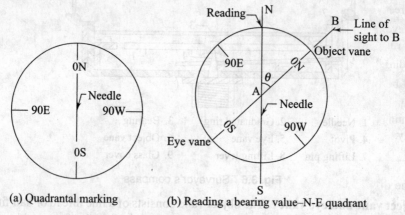

(a) Quadrantal marking   (b) Reading a bearing value–N-E quadrant

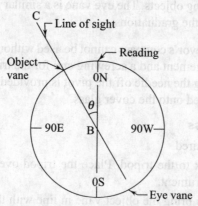

(c) Reading a bearing value–N-W quadrant

Fig. 3.7 Surveyor's compass–Graduations on the ring

3.2.3 Comparison Between Prismatic and Surveyor's Compasses

The prismatic compass and the surveyor's compass are both based on the same principle of orientation of a magnetic needle along the north–south direction. Both the instruments measure magnetic bearings.

Table 3.1 Differences between the prismatic compass and surveyor's compass

Prismatic compass	Surveyor's compass
It has a broad needle.	It has an edge bar needle.
The graduated ring moves with the needle.	The graduated ring is fixed to the box and is independent of needle.
The graduations are from 0° to 360° in the clockwise direction, with zero at the south end.	The graduations are from 0° to 90° in four quadrants. North and south are marked with 0° east and west are marked with 90°. East and west are also interchanged.
Measures whole circle bearings.	Measures quadrantal bearings.
Degrees are numbered inverted.	Degrees are numbered erect.
The eye vane has a prism to read the graduated ring.	The eye vane has no prism and is not used for reading.
Readings are taken from the south end.	Readings are taken from the north end.
Can be used in the hand-held position also.	Has to be used with a tripod only.
Sighting and reading are done simultaneously.	The object is sighted first. The observer then has to move to the object vane side to take the reading.

The differences in the construction and use of the prismatic compass and surveyor's compass are given in Table 3.1.

3.3 Designation of Bearings

We have already discussed bearings. In this section we will deal with the designation of bearings, their interconversion, and the calculation of angles.

Meridians

Meridians are reference directions with reference to which bearings are measured. There are three different types of meridians which can be used as reference directions.

True meridian The true or geographic meridian at a point is the line of intersection of a plane passing through the north and south poles and the point with the surface of the earth. Since the earth is approximately a sphere, it is clear that the meridians through different points meet at the north and south poles. Obviously, the true meridians through different points are not parallel. However, in the case of small surveys, they can be assumed to be parallel without making any significant error. The true meridian at a place can be established through astronomical observations. The direction of the true meridian at a place does not change and hence bearings taken with respect to the true meridian remain constant. If the magnetic bearing of the sun is taken at noon, the location of the true meridian at the point can be found. The sun at noon is on a plane passing through the north and south poles at a place. The *true bearing* of a survey line is the horizontal angle that the line makes with the true meridian passing through one of its ends.

Magnetic meridian The magnetic meridian through a point on the ground is the direction taken by a freely suspended magnetic needle placed at that point. The magnetic needle takes the direction of the magnetic north and south poles and is different from the true meridian. Since it is determined by a magnetic needle, the magnetic meridian can be affected by any serious magnetic interference, such as an overhead electric cable or the presence of magnetic substances. These will be explained later. The *magnetic bearing* of a survey line is the horizontal angle between the line and the magnetic meridian passing through one of its ends. A compass measures the magnetic bearing of a line.

Arbitrary meridian The arbitrary meridian at a point is any well-defined direction between any two points, such as the spire of a church, a well-defined point on the ground, or a tower. Such meridians can be used for local surveys as they will serve the purpose of a reference direction, and the required computations are possible with such data. The *arbitrary bearing* of a line is the horizontal angle between the line and the direction of the arbitrary meridian through one end of the line.

Fore bearing and back bearing

Consider the line AB shown in Fig. 3.8. If we take the direction of the meridian as upwards, as shown in the figure, and the bearing is measured clockwise from this direction, then let the bearing of AB as measured at A be θ and the bearing of BA as measured at B be β. It is clear that these two bearing angles are not equal. The bearing of AB measured at A is known as the fore bearing of AB and the bearing of BA as measured at B is known as the back bearing of AB. Therefore, when we designate a bearing as that of AB, it means that it is measured at A from the reference direction through A. When a bearing is mentioned as that of BA, it means that it is measured at B for the line BA from the meridian direction through B. It is clear from Fig. 3.8 that the two whole circle bearings, measured at A and B, differ by 180°; that is, in the case shown, $\beta - \theta = 180°$.

In all further discussion on bearings, it will be assumed that we are referring to magnetic bearings.

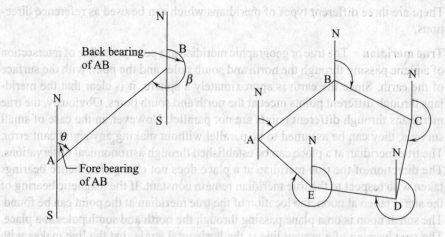

Fig. 3.8 Fore bearing and back bearing   **Fig. 3.9** Whole circle bearings

Whole circle bearing

It was mentioned earlier that the whole circle bearing is measured from the north direction as the horizontal angle between that direction and the line. The whole circle bearing can have a value between 0° and 360°. Referring to Fig. 3.9, the whole circle bearings of the lines will be recorded as shown in Table 3.2.

Table 3.2 Whole circle bearings of lines

Line	Length	Bearing
AB	150.65	62° 30′
BC	142.85	120° 15′
CD	168.90	210° 30′
DE	121.55	287° 45′
EA	124.02	333° 23′

The advantage of the WCB system of designating bearings is that only one value (that of the angle) needs to be recorded. The bearing is always measured from the magnetic north direction, clockwise. The disadvantage of this system is that the angles need to be converted to acute values for calculating trigonometric ratios, if one is using logarithmic tables.

Reduced or quadrantal bearings

At any point on the ground, the north–south direction and the east–west direction divide the horizontal plane containing that point into four quadrants. Figure 3.10(a) shows the cardinal directions and the four quadrants. The quadrants are numbered I, II, III, and IV arbitrarily.

Quadrantal bearings are always measured from the north or south direction as acute angles. Thus, in quadrants I and II the reference direction is north and the angle is measured either to the east or to the west. In quadrants III and IV, the angle is measured from the south, either to the west or to the east. This is shown in Fig. 3.10(b).

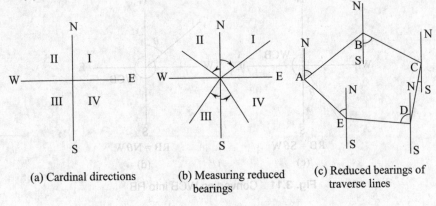

(a) Cardinal directions	(b) Measuring reduced bearings	(c) Reduced bearings of traverse lines

Fig. 3.10 Reduced bearings

Table 3.3   Quadrantal bearings of lines

Line	Length	Bearing
AB	150.65	N62° 30′E
BC	142.85	S59° 45′E
CD	168.90	S30° 30′W
DE	121.55	N72° 15′W
EA	124.02	N26° 37′W

The quadrantal bearings of lines are recorded as shown in Table 3.3 for the traverse shown in Fig. 3.10(c).

Conversion of bearings

If the whole circle bearings are given, it will be necessary to convert them to quadrantal or reduced bearings. Similarly, quadrantal bearings can also be converted to whole circle bearings.

Whole circle bearing to reduced bearing   To convert WCB (measured clockwise from the north direction) to reduced bearings, the following simple rules are followed (Fig. 3.11).

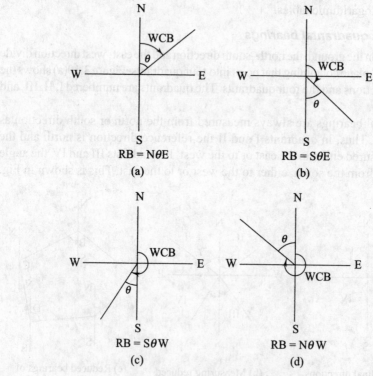

Fig. 3.11   Converting WCB into RB

(a) If the WCB is less than 90°, the RB is numerically equal to the WCB. The quadrant designation is N-E.

(b) If the WCB is between 90° and 180°, the RB is equal to 180° – WCB. The quadrant designation is S-E.

(c) If the WCB is between 180° and 270°, the RB is equal to WCB – 180°. The quadrant designation is S-W.

(d) If the WCB is between 270° and 360°, the RB is equal to 360° – WCB. The quadrant designation is N-W.

Quadrantal bearing to whole circle bearing To convert given quadrantal bearings to WCB, the following simple rules are to be followed (Fig. 3.12).

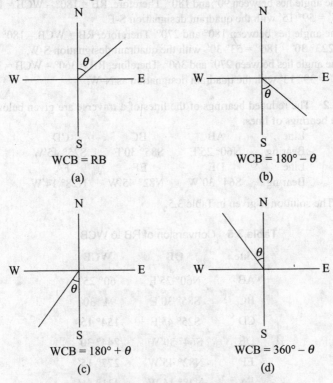

Fig. 3.12 Converting RB into WCB

(a) If the quadrant designation is N-E, the WCB is numerically equal to the reduced bearing.

(b) If the quadrant designation is S-E, the WCB is equal to 180° – QB.

(c) If the quadrant designation is S-W, the WCB is equal to 180° + QB.

(d) If the quadrant designation is N-W, the WCB is equal to 360° – QB.

The following examples illustrate the converson of bearings.

Example 3.1 The WCBs of a traverse are recorded as given below. Convert these to reduced bearings.

Line	AB	BC	CD	DA
Bearing	70° 30′	120° 45′	223° 30′	320° 47′

Solution The solution is given in Table 3.4.

Table 3.4 Conversion of WCB to RB

Line	WCB	RB
AB	70° 30′	N70° 30′E
BC	120° 45′	S59° 15′E
CD	223° 30′	S53°30′W
DA	320° 47′	N39° 13′W

AB: The angle being < 90°, the RB is equal to the WCB with the quadrant designation N-E.

BC: The angle lies between 90° and 180°. Therefore, RB = 180° − WCB = 180° − 120° 45′ = 59° 15′ with the quadrant designation S-E.

CD: The angle lies between 180° and 270°. Therefore, RB = WCB − 180° = 223° 30′ − 180° = 53° 30′ with the quadrant designation S-W.

DA: The angle lies between 270° and 360°. Therefore, RB = 360° − WCB = 360° − 320° 47′ = 39° 13′ with the quadrant designation as N-W.

Example 3.2 The reduced bearings of the lines of a traverse are given below. Find the whole circle bearings of lines.

Line	AB	BC	CD
Bearing	N60° 25′E	S85° 30′E	S25° 45′W
Line	DE	EF	FA
Bearing	S64° 30′W	N82° 45′W	N28° 14′W

Solution The solution is given in Table 3.5.

Table 3.5 Conversion of RB to WCB

Line	QB	WCB
AB	N60° 25′E	60° 25′
BC	S85° 30′E	94° 30′
CD	S25° 45′E	154° 15′
DE	S64° 30′W	244° 30′
EF	N82° 45′W	277° 15′
FA	N28° 14′W	331° 46′

AB: The quadrant designation being N-E, WCB = RB.

BC: The quadrant designation is S-E. Therefore, WCB = 180° − RB = 180° − 85° 30′ = 94° 30′.

CD: The quadrant designation is S-E. Therefore, WCB = 180° − 25° 45′ = 154° 15′.

DE: The quadrant designation is S-W. Therefore, WCB = 180° + RB = 180° + 64° 30′ = 244° 30′.

EF: The quadrant designation is N-W. Therefore, WCB = 360° − RB = 360° − 82° 45′ = 277° 15′.

FA: The quadrant designation is N-W. Therefore, WCB = 360° − 28° 14′ = 331° 46′.

3.4 Calculation of Included Angles from Bearings

At the point where two survey lines meet, two angles are formed—an exterior angle and an interior angle. The interior angle or included angle is generally the smaller angle (< 180°) but this need not necessarily be so always. This is illustrated by the traverse shown in Fig. 3.13(b), where the interior angle at point E is more than 180°.

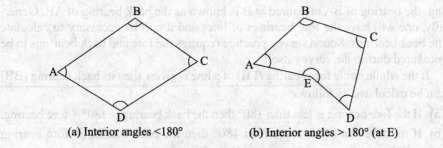

(a) Interior angles <180°   (b) Interior angles > 180° (at E)

Fig. 3.13 Interior angles

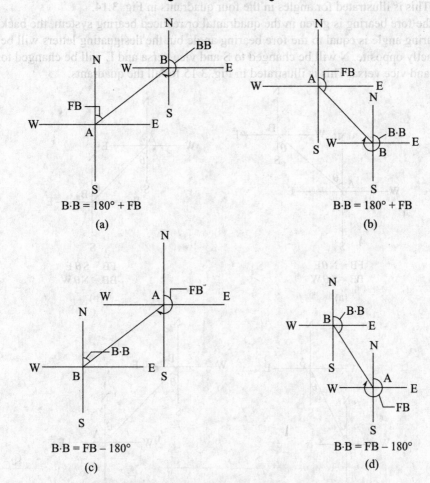

B·B = 180° + FB   B·B = 180° + FB

(a)   (b)

B·B = FB − 180°   B·B = FB − 180°

(c)   (d)

Fig. 3.14 Fore bearings and back bearings (WCB)

Given the bearings of the lines of a traverse, it is possible to calculate the included angles, which can be used for plotting the traverse. While some rules will be given for this purpose, it is strongly recommended that a rough sketch of the traverse be drawn for the purpose of calculating the interior angles or bearings from included angles. A sketch always gives a better idea for calculations.

3.4.1 Fore Bearings and Back Bearings of Lines

As mentioned earlier, if the bearing of AB measured at A is taken as the fore bearing, the bearing of BA measured at B is known as the back bearing of AB. Generally, one will have the fore bearings of lines and it will be necessary to calculate the back bearings. Sound survey practice requires the fore and back bearings to be measured during the survey itself.

If the whole circle fore bearing (FB) of a line is given, then its back bearing (BB) can be calculated as follows.

(a) If the fore bearing is less than 180° then the back bearing is 180° + fore bearing.

(b) If the fore bearing is more than 180° then the back bearing is fore bearing − 180°.

This is illustrated for angles in the four quadrants in Fig. 3.14.

If the fore bearing is given in the quadrantal or reduced bearing system, the back bearing angle is equal to the fore bearing angle but the designating letters will be exactly opposite. N will be changed to S and vice versa and E will be changed to W and vice versa. This is illustrated in Fig. 3.15 for all the quadrants.

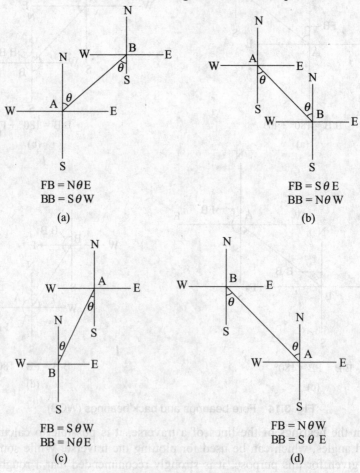

(a)

FB = N θ E
BB = S θ W

(b)

FB = S θ E
BB = N θ W

(c)

FB = S θ W
BB = N θ E

(d)

FB = N θ W
BB = S θ E

Fig. 3.15 Fore bearings and back bearings (RB)

3.4.2   Included Angles from Bearings

Included angles from WCBs

Assume that the whole circle bearings are given and at any point P, two lines meet, say, PO and PQ (shown in Fig. 3.16) in a traverse MNOPQRM. In this order, the WCB of PO is the back bearing of OP and the WCB of PQ is the fore bearing of PQ. Knowing the fore bearing and back bearing of lines meeting at a point, the included angle between PO and PQ (or ∠OPQ) = difference between the fore and back bearings. If it is less than 180°, it is the interior angle generally. If the difference is more than 180°, then (360° – difference) is the interior angle. As mentioned earlier, one should draw a rough sketch of the traverse to apply any rule correctly. Different causes are shown in Fig. 3.16 explaining the rule.

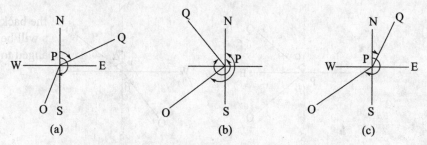

(a)   (b)   (c)

Fig. 3.16   Included angles from WCBs of lines

Included angles from RBs

If the reduced bearings are given, then the included angles can be obtained by applying the following rules.

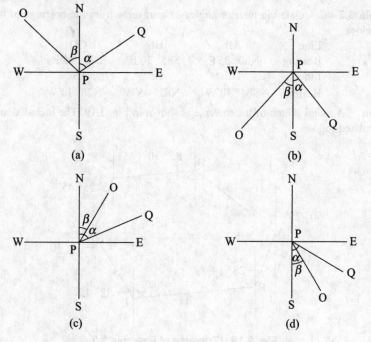

(a)   (b)

(c)   (d)

Fig. 3.17   Included angles from RBs of lines

Let the fore bearing of line PQ be equal to α and the back bearing of line PO be β.

(a) If both the bearings are from the same direction (north or south), then the included angle = $\alpha + \beta$ if the second letters of both the bearings are not the same (E or W) and $\alpha - \beta$ if the second letters of both the bearings are the same (both bearings are N-E or N-W or S-E or S-W). This is illustrated in Fig. 3.17.

(b) If the two bearings are measured from two different directions (e.g., α from N and β from S or vice versa), then the interior angle depends upon the second letter, W or E.

(c) If the second letters are the same, then the included angle = $180° - (\alpha + \beta)$.

(d) If the second letters are different, then the included angle = $180° - (\alpha - \beta)$.

All these cases are illustrated in Fig. 3.18.

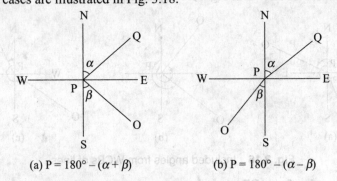

(a) P = $180° - (\alpha + \beta)$ (b) P = $180° - (\alpha - \beta)$

Fig. 3.18 Included angles from RBs of lines

The following examples illustrate the application of these rules.

Example 3.3 Calculate the interior angles of a traverse from the bearings of the lines given below.

Line	AB	BC	CD
Bearing	N60° 25′E	S85° 30′E	S25° 45′E
Line	DE	EF	FA
Bearing	S64° 30′W	N82° 45′W	N28° 14′W

Solution A rough sketch of the traverse is shown in Fig. 3.19. The included angles as are calculated follows.

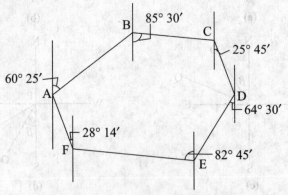

Fig. 3.19 Traverse of Example 3.3

∠A: Back bearing of FA = S28° 14′E; fore bearing of AB = N60° 25′E; included angle = 180° − (28° 14′ + 60° 25′) = 91° 21′. [Rule: Bearings measured from different directions; second letter designation the same; 180° − ($\alpha + \beta$).]

∠B: Bearing of BA = S60° 25′W; fore bearing of BC = S85° 30′E; included angle at B = 60° 25′ + 85° 30′ = 145° 55′. [Rule: Bearings measured from the same irection (S); second letter designation different; included angle = ($\alpha + \beta$).]

∠C: Bearing of CB = N85° 30′W; bearing of CD = S25° 45′E; ∠C = 180° − (85° 30′ − 25° 45′) = 120° 15′. (Rule: Bearings measured from different directions; second letter is different; included angle = 180° − $\alpha - \beta$).]

∠D: Bearing of DC = N25° 45′W; bearing of DE = S64° 30′W; ∠D = 180° − (25° 45′ + 64° 30′) = 89° 45′. (Rule: Same as for ∠A.)

∠E: Bearing of ED = N64° 30′E; bearing of EF = N82° 45′W; ∠E = 64° 30′ + 82° 45′ = 147° 15′. (Rule: Same as for ∠B.)

∠F: Bearing of FE = S82° 45′E; bearing of FA = N28° 14′W; ∠F = 180° − (82° 45′ − 28° 14′) = 125° 29′. (Rule: Same as for ∠C.)

Check: Sum of interior angles = ($2n − 4$) right angles = ($2 \times 6 − 4$) × 90° = 720°; sum of angles = 91° 21′ + 145° 55′ + 120° 15′ + 89° 45′ + 147° 15′ + 125° 29′ = 720°.

Example 3.4 Whole circle bearings of the lines of a traverse ABCDA are given below. Find the interior angles at A, B, C, and D.

Line	AB	BC	CD	DA
Bearing	70° 30′	120° 45′	223° 30′	320° 47′

Solution The angles can be calculated as the difference of the bearings measured at a point. The sketch of the traverse is shown in Fig. 3.20.

∠A = bearing of AD − bearing of AB = (320° 47′ − 180°) − 70° 30′ = 70° 17′

∠B = bearing of BA − bearing of BC = (180° + 70° 30′) − 120° 45′ = 129° 45′

∠C = bearing of CB − bearing of CD = (180° + 120° 45′) − 223° 30′ = 77° 15′

∠D = Bearing bearing of DA − bearing of DC = 320° 47′ − (223° 30′ − 180°) = 277° 17′ >180°

∠D = 360° − 277° 17′ = 82° 43′

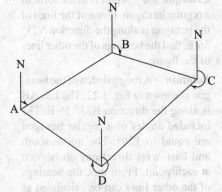

Fig. 3.20 Traverse of Example 3.4

Check: Sum of interior angles = ($2n − 4$) right angles = 4 × 90° = 360°. Sum of interior angles = 70° 17′ + 129° 45′ + 77° 15′ + 82° 43′ = 360°.

3.4.3 Bearings from Internal Angles

Sometimes it becomes necessary to find the bearings of lines from internal angles and the known bearing of a line. This is just the reverse of determining included angles from bearings, and with a neat sketch of the traverse to help, it is a simple matter to calculate the bearings of other lines. It is easier to work with whole circle bearings, which can be later converted to reduced bearings if required. The following examples illustrate the procedure.

Example 3.5 The interior angles of a traverse are given below. If the bearing of line AB is N36° 45′W, find the bearings of the remaining lines of the traverse. ∠A = 114° 51′, ∠B = 114° 30′, ∠C = 108° 15′, ∠D = 90° 45′, ∠E = 111° 39′.

Solution A rough sketch of the traverse is shown in Fig. 3.21. Draw the north–south and east–west lines through all the points. From the rough sketch, the bearings of the other lines can be calculated as follows.

Bearing of BC = 180° – (∠B + bearing of AB) = 180° – 114° 30′ – 36° 45′ = 28° 45′; the quadrant is N-E; therefore, bearing of BC = N28° 45′E.

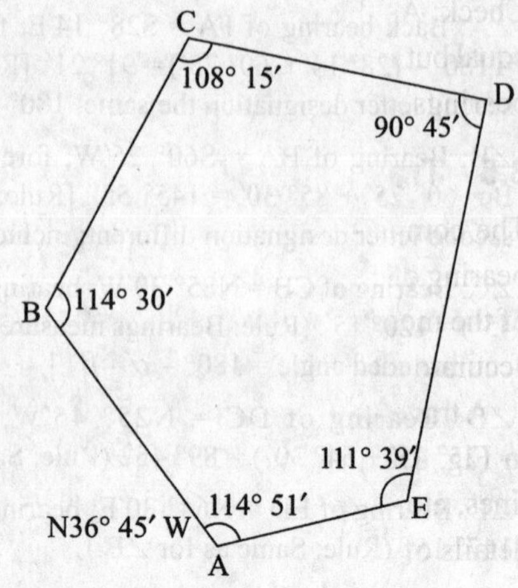

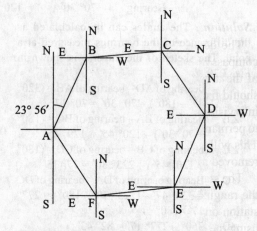

Fig. 3.21 Traverse of Example 3.5

Bearing of CD = ∠C – bearing of BC = 108° 15′ – 28° 45′ = 79° 30′; the quadrant designation is S-E; bearing of CD = S79° 30′E.

Bearing of DE = 180° – ∠D – bearing of CD = 180° – 90° 45′ – 79° 30′ = 9° 45′; the quadrant designation is S-W; bearing of DE = S9° 45′W.

Bearing of EA = 180° – (∠E – bearing of ED) = 180° – 111° 39′ + 9° 45′ = 74° 54′; the quadrant designation is S-W; bearing of EA = S74° 54′W.

Example 3.6 A plot is in the form of a regular hexagon. If one of the lines of the hexagon is along the direction N23° 56′E, find the bearings of the other lines of the figure.

Solution A rough sketch of the hexagon is shown in Fig. 3.22. The line AB is along the direction N23° 56′E. The included angles of a regular hexagon are equal to 120°. The north–south and east–west directions are shown at each point. From this, the bearings of the other lines can be calculated as follows.

Bearing of AB = N23° 56′E.

Fig. 3.22 Hexagon of Example 3.6

Bearing of BC = 180° – ∠B + bearing of AB = 180° – 120° + 223° 56′ = 83° 56′, the quadrant is N-E; the bearing of BC = N83° 56′E.

Bearing of CD = ∠C – bearing of BC = 120° – 83° 56′ = 36° 04′; the quadrant is S-E; the bearing of CD = S36° 04′E.

Bearing of DE = 180° – ∠D – bearing of CD = 180° – 120° – 36° 04′ = 23° 56′, the quadrant is S-W; bearing of DE = S23° 56′W.

Bearing of EF = 180° – 120° + bearing of DE = 180° – 120° + 23° 56′ = 83° 56′; the quadrant is S-W; bearing of EF = S83° 56′W.

Bearing of FA = ∠F – bearing of FE = 120° – 83° 56′ = 36° 04′; the quadrant is N-W; the bearing of FA = N36° 04′W.

Check: As the opposite sides of the hexagon are parallel, the bearings should be equal but in different quadrants. It can be seen that this is the case with the calculated bearings.

3.5 Traversing with Chain and Compass

The compass is only an angle-measuring instrument that measures the magnetic bearing of survey lines. The measurement of magnetic bearing can be used in place of the measurement of chain angles as was described in Chapter 2. It will be more accurate as far as the directions of lines are concerned.

A traverse in chain survey can be an open traverse or a closed traverse as shown in Fig. 3.23. An open traverse is generally run along river banks, roads, railway lines, etc. A closed traverse is run to obtain the map of a small area and fill in the details of the features in that area.

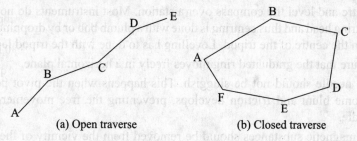

| (a) Open traverse | (b) Closed traverse |

Fig. 3.23 Types of traverses

The details of fieldwork outlined in Chapter 2 for chain surveying apply to compass and chain traversing too. The surveyor should conduct a reconnaissance of the area to identify the topographical features of the area to be surveyed. He should fix survey stations as far away as possible, ensuring at the same time their intervisibility and ease of chaining. The starting station must be fixed with reference to permanent features such as building walls and similar structures near the station. This will be useful to locate the point at a later time when the pegs or arrows are removed after the survey.

The stations should be marked with pegs with a cross mark over them to place the ranging rod. The fieldwork consists of placing the compass over the survey station on a tripod. The compass should be centred over the station and levelled using the ball and socket arrangement. In the case of an open traverse, the bearing of the first line is observed. In the case of a closed traverse, the fore bearing of the first line and the back bearing of the last line are observed when the instrument is set up at the first station. In the field book, bearings are recorded separately by naming the stations in clockwise or counter clockwise order. Chaining is done for the first line and offset measurements for the details are carried out as for chain surveying. These details are recorded as explained in Chapter 2. Each survey line should be recorded in one page and necessary notes should be made on that page for any future reference.

The compass is then shifted to the next station and placed over it. The back bearing of the first line is observed after the temporary adjustments are completed. One should note whether or not the back bearing and fore bearing of the same line differ by 180° in the case of the prismatic compass. If there is any difference and the difference is small, a second reading should be made. If the difference still remains, it may be due to an external magnetic influence, which can be corrected later. If the difference in the readings is very large, it will be necessary to take the fore bearing and back bearing again to verify and eliminate the error.

The fore bearing of the next line is then taken and the line is measured with a chain or tape. The process is then repeated for all other lines. The field notes should be prepared in detail for future reference. The prismatic compass is a fine instrument and is used for rough surveys where speed is important rather than precision. It is ideal for such surveys.

3.5.1 Precautions in Using Prismatic Compass

The following precautions should be taken in using the compass during traversing.

1. Centre and level the compass over a station. Most instruments do not have a centring head and thus centring is done with a plumb bob or by dropping a stone from the centre of the tripod. Levelling has to done with the tripod legs.
2. Ensure that the graduated ring moves freely in a horizontal plane.
3. The needle should not be sluggish. This happens when the pivot point has become blunt and friction develops, preventing the free movement of the needle.
4. All magnetic substances should be removed from the vicinity of the needle. Metallic substances such as key chains can affect the readings.
5. Always take the fore bearings and back bearings of the lines as a check. If there is a discrepancy, repeat the readings. If the discrepancy still remains, make a note of the presence of local external magnetic influences.
6. Readings should be taken with care, noting down the figures along with the correct direction. In the case of the surveyor's compass, the directions should also be noted.
7. With the prismatic compass, readings should be taken straight along the needle, avoiding parallax.
8. Take double readings for important lines.
9. The intervisibility of stations should be taken care of while selecting stations. In case the stations are not intervisible, carefully lay out a parallel line using offsets and take the bearing of this line.

3.5.2 Errors in Compass Surveying

The following errors are common in surveying with a compass.

Instrumental errors

(a) The needle may not be straight, giving wrong readings.
(b) The pivot point may have become blunt and the needle may not move freely.
(c) The line of sight may not pass through the centre of the graduated ring.
(d) The ring may not move in a horizontal plane due to the dip of the needle as a result of the wrong adjustment of the balancing weight.

(e) The cross hair in the objective vane may not be straight or may have become loose.

Personal errors

(a) Reading the graduations in the wrong direction or reading the quadrants wrongly
(b) Improper centring of the compass over the station
(c) Not levelling the compass properly
(d) Not bisecting the signal at a station properly

Other errors

(a) Variation in declination during the day, when the survey is carried out over a long duration during the day
(b) Local attraction due to the proximity of external magnetic influences at one or more stations
(c) Other variations due to magnetic storms, cloud cover, etc., which affect the magnetic needle

3.6   Corrections to Measured Bearings—Local Attraction

As indicated earlier, the compass measures the magnetic bearing of lines based on the principle that the magnetic needle aligns itself along the magnetic lines of force due to the earth's magnetism. It is clear that the presence of any external magnetic influence, such as iron ore mines, overhead electric cables, electric poles, and large quantities of magnetic material, will affect the magnetic bearing. This effect is known as *local attraction*. The term local attraction is also used for errors in bearings.

The presence of local attraction at a station is indicated if the fore bearing and back bearing of a line do not agree with the rules mentioned earlier. The prismatic compass measures the whole circle bearings of lines and the fore and back bearings should differ by 180°. This should be checked during fieldwork itself and if the difference is small and remains irrespective of checking again, then a note should be made of the fact and the likely feature nearby that may be the cause. Once local attraction is detected, corrections should be made to the bearings before plotting.

3.6.1   Methods to Correct Bearings for Local Attraction

In general, it will be observed that the fore and back bearings of at least one line will be correct, and the difference is 180° in the case of whole circle bearings and numerically equal in the case of reduced bearings. If the observed bearings of no line differ by 180° or are numerically equal, then the bearing of a line with the least difference should be identified and the corrections applied from that line.

Two methods are employed to correct the observed bearings. In the first method, included angles are calculated first. The principle used is that even if a station is affected by local attraction, the included angle calculated there should be correct, as both the lines are equally affected. Then, starting from the bearing of a line not affected by local attraction, all the other bearings are calculated using the included angles. In the case of a closed traverse, it is clear that the sum of included angles can be known from the number of sides. The advantage of this method is that there is a check on the included angles also while the corrections are made. It should be observed that since local attraction, if it exists, affects all the bearings measured at a station equally, the included angles are correct even at a station affected by local attraction. The disadvantage of this method is that it requires more effort for calculating the included angles.

In case the included angles do not add up to the value required from the number of sides, the method adopted is to distribute the error in the sum of included angles equally amongst all the angles and find the corrected included angles. These included angles are then used to correct the bearings, starting from the line not affected by local attraction. The following procedure is to be adopted for correcting bearings affected by local attraction by the method of included angles.

1. By inspecting the observed bearings, find the line not affected by local attraction. If no line is found to be free of local attraction, find the line having the least difference between the fore and back bearings to start the corrections. Draw a rough sketch of the traverse to help in the calculation.

2. Find the included angles of the closed traverse by the methods outlined earlier.

3. Check that the sum of the included angles is equal to $(2n - 4)$ right angles. If it is not so, find the error and correct the included angles by distributing the error equally among all the angles. This will give us correct included angles that add up to the value required based on the number of sides of the traverse.

4. From the unaffected or the least affected line, start the corrections by finding bearings from included angles. The method to find bearings from included angles has already been outlined earlier. This will give us the correct bearings of lines.

The second method for correcting bearings for local attraction is based on the principle that the effect of local attraction is the same for all the bearings taken at a station affected by local attraction. The magnetic needle at the affected station is deflected by the same amount and in the same direction for all bearings taken at that station. As an example, in a traverse PQRSTUP, if it is found that the fore and back bearings of line RS are correct, then it can be assumed that stations R and S are not affected by local attraction. The bearings of lines TS and QR can be corrected to match the bearings of ST and RQ. The quantum of local attraction and its direction will then be known for stations T and Q. The bearings of TU and QP can then be corrected by applying the same amount of correction to the bearings observed for stations T and Q. Similar corrections can be made to bearings taken at other stations. The procedure for correcting the bearings using this method entails the following steps.

1. By inspecting the observed bearings, find the line not affected by local attraction. Assume that the two end stations of that line are not affected by local attraction.

2. The forward and backward bearings taken from that station are then correct. The bearing to this station taken from the next station is then corrected as required. This will provide the amount of correction to be applied to the bearings taken at the next station.

3. Correct the other bearings taken at that station by the same amount.

4. Proceeding to the next station, correct the other bearings to match the corrected bearings of step 3. The quantum of local attraction at that station is thus determined.
5. Proceed similarly to all the stations of the traverse and correct the bearings.

The following examples illustrate the procedure for correcting bearings for local attraction.

Example 3.7 The bearings observed at the stations of a closed traverse are given below. Check whether the bearings are correct. If not, correct the bearings by the method of included angles.

AB: 122° 15′	BA: 302° 15′
BC: 66° 00′	CB: 243° 45′
CD: 308° 15′	DC: 133° 00′
A: 198° 00′	AD: 15° 30′

Solution A rough sketch of the traverse is drawn as shown in Fig. 3.24.

From an inspection of the bearings, it is seen that the bearings of lines AB and BA differ by 180°. Thus A and B can be assumed to be free of local attraction. The included angles of the traverse can be calculated as follows.

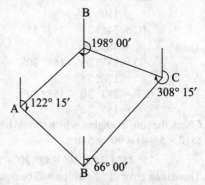

∠A = bearing of AB – bearing of
AD = 122° 15′ – 15° 30′ =
106° 45′

∠B = bearing of BA – bearing of BC
= 302° 15′ – 66° 00′ = 236°
15′; this is more than 180° and

Fig. 3.24 Traverse of Example 3.7

hence the exterior angle ∠B = 360° – 236° 15′ = 123° 45′
∠C = bearing of CD – bearing of CB = 308° 15′ – 243° 45′ = 64° 30′
∠D = bearing of DA – bearing of DC = 198° 00′ – 133° 00′ = 65° 00′
Sum of interior angles = 106° 45′ + 123° 45′ + 64° 30′ + 65° 00′ = 360° [Check:
(2 × 4 – 4) right angles = 360°]
Bearing of BC = 123° 45′ – (360° – bearing of BA) (refer to Fig. 3.24)
= 123° 45′ – (360° – 302° 15′) = 66° 00′
Bearing of CB = 246° 00′
Bearing of CD = bearing of CB + ∠C = 246° 00′ + 64° 30′ = 310° 30′
Bearing of DC = 310° 30′ – 180° = 130° 30′
Bearing of DA = bearing of DC + ∠D = 130° 30′ + 65° = 195° 30′
Bearing of AD = 195° 30′ – 180° = 15° 30′

The corrected bearings are given in Table 3.6.

Table 3.6 Corrected bearings of Example 3.7

Line	Fore bearing	Back bearing
AB	122° 15′	302° 15′
BC	66° 00′	246° 00′
CD	310° 30′	130° 30′
DA	195° 30′	15° 30′

Example 3.8   The following bearings are observed while traversing with a compass and tape. Check the bearings for local attraction. Correct the bearings by the method of included angles.

AB: 188° 45'   BA: 7° 45'
BC: 118° 15'   CB: 298° 15'
CD: 346° 35'   DC: 166° 30'
DE: 337° 05'   ED: 158° 10'
EA: 293° 30'   AE: 113° 00'

Solution   A rough sketch of the traverse is shown in Fig. 3.25. The included angles are calculated as follows:

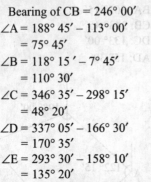

Bearing of CB = 246° 00'
∠A = 188° 45' – 113° 00'
   = 75° 45'
∠B = 118° 15' – 7° 45'
   = 110° 30'
∠C = 346° 35' – 298° 15'
   = 48° 20'
∠D = 337° 05' – 166° 30'
   = 170° 35'
∠E = 293° 30' – 158° 10'
   = 135° 20'

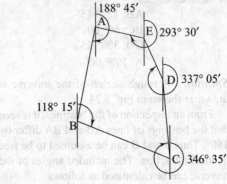

Fig. 3.25   Traverse of Example 3.8

Check the sum of angles, which should be equal to $(2 \times 5 - 4) \times 90° = 540°$:

75° 45' + 110° 30' + 48° 20' + 170° 35' + 135° 20' = 540° 30'

There is an error of 30', which will be equally distributed among the angles. As there are five angles, and the sum is more, 6' will be deducted from each angle to get the correct interior angles. The interior angles will then be ∠A = 75° 39', ∠B = 110° 24', ∠C = 48° 14', ∠D = 170° 29', and ∠E = 135° 14'. We note that the bearing of BC and CB differ by 180°. Stations B and C are therefore free of local attraction. Taking the bearing of CB as correct,

Bearing of CD = bearing of CB + ∠C = 298° 15' + 48° 14' = 346° 29'
Bearing of DC = 346° 29' – 180° = 166° 29'
Bearing of DE = bearing of DC + ∠D = 166° 29' + 170° 29' = 336° 58'
Bearing of ED = 156° 58'
Bearing of EA = bearing of ED + ∠A = 156° 58' + 135° 14' = 292° 12'
Bearing of AE = 112° 12'
Bearing of AB = bearing of AE + ∠A = 112° 12' + 75° 39' = 187° 51'
Bearing of BA = 7° 51'
Bearing of BC = bearing of BA + ∠B = 7° 51' + 110° 24' = 118° 15'

The corrected bearings are listed in the Table 3.7.

Table 3.7   Corrected bearings of Example 3.8

Line	Fore bearing	Back bearing
AB	187° 51'	7° 51'
BC	118° 15'	298° 15'
CD	346° 29'	166° 29'
DE	336° 58'	156° 58'
EA	292° 12'	112° 12'

Example 3.9 The observed bearings of a closed traverse are given below. Find the stations affected by local attraction and correct the bearings by finding the local attraction at the affected stations.

AB: 36° 00'   BA: 216° 45'
BC: 98° 15'   CB: 276° 00'
CD: 201° 45'   DC: 23° 15'
DA: 322° 45'   AD: 142° 45'

Solution A rough sketch of the traverse is shown in Fig. 3.26. It must be noted that the bearings of AD and DA differ exactly by 180°. Stations A and D are thus not affected by local attraction. Thus the bearing of AB can be taken to be correct.

The bearing of BA should be = 36° 00' + 180° = 216° 00'. The needle at B has thus rotated clockwise by 45' due to local attraction. The bearing of BC, therefore, should be 98° 15' − 00° 45' = 97° 30'.

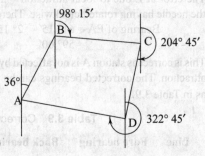

Fig. 3.26 Traverse of Example 3.9

The bearing of CB, therefore, should be equal to 180° + 97° 30' = 277° 30'. This indicates that the needle has rotated counter-clockwise by 1° 30' at station C.

The bearing of CD should therefore be equal to 201° 45 + 1° 30' = 203° 15' and the bearing of DC = 203° 15' + 180° = 23° 15', which is correct. The corrected bearings are listed in Table 3.8.

Table 3.8 Corrected bearings of Example 3.9 [modified for illustration]

Line	Fore bearing	Back bearing	Station	Local attraction
AB	36° 00'	216° 00'	A	Not affected
BC	97° 30'	277° 30'	B	Needle has turned clockwise by 00° 45'
CD	203° 15'	23° 15'	C	Needle has turned counter-clockwise by 1° 30'
DA	322° 45'	142° 45'	D	Not affected

Example 3.10 The bearings in the quadrantal system observed while traversing with a compass and chain are given below. Find the local attraction at the affected stations and also the corrected bearings.

AB: S36° 15'E   BA: N36° 15'W
BC: S44° 30'W   CB: N45° 30'E
CD: N71° 45'W   DC: S71° 00'E
DE: N14° 00'E   ED: S14° 30'W
EA: N61° 15'E   AE: S59° 00'W

Solution While dealing with reduced bearings, remember that RBs are measured from North and South directions and draw sketches to find corrections. A rough sketch of the traverse is shown in Fig. 3.27. It is observed that the bearings of AB and BA are numerically equal and the quadrants are reversed, which are correct. Stations A and B are thus not affected by local attraction.

Taking the bearing of BA to be correct, the bearing of BC is also correct. The bearing of CB should be 44° 30'. The local attraction at station C = 1°, the needle having rotated clockwise. This correction is applied to the bearing of CD. Therefore,

Bearing of CD = 71° 45' + 1° = 72° 45'

The bearing of DC should be equal to S72° 45′E. Local attraction at D = 1° 45′, the needle having rotated clockwise. Therefore

$$\text{Bearing of DE} = 14° 00′ − 1° 45′$$
$$= 12° 15′$$
$$\text{Bearing of ED} = \text{S}12° 15′ \text{W}$$

The error at E due to local attraction = 2° 15′, the needle having rotated clockwise. Therefore,

$$\text{Bearing of EA} = 61° 15′ − 2° 15′$$
$$= 59° 00′$$

This is correct as station A is not affected by local attraction. The corrected bearings can be listed as in Table 3.9.

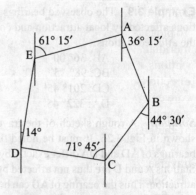

Fig. 3.27   Traverse of Example 3.10

Table 3.9   Corrected bearings of Example 3.10

Line	Fore bearing	Back bearing	Station	Local attraction
AB	S36° 15′E	N36° 15′W	A	Nil
BC	S44° 30′W	N44° 30′E	B	Nil
CD	N72° 45′W	S72° 45′E	C	1° clockwise
DE	N12° 15′E	S12° 15′W	D	1° 45′ clockwise
EA	N59° 00′E	S59° 00′W	E	2° 15′ clockwise

3.7   Plotting and Adjusting a Traverse

Once the fieldwork is over, the traversing data is to be transferred to a drawing sheet by plotting the traverse to some standard scale. There are many methods of plotting a traverse. Some of the methods are outlined in the following. The data given in Table 3.10 will be used to illustrate the procedure in each case.

A rough sketch of the traverse should always be prepared as an aid to visualizing the plan and placing the drawing properly on the drawing sheet. This is shown in Fig. 3.28.

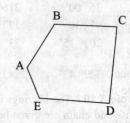

Fig. 3.28   Rough sketch of traverse (Table 3.10)

Table 3.10   Traverse data

Line	Length	Bearing
AB	402	60° 45′
BC	398	109° 30′
CD	430	202° 30′
DE	490	278° 15′
EA	274.5	343° 38′

3.7.1 Plotting by Bearings

To plot a traverse using bearings, parallel meridians have to be drawn through all stations (refer to Fig. 3.29). The north direction should be fixed arbitrarily and conveniently on paper and also marked separately on the right-hand top corner of the sheet by drawing a parallel line. The following steps are used to plot the traverse.

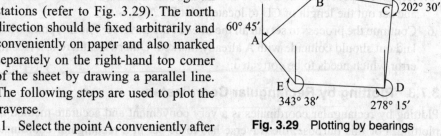

Fig. 3.29 Plotting by bearings

1. Select the point A conveniently after fixing the scale of the drawing. It should be ensured that the drawing is placed reasonably in the centre of the sheet and well within the borders of the sheet.
2. Draw the north direction through A. Set out the bearing of AB and draw the line AB in that direction. In the case shown, the line AB is along a direction 60° 45′ clockwise from the North meridian. Set out the length of AB (402 m) to scale and mark point B.
3. Draw the north direction through B, parallel to the north direction through A.
4. Set out the line BC at an angle of 109° 30′ from the north direction. Set out the length BC = 398 m to the scale chosen. This fixes the position of point C.
5. Draw a parallel north direction (meridian) through C and repeat the procedure till all the stations are covered.
6. The closed traverse given by the data of Table 3.10 in such a way should close, that A of line EA when drawn should coincide with the point A already marked. If this does not happen, the difference is known as the *closing error*. The procedure for correcting this error will be outlined later.

3.7.2 Plotting by Included angles

In place of the bearings, the included angles can be used for plotting the traverse (refer to Fig. 3.30). The included angles can be calculated by the rules given earlier. The following steps are taken to plot a traverse using included angles.

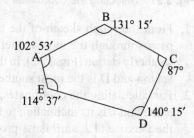

Fig. 3.30 Plotting by included angles

1. Calculate the included angles of the traverse. From the data given in Table 3.10, the included angles can be calculated as ∠A = 102° 53′, ∠B = 131° 15′, ∠C = 87°, ∠D = 104° 15′, and ∠E = 114° 37′.
2. Check that the sum of the included angles is equal to $(2n-4)$ right angles. This works out to $6 \times 90° = 540°$ in the case being given as an example. In this case the sum of included angles = 540°.
3. Pick a convenient point on paper and mark it 'A'. The point is so chosen that the whole traverse can be adjusted properly and centrally on the sheet. Mark the north direction through A. Set out AB in the direction of its bearing and measure the length AB to scale to set point B.

4. Set out the included angle at B (= 114° 30′) to obtain the direction of BC. Set out the length of BC to scale to locate point C.
5. Repeat the procedure at C by measuring the included angle at C (= 108° 15′) and set out the length of CD to locate point D.
6. Continue the process to set out all the lines of the traverse. The last point when laid out should coincide with A already marked. Otherwise, there is a closing error which needs to be corrected.

3.7.3 Plotting by Rectangular Coordinates

Plotting by rectangular coordinates is a very convenient and accurate method of plotting. For this purpose, the traverse has to be set in a rectangular coordinate system with a conveniently selected origin. The origin can be selected anywhere, but to make plotting easier, it will be convenient if all the coordinates are positive. This can be achieved by making the y-axis pass through the most westerly station and the x-axis pass through the most southerly station. The most westerly station will have a zero x-coordinate and the most southerly station will have a zero y-coordinate. The procedure is as follows (refer to Fig. 3.31).

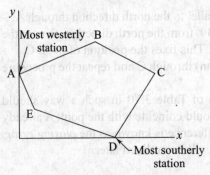
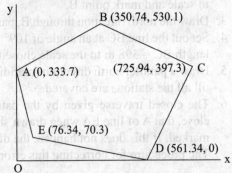

Fig. 3.31 Selecting the coordinate sytem   **Fig. 3.32** Plotting by coordinates

1. From the rough sketch of the traverse, select the origin such that the y-axis passes through the most westerly station and x-axis passes through the most southerly station (Fig. 3.32). In the case being considered, A is the most westerly station and D is the most southerly station.
2. For calculating the coordinates, it will be easy to see that if L is the length of a line and θ is its inclination to the vertical, then $L \cos\theta$ is its projection along the y-axis and $L \sin\theta$ is its projection along the x-axis. Use these projected distances to calculate the coordinates.
3. It is easy to see that the origin is 333.7 m below station A and 561.34 m to the left of D. This makes the coordinates of A (0, 333.7) and those of D (561.34, 0). Calculate the coordinates of the stations as shown in Table 3.11 below. Knowing the coordinates of A, obtain the x-coordinate of B by algebraically adding $L \sin\theta$ to the x-coordinate of A, and the y-coordinate of B by adding $L \sin\theta$ to the y-coordinate of A. Obtain the coordinates of the other points similarly. Thus, the x-coordinates of the points are the following:

$$B = 0 + 350.74 = 350.74, \quad C = 350.74 + 375.2 = 7225.94$$

$$D = 725.94 - 164.6 = 561.34$$
$$E = 561.34 - 485 = 76.34 \quad (\text{Check: } A = 76.34 - 76.34 = 0)$$

Table 3.11 Calculation of coordinates

Station	L cosθ	L sinθ	x	y
A	196.4	350.74	0	333.7
B	- 132.8	375.20	350.74	530.1
C	- 397.3	- 164.60	725.94	397.3
D	70.3	- 485.00	561.34	0
E	263.4	- 76.34	76.34	70.3

The *y*-coordinates of the points are the following:

$A = 333.37, \quad B = 333.7 + 196.4 = 530.1$

$C = 530.1 - 132.8 = 397.3 \quad D = 397.3 - 397.3 = 0$

$E = 0 + 70.3 = 70.3 \quad$ (Check: $70.3 + 263.4 = 333.7$, *y*-coordinate of A)

4. With these coordinates, plot each station independently of the other stations.
5. To plot the traverse, select the origin conveniently on paper and draw the rectangular axis through it. Plot each point by measuring the (x, y) coordinates of the point from the origin, to the scale selected. After all the points are plotted, join them to get the traverse.
6. Check the plots by measuring the lengths of the line to scale from the plots and comparing with the measured values.

3.7.4 Plotting by Deflection Angles

A traverse can also be plotted using the deflection angles of lines from the previous line. The deflection angle is that angle by which a traverse line deflects from the direction of the previous line (refer to Fig. 3.33). The deflection angle can be to the right or left of the previous line. This method is very suitable for open and closed traverses.

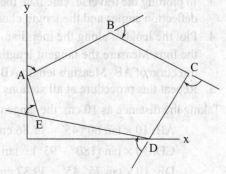

The deflection angle can be calculated from a rough sketch of the traverse and the bearings. The following steps have to be taken to plot the traverse using deflection angles.

Fig. 3.33 Plotting by deflection angles

1. Select one line as the reference line. In the case being described, the meridian through station A has been selected as the reference line. The deflection angle of line AB = the bearing of line AB = 60° 45′.
2. Calculate the deflection angles of the remaining lines. The deflection angles of lines BC, CD, DE, EA are as follows. B: (109° 30′ – 60° 45′) = 48° 45′, C: (202° 30′ – 109° 30′) = 93° 00′, D: (278° 15′– 202° 30′) = 75° 45′, E: (343° 49′ – 278° 15′) = 65° 23′.
3. Mark the line AB conveniently on paper to the scale selected.
4. From the end of line AB at B, mark the deflection angle of BC. Set out the length BC according to scale to locate point C.

5. Repeat the procedure at C, D, and E.
6. When the last line is plotted to scale, it should coincide with point A.

3.7.5 Plotting by Tangent Lengths

Plotting by tangent lengths is a graphical method of plotting based on deflection angles. The procedure is as follows.

1. Fix the position of the starting point suitably on paper. Draw the direction of the meridian through A.
2. From Fig. 3.34(a), it is clear that if the bearing of AB is θ, then for any length *l along the meridian*, the perpendicular distance to the direction of AB = *l* tanθ. This is known as the *tangent length*. *l* can be any convenient length.

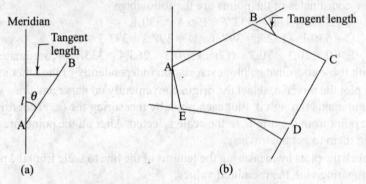

(a)   (b)

Fig. 3.34 Plotting by tangent lengths

3. In plotting the traverse, calculate the tangent lengths of all the lines from their deflection angles and the length chosen, which can be 10 cm or 20 cm.
4. Plot the length *l* along the meridian. From this point, draw a perpendicular to the line. Measure the tangent length along this perpendicular. This gives the direction of AB. Measure length AB to scale along this direction.
5. Repeat this procedure at all stations for plotting all the points of the traverse.

Taking the distance as 10 cm, the tangent distances are calculated as follows:

> AB: $10 \times \tan 60° \, 45' = 17.86$ cm,   BC: $10 \times \tan 48° \, 45' = 11.4$ cm
>
> CD: $10 \times \tan (180° - 93°) = \tan 87° = -190.8$ cm
>
> DE: $10 \times \tan 75° \, 45' = 39.37$ cm
>
> EA: $\tan 65° \, 23' \times 10 = 21.82$ cm

(It will be noticed that the tangent lengths for the directions of CD and DE are very high due to the large value of the deflection angle. It will be advisable to use the deflection angle directly in such cases.)

3.7.6 Plotting by Chords

Plotting by chords is another graphical method using deflection angles. As shown in Fig. 3.35(a), the chord length for a convenient length (say, 10 cm) is used for plotting the deflection angle. If the length is *l* and the deflection angle is θ, $2l \sin(\theta/2)$ is the chord length. The procedure is the following.

1. Select a convenient point to plot the starting station A. Draw the meridian through A.

2. Using the bearing of AB and a conveniently chosen length (10 cm), calculate the chord length. The chord length for AB is $2 \times 10 \sin (60.75°/2) = 10.1$ cm. The chord lengths for the other lines are BC = 8.25 cm, CD = 14.51 cm, DE = 12.28 cm, and EA = 10.8 cm.

3. With A as the centre and 10 cm as the radius, draw an arc. With a point on the meridian 10 cm from A, draw an arc with the chord length to obtain the direction of AB.

4. Mark the length of AB along this line to the scale chosen for the plot.

5. Repeat the procedure at other stations to get all the points.

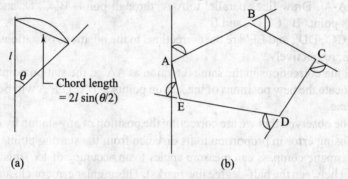

Fig. 3.35 Plotting by chord length

3.7.7 Adjusting a Traverse

In Fig. 3.36, the distance from A to A′ is the closing error, as explained before. If the closing error is small, it can be adjusted by a simple graphical method.

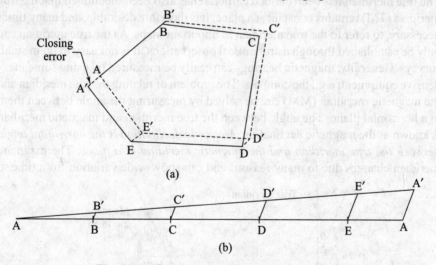

Fig. 3.36 Graphical correction of closing error

Figure 3.36 shows a traverse ABCDEA. When the traverse is plotted, let the last line EA drawn end at a point A′. The traverse does not close and the closing error or *error of closure* is AA′. The closing error being AA′ in magnitude and direction, one method is to move the points B, C, D, and E by some amount in the same direction as A'A. This will make A' come to A while the plotted points B, C, D, and E will move to new positions, thus giving a closed traverse.

The graphical method of adjusting the traverse will only be illustrated here. This method is based on *Bowditch's rule*, which will be explained in Chapter 6. The procedure is as follows and is illustrated in Fig. 3.36(b).

1. Plot the traverse and obtain the magnitude and direction of the closing error.
2. Draw a straight line ABCDEA to some suitable scale representing the lengths of the traverse lines.
3. Mark A' equal in magnitude and direction (A'A) to the closing error from A.
4. Join A-A'. Draw lines parallel to A-A' through points B, C, D, and E. This yields points B', C', D', and E'.
5. BB', CC', DD', and EE' are the corrections to the positions of stations B, C, D, and E, respectively.
6. Mark the corrections in the same direction as AA' at the station points plotted and locate the new positions of the station points. AB'C'D'E'A will be a closed traverse.

It will be observed that we are correcting the position of any station by distributing the closing error in proportion to its distance from the starting point.

The prismatic compass can measure angles to an accuracy of 15' (this too is by judgement between the half-degree line marks). The angular error of closure in such a case is limited to $15(n)^{1/2}$ minutes, where n is the number of sides of the traverse. The relative error of closure in linear measure is limited to 1/300 to 1/600.

3.8 Magnetic Declination

The true meridian has been defined earlier. It has also been mentioned that the true meridian (TM) remains constant at a place. It is therefore desirable, and many times necessary, to refer to the true meridian in important maps. As the true meridian can only be established through astronomical observations, it is not easy to do in small surveys. Generally, magnetic bearings can easily be measured by using some inexpensive equipment, e.g., the compass. The problem of relating the true meridian and the magnetic meridian (MM) can be solved by measuring the angle between them in a horizontal plane. The angle between the true meridian and magnetic meridian is known as the magnetic declination. *Magnetic declination is the horizontal angle between the true meridian and the magnetic meridian at a place.* The magnetic meridian changes due to many reasons and can show wide variation from time to

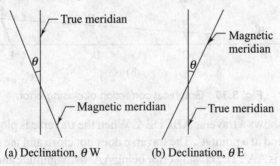

(a) Declination, θ W   (b) Declination, θ E

Fig. 3.37 Magnetic declination

time. Thus it can lie either to the west of the true meridian or to the east of the true meridian. Thus the declination is written as θE or θW. The declination θE means that the magnetic meridian lies $\theta°$ to the east of the true meridian (Fig. 3.37).

There are four types of variation in declination—diurnal, annual, secular, and irregular.

Diurnal variation   The daily variation of the position of the magnetic needle as it points to the north is known as diurnal variation. The variation can be as much as 12′ a day. Diurnal variation occurs with (a) the time of the day, (b) locality, less at the equator compared to places of high latitude, and (c) season of the year, being more in summer than in winter. It changes from year to year as well.

Annual variation   If the location of the magnetic meridian is observed regularly over a period of one year, it will be observed that there is a variation in the position of the magnetic meridian. This variation can be as much as 2 min.

Secular variation   In general, the magnetic meridian swings like a pendulum, swinging in one direction for a long period of time (200 years or more) and then swinging in the reverse direction. This is known as secular variation.

Irregular variation   Magnetic declination is subjected to irregular variation due to magnetic disturbances or storms due to sun spots, volcanic eruptions, seismic shocks, and similar causes. This variation can be as much as $2°$.

Of these variations, secular variation is the most important as it continuously increases in magnitude over a long period of time. Thus, magnetic bearings are not very reliable measurements over a period of time. In all maps, if the magnetic declination at the time of survey is mentioned, indicating the annual variations at that place, then it will be possible to relate the map with the ground even after many years. The magnetic declination at the time of survey enables us to determine the true bearing of lines. Knowing the magnetic declination at the present time, we can determine the magnetic bearing of a line at present. This will enable us to relate old maps with the magnetic bearings of lines to the present times.

Isogonic charts   In order to help surveyor's determine declinations at a place, isogonic charts are published. These charts show (a) isogonic lines connecting places of equal declinations, (b) agonic lines showing places having zero declination, and (c) lines connecting places having equal variations in declination.

Determination of true bearings   Knowing the declination at a place, the true bearing of a line can be determined from its magnetic bearing. In the case of whole circle bearings, true bearing = magnetic bearing ± declination. The plus sign is used when the declination is towards the east and the minus sign when the declination is towards the west.

In the case of reduced bearings, true bearing = magnetic bearing ± declination. If the declination is towards the east (magnetic meridian to the east of true meridian), use the plus sign if the bearing is N-E or S-W (first and third quadrants) and the minus sign if the bearing is N-W or S-E (second and fourth quadrants). If the declination is towards the west, use the plus sign for the N-W and S-E bearings and the minus sign for the N-E and S-W bearings. It is always desirable to draw a sketch and compute the true bearings. It will sometimes be necessary to determine the magnetic bearing from the true bearing. This can be done similarly. Always draw a sketch instead of remembering the formula.

Example 3.11 The following bearings were observed from a station during traversing: 184° 35′ and 124° 30′. If the declination at the place is known to be 1° 45′E, find the true bearings of the lines.

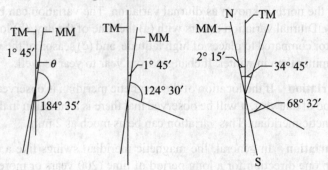

Fig. 3.38 Example 3.11   **Fig. 3.39** Example 3.12

Solution The given bearings are whole circle bearings. From Fig. 3.38, it is clear that

True bearing of line = magnetic bearing + declination

Magnetic bearing = 184° 35′,   true bearing = 184° 35′ + 1° 45 = 187° 20′

Magnetic bearing = 124° 30′,   true bearing = 124° 30′ + 1° 45′ =126° 15′

Example 3.12 The following were observed as the fore and back bearings of lines from a station: N34° 45′E and S68° 32′E. If the magnetic declination at the place is known to be 2° 15′W, find the true bearings of the lines.

Solution Figure 3.39 shows the line and the bearings. For a magnetic bearing with designation N-E, the declination towards the west has to be subtracted to get the true bearing.

Magnetic bearing = N34° 45′E

True bearing = 34° 45′ − 2° 15 = N32° 30′E

For a magnetic bearing with the designation S-E, the declination has to be added to get the true bearing.

Magnetic bearing = S68° 32′E

True bearing = 68° 32′ + 2° 15′ = S70° 47′E

Example 3.13 On an old map, the bearing of a line is given as 148° 30′. The declination at the time of survey was recorded as 3° 45′E [Fig. 3.40(a)]. If the present declination is 2° 15′E, find the magnetic bearing to which this line has to be set now.

Solution We first find the true bearing of the line. With east declination, a whole circle bearing is converted to the true bearing by adding the declination.

Magnetic bearing = 148° 30′, true bearing = 148° 30′ + 3° 45′

= 152° 15′ [Fig. 3.40(b)]

The present declination is also east. Therefore,

Magnetic bearing = true bearing − declination = 152° 15′ − 2° 15′ = 150°

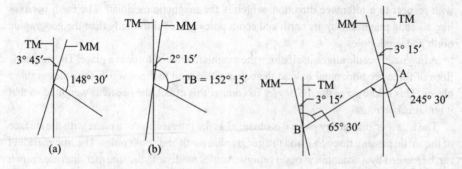

Fig. 3.40 Example 3.13   **Fig. 3.41** Example 3.14

Example 3.14 The fore and back bearings of a line AB were observed to be 245° 30′ and 65° 45′. It is known that station A is free of local attraction. The declination at the place was found to be 3° 15′W. Find the true bearing of line BA.

Solution Refer to Fig. 3.41. First we correct the bearing for local attraction. The back bearing of line BA should be equal to 245° 30′ – 180° = 65° 30′.

> Magnetic bearing of BA = 65° 30′
> True bearing = magnetic bearing – west declination
> = 65° 30′ – 3° 15′ = 62° 15′

3.9 Upkeep and Maintenance of Prismatic Compass

The prismatic compass is a fine instrument that is easy to set up and use. It is ideally suited for rough, speedy survey work. The following points are important for maintaining the compass in good condition. The surveyor's compass is not discussed here as it is no longer in use.

1. The compass comes in a leather cover. The compass should be kept in its cover when not in use.
2. The compass should be tested frequently, before using it for surveying. An important test is to see whether the magnetic needle is freely moving or sluggish. For this purpose, set up the compass at a point and take the bearing of a line connecting to a well-defined object. Having taken the reading, rotate the compass and immediately bring it back to the same point, bisect, and take the reading. The reading should be the same. If not, the needle is not moving properly due to a blunt pivot.
3. Check whether the needle (or the graduated ring) moves in a horizontal plane. This can be due to the dip of the needle, in which case the rider weight can be adjusted.
4. Keep the instrument free of dust and clean the glass cover with a fine cloth.
5. There are not many adjustments that can be done by the user; a qualified mechanic needs to do the servicing or repair.

Summary

The directions of the lines of a traverse can be measured either as angles between the lines or as angles with respect to a reference direction. A magnetic compass measures the angle with respect to a reference direction, which is the magnetic meridian. The earth behaves like a strong magnet, with its north and south poles placed differently than the geographic north and south poles.

A magnetic needle aligns itself along the magnetic lines of force at a place. The magnetic lines of force are horizontal only at the equator and tend to dip towards the poles at other places. This is known as the *dip angle*. To counter this effect, the needle is weighted so that it remains horizontal.

The true meridian at any place P is obtained as the intersection of a plane with the surface of the earth, passing through P and the geographic north and south poles. The true meridian can be located by astronomical observations. Angles made with the true meridian are known as *true bearings* and do not change. Angles made with the magnetic meridian, the directions taken by the magnetic needle, are known as *magnetic bearings*. They are subject to change and do not remain the same at all times.

Of the two types of compasses, prismatic and surveyor's, the prismatic compass is commonly used. Surveyor's compass is no longer used. The prismatic compass measures the whole circle bearing of a line, which is the angle made by the line with the north direction. The whole circle bearing can have values between 0° and 360°. Surveyor's compass on the other hand measures the acute angle (< 90°), either from the north or south direction towards the east or west. These angles are known as *magnetic bearings*. In the prismatic compass a prism is used to take the readings simultaneously with sighting. The compass is graduated to half-degrees and it is possible to take readings up to 15′ by judgement.

A compass is used with a chain for traversing. Knowing the bearings of lines, it is possible to find included angles. If the included angles and bearing of one line are known, the bearings of other lines can be similarly computed.

The orientation of a magnetic needle is affected by the presence of magnetic substances, such as iron ore mines and overhead electric cables. This is known as *local attraction*. The bearings can be corrected for local attraction.

A traverse, closed or open, can be plotted by many methods—using bearings, included angles, coordinates, deflection angles, or graphical methods. A closed traverse that does not close on plotting can be corrected by distributing the error among the lines proportionately.

Magnetic declination is the angle between the true meridian and the magnetic meridian at a place. The declination can be towards the east or west, depending upon whether the magnetic meridian lies to the east or the west of the true meridian. Since the magnetic meridian changes due to various reasons, if a map is prepared using the magnetic meridian, it is necessary to indicate the declination at the time of the survey on the map. This will help to relocate the line later on the ground from the true bearing.

Exercises

Multiple-Choice Questions

1. Prismatic compass is an instrument used to measure
   (a) distances between points
   (b) vertical angles
   (c) angles between two lines
   (d) magnetic bearings of lines
2. Magnetic bearing of a line is
   (a) the horizontal angle between any two lines
   (b) the vertical angle of a line from the horizontal plane

    (c) the horizontal angle between the magnetic meridian and the line

    (d) the vertical angle between the magnetic meridian and the line

3. The graduations in the ring of a prismatic compass are in the
    (a) whole circle system from 0 to 360 degrees
    (b) quadrantal system from 0 to 90 degrees in the four quadrants
    (c) semicircle system from 0 to 180 degrees from the north in clockwise and counter clockwise directions
    (d) semicircle system from 0 to 180 degrees from the south in clockwise and counter-clockwise directions

4. The zero graduation in a prismatic compass is marked in the
    (a) north end of the circle   (c) in the south end of the circle
    (b) in the east end of the circle   (d) in the west end of the circle

5. The prismatic compass gives the
    (a) quadrantal bearing of lines
    (b) whole circle bearing of lines
    (c) angle between the previous line and the forward line
    (d) the deflection angle between the lines meeting at the station

6. If the whole circle bearing of a line is $237° 45' 30''$, its quadrantal bearing is
    (a) S $57° 45' 30''$W   (c) S $32° 14' 30''$ W
    (b) S $47° 45'30''$ E   (d) S$32° 14' 30''$ E

7. If the quadrantal bearing of a line is S $25° 32' 00''$E its whole circle bearing is
    (a) $115° 32' 00''$ (b) $205° 32' 00''$ (c) $154° 28' 00''$ (d) $64° 28' 00''$

8. The whole circle bearing of a line AB is $132° 35' 45''$. The whole circle bearing of line BA is
    (a) $42° 35'45''$ (b) $312° 35'45''$ (c) $47° 24'15''$ (d) $227° 24'15''$

9. The whole circle bearing of line AB is $210° 25' 35''$. The quadrantal bearing of line BA is
    (a) N $30° 25' 35''$E (b) S $30° 25' 35''$ E (c) N$59° 34' 25''$W(d) N$59° 34' 25''$E

10. If the magnetic bearing of a line is designated as SW in the quadrantal system, then the whole circle bearing of the line will be between
    (a) $0°$ to $90°$ (b) $90°$ to $180°$ (c) $180°$ to $270°$ (d) $270°$ to $360°$

11. If the whole circle bearing of a line is between $90°$ and $180°$, then the line lies in the
    (a) NE quadrant   (b) NW quadrant (c) SE quadrant   (d) SW quadrant

12. If the bearing of line AB is $185° 34' 20''$ and that of BC is $235° 46' 40''$, then the included angle between the lines is
    (a) $50° 12' 20''$ (b) $61° 30' 00''$ (c) $124° 13' 20''$ (d) $129° 47' 40''$

13. If the bearing of line AB is N$87° 32'00''$E and that of line BC is S $65° 35'20''$E, then the included angle at B is
    (a) $21° 56'40''$ (b) $86° 52'40''$ (c) $115° 6'40''$ (d) $153° 07'20''$

14. If the bearing of line AB is N$87° 32'00''$E and that of line BC is S $65° 35'20''$E, then the deflection angle between the lines at B is
    (a) $26° 52'40''$ (b) $92° 28'00''$ (c) $114° 24'40''$ (d) $153° 07'20''$

15. Two lines AB and BC are part of a closed traverse. The bearing of BC is $173° 25' 00''$ and the included angle between the lines if $122° 25'00''$. The bearing of line AB is
    (a) $115° 50'$ (b) $186° 35'$ (c) $263° 25'$ (d) $295° 25'$

16. If the bearing of line AB is N$65° 35' 00''$E and the deflection angle, clockwise, between the lines AB and BC is $34° 45'00''$, then the bearing of line BC is
    (a) N $10° 20'$E   (b) S $79° 40'$E   (c) N $34° 45'$E (d) N $30° 50'$E

17. The whole circle bearing of side AB of an equilateral triangle ABC is $38° 45'$. Then, the bearing of the third side CA of the triangle is
    (a) $278° 45'$ (b) $197° 30'$ (c) $98° 45'$ (d) $81° 15'$

18. If the bearing of one side AB of a regular hexagon ABCDEF is 36° 45′, the bearing of the adjacent side BC of the hexagon is
    (a) 83° 15′   (b) 96° 45′   (c) 156° 45′   (d) 216° 45′

19. ABCDE is a regular pentagon. If the bearing of AB is N 36° 00′E, bearing CD is
    (a) S90° 00′E   (b) S90° 00′W   (c) S 00° 00′E   (d) N 00° 00′E

20. ABCD is a square. If the bearing of AB is N 30° 30′ E, the bearing of diagonal BD is
    (a) S 59° 30′E   (b) S 84° 30′E   (c) S 14° 30′E   (d) N 59° 30′E

21. ABCD is a square. If the bearing of AB is 30° 30′, the bearing of diagonal CA is
    (a) N 59° 30′ W   (b) S 75° 30′ W   (c) S 59° 30′ E   (d) S 14° 30′ E

22. ABC is an equilateral triangle. If the bearing of AB is N 45° 30′ E, the bearing of altitude BD is (BD is perpendicular to AC)
    (a) S 15° 30′ E   (b) S 15° 30′ W   (c) N 15° 30′ W   (d) S 75° 30′ W

23. In an isosceles triangle, AB is the base and the bearing of AB is N 75° 45′E. If the bearing of AC is N 30° 45′E, the bearing of BC is
    (a) S 45° 00′W   (b) N 59° 15′W   (c) S 34° 45′W   (d) S 64° 15′ E

24. ABCDEF is a regular hexagon. If the bearing of AB is 53° 45′, the bearing of EF is
    (a) 173° 45′   (b) 186° 15′   (c) 126° 15′   (d) 113° 45′

25. In a right angled triangle, BC is the hypotenuse. If AB: AC is 3:4 and bearing of AC is 65° 35′, the bearing of BC is
    (a) 24° 25′   (b) 36° 52′   (c) 53° 08′   (d) 102° 27′

26. Magnetic declination is
    (a) the deflection of magnetic needle due to external magnetic sources
    (b) the error in the bearings due to external magnetic influences
    (c) the angle between the true meridian and the magnetic meridian at a place
    (d) the dip of the needle to earth's magnetic field

27. Local attraction is
    (a) the effect on the magnetic needle due to external magnetic influences
    (b) the dip of the needle due to earth's magnetic field
    (c) the amount by which the compass is kept off-centre from the station
    (d) the parallax error in reading the graduations on the ring

28. Magnetic declination enables us to
    (a) find true bearings of lines
    (b) determine the correct functioning of magnetic needle
    (c) test and adjust a prismatic compass
    (d) find local attraction at a place

29. Magnetic bearing of a line is found as 35° 45′. If the declination is 3° 45′ E, true bearing is
    (a) 32° 00′   (b) 39° 30′   (c) 35° 45′   (d) 3° 45′

30. The magnetic bearing of a line AB in a map is given as 67° 30′. The magnetic declination at the time is mentioned as 3° 30′ W. If the present declination is 1° 30′E, the magnetic bearing of the line at present is
    (a) 62° 30′   (b) 69° 00′   (c) 70° 30′   (d) 72° 30′

Review Questions

1. State the principle on which a compass works.
2. Draw a neat sketch of a prismatic compass and label its various parts. State the function of each part.
3. Explain with the help of sketches the relation between the whole circle and quadrantal bearings.
4. Explain the method of calculating the internal angles between the lines of a traverse

with an example when the bearings are given in both the whole circle system and the quadrantal system.

5. Explain the various types of meridians used in compass surveys and their advantages and disadvantages.
6. Define and explain the term dip angle. How does it affect the movement of a magnetic needle? How is it remedied?
7. Define and explain the term declination. Enlist the causes of declination.
8. Describe briefly the different types of variations in declination.
9. Explain the differences between a prismatic compass and the surveyor's compass.
10. Briefly explain the method of traversing with a chain and compass.
11. Explain briefly the likely errors in compass surveying.
12. Briefly describe the care and maintenance of a compass.
13. Explain the different methods of plotting a compass traverse survey.
14. Explain briefly the phenomenon of the earth's magnetism.
15. Explain the term closing error. Which principle is used to adjust it?

Problems

1. Convert the following whole circle bearings to reduced bearings: 67° 30′, 278° 45′, 123° 55′, 270° 00′, 326° 30′, 180° 00′.
2. Convert the following reduced bearings to whole circle bearings: N45° 30′W, S67° 30′W, S28° 45′E, S86° 32′W, N48° 34′E.
3. The following pairs of bearings are the fore bearings of consecutive lines of a traverse. Find the angles between the lines: (i) 24° 45′, 320° 55′, (ii) 98° 36′, 254° 48′, (iii) 194° 45′, 84° 55′, (iv) 320° 48′, 264° 24′, (v) 64° 35′, 172° 45′, (vi) 202° 36′, 324° 56′.
4. The following pairs of reduced bearings were taken as the fore bearings of two consecutive lines of a traverse. Find the angles between the lines: (i) N36° 45′W, S22° 10′E, (ii) S26° 45′E, N38° 24′E, (iii) N48° 24′E, N32° 48′W, (iv) S22° 48′W, S64° 42′E, (v) S12° 48′W, N82° 56′E, (vi) N20° 48′W, S40° 54′W.
5. The bearing of line AB of a closed traverse ABCDEA is 26° 35′. The interior angles of the traverse are ∠A = 68° 45′, ∠B = 138° 30′, ∠C = 131° 30′, ∠D = 90°, ∠E = 111° 15′. Find the whole circle bearings of the lines of the traverse.
6. The bearing of the line AB of an open traverse ABCDE is N62° 45′E. The deflection angles at other points are the following. B: 36° 45′ (R), C: 48° 25′ (L), D: 68° 45′ (R), E: 42° 36′ (L). Find the bearings of the remaining lines of the traverse. (R and L indicate that the deflection angle is to the right and left, respectively.)
7. The bearing of the line AB of a traverse ABCDEFA is N75° 15′E. The included angles of the traverse are ∠A = 52° 14′, ∠B = 165° 15′, ∠C = 90°, ∠D = 90°, ∠E = 90°, and ∠F = 232° 31′. Find the bearings of the remaining lines of the traverse.
8. One side of a plot in the form of an equilateral triangle has the bearing 84° 55′. Find the bearings of the remaining lines of the triangle.
9. A plot has to be laid out in the form of a regular pentagon. If the bearing of one of the lines of the pentagon is N37° 45′E, find the bearings of the remaining lines.
10. If, in a regular octagon, one line lies along the bearing N34° 55′E, find the bearings of the remaining lines of the figure.
11. The lengths and bearings of the lines of a traverse are given in Table 3.12. Select the origin such that all the coordinates are positive. Find the coordinates of the stations for plotting.

Table 3.12 Traverse data

Line	Length	Bearing
AB	330	90° 00′
BC	235	131° 30′
CD	250	180° 00′
DE	348	270° 00′
EA	435.3	338° 45′

12. The included angles of a traverse are given below. Find the bearings of the other lines and the deflection angles at the stations for plotting. Bearing of AB = S45° 00′E, ∠A = 101° 48′, ∠B = 95° 30′, ∠C = 75° 15′, ∠D = 87° 27′.

13. The whole circle bearings of the lines of a closed traverse are given below. Determine which stations, if any, are affected by local attraction and correct the bearings by calculating the included angles.

    PQ: 41° 20′   QP: 221° 20′
    QR: 114° 30′   RQ: 293° 50′
    RS: 164° 40′   SR: 346° 20′
    SP: 275° 30′   PS: 94° 30′

14. The bearings of the lines of a traverse are given below. Find the included angles and correct the bearings for local attraction, if any.

    PQ: 73° 40′   QP: 252° 30′
    QR: 113° 50′   RQ: 295° 20′
    RS: 164° 20′   SR: 344° 20′
    ST: 223° 40′   TS: 43° 00′
    TP: 303° 50′   PT: 123° 45′

15. The bearings of the lines of an open traverse are given below. Find the local attraction, if any, at the stations and correct the bearings.

    AB: 36° 10′   BA: 216° 10′
    BC: 109° 20′   CB: 288° 40′
    CD: 159° 30′   DC: 341° 10′
    DA: 270° 20′   AD: 89° 20′

16. The bearings of the lines of a traverse are given below. Find the local attraction, if any, at the stations and correct the bearings.

    PQ: 60° 30′   QP: 240° 30′
    QR: 99° 15′   RQ: 278° 00′
    RS: 218° 30′   SR: 38° 45′
    SP: 320° 45′   PS: 141° 45′

17. The bearings of the lines of a closed traverse are given below. Determine the stations affected by local attraction and the magnitude of the local attraction. Correct the bearings.

    AB: 94° 30′   BA: 276° 30′
    BC: 174° 30′   CB: 353° 15′
    CD: 231° 00′   DC: 51° 00′
    DA: 339° 30′   AD: 158° 45′

18. The bearings of the lines of a traverse are given below. At which station do you suspect local attraction? Correct the bearings for local attraction.

    PQ: S47° 00′ E   QP: N47° 00′ W

PQ: S47° 00′ E   QP: N47° 00′ W

QR: S61° 30′ E   RQ: N62° 10′ W

RS: N7° 00′ W   SR: S4° 50′ E

SP: S61° 30′ W   PS: N60° 0′ E

19. The bearings taken for four lines of a traverse are given below. Find the true bearings of the lines if the declination at the time of the survey was 2° 45′W. N57° 30′E, N22° 45′E, S78° 30′E, S20° 15′W.

20. The bearing of a line is given in an old map as N56° 30′W. The declination at the time of the survey is recorded as 1° 45′E. If the present declination is 2° 30′W, find the magnetic bearing to which the line has to be set now.

21. The bearing of the sun at noon was measured with a compass and found to be 3° 30′. If the magnetic bearing of a line AB was also measured and found to be S56° 30′W, find the true bearing of this line.

22. The bearing of a line was measured as S54° 35′W. It is known from other measurements that the needle at A has rotated clockwise by 1° 30′ due to local attraction. If the declination at the place is known to be 2° 15′W, find the true bearing of the line.

Measurement of Directions— Theodolite Surveying

Learning Objectives

After going through this chapter, the reader will be able to

- describe the main parts of a theodolite and their functions
- explain the relationship between the fundamental lines of a theodolite
- define and explain the basic terms connected with the instrument
- explain the temporary adjustments of the instrument
- explain the methods to measure horizontal angles between lines
- explain the method to measure vertical angles
- explain the method of theodolite traversing
- explain the likely errors and the precautions to be taken while working with a theodolite

Introduction

In the previous chapter, we discussed compass surveying. A compass measures the direction by measuring the angle between the line and a reference direction, which is the magnetic meridian. A compass can measure angles up to an accuracy of 30′ and by judgement up to an accuracy of 15′. The principle of working of the compass is based on the property of the magnetic needle, which when freely suspended, takes the north–south direction. Compass measurements are thus affected by external magnetic influences and therefore a compass is unsuitable in some areas. In this chapter, we will discuss another method of measuring directions of lines; a theodolite is very commonly used to measure angles in survey work.

There are a variety of theodolites—vernier, optic, electronic, etc. The improvements (from one form to the other) have been made to ensure ease of operation, better accuracy, and speed. Electronic theodolites display and store angles at the press of a button. This data can also be transferred to a computer for further processing. We start our discussion with the simplest theodolite—the vernier theodolite. Other forms of theodolites are discussed at the end of this chapter.

The vernier theodolite is a simple and inexpensive instrument but very valuable in terms of measuring angles. The common vernier theodolite measures angles up to an accuracy of 20″, which is sufficient for most of the normal work done. Unlike in a compass, where the line of sight is simple, restricting its range, theodolites are provided with telescopes which provide for much greater range and better accuracy in sighting distant objects. It is, however, a delicate instrument and needs to be handled carefully. The theodolite measures the horizontal angles between lines and can also measure vertical angles. The horizontal angle measured can be the included angle, deflection angle, or exterior angle in a traverse. The vertical angle is the angle in a vertical plane between the inclined line of sight of the instrument and the horizontal. In the following sections we will discuss the vernier theodolite as well as its applications in surveying.

4.1 Vernier Theodolite

The vernier theodolite is also known as a *transit*. In a transit theodolite or simply transit the telescope can be rotated in a vertical plane. Earlier versions of theodolites were of the non-transit type and are obsolete now. Only the transit theodolite will be discussed here.

Two different views of a vernier theodolite are shown in Figs 4.1(a) and (b). The instrument details vary with different manufacturers but the essential parts remain the same. The main parts of a theodolite are the following.

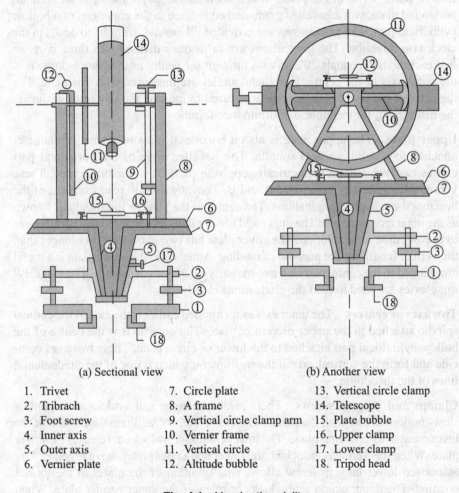

(a) Sectional view (b) Another view

1. Trivet	7. Circle plate	13. Vertical circle clamp
2. Tribrach	8. A frame	14. Telescope
3. Foot screw	9. Vertical circle clamp arm	15. Plate bubble
4. Inner axis	10. Vernier frame	16. Upper clamp
5. Outer axis	11. Vertical circle	17. Lower clamp
6. Vernier plate	12. Altitude bubble	18. Tripod head

Fig. 4.1 Vernier theodolite

Levelling head The levelling head is the base of the instrument. It has the provision to attach the instrument to a tripod stand while in use and attach a plumb bob along the vertical axis of the instrument. The levelling head essentially consists of two triangular plates kept a distance apart by levelling screws. The upper plate of the levelling head, also known as the *tribrach*, has three arms, each with a foot screw. Instruments with four foot screws for levelling are also available. In terms of wear and tear, the three-foot-screw instrument is preferable. The lower plate, also known as the *trivet*, has a central hole and a hook to which a plumb bob can be attached. In modern instruments, the base plate of the levelling head has two plates which can move relative to each other. This allows a slight movement of the levelling head relative to the tripod. This is called a *shifting head* and helps in centring the instrument over the station quickly. The functions of the levelling head are to support the upper part of the instrument, attach the theodolite to a tripod, attach a plumb bob, and help in levelling the instrument with the foot screws.

Lower plate The lower plate, also known as the *circle plate*, is an annular, horizontal plate with a bevelled graduated edge fixed to the upper end of a hollow cylindrical part. The graduations are provided all around, from 0° to 360°, in the clockwise direction. The graduations are in degrees divided into three parts so that each division equals 20'. An axis through the centre of the plate is known as the outer axis or the centre. Horizontal angles are measured with this plate. The diameter of the lower plate is sometimes used to indicate the size of or designate the instrument; for example, a 100-mm theodolite.

Upper plate The upper plate is also a horizontal plate of a smaller diameter attached to a solid, vertical spindle. The bevelled edge of the horizontal part carries two verniers on diametrically opposite parts of its circumference. These verniers are generally designated A and B. They are used to read fractions of the horizontal circle plate graduations. The centre of the plate or the spindle is known as the inner axis or centre. The upper and lower plates are enclosed in a metal cover to prevent dust accumulation. The cover plate has two glass windows longer than the vernier length for the purpose of reading. Attached to the cover plate is a metal arm hinged to the centre carrying two magnifying glasses at its ends. The magnifying glasses are used to read the graduations clearly.

Two axes or centres The inner axis as mentioned earlier is the axis of the conical spindle attached to the upper or vernier plate. The outer axis is the centre of the hollow cylindrical part attached to the lower or circle plate. These two axes coincide and form the vertical axis of the instrument, which is one of the fundamental lines of the theodolite.

Clamps and tangent screws There are two clamps and associated tangent or slow-motion screws with the plates. The clamp screws facilitate the motion of the instrument in a horizontal plane. The lower clamp screw locks or releases the lower plate. When this screw is unlocked, the lower and upper plates move together. The associated lower tangent screw allows small motion of the plates in the locked position. The upper clamp screw locks or releases the upper vernier plate. When this clamp is released (with the lower clamp locked), the lower plate does not move but the vernier plate moves with the instrument. This causes a change in the reading. The upper tangent screw allows for a small motion of the vernier plate for fine adjustments. When both the clamps are locked, the instrument cannot move in the horizontal plane. The construction of the clamp and tangent screws is shown in Fig. 4.2.

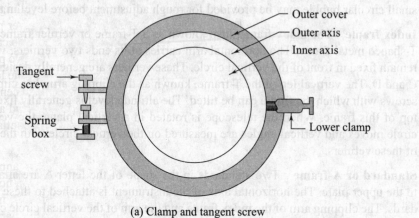

(a) Clamp and tangent screw

(b) Vertical circle graduations (20′ main scale)

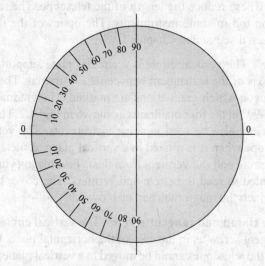

Quadrants 0–90°

(c) Main circle plate graduations

Fig. 4.2 Parts of a theodolite

Plate level   The plate level is a spirit level with a bubble and graduations on the glass cover. A single level or two levels fixed in perpendicular directions may be provided. The spirit level can be adjusted with the foot screw of the levelling head. The bubble of the spirit level can be moved with the foot screws of the levelling head, which is a very fundamental adjustment required for using the theodolite. A small circular bubble may be provided for rough adjustment before levelling.

Index frame   The index frame, also known as a T-frame or vernier frame, is a T-shaped metal frame. The horizontal arm carries at its ends two verniers, which remain fixed in front of the vertical circle. These verniers are generally designated C and D. The vertical leg of the T-frame, known as the clipping arm, has clipping screws with which the frame can be tilted. The altitude level is generally fixed on top of this frame. When the telescope is rotated in a vertical plane, the vertical circle moves and vertical angles are measured on the vertical circle with the help of these verniers.

Standard or A-frame   Two standards in the shape of the letter A are attached to the upper plate. The horizontal axis of the instrument is attached to these standards. The clipping arm of the index frame and the arm of the vertical circle clamp are also attached to the A-frame. The A-frame supports the telescope and the vertical circle.

Telescope   The telescope is a vital part of the instrument. It enables one to see stations that are at great distances. The essential parts of a telescope are the eyepiece, diaphragm with cross hairs, object lens, and arrangements to focus the telescope. A focusing knob is provided on the side of the telescope. Earlier, external focusing telescopes were used. Today, only internal focusing telescopes are used in theodolites. These reduce the length of the telescope. The telescope may carry a spirit level on top in some instruments. The optics of the telescope and other details have been discussed in Chapter 1.

Vertical circle   The vertical circle is a circular plate supported on the trunnion or horizontal axis of the instrument between the A-frames. The vertical circle has a bevelled edge on which graduations are marked. The graduations are generally quadrantal, 0°–90° in the four quadrants as shown in Fig. 4.2. The full circle system of graduations can also be seen in some instruments. The vertical circle moves with the telescope when it is rotated in a vertical plane. A metal cover is provided to protect the circle and the verniers from dust. Two magnifying glasses on metal arms are provided to read the circle and verniers. The cover has glass or plastic windows on which the magnifiers can be moved.

Vertical circle clamp and tangent screw   The vertical circle is provided with a clamp and tangent screw as in the case of the horizontal plate. Upon clamping the vertical circle, the telescope cannot be moved in a vertical plane. The tangent screw allows for a slow, small motion of the vertical circle.

Altitude level   The altitude level is fixed on the T-frame. A highly sensitive bubble is used for levelling, particularly when taking vertical angle observations.

Compass   A circular or trough magnetic compass is generally fitted to the theodolite for measuring the magnetic bearings of lines. It is fitted on the cover of the horizontal plates. Two plates with graduations are provided in the compass box for ensuring that the needle ends are centred. The needle can be locked or released by a pin. When released, the telescope can be turned in azimuth to make the north end of the needle point to the north by making it read zero.

Tripod   One accessory essential with the theodolite is the tripod on which it is mounted when it has to be used. The tripod head is screwed onto the base or the lower part of the levelling head. Its legs should be spread out for stability. The legs of the tripod are also used for rough levelling.

Plumb bob   A heavy plumb bob on a good string with a hook at the end is required for centring the theodolite over a station. The plumb bob is fixed to the hook or other device projecting from the centre of the instrument in a central opening in the levelling head.

Main circle and vernier graduations   In most of the instruments, the vernier enables readings up to $20''$ of the arc. This is made possible by marking the graduations on the circle and the vernier suitably as follows. As shown in Fig. 4.2(b), the main circle is graduated into degrees and each degree is divided into three parts. Each main scale division thus represents $2'$. For the vernier, 59 main scale divisions are taken and divided into 60 parts. 59 main scale divisions form $59 \times 20'$. Therefore, each vernier scale division represents $59 \times 20/60$ minutes. As you would have studied earlier, least count of the vernier = difference between a main scale division and a vernier scale division = main scale division – vernier scale division. Hence, in this case,

$$\text{Least count} = 20' - 59 \times 20/60 = 1/3 = (1/3) \times 60'' = 20''$$

Thus the least count of the vernier in common theodolites is $20''$.

4.2 Terminology

It is important to clearly understand the terms associated with the theodolite and its use and meaning. The following are some important terms and their definitions.

Vertical axis   It is a line passing through the centre of the horizontal circle and perpendicular to it. The vertical axis is perpendicular to the line of sight and the trunnion axis or the horizontal axis. The instrument is rotated about this axis for sighting different points.

Horizontal axis   It is the axis about which the telescope rotates when rotated in a vertical plane. This axis is perpendicular to the line of collimation and the vertical axis.

Telescope axis   It is the line joining the optical centre of the object glass to the centre of the eyepiece.

Line of collimation   It is the line joining the intersection of the cross hairs to the optical centre of the object glass and its continuation. This is also called the line of sight.

Axis of the bubble tube   It is the line tangential to the longitudinal curve of the bubble tube at its centre.

Centring   Centring the theodolite means setting up the theodolite exactly over the station mark. At this position the plumb bob attached to the base of the instrument lies exactly over the station mark.

Transiting   It is the process of rotating the telescope about the horizontal axis through 180°. The telescope points in the opposite direction after transiting. This process is also known as *plunging* or *reversing*.

Swinging   It is the process of rotating the telescope about the vertical axis for the purpose of pointing the telescope in different directions. The right swing is a rotation in the clockwise direction and the left swing is a rotation in the counter-clockwise direction.

Face-left or normal position   This is the position in which as the sighting is done, the vertical circle is to the left of the observer.

Face-right or inverted position   This is the position in which as the sighting is done, the vertical circle is to the right of the observer.

Changing face   It is the operation of changing from face left to face right and vice versa. This is done by transiting the telescope and swinging it through 180°.

Face-left observation   It is the reading taken when the instrument is in the normal or face-left position.

Face-right observation   It is the reading taken when the instrument is in the inverted or face-right position.

4.3   Temporary Adjustments

Theodolite has two types of adjustments—temporary and permanent. Temporary adjustments are to be done at every station the instrument is set up. Permanent adjustments deal with the fundamental lines and their relationships and should be done once in a while to ensure that the instrument is properly adjusted. The fundamental lines and their desired relationships are explained later in this chapter and the permanent adjustments are explained in detail in Chapter 12. In this section we will discuss temporary adjustments.

The temporary adjustments are the following: (a) setting up and centring, (b) levelling, (c) focusing the eyepiece, and (d) focusing the objective.

4.3.1   Setting Up and Centring

The following procedure is adopted for this operation.
1. Remove the theodolite from its box carefully and fix it onto a tripod kept over the station where the instrument is to be set up. The tripod legs should be well apart and the telescope should be at a convenient height for sighting.
2. Tie a plumb bob onto the hook provided at the base. If there is no shifting head in the instrument, centre it by adjusting the tripod legs and shifting the instrument as a whole to bring the plumb bob over the station mark.

3. To centre the plumb bob, shift the tripod legs radially as well as circumf-erentially. *Moving any leg radially shifts the plumb bob in the direction of the leg*. This does not affect the level status of the instrument. *Moving any leg circumferentially does not appreciably shift the plumb*. However, this movement tilts the instrument and affects the level of the plate bubbles. By moving the legs the plumb bob is brought over the station mark at the same time ensuring that the instrument is approximately level. This saves a lot of time for the next operation of levelling.
4. If the instrument has a shifting head with a clamp, first centre the instrument using the legs. Make the final adjustment by loosening the clamp and shifting the head (or the instrument as a whole) to bring the plumb bob over the station mark.

In all operations, the starting step should be to first bring the plumb bob very close to the mark and then make the final adjustment using the legs or the shifting head.

4.3.2 Levelling

After setting up and centring the instrument, levelling is done. Levelling has to be done at every station the instrument is set up. By levelling the instrument, it is ensured that as the instrument is swung about the vertical axis, the horizontal plate moves in a horizontal plane. The instrument may have a three-screw or a four-screw levelling head. The levelling operations differ slightly in these two cases as detailed in the following sections. Most instruments have only one bubble tube, but some instruments have two bubble tubes set at right angles over the plates.

Three-screw levelling head

When the theodolite has a three-screw levelling head, the following procedure is adopted.

1. Swing the theodolite and bring the plate bubble parallel to any two of the foot screws. Centre the bubble by rotating the foot screws. To do this, hold the foot screws by the thumb and forefinger of each hand and *rotate both either inwards or outwards* [see Fig. 4.3(a)]. Also note that the bubble moves in the direction of movement of the left thumb during this operation.
2. Once the bubble traverses (or comes to the central position from the gradua-tion of the tube), swing the instrument and bring the bubble over the third foot screw. In this position, the bubble tube is at right angles to the earlier position. Centre the bubble by rotating the third foot screw alone.

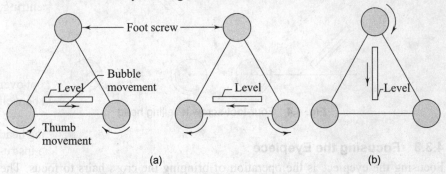

Fig. 4.3 Three-foot-screw levelling head

3. Bring the plate bubble to its previous position by swinging the instrument back. Check whether the bubble traverses. If it does not traverse, bring the bubble to the centre using the two foot screws as before.
4. Repeat the procedure till the bubble traverses in both these positions.
5. Swing the instrument through 180° and check whether the bubble traverses. The bubble should traverse in all positions if the instrument has been properly adjusted.

If two plate bubbles are provided [see Fig. 4.3(b)], the procedure is the same except that swinging the instrument through 90° is not required. When one plate level is kept parallel to a pair of foot screws, the other plate level is over the third foot screw (in a perpendicular direction). The third foot screw is adjusted alternately by the same process using the foot screws over which they are parallel.

Four-screw levelling head

When the theodolite has a four-screw levelling head, the following procedure is adopted.
1. After setting up and centring the theodolite, bring the plate level parallel to any one pair of diagonally opposite foot screws. Operate these foot screws to centre the bubble (Fig. 4.4).
2. Swing the instrument to bring the plate level parallel to the other pair of foot screws. Centre the bubble.
3. Swing it back to the previous position. Check whether the bubble traverses. If it does not, centre it with the foot screws to which the level is parallel.
4. Swing it back, check the position of the bubble, and repeat the procedure.
5. Once the bubble traverses in the two orthogonal positions, swing it through 180°. The bubble should traverse in this position or in any other position.

If two plate levels are provided, the procedure is the same. Bring one plate level parallel to a pair of opposite foot screws. The other pair will be parallel to the remaining pair of foot screws. There is no need to swing the instrument. Bring the bubble to the central position alternately and check in the other positions.

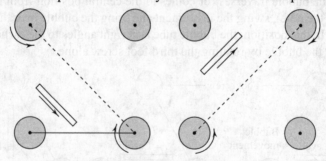

Fig. 4.4 Four-foot-screw levelling head

4.3.3 Focusing the Eyepiece

Focusing the eyepiece is the operation of bringing the cross hairs to focus. The focusing position varies with the eyesight of the observer. If the same observer is taking the readings, this has to be done only once. To focus the eyepiece, use the following procedure.

1. Keep a piece of white paper in front of the telescope or direct the telescope towards a clear portion of the sky.
2. Looking through the telescope, adjust the vision by rotating the eyepiece till the cross hairs come into sharp and clear view.
3. If the eyepiece has graduations, note the graduation at which you get a clear view of the cross hairs. This can help in later adjustment if required.

4.3.4 Focusing the Objective

The objective lens has to be focused whenever an object is sighted, as this depends upon the distance between the instrument and the object. A focusing screw on the side of the telescope is operated to focus the objective. This operation brings the image of the object in the plane of the cross hairs. This helps to exactly bisect the object, be it a ranging rod or an arrow. To focus the objective, swing the instrument to bring the object into view by looking over the telescope. Rotate the focusing knob till the object is in sharp view along with the cross hairs.

4.3.5 Using the Theodolite

The theodolite is mainly used to measure horizontal and vertical angles, even though many other operations can be done with the instrument. It is a delicate and sensitive instrument and needs to be handled carefully. The following points should be noted while using the instrument.

1. The theodolite should be set up and levelled at every station. This is a fundamental, necessary operation and should be carried out carefully.
2. In measuring horizontal angles, the inclination of the telescope is not significant. The line of sight is arranged to bisect the object clearly.
3. The graduated circle plate gives the outer axis and the vernier plate provides the inner axis. Both the axes coincide if the instrument is properly adjusted and form the vertical axis.
4. There are three clamp screws each with its own tangent screw. The *lower clamp* screw releases the lower plate, the *upper clamp* screw releases the upper vernier plate, and the third vertical circular clamp releases the vertical circle. One should be familiar with the location of the clamp screws and the corresponding tangent screws.
5. Each clamp screw releases one plate. The lower plate is released by the lower clamp screw. When this plate is released, swinging the instrument or rotating it in a horizontal plane causes no change in the reading of the circle, as both the plates move together. This is used when an object has to be sighted with the zero setting of the circle or with any other reading without changing the reading.
6. Both the clamp screws should not be released together. When the lower clamp is tight and the upper clamp screw is released, the upper plate moves relative to the lower plate and the reading changes. This is done when one has to measure an angle.
7. The clamp screws should be tightened very near to their final position so that only a very small movement has to be effected by the tangent screw. For each clamp screw, the corresponding tangent screw should be used for final adjustment.

8. To set the instrument to zero at the plate circle, release the upper clamp and rotate the instrument about the vertical axis. On the vernier A, make the zero of the circle coincide with the zero of the vernier. Tighten the upper clamp and using the upper tangent screw, make the zeros exactly coincide. This can be verified by looking through the magnifying glasses and seeing that the graduations on either side are symmetrical. Verify the condition on vernier B as well, where the 180° graduation should coincide with the zero of the vernier.

9. While bisecting the signals or setting the zero reading, keep the line of sight in such a position that the tangent screw moves the sight in the same direction as the movement of the instrument. If the movement is clockwise, then the tangent screw is adjusted to move the cross hairs from left to right.

10. Operate a tangent screw only after clamping the corresponding clamp screw.

11. The magnifying glasses are so fixed that they can be moved along the circle. Read the circle by bringing the glass over the reading and looking directly over the reading to avoid any parallax error.

12. While bisecting stations with the theodolite, the station mark should be very clear and must be a point. Bisect either the cross marks on pegs at their intersection or the ranging rod and arrow at their lowest pointed end.

13. Clamp screws and tangent screws need careful handling. Do not apply great force on these screws and handle them delicately during survey work.

4.4   Measuring Horizontal Angles

To measure the horizontal angle between two lines, the following procedure is adopted.

1. Referring to Fig. 4.5, the angle POQ is to be measured. Set up the theodolite at O.

2. Set the instrument to read 0° 00′ 00″. This is not strictly required, as the angle can be determined as the difference between the initial and final readings. However, it is convenient to make the initial reading zero. For this, release the upper clamp and rotate the instrument to make the reading approximately zero. Clamp the upper plate and using the upper tangent screw, make the reading exactly zero. Vernier A reads zero and vernier B reads 180° 00′ 00″.

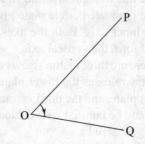

Fig. 4.5 Measuring a horizontal angle

3. Release the lower plate and rotate the instrument to bisect the station P. After approximately bisecting it, clamp the lower plate and using the lower tangent screw, bisect the signal exactly. The readings on the plates do not change as both the plates move together in this operation. Check that the readings on vernier A and B are zero and 180°, respectively.

4. Release the upper plate by loosening the upper clamp. Rotate the instrument to bisect the station signal at Q. Tighten the upper clamp. Using the upper tangent screw, exactly bisect the signal at Q.

5. Read both the verniers A and B. The reading at A will give the angle directly. The reading at B will be 180° + ∠POQ.
6. If there is any difference, take the average of the two values as the correct angle.

Horizontal angles are measured this way for ordinary work. The accuracy can be improved by reading the angles with face-left and face-right observations and taking the average of the two. For more precise work, the angles are repeatedly measured with both the faces and the average taken. This method is known as the *repetition method* and is described below.

4.4.1 Method of Repetition

In the method of repetition, the horizontal angle is measured a number of times and the average value is taken. It is usual to limit the number of repetitions to three with each face except in the case of very precise work. With large number of repetitions, errors can also increase due to bisections, reading the verniers, etc. Very large number of repetitions necessarily do not lead to a more precise value of the angle. However, a number of errors are eliminated by the repetition method. The procedure is as follows (Fig. 4.6).

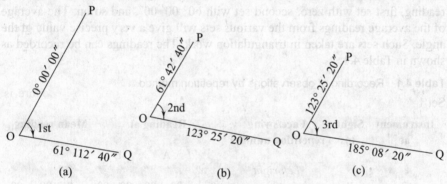

Fig. 4.6 Repetition method

1. Angle POQ is to be measured. Set up, centre, and level the theodolite at O. Ensure that the instrument is in the normal position, i.e., face left.
2. Set the instrument to read 0° 00′ 00″. For this release the upper clamp and bring the zero of the vernier (at vernier A) very close to the zero of the circle. Clamp the upper plate and using the upper tangent screw, coincide the two zeros exactly.
3. Loosen the lower clamp and rotate the instrument so that the left signal at P is approximately bisected. Tighten the lower clamp and using the lower tangent screw, bisect the signal at P exactly. Read the verniers at A and B. The reading should not change and they should read zero and 180°.
4. Loosen the upper clamp and rotate the instrument clockwise to approximately bisect the right signal at Q. Using the upper tangent screw, bisect the signal at Q exactly.
5. Read the verniers at A and B. The reading at A gives the value of the angle directly. The reading on the vernier at B will be 180° + the angle. Record both the readings.
6. Release the lower clamp and rotate the instrument clockwise to bisect the signal at the left station P again. Using the lower tangent screw, bisect the signal

exactly. Check the readings on the verniers at A and B. They should remain the same as recorded.

7. Release the upper clamp and rotate the instrument to bisect the signal approximately. Lock the upper clamp and bisect the signal exactly with the upper tangent screw. Read the verniers A and B. The readings should be twice the angle. Record the readings from both the verniers.

8. Repeat the procedure once more. We thus have three repetitions of the value. Record the readings of both the verniers.

9. Change the face of the instrument. Invert the telescope and make it face right. Repeat the above procedure to have three more readings of the angle.

10. The average of the face-left readings and the average of the face-right readings are averaged to get a very precise value of the angle.

Sets for precision work

In works requiring greater precision, sets of readings are taken. One set may consist of six face-left readings and an equal number of face-right readings. The readings may be taken in clockwise and counterclockwise directions with different faces. A number of similar sets may be taken. Sets may also start with a different initial reading, first set with zero, second set with 60° 00′ 00″, and so on. The average of the average readings from the various sets will give a very precise value of the angle. Such sets are taken in triangulation work. The readings can be recorded as shown in Table 4.1.

Table 4.1   Recording of observations by repetition method

Set 1

Instrument at	Sight to	Face/swing repetition number		Reading at					Mean reading		
					A		B				
		Left/right	°	′	″	′	″	°	′	″	
O	P		0	00	00	00	00	0	00	00	
	Q	1	61	42	40	42	40	61	42	45	
	P		61	42	40	42	40				
	Q	2	123	25	20	25	20	123	25	20	
	P		123	25	40	26	40	123	26	40	
	Q	3	185	08	20	08	20	185	08	20	

Mean value of the angle = 61° 42 ′ 47″

Set 2

Instrument at	Sight to	Face/swing repetition number		Reading at					Mean reading		
					A		B				
		Right/left	°	′	″	′	″	°	′	″	
O	P		00	00	00	00	00				
	Q	1	61	42	40	42	40				
	P										
	Q	2	123	25	40	25	40				
	P										
	Q	3	185	08	00	08	00	185	08	00	

Mean value of the angle = 61° 42′ 40″

The method of repetition helps to eliminate the following errors.

(a) Errors caused by the eccentricity of the centres and verniers, by reading both the verniers and averaging.

(b) Graduation errors by reading from different parts of the circle.

(c) Imperfect adjustment of the line of collimation and horizontal axis by face-left and face-right observations.

(d) Observational errors and other errors tend to be compensated by the large number of readings.

However, the errors due to levelling cannot be compensated. This has to be done by permanent adjustment. Also a large number of repetitions tend to increase the wear of clamp and tangent screws.

Therefore, from the two sets,

$$\text{Mean value of the angle} = (1/2)(61° 42' 47'' + 61° 42' 40'') = 61° 42' 44''$$

4.4.2 Method of Reiteration

The method of reiteration is another method of measuring horizontal angles. This method is useful when a number of angles are to be measured at one point. In Fig. 4.7, let O be the point where the instrument is set up and P, Q, R, and S be the stations. Angle POQ, QOR, and ROS are to be measured. In the retiration method, each angle is measured in succession and finally the line of sight is brought back to P, i.e., the line of sight is made to close the horizon. The instrument is

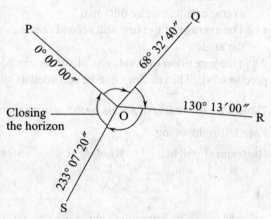

Fig. 4.7 Reiteration method

turned through 360°. Obviously, the instrument should read, upon closing the horizon, the same reading set initially at P. The procedure is as follows.

1. Set up and level the theodolite at O. Keep the instrument in the normal position, i.e., face left. Set the vernier at A to read zero using the upper clamp and upper tangent screw. Check that the vernier at B reads 180°.

2. Loosen the lower clamp and swing the instrument to bisect the station mark at P. Tighten the screw and using the lower tangent screw, finely bisect the signal at P. Check that the verniers at A and B read zero and 180°, respectively.

3. Release the upper plate with the upper clamp, swing the instrument clockwise to bisect the signal at Q. Tighten the clamp and using the upper tangent screw, bisect the mark at Q exactly.

4. Read the verniers at A and B and record both the readings.

5. Release the upper clamp screw, bisect the signal at R. Tighten the clamp and bisect the mark at R exactly with the upper tangent screw. Read the verniers at A and B and record the readings.

6. Continue the procedure with other stations.

7. After the last angle has been recorded, release the upper clamp and swing the instrument to close the horizon and bisect the station mark at P. Check that the verniers at A and B now read the initial reading set while starting. If there is any discrepancy, note that down for the final adjustment.

8. This completes the first readings of angles with the left face. Note that *the lower clamp and the lower tangent screw should not be touched during the entire process* after bisecting the station mark at P.

9. Each angle is calculated from the mean value of the readings of verniers A and B at each station. The difference between these mean values gives the value of the required angle.

10. Now repeat the procedure by changing the face. The telescope will be in an inverted position (right face). To be more accurate, the second set of readings may be taken with the initial value of the reading different from zero. The instrument can be set to read 30° or 60° initially. The reading may also be taken in an counterclockwise direction along with the change of face. A second average value can be obtained.

11. The average of the first and second readings will give a very accurate value of the angle.

As in the repetition method, sets of values can be obtained and averaged for more precise work. The readings can be recorded as shown in Table 4.2.

Table 4.2   Recording of angles by the reiteration method

Face left/right swing

Instrument at	Sight to	Reading at A			B		Mean reading			Angle	Value		
		°	′	″	′	″	°	′	″		°	′	″
O	P	00	00	00	00	00	00	00	00				
	Q	68	32	40	32	40	68	32	40	POQ	68	32	40
	R	130	13	00	13	00	130	13	00	QOR	61	40	20
	S	233	07	20	07	20	233	07	20	ROS	102	54	20
	P	00	00	00	00	00	00	00	00				

Face right/right swing

Instrument at	Sight to	Reading at A			B		Mean reading			Angle	Value		
		°	′	″	′	″	°	′	″		°	′	″
O	P	00	00	00	00	00	00	00	00				
	Q	68	32	20	32	20	68	32	20	POQ	68	32	20
	R	130	13	20	13	20	130	13	20	QOR	61	41	00
	S	233	07	40	07	40	233	07	40	ROS	102	54	20
	P	00	00	00	00	00	00	00	00				

The mean values of the angles are

∠POQ = 68° 32′ 30″,   ∠QOR = 61° 41′ 10″,   ∠ROS = 102° 54′ 20″

4.5 Measuring Vertical Angles

A vertical angle is made by an inclined line of sight with the horizontal. The line of sight may be inclined upwards or downwards from the horizontal. Thus one may have an angle of elevation or depression. See Fig. 4.8. For measuring vertical angles, the theodolite is levelled with respect to the altitude bubble.

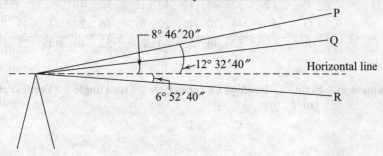

Fig. 4.8 Measuring vertical angles

The procedure for measuring vertical angles is as follows.
1. Set up the theodolite at the station from where the vertical angle is to be measured. Level the instrument with reference to the plate bubble.
2. Further level the instrument with respect to the altitude level fixed on the index arm. This bubble is generally more sensitive. The procedure for levelling is the same. Bring the altitude level parallel to two foot screws and level till the bubble traverses. Swing through 90° to centre the bubble again with the third foot screw. Repeat till the bubble traverses.
3. Swing the telescope to approximately direct the line of sight towards the signal at P. Loosen the vertical circle clamp screw and incline the line of sight to bisect P. Clamp the vertical circle and bisect the signal exactly with the horizontal cross hair.
4. Read the verniers C and D. The average of these readings gives the value of the angle.

This procedure assumes that the instrument is properly adjusted. If there is an index error, the instrument does not read zero when the bubble is in the centre and the line of sight is horizontal, the adjustment is done by the clip screw. There may be a small index error, which can be accounted for in the value of angle. The readings can be recorded as shown in Table 4.3.

4.5.1 Measuring Vertical Angle Between Two Points

The two points may be above the horizontal or below the horizontal or one may be above and the other below. In all cases, the vertical angles between the instrument and the points are measured. If the points lie on the same side of the horizontal, the vertical angle between the points is the difference between the measured angles. If they lie on either side of the horizontal through the instrument, the vertical angle between the points is the sum of the angles measured.

Table 4.3   Recording of vertical angles

Face left

Instrument at	Sight to	Reading on vernier					Mean angle			Vertical angle		
			C		D							
		°	′	″	′	″	°	′	″	°	′	″
O	P	12	32	40	32	40	12	32	40			
	Q	8	46	20	46	20	8	46	20	3	46	20
	R	(–)6	52	40	52	40	(–)6	52	40	19	25	20

Face right

Instrument at	Sight to	Reading on vernier					Mean angle			Vertical angle		
			C		D							
		°	′	″	′	″	°	′	″	°	′	″
O	P	12	32	20	32	20	12	32	20			
	Q	8	46	20	46	20	8	46	20	3	46	00
	R	(–) 6	52	20	52	20	(–)6	52	20	19	24	40

The mean vertical angles are

Between P and Q = 3° 46′ 10″,   between P and R = 19° 25′ 00″

4.6   Other Theodolite Operations

The theodolite is a very versatile instrument and can be used effectively for many accurate operations. Some of these are discussed below. Traversing with the theodolite is discussed in Section 4.7.

4.6.1   Measuring the Magnetic Bearing of a Line

The following procedure is adopted to measure the magnetic bearing of a line.

1. To determine the magnetic bearing of line PQ, set and level the theodolite at P (Fig. 4.9).

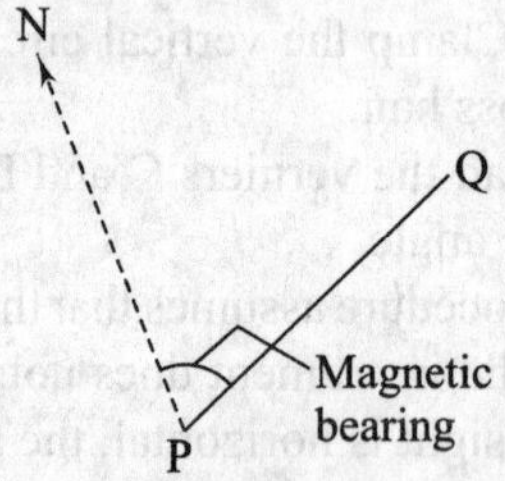

Fig. 4.9 Measuring magnetic bearing

2. Use the upper tangent screw to release the upper plate and swing the instrument to set the reading on vernier A to read zero approximately. Clamp the upper plate and with the upper tangent screw, make the zero of the vernier and the circle coincide exactly.

3. Release the lower plate and the magnetic needle of the compass. Swing the instrument so that the magnetic needle is nearly at the centre of its run, with the north end of the needle pointing to the zero of the graduations on the compass. Tighten the lower clamp and using the lower tangent screw, bring the magnetic

needle to read exactly zero at the north end. Check the verniers at A and B. They should read zero and 180° as set earlier.

4. The line of sight of the instrument is in the direction of the magnetic meridian at P. Release the upper clamp and swing the instrument to bisect the signal at Q. Using the upper tangent screw, bisect the signal exactly at Q.
5. Read both the verniers. The average of the two readings gives the magnetic bearing of the line PQ.

4.6.2 Prolonging a Given Line

Let PQ be the line given. It is desired to prolong the line to another point T in line with it and establish intermediate points R and S (Fig. 4.10). There are many ways to do this. The following three methods are common.

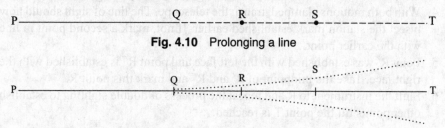

Fig. 4.10 Prolonging a line

Fig. 4.11 Back sighting method

Direct method

The procedure is as follows.
1. Set up and level the theodolite at P.
2. Swing the instrument and bisect the signal at Q. Exact bisection can be done with the lower clamp and the lower tangent screw.
3. Establish a point along the line of sight at R.
4. Shift the instrument to Q. Level the instrument and sight the signal or station mark at R. Establish a third point at S along the line of sight.
5. Shift the instrument and set it up at R. Sight the station mark at S. Establish another station along the line of sight.
6. Continue the process until the point T is reached.

The method is suitable when the instrument is aligned properly.

Back sighting

The following procedure is adopted (Fig. 4.11).
1. Set up and level the theodolite at Q. Swing the instrument to take a back sight on P. Clamp both the motions (upper and lower plates) and transit the telescope.
2. Set a point R along the line of sight at a convenient distance.
3. Shift the instrument to R. Set up and level the instrument at R. Swing the instrument and take a back sight on the station mark at Q. Clamp both the motions and transit the telescope. Fix a point S along the line of sight.
4. Now shift the instrument to S and repeat the procedure till the point T is reached.

This method also assumes that the instrument is properly aligned. If not, when the instrument is transited, the line of sight may not be along PQ and dotted lines will be the result.

Double sighting

The double sighting method is used when the instrument is not properly aligned and does not give precise results. The procedure is as follows (Fig. 4.12).

1. Set up the theodolite at Q and level the theodolite. Release the lower plate, swing the instrument, and take a back sight on P.
2. With both the plates clamped, transit the telescope and establish a point in line with the line of sight.
3. Change the face of the instrument and again take a back sight on P.
4. With both motions clamped, transit the telescope. The line of sight should now bisect the station mark established earlier. If not, mark a second point in line with the earlier point.
5. Point R′ was established with the left face and point R″ is established with the right face. Take the midpoint of R′ and R″ and mark this point R.
6. Shift the instrument to R and repeat the process of double sighting to establish other points till the point T is reached.

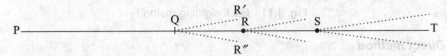

Fig. 4.12 Double sighting

Laying a straight line between two given points

There are three methods for laying a straight line between two given points depending upon whether the two stations are intervisible, intervisible from an intermediate point, or are not intervisible from any intermediate point.

When the stations are intervisible   When the two given stations are intervisible, the procedure is simple, as follows (Fig. 4.13).

1. Set up and level the theodolite at P.
2. Sight the signal at Q and clamp both the motions. Finely bisect the signal at Q using the lower tangent screw.
3. Establish points R, S, T, etc. along the line of sight and fix station marks.
4. PRSTQ is a straight line.

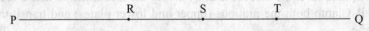

Fig. 4.13 Laying out a straight line

When the stations are visible only from an intermediate point   In this case a method known as *balancing-in* is used. The procedure is as follows (Fig. 4.14).

1. Determine a point midway between P and Q from which both P and Q are visible.
2. By judgement, establish a point R in line with P and Q.

3. Set up and level the theodolite at R.
4. Swing the instrument and take a back sight on P. Transit the telescope and sight the station mark at Q. It may not be in line as the position of R is only set by estimation.
5. Estimate a new position of R to be in line with P and Q.
6. Repeat the procedure till, after back sighting on P and plunging the telescope, the signal at Q is bisected.
7. R is then a point in line with P and Q.

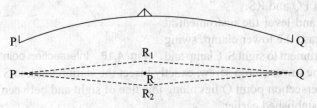

Fig. 4.14 Balancing-in method

When the stations are not visible from any intermediate point   In this case the *random line method* is used. The procedure is as follows (Fig. 4.15).

1. Run a random line at any angle to one side of the line joining the two stations P and Q.
2. To do this, set up the theodolite at P. Centre and level the instrument.
3. Take a line of sight along PQ′ and establish point Q′ such that Q is visible from Q′ and the distance QQ′ can be measured.
4. Set up a sufficient number of intermediate points R′, S′, etc. along the line of sight of PQ′.
5. Measure QQ′, PQ′, PR′, PS′, etc. accurately with a steel tape.
6. Set up and level the instrument at Q′. Measure angle PQ′Q = α accurately.
7. Set up the instrument at R′ and level it. Swing the instrument and lay off an angle equal to α. The point R on line PQ can be obtained by measuring a distance R′R on the line laid at angle α. (The method to lay off an angle is given later.)
8. The length of this line can be determined using similar triangles PQ′Q and PR′R. R′R is equal to QQ′ × PR′/ PQ′.
9. Repeat the same procedure with points S′, etc. PRSQ is a straight line between P and Q.

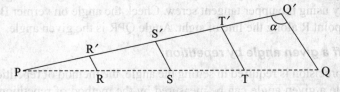

Fig. 4.15 Random line method

Locating the point of intersection of two given lines

Consider four points P, Q, R, and S as shown in Fig. 4.16. The objective is to find the point of intersection of the two lines. The procedure is as follows.

1. Set up and level the theodolite at P.
   Release the lower clamp, swing the
   instrument, and bisect the signal at
   Q. Using the lower tangent screw,
   exactly bisect the station mark at Q.
2. Along the line of sight, set up two
   points 1 and 2, roughly on either
   side of an estimated intersection
   point of PQ and RS.
3. Set up and level the instrument at
   R. Release the lower clamp, swing
   the instrument to sight S. Clamp and

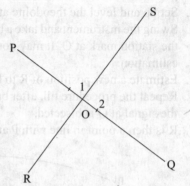

Fig. 4.16 Intersection point of two lines

using the lower clamp screw, exactly bisect the station mark at S.
4. The intersection point O lies along this line of sight and between the points 1
   and 2 established earlier.
5. To locate O, establish a line between 1 and 2 with a fine string. On this string,
   sighting from R to S, obtain the point O and mark it accurately.
6. The point can be checked by setting up the instrument at O and checking
   whether PQ and RS are in line with the station point.

Laying off a given angle

The procedure for setting an angle on the ground is as follows (Fig. 4.17).
1. Let the angle to be set be 38° 47′ 20″ with
   a line PQ.
2. Set up the theodolite at P, centre it, and level
   it with reference to the plate level.
3. Release the upper clamp, swing the instrument to make vernier A read 0° 00′ 00″. Set
   the reading exactly to zero with the upper

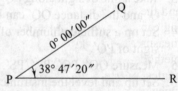

Fig. 4.17 Laying out an angle

tangent screw. Read vernier B, which should register 180° 00′ 00″.
4. Release the lower clamp and swing the instrument to bisect the signal at Q.
   Exactly bisect the signal at Q with the lower tangent screw. Check the reading
   on both the verniers, which should be the same as set before (0° 00′ 00″).
5. Release the upper clamp and swing the instrument clockwise. Set vernier A to
   read approximately the given angle and clamp. Set the angle to the given value
   exactly using the upper tangent screw. Check the angle on vernier B.
6. Set a point R along the line of sight. Angle QPR is the given angle.

Laying off a given angle by repetition

If greater precision is required in setting the angle, the method of repetition can be
used. While a given angle can be measured by the method of repetition directly,
there is no corresponding direct method to set an angle. An indirect method is
employed. The procedure is as follows (Fig. 4.18).
1. Let the angle to be set off be 54° 36′ 35″ with the line OP.
2. Set up and level the theodolite at O. Set the instrument to read zero using the
   upper clamp and the upper tangent screw. Check vernier B also.

3. Release the lower clamp, bisect the station mark at P, and exactly bisect the mark with the lower tangent screw. Check the readings on both the verniers; they should remain the same.

4. Release the upper clamp, swing the instrument clockwise, and set the instrument to read 54° 36′ 00″. With the upper tangent screw, make the readings exactly 54° 36′ 00″. Make a station mark along the line of sight. Point Q′ is established.

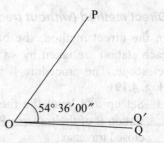

Fig. 4.18 Laying out an angle by repetition

5. Measure the angle POQ′ by the repetition method. Take three readings with face left and three readings with face right. Find the average value of the angle from these observations. Let the average value be equal to 54° 36′ 12″.

6. The difference between the angle to be set and this angle is 23″. The angle is corrected by linear measurements, as the difference in angles is too small.

7. Let the length of the line OQ′ be 100 m. Then Q′Q = 100 tan (23″) = 0.011 m.

8. Measure this distance accurately along the perpendicular with a steel tape to get point Q. Angle POQ is the required angle.

9. Measure the angle POQ again by repetition to check.

4.7 Theodolite Traversing

We have seen that chain traversing or chain triangulation is done using only a chain and the details are filled in by taking offsets. With a compass and chain, traversing can be done with the directions of lines coming from bearings and distances measured with a chain or tape. With the theodolite, traversing can be done by many methods—measuring bearings of lines, measuring the included angles, deflection angles, or external angles of a traverse, with the distances measured using a tape. Open and closed traverses can be surveyed accurately with a theodolite.

Traversing includes the survey of the framework or skeleton and the filling up of details with offsets. The skeleton is not restricted to a framework of triangles and can have any shape. The survey lines are arranged to suit the terrain and the details to be located.

The basic procedure for theodolite traversing is the same as that in any other method of traversing. First reconnaissance has to be conducted with a sketch drawn of the terrain using approximate locations of traverse stations, then the important details are to be picked up, the intervisibility of stations to be checked, and the basic equipment required for the survey to be collected. Theodolite traversing requires station marking tools such as pegs, arrows, a theodolite with its stand, and a steel tape. The decision regarding which method of surveying will be used, by angles or bearings, should also be taken in advance including the scale of the plot to be prepared finally.

4.7.1 Traversing with Magnetic Bearings

The method of traversing with magnetic bearings is suitable in places where local attraction due to external magnetic influences is absent. There are many methods of measuring angles. This method is also known as the *fast needle method of traversing*.

Direct method (without transiting)

In the direct method, the bearings at each station are taken by swinging the telescope. The procedure is as follows (Fig. 4.19).

1. Set up and level the theodolite at station P of the traverse PQRSTP, a closed traverse.
2. Using the upper clamp and upper tangent screw, set vernier A to read zero.
3. Loosen the magnetic needle. Release the lower clamp and point the telescope in the direction of the magnetic meridian till the magnetic needle comes to rest at the zero position. Using the lower tangent screw, the north end of the magnetic needle to read exactly zero.

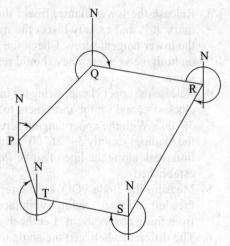

Fig. 4.19 Magnetic bearings– Direct method (without transiting)

4. Release the upper plate and swing the instrument to bisect the signal at Q. With the upper tangent screw, bisect the station mark exactly.
5. Read vernier A, which gives the bearing of the line PQ.
6. Keeping both the clamps tight, shift the instrument to Q. Set up and level the instrument. Check the reading on vernier A. It should be the same as the magnetic bearing of the line PQ (if not, this can be corrected and the bearing value noted earlier be set on vernier A).
7. Release the upper clamp. Swing the instrument clockwise to bisect the station mark at R. Using the upper tangent screw, bisect the station mark R exactly. Read the vernier at A and note down the reading.
8. With both clamps tight, shift the instrument to R and repeat the procedure. The work is continued at all stations in a similar manner.

The following points should be carefully noted.

(a) We get the correct reading of the bearing of PQ at P on vernier A.
(b) At Q, as the theodolite is swung to sight R, the instrument is 180° out of orientation and a correction has to be applied to the reading on vernier A. In this position, vernier B gives the correct reading of the magnetic bearing of line QR. This can be seen from the sample readings given in Table 4.4.
(c) It is more convenient to read one vernier only throughout the operation. In such a case, the bearings at Q and S (or the second, fourth, sixth, etc.) stations need to be corrected by 180°. The bearings at P, R, and T (or the first, third, fifth, etc.) can be read directly on vernier A.
(d) Alternately, read vernier A for the magnetic bearings of PQ, RS, and TP (or the first, third, fifth, etc.) and vernier B for the magnetic bearings of QR and ST (or the second, fourth, sixth, etc.).

(e) At station P, the first station, observe the magnetic bearing of the last line PT before measuring the bearing of PQ. This can be used for checking.

Table 4.4 Recording magnetic bearings

Instrument at	Sight to	Reading on vernier A			Reading on vernier B			Line	Bearing		
		°	′	″	°	′	″		°	′	″
P		00	00	00	00	00	00				
	Q	62	42	20	242	42	20	PQ	62	42	20
	T	201	59	40	21	59	40				
Q	R	334	53	40	154	53	40	QR	154	53	40
R	S	202	32	40	22	32	40	RS	202	32	40
S	T	101	43	20	281	43	20	ST	281	43	20
T	P	21	59	40	201	59	40	TP	21	59	40

Direct method (with transiting)

In this method, instead of swinging the instrument, the telescope is transited at the second station. This gives the correct reading of the bearings at the station. The procedure is as follows (Fig. 4.20).

1. Set up and level the theodolite at P. Set the reading on the vernier to zero with the upper clamp and the upper tangent screw.

2. Release the lower clamp and swing the instrument towards the magnetic meridian. Release the needle of the compass. With the lower tangent screw, set the magnetic needle to read exactly zero.

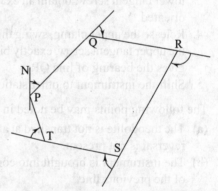

Fig. 4.20 Magnetic bearings–Direct method (with transiting)

3. Release the upper clamp and swing the instrument to bisect the signal at Q. With the upper tangent screw, exactly bisect the station mark Q. Read vernier A and record the reading.

4. With both plates clamped, shift the instrument to Q, and set up and level the instrument.

5. Release the lower clamp and swing the instrument to take a back sight on P. Exactly bisect the signal with the lower tangent screw. Check that the reading on vernier A is the bearing of PQ.

6. Transit the telescope. The line of sight is now directed in the direction of the extension of line PQ. The reading of the bearing is correct in this position.

7. Release the upper clamp and swing the instrument to bisect the station mark at R. Exactly bisect the signal with the upper tangent screw. Read vernier A and record. This will be the correct bearing of line QR.

8. Shift the instrument to the next station and repeat the procedure.

The following point may be noted in this method: the instrument is transited at every station to orient it with the line of sight.

Back bearing method

In this method the instrument is set to the back bearing of the line to get the correct orientation of the line. The procedure is as follows.

1. Set up and level the theodolite at P. With the upper clamp and upper tangent screw, set vernier A to read zero. With the lower clamp and lower tangent screw, bring the line of sight to the magnetic meridian by releasing the magnetic needle. When the needle reads zero, clamp the instrument.
2. Release the upper clamp, swing the instrument to bisect the signal at Q, and bisect the signal exactly with the upper tangent screw. Read vernier A and record the reading as the bearing of PQ.
3. Shift the instrument to Q. Calculate the back bearing of line PQ. With the upper clamp and the upper tangent screw, set the reading of vernier A to the back bearing of PQ. Release the lower clamp and take a back sight on P. With the lower tangent screw, obtain an exact bisection. Now the instrument is correctly oriented.
4. Release the upper clamp, swing the instrument, and bisect the signal at R. With the upper tangent screw, exactly bisect the signal. Read vernier A. This reading gives the bearing of line QR.
5. Shift the instrument to other stations and repeat the procedure.

The following points may be noted in this procedure.

(a) The theodolite is not transited at any station. Thus the error due to line of sight reversal is not present.
(b) The instrument is brought into correct orientation by setting the back bearing of the previous line.
(c) The instrument is swung and rotated in a horizontal plane at every station to bisect the signal at the next station.
(d) The reading on vernier A gives the correct bearings of the lines at every station.

4.7.2 Traversing by Angles

Instead of bearings of lines, the angles of a traverse can be measured with the theodolite. The angles measured can be included (interior) angles, exterior angles, or deflection angles. In the case of a closed traverse, any of the angles can be measured. In the case of open traverses, deflection angles or angle between lines can be measured. Whichever angle is measured, the bearing of the first line is always measured to get the direction of the magnetic meridian at that place.

Traversing by interior or exterior angles

Figure 4.21 shows the interior and exterior angles of a traverse. If the traverse is covered in the clockwise direction, the angles measured are interior angles. If we go counterclockwise over the traverse, we measure the exterior angles. The general procedure in either case is as follows.

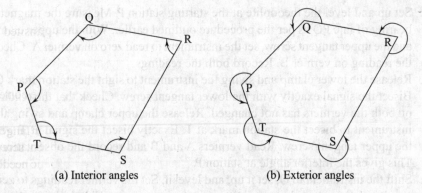

(a) Interior angles **(b) Exterior angles**

Fig. 4.21 Traversing by angles

Table 4.5 Recording of interior angles

Face left/swing right; traversing clockwise

Instrument at	Sight to	Reading on vernier					Angle	Horizontal angle		
		A			B					
		°	′	″	′	″		°	′	″
P	Q	00	00	00	00	00				
	T	139	17	20	17	20	TPQ	139	17	20
Q	R	00	00	00	00	00				
	P	87	48	40	48	40	PQR	87	48	40
R	S	00	00	00	00	00				
	Q	132	21	00	21	00	QRS	132	21	00
S	T	00	00	00	00	00				
	R	100	49	00	49	00	RST	100	49	00
T	S	00	00	00	00	00				
	P	79	44	00	44	00	STP	79	44	00

Face right/swing right; traversing clockwise

Instrument at	Sight to	Reading on vernier					Angle	Horizontal angle			Mean angle		
		A			B								
		°	′	″	′	″		°	′	″	°	′	″
P	T	00	00	00	00	00							
	Q	139	17	00	17	00	TPQ	139	17	00	139	17	10
Q	P	00	00	00	00	00							
	R	87	48	20	48	20	PQR	87	48	20	87	48	30
R	S	00	00	00	00	00							
	Q	132	21	20	21	20	QRS	132	21	20	132	21	10
S	T	00	00	00	00	00							
	R	100	49	20	49	20	RST	100	49	20	100	49	10
T	P	00	00	00	00	00							
	S	79	43	40	43	40	STP	79	43	40	79	43	50

1. Set up and level the theodolite at the starting station P. Measure the magnetic bearing of line PQ as per the procedure outlined earlier. With the upper clamp and the upper tangent screw, set the instrument to read zero on vernier A. Check the reading on vernier B. Record both the readings.

2. Release the lower clamp and swing the instrument to sight the station mark Q. Bisect the signal exactly with the lower tangent screw. Check that the reading on both the verniers has not changed. Release the upper clamp and swing the instrument to bisect the station mark at T. Exactly bisect the signal at T with the upper tangent screw. Read verniers A and B and record the observations. This gives the interior angle at station P.

3. Shift the instrument to Q, set it up, and level it. Set the vernier readings to zero again with the upper clamp and the upper tangent screw. With the lower clamp and lower tangent screw, bisect exactly the station mark at R. Check the readings on the verniers, which should read zero and 180°.

4. Release the upper clamp and swing the instrument to bisect the signal at P. Read verniers A and B and record the readings. This will give the interior angle at Q.

5. Repeat the procedure at all stations of the traverse.

The readings can be recorded as shown in Table 4.5. To achieve greater accuracy, the angles can be measured at a station with both faces, face left and face right. For high, precision work, angles can be read by the method of repetition with three repetitions for each face. When you go counterclockwise and measure the angles from the previous station to the forward station, you measure the exterior angles. The procedure is exactly the same.

Traversing by deflection angles

Traversing by deflection angles is the preferred method for open traverses. The method can be used for closed traverses as well. The procedure is as follows (Fig. 4.22).

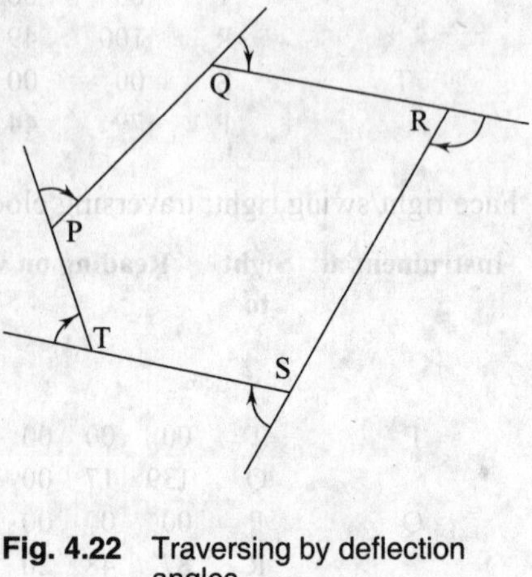

Fig. 4.22 Traversing by deflection angles

1. For the closed traverse PQRSTP shown, set up and level the theodolite at P. Observe and record the magnetic bearing of the line PQ by the method outlined above.

2. Using the upper clamp and upper tangent screw, set vernier A to read zero. Check vernier B. Record the readings of both the verniers.

3. Using the lower clamp and lower tangent screw, bisect the signal at the last station T. Transit the telescope. Release the upper clamp, and swing the instrument to bisect the signal at the next station Q. With the upper tangent screw, bisect the signal exactly. Read the verniers. This will give the deflection angle at P. While line PQ can be plotted with the bearing, this can help as a check.

4. Shift the instrument to Q, set it up, and level it. With the upper clamp and tangent screw, set vernier A to read zero. With the instrument reading zero, release the lower clamp and swing the instrument to bisect the signal at P. Exactly bisect the signal with the lower tangent screw.

5. Transit the telescope. Release the upper clamp and swing the instrument to bisect the signal at R. Clamp the upper plate and, with the upper tangent screw, exactly bisect the signal. Read both the verniers and record.

6. Repeat the procedure at all the stations.

The observations can be recorded as shown in Table 4.6. It should be noted that deflection angles may be left or right as shown in the table. A deflection angle is right, when the next line is to the right of the continuation of the previous line. A deflection angle is left, when the next line is to the left of the continuation of the previous line. It should be recorded simultaneously with the observed angles.

Table 4.6 Recording of deflection angles

Face left

Instrument at	Sight to	Reading on vernier					Mean value			Deflection angle		
			A			B						
		°	′	″	′	″	°	′	″	°	′	″
P	North	00	00		00	00 00						
	Q	61	46		20	46 20	61	46	20			
Q	P	00	00		00	00 00						
	R (R)	15	36		40	36 20	15	36	30	15	36	30
R	Q	00	00		00	00 00						
	S (R)	44	32		20	36 40	44	36	30	44	36	40
S	R	00	00		00	00 00						
	T (R)	54	28		40	28 40	54	28	40	54	28	40

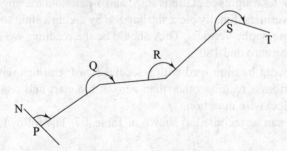

Fig. 4.23 Open traverse—Direct angles

Open traverse with direct angles

Instead of deflection angles, direct angles between the lines can be measured in an open traverse. The procedure is the following (Fig. 4.23).

1. Set up the theodolite at station P and level it. Record the magnetic bearing of line PQ.

2. Shift the instrument to Q. Set up and level it. Set vernier A to read zero using the upper clamp and upper tangent screw. Check vernier B also.

3. Release the lower clamp and swing the instrument to take a sight on the station mark at P. Exactly bisect the signal with the lower tangent screw.

4. Release the upper plate with the upper clamp and swing the instrument clockwise to bisect the signal at R. This reading gives the angle between the lines QP and QR.

5. Shift the instrument to other stations and repeat the procedure.

Closed traverse with central angles

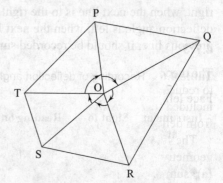

Fig. 4.24 Closed traverse by central angles

In the case of a closed traverse, another procedure can be adopted to survey the skeleton of the traverse. The method of reiteration can be conveniently employed here. The procedure is as follows (Fig. 4.24).

1. Select a convenient point O within the traverse from which all the points of the traverse are visible and chaining to those points is possible.

2. Set up and level the theodolite at O. Set the instrument to read zero with the upper clamp and upper tangent screw. Vernier A should read zero and vernier B should read 180°.

3. Release the lower clamp and swing the instrument to sight the first station P. Exactly bisect it with the lower tangent screw.

4. Check the reading on the verniers. They should be the same as set. Release the upper clamp and swing the instrument to bisect the signal at Q. Bisect it exactly with the upper tangent screw. Note the reading on both the verniers.

5. Similarly, take sights to stations R, S, and T and note the angle readings from both the verniers. Finally close the horizon by taking a sight to station P. Check the reading on the verniers. They should be the readings we started with at P, in this case zero and 180°.

6. Accuracy can be improved with a second set of readings with changed face, starting from a reading other than zero at the start and reading angles in a counterclockwise direction.

The readings can be recorded as shown in Table 4.7. The following points should be noted.

(a) Along with the angles, the distances OP, OQ, etc. are measured with a tape. These are required for plotting the traverse.

(b) As a check, the distances PQ, QR, etc. are also measured. This is not extra work, as these distances are measured for the purpose of taking in details with offsets.

(c) There is no direct measurement of the traverse angles. They can be calculated from the properties of triangles, knowing the three sides.

Table 4.7 Recording of observations

Face left

Instrument at	Sight to	Reading on vernier					Angle	Horizontal angle		
		A °	**B** ′	″	′	″		°	′	″
O	P	00	00	00	00	00				
	Q	62	43	40	43	40	POQ	62	43	40
	R	120	38	00	38	00	QOR	57	54	20
	S	192	53	40	53	40	ROS	72	15	40
	T	273	15	00	15	20	SOT	80	21	10
	P	359	59	40	59	40	TOP	86	44	50

4.7.3 Interconversion of Angles

The theodolite measures the whole circle bearings of lines. These can be converted to reduced bearings by the methods discussed in Chapter 3. Also, one can calculate included angles from bearings and vice versa. Included angles can also be calculated from deflection angles and vice versa.

The following relationships of the angles of a closed traverse are known from geometry:
(a) sum of the interior angles = $(2n - 4)$ right angles
(b) sum of exterior angles = $(2n + 4)$ right angles
(c) sum of the deflection angles = 4 right angles
It is desirable to draw a rough sketch of the traverse before attempting to solve problems. The following examples illustrate these principles.

Example 4.1 The whole circle bearings of the lines of a closed traverse are given below. Find the included angles of the traverse.

Line	AB	BC	CD	DA
Bearing	78° 40′ 20″	152° 31′ 40″	251° 18′ 40″	3° 44′ 15″

Solution It is desirable to draw a rough sketch of the traverse as shown in Fig. 4.25. The solution is shown in Table 4.8.

Table 4.8 Solution to Example 4.1

Line	Fore bearing	Back bearing	Angle at	°	′	″
AB	78° 40′ 20″	258° 40′ 20″	A	105	03	55
BC	152° 31′ 40″	332° 31′ 40″	B	106	08	40
CD	251° 18′ 40″	71° 18′ 40″	C	81	13	00
DA	3° 44′ 15″	183° 44′ 15″	D	67	34	25

First we calculate the back bearings of the lines. The included angle is the difference between the back bearing of a line and the fore bearing of the next line. Thus,

∠DAB = bearing of DA − bearing of AB = 183° 44′ 15″ − 78° 40′ 20″
= 105° 03′ 55″
∠ABC = bearing of BA − bearing of BC = 258° 40′ 20″ − 152° 31′ 40″
= 106° 08′ 40″

∠BCD = Bearing of CB – bearing of CD = 332° 31′ 40″ – 251° 18′ 40″
= 81° 13′ 00″

∠CDA   = bearing of CD – bearing of DA = 71° 18′ 40″ – 3° 44′ 15″
= 67° 34′ 25″

Check: Sum of included angles of a traverse = (2*n* – 4) right angles = 360 or a four-sided figure.

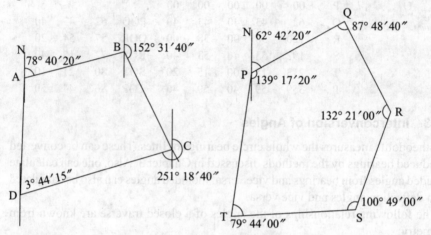

Fig. 4.25   Traverse of Example 4.1   **Fig. 4.26**   Traverse of Example 4.2

Example 4.2   The bearing of the first line PQ of a traverse is 62° 42′ 20″. The included angles of the traverse are given below. Calculate the whole circle bearings of the lines.

Station	P	Q	R	S	T
Angle	139° 17′ 20″	87° 48′ 40″	132° 21′ 00″	100° 49′ 00″	79° 44′ 00″

Solution   A rough sketch of the traverse is shown in Fig. 4.26. The solution is given in Table 4.9.

Table 4.9   Solution to Example 4.2

Station	Included angle	Line	Fore bearing	Back bearing
P	139° 17′ 20″	PQ	62° 42′ 20″	242° 42′ 20″
Q	87° 48′ 40″	QR	154° 53′ 40″	334° 53′ 40″
R	132° 21′ 00″	RS	202° 32′ 40″	22° 32′ 40″
S	100° 49′ 00″	ST	281° 43′ 40″	101° 43′ 40″
T	79° 44′ 00″	TP	21° 59′ 40″	201° 59′ 40″

Example 4.3   A closed traverse ABCDEA was conducted and the included angles measured. The bearing of the first line AB was 78° 40′ 20″. Find the deflection angles at each vertex and the bearings of the lines.

Station	A	B	C	D
Angle	105° 03′ 55″	106° 08′ 40″	81° 13′ 00″	67° 34′ 25″

Solution   A rough sketch of the traverse is shown in Fig. 4.27. The solution is given in Table 4.10.

Table 4.10 Solution to Example 4.3

Station	Included angle			Deflection angle			Line	Bearing		
	°	′	″	°	′	″		°	′	″
A	105	03	55	74	56	05	AB	78	40	20
B	106	08	40	73	51	20	BC	152	31	40
C	81	13	00	98	47	00	CD	251	18	40
D	67	34	25	112	25	35	DA	3	44	15

Example 4.4 The deflection angles measured in a closed traverse ABCDEA are given below. Find the included angles of the traverse.

Station	Defl. angle
A	81° 32′ 40″
B	65° 54′ 20″
C	72° 14′ 30″
D	68° 25′ 50″
E	NA

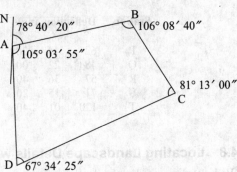

Fig. 4.27 Traverse of Example 4.3

Solution The deflection angle at E either has not been measured or is not available. The sum of deflection angles in a closed traverse is equal to 360°, irrespective of the number of sides. The sum of the four deflection angles given = 288° 07′ 20″. Deflection angle at E = 360°− 288° 07′ 20″. The included angle is equal to 180° − deflection angle. The solution is as shown below.

Station	Defl. angle	Incl. angle
A	81° 32′ 40″	98° 27′ 20″
B	65° 54′ 20″	114° 05′ 40″
C	72° 14′ 30″	107° 45′ 30″
D	68° 25′ 50″	111° 34′ 10″
E	71° 52′ 40″	108° 07′ 20″

As a check, sum of the included angles = 540° 00′ 00″.

Example 4.5 A traverse was conducted and the exterior angles were measured. Find the deflection angles from the data given below.

Station	A	B	C	D
Exterior angle	254° 56′ 05″	253° 51′ 20″	278° 47′ 00″	NA

Solution The sum of the exterior angles of a closed traverse = $(2n + 4)$ right angles. This condition can be used to evaluate the unknown exterior angle.

$(2n + 4)$ right angles = $(2 \times 4 + 4) \times 90$ = $12 \times 90°$ = 1080°

Sum of given angles = 787° 34′ 25″

Exterior angle at D = 1080° 00′ 00″ − 787° 34′ 25″ = 292° 25′ 35″

Deflection angle = exterior angle − 180°; therefore, the deflection angles are

A: 74° 56′ 05″, B: 73° 51′ 20″, C: 98° 47′ 00″, D: 112° 25′ 35″

Example 4.6 The deflection angles of a closed traverse were measured with a theodolite. Find the bearings of the lines of the traverse from the deflection angles given below. The bearing of the first line PQ was 54° 25′ 30″.

Station	Defl. angle
P	62° 44′ 20″
Q	48° 22′ 00″
R	57° 36′ 40″
S	71° 15′ 20″
T	120° 01′ 40″

Solution   The solution is shown in Table 4.11.

Table 4.11   Solution to Example 4.6

Station	Deflection angle			Line	Bearing		
	°	′	″		°	′	″
P	62	44	20	PQ	54	25	30
Q	48	22	00	QR	102	47	30
R	57	36	40	RS	160	24	10
S	71	15	20	ST	231	39	30
T	120	01	40	TP	351	41	10

4.8   Locating Landscape Details with the Theodolite

We have discussed so far methods to survey the main frame or the skeleton of the survey. In most surveys, it is necessary to locate details such as buildings, railway lines, canals, and other landmarks along with the survey. A transit with a steel tape is used to locate details, and many methods are available, as the transit is an angle-measuring instrument. The following methods can be used.

Angle and distance from a single station

A point can be located with an angle to the station along with the distance from that station as shown in Fig. 4.28(a). The angle is preferably measured from the same reference line to avoid confusion. A sketch with the line and the distance and angle measured will help in plotting later. A road can be located as shown in Fig. 4.28(b). Angles to a number of points are measured and with each angle two distances are measured to locate the road.

Angle from one station and distance from another

If for any reason, it is not possible to measure the angle and distance to an object from the same point, it may be possible to locate the point by measuring angles from one station and distances from the other. The recorded data should clearly indicate the stations from which the angle and distance are measured. Figure 4.28(c) shows this method of measuring. The angle is measured from station A to point P. When the instrument is shifted to B, the distance to point P is measured from B with a steel tape.

Angles from two stations

If for some reason, it is not possible to measure distances, then angles from two stations are enough to locate a point. As shown in Fig. 4.28(d), the point P is located by measuring angles to point P from stations A and B.

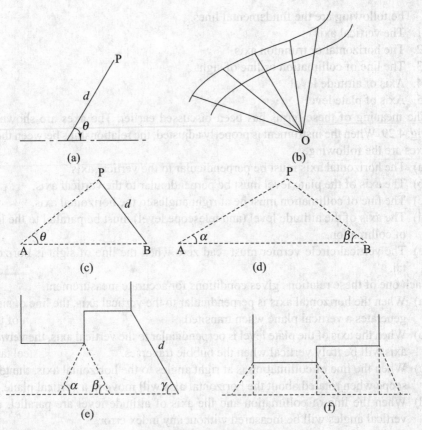

Fig. 4.28 Filling-in details

Intermediate points on lines

If the stations are far away, any of the above methods can be used to locate details from any one or two selected points on the main line. The angle and distance or two angles from two different points can be used to locate points. The end of a chain length or other convenient chainages can be selected. The field book should clearly indicate the points and the angles and distances from the points.

Offsets

As in the case of chain surveying, details can be located by offsets with the offset rod or tape as described earlier from the main chain lines. This is suitable when a number of points have to be located along the chain line.

4.9   Fundamental Lines and Desired Relationships

The theodolite has to be properly adjusted before it is used. All the methods described above assume that the instrument is properly adjusted, though some techniques whereby certain errors due to bad adjustment of the instrument can be eliminated were discussed. However, the instrument should be in adjustment for survey work. Here we discuss only the fundamental lines and their relationships for getting accurate results using the theodolite. The actual procedure for adjustment is discussed in Chapter 12.

The following are the fundamental lines.
1. The vertical axis
2. The horizontal or trunnion axis
3. The line of collimation or line of sight
4. Axis of altitude level
5. Axis of plate level

The meaning of these terms has been discussed earlier. The axes are shown in Fig. 4.29. When the instrument is properly adjusted, the relationships between these axes are the following.

(a) The horizontal axis must be perpendicular to the vertical axis.
(b) The axis of the plate level must be perpendicular to the vertical axis.
(c) The line of collimation must be at right angles to the horizontal axis.
(d) The axis of the altitude level (and telescope level) must be parallel to the line of collimation.
(e) The vertical circle vernier must read zero when the line of sight is horizontal.

Each one of these relations gives conditions for accurate measurement.

(a) When the horizontal axis is perpendicular to the vertical axis, the line of sight generates a vertical plane when transited.
(b) When the axis of the plate level is perpendicular to the vertical axis, the vertical axis will be truly vertical when the bubble traverses.
(c) When the line of collimation is at right angles to the horizontal axis, the telescope when rotated about the horizontal axis will move in a vertical plane.
(d) When the line of collimation and the axis of altitude level are parallel, the vertical angles will be measured without any index error.
(e) The index error due to the displacement of the vernier is eliminated when the vernier reads zero with the line of collimation truly horizontal.

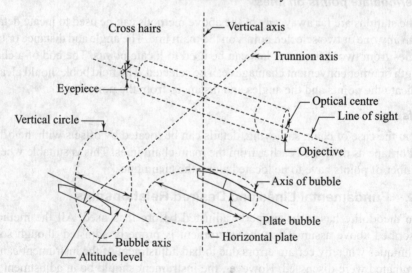

Fig. 4.29 Fundamental lines of a theodolite

4.10 Errors in Theodolite Work

The theodolite is a very versatile but delicate instrument. The fundamental lines retain the desired relationships just for a few uses. Thus, errors creep into the measured values. Other errors are due to incorrect use and operations by personnel. Still other errors may be attributed to natural causes. These are discussed below.

4.10.1 Instrumental Errors

Instrumental errors are those due to the maladjustment of or faults in the instrument. The following errors are common.

Maladjustment of plate level

When the axis of the plate level, which is a line tangential to the bubble tube at its centre, is not perpendicular to the vertical axis, the horizontal circle is inclined when the bubble traverses. The angles measured are not in a horizontal plane and an error will be introduced in the measurements. This error becomes serious when the angle is measured between two points at different elevations. A permanent adjustment has to be carried out to set this right. The instrument can also be levelled with respect to the altitude level, which is a more sensitive bubble tube.

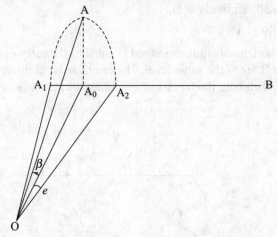

Fig. 4.30   Line of collimation not at right angles to the trunnion axis

Line of collimation not being at right angles to trunnion axis

When the line of collimation is not perpendicular to the horizontal or trunnion axis, it will not move in a vertical plane when the telescope is rotated in a vertical plane. The trace of the movement of the telescope is the surface of a cone. As shown in Fig. 4.30, let A and B be two point at different elevations, A being higher. When the point A is sighted and the telescope is brought down to sight point B, the telescope traces a curve and comes to points A_1 or A_2, depending upon whether the observation is face right or face left. A_0 is the point directly below A. Angles A_0OA_1 and A_0OA_2 are the angles by which the line of sight is misaligned. If these angular errors are equal to e, then $e = \beta \sec\alpha_1$, where α_1 is the angle of elevation of A with the horizontal. A similar error $e = \beta \sec\alpha_2$, where β is the error in the line of collimation. It is clear that when face-left and face-right observations are taken and averaged, this error is eliminated. A small error of $\beta(\sec\alpha_1 - \sec\alpha_2)$ remains when A has an elevation of α_1 and B has an elevation of α_2 with the horizontal. This will be zero when A and B are at the same level.

Horizontal axis not perpendicular to vertical axis

When the horizontal axis is not perpendicular to the vertical axis, an error is introduced, as the line of sight will not move in a vertical plane but along an inclined plane. This error is again significant when the two points between which the angles are measured are at considerable difference in elevation.

As shown in Fig. 4.31, let A and B be two points at a considerable difference in elevation. A_0 is a point directly below A in a vertical plane. A_1 and A_2 are points, in a horizontal plane with B, obtained when the telescope is lowered with face left and with face right to sight Q. a_1 is the vertical angle to A and α_2 is the vertical angle to B. In Fig. 4.31,

$$\tan\alpha = A_0A/OA_0, \quad \tan\beta = A_0A_1/AA_0, \quad \tan e = A_0A_1/OA_0$$

Substituting,

$$\tan e = \tan\alpha_1\tan\beta = \beta\tan\alpha_1, \quad e = \beta\tan\alpha_1$$

as β and e are small. Similarly at B,

$$e = \beta\tan\alpha_2$$

The error in the horizontal angle measured is equal to $\beta(\tan\alpha_1 - \tan\alpha_2)$ and is zero when A and B are at the same level. The error can be eliminated by recording observations with both the faces and averaging the results.

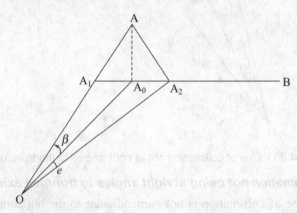

Fig. 4.31 Horizontal axis not perpendicular to vertical axis

Eccentricity of inner and outer axes

In a perfectly adjusted instrument, the inner and outer axes coincide and form the vertical axis. If the centres of the upper (vernier) plate and the circle plate do not coincide, an error is introduced in the measurement. As shown in Fig. 4.32, let the centre of the circle plate be O and that of the vernier plate be O′. The position of the vernier is P before the telescope is rotated to measure the angle. After the telescope is turned, the position of the vernier is Q. The corresponding points on the diametrically opposite verniers are R and S. As the vernier is rotated to measure the angle between two targets P and Q, the vernier rotates about O′. The arc through which the instrument turns measures the angle POQ and not the true angle PO′Q. If T is the intersection of lines OP and O′Q, then from the property of triangles that the exterior angle is equal to the sum of interior opposite angles,

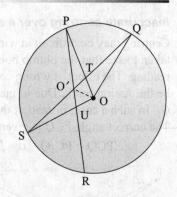

$$PTQ = PO′Q + OPO′$$

from triangle PO′T. Also

Fig. 4.32 Inner and outer axes not aligned

$$PTQ = POQ + O′QO$$

from triangle OTQ. Therefore,

$$PO′Q = POQ + O′QO – OPO′$$

If R and S are points opposite to P and Q on the circle, then on similar lines,

$$RU′S = RO′S + O′SO \quad \text{and} \quad RUS = ROS + O′RO$$

From this,

$$RO′S = O′RO + ROS – O′SO$$

From the figure, it is clear that

$$RO′S = PO′Q, \quad OPO′ = O′RO, \quad \text{and} \quad O′SO = O′QO$$

Adding the two equations for PTQ and RO′S, we get

$$2PO′Q = POQ + ROS \quad \text{or} \quad PO′Q = (1/2)(POQ + ROS)$$

The true angle is the mean of the measured value at the two verniers. Measuring at verniers A and B and averaging the results gives the correct angle.

Axis of altitude level not parallel to line of collimation

This condition results in error in the measurement of vertical angles, as the zero line of the verniers does not act as a true reference line for measuring vertical angles. Taking observations with both the faces and averaging the results eliminate the error.

Faulty circle graduations

The graduations of the circle may be faulty, not being equidistant in different parts of the circle. The error can be minimized by taking the readings for an angle several times with different parts of the circle and averaging the results.

Verniers being eccentric

This condition arises when the zeroes of the verniers are not at 180° or diametrically opposite. Taking readings at verniers A and B and calculating the average eliminates this error.

4.10.2 Personal Errors

Personal errors arise due to faulty manipulation of the instrument controls and errors in sighting, reading, and recording.

Inaccurate centring over a station

Centring may be difficult in windy conditions. Also proper care may not have been taken to see that the plumb bob is exactly over the station mark before taking the reading. The effect of wrong centring can be derived as follows. In Fig. 4.33, let C be the station mark. Due to inaccurate centring, the centre of the instrument is at C_1. In such a case, instead of the correct angle PCQ, the angle measured is PC1Q. The correct angle PCQ is given by

$$\angle PCQ = PC_1Q - \alpha - \beta = PC_1Q - (\alpha + \beta)$$

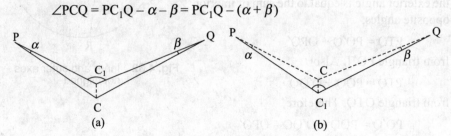

(a)   (b)

Fig. 4.33   Innacurate centring

Table 4.12   Variation of centring error with the length of sight

Centring error (mm)	Length of sight (m)	Angular error
5	25	41″
	50	21″
10	25	1′ 22″
	50	41″
15	25	2′
	50	1′
20	25	2′ 45″
	50	1′ 22″

On the other hand, if the centring is done over C_2 on the other side, then

Correct $\angle PCQ = \angle PC_2Q + (\alpha + \beta)$

The error due to centring, $\pm (\alpha + \beta)$, depends upon the length of sights and the error in centring. Table 4.12 shows the error due to centring and how it varies wih the length of sights.

Slip

Errors due to slip may occur when the instrument is not put firmly on a tripod. The error can also result from the clamps not being properly tightened or the shifting head not being properly clamped.

Wrong manipulation of controls

This is a common error. Clamping the instrument and operating the wrong tangent screw can result in erroneous zero setting or reading of the angle. The remedy is to familiarize oneself with the controls, operate them carefully, and verify the readings frequently.

Inaccurate levelling

The levelling operation is fundamental to the theodolite. Levelling should be done carefully and verified by turning the instrument through 180°. This error is similar to the instrumental error of the plate level not being perpendicular to the vertical axis.

Inaccurate bisection of target

As angles are to be measured accurately, ideally a point target is required. Many often, it is not possible to observe the cross mark on the peg over the station. Care must be taken to ensure that if a ranging rod is sighted, the lowest point is observed and an appropriate mark is made for bisection.

Non-verticality of ranging rod

The ranging rod or other object held over a station has to be truly vertical. Otherwise an error is introduced in the reading. The error is inversely proportional to the length of sight. The error is given by tan e = error in verticality/length of sight.

Displacement of target

All stations should be properly marked. If the target is misplaced or shifted during surveying, serious error can occur.

Parallax

To avoid parallax, the eyepiece should be focused properly. The focusing of the objective should be done at every sight to get a sharp and clear image.

Mistakes in setting and reading

Mistakes in reading the main and vernier scales are common. These could be due to reading the vernier in the wrong direction, inaccurate adjustment of the vernier, or reading the main and vernier scales at an angle instead of directly over the graduations.

4.10.3 Errors due to Natural Causes

Errors due to natural causes include errors due to (i) settlement of the tripod due to soft soil (ii) wind, causing vibrations and turning, and (iii) high temperature, causing faults in reading due to refraction, differential expansion of different parts, and direct sunlight on the instrument making sighting and reading difficult.

4.11 Precision in Linear and Angular Measurements

In measuring angles and distances with a theodolite and a tape, one must maintain consistency between the linear and angular measurements. As shown in Fig. 4.34, let AB and AC be two lines. The angle between the two lines is α and the length of line AC is l. Let the angular error in measuring the angle be $d\alpha$ and the linear error in measuring the length be n/l. The true position of the required point is C, but due to linear and angular errors, the position shifts to C_1. C_2C_1 is the linear error n.

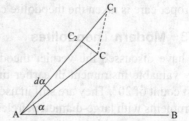

Fig. 4.34 Precision in linear and angular measurements

$CC_2 = l \tan d\alpha$; $C_1C_2 = n$. Equating the two, $\tan d\alpha = l/n$. Linear error is usually expressed as 1 in L. Table 4.13 below verifies the relationship derived above for various angles.

Table 4.13 Relationship between linear and angular error

Angular error	1/L	Linear error	Angular precision
10″	1/206,26	1 in 500	6′ 50″
20″	1/103,13	1 in 3000	1′ 8″
30″	1/6875	1 in 5000	41″
45″	1/4584	1 in 10,000	21″
60″	1/3438	1 in 25,000	8″

To illustrate, an angular error of 10″ gives $\tan 10″ = 0.00004848$. The inverse of this value is 1/20626 which is the corresponding linear error. If the linear error is 1 in 500 (1/500), $\tan^{-1}(1/500) = 0.11459°$ or $0°06′52″$.

4.12 Upkeep and Maintenance of Theodolite

The vernier theodolite is a very fine instrument used to measure angles. The following points should be taken care of to maintain the instrument in good condition.

(a) The instrument should be properly kept in the box supplied with the equipment. The clamp screws are loosened to enable movement and the instrument is then kept in the proper position in the box. The instrument is kept in a tight position within the box by appropriate holding knobs or plates. The instrument should not move around when it is carried in the box.

(b) It should be kept free from dust and water. It is preferable to have a field umbrella to protect the instrument from direct sunshine during fieldwork.

(c) It should be protected by a plastic cover or any other similar arrangement when kept at a station while taping or locating details.

(d) Depending upon the extent of usage, the instrument should undergo permanent adjustments frequently.

(e) In windy conditions and on slopes, the instrument should be held firmly to prevent it from falling off the tripod.

(f) Anti-moisture gel must be kept in the box to prevent damage to the optical parts due to excessive humidity.

(g) The instrument should be dusted to remove dust and moisture before keeping it in the box after the day's work.

If proper care is taken the theodolite can provide good service for a long time.

4.13 Modern Theodolites

We have discussed the vernier theodolite earlier. The vernier theodolite was a very valuable instrument in earlier times. Most vernier-type instruments have a least count of 20″. They are still in use in many places. For geodetic work, vernier instruments with large-diameter circles were used with a small least count in early days. To improve this instrument and measure angles with greater precision, many modifications were made to the design of angle-measuring instruments. Optic and electronic theodolites, which need less maintenance, were developed for ease of operation and greater precision. In these theodolites, permanent adjustments remain for a longer period of time. Angles can be read to an accuracy of 1″ or even less. In the following sections, we will discuss these instruments briefly. The details vary with the manufacturers but the essential principles remain the same.

4.13.1 Micro-optic Theodolites

Micro-optic theodolites can read angles to an accuracy of 10″ or even less. The essential principle is illustrated in Fig. 4.35. The special features of such theodolites are as follows.

(a) Conventional metal circles are replaced by glass circles on which the graduations are etched by photographic methods. The graduations can be made finer and sharper by this technique. Both the horizontal and vertical circles are made of glass and generally graduated to 10′.

(b) Light passing through the circle at the point of the reading is taken through a set of prisms to the field of view of the observer. For passing light through glass circles, sunlight is reflected through a reflecting prism and passed through the circle. In case night operation is required, the battery-operated light provided in the instrument can be used.

(c) Both the horizontal and vertical circles are seen at the same time in the field of view. This is an advantage, as the readings of both the circles can be taken at the same time. Some manufacturers make a switching arrangement so that the horizontal or vertical circle reading can be seen along with the micrometer reading.

(d) The optical micrometer is used to read fractions of the main scale division. Depending upon the reading system, angles can be read up to 10′ or less.

(e) The circles are generally graduated to 10′ or 20′ of the arc. The micrometer can be read after coinciding the index with the nearest main scale division. The fractions are then read from the micrometer scale, which is also seen in the field of view.

(f) A small, separate reading telescope is provided besides the main telescope. It eliminates the need to move while bisecting an object and taking the reading.

(g) In most instruments, diametrically opposite ends of the circle are brought together in the field of view.

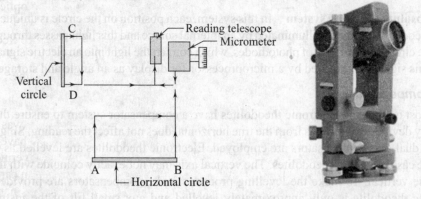

Fig. 4.35 Micro-optic theodolite, also see Plate 1

4.13.2 Electronic Theodolites

The electronic theodolite (Fig. 4.36) is a recent development in the manufacture of theodolites. The working principle remaining the same, the graduated circle is made differently and the method of taking readings is also different. At the press of a button, the angles are displayed and can be stored in memory. The following are the characteristic features of electronic theodolites.

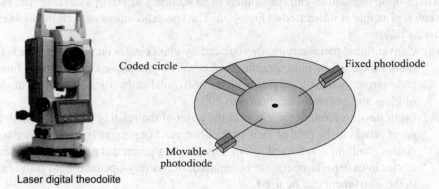

Laser digital theodolite

Coded circle — Fixed photodiode

Movable photodiode

Fig. 4.36 Incremental encoding

Graduated circle

The graduated circle is made of glass as in an optical theodolite, but the graduations are made in a special code to be read by photodiodes. The photodiodes convert the readings into electrical signals that are processed by a microprocessor into angles that are digitally displayed on an LCD or LED panel. The graduated circle can be encoded in two ways—the incremental system and the absolute encoding system.

Incremental system In this system, two photodiodes are placed at diametrically opposite positions over the glass circle. The circle is encoded by dividing it into a large number of parts. One of the photodiodes is fixed in position and indicates the zero of the circle. As the telescope is turned in azimuth, the other photodiode moves along with the telescope. The circle does not move. The photodiode measures the movement of the alidade from the code etched on the circle and the signal is sent to a processor for converting the measurement into an angle for display and storage.

Absolute encoding system In this system, each position on the circle is uniquely encoded. The circle is illuminated by an internal source and this light passes through the circle onto an array of photodiodes, which convert the light into an electric signal. This signal is processed by a microprocessor for display as an angle and storage.

Compensator

Most optical and electronic theodolites have a compensator system to ensure that any deviation of the axis from the true horizontal does not affect the reading. Single or dual-axis compensators are employed. Electronic theodolites are levelled as in the case of vernier theodolites. The vertical axis may not exactly coincide with the true vertical. To make the levelling process simple, compensators are provided. The theodolite is only approximately levelled and any small tilt of the instrument is taken care of by the compensator. The compensator systems are shown in Fig. 4.37.

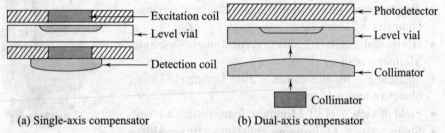

(a) Single-axis compensator   (b) Dual-axis compensator

Fig. 4.37 Compensator systems

The single-axis compensator measures the tilt of the vertical axis and compensates by adjusting the vertical angle. The dual-axis compensator measures the tilt of the vertical axis as well as the trunnion axis. The horizontal angle is also adjusted for any tilt in the trunnion axis. Single-axis compensator systems are of liquid or magnetic type. Current is passed through the central coil known as the *excitation coil*. This generates a current in the other two coils. The current values will be equal if the instrument is level and different if there is any tilt in the vertical axis. This difference is measured and converted into a tilt angle. The measured angle is also compensated for this tilt by the processor before displaying the angle.

In dual-axis compensators, light from an internal source is passed onto a lens for evenly distributing the light onto a liquid vial. A photodiode divided into four sections receives this light. The light falling on the various sections of the photodiode will be even if there is no tilt in the axes and uneven if there is any tilt in the axes. The difference in light intensity is converted into tilt angles for both the axes by the processor. The horizontal and vertical angles are also corrected for tilt before displaying.

Display consoles may be provided either on one side or on both sides. Most electronic theodolites have a zero set facility—changing the direction of the reading from clockwise to counterclockwise, a display of angles, a display of grades, etc. The angles are measured and displayed within a few seconds of pressing a button. The batteries last for about 75 to 100 hours of operation. The data from the theodolite can be transferred to a computer for further processing.

Digital theodolites

Digital theodolites are very fine instruments for angle and distance measurements. The instruments are light weight and are similar to electronic theodolites in construction.

The instrument is set up over a station as in the case of normal theodolites. They will have extendable tripod legs which can be adjusted for comfortable viewing. The centring and levelling operations are done with a circular vial for coarse setting and the plate vial for fine levelling of the instrument. Once the target is sighted, one has to press only a measure button to get the readings of angles and distances. Some models also have a laser pointer for easy alignment in critical cases and for staking out operations. With the arrival of total stations, these theodolites have less demand though they are cheaper compared to a total station.

The following are typical features in a digital theodolite:

- Angle measurement – by absolute encoding glass circle;
  Diameter – 71 mm
- Horizontal angle – 2 sides; vertical angle–one side;
  Minimum reading -1″/5″
- Telescope – Magnification – 30x; Length–152 mm;
  objective lens – 45 mm
- Field of view – 1°30′ Minimum focus distance – 1m
- Stadia values: Multiplying constant – 100; additive
  constant – 0

Also see Plate 1

- Laser pointer – coaxial with telescope; 633 nm class II laser; Method – focus-
  ing for alignment and stake out operations
- Display on both sides; 7-segment LCD unit
- Display and reticle illuminated
- Compensator – tilt sensor; vertical tilt sensitivity ± 3′
- Optical plummet – magnification – 3x; field of view – 3°; focusing from 0.5 m
  to infinity
- Level sensitivity – Plate vial – 40″/2mm; circular vial – 10′/2 mm
- Power supply – 4 AA size batteries; Operating times – Theodolite only – 140
  hours
- Laser only – 80 hours; Theodolite + laser – 45 hours
- Weight – 4.2 kg

Optical plummet

Most instruments are provided with an optical or laser plummet for accurate centring.
The plummet system is shown in Fig. 4.38. A small telescope is provided at the
tribrach level. The station point can be seen through the telescope very accurately.
The image of the station point is brought to the view of the telescope through a
reflecting prism along the centre of the instrument.

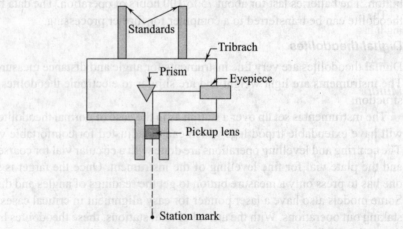

Fig. 4.38 Optical plummet

Summary

A theodolite is commonly used to measure horizontal and vertical angles. The common vernier theodolite consists of a levelling head for levelling the instrument, a horizontal graduated circle with two verniers fixed diametrically opposite, a supporting A-frame, an index frame, a vertical circle and verniers, and a telescope. The instrument is attached to a tripod while in use. The theodolite can generally measure angles up to 20″ accuracy. The horizontal circle is graduated from 0° to 360° while the vertical circle may be graduated in quadrants from 0° to 90°.

Temporary adjustments of a theodolite are those adjustments that are to be done at every station it is set up. These are centring and levelling. Levelling can be done using spirit levels fitted on the horizontal plates. It is done by centring the bubble in two directions at right angles repeatedly till the bubble traverses in all positions. Centring is done using a plumb bob attached to the centre of the theodolite base.

Horizontal angles are measured by the repetition method, in which the angle is measured several times and averaged. Equal number of repetitions are done with both the left and right faces. The reiteration method is very useful when several angles are to be measured at a single station. This is done by measuring the angles in the clockwise or counterclockwise direction sequentially and closing the horizon finally by sighting the first station again. Left-face and right-face observations are taken to ensure more accuracy.

Vertical angles are measured using the vertical circle. In this case, levelling is done using the index or altitude level or telescope level. The angle measured is the angle between the horizontal plane through the horizontal axis and the inclined line of sight.

The theodolite can be used to measure bearings of lines using the compass attached to the instrument. Other operations that can be done using the instrument include prolonging a line, setting a line between given points, and finding the intersection of lines.

The theodolite is also commonly used for traversing. Traversing is done by measuring included angles, exterior angles, bearings, or deflection angles. Details are located by a combination of lengths and angles or angles alone.

The fundamental lines of a theodolite are the vertical axis, horizontal or trunnion axis, axis of plate level, axis of altitude level, and line of collimation. These lines are either parallel or perpendicular to each other in an adjusted instrument.

Common errors in theodolite work include instrumental errors due to maladjustments of the instrument; errors of manipulation, sighting, and recording; and errors due to natural causes.

The theodolite is a very fine and delicate instrument and should be taken care of by keeping it in adjustment and ensuring that dust and moisture do not spoil its sensitive parts.

Exercises

Multiple-Choice Questions

1. A theodolite is often designated by
   - (a) diameter of the objective lens
   - (b) diameter of the horizontal circle
   - (c) diameter of the eye piece lens.
   - (d) length of the telescope
2. To change the reading on the circle while measuring an angle,
   - (a) upper clamp is tightened and lower clamp is loosened
   - (b) upper clamp is loosened and lower clamp is tightened
   - (c) both upper and lower clamps are loosened
   - (d) both upper and lower clamps are tightened

3. The horizontal circle in a theodolite is graduated in
   (a) the quadrantal system from 0 to 90 in the four quadrants
   (b) the whole circle system from 0 to 360
   (c) the semi-circle system from 0 to 180 in the right and left halves
   (d) a way similar to that in a prismatic compass
4. The vertical circle of a theodolite is generally graduated in
   (a) the quadrantal system from 0 to 90 in the four quadrants
   (b) the whole circle system from 0 to 360
   (c) the semi-circle system from 0 to 180 in the right and left halves
   (d) a way similar to that in a prismatic compass
5. When you transit the telescope, you rotate the telescope about
   (a) the vertical axis   (c) the optical axis of the telescope
   (b) the trunnion axis   (d) the line of collimation
6. A face-left or normal position of observation means
   (a) the vertical circle of the instrument is on the left of the observer
   (b) the vertical circle is on the right side of the observer
   (c) the station being observed is to the left of the observer
   (d) the left side vernier is read for the angle value
7. Focusing the objective lens is an operation done
   (a) to get the cross hairs to clear view
   (b) every time to focus the object being sighted
   (c) for reading the vernier clearly
   (d) for reading vertical angles
8. In the method of repetition for measuring horizontal angles, to rotate the instrument without changing the reading,
   (a) lower clamp screw is tightened and upper clamp is loosened
   (b) lower clamp screw is loosened and upper clamp screw is tightened
   (c) any one of the clamp screw is loosened
   (d) both the clamp screws are loosened
9. In the method of reiteration of measuring horizontal angles,
   (a) the same angle is measured three times
   (b) the same angle is measured by face-left and face-right observations
   (c) the angle is measured and the instrument turned to close the horizon
   (d) the angle is measured three times each using face-left and face-right observations
10. The inner axis of a theodolite is
    (a) the plate bubble axis on the horizontal circle
    (b) the axis passing through the centre of the vertical circle
    (c) the trunnion axis
    (d) the centre of the spindle carrying the vernier circle
11. The outer axis of a theodolite is
    (a) the axis of the altitude level
    (b) the trunnion axis
    (c) the axis passing through the centre of the horizontal graduated circle
    (d) the line of collimation of the theodolite
12. The index frame or vernier frame is
    (a) the A- frame attached to the telescope
    (b) the verner circle of the horizontal circle
    (c) the T-shaped frame carrying the vernier of the vertical circle
    (d) the base of the theodolite having the leveling head

13. A micro-optic theodolite has horizontal and vertical circles made of
    (a) steel   (c) aluminium
    (b) glass   (d) horizontal and vertical circles made of light plastic
14. In electronic theodolites, the horizontal and vertical circles are made of
    (a) glass having specially coded graduations read by photo diodes
    (b) special metal read electronically
    (c) steel plates with specially coded graduations
    (d) plastic with coded graduations
15. Optical plummet is used
    (a) in optic theodolites for measuring angles
    (b) in electronic theodolites for compensating any tilt in the axis
    (c) for accurate centrring of the theodolite over a station
    (d) for accurate levlling of the theodolite

Review Questions

1. Draw a neat sketch of a vernier theodolite. Describe its main parts and their functions.
2. List the fundamental lines of a theodolite. Explain briefly the desired relationships between these lines and the effects if such relationships are not maintained.
3. With a neat sketch, explain how the main scale and vernier scale are graduated to obtain a least count of 20″.
4. Explain the temporary adjustments of a theodolite.
5. Explain the steps involved in measuring horizontal angle with a theodolite.
6. Explain the repetition method to measure horizontal angles and how readings are recorded.
7. Explain the procedure for the reiteration method of measuring horizontal angles.
8. Explain the step-by-step procedure to measure the vertical angle between two points.
9. Explain briefly the different methods of traversing with a theodolite.
10. Explain the procedure to measure the bearing of a line with a theodolite.
11. Describe briefly the methods used to locate details with a theodolite.
12. Describe a method to range a line between two points when the stations are not intervisible from an intermediate point.
13. Explain the different methods to prolong a line.
14. Explain briefly the possible instrumental errors in theodolite work and the precautions that should be taken to eliminate them.
15. List the possible personal errors in theodolite work and the precautions that should be taken to eliminate them.
16. Describe how consistency in linear and angular measurements can be achieved.

Problems

1. The following included angles were measured in a closed traverse. If the bearing of the line AB is 70° 40′ 30″, find the bearings of the remaining lines.

Station	A	B	C	D	E
Angle	78° 42′ 30″	104° 15′ 20″	92° 44′ 40″	112° 36′ 10″	131° 41′ 20″

2. The bearings of lines measured in a theodolite survey are given below. Find the included angles of the traverse.

Line	PQ	QR	RS
Bearing	160° 33′ 20″	58° 47′ 40″	320° 46′ 30″

Line	ST	TU	UP
Bearing	280° 27′ 30″	225° 52′ 40″	114° 58′ 20″

3. The included angles of a traverse are given below. Find the deflection angles at each vertex.

Station	A	B	C	D	E
Angle	102° 30′ 40″	124° 10′ 40″	96° 08′ 20″	131° 00′ 40″	86° 09′ 40″

4. The deflection angles measured in a traverse are given below. Find the included angles of the traverse.

Station	A	B	C
Defl. angle	22° 36′ 20″	70° 18′ 40″	88° 20′ 20″
Station	D	E	
Defl. angle	96° 13′ 10″	82° 31′ 30″	

5. In a traverse survey, deflection angles were measured. If the bearing of the line PQ is 63° 45′ 20″, find the bearings of the remaining lines of the traverse.

Station	P	Q	R
Defl. angle	32° 10′ 40″	68° 30′ 20″	81° 18′ 30″
Station	S	T	
Defl. angle	84° 00′ 10″	94° 00′ 20″	

6. The exterior angles of a traverse were measured during the survey of an area. If the bearing of the line PQ is 45° 32′ 40″, find the bearings of the remaining lines.

Station	P	Q	R
Ext. angle	202° 36′ 20″	250° 18′ 40″	268° 20′ 20″
Station	S	T	
Ext. angle	276° 13′ 10″	262° 31′ 30″	

Plane Table Surveying

Learning Objectives

After going through this chapter, the reader will be able to
- describe the plane table and the accessories used along with it
- explain the temporary adjustments of a plane table
- explain the different methods of locating points with the plane table
- explain the two-point problem and its solution
- explain the three-point problem and its solution
- list the likely errors and the precautions to be taken in plane table surveying
- explain the advantages and disadvantages of plane table surveying

Introduction

We have seen the methods used to measure distances with a chain or tape, directions with the compass, and horizontal and vertical angles with the theodolite, which is another way of measuring directions. After the data has been collected through survey with such instruments, paperwork of correcting the measurements, plotting, computing areas, etc. starts. Plane table survey is a technique where the fieldwork and a part of the paperwork are done simultaneously in the field.

Plane tabling is ideally suited to filling in details on a map already prepared and available on the drawing sheet. This is done along with a chain or tape. Plane table survey can also be used to prepare a map, with the linear measurements being done with a chain or tape. When a map has already been prepared, the table has to be oriented in the field according to the map plotted on the drawing sheet.

In this chapter we will discuss with the plane table and its accessories, methods of plane tabling, and the methods of orienting the plane table for the two- and three-point problems.

5.1 Plane Table and Accessories

Plane table survey equipment usually consists of the following.

Plane table

The plane table is a well-seasoned, good-quality drawing board, varying in size from 45 cm to 75 cm. A drawing sheet can be placed and fixed on the table using pins, tape, or other similar arrangements [Fig. 5.1(a)].

A simple plane table has a drawing board provided with a ball and socket arrangement for levelling the table, with an arrangement to fix the table to a tripod. The table can be fixed in position, after levelling it with the tripod legs, with a clamp. This type of table is normally used in surveying.

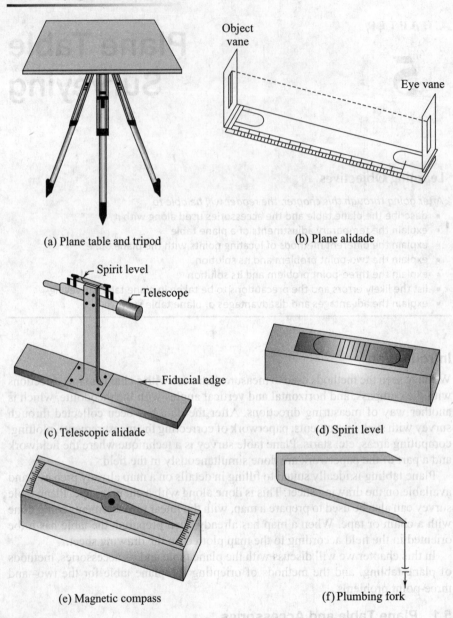

(a) Plane table and tripod

(b) Plane alidade

Spirit level

Telescope

Fiducial edge

(c) Telescopic alidade

(d) Spirit level

(e) Magnetic compass

(f) Plumbing fork

Fig. 5.1 Plane table and accessories

The second type, called *Johnson's table*, also comes with a ball and socket arrangement. There are two thumb screws fitted onto a vertical spindle. The upper thumb screw, when loosened, allows the table to be tilted for levelling. The lower thumb screw is used for rotating the table for sighting.

The third type of plane table comes with a levelling head with three foot screws, as in a level or theodolite. With the help of a spirit level placed in two perpendicular positions, the table can be accurately levelled using the foot screws.

Alidade

An alidade is a sighting instrument that can be of two types: plane alidade and telescopic alidade. A plane alidade consists of a metallic (brass or gunmetal) or good-quality wooden rule [Fig. 5.1(b)] with a fine bevelled edge called the *fiducial edge*. The fiducial or ruling edge is used for drawing lines and for sighting by placing the ruling edge along the points or lines marked on the sheet. Two frames are attached vertically at its two ends. The eye vane part is a small metal frame with a slit for sighting with the eye. The object vane, the part directed towards the object, is a frame with a fine thread or hair placed vertically for bisecting objects. The centre of the eye vane and the thread provides the line of sight. Sometimes, a string is tied between the two vanes at the top to help as a line of sight.

A telescopic alidade [Fig. 5.1(c)] consists of a telescope which provides a very accurate line of sight. The telescope is fitted onto an A-frame, which is fitted onto a heavy rule. The telescope may be provided with a spirit level to help level the table. There may also be a vertical graduated arc to measure the angle of an inclined sight. The rule has a bevelled edge, which is used as a fiducial edge. The telescopic alidade is used for precision work, has a wide range for the line of sight, and can easily take inclined sights. The table can easily be levelled. The telescopic alidade can also be used to measure horizontal distances based on tachometric principles, to which will be discussed later.

Spirit level

With a simple plane table, it is necessary to have a spirit level [Fig. 5.1(d)]. The spirit level can be used to level the plane table. It can be a tubular level, which can be placed in two perpendicular positions and levelled. A circular level can also be used; when the table is level, the bubble is in the centre.

Compass

A trough compass [Fig. 5.1(e)] is used for orienting the table when it has to be used in more than one station. The trough compass should have two, long parallel edges, either of which can be used for drawing the magnetic meridian at the station. When the table is shifted to another station, the trough compass is placed along the meridian previously drawn and rotated to make the needle read zero. The table is then oriented to the same position it had occupied at the previous station.

Plumb bob

The plumb bob is fitted onto a folded frame [Fig. 5.1(f)] so that the straight edge can be placed on the table and the bottom leg of the frame is under the table. The top leg has a fine point, which can be kept on a plotted point (or a point can be plotted against it). The plumb bob remains vertically below this point. The bob can be adjusted to be over the station mark for centring by moving the tripod legs.

Drawing paper

Good-quality drawing sheet, tinted off-white to reduce strain on the eyes, should be used. The paper should be seasoned to reduce the effects of temperature changes and humidity. It should be fixed onto the board with pins or tape. Note that pins are likely to come in the way of the alidade while moving it around. Along with the sheet, good-quality pencils and erasers should be carried for the fieldwork.

5.2 Setting up the Plane Table in the Field

Before working on the plane table, the following adjustments have to be made.

Setting up over the station

The plane table is screwed onto the tripod and the tripod is set, adjusting the table in such a way that the table is approximately level. The drawing sheet, either with the stations already marked or fresh, is placed on the table and fixed with drawing pins or adhesive tape.

Levelling the table

The plane table is levelled using the tripod legs and the spirit level. The tubular level is placed in two perpendicular directions and the tripod legs adjusted to bring the level to the centre. This can be achieved with a few trials.

Centring the table

This becomes important if the drawing sheet already has a point plotted and the station occupied by the table is marked on the sheet. Using the plumb bob, the marked station has to be brought over the point plotted on the sheet. To do this, place the plumb bob with its upper leg on the sheet and its pointed end over the station mark. The lower end of the plumb bob will show where the point plotted on the sheet lies on the ground. If it is not over the station mark, adjust the legs of the tripod to bring the plumb bob over the station mark.

If the work is being started with a fresh sheet, it will only be necessary to place the lower end of the plumb bob over the station mark on the ground and mark a point against the pointed end of the upper end on the sheet. This will give the position of the ground point on the sheet. It will be shown later that centring is not very critical except in the case of large-scale drawings.

Orienting the plane table

Orienting the table is an important operation if the table has to be set up at more than one station. *Orienting essentially means to keep the table in parallel positions at all stations.* If this is not done, the plotted plan will be distorted. The table can be oriented in two ways (refer Fig. 5.2).

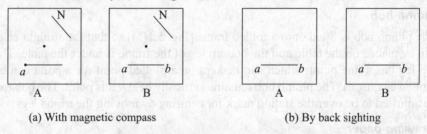

(a) With magnetic compass   (b) By back sighting

Fig. 5.2   Orienting the plane table

Using the trough compass   When the plane table is set up at a station, after centring and levelling the table, a trough compass is kept on the table at a suitable place. The compass is rotated so that the needle is in the centre or reads zero. A line is marked alongside the compass box to indicate the direction of the meridian at the place. When the plane table is shifted to another station, the trough compass is kept along the magnetic meridian previously marked. The table is rotated till the needle is in the centre or reads zero. The table is clamped in that position. The table is now in a position parallel to the position occupied at the first station.

By back sighting   To understand how the plane table is oriented by back sighting, assume that the plane table is kept at point A. The table is centred and levelled. The point occupied by the table is marked on the sheet using the plumb bob. Point *a* is thus marked on the sheet. Using the alidade, the signal at B is sighted and a line drawn along the alidade to mark its direction. The length AB is measured and marked to a suitable scale to obtain point *b*. When the table is shifted to B, the table is oriented by keeping the alidade along *b-a* and sighting the signal at A, keeping *a* towards A. When the signal at A is sighted, the table is clamped. Now A-B on the ground is parallel to *a-b* on the sheet.

There can be situations in which the table has to be oriented with points marked on the table, but the table is kept at a different station and oriented. With two points, this becomes the two-point problem, and with three points marked on the sheet, this becomes the three-point problem. Both are problems of orientation and are dealt with in detail later.

5.3   Plane Tabling Methods

The plane table can be used in four ways: (i) radiation, (ii) intersection, (iii) traversing, and (iv) resection. As mentioned earlier, the plane table is generally used with the traverse drawn on a sheet for filling in details. It can also be used for doing fresh surveys without requiring too much precision. The above-mentioned methods cover both these aspects of surveying with the plane table. The following notation will be used during this discussion: capital letters will be used for stations on the ground and lower case letters will be used for the plotted positions on the sheet. Thus A, B, C, ... will be ground stations while *a*, *b*, *c*,... will be the corresponding positions on the sheet or plan.

5.3.1   Radiation

The method of radiation is illustrated in Fig. 5.3. It requires the plane table to occupy a single station. Orientation of the table is not required in such a case. To conduct the survey of an area, the table is kept at a convenient station P commanding a full view of the area to be surveyed. The following procedure is adopted.

1. The plane table is kept at station P, centred, and levelled. Using the trough compass, mark the magnetic meridian on the sheet at a convenient place.
2. Mark the station point P on the sheet using the plumb bob. This will be point *p* on the sheet.
3. Stations A, B, C, D, E, ... are then conveniently marked on the ground so that they are visible from P.
4. The alidade is then kept over *p* and the flag at point A is sighted. Once the signal at A is bisected, a ray/line is drawn along the edge of the alidade to represent the line *p-a*. The distance from P to A is measured with a chain or tape. This distance is marked to a chosen scale along *p-a* to obtain point *a*.

5. Similarly, points B, C, D, E, etc. are represented on the sheet by *b, c, d, e*, etc. by the radial lines sighted from P.
6. The table should be kept clamped throughout this process. The lines AB, BC, CD, etc. can be measured in the field and can be checked with the marked distances on the sheet as per scale.

This method has great advantages in small-area surveying. There is no need for orienting the table, as the table is kept at one station only. The details can be filled in once the main stations are marked on the sheet.

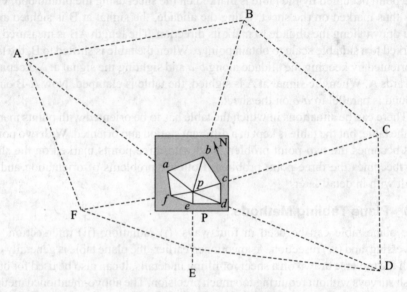

Fig. 5.3 Radiation method

5.3.2 Intersection

The intersection method is illustrated in Fig. 5.4. As can be seen, the method requires setting the table up at a minimum of two stations. Orientation is then a necessity and can be done by back sighting. Two stations A and B are selected so that they command a full view of the area to be surveyed. The method requires the following steps.

1. Set up the plane table at A. Level and centre it. Using the trough compass, mark the magnetic meridian on the sheet. This should be done for reference even though it may not be used for any purpose during the survey.
2. Transfer the point A on the ground to the sheet by the plumbing fork as point *a*. Keep the alidade over *a* and bisect the signal at B. Draw a line along the edge of the alidade to represent line AB. Then measure the distance AB and mark this distance to the chosen scale along *a-b* to get point *b*. This is the base line on which the survey is based. No other linear measurement is required.
3. Now sight the other stations, P, Q, R, etc. keeping the alidade at point *a*. The rays drawn towards these stations are now available on the sheet.
4. Shift the table to B; centre and level it. Now orient the table by back sighting along *b-a* to the signal kept at A. One may have to repeat centring and orienting

to get the orientation correct. The line BA on the ground is now parallel to the line b-a on the sheet.

5. Keep the alidade at *b* sight the signals at P, Q, R, etc. and draw rays from *b*. These rays intersect the previously drawn rays from *a* at points *p*, *q*, *r*, etc.

6. The distances need not be measured in this method. However, one may measure one or two distances for checking purpses.

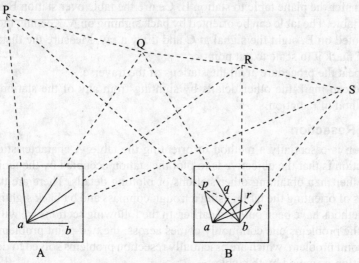

Fig. 5.4 Intersection method

5.3.3 Traversing

Traversing is a method of surveying whereby a series of lines are surveyed. The traverse may be an open traverse or a closed one. The following procedure is adopted in traverse surveying using the plane table (refer to Fig. 5.5).

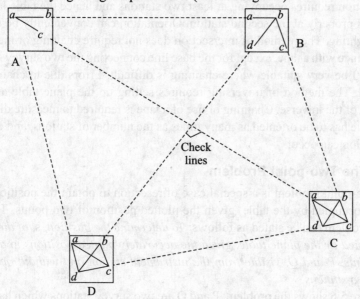

Fig. 5.5 Traversing

1. Keep the plane table at station A; centre and level it. Transfer the ground point A to the sheet with the plumbing fork and mark point *a* on the sheet.
2. Using the alidade, sight the signal at station B and draw the ray. Measure the distance AB and mark it to scale on this ray to get point *b*. In the case of a closed traverse, the last station D can also be sighted from *a* and *d* marked on the sheet.
3. Transfer the plane table to station B. Centre the table over station B and level the table. The table can be oriented by back sighting on A. Keeping the alidade pivoted on B, sight the signal at C and draw a ray. Measure the distance BC and mark it to scale to get point *c*.
4. Repeat the procedure at all the stations of the traverse.
5. One can mark the other details by sighting from any of the stations by the method of radiation.

5.3.4   Resection

Resection is essentially a method of orienting the table. A characteristic feature of resection is that the objective is to plot the station occupied by the table on the sheet rather than obtaining other stations or plotting details. There are two basic methods of orienting the table: using a trough compass and by back sighting. Both these methods have been outlined earlier. In the following sections we will discuss two of the problems one commonly comes across: the two-point problem and the three- point problem, which are essentially resection problems solved to determine the position occupied by the table.

5.3.5   Comparison of the Methods

The method of radiation is the simplest way to use the plane table, in that it requires occupying only one station. There is no need to orient the table. For future reference, the magnetic meridian should be marked on the sheet. The method of intersection requires occupying at least two stations and hence the table has to be oriented properly at the second station. Orientation can conveniently be done by back sighting. The method of intersection does not require chaining or measuring the distance with a tape, except for the base line connecting the two survey stations. This will be very suitable where chaining is difficult. Errors due to chaining are avoided. The method of traversing requires setting up the plane table at all the stations of the traverse. Chaining or use of a tape is required to measure distances. The table has to be oriented as many times as the number of stations, and errors of orientation can occur.

5.4   The Two-point Problem

The two-point problem is a special case of resection to obtain the position of the station occupied by the table, given the plotted position of two points. The two-point problem can be stated as follows: *To determine the position, s, of the station S occupied by the plane table, given the accurately plotted positions, p and q, of two points, P and Q, visible from the instrument station and without occupying these two stations.*

   Figure 5.6 shows the problem. P and Q are two survey stations which have been surveyed earlier and their positions plotted as *p* and *q* on the sheet. S is the instrument station (where the plane table is to be set up) and it is required to get the position of S on the sheet as s when the table is oriented correctly, i.e., when P-Q is parallel to *p-q*. The following procedure can be adopted to solve the two-point problem.

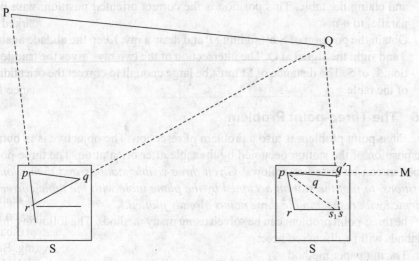

Fig. 5.6 Two-point problem

1. Select an auxiliary station R such that the signals at P, Q, and S are clearly visible from R and the angles formed by P, Q, and S are not very acute.
2. Set up the plane table at R. Level the table. Approximately orient the table so that *p-q* is nearly parallel to P-Q. Clamp the table in this position.
3. Plot the position of R on the table by sighting to P and Q. For this, keep the alidade against *p* and sight the signal at P. Draw a line along the ruling edge of the alidade. Similarly, sight the signal at Q by pivoting the alidade against *q* and draw a ray. The intersection of the two rays drawn gives the position of the station occupied by the table. Label this point *r*. Point *r* is obtained as the position of the station occupied and is accurate to the extent the line *p-q* is parallel to P-Q.
4. Transfer the point *r* on the table to the ground as R using the plumbing fork. A peg can be driven to locate the ground station.
5. With the alidade kept against *r*, sight the ranging rod or other signal at S and draw a line. Mark the distance S by approximation or rough chaining. Point s_1 is thus obtained.
6. Shift the table to S. Level and centre the table over s_1. Orient the table by back sighting at R. For this, keep the alidade against s_1 and sight the signal at R by rotating the table. Clamp the table in this position.
7. With the alidade kept against *p*, sight the station P and draw a ray. This ray intersects the line *r-s* at *s*, giving the station *s*. Keep the alidade against *s*, sight the signal at Q, and draw a ray. This ray will intersect the ray *r-q* not at *q* but at *q'*, as the orientation of the table is only approximate.
8. *p-q'* is the representation obtained of *p-q* due to the error in orientation. The angle *q-p-q'* is the angular error in orientation.

9. To remove this error, place the alidade against p-q' and keep a ranging a rod at a large distance M.
10. Keep the alidade against p-q and rotate the table until the signal at M is sighted and clamp the table. This position is the correct oriented position, with P-Q parallel to p-q.
11. Obtain the position of S by sighting P and draw a ray. Keep the alidade against q and sight the signal at Q. The intersection of the two rays gives the true position, s, of S. The distance of M must be large enough to correct the orientation of the table.

5.5 The Three-point Problem

The three-point problem is also a problem of resection. The objective is to obtain the position of the station occupied by the table after orientation. The three-point problem can be stated as follows: *Given three visible stations and their plotted positions, to plot the station occupied by the plane table with the table correctly oriented with respect to the three points already plotted.*

The three-point problem can be solved using many methods. The following three methods will be discussed here:

1. Tracing paper method
2. Graphical method
3. Trial and error method

5.5.1 Tracing Paper Method

The following procedure is adopted (refer to Fig. 5.7) in the tracing paper.

1. Let P, Q, and R be the three stations whose plotted positions p, q, and r are available on the plan. Let S be the station where the plane table is to be set up. The objective is to orient the table with respect to P, Q, and R and then plot the position of S.
2. Set up the plane table at S and level it. Fix the sheet with the plotted positions of P, Q, and R on the table. Rotate the table and approximately orient it visually so that A-B or B-C is parallel to a-b or b-c.
3. Fix a tracing paper on the table over the sheet. Mark the position of S on the tracing sheet as s_1 using a plumbing fork. This is only approximate, as the table is not oriented.
4. With the alidade centred on s_1, sight the signals at P, Q, and R and draw rays s_1p_1, s_1q_1, and s_1r_1. These rays will not pass through the point p, q, and r plotted earlier, as the orientation is only approximate.
5. Loosen the tracing paper from the table. Rotate the tracing paper so that rays s_1p_1, s_1q_1, and s_1r_1 pass through p, q, and r. Fixing the sheet in such a position, transfer the position of s_1 to the drawing sheet as s.
6. Remove the tracing paper. Place the alidade along s-p. It will not pass through the signal at P, as the table is not oriented. Rotate the table so that the signal at P is sighted when the alidade is kept along s-p and clamp it. The table is now oriented.
7. Check the orientation by sighting the signals at Q and R by placing the alidade along s-q and s-r, respectively. If the orientation is correct, these signals must be seen in these positions.

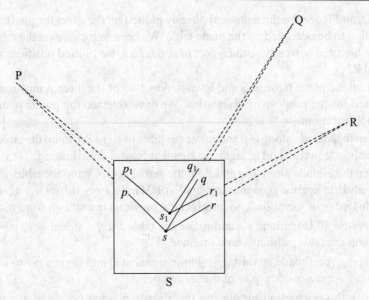

Fig. 5.7 Tracing paper method

5.5.2 Graphical Method

Of the many graphical methods, the Bessel's graphical solution is the more commonly employed method. It is based upon the fact that the three plotted positions and the plot of the station occupied by the table lie on the circumference of a circle when the table is oriented. The procedure is as follows and is illustrated in Fig. 5.8.

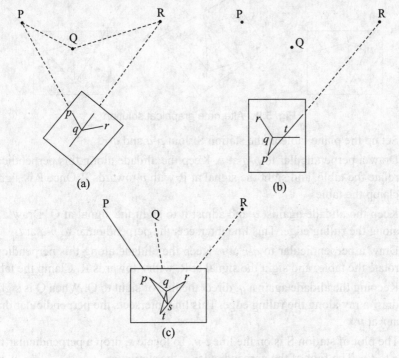

Fig. 5.8 Bessel's graphical method

1. P, Q, and R are the three stations already plotted on the sheet (or plan). S is the station to be occupied by the plane table. We need to locate *s* on the table after it is oriented correctly with respect to *p*, *q*, and *r*, the plotted positions of P, Q, and R.

2. Set up the plane table at S and level it. Any two of the three points can be selected for the purpose of orientation. We have selected the points *p* and *q* for the illustration.

3. Keep the alidade along *q-p* and rotate the table to sight P. Clamp the table. Keep the alidade pivoted on *q*, sight the signal at R, and draw line *q-t*.

4. Keep the alidade along *p-q* and sight the signal at Q. Clamp the table. Keeping the alidade against *p*, sight the signal at R. Draw a ray through *p* against the ruling edge of the alidade to intersect the previous line drawn from *q* at *t*.

5. Keep the alidade along *t-r* and rotate the table till the signal at R is sighted. Clamp the table, which is now oriented.

6. Keeping the alidade against *q*, sight the signal at Q and draw a ray to intersect the ray *t-r* at *s*. *s* is the plot of the station point.

7. Check the orientation by placing the alidade pivoted on *p* and sighting the signal at P. The line should pass through *s* if the table is oriented.

The same result can be achieved by the following alternative procedure (Fig. 5.9).

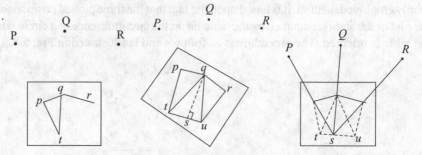

Fig. 5.9 Alternate graphical solution

1. Set up the plane table at the station S. Join *p-q* and *q-r*.

2. Draw a perpendicular to *p-q* at *p*. Keep the alidade along this perpendicular, rotate the table, and sight the signal at P, with *p* towards P. Once P is sighted, clamp the table.

3. Keep the alidade against *q* and adjust it to sight the signal at Q. Draw a line along the ruling edge. This line intersects the perpendicular to *p-q* at *t*.

4. Draw a perpendicular to *q-r* at *r*. Keep the alidade along this perpendicular, rotate the table, and sight the signal at R, with *r* towards R. Clamp the table.

5. Keeping the alidade against *q*, direct the line of sight to Q. When Q is sighted, draw a ray along the ruling edge. This line intersects the perpendicular drawn at *r* at *u*.

6. The plot of station S is on the line *t-u*. To locate *s*, drop a perpendicular from *q* to *t-u*. The foot of the perpendicular is the station *s*.

7. To orient the table, keep the alidade along *s-p* and sight the signal at P by rotating the table. Clamp the table, which is now oriented.

8. To check the orientation, sights can be taken to Q and R, both of which should pass through *s* if the orientation is correct.

5.5.3   Trial and Error Method

As the name implies, the position of the station occupied is obtained by trial and error. This is also known as *Lehmann's method*. Let P, Q, and R be the ground stations and *p*, *q*, and *r* be the corresponding plotted positions on the plan. If rays *p*-P (keeping the alidade against *p*, sighting the signal at P, and drawing a line along the ruling edge), *q*-Q, and *r*-R are drawn, they will not coincide at a point as they should if the table is oriented. The rays will form a small triangle known as the *triangle of error* as shown in Fig. 5.10(a). This triangle can be made smaller and smaller and finally eliminated by trial and error. Some rules known as Lehmann's rules are used to select a new position of the occupied station as a better choice than that obtained earlier. The following procedure is adopted in this method [Fig.5.10b)]

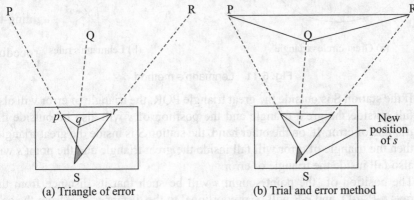

(a) Triangle of error   (b) Trial and error method

Fig. 5.10   Trial and error method

1. Set up and level the table at S, the station to be plotted. The three stations, P, Q, and R, must be visible and should not subtend very small angles at S.
2. Orient the table by keeping *p-q* approximately parallel to P-Q and clamp the table.
3. Keeping the alidade against *p*, set it to sight the signal at P, and draw a ray. Keep the alidade against *q*, set the alidade to sight the signal at Q, and draw a ray. Keeping the alidade against *r*, set the alidade to sight the signal at R and draw a ray. The three rays drawn will not meet at one point, as the orientation is only approximate. They will form a small triangle of error.
4. To eliminate the triangle of error, select another position of *s*, the plan position of the station occupied. Draw the rays *s-p*, *s-q*, and *s-r*. The three rays will again not meet at a point but will form a smaller triangle of error.
5. The process can be repeated to eliminate the triangle of error.

Lehmann's rules

In step 4 above, we selected a position of *s* to start the trial. The selection of the position of *s* depends upon the three rays and the triangle of error. Lehmann formulated three basic rules which help us to select the position of s. s is known as the point sought, as we are interested in locating the position of the station occupied. The three ground stations P, Q, and R form a triangle known as the great triangle and the circle passing through P, Q, and R is known as the great circle (Fig. 5.11).

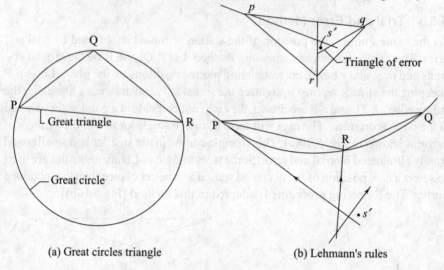

(a) Great circles triangle　　　　　　　　(b) Lehmann's rules

Fig. 5.11　Lehmann's method

(a) If the station S is outside the great triangle PQR, the triangle of error will also fall outside the great triangle and the position of s will also be outside the triangle of error. If, on the other hand, the station S is inside the great triangle, then the triangle of error will fall inside the great triangle and the point s will also fall inside the triangle of error.

(b) The position of the point sought s will be such that its distance from the rays p-P, q-Q, and r-R will be proportional to the distances of S from the corresponding points P, Q, and R.

(c) The point sought s will lie to the same side of the three rays, either to the left or to the right of p-P, q-Q, and r-R.

These three rules are sufficient to proceed with the trials. The following additional subrules will facilitate the selection of trial positions and the station S.

(i) If the station S is outside the great circle, select the point s such that the intersection of the rays to two nearer points lies midway between the ray to the most distant point and the point s [Fig. 5.12(a)].

(ii) If the station S lies outside the great triangle but inside the great circle, then select s such that the ray to the middle point lies midway between the point s and the intersection of the rays to the extreme points [Fig. 5.12(b)].

(iii) If the station S lies very close to one of the sides of the great triangle as in Fig. 5.12(c), the point s should lie on the same side of the three rays and between the very nearly parallel rays drawn to the extremities of the line.

(iv) If the points P, Q, and R lie on a straight line, PQR is a chord of the great circle of radius equal to infinity. The point s, in such a case, should be chosen such that the ray to the middle point lies between the intersection of the rays to the other two points and the point s.

(v) The position of *s* is indeterminate if the station S lies on the great circle passing through the three points P, Q, and R, as the three rays will intersect at one point even if the table is not oriented.

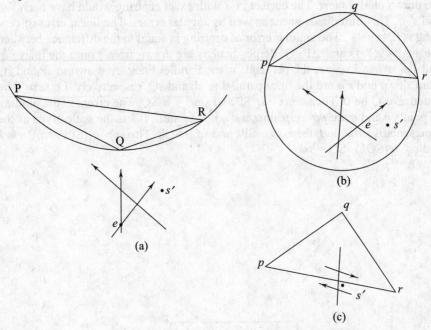

Fig. 5.12 Lehmann's additional rules

Strength of fix

The accuracy with which the point sought s can be determined depends upon the relative positions of the four points P, Q, R, and S. This is known as the strength of fix. The strength of fix is good when

(a) S lies within the triangle PQR,

(b) the middle station is much nearer compared to the other two stations, and

(c) the angle subtended by PQ and QR at S is such that one is large and the other is small, provided the distance between the points subtending the smaller angle is not very small.

The strength of fix is poor when

(a) S is near the great circle passing through P, Q, and R and

(b) the angles subtended by the two lines PQ and QR are both very small.

5.6 Errors due to Centring

Centring is the process of ensuring that the plane table is kept at a station in such a way that the plotted position of the point is exactly over the station point. Since the process of orienting the table requires rotating the table, centring gets affected even after proper orientation is done. Thus the steps of orientation and centring are to be carried out repeatedly to get both correct. We will now examine the error introduced due to incorrect centring.

Referring to Fig. 5.13, let s be the plotted position of the station occupied by the table; the actual position with accurate centring should have been s'. If we consider two stations P and Q which were sighted and rays drawn, rays s-p and s-q are drawn on the sheet. The correct rays with exact centring would have been s'-p and s'-q. This introduces linear as well as angular errors. The linear error of centring is $E_c = s - s'$. The angular error of centring is equal to the difference between the angles $\angle PsQ$ and $\angle Ps'Q$. Perpendiculars are drawn from s onto the lines $s'q'$ and $s'p'$. s-t and s-r are the perpendiculars. Parallel lines are drawn to sP and sQ from s'; $s'p$ and $s'q$ are the lines parallel to sP and sQ, respectively. Let $\angle ps'P$ be α and $\angle qs'Q$ be β. Then $\sin\alpha = sr/SP$ and $\sin\beta = st/SQ$. The errors in the plotting of points p and q are $pp' = sp \sin\alpha$ and $qq' = sq \sin\beta$. If k is the scale given as the representative fraction, then, $sp = SPk$ and $sq = SQk$. Then $pp' = (SPk)sr/SP = srk$ and $qq' = (SQk)s/SQ = stk$.

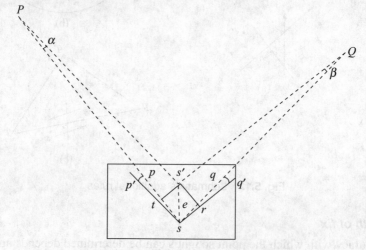

Fig. 5.13 Error due to centring

Generally, 0.25 mm is taken as the unit of precision for plotting. If ksr or kst is to remain insignificant, sr or st must be equal to or less than $0.25/k$ mm or $0.00025/k$ m. The following are the values for different scales:

Scale	k	st/sr
1 mm = 0.1 m	1/100	0.025 m or 25 mm
1 mm = 1m	1/1000	0.25 m or 250 mm
1 mm = 5 m	1/5000	1.25 m or 1250 mm
1 mm = 10 m	1/10,000	2.5 m or 2500 mm

Thus, one can easily see that centring is not significant for small-scale work. The station can be anywhere below the board and it will not be reflected in the drawing sheet. Effort should be put into centring the table only for large-scale work.

Example 5.1 In setting up the plane table at a station, the centring was not done with care and it was found that the displacement of the plotted point at right angles to the ray was 25 cm. Determine the error in the plotted positions if the scale was (a) 1 mm = 0.1 m (b) 1 mm = 1 m, and (c) 1 mm = 2 m.

Solution When the scale is 1 mm = 0.1 m, the representative fraction $k = 1/(0.1 \times 1000) = 1/100$. Displacement of the plotted point = 25 cm × 1/100 = 0.25 cm or 2.5 mm (this is significant error). When the scale is 1 mm = 1 m, $k = 1/1000$. Displacement of the plotted point = 25 cm × 1/1000 = 0.025 cm or 0.25 mm (this is a limiting case). When the scale is 1 mm = 2 m, $k = 1/2000$. Displacement of the plotted point = 25 cm/2000 = 0.0125 cm or 0.125 mm (this is not significant).

5.7 Adjustments of the Plane Table

The following adjustments are to be made to the plane table at intervals of time to obtain accurate results from the plane table survey.

1. The surface of the board should be a perfect plane. This can be checked by placing a straight edge over the table in many directions. If the board is made of well-seasoned wood, the board generally will not twist or warp. By using a planer or sand paper, the table top can be levelled to some extent.

2. The vertical axis of the instrument is the line through the centre of the ball and socket arrangement or the levelling head or any other arrangement provided. When this axis is vertical, the plane of the board should be horizontal. This can be checked using a spirit level. After setting up the table, keep a spirit level on the table and centre it. Turn the table through 180° and observe whether the table is centred or not. Half the apparent error is corrected by placing a packing on the underside of the table between the table and its support. Repeat the check till the bubble remains centred in all directions.

3. The ruling or fiducial edge of the alidade should be straight. This can be checked by drawing a line on the sheet along the ruling edge and then reversing the alidade end for end and drawing another line along the same line. If the two lines are coincident all along the line, then the edge is straight. Otherwise, the edges need to be filed and made straight.

4. The sight vanes of the alidade should be perpendicular to the base of the ruler. If the hair is loose or the vane is tilted, it may not provide a vertical line to bisect objects. To test this, a plumb line can be bisected to check the verticality of the sight vane. The correction can be done by providing packing or by filing the base.

5.8 Errors in Plane Tabling

The following are some of the chief sources of error.

(a) The board not being perfectly levelled
(b) The board not being properly oriented
(c) The alidade not being correctly pivoted against the plotted point while sighting
(d) The objects not being sighted accurately
(e) The rotation of the table between sights due to insufficient clamping
(f) The rays not being drawn accurately through the point
(g) Inaccurate scaling and plotting
(h) Expansion and contraction of paper

The orientation should be checked by sighting distant and prominent objects as frequently as possible. A sharp pencil must be used for drawing. Before drawing a ray the alidade should be placed correctly against the plotted point. The table should be properly clamped before starting to use it at a station.

5.9   Advantages and Disadvantages of Plane Tabling

The advantages of plane tabling are the following.
(a) Survey work can be done with high speed.
(b) A field book is not required, as the plotting is done along with the fieldwork.
(c) Errors of booking in the field book are avoided.
(d) Plane tabling is very suitable for preparing small-scale maps.
(e) Measurement errors and errors of plotting can be readily checked.
(f) Many features can be accurately represented, as the surveyor has the objects or features in view while plotting.
(g) The method is suitable in magnetic areas, where a compass survey will not be reliable.
(h) It is inexpensive.
(i) As the instruments used are very simple, most people can handle the job without special skills.

The disadvantages of plane tabling are the following.
(a) Plane tabling requires many accessories, and the chances of losing one or the other are high.
(b) The equipment as a whole is heavy and cumbersome to carry during field-work.
(c) It is not suitable when accuracy is required.
(d) Absence of field notes becomes a great disadvantage if some calculations are required.
(e) Survey cannot be done in the rainy season; essentially a tropical method.

Summary

Plane table surveying is a technique where part of the paperwork is combined with the fieldwork. It is ideally suited for filling in details with survey lines already plotted by some other method.

A plane table consists of a board with an arrangement to fix it to a tripod. The board can also be rotated with a ball and socket arrangement. An alidade with a fine ruling edge is an essential part of the plane table. A plane or telescopic alidade is used for sighting objects. Generally, a spirit level, magnetic (trough) compass, and plumb bob on a U-frame also form a part of the plane tabling equipment. A fresh drawing sheet or a sheet with plotted points is used on the board.

The plane table has to be kept at a station, centred, and levelled. Centring is generally done by adjusting the legs of the tripod. Levelling is done using the spirit level, keeping it in two perpendicular positions. The direction of the magnetic meridian is generally marked with the trough compass. If the sheet contains plotted points, the table needs to be oriented. Centring is not very important for plane tabling when the scale gives a small-sized plot.

Plane tabling can be done by the radiation method, in which a single station is occupied by the table and other points or features are fixed by radial rays and distances measured using a chain or tape. In the intersection method, two stations are occupied by the table and other points or features are marked by intersecting rays from the two stations. The measurement of distance is not required in this case, even though distances are measured for the purpose of checking. The plane table can also be used for traversing, open or closed. The table is kept at each of the stations of the traverse. Distances are measured with a chain or tape. Resection is another technique where the objective is to orient the table and plot the position of the station occupied by the plane table. Simple resection techniques use the magnetic meridian for orienting or the back sighting method. Two- and three-point problems are essentially resection problems. The objective in these cases is to orient the table and mark the position of the station occupied by the table.

Plane tabling has many advantages and disadvantages. The major advantages are that no field notes are required, plotting is done as the surveyor sees the features, and many common errors are eliminated. Lack of field notes can be a disadvantage in case further work is to be done in the future and it is not suitable for high-precision work.

Exercises

Multiple-Choice Questions

1. An alidade used with the plane table is used for
   - (a) centring the plane table
   - (b) sighting objects
   - (c) levelling the plane table
   - (d) determining distances of objects
2. A plane table is oriented by the
   - (a) method of radiation
   - (b) method of intersection
   - (c) method of back sighting
   - (d) use of plumb bob
3. Orienting the plane table is the operation of
   - (a) centring the table
   - (b) plotting points on the paper placed on the plane table
   - (c) keeping the plane table in parallel position at all stations
   - (d) keeping the plane parallel to the ground at a station
4. The radiation method of plane tabling involves
   - (a) keeping the plane table at one station
   - (b) keeping the plane table on at least two stations
   - (c) keeping the plane table at many stations
   - (d) obtaining points on the ground by intersection of two lines
5. The two-point problem in plane tabling involves
   - (a) locating two points from the given plane table station
   - (b) locating the plane table station and a given point
   - (c) given location of two points, locating the station occupied by the plane table
   - (d) given the location of plane table station, locating two previous stations occupied by the plane table
6. The two-point problem is essentially a problem of
   - (a) orienting the plane table
   - (b) centring the table
   - (c) finding the line joining two points
   - (d) solving a triangle
7. The three-point problem in plane tabling involves
   - (a) determining the position of three points
   - (b) locating the station occupied by plane table given the position of three points
   - (c) locating the position of two points given the position of one point
   - (d) surveying a triangular area
8. In using trial and error method to solve the three point problem, the strength of fix is good when
   - (a) the station occupied by the plane table lies within the triangle formed by the three points

(b) the station occupied by the plane table lies near the circle passing through the three points

(c) the station occupied by the plane table lies on the circle passing through the three points

(d) the angle subtended by the two lines obtained by joining the three points are very small

9. Centring the plane table over the station occupied is
   (a) a very important step in plane tabling
   (b) not significant at all
   (c) significant only for large-scale work
   (d) significant for small scale work

10. If the displacement of a plotted point perpendicular to the ray was 25 cm, the scale at which the error centring becomes significant is, taking the plotting accuracy as 0.025 cm, is
    (a) 1/10000   (b) 1/5000   (c) 1/1000   (d) 1/100

Review Questions

1. Enlist and explain the function of each of the instruments required for plane table surveying.
2. What are the adjustments required at a station before starting a plane table survey? How are these adjustments done?
3. Enlist and explain briefly the different methods of plane table surveying.
4. Explain the advantages and disadvantages of the different methods of plane table surveying.
5. What is resection? How is it different from the other methods?
6. Explain the method of traversing with the plane table.
7. Explain the two-point problem and the method of solution.
8. Explain the three-point problem and the different methods of solving it.
9. Show that approximate centring of the table over a station is sufficient in plane table surveying.
10. Explain the advantages and disadvantages of plane table surveying.
11. What are the likely errors in plane table surveying and what precautions should be taken to eliminate them?

Problems

1. The plane table at a station S was not centred properly. The ray, *sa*, to a station A was drawn from S. If the ray was displaced at right angles by 250 mm, find the displacement of A from its true position if (a) the scale used is 1 mm = 2 m and the distance of station A from S is 20 m and (b) if the scale used is 1 mm = 20 m and the distance of station A from S is 2000 m.
2. A plane table is of size 600 × 750 mm and was set up over a station. Assuming that the error in centring was maximum with the station point lying within the boundaries of the board, find the maximum scale that can be used so that the error in displacement of points plotted at a maximum distance of 1000 m is not significant.

CHAPTER

6

Traverse Computations

Learning Objectives

After going through this chapter, the reader will be able to
- compute the latitudes and departures of a survey line
- apply the necessary checks for open and closed traverses
- compute the coordinates of a closed traverse
- determine the error of closure and balance the traverse by different methods
- calculate the missing measurements (angles and lengths) of a closed traverse

Introduction

In Chapters 2, 3, and 4, we discussed the methods of traversing. While traversing can be done by a chain or tape alone, invariably an angle-measuring instrument is also used for traversing. For works that do not require high precision, a compass survey can be done. The angles (strictly the bearings) can be measured to an accuracy of 15′ in a compass survey. A normal vernier theodolite, on the other hand, measures angles up to an accuracy of 20″. For angles measured with such high precision (even higher precision is possible with electronic theodolites), the length needs to be measured to a greater level of precision.

Whatever be the method of surveying, the survey data is used for plotting traverses and making further calculations. Some methods have already been outlined in Chapter 3. In this chapter we will focus on the traverses conducted with a theodolite.

6.1 Traverse Parameters

A traverse is a series of survey lines connected to each other. In an open traverse the end point is not the same as the starting point. This is shown in Fig. 6.1(a). The traverse starts at point A and goes along a direction that does not return it to point A. A, B, C, etc. are the survey stations. The lengths of the lines AB, BC, CD, etc. are measured along with either the bearings of these lines or the angles between the lines. Even when the angles are measured, the bearing of at least one line is measured to get the meridional direction. The angles measured can be the bearings, the angle between the lines, or the deflection angles, as shown in Fig. 6.1(a).

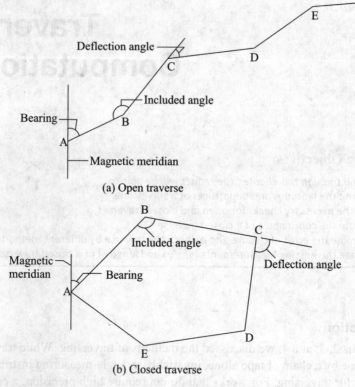

(a) Open traverse

(b) Closed traverse

Fig. 6.1 Traverses

A closed traverse [Fig. 6.1(b)] has the same starting and end points. ABCDEA is thus a closed traverse. Starting at station A, the path can be traversed in either the clockwise or the counterclockwise direction. The lengths and bearings of the lines, the lengths and angles between the lines, or the lengths and the deflection angles are measured. More checks are possible in a closed traverse than in an open traverse.

The traverse parameters are the lengths and directions of the lines. The directions can be bearings, angles between lines, or deflection angles.

6.1.1 Open Traverse

In the case of an open traverse, there are no checks possible for the accuracy of the work done. To check the angles measured, some check lines are chosen and their bearings measured during the survey work as described earlier. However, it can be seen that if the bearing of the first line is measured or known, then the bearing of the last line can be related to the bearing of the first line. This is done by measuring or calculating the deflection angles of the lines. The deflection angles can be towards the left or towards the right. If the right deflection angles (clockwise) are considered positive and the left deflection angles (counterclockwise) considered negative, then the bearing of the last line = bearing of the first line + sum of right deflection angles – sum of left deflection angles. In other words, bearing of the last line = bearing of the first line + algebraic sum of deflection angles. This can be a check on the calculations. The following examples illustrate this principle.

Example 6.1 The bearing of line AB of an open traverse ABCDEF is 36° 45′. The deflection angles between the lines were measured with a theodolite and were as follows: 26° 37′ (R) at B, 66° 45′ (L) at C, 20° 56′ (R) at D, 33° 54′ (R) at E, and 26° 54′(L) at F. If the bearing of the last line observed was 24° 33′, check whether the observations are correct.

Solution   A rough sketch of the traverse is shown in Fig. 6.2(a).

Sum of positive (right) deflection angles   = 26° 37′ + 20° 56′ + 33° 54′
                                            = 81° 27′

Sum of negative deflection angles = 66° 45 + 26° 54′ = 93° 39′

Bearing of the last line EF = 36° 45′ + 81° 27′ – 93° 39′ = 24° 33′

The observed bearing and calculated bearing of the last line are correct. The observations can therefore be considered to be correct.

Example 6.2   In an open traverse ABCDE, the bearing of the first line observed is not clearly written. The bearing of the last line DE is recorded as S36°45′ E. If the deflection angles are 33° 54′ (R) at D, 42° 34′ (R) at C, and 18° 24′ (L) at B, calculate the bearing of the first line.

Solution   A rough sketch of the traverse is shown in Fig. 6.2(b). It will be easier to work with whole circle bearings (WCBs). We convert the bearing of line DE as 180° – 36° 45′ = 143° 15′. We then apply the condition that the bearing of last line = bearing of first line + algebraic sum of deflection angles. Thus,

143° 15′ = bearing of AB + 42° 34′ + 33° 54′ – 18° 24′

Bearing of AB = 85° 11′

Bearing of first line = N85° 11′E

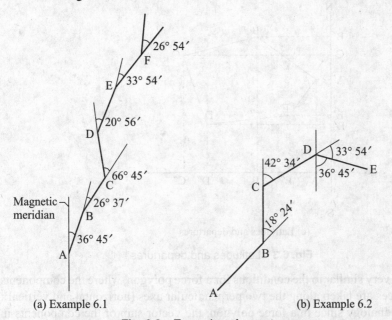

(a) Example 6.1   (b) Example 6.2

Fig. 6.2   Examples of open traverses

6.1.2   Closed Traverse

In the case of a closed traverse, as in traverse ABCDEA, the traverse returns to the starting point. A closed traverse is mathematically a closed polygon. Obviously it has to satisfy the conditions that apply to a closed figure. If we consider two directions at right angles, the north–south and the east–west [Fig. 6.3(a)], each of the lines of the polygon ABCD has projections on them as shown in Fig. 6.3(b). It is easy to see that the algebraic sum of the projections (taking projections upward and towards the right as positive) in the north–south and east–west directions should be zero. There are, thus, two independent conditions for a closed traverse. To elaborate further, in the traverse shown in Fig. 6.3(c), if we go around the traverse in a clockwise direction, the projections on the north–south and east–west directions and their signs are given in Table 6.1.

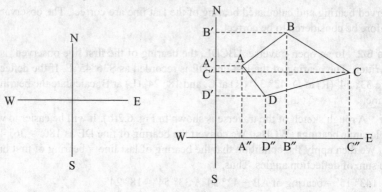

(a) Cardinal directions   (b) Projections on the N-S and W-E directions

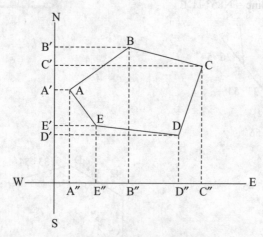

(c) Latitudes and departures

Fig. 6.3 Latitudes and departures

   This is very similar to the conditions for a force polygon, where the components of the forces are taken along the two perpendicular axes (horizontal and vertical). With this analogy since in a force polygon, the vector sum of the components in the two perpendicular directions is zero, it is easy to see that the algebraic sum of the projections in the second and third columns of Table 6.1 should be zero. These components are known as *latitudes* and *departures* of traverse lines. We will discuss these projected lengths in detail, leading us to the conditions to be satisfied by a closed traverse.

Table 6.1 Assigning signs to projected lengths

Line	North–south projection	East–west projection
AB	$+ A'B'$	$+ A''B''$
BC	$- B'C'$	$+ B''C''$
CD	$- C'D'$	$- C''D''$
DE	$+ D'E'$	$- D''E''$
EA	$+ E'A'$	$- E''A''$

6.2 Latitudes and Departures of Lines

The latitude of a line is the projection of the line on the north–south meridian. If the bearings of a line are measured, the whole circle bearings may have to be converted to reduced bearings if one is working with logarithmic tables. Then the bearings are either from the north or from the south direction. Hence the projection on the meridian is the length of the line multiplied by the cosine of the bearing. This is shown in Fig. 6.4(a). Thus, if l is the length of the line and θ is its reduced bearing, then $l \cos \theta$ is the projection on the meridian or the latitude of the line.

The latitude of the line can be positive or negative. It is positive when the bearing is measured from the north direction and negative when the bearing is measured from the south direction. The positive latitude is also called *northing*, as it is measured towards the north. Similarly, the negative latitude is called *southing*, as it is measured towards the south. If the bearing of the line is N-E or N-W, the latitude is positive. If the bearing of the line is S-W or S-E, the latitude is negative.

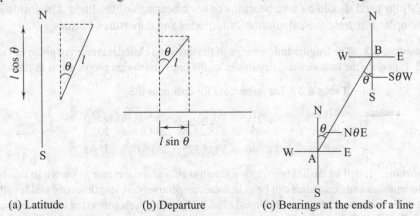

(a) Latitude (b) Departure (c) Bearings at the ends of a line

Fig. 6.4 Computing latitudes and departures

The departure of a line is the projection of the line on the east–west direction. This is shown in Fig. 6.4(b). As in the case of the latitude, since the reduced bearings are measured from the north or south directions only, the departure is equal to the length of the line multiplied by the sine of the angle. Thus, if l is the length of the line and θ is its reduced bearing, then the departure of the line is equal to $l\sin\theta$.

The departure of the line can be positive or negative. The departure is positive when the bearing is measured towards the east and negative when the bearing is measured towards the west. A positive departure is called *easting* and a negative departure is called *westing*. The departure of a line is positive if the bearing is N-E or S-E. The departure is negative if the bearing is N-W or S-W.

The rules for assigning signs to latitudes and departures based on their bearings are given in Table 6.2.

Table 6.2 Signs of latitudes and departures

Whole circle bearing (θ)	Quadrantal bearing quadrant	Sign of latitude	Sign of departure
$0 < \theta < 90$	N-E I quadrant	+	+
$90 < \theta < 180$	S-E IV quadrant	–	+
$180 < \theta < 270$	S-W III quadrant	–	–
$270 < \theta < 360$	N-W II quadrant	+	–

Latitudes and departures of lines are also known as dependent or consecutive coordinates. This is so because they can be considered as coordinates with the origin at one end of the line.

An important point to remember is that for a line AB [Fig. 6.4(c)], the bearing measured at A is NθE, while the bearing measured at B is SθW. The latitude and departure of line AB calculated with the bearing measured at A will be equal in magnitude but opposite in sign to those calculated for line BA with the bearing measured at B. Therefore, it is important to go either in the clockwise or the counterclockwise direction while calculating the latitudes and departures of lines. All the bearings used should be fore bearings or back bearings of the lines. The following examples illustrate the calculation of latitudes and departures of lines.

Example 6.3 The lengths and bearings of the lines of a closed traverse are given in Table 6.3. Calculate the latitudes and departures of the lines and assign proper signs to them.

Table 6.3 Traverse data for Example 6.3

Line	AB	BC	CD	DA
Length (m)	156.5	178.2	234.8	203.1
Bearing	78° 40′	152° 32′	251° 18′	3° 45′

Solution It will be desirable to draw a rough sketch of the traverse as shown in Fig. 6.5. The latitudes and departures can be calculated, respectively, as length × cosθ and length × sin θ. The values are shown in Table 6.4. The WCBs have been converted to reduced bearings for calculating the sine and cosine values of the bearings.

Table 6.4 Latitudes and departures calculated from data in Table 6.3

Line	Length (m)	Bearing	Latitude	Departure
AB	156.5	N78° 40′E	+ 30.74	+153.45
BC	178.2	S27° 28′E	– 158.11	+82.19
CD	234.8	S71° 18′W	–75.28	– 222.4
DA	203.1	N3° 45′E	+202.64	–13.25

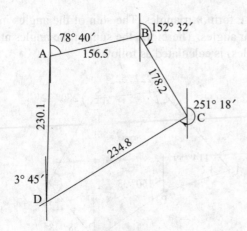

Fig. 6.5

Example 6.4 The lengths and bearings of the lines of a traverse are given in Table 6.5. Calculate the northings, southings, eastings, and westings of the lines.

Table 6.5 Traverse data for Example 6.4

Line	PQ	QR	RS	ST	TU	UP
Length (m)	151.24	258.52	270.38	121.84	203.73	174.20
Bearing	160° 33′	58° 48′	320° 47′	280° 27′	222° 52′	114° 59′

Solution A rough sketch of the traverse is shown in Fig. 6.6. The values can be calculated and tabulated as shown in Table 6.6.

Table 6.6 Northings, southings, eastings, and westings (Example 6.4)

Line	Length	Bearing	Northing	Southing	Easting	Westing
PQ	151.24	S19° 27′E		142.61	50.36	
QR	258.52	N58° 48′E	133.92		221.13	
RS	270.38	N39° 13′W	209.48			170.95
ST	121.84	N79° 33′W	22.1			119.82
TU	203.73	S42° 52′W		149.32		138.59
UP	174.2	S65° 01′E		73.57	157.87	

6.3 Checks for a Closed Traverse

The following checks are available for the angles, depending upon which angles are being considered, in a closed traverse (refer to Fig. 6.7).

1. The sum of the interior angles of a closed traverse is equal to $(2n - 4)$ right angles, where n is the number of sides of the traverse. This can be proved easily. Take a point 'O' within the traverse [Fig. 6.7(a)]. Join O to the vertices A, B, C, D, and E. The sum of the angles at O is $\angle AOB + \angle BOC + \angle COD + \angle DOE + \angle EOA = 360°$ (angle around a point) = 4 right angles. By joining

the vertices, we form *n* triangles. The sum of the angles in *n* triangles = *n* × 180° = 2*n* right angles. Therefore, the sum of the angles at the vertices (sum of interior angles) is calculated as follows:

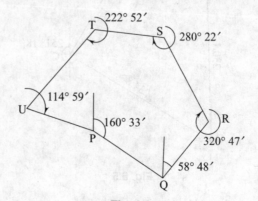

Fig. 6.6

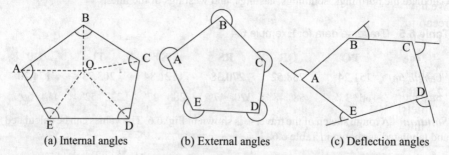

| (a) Internal angles | (b) External angles | (c) Deflection angles |

Fig. 6.7 Traverse angles

Sum of interior angles= sum of angles in *n* triangles – sum of angles at O
$$= 2n \text{ right angles} - 4 \text{ right angles}$$
$$= (2n - 4) \text{ right angles}$$

Therefore, for a four-sided traverse, the sum is equal to 360°[(2 × 4 – 4) × 90°]; for a five-sided traverse, the sum is equal to 540° [(2 × 5 – 4) × 90°]; for a six-sided traverse, the sum is equal to 720°[(2 × 6 – 4) × 90°]; and so on.

2. The sum of the external angles of a traverse is equal to (2*n* + 4) right angles, where *n* is the number of sides of the traverse. This can be proved easily. In an *n*-sided traverse, there are *n* vertices. At each vertex, the sum of the interior and exterior angles = 360° = 4 right angles. Therefore, the sum of the interior and exterior angles at *n* vertices = 4*n* right angles.

Sum of the interior angles = (2*n* – 4) right angles
as proved earlier. Therefore,

Sum of exterior angles = [4*n* – (2*n* – 4)] right angles
$$= (2n + 4) \text{ right angles}$$

Therefore, for a four-sided traverse, this sum is equal to 1080°.

3. If the deflection angles of a traverse are measured, the sum of the deflection angles is equal to 360° (irrespective of the number of sides). This can be verified by drawing the deflection angles of all the traverse lines from one point and noting

that they add up to 360°. This is so because as the traverse lines deflect from the previous line by a certain amount and the traverse ends at the starting point, it is equivalent to making a complete revolution around a point. At any vertex of the polygon, sum of interior angle + deflection angle = 180° = 2 right angles. At n vertices, this sum = $2n$ right angles. Therefore, n interior angles + n deflection angles = $2n$ right angles; $(2n - 4)$ right angles + n deflection angles = $2n$ right angles; n deflection angles = 4 right angles.

As mentioned earlier, the two parameters of a traverse are the lengths and bearings of the lines forming the traverse. A check on the sum of angles by any of the methods given above ensures that the angles are correct. The traverse may still have an error and may not close if plotted. This will be due to the errors in the lengths of the lines. During the survey, the lines need to be measured in both the directions, at different times, and preferably by different persons. This will ensure that the errors, if any, are small and can be adjusted.

6.4 Plotting a Traverse and Determining the Closing Error

We have already discussed the methods of plotting a traverse in Chapter 3. We just recapitulate these methods in this section (refer to Fig. 6.8).

1. **Plotting by bearings** This is done when the bearings of the survey lines are measured (or calculated from other values). It requires parallel meridian lines to be drawn through all the points of the traverse.

2. **Plotting by included angles** When the included angles are measured in the field (or calculated from other values), the traverse can be plotted using the interior angles. The bearing of at least one line is required to orient the traverse properly.

3. **Plotting by deflection angles** This is done when the deflection angles are measured (or calculated from other values). The deflection angle is the angle by which a line deflects or deviates in direction from the previous line.

4. **Plotting by tangent lengths** This is a graphical method of plotting deflection angles. Taking any convenient measure (10 cm), the tangent length is $10\tan\theta$, where θ is the deflection angle. The 10 cm length is measured on the extension of the line and the tangent length is measured at right angles to it. If the deflection angle is nearer to 90°, the tangent lengths will be very large and may not be convenient for plotting.

5. **Plotting by chord lengths** This is another graphical method. To plot the traverse by chord lengths, a convenient length l, say 10 cm or 20 cm, is taken. The chord length is then calculated as $2l \sin(\theta/2)$. With the starting point A of the line as the centre, an arc equal in length to l is drawn which cuts the extension of the previous line at, say,1. With 1 as the centre, another arc equal in length to the calculated chord length is drawn, cutting the previous arc drawn at, say, 2. A-2 then gives the direction of the line.

6. **Plotting by coordinates** The traverse can be set in a departure–latitude or x-y coordinate system. The origin can be selected at the most westerly station, which makes all the x-coordinates positive. As a better choice, the origin can be so selected that the y-axis passes through the most westerly station and the x-axis passes through the most southerly station. The x- and y-coordinates of

all the points will be positive in such a case. Knowing the coordinates, each station is plotted independently of the other points, thus eliminating cumulative errors. Plotting by coordinates is the most accurate and reliable method of plotting a traverse.

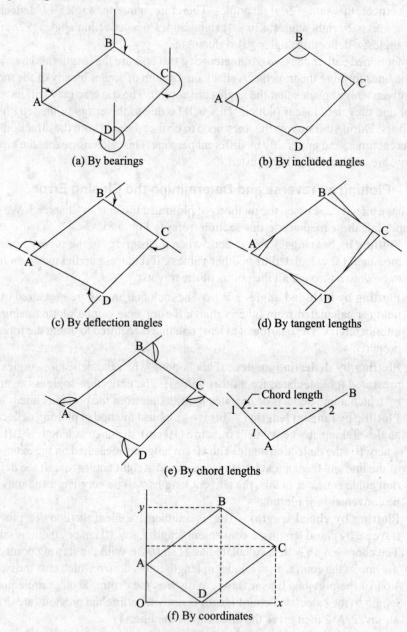

(a) By bearings

(b) By included angles

(c) By deflection angles

(d) By tangent lengths

(e) By chord lengths

Chord length

(f) By coordinates

Fig. 6.8　Plotting a traverse

The following example illustrates the calculation of quantities required for all the methods of plotting.

Example 6.5 From the traverse data given in Table 6.7, calculate the quantities required for plotting the traverse by (i) bearings, (ii) included angles, (iii) deflection angles, (iv) tangent lengths, (v) chord lengths, and (vi) coordinates.

Table 6.7 Traverse data for Example 6.5

Line	Length (m)	Bearing
PQ	201.54	62° 42′
QR	189.68	154° 54′
RS	231.94	202° 32′
ST	272.55	281° 44′
TP	256.83	22° 00′

Solution A rough sketch of the traverse is drawn as in Fig. 6.9

Plotting by bearings: The traverse data is given in terms of bearings. Therefore, no further calculation is required. The given lengths and bearing angles are sufficient for plotting.

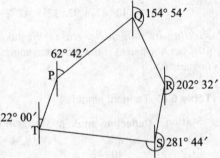

Plotting by included angles: The included angles between the lines have to be calculated. With whole circle bearings, the included angle is the difference between the fore bearings and back bearings of the lines meeting at a station. The included angles can be calculated as follows.

Fig. 6.9 Traverse for Example 6.5

At station P: ∠TPQ = difference between the bearings of PQ and PT. Bearing of PQ = 62° 42′, bearing of PT = 22° + 180° = 202° 00′. Therefore,

$$\angle TPQ = 202° \ 00′ - 62° \ 42′ = 139° \ 18′$$

At station Q: ∠PQR = difference in the bearings of QP and QR. Bearing of QP = 62° 42′ + 180° = 242° 42′, bearing of QR = 154° 54′. Therefore,

$$\angle PQR = 242° \ 42′ - 154° \ 54′ = 87° \ 48′$$

At station R: ∠QRS = difference between the bearings of RQ and RS. Bearing of RQ = 154° 54′ + 180° = 334° 54′, bearing of RS = 202° 32′. Therefore,

$$\angle QRS = 334° \ 54′ - 202° \ 32′ = 132° \ 22′$$

At station S: ∠RST = difference between the bearings of SR and ST. Bearing of SR = 202° 32′ − 180° = 22° 32′, bearing of ST = 281° 44′. Therefore,

$$\angle RST = 281° \ 44′ - 22° \ 32′ = 259° \ 12′ > 180°$$
$$= 360° \ 00′ - 259° \ 12′ = 100° \ 48′$$

At station T: ∠STP = difference between the bearings of TS and TP. Bearing of TS = 101° 44′, bearing of TP = 22° 00′. Therefore,

$$\angle STP = 101° \ 44′ - 22° = 79° \ 44′$$

Check: Sum of included angles = 139° 18′ + 87° 48′ + 132° 22′ + 100° 48′ + 79° 44′ = 540° 00, which is equal to $(2 \times 5 - 4) \times 90°$.

Plotting by deflection angles: The deflection angle Δ is the difference between the fore bearings of the two lines meeting at a point. The deflection angles can be calculated as follows:

At P:

$$\Delta_P = \text{bearing of PQ} - \text{bearing of TP} = 62° \ 42' - 22° \ 00' = 40° \ 42'$$

At Q:

$$\Delta_Q = \text{bearing of QR} - \text{bearing of PQ} = 154° \ 54' - 62° \ 42' = 92° \ 12'$$

At R:

$$\Delta_R = \text{bearing of RS} - \text{bearing of QR} = 202° \ 32' - 154° \ 54' = 47° \ 38'$$

At S:

$$\Delta_S = \text{bearing of ST} - \text{bearing of RS} = 281° \ 44' - 202° \ 32' = 79° \ 12'$$

At T:

$$\Delta_T = \text{bearing of TP} - \text{bearing of ST} = 382° \ 00' - 281° \ 44' = 100° \ 16'$$

In the last case, note that the bearing of 22° 00′ is converted to 22° + 360° = 382° 00′ to obtain the deflection angle. Checking the sum of deflection angles:

$$\text{Sum} = 40° \ 42' + 92° \ 12' + 47° \ 38' + 79° \ 12' + 100° \ 16' = 360°$$

Plotting by tangent lengths: We take a length of 10 cm. The tangent length is $10 \times \tan \Delta$, where Δ is the deflection angle. The tangent lengths can be calculated as shown in Table 6.8.

Table 6.8 Tangent lengths

Station	Deflection angle Δ	Tangent length = 10 tan Δ (cm)	Remarks
P	40° 42′	8.6	
Q	92° 12′	260.3	The high value is due to Δ being very close to 90°.
R	47° 38′	10.96	
S	79° 12′	52.42	Δ closer to 90°.
T	100° 16′	55.2	Δ closer to 90°.

Plotting by chord lengths: If we take a length of 10 cm, the chord length is $= 2 \times 10 \times \sin(\Delta/2) = 20 \sin(\Delta/2)$, where Δ is the deflection angle. These are calculated for the station points as shown in Table 6.9.

Table 6.9 Chord lengths

Station	Deflection angle (Δ)	Chord length = 20 sin(Δ/2) (cm)
P	40° 42′	20 sin (20° 21′) = 6.95
Q	92° 12′	20 sin (46° 06′) = 13.8
R	47° 38′	20 sin (23° 49′) = 8.1
S	79° 12′	20 sin (39° 36′) = 12.8
T	100° 16′	20 sin (50° 08′) = 15.35

Plotting by coordinates: From the rough sketch of the traverse, it can be seen that T is the most westerly station and S is the most southerly station. We take the x-axis through S and the y-axis through T. By calculating the latitudes and departures of the lines shown in Table 6.10, we can see that the origin is 55.43 m below T and 266.85 m to the left of S. This gives the coordinates of T as (0, 55.43) and those of S as (266.85, 0). The coordinates (Table 6.11) of the remaining stations can be calculated by algebraically adding the departures to the x-coordinates and the latitudes to the y-coordinates. Thus,

Station P:

  x-coordinate   = 0 + 96.18 = 96.18

  y-coordinate   = 55.43 + 238.13 = 293.56

Station Q:

  x-coordinate   = 96.18 + 179.09 = 275.27

  y-coordinate   = 293.56 + 92.44 = 386.00

Station R:

  x-coordinate   = 275.27 + 80.46 = 355.73

  y-coordinate   = 386.00 − 171.77 = 214.23

Table 6.10   Calculation of latitudes and departures

Line	Length (m)	Bearing	Latitude	Departure
PQ	201.54	62° 42′	92.44	179.09
QR	189.68	154° 54′	− 171.77	80.46
RS	231.94	202° 32′	− 214.23	− 88.58
ST	272.55	281° 44′	55.43	− 266.85
TP	256.83	22° 00′	238.13	96.18

Table 6.11   Coordinates (Example 6.5)

Station	x-coordinate	y-coordinate
P	96.18	293.56
Q	275.27	386.00
R	355.73	214.23
S	266.85	0
T	0	55.43

Station S:

  x-coordinate = 355.73 − 88.58 = 266.85

  y-coordinate = 214.23 − 214.23 = 0

Check: At station T,

  x-coordinate = 266.85 − 266.85 = 0

  y-coordinate = 0 + 55.43 = 55.43

6.4.1 Closing Error

When a closed traverse is plotted by any of the above methods, the end point of the last line may not coincide with the starting point, as shown in Fig. 6.10. The traverse ABCDEA is plotted starting at A and the last line comes out to be EA′. The distance AA′ is known as the closing error. This error can occur due to a combination of errors in angular measurements and linear measurements. Before a closed traverse is plotted, the following checks should be applied.

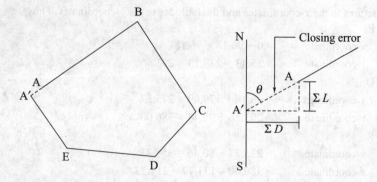

Fig. 6.10 Closing error

- Verify that the sum of the interior angles is equal to $(2n - 4)$ right angles.
- Calculate the latitudes and departures of the lines.
- Check that the sum of latitudes $\Sigma L = 0$ and the sum of departures $\Sigma D = 0$.
- The closing error $e = [(\Sigma L)^2 + (\Sigma D)^2]^{1/2}$.
- The direction of the closing error is given by $\tan\theta = \Sigma D/\Sigma L$, θ being the angle made with the north or south direction. The quadrant of the error can be determined from the signs of ΣD and ΣL.

Correcting angles

If the interior angles of the traverse are measured and they do not satisfy the geometric condition that their sum should be equal to $(2n - 4)$ right angles, then the angular error can be distributed equally amongst all the angles. It is also possible that if the surveyor knows which angles are likely to be in error, he may then distribute the error among those angles only.

Correcting bearings

If the bearings of a traverse are measured and it is found that the observed back bearing of the last line and the fore bearing of the same line do not agree, then the error can be distributed among the angles. Here again, it is possible to correct only a few angles if the surveyor is sure about only some of the angles being affected. The error in the bearing, e, can be distributed as follows.

> Error in the bearing of first line = e/N
> Error in the bearing of second line = $2e/N$
> Error in the bearing of last line = $Ne/N = e$

This is equivalent to applying the correction by distributing it equally amongst the observed angles.

Example 6.6 A traverse survey was conducted and the data obtained is given in Table 6.12. Find the magnitude and direction of the closing error if any.

Solution A rough sketch of the traverse is shown in Fig. 6.11. The latitudes and departures can be conveniently calculated in a tabular form as shown in Table 6.13.

Table 6.12 Traverse data

Line	AB	BC	CD	DA
Length	156.5	178.2	234.8	202.6
Bearing	78° 40′	152° 32′	251° 18′	356° 15′

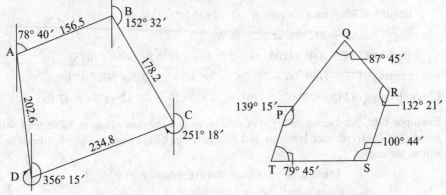

Fig. 6.11　　　　　　　**Fig. 6.12**

Table 6.13 Calculation of closing error (Example 6.6)

Line	Length	Bearing	Latitude	Departure
AB	156.5	78° 40′	30.77	153.44
BC	178.2	152° 32′	– 158.11	82.19
CD	234.8	251° 18′	– 75.28	– 222.4
DA	202.6	356° 15′	202.16	– 13.25

Sum of latitudes = 30.77 – 158.11 – 75.28 + 202.16 = – 0.46 m

Sum of departures = 153.44 + 82.19 – 222.4 + 13.25 = – 0.02 m

Closing error = $\sqrt{[(-0.46)^2 + (-0.02)^2]}$ = 0.46 m

Direction of closing error:

tan θ = (– 0.02)/(– 0.46) = 0.0438,　θ = 2.5°

As the latitude and departure are both negative, the closing error lies in the S-W quadrant.

Example 6.7 A traverse survey was conducted with a theodolite and the interior angles were measured. From the data given below, determine the angular error and correct the angles assuming that all the angles were measured with equal precision. If the bearing of line PQ is N62° 42′E, find the bearings of the remaining lines.

Station	P	Q	R	S	T
Included angle	139° 15′	87° 45′	132° 21′	100° 44′	79° 45′

Solution A rough sketch of the traverse is shown in Fig. 6.12. The sum of the included angles should be equal to (2 × 5 – 4) right angles = 6 right angles = 540°. The sum of the interior angles of the traverse is equal to

Sum = 139° 15′ + 87° 45′ + 132° 21′ + 100° 44′ + 79° 45′

= 539° 50′

The difference of 10′ is distributed equally among all the angles. The corrected angles are ∠P = 139° 17′, ∠Q = 87° 47′, ∠R = 132° 23′, ∠S = 100° 46′, and ∠T = 79° 47′, by adding 2′ to each angle. The bearings can be calculated as follows:

Bearing of PQ = N62° 42′E, back bearing of PQ (WCB) = 242° 42′

Bearing of QR = 242° 42′ − 87° 45′ = 154° 57′ or S25° 03′E

Bearing of RS = back bearing of QR − 132° 21′ = 334° 57′ − 132° 21′

= 202° 36′ or S22° 36′W

Bearing of ST = 360° − (100° 44′ − 22° 36′) = 281° 52′ or N78° 08′W

Bearing of TP = (180° 00′ − 78° 08′) − 79° 45′ = 22° 07′ or N22° 07′E

Check: Bearing of PQ = (22° 07′ + 180° 00′) − 139° 15′ = 62° 42′ or N62° 42′E.

Example 6.8 A traverse survey was conducted and the data given in Table 6.14 was noted. The observed back bearings and fore bearings of the lines are given. Check and correct the angles.

Table 6.14 Traverse data (Example 6.8)

Line	Fore bearing	Back bearing
AB	160° 33′	340° 33′
BC	58° 48′	238° 48′
CD	320° 47′	140° 47′
DE	280° 27′	100° 27′
EF	222° 52′	42° 52′
FA	114° 53′	294° 59′

Solution A rough sketch of the traverse is shown in Fig. 6.13. It can be noted from the values of the bearings that the fore and back bearings of FA do not agree with the rule that the fore and back bearings should differ by 180°.

The back bearing of FA, taken at A is 294° 59′. This should be taken as correct. The bearing of FA should be 114° 59′. This is an error of 6′ in the bearings. This error will be distributed equally among all the angles. The correction for AB is 1′ (e/N), for BC 2′ ($2e/N$), and so on. The corrected fore and back bearings are given in Table 6.15.

Table 6.15 Corrected bearings

Line	Corrections	Fore bearing	Back bearing
AB	1′	160° 34′	340° 34′
BC	2′	58° 50′	238° 50′
CD	3′	320° 50′	140° 50′
DE	4′	280° 31′	100° 31′
EF	5′	222° 57′	42° 57′
FA	6′	114° 59′	294° 59′

We calculate the interior angles:

∠A = 294° 59′ − 160° 34′ = 134° 25′

From the figure, it is clear that this is the exterior angle. Therefore,

∠A = 360° – 134° 25′ = 225° 35′

∠B = 340° 34′ – 58° 50′ = 281° 44′ >180°

∠B = 360°– 281° 45′ = 78° 16′

∠C = 320° 50′ – 238° 50′ = 82° 00′

∠D = 280° 31′ – 140° 50′ = 139° 41′

∠E = 222° 57′ – 100° 31′ = 122° 24′

∠F = 114° 59′ – 42° 57′ = 72° 02′

Sum of interior angles = 225° 35′ + 78° 16′
+ 82° 00′ + 139° 41′

+ 122° 24′ + 72° 02′

= 720°

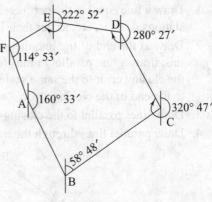

Fig. 6.13

6.5 Balancing a Traverse

The two conditions to be satisfied by a closed traverse are $\Sigma L = 0$ and $\Sigma D = 0$. This means that the algebraic sum of the latitudes is equal to 0 as is the algebraic sum of the departures. This also means that the sum of the northings is equal to the sum of the southings, and the sum of the eastings is equal to the sum of the westings.

Depending upon whether greater reliance is placed on angular measurements or linear measurements, or equal reliance is placed on both of them, there are different methods of adjusting a traverse. These methods are discussed as follows..

6.5.1 Bowditch's Rule

Bowditch's rule also known as the *compass rule* is applied when the angular and linear measurements are deemed to have equal precision. From the traverse data, the angles are first corrected to satisfy the geometrical condition of the traverse. Then the latitudes and departures of the lines are calculated. Finally, the algebraic sum of the latitudes, ΣL, and the algebraic sum of the departures, ΣD, are calculated. In a closed traverse, both these sums should be zero. If one or both are not zero, then corrections are applied as follows:

Correction to latitude of any side = error in latitude × (length of that side/perimeter)

Correction to departure of any side = error in departure × (length of that side/perimeter)

If E_L and E_D are the errors in latitude and departure of the traverse, P is the perimeter, and L is the length of any side, then

Correction to latitude of any side = $(- E_L)(L/P)$

Correction to departure of any side = $(- E_D)(L/P)$

It is evident that the error is distributed among the sides in proportion to their length.

6.5.2 Graphical Method

A graphical method of applying the Bowditch's rule is sometimes employed to avoid calculating latitudes and departures. The method was discussed in Chapter 3. We recapitulate only the procedure here [Fig. 6.14].

1. Draw a line equal to the perimeter of the traverse to a suitable scale. Mark the stations A, B, C, etc. as per their length and the scale chosen.
2. Draw at the end of the line the closing error in magnitude and direction. For this, draw a line parallel to the closing error and mark on it the magnitude of the closing error to the same scale as the traverse scale. Join the starting point to the end of the closing error.
3. Draw lines parallel to the closing error through points B, C, D, etc.
4. Draw parallel lines through the traverse points to the closing error.

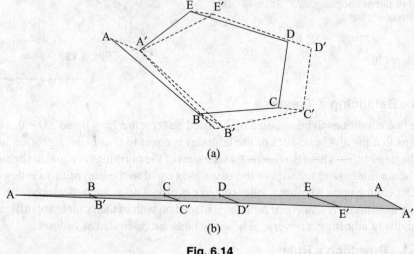

(a)

(b)

Fig. 6.14

5. Mark on these lines corrections BB′, CC′, etc. taken from Fig. 6.14(b).
6. Join A′, B′ C′, D′, E′, A′ to get the corrected taerse.

6.5.3 Transit Method

The transit method assumes that the angles are measured to greater precision than the lengths. The method can be applied in two ways. In the first method, latitudes and departures are corrected as follows:

Correction to latitude of a line = total error in latitude × (latitude of the line/arithmetic sum of latitudes)

Correction to departure of a line = total error in departure × (departure of the line/arithmetic sum of departure of the lines)

Note that the arithmetic sum is taken, which means that the sum of latitudes/departures is determined without taking their signs into account.

The second method corrects the northings, southings, eastings, and westings. Thus the corrections can be applied as follows:

Correction to northing of a line = (1/2) total error in latitude × (northing of the line/sum of northings)

Correction to southing of any line = $(1/2)$ total error in latitude $\times$ (southing of the line/sum of southings)

Correction to easting of any line = $(1/2)$ total error in departure $\times$ (easting of the line/sum of eastings)

Correction to westing of any line = $(1/2)$ total error in departure $\times$ (westing of the line/sum of westings)

6.5.4 Axis Correction Method

In the axis correction method, it is assumed that the angles are measured very precisely. In the adjusted traverse, the lengths are altered but the lines are kept parallel to their initial positions. The procedure for correction by the axis correction method is as follows (refer to Fg. 6.15).

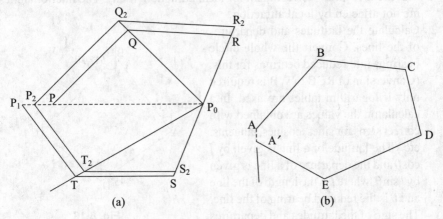

Fig. 6.15

1. Plot the traverse to a suitable scale and find the closing error. The plot obtained is PQRSTP$_1$. The closing error is PP$_1$.
2. Extend the closing error into the traverse; the line may cut the traverse approximately into two equal parts as in Fig. 6.15(a). In such a case, carry out the correction as per the procedure given in step 4 and further.
3. In case the closing error line, when extended, does not pass through the traverse at all or divides the traverse into two very unequal parts, as in Fig. 6.15(b), it will be necessary to make some adjustments before proceeding with the correction. In such a case, transfer the closing error to some other suitable station so that the closing error line when extended divides the traverse into two approximately equal parts. This will be illustrated with an example later.
4. Extend the closing error line to cut a traverse line at a point P$_0$. This line is known as the *axis for correction*. This line should divide the traverse into two nearly equal parts.
5. Divide the closing error P-P$_1$ into two equal parts with the point P$_2$. Thus, PP$_2$ = P$_1$P$_2$.
6. Carry out the corrections taking P$_2$ as the correct location of P. Connect the axis point PO to points Q, R, S, and T of the traverse and extend if necessary.

7. Draw lines parallel to the existing lines of the traverse. Thus, draw a line parallel to PQ through P_2 to get the point Q_2 on the line P_0Q or its extension.

8. Repeat this with all the stations to get the corrected traverse P_2-Q_2-R_2-S_2-T_2-P_2. This traverse closes keeping the directions of the lines the same as before.

Gale's traverse table

Angles are corrected and a traverse is balanced generally in tabular form. The most common form of the table is known as Gale's traverse table. The preparation of Gale's traverse table will require the following steps.

1. If the interior angles of the traverse are measured, check the sum of the interior angles to see whether this matches with the theoretical sum of $(2n - 4)$ right angles. If not, correct by distributing the error among all the angles. If the bearings are measured, correct them for local attraction, if any. The interior angles are not affected by local attraction.

2. Calculate the latitudes and departures of the lines. Convert the whole circle bearings into reduced bearings for this. (Conversion of RCB to WB is required only if logarithm tables are used. In a calculator, the values are obtained with correct signs for sines, cosines, tangents, etc.) The latitude of a line is given by $l\cos\theta$ and the departure of a line is given by $l\sin\theta$, where l is the length of the line and θ is the reduced bearing of the line. The sign of the latitudes and departures are known from the quadrants in which the lines lie.

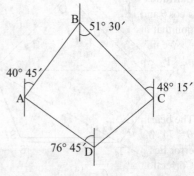

Fig. 6.16

3. Find the algebraic sum of latitudes and departures. For a closed traverse, each of these sums should be zero.

4. If not, correct by applying the methods outlined above. Use the Bowditch's rule for a compass traverse and the transit rule for a theodolite traverse.

5. Once the corrections are done, plot the traverse.

The preparation of Gale's traverse table and corrections by other methods are illustrated through the examples given below.

Example 6.9   A traverse was run with a prismatic compass and the lengths and bearings of the lines observed are given below. Check whether or not the traverse closes. If not, balance it using (i) Bowditch's rule and (ii) the graphical method.

Line	AB	BC	CD	DA
Length	105.8	142.5	188.8	188.9
Bearing	N40° 45′W	N51° 30′E	S48° 15′E	S76° 45′W

Solution   A rough sketch of the traverse is shown in Fig. 6.16. As the back bearings are not available, we assume that there is no local attraction. We first calculate the included angles to check whether their sum is 360°.

$$\angle A = 40° 45' + 76° 45' = 117° 30'$$

$$\angle B = 180° - (40° \ 45' + 51° \ 30') = 87° \ 45'$$

$$\angle C = 51° \ 30' + 48° \ 15' = 99° \ 45'$$

$$\angle D = 180° - (48° \ 15' + 76° \ 45') = 55° \ 00'$$

$$\angle A + \angle B + \angle C + \angle D = 117° \ 30' + 87° \ 45' + 99° \ 45' + 55° \ 00' = 360°$$

We now calculate the latitudes and departures of lines (see Table 6.16).

Table 6.16 Calculation of latitudes and departures

Line	Length	Bearing	Latitude	Departure	Lat. corr.	Dep. corr.
AB	105.8	N40° 45'W	80.15	– 69.06	0.0271	0.0946
BC	142.5	N51° 30'E	88.71	111.52	0.0364	0.1275
CD	188.8	S48° 15'E	– 125.72	140.85	0.0482	0.1689
DA	188.9	S76° 45'W	– 43.3	– 183.87	0.0483	0.1690

Latitudes and departures are calculated from the lengths and bearings. Latitude = $l \cos \theta$ and departure = $l \sin \theta$. Signs are assigned to latitudes and departures by the convention that north latitudes and east departures are positive.

We now find the sum of columns 4 and 5 of Table 6.16. We get

Sum of latitudes (column 4) = – 0.16

Sum of departures (column 5) = – 0.56

Closng error = $\sqrt{(-0.16)^2 + (-0.56)^2}$ = 0.5824 m

The bearing of the closing error is given by $\tan \theta = (-0.56)/(-0.16) = 3.5$. This gives angle $\theta = 74° \ 03'$. As the latitude and departure of the closing error are both negative, the closing error lies in the S-W quadrant.

Corrections to latitudes and departures are done as per Bowditch's rule. Correction to latitude of a line = total error in latitude × (length of line/perimeter of traverse). Correction to departure = total error in departure × (length/perimeter).

Perimeter = 105.8 + 142.5 + 188.8 + 188.9 = 626 m

These corrections are shown in columns 6 and 7 of Table 6.16. The corrected latitudes and departures are shown in Table 6.17.

Table 6.17 Corrected latitudes and departures

Line	Length	Bearing	Latitude	Departure
AB	105.8	N40° 45'W	80.1771	– 68.9654
BC	142.5	N51° 30'E	88.7464	111.6475
CD	188.8	S48° 15'E	– 125.6718	141.0189
DA	188.9	S76° 45'W	– 43.2517	– 183.7010

Table 6.18 Traverse data for Example 6.10

Line	Length	Station	Included angle
PQ	102.8	P	131° 14' 30"
QR	98.4	Q	84° 19' 25"
RS	110.8	R	116° 35' 25"
ST	82.8	S	119° 58' 05"
TP	113.29	T	87° 54' 05"

Example 6.10 The traverse data given in Table 6.18 contain the lengths and interior angles of a traverse PQRSTP. The bearing of line PQ was observed and recorded as S36° 12′ 30″E. Check the traverse for angles and closing errors, if any. Find the correct latitudes and departures by the transit method.

Solution   A rough sketch of the traverse is shown in Fig. 6.17.

Sum of included angles = 131° 14′ 30″ + 84° 19′ 25″ + 116° 35′ 25″

$$+ 119° 58′ 05″ + 87° 54′ 05″$$

$$= 540° 01′ 30″$$

The error of 1′ 30″ is distributed equally among all angles. Each angle will be reduced by 90″/5 = 18″. The corrected included angles and the bearings of the lines calculated from these angles are shown in Table 6.19.

Table 6.19   Corrected included angles and bearings

Station	Included angle	Line	Bearing	Length	Latitude	Departure
P	131° 14′ 12″	PQ	S36° 12′ 30″E	102.8	− 82.7853	60.6081
Q	84° 19′ 07″	QR	N48° 06′ 37″E	98.4	65.7015	73.2521
R	116° 35′ 07″	RS	N15° 18′ 16″W	110.8	106.8707	− 29.2453
S	119° 57′ 47″	ST	N75° 20′ 29″W	82.8	20.9533	− 80.1049
T	87° 53′ 47″	TP	S12° 33′ 18″W	113.29	− 110.5810	− 24.6266
			Total		0.1592	− 0.1166

Calculation of bearings   From the corrected included angles and the known bearing of PQ, the bearings of other lines can be calculated. Thus,

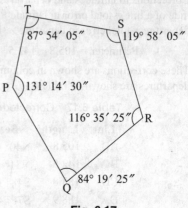

Fig. 6.17

Bearing of PQ = S36° 12′ 30″E

Bearing of QR = 84° 19′ 07″ − 36° 12′ 30″
= N48° 06′ 37″E

Bearing of RS = 180° − 116° 35′ 07″ − 48° 06′ 37″ = N15° 18′ 16″W

Bearing of ST = 180° − 119° 57′ 43″ − 15° 18′ 16″ = N75° 20′ 29″W

Bearing of TP = 87° 53′ 47″ − 75° 20′ 29″
= S12° 33′ 18″W

Check bearing of PQ: 180° − 131° 14′ 12″ − 12° 33′ 18″ = S36° 12′ 30″E

As mentioned earlier, there are two methods of applying the corrections, one in terms of latitudes and departures and the other in terms of northings, southings, eastings, and westings.

In terms of latitudes and departures   In this method,

Correction to latitude of a line = total error in latitude × latitude of the line/ arithmetic sum of latitudes

Correction to departure of a line = total error in departure × departure of the line/
arithmetic sum of departures

Arithmetic sum of latitudes   = 387.0329
Arithmetic sum of departures   = 267.6893

The corrections based on this are shown in Table 6.20.

Table 6.20 Corrected latitudes and departures

Line	Length	Corrections to latitude	Corrections to departure	Corrected latitude	Corrected departure
AB	102.8	− 0.0341	0.0264	− 82.8194	60.6345
BC	98.4	− 0.0270	− 0.0319	65.6745	73.1284
CD	110.8	0.044	− 0.0127	106.8267	− 29.2326
DE	82.8	0.0086	− 0.0349	20.9447	− 80.0700
EA	113.29	− 0.0455	0.0107	− 110.6265	− 24.6159
			Total	0.0000	0.0000

In terms of northings, southings, etc.   According to this method,

Correction to northing = (1/2) total error in latitude × northing of line/
sum of northings

Corrections to southing = (1/2) total error in latitude × southing of line/
sum of southings

Correction to easting = (1/2) total error in departure × easting of line/
sum of eastings

Correction to westing = (1/2) total error in departure × westing of line/
sum of westings

These corrections are shown in Table 6.21.

Table 6.21 Corrections to northing, southing, easting, and westing

Line	Northing	Southing	Easting	Westing	Corr. to northing	Corr. to southing	Corr. to easting	Corr. to westing
AB		82.7853	60.6081			0.0341	0.0264	
BC	65.7015		73.2521		0.027		0.0319	
CD	106.8707			29.2453	0.0440			0.0127
DE	20.9533			80.1049	0.0086			0.0348
EA		110.5810		24.6266		0.0455		0.0107

The corrected latitudes and departures are shown in Table 6.22.

Table 6.22 Corrected latitudes and departures

Line	Length	Bearing	Corrected latitude	Corrected departure
AB	102.8	S36° 12′ 30″E	82.8194	60.6345
BC	98.4	N48° 06′ 47″E	− 65.6744	73.2840
CD	110.8	N15° 18′ 06″E	− 106.8268	− 29.2325
DE	82.8	N75° 20′ 19″W	− 20.9447	80.0701
EA	113.29	S12° 33′ 28″W	110.6265	24.6159
	Total		0.0000	0.0000

A combination of Tables 6.20–6.22 gives Gale's traverse table and is shown in Table 6.23.

Example 6.11 From the traverse data given below, determine the closing error, if any, and correct the traverse by the axis correction method.

Line	AB	BC	CD	DE	EA
Length (m)	201.54	189.68	231.94	272.55	257.15
Bearing	62° 42′	154° 54′	202° 32′	281° 44′	22° 15′

Solution A rough sketch of the traverse is shown in Fig. 6.18(a). This is a graphical method and the solution is shown Fig. 6.18(b). ABCDEA′ is the plot of the traverse with the given data. AA′ is the closing error. We extend AA′ and let it meet the line CD at point P. PA′ is the axis for correction and divides the traverse approximately into two equal parts. The procedure for correction by the axis correction method is as follows.

1. Join PB, PC, PD, PE.
2. Bisect AA′ to get point A_1.
3. Draw line A_1B_1 parallel to AB to intersect PB at B_1. A_1B_1 is the corrected line for AB. The line retains its direction but has reduced in length.

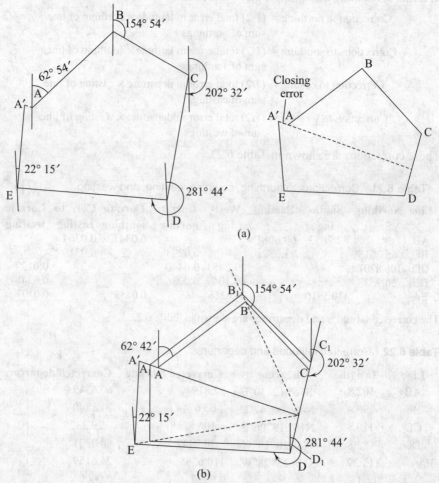

(a)

(b)

Fig. 6.18

Table 6.23 Gale's traverse table

Station	Included angle	Corrected included angles	Line	Length	Bearing	Latitude	Departure	Corrections to latitude	Corrections to departure	Corrected latitude	Corrected departure
P	131° 14' 30"	131° 14' 12"	PQ	102.8	S36° 12' 30"E	−82.7853	60.6081	−0.0341	0.0264	−82.8194	60.6345
Q	84° 19' 25"	84° 19' 07"	QR	98.4	N48° 06' 37"E	65.7015	73.2521	0.0270	−0.0270	65.6745	73.284
R	116° 35' 25"	116° 35' 07"	RS	110.8	N15° 18' 16"W	106.8707	−29.2453	0.044	−0.0127	106.8267	−29.2326
S	119° 58' 05"	119° 57' 47"	ST	82.8	N75° 20' 29"W	20.9533	−80.1049	0.0086	−0.0349	20.9447	−80.070
T	87° 54' 05"	87° 53' 47"	TP	113.29	S12° 33' 18"W	−110.5810	−24.6266	−0.0455	0.0107	−110.6265	−24.6159

Line	Northing	Southing	Easting	Westing	Corr.	Corr. to southing	Corr. to easting	Corr. to westing	Corrected latitude	Corrected departure
PQ		82.7853	60.6081			0.0341	0.0264		82.8194	60.6345
QR	65.7015		73.2521		0.027		0.0319		65.674	73.1284
RS	106.8707			29.2453	0.044			0.0127	106.8267	−29.2326
ST	20.9533			80.1049	0.0086			0.0349	−20.9447	80.07
TP		110.5810		24.6266		0.0455		0.0107	110.6265	24.6159

4. Through B_1, draw a line parallel to BC to intersect PC or its extension at C_1.
5. Through A_1, draw a line parallel to AE to intersect PE at E_1.
6. From E_1, draw a line parallel to ED to intersect PD or its extension at D_1.
7. $A_1B_1C_1D_1E_1A_1$ is the corrected traverse, which now closes.

Example 6.12 The data recorded for a traverse is given below. Determine the closing error, if any. Balance the traverse by the axis correction method.

Line	PQ	QR	RS	ST	TP
Length (m)	105.8	115.4	122.3	98.4	78.1
Bearing	N30° 15′W	N60° 30′E	S58° 45′E	S50° 30′W	S73° 35′W

Solution The traverse drawn to scale is shown in Fig. 6.19. It is seen that the closing line does not intersect the traverse when produced. It is therefore necessary to transfer the closing error to some other point. From P draw a line PT′ parallel and equal to P′T. From this new position of T, draw a line T′S′ parallel and equal to TS. SS′ is the closing error and when extended approximately cuts the traverse into two equal parts. The corrections are then done with reference to the axis SA. A is the point where the closing line when extended cuts the line PQ. The procedure then is as follows.

1. Take the middle point of SS′ and mark it S_1.
2. Join AT′. Draw a line parallel to ST′ through S_1 to cut AT′ at T_1.
3. Join AR. Draw a line parallel to RS through S_1 to intersect AR at R_1.
4. Draw a line through R_1 parallel to QR to cut AQ or its extension at Q_1.
5. Draw a line through T_1 parallel to QP to cut AP or its extension at P_1.
6. $P_1Q_1R_1S_1T_1P_1$ is the adjusted traverse.

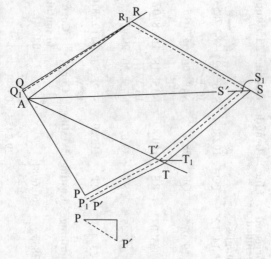

Fig. 6.19

6.6 Omitted Measurements

Often during fieldwork, it may not be possible to measure some parameter of a traverse line. This can also be caused by carelessness or omission by the surveyor. If the traverse is a closed one, it is possible to calculate any two quantities if they are missing. This is done using the basic conditions of a closed polygon.

For a closed traverse, the algebraic sum of latitudes and that of departures should each be zero. Thus, we have two conditions for a closed traverse: $\Sigma L = 0$ and $\Sigma D = 0$. ΣL is the algebraic sum of latitudes and ΣD is the algebraic sum of departures. If l_1, l_2, l_3, ... are the lengths of the sides of the traverse and θ_1, θ_2, θ_3, ... their reduced bearings, then

$$\Sigma L = l_1 \cos\theta_1 + l_2 \cos\theta_2 + l_3 \cos\theta_3 + \cdots = 0$$

$$\Sigma D = l_1 \sin\theta_1 + l_2 \sin\theta_2 + l_3 \sin\theta_3 + \cdots = 0$$

These two conditions enable us to evaluate any two unknown quantities. The unknown quantities can be the lengths and bearings of lines. If more than two quantities are missing, then the problem is indeterminate. Depending upon the parameters missing, the following types of problems can arise and can be solved using the two conditions stated above.

1. A single line affected:
   - (a) Length of a line missing
   - (b) Bearing of a line missing
   - (c) Length and bearing of a line missing

2. More than one line affected:
   - (a) Lengths of two adjacent lines missing
   - (b) Bearings of two adjacent lines missing
   - (c) Length of one line and bearing of an adjacent line missing

In case 2, the affected sides may be adjacent or separated by other lines.

Case 1: A single line affected

(a) Length of a line missing Let l be the unknown length of the line. Its bearing q is known. We will determine the algebraic sum of the latitudes and departures of the lines whose lengths and bearings are known. Let these sums be ΣL_1 and ΣD_1. Then we have the condition that

$$\Sigma L_1 + l \cos\theta = 0 \qquad \text{and} \qquad \Sigma D_1 + l \sin\theta = 0$$

From these equations, we get

$$l \cos\theta = -\Sigma L_1 \qquad \text{and} \qquad l \sin\theta = -\Sigma D_1$$

The length can be evaluated from any of these expressions and checked with the other. It may be noted that we have two conditions and only one unknown.

(b) Bearing of a line missing This is similar to case (a). As in case (a), we have

$$\Sigma L_1 + l \cos\theta = 0 \qquad \text{and} \qquad \Sigma D_1 + l \sin\theta = 0$$

From these, we get

$$l \cos\theta = -\Sigma L_1 \qquad \text{and} \qquad l \sin\theta = -\Sigma D_1$$

Any of these equations can be used to find θ. Alternatively, dividing one by the other, we get $\tan\theta = (-\Sigma D_1 / -\Sigma L_1)$. The bearing is thus known and its quadrant can be determined from the signs of $-\Sigma L_1$ and $-\Sigma D_1$.

(c) Length and bearing of a line missing   In this case there are two unknowns and both the equations are needed for solution. Let ΣL_1 and ΣD_1 be the algebraic sum of the latitudes and departures, respectively, of the remaining lines whose lengths and bearings are known. We have, as in the previous two cases,

$$\Sigma L_1 + l\cos\theta = 0 \quad \text{and} \quad \Sigma D_1 + l\sin\theta = 0$$

Here l is the unknown length of the line and θ its reduced bearing. We have two equations, which can be written as

$$l\cos\theta = -\Sigma L_1 \quad \text{and} \quad l\sin\theta = -\Sigma D_1$$

Dividing one by the other, we get

$$\tan\theta = (-\Sigma D_1 / -\Sigma L_1)$$

The reduced bearing can be found from this equation. The quadrant of the bearing is found from the signs of $-\Sigma L_1$ and $-\Sigma D_1$. The length of the line is found from any of the relations

$$\Sigma L_1 + l\cos\theta = 0 \quad \text{and} \quad \Sigma D_1 + l\sin\theta = 0$$

Case 2: Two lines affected

(a) Lengths of two adjacent lines missing   Consider the case shown in Fig. 6.20. In traverse PQRSTP, let the lengths of lines PQ and PT (or any two adjacent sides) be missing. We consider the traverse QRSTQ, which is a closed traverse; QT is the closing line. The length and bearing of QT can be calculated from the conditions that $\Sigma L = 0$ and $\Sigma D = 0$ for this traverse.

We now consider the triangle QTP. Since the bearings of all the lines of the triangle are known, the included angles of the triangle can be calculated. Let $\angle QTP = \beta$, $\angle PQT = \alpha$, $\angle QPT = \gamma$. We have the relation between the sides and angles of a triangle that the sides are proportional to the sine of the opposite angle. This gives $PQ/\sin\beta = PT/\sin\alpha = QT/\sin\gamma$. PQ and PT, the unknown lengths, can be evaluated from these equations as

$$PQ = QT(\sin\alpha/\sin\gamma) \quad \text{and} \quad PT = QT(\sin\beta/\sin\gamma)$$

Alternatively, let l_1 and l_2 be the lengths of sides PQ and PT and θ_1 and θ_2 be their bearings (bearings are known). We can write the equations $\Sigma L = 0$ and $\Sigma D = 0$ with these values and solve for l_1 and l_2.

(b) Bearings of two adjacent lines missing   In this case the lengths of all the sides are known. The following procedure can be adopted (refer to Fig. 6.20).

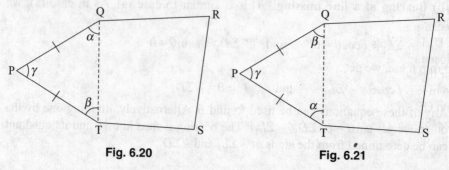

Fig. 6.20   Fig. 6.21

1. Let the bearings of PQ and PT be the unknowns. Close the traverse QRSTQ and find the length and bearing of the line QT. This is done by the two conditions $\Sigma L = 0$ and $\Sigma D = 0$ for this traverse.

2. When the length QT is calculated, in triangle QTP, the lengths of all the sides are known. Calculate its area using the Hero's formula: area $= \sqrt{s(s-a)(s-b)(s-c)}$, where s is the half-perimeter of the triangle QTP $= (1/2)(QT + TP + PQ)$, and $a = QT$, $b = TP$, and $c = PQ$.

3. Also calculate the area of the triangle QTP from the sine formula, which is area of a triangle = product of two sides × the sine of the angle between them. If α, β, and γ are the angles as shown in Fig. 6.20, then the area of the triangle QTP is also given by area $= (1/2)PT \times TQ \times \sin\alpha = (1/2) TQ \times QP \times \sin\beta = (1/2) PQ \times PT \sin\gamma$.

4. Equating the area calculated in step 2 to these expressions, calculate the angles α, β, and γ. Only two angles need to be calculated. The third angle can be calculated from the condition that the sum of the angles of a triangle $= 180°$.

5. Knowing the angles α, β, and γ, calculate the bearings of the lines QP and PT.

(c) Length of one line and bearing of an adjacent line missing Referring to Fig. 6.21, in the traverse PQRSTP, let the two missing quantities be the length of PQ and bearing of TP. The general procedure for solution is as follows.

1. As before, close the traverse QRSTQ and find the length and bearing of QT.

2. The triangle QTP has to be solved for finding the missing quantities.

3. In triangle QTP, the lengths of sides QT and TP are known and the bearings of PQ and QT are known.

4. Find the angle PQT (angle β) from the bearings of PQ and QT.

5. If a, b, c are the sides of a triangle and A, B, C are the angles opposite them, then we have two equalities from the properties of a triangle:
$a/\sin A = b/\sin B = c/\sin C$

6. Rewrite this relation for triangle QTP as PQ/$\sin\alpha$ = PT/$\sin\beta$ = QT/$\sin\gamma$.

7. Determine the length of PQ and angle γ from these equations and the condition that the sum of the angles in a triangle $= 180°$. The bearing of the line PT can then be calculated.

When the lines with missing parameters are not adjacent

Consider the traverse PQRSTUP shown in Figs 6.22(a) and (b). In case (a), the missing measurements are of the lines UT and RS. Through one end of one of the affected lines, in this case T, draw a line parallel and equal to RS. Through R, draw a line equal and parallel to ST. The two affected lines are brought adjacent because RS = R'T. The traverse PQRR'U is a closed traverse with the closing line R'U. The length and bearing of R'U can be determined from the conditions of the closed traverse. The triangle R'TU can then be solved to get the unknown values for the lines UT and R'T (same as RS).

In Fig. 6.22(b) the affected lines (whose length or bearing is unknown) are UT and QR. US' is drawn parallel and equal to TS and RQ' is drawn parallel and equal to RS. Join S'Q'. It is clear that SS' = UT and Q'S = QR. The affected sides are thus brought adjacent. From the closed traverse UPQQ'S'U, the length and bearing of line Q'S' can be determined. The triangle Q'S'S can then be solved to find the missing parameters of these lines.

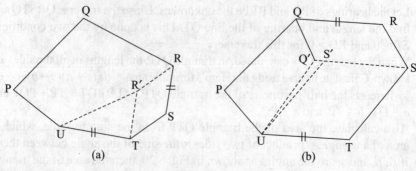

Fig. 6.22

In solving some of these problems, there can be ambiguity about the answer. This can be taken care of by drawing a neat sketch and finding the correct solution.

Area of a traverse

The area of a closed traverse can be determined using many methods. These are detailed in Chapter 8 on computation of areas.

Example 6.13   The survey data of a traverse is given in Table 6.24. The length and bearing of one side were not recorded during the survey. Find the missing measurements.

Table 6.24   Traverse data for Example 6.13

Line	Length	Bearing
AB	201.8	N45° 00′W
BC	288.4	N60° 30′E
CD	192.6	S34° 45′E
DA	Missing	Missing

Solution   A rough sketch of the traverse is shown in Fig. 6.23. The latitudes and departures of the first three lines can be calculated as given in Table 6.25.

Algebraic sum of latitudes = 142.69 + 142.01 − 158.25 = 126.5

Algebraic sum of departures = − 142.69 + 251.01 + 109.2 = 218.1

If L is the latitude of DA and D the departure of DA, we have

$126.5 + L = 0$, $L = -126.5$ and $218.1 + D = 0$, $D = -218.1$

Fig. 6.23

Table 6.25   Calculation of latitudes and departures

Line	Length	Bearing	Latitude	Departure
AB	201.8	N45° 00′W	142.69	− 142.69
BC	288.4	N60° 30′E	142.01	251.01
CD	192.6	S34° 45′E	− 158.25	109.78
DA	–	–	–	–

Legth of DA $= \sqrt{(L^2 + D^2)} = \sqrt{[(-126.5)^2 + (-218.1)^2]} = 252.13$ m

$\tan\theta = (-218.1)/(-126.5) = 1.7241, \qquad \theta = 59° 53' 09''$

As both the latitude and departure of DA are negative, the quadrant is S-W.

Length of DA $= 252.13$ m,   bearing of DA $= S59° 53' 09''W$

Example 6.14 From the records available from survey notes, it is observed that the lengths of two lines were not readable. From the available data given below, find the lengths of the two sides.

Line	PQ	QR	RS	ST	TP
Length (m)	178.6	228.4	Missing	Missing	238.8
Bearing	S52° 30'E	N48° 45'E	N18° 15'W	S78° 30'W	S32° 30'W

Solution A rough sketch of the traverse is shown in Fig. 6.24. The latitudes and departures of the lines can be calculated as in Table 6.26. Let L_1 and L_2 be the lengths of RS and ST. Since the bearings of these lines are known, two simultaneous equations with unknowns L_1 and L_2 will evolve out of the conditions $\Sigma L = 0$ and $\Sigma D = 0$.

Table 6.26 Calculation of latitudes and departures (Example 6.14)

Line	Length	Bearing	Latitude	Departure
PQ	178.6	S52° 30'E	− 108.72	141.69
QR	228.4	N48° 45'E	150.59	171.72
RS	L_1	N18° 15'W	$0.9497L_1$	$- 0.3132L_1$
ST	L_2	S78° 30'W	$- 0.1994L_2$	$- 0.9799L_2$
TP	238.8	S32° 30'W	− 201.4	− 128.31

Table 6.27 Latitudes and departures of traverse PQRTP

Line	Length	Bearing	Latitude	Departure
PQ	178.6	S52° 30'E	− 108.72	141.69
QR	228.4	N48° 45'E	150.59	171.72
RT	L	θ	$L\cos\theta$	$L\sin\theta$
TP	238.8	S32° 30'W	− 201.4	− 128.31

$\Sigma L = 0,$ $- 108.72 + 150.59 + 0.9497L_1 - 0.1994L_2 - 201.4 = 0$

$\Sigma D = 0,$ $141.69 + 171.72 - 0.3132L_1 - 0.9799L_2 \; 128.31 = 0$

Solving,

$L_1 = 194.6$ m, $L_2 = 126.7$ m

Length of line RS $= 194.6$ m

Length of line ST $= 126.7$ m

Alternative solution Close the traverse with the line RT, thus eliminating the lines with unknown parameters. The length and bearing of RT can be calculated from the data of Table 6.27.

$\Sigma L = 0,$ $- 159.53 + L\cos\theta = 0,$ $L\cos\theta = 159.53$

$\Sigma D = 0,$ $185.10 + L\sin\theta = 0,$ $L\sin\theta = - 185.10$

Legth of RT $= \sqrt{[(159.53)^2 + (-185.10)^2]} = 244.36$ m

The bearing of RT is given by

$$\tan\theta = -185.10/159.53 =$$
$$1.1602, \qquad \theta = 49° \ 14' \ 36''$$

As the departure of line RT is negative and latitude positive, the quadrant is N–W. We now solve the triangle RTS shown in Fig. 6.24. RS and ST are the unknowns. The bearings of these two lines are known. From these, the angles can be calculated as follows:

$$\alpha = 180° - (78° \ 30' + 49° \ 14' \ 36'') =$$
$$52° \ 15' \ 24''$$

$$\beta = 49° \ 14' \ 36'' - 18° \ 15' = 30° \ 59' \ 36''$$

$$\gamma = (78° \ 30' + 18° \ 15') = 96° \ 45'$$

We can solve for the unknown lengths using the sine rule:

$$\text{RT}/\sin 96° \ 45' = \text{RS}/52° \ 15' \ 24'' = \text{ST}/\sin 30° \ 59' \ 36''$$

Solving,

$$\text{RS} = 194.6 \text{ m} \qquad \text{and} \qquad \text{ST} = 126.7 \text{ m}$$

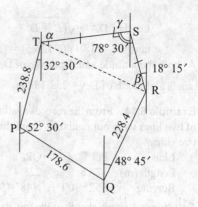

Fig. 6.24

Example 6.15 Due to some problems with equipment, the bearings of two sides were not taken for a closed traverse ABCDEA. From the available data, compute the bearings of the two sides.

Line	AB	BC	CD	DE	EA
Length (m)	230.5	250.2	210.8	240.3	265.4
Bearing	N36° 45′E	S82° 48′E	S10° 15′E	Missing	Missing

Solution A rough sketch of the traverse is shown in Fig. 6.25. We join the points D and A to get a closed traverse ABCDA. The length and bearing of DA can be found using the two conditions. The calculation of latitudes and departures can be done as shown in Table 6.28.

Table 6.28 Calculation of length and bearing of DA (Example 6.15)

Line	Length	Bearing	Latitude	Departure
AB	230.5	N36° 45′E	184.69	137.91
BC	250.2	S82° 48′E	−31.36	248.23
CD	210.8	S10° 15′E	−207.43	37.51
DA	–	–	–	–

The sums of the latitudes and departures in columns 4 and 5 of Table 6.28 are the following:

Sum of latitudes $= -54.1$,   sum of departures $= +423.65$

Latitude of DA $= 54.1$,   departure of DA $= -423.65$

(from closed traverse ABCDA)

Length of DA $= \sqrt{(54.1)^2 + (-423.65)^2} = 427.1$ m

Bearing of DA: $\tan\theta = -423.65/54.1 = 7.8309$, $\theta = $ N82° 43′ 12″W

Next we consider triangle DAE. The lengths of all the sides are known. The angles are not known.

DA $= 427.1$ m,   AE $= 265.4$ m,   ED $= 240.3$ m

Perimeter = 932.8 m

$S = (1/2)$ perimeter = 466.4 m

Area of $\triangle DAE = \sqrt{[466.4(466.4-427.1)(466.4-265.4)(466.4-240.3)]}$
$= 28862$ sq. m

Area is also given by

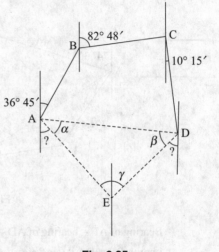

Fig. 6.25

$(1/2)AE \times AD \times \sin\alpha = (1/2)AD \times DE \times \sin\beta$

$(1/2) \times 265.4 \times 427.1 \times \sin\alpha = 28{,}862$ sq. m, $\alpha = 30° 36' 36''$

$(1/2)AD \times DE \times \sin\beta = 28{,}862$, $= 34° 13' 28''$

Bearing of DE $= 180° - (82° 43' 12'' + 34° 13' 28'') = 63° 03' 20''$

$= S63° 03' 20''W$

Bearing of AE $= 82° 43' 12'' - 30° 36' 36'' = 52° 06' 36''$

$= S52° 06' 36''E$

Example 6.16 A traverse ABCDEA was conducted and due to difficulties in the field, the bearing of line EA and length of line DE could not be measured. From the remaining data of the traverse, find the missing quantities.

Line	AB	BC	CD	DE	EA
Length (m)	301.5	288.4	199.5	Missing	201.4
Bearing	N74° 30′E	S60° 15′E	S30° 45′W	N82° 15′W	Missing

Solution The two affected lines are adjacent. We close the traverse with the line DA. The length and bearing of line DA can be calculated. The latitudes and departures of the traverse ABCDA are calculated as shown in Table 6.29.

Table 6.29 Calculation of latitudes and departures (Example 6.16)

Line	Length	Bearing	Latitude	Departure
AB	301.5	N74° 30′E	80.57	290.53
BC	288.4	S60° 15′E	– 143.11	250.39
CD	199.5	S30° 45′W	– 171.45	– 102
DA	–	–	L	D

We get from columns 4 and 5 of Table 6.29,

$L - 233.99 = 0$, $\quad D + 438.92 = 0$, $\quad L = 233.99$, $\quad D = -438.92$

Length of DA $= \sqrt{(233.99)^2 + (-438.92)^2} = 497.4$ m

Bearing of DA: $\tan\theta = (-438.92/233.99)$, $\quad \theta = 61.94° = 61° 56' 24''$

$\alpha = 82.25 - 61.94 = 20.31°$

We solve triangle DAE to get the length of DE and bearing of AE. The sine rule is applied (Fig. 6.26).

$$\sin\gamma = \frac{DA}{AE} \sin\alpha \quad (\alpha = 82° 15' - 61° 56' 24'' = 20°18' 36'')$$

$$= \frac{497.4 \times \sin(20°18'36'')}{201.4} = 0.8572, \quad \gamma = 59°$$

From the figure it is clear that the angle is $180° - 59° = 121°$. Hence,

$\beta = 180° - 121° - 20.31° = 38.69°$

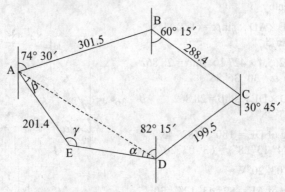

Fig. 6.26

Bearing of AE = bearing of AD – β = 61.94° – 38.69° = 23.25°

Length DE = DA × (sinβ/sinγ) = 497.4 × sin 38.69°/sin 59° = 362.74 m

Example 6.17 In conducting a traverse PQRSTP, the lengths of the lines QR and TP could not be measured correctly. Find the lengths of these lines from the remaining data given below:

Line	PQ	QR	RS	ST	TP
Length (m)	357.2	Missing	389.2	253.4	Missing
Bearing	S52° 30′E	N48° 45′E	N18° 15′W	S78° 30′W	S32° 30′W

Solution A rough sketch of the traverse is given in Fig. 6.27. In this case the two lines with missing parameters are not adjacent. Hence, it is necessary to bring these two lines adjacent by transferring the lines as follows. Draw a line parallel to PT through Q and a line parallel to PQ through T to meet at T′. QT′ = PT and has been brought near QR, the other affected side. We find T′R by solving the traverse RSTT′R. The calculations are shown in Table 6.30.

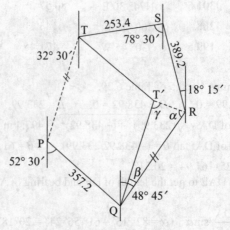

Fig. 6.27

Sum of latitudes = 369.2 – 50.52 – 217.45 = 101.23

Sum of departures = – 121.88 – 248.31 + 283.38 = – 86.81

From the closed traverse RSTT′R,

Latitude of T′R = – 101.23,   departure of T′R = + 86.81

Length of T′R = $\sqrt{(-101.23)^2 + (86.81)^2}$ = 133.35 m

Bearing of T′R: $\tan\theta$ = 86.81/(– 101.23), θ = 40.61 = S40° 36′ 36″E

Table 6.30 Calculation of the length and bearing of T′R (Example 6.17)

Line	Length	Bearing	Latitude	Departure
RS	389.2	N18° 15′W	369.62	– 121.88
ST	253.4	S78° 30′W	– 50.52	– 248.31
TT′	357.2	S52° 30′E	– 217.45	283.38
T′R	–	–	–	

Next we solve the triangle T′QR. The bearings of all the sides are known. The lengths of sides QT′ and QR are to be found. The sine rule can be used for this. The angles α, β, and γ are the following:

α = 180° – 48° 45′ – 40° 36′ 36″ = 90° 38′ 24″

β = 48° 45′ – 32° 30′ = 16° 15′

γ = 32° 30′ + 40° 36′ 36″ = 73° 06′ 36″

We have from the triangle

T′R/sin 16° 15′ = QT′/sin 90° 38′ 24″ = QR/sin 73° 06′ 36″

Solving,

QT′ = 476.51 m,   QR = 456 m

The missing sides are PT = QT′ = 476.51 m and QR = 456 m.

Example 6.18 From the traverse data given below, find the length of CD and bearing of EA.

Line	AB	BC	CD	DE	EA
Length (m)	178.6	228.4	–	126.7	238.8
Bearing	S52° 30′E	N48° 45′E	N18° 15′W	S78° 30′W	–

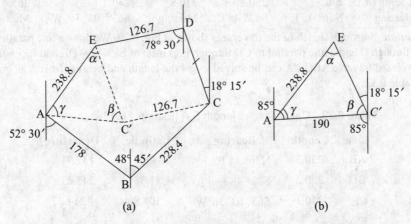

(a)   (b)

Fig. 6.28

Solution A rough sketch of the traverse is drawn as shown in Fig. 6.28. Lines CD and EA are to be brought adjacent. For this, draw a line parallel to CD through E and a line parallel to DE through C. EC' = CD and the two lines with missing measurements are brought adjacent. CC" = DE. The traverse ABCC'A is a closed traverse and the length and bearing of C'A can be found. The calculations are shown in Table 6.31.

Table 6.31 Calculation of length and bearing of C'A (Example 6.18)

Line	Length	Bearing	Latitude	Departure
AB	178.6	S52° 30′E	− 108.73	141.69
BC	228.4	N48° 45′E	150.60	171.72
CC"	126.7	S78° 30′W	− 25.26	− 124.15
C'A	–	–	–	–

Latitude of C'A = − (− 108.75 + 150.60 − 25.26) = −16.61

Departure of C'A = − (141.69 + 171.72 − 124.15) = − 189.26

Length of C'A = $\sqrt{(-16.61)^2 + (-189.26)^2}$ = 190 m

Bearing of C'A: tan θ = (− 189.26)/(− 16.61), θ = S48° 00′ 00″W

We can now solve the triangle C'AE to obtain the length of CD and the bearing of EA. In triangle C'AE, shown enlarged in Fig. 6.28(b),

β = 180° − 18° 15′ − 85° 00′ = 76° 45′

Applying the sine rule,

190/sinα = 238.8/sinβ, α = 50° 45′, γ = 180° − α − β = 52° 30′

EC'/sin 52° 30′ = 238.8/sin 76° 45′

EC' = CD = 238.8 × sin 52° 30′/sin 76° 45′ = 194.6 m

Length of CD = 194.6 m; bearing of AE = 85° − γ = 85° − 52° 30′ = 32° 30′; bearing of EA = S30° 30′W.

Example 6.19 In a closed traverse ABCDEA, the bearings of two lines could not be observed due to difficulties in the field. Compute the bearings of the two lines from the data of the traverse given below.

Line	AB	BC	CD	DE	EA
Length (m)	230.5	250.2	210.8	240.3	265.4
Bearing	N36° 45′E	S82° 48′E	Missing	S63° 03′ 36″W	Missing

Solution A rough sketch of the traverse is shown in Fig. 6.29. We draw a line parallel to DE through C and a line parallel to CD through E to meet at E'. CE' = DE and EE' = CD. The closed traverse ABCE'A can be solved to get the length and bearing of E'A as given in Table 6.32.

Table 6.32 Calculation of length E'A (Example 6.19)

Line	Length	Bearing	Latitude	Departure
AB	230.5	N36° 45′E	184.69	137.91
BC	250.2	S82° 48′E	− 31.38	248.23
CE'	240.3	S63° 03′ 36″W	− 108.86	− 214.22
E'A				

Latitude of E′A = – 44.45, departure of E′A = – 171.92

Length of E′A = $\sqrt{(-44.45)^2 + (-171.92)^2}$ = 177.57 m

Bearing of E′A: $\tan\theta$ = (– 171.92)/(– 44.45), θ = 75° 30′

Therefore, the bearing is S75° 30′W.

Triangle E′EA can be solved to determine the unknown bearings. All the sides of the triangle are known.

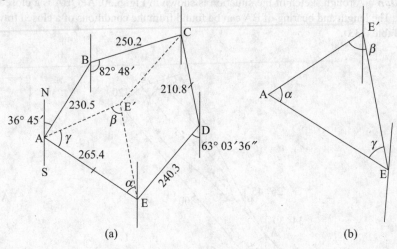

Fig. 6.29

Perimeter = 177.57 + 265.4 + 210.8 = 653.77

Half-perimeter s = 326.89

Area of triangle = $\sqrt{326.89(326.89-177.57)(326.89-210.8)(326.89-265.4)}$

= 18,666.33 sq. m

Area is also given by (1/2)(product of two sides × sine of angle between them). From Fig. 6.29(b),

Area = (1/2) AE × EE′ × sin γ = (1/2) AE × AE′ × sinα

= (1/2) AE′ × EE′ × sinβ

$\sin\gamma$ = 2 × 18,666.33/(265.4 × 210.8), $\sin\gamma$ = 0.6673, γ = 41.84 = 41° 50′ 24″

$\sin\alpha$ = 2 × 18,666.33/(265.4 × 177.57), $\sin\alpha$ = 0.7922, α = 52.36 = 52° 21′ 36″

$\sin\beta$ = 2 × 18,666.33/(177.57 × 210.8), $\sin\beta$ = 0.9973, β = 85.8 = 85° 48′ 00″

From these values,

Bearing of CD = S10° 18′E, bearing of EA= N52° 08′ 24″W

The following examples illustrate the application of the principles given for the solution of traverse problems.

Example 6.20 In conducting the survey of an area, a small hillock came in the way of a survey line, obstructing vision and making chaining impossible. To overcome this, from one side of the hillock, a right angle was laid out at station A. A length of 120 m was measured. Another right angle was measured and a length of 800 m was measured when the other side of the hillock was reached. Still another right angle was measured and a length of 120 m was measured to locate station B. To check the accuracy of this work, a traverse was surveyed on the other side. Details of the traverse APQB (see Fig. 6.30) are given below. Verify the length of AB and check whether AB is along the same bearing as line OA, if the bearing of OA = N40° 45′E.

Line	AP	PQ	QB
Length (m)	165.8	600	269.63
Bearing	N34° 30′W	N26° 45′E	S79° 08′E

Solution A rough sketch of the situation is shown in Fig. 6.30. APQBA is a closed traverse. The length and bearing of BA can be found from the conditions of a closed traverse see (Table 6.33).

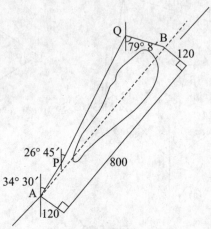

Fig. 6.30

Table 6.33 Calculation of length and bearing of BA (Example 6.20)

Line	Length	Bearing	Latitude	Departure
AP	165.8	N34° 30′W	136.64	– 93.91
PQ	600	N26° 45′E	535.78	270.06
QB	269.63	S79° 08′E	– 66.37	346.06
BA	–	–	–	–

Latitude of BA = – (136.64 + 535.78 – 66.37) = – 606.05

Departure of BA = – (– 93.91 + 270.06 + 346.06) = – 522.21

Length of BA = $\sqrt{(-606.05)^2 + (-522.21)^2}$ = 799.999 m

Bearing of BA: $\tan\theta$ = (– 522.21/– 606.05), θ = 40.7502

Bearing of BA = S40° 45′W

Both the length and bearing of BA calculated from the traverse agree with the earlier rectangle set out.

Example 6.21 OP is a survey line coming from one side of a lake. P and Q are two stations on either side of the lake. P and Q are intervisible but the length of PQ cannot be measured. A traverse PABCQ was run around the lake and the length of line CQ could not be measured due to some obstacles. Find the length CQ for Q to be on the extension of OP and the length PQ from the following:

Line	PA	AB	BC	CQ
Length	150	500	300	–
Bearing	N10° 15′E	N32° 45′E	N65° 30′E	S50° 30′E

Solution A rough sketch of the situation is shown in Fig. 6.31. PABCQP is a closed traverse. Lengths of lines CQ and QP are to be determined. We calculate the latitudes and departures as shown in Table 6.34. Assume that the length of CQ = L_1 and that of QP = L_2.

Table 6.34 Calculation of latitudes and departures of traverse PABCQP

Line	Length	Bearing	Latitude	Departure
PA	150	N10° 15′E	147.6	26.69
AB	500	N32° 45′E	420.52	270.49
BC	300	N65° 30′E	124.41	272.99
CQ	L_1	S50° 30′E	$0.6361L_1$	$0.7716L_1$
QP	L_2	S68° 45′W	$0.3624L_2$	$0.932L_2$

$$\Sigma L = 0, \quad 0.6361L_1 + 0.3624L_2 = 692.53$$
$$\Sigma D = 0, \quad -0.7716L_1 + 0.932L_2 = 570.17$$

Solving,

$$L_1 = 502.937 \text{ m}, \quad L_2 = 1028.214 \text{ m}$$

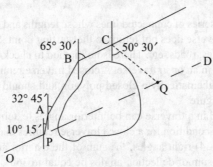

Fig. 6.31

Length of CQ = 502.937 m (so that OPQ is a straight line)

Length of PQ = 1028.214 m

Example 6.22 A traverse ABCDA was conducted and the following data was obtained. It was required to connect the midpoint E of CD to the midpoint F of AB. Determine the length and bearing of EF.

Line	AB	BC	CD	DA
Length (m)	612	502	1124	452
Bearing	N80° 06′E	N12° 45′E	S80° 06′W	S15° 45′E

Table 6.35 Calculations for Example 6.22

Line	Length	Bearing	Latitude	Departure
AD	488.7	N50° 06′W	313.74	– 374.69
DE	562.34	N75° 30′E	140.8	544.43
EF	–	–	–	–
FA	306.18	S80° 06′W	– 52.64	– 301.62

Solution A rough sketch of the traverse is shown in Fig. 6.32. Traverse AFED can be solved to get the length and bearing of EF. The calculations are shown in Table 6.35.

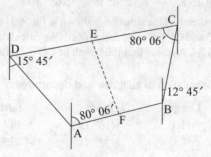

Fig. 6.32

From Table 6.35,

Latitude of EF = – 401.9,   departure of EF = 131.88

Length of EF = 422.99 m

Bearing of EF: tan θ = 131.88/(– 401.9) = 0.3281, θ = 18.16°

Bearing of EF = S18° 09′ 36″E

Summary

A traverse consists of a series of connected lines whose lengths and directions are measured in the field. An open traverse does not return to the starting point. Such a traverse is taken along rivers, railway lines, roads, etc. There is no method to check the accuracy of angular or linear measurements in an open traverse. A closed traverse returns to the starting point. A closed traverse is mathematically a closed polygon and should satisfy the geometrical conditions of a closed figure.

The angles measured in a traverse can be interior angles, external angles, or deflection angles. The geometrical condition for a closed traverse requires that the sum of the interior angles be equal to $(2n - 4)$ right angles, the sum of the exterior angles be equal to $(2n + 4)$ right angles, and the sum of deflection angles be equal to 360°, where n is equal to the number of sides of the traverse.

Closed traverse problems are essentially solved by determining the latitudes and departures of lines, which are projections of the lines on two mutually perpendicular directions. The latitude of a line is its projection on the meridian (north–south direction) and the departure of a line is its projection on the east–west direction. Latitudes and departures are calculated from the lengths and reduced bearings of lines. If l is the length and θ the reduced bearing, then latitude = $l \cos\theta$ and departure = $l \sin\theta$.

The condition for a closed traverse in terms of latitudes and departures is that the algebraic sum of the latitudes and the algebraic sum of the departures should each be zero. This gives two conditions for checking the accuracy of survey work or for finding missing quantities.

A traverse can be plotted using (a) bearings and lengths, (b) interior angles and lengths, (c) deflection angles and lengths, (d) tangent lengths, (e) chord lengths, or (f) coordinates. Plotting by coordinates is the most accurate method. When a closed traverse is plotted, it may not close, giving rise to what is known as the closing error. The closing error can be determined without plotting as $[(\Sigma L)^2 + (\Sigma D)^2]^{1/2}$.

A traverse is balanced and adjusted before plotting. Balancing is done using the following.

(a) Bowditch's (or compass) rule, in which the error is distributed in proportion to the length of the line, assuming that the lengths and angles are measured with equal precision.

(b) A graphical method equivalent to Bowditch's rule.

(c) Transit rule, which assumes that angles are measured more accurately than distances. In this method, the error in latitude is distributed proportionately to the latitude of lines and similarly for departures.

(d) Axis correction method, in which angles are again assumed to be measured more accurately. The directions of the lines are maintained while adjusting using this graphical method.

The two conditions that $\Sigma L = 0$ and $\Sigma D = 0$ for a closed traverse enable us to calculate any two unknown quantities (lengths or angles). This often helps a surveyor to calculate missing measurements or measurements that could not be taken due to difficulties in the field.

Exercises

Multiple-Choice Questions

1. An open traverse
   (a) is conducted in an open area
   (b) does not come back to the starting point while closing
   (c) is done with a prismatic compass
   (d) is done as a part of chain traversing

2. The latitude of a line is
   (a) the average latitude in which the line lies
   (b) the projection of the line on the east–west direction
   (c) the projection of the line on the north–south meridian
   (d) the projection of the line on any reference direction

3. The departure of a line of a traverse is
   (a) the deviation in the alignment of the line
   (b) the projection of the line on the east–west direction
   (c) the projection of the line on the north–south meridian
   (d) the projection of the line in any reference direction

4. If the latitude of a line is 123.65 m and its bearing is 155° 30′, then the length of the line is
   (a) 135.88 m   (b) 271.32 m   (c) 297.81 m   (d) 302.65 m

5. If the departure of a line is 87.75 m and its bearing is 283° 45′, then the length of the line is
   (a) 88.65 m   (b) 90.33 m   (c) 283.45 m   (d) 369.18 m

6. If the latitude and departure of a line are 135.8 m and 98.75 m, then the length of the line is
   (a) 145 m   (b) 167.9 m   (c) 198.75 m   (d) 203.25 m

7. If the latitude and departure of a line are 135.65 m and – 75.65 m respectively, then the bearing of the line is
   (a) 119° 09′   (b) 123° 54′   (c) 146° 06′   (d) 330° 51′

8. If the latitude of a line is 75.35 m and its length is 122.8 m, then its departure is
   (a) 65.35 m   (b) 81.17 m   (c) 96.96 m   (d) 104.66 m

9. If the departure of a line is 134.8 m and its bearing is 138° 45′, then its latitude is
   (a) –204.4 m   (b) –153.71 m   (c) +153.71 m   (d) +204.4 m

10. The latitude of a line with a bearing of 235° 30′ is – 102.65 m. The departure of the line is
    - (a) –181.23 m
    - (b) –149.35 m
    - (c) +149.35 m
    - (d) +181.23 m
11. The sum of included angles of a closed traverse of 5 sides is
    - (a) 360°
    - (b) 540°
    - (c) 900°
    - (d) 1260°
12. The sum of external angles of an *n*-sided traverse is
    - (a) *n* right angles
    - (c) 2*n* right angles
    - (b) (2*n* – 4) right angles
    - (d) (2*n* + 4) right angles
13. In a closed traverse of *n* sides, the sum of deflection angles is
    - (a) *n* right angles
    - (c) (2*n* – 4) right angles
    - (b) 2*n* right angles
    - (d) 4 right angles
14. For a closed traverse of 8 sides, the sum of external angles is
    - (a) 360°
    - (b) 720°
    - (c) 1440°
    - (d) 1800°
15. If we take a length of 20 cm and the deflection angle at a point is 30° 45′, then the chord length for plotting is
    - (a) 5.3 cm
    - (b) 10.6 cm
    - (c) 11 cm
    - (d) 38.5 cm
16. The bearing of line AB is 137° 30′ and its length is 125.8 m. If the coordinates of A are (0, 123.45), X-coordinates of B is
    - (a) –84.99
    - (b) +84.99
    - (c) + 92.74
    - (d) +208.44
17. The coordinates of point A of a traverse are (– 34.5, 83.7). If the length of line AB is 128 m and its bearing is S45° 30′E, then the coordinates of B are
    - (a) (55.21, 7.59)
    - (c) (125.79, –6.01)
    - (b) (56.79, –6.01)
    - (d) (55.21, 173.41)
18. If we take a length of 10 cm and the deflection angle at the point is 30° 45′, then the tangent length for plotting the line is
    - (a) 2.75 cm
    - (b) 5.3 cm
    - (c) 5.94 cm
    - (d) 11.9 cm
19. If measurements in a closed traverse are missing, then the conditions used for evaluating any two unknown quantities are
    - (a) sum of external angles and internal angles at all joints must be 4 right angles
    - (b) sum of deflection angles is 4 right angles
    - (c) algebraic sum of northings is equal to sum of westings
    - (d) algebraic sum of latitudes and algebraic sum of departures each must be zero
20. If some parameters of a closed traverse are missing, it is possible to determine
    - (a) any three quantities using the conditions of a closed traverse
    - (b) any number of quantities using the conditions of a closed traverse
    - (c) any two quantities if they are of adjacent lines
    - (d) any two quantities of a closed traverse
21. Bowditch's rule is especially applicable
    - (a) where linear measurements are more precise than angular measurements
    - (b) where angular measurements are more precise than linear measurements
    - (c) where linear and angular measurements are of equal precision
    - (d) in the case of all traverse adjustments
22. Bowditch's rule is applied as

    (a) Correction to any side $= \left(\dfrac{\text{closing error} \times \text{legnth of the side}}{\text{perimeter}} \right)$

    (b) Correction to latitude of a side $= \left(\dfrac{\text{total error in latitude} \times \text{latitude of line}}{\text{perimeter of the traverse}} \right)$

    (c) Correction to latitude of a line $= \left(\dfrac{\text{total error in latitude} \times \text{latitude of line}}{\text{arithmetic sum of latitudes}} \right)$

(d) Correction to latitude of any side $= \left(\dfrac{\text{error in latitude} \times \text{ length of the side}}{\text{perimeter of traverse}} \right)$

23. The transit rule for adjustment of traverse is eminently suited
    (a) where linear measurements are more precise than angular measurements
    (b) where angular measurements are more precise than linear measurements
    (c) where angular and linear measurements are of equal precision
    (d) irrespective of precision in linear and angular measurements

24. The transit method of adjusting a traverse applies the corrections as

    (a) Correction to any side $= \left(\dfrac{\text{closing error } \times \text{ legnth of the side}}{\text{perimeter}} \right)$

    (b) Correction to latitude of a side $= \left(\dfrac{\text{total error in latitude} \times \text{ latitude of line}}{\text{perimeter of the traverse}} \right)$

    (c) Correction to latitude of a line $= \left(\dfrac{\text{total error in latitude} \times \text{ latitude of line}}{\text{arithmetic sum of latitudes}} \right)$

    (d) Correction to latitude of any side $= \left(\dfrac{\text{error in latitude} \times \text{ length of the side}}{\text{perimeter of traverse}} \right)$

25. The axis correction method of adjusting a traverse is applicable where
    (a) linear measurements are more accurate than angular measurements
    (b) angular measurements are more accurate than linear measurements
    (c) both angles and lines are measured with equal precision
    (d) no assumption needs to be made of precision of measurements

Review Questions

1. Define the terms latitude and departure of a survey line. How are they calculated?
2. What are the geometrical conditions used to check the accuracy of angular measurements in a closed traverse?
3. Explain the different methods of plotting a traverse.
4. Explain the term closing error. How do you find its magnitude and bearing?
5. Explain with neat sketches the different methods of balancing a traverse.
6. Explain the difference between Bowditch's rule and the transit rule.
7. If the sum of the interior angles of a closed traverse is equal to $(2n - 4)$ right angles, is the traverse balanced? Explain.
8. Explain the procedure for evaluating missing quantities in a closed traverse.
9. If the bearings of two adjacent lines are missing, explain how you will determine these.
10. If the lines with missing parameters are not placed adjacently, how does one proceed to evaluate the unknown values?

Problems

1. If the latitude of a line is 184.96 m and its departure is − 214.44 m, find its length and bearing.
2. The length of a line is 358.62 m and its latitude is + 185.32. What are the possible bearings of the line?
3. The (x, y) coordinates of two points A and B are (−130.82, 234.92) and (62.98, 284.68), respectively. Find the length and bearing of the line.
4. The fore and back bearings of the lines of a traverse are given below. Correct the bearings and check the geometrical condition of interior angles.

Line	AB	BC	CD	DE	EA
Fore bearing	61° 12′	153° 24′	201° 02′	280° 14′	20° 30′
Back bearing	241° 12′	333° 24′	21° 02′	100° 14′	200° 20′

5. In an open traverse ABCDE, the bearing of line AB was measured as N45° 30′E. The deflection angles measured at B, C, D and E were 10° 12′ (R), 18° 10′ (L), 14° 45′ (L), respectively. If the bearing of line DE was measured to be N22° 47′E, check whether the observations agree with this.

6. From the traverse data given below, find the data required for plotting by (i) bearings, (ii) included angles, (iii) deflection angles, (iv) tangent lengths, and (v) coordinates.

Line	AB	BC	CD	DA
Length (m)	86.8	100.2	93.6	22.54
Bearing	320° 30′	64° 45′	201° 15′	183° 50′

7. From the traverse data given below, find the closing error, if any, and its bearing.

Line	PQ	QR	RS	SP
Length (m)	340.2	350.6	440.8	423.2
Bearing	70° 30′	120° 45′	223° 30′	320° 47′

8. From the traverse data given below, check whether the traverse closes. If not, balance the traverse using Bowditch's rule, both analytically and graphically.

Line	AB	BC	CD	DA
Length (m)	310.5	340.8	405.2	279.2
Bearing	S45°E	N50° 30′E	N54° 15′W	S33° 18′W

9. Prepare Gale's traverse table using the data of the closed traverse given below, after checking and balancing the traverse. The bearing of line AB observed was 222° 01′ 30″.

Line	AB	BC	CD	DA
Length	155.25	170.4	202.6	139.4
Station	A	B	C	D
Included angle	101° 39′ 30″	95° 32′ 50″	75° 15′ 30″	87° 32′ 50″

10. Find the length and bearing of line BC from the partial data available for traverse ABCDA.

Line	AB	BC	CD	DA
Length (m)	156.5	–	234.8	203.1
Bearing	78° 40′	–	251° 18′	3° 45′

11. Find the lengths of lines RS and ST of a traverse PQRSTP from the data given below:

Line	PQ	QR	RS	ST	TP
Length (m)	201.54	189.68	–	–	256.83
Bearing	62° 42′	154° 54′	202° 32′	281° 44′	22°

12. The bearings of two lines could not be measured due to field problems. Find these bearings from the remaining data available.

Line	AB	BC	CD	DE	EF	FA
Length (m)	151.24	258.52	270.32	121.8	203.7	174.2
Bearing	160° 33′	58° 48′	320° 47′	–	–	114° 59′

13. In conducting a traverse ABCDEA, the length of line DE and the bearing of line EA could not be measured. Find these missing measurements from the data recorded.

Line	AB	BC	CD	DE	EA
Length (m)	100.77	94.84	115.97	136.27	–
Bearing	N62° 42′E	S25° 06′E	S22° 32′W	–	N22°E

14. In conducting a traverse ABCDEA, the lengths of two lines BC and DE could not be measured. From the available data, find the lengths of these two lines.

Line	AB	BC	CD	DE	EA
Length (m)	186.8	–	193.6	–	151.0
Bearing	N39° 30′W	N64° 45′E	S22° 45′E	S3° 48′W	N58° 18′W

15. In a traverse survey notebook, the bearings of two lines were not recorded. Find the bearings of these lines from the data given below.

Line	PQ	QR	RS	ST	TP
Length (m)	302.31	284.52	347.91	408.81	385.26
Bearing	52° 12′	–	192° 02′	271° 14′	–

16. Find the length of BC and bearing of EA from the data given below.

Line	AB	BC	CD	DE	EA
Length (m)	282.2	–	324.7	381.6	359.6
Bearing	61° 30′	151° 24′	201° 02′	280° 14′	–

17. In surveying a triangle ABC, only the lengths of three sides, one included angle at A, and the bearing of AB could be recorded. Find the bearings of the remaining two lines from the available data.

Line	AB	BC	CA
Length (m)	102.5	87.4	154.8

The bearing of AB is N42° 45′E and the angle BAC = 32° 30′.

18. In order to continue the survey line beyond a hillock, two traverses were conducted, one from each side. The left traverse ABCD, starting from point A of the line OA, was completed giving point E in continuation of OA. The right traverse APQR was similarly conducted. Determine whether one or both of these traverses give the correct location of the point in continuation of OA. Bearing of OA = N62° 45′E.

Traverse ABCD.

Line	AB	BC	CD
Length (m)	150	600	186.2
Bearing	N6° 30′E	N60° 30′E	S66° 13′E

Traverse APQR

Line	AP	PQ	QR
Length (m)	160	700	110.5
Bearing	S61° 40′E	N60° 30′E	N39° 45′W

19. A survey line AC was laid out with a bearing of N36° 45′E. AC = 1000 m. Another survey line AD, 500 m long, was laid out with a bearing of N87° 30′E. Determine the length and bearing of DB and DC such that B is the midpoint of AC.

20. A and C are survey stations on either side of a wooded area. C is not visible from A and the distance AC cannot be measured. The survey line OA (bearing N48° 30′E) has to be continued across the area as CD. For this, a line AB was laid in the east direction for a length of 300 m. Line BC with a bearing of N10° 15′E was to be measured to locate C. Find the length BC so that C is on the continuation of line OA. If the coordinates of A are (3230.54, 6730.76) with reference to a general coordinate system, find the coordinates of C.

Measurement of Elevations—Levelling

Learning Objectives

After going through this chapter, the reader will be able to

- explain the basic principles of spirit levelling
- draw a neat sketch and explain the different parts of a levelling instrument
- explain the temporary adjustments of a level
- describe methods to fill data into a field book as well as the reduction of levels by the height of collimation and rise and fall methods
- describe the different methods of levelling, such as fly levelling, check levelling, and reciprocal levelling
- explain the basic principles of hypsometry and barometric levelling
- explain the effects of curvature and refraction on levelling
- explain methods to overcome the obstacles encountered during levelling
- explain the different types of errors and the avoidance of such errors

---

Introduction

In the last few chapters, we discussed the measurement of distances, magnetic bearings, and horizontal and vertical angles. The distances were measured as horizontal distances for plotting purposes and the angles were measured as angles in a horizontal plane. The vertical angles were measured as angles in a vertical plane to a point or as the difference in the vertical angles between points.

The points on the surface of the earth do not lie in a plane and there is a difference in the heights of the different points. A complete relief map of the surface will require the determination of such differences in heights and representing them on the plan to get a complete picture of the topography. The height of a point from a datum is known as the *elevation* and is represented by what are called *reduced levels*. The instrument for measuring the elevations is known as a *level* and the process of determining elevations is called *levelling*. A theodolite can be used as a level as well, as the essential parts of a level are built in the theodolite. However, there are separate levelling instruments which are used only for levelling. In this chapter, we will discuss the instruments and methods used to determine the elevations of points.

7.1 Terminology

The following terms should be understood clearly before we can take up the instruments and method of levelling (refer to Fig. 7.1).

Level surface   It is a surface parallel to the mean spheroidal surface of the earth at every point. The water surface in a lake when it is still is a perfect example of a level surface. The level surface does not lie in a horizontal plane. Because of the distances involved, a horizontal surface is considered a level surface in ordinary survey work. Only when the distances are large, we take into consideration the curvature of the surface.

Level line   It is a line lying on the level surface.

Horizontal plane   A horizontal plane through a point is a plane tangential to the level surface passing through that point. It is perpendicular to the plumb line or gravity line.

Horizontal line   It is a line lying in the horizontal plane. It is tangential to the level surface and perpendicular to the plumb line.

Vertical line   It is a line perpendicular to the level surface and lies along the plumb line through that point.

Vertical plane   It is a plane containing the vertical line.

Datum surface   It is a level surface whose elevation is known or assumed. In the levelling operation, to get the elevations of points, heights are taken from the datum surface in a vertical plane.

Elevation   The elevation of a point is the height in a vertical plane from the datum surface.

Difference in elevation   The difference in elevation between two points is the vertical distance between the level surfaces through the two points.

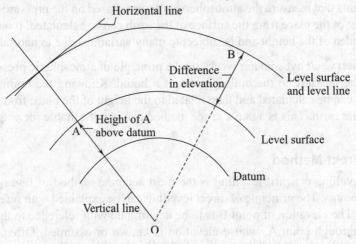

Fig. 7.1   Levelling terminology

Reduced level   The reduced level of a point is the height of the point obtained by adding the known or assumed datum surface elevation and the elevation of the point from the datum surface.

Benchmark It is a fixed reference point through a level surface of known elevation. In ordinary survey work, a benchmark can be arbitrarily chosen and its elevation assumed. The GTS benchmark refers to the mean sea level at a point by collecting data of sea level elevations over long periods of time. Such benchmarks are available at many places in India for reference purposes.

7.2 Methods of Finding Elevations

There are many methods of determining the elevations of points. In survey work the direct method of finding heights is commonly used. The following methods are available based on different principles:

Direct method The direct method of levelling is also known as *spirit levelling*. A spirit level is a level tube with a bubble which when centred makes for a horizontal surface tangential to the level surface at a point. A telescope attached to the spirit level then gives a horizontal plane as it moves in different directions. The horizontal plane is then taken as a reference from which heights can be measured with a graduated rod known as a *levelling staff*. When the instrument is set up at different stations, horizontal parallel planes are obtained at instrument stations. These can be related to each other by taking common readings to the same point from two different instrument stations. This is the most common method used in surveying and is very accurate compared to other methods. In the case of long sights, the curvature of the level surface can also be taken into account and the level measurements from the horizontal surface corrected to get measurements from the true level surface.

Barometric levelling This method works on the principle that the atmospheric pressure decreases as we go higher from the surface of the earth. Barometers are instruments that measure the atmospheric pressure. Based on the pressure reading, the height of the place from the surface of the earth can be calculated. It only gives a rough idea of the height and is subject to many variations due to natural causes.

Hypsometry A hypsometer works on the principle of atmospheric pressure, but relates the pressure to the boiling point of a liquid. Knowing the boiling point, pressure can be calculated and then related to the height of the place from the surface of the earth. This is again a crude method and is not suitable for engineering surveys.

7.3 Direct Method

Direct levelling or spirit levelling is the most accurate method of finding elevations of points. The principle of direct levelling can be explained with reference to Fig. 7.2. The elevation of point B is to be determined with reference to the datum surface through point A, whose elevation is known or assumed. Often we will be interested only in the difference in elevations between points and not in their absolute elevation with reference to the nationally accepted mean sea level. The following procedure is adopted.

1. Set up a levelling instrument (essentially consisting of a spirit level and a telescope) at point C. It must be clearly understood that *the instrument station, or where the instrument is set up, is not the point whose elevation is*

determined. Set up the instrument at a station for convenience of observation to the points whose elevations are to be determined.

2. Centre the spirit level so that the bubble traverses (remains at the centre). In this position, the telescope, when rotated horizontally, traces a horizontal plane and the line of sight remains in that plane in all directions.

3. Keep a measuring rod at A and take a reading on the measuring rod with the horizontal hair of the telescope.

4. Now turn the telescope to sight point B. Keep the measuring rod vertically at B and take a reading corresponding to the horizontal hair of the telescope.

5. Knowing the two readings, determine the elevation of point B with respect to point A.

The methods for determining the elevations of points will be discussed in detail later. Only an idea of the technique is given here. There are two methods to find the elevation of point B. Let the assumed elevation of point A be 100 and the readings on the vertical measuring rod at points A and B as observed be 2.345 m and 1.368 m, respectiely.

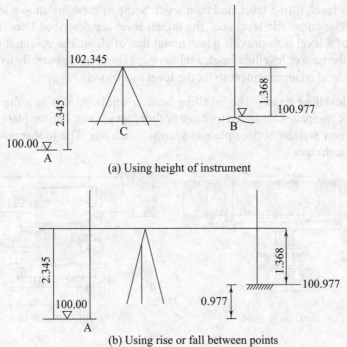

(a) Using height of instrument

(b) Using rise or fall between points

Fig. 7.2 Basic principle of spirit levelling

In the first method, we find the height of collimation or the elevation of the line of sight. *The height of collimation provides a horizontal plane to which all readings taken from that instrument station can be related.* The reading on point A being 2.345 m, the line of sight of the telescope has an elevation of $100 + 2.345 = 102.345$ m. This is evident from Fig. 7.2. The reading on B being 1.368 m, point B is 1.368 m below the line of sight of the telescope. The elevation of point B is, therefore, $102.345 - 1.368 = 100.977$ m. The elevation of point B is thus determined.

The second method uses the principle that *when readings are taken from the same line of sight to two points, one point is above or below the other depending upon whether the reading is less or more*. The reading at A is 2.345 m and at B is 1.368 m. Point B is higher because the reading is less than that at A *with the line of sight in the same horizontal plane*. Figure 7.2(b) makes it clear. This is known as the rise of point B above A. The rise is 2.345 – 1.368 = 0.977 m. Thus, point B is 0.977 m above A and its elevation is 100 + 0.977 = 100.977 m, as before. This is the principle of direct levelling. As we have seen, two instruments are required for this operation—(i) An instrument that provides the horizontal line of sight in all directions and (ii) a vertical measuring rod. One is the levelling instrument or simply the level and the other is the levelling rod or levelling staff. These will now be discussed in detail.

7.4 Levelling Instruments

There are many types of levelling instruments each with some minor modification for ease of operation. The main types of levels are dumpy level, reversible level, Cushing's level, tilting level, and auto level. Some of these are shown in Figs 7.3 and 7.4. The automatic level and the digital level are described later. The basic purpose of a level is to provide a horizontal line of sight. The essential parts of a level are the base or levelling head, a telescope, a spirit level generally fixed on the telescope, and an arrangement to fix the level to a tripod.

Base or levelling head   The levelling head is similar to that in a theodolite. It consists of two plates separated by three or four foot screws. The top plate is known as a *tribrach* and the bottom plate is known as a *trivet*. The foot screws help to level the instrument.

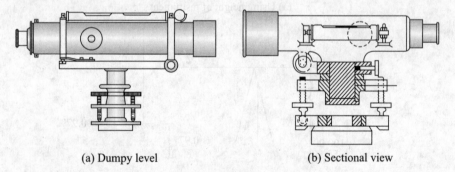

(a) Dumpy level   (b) Sectional view

Fig. 7.3   Dumpy level

Telescope   The telescope has been described in Chapter 1. With improvement in optics, telescopes have become shorter and compact in size and are more powerful with a higher range. The telescope is used to sight the levelling staff to take readings.

Altitude bubble   The bubble is generally fixed onto the telescope. The spirit level can be made to traverse, or remain in the centre of its run, by the foot screws. Initial adjustment is done with the tripod legs for approximate levelling. The base plate or trivet has arrangements to fix the level onto a tripod.

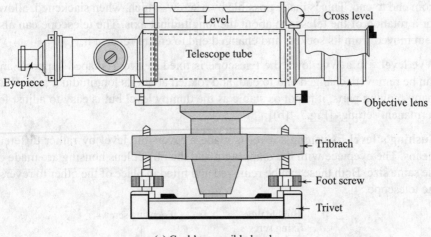

(a) Cook's reversible level

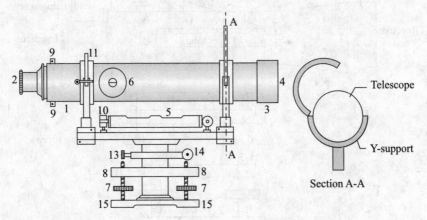

(b) Wye level

1. Telescope
2. Eyepiece
3. Ray shade
4. Objective end
5. Bubble tube
6. Focusing screw
7. Foot screw
8. Tribrach
9. Diaphragm adjusting screws
10. Bubble tube adjusting screws
11. Wye clip
12. Clip half open
13. Clamp screws
14. Tangent screw
15. Trivet stage

Fig. 7.4 Other forms of levels

Dumpy level A dumpy level (Fig. 7.3) is a very solid and old type of instrument with very few moving parts and is considered a stable instrument. In this instrument, the telescope is fixed to a vertical spindle and cannot be tilted or otherwise moved. In a solid dumpy level, the telescope tube and the vertical spindle are cast in one piece.

Cook's reversible level In Cook's reversible level [Fig. 7.4(a)], the essential parts remaining the same, a provision is made to rotate and change the telescope from end to end. This is made possible by a screw, which, when slackened, allows for a rotation of the telescope about the longitudinal axis. The telescope can also be removed from its sockets and changed end to end to reverse it.

Wye level   In a Wye level the telescope is fixed to two Y-shaped supports and can be removed, changed end to end, and rotated about its longitudinal axis. With many moving parts, it is not as stable as the dumpy level but is easy to adjust for permanent settings [Fig.7.4(b)].

Cushing's level   Cushing's level is made a reversible level by rather different means. The eyepiece with the diaphragm and the object lens housing are made of the same size. Both these can be removed and fitted in place of the other to reverse the telescope.

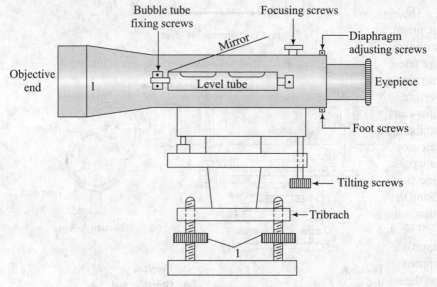

Fig. 7.5   Tilting level

Tilting level   The tilting level (Fig. 7.5) is a later addition to the family of levels. In a tilting level, the fundamental conditions that the vertical axis should be vertical and the line of collimation perpendicular to it are done away with. This is made possible by making a provision to tilt the telescope about the horizontal axis with a fine-pitched screw. Even if the vertical axis is not vertical, the line of collimation can be made horizontal with this screw. Levelling takes less time, as only approximate levelling of the instrument is required either with a circular bubble or with the telescope bubble. Whenever a reading is taken, the bubble is centred to make the line of collimation horizontal. If many readings are to be taken from one station, the advantage is not apparent. However, the operation can be completed speedily when precise levelling is required to transfer benchmarks or to determine the reduced level of a point by fly levelling.

7.4.1   Modern Levelling Instruments

We have discussed many conventional levels in the foregoing paragraphs. Levelling instruments have undergone constant development over the years. The objective of such development is to speed up the operation of levelling and reduce the operational fatigue. The modern instruments that are a result of this progress provide far more accurate data, acquired with greater ease. In this section, we will discuss two such instruments.

Automatic levels

As the names suggests, automatic levels [Fig. 7.6(a)], though similar to dumpy levels in construction, eliminate the need for the manual adjustment required in the case of dumpy levels to ensure accurate levelling. These instruments use a compensator mechanism to keep the line of sight horizontal even if it is not perpendicular to the vertical axis. The compensator mechanism made of mirrors or prisms is built into the telescope tube and can be activated if the line of sight is horizontal within 15′ to 30′ of the true horizontal.

The compensator mechanism is shown in Fig. 7.6(b). It consists of two mirrors or prisms (F) fixed to the barrel of the telescope. Another prism or mirror (S) is suspended by two non-magnetic wires and moves by the action of gravity when the telescope is tilted. A ray of light gets reflected from the fixed and suspended mirrors or prisms and always remains horizontal. A damping device with counterweight is provided to prevent excessive oscillations of the system. The mechanism also consists of a circular bubble, which is brought approximately to the centre by adjusting the foot screws. This is required to activate the compensator when the telescope is not perfectly level. One can also check whether or not the compensator is working either by rotating the foot screws or by pressing a button provided on one side of the telescope. When the compensator is working, the action of tilting using the foot screws or the button brings the line of sight back to the same reading, after a momentary shift.

The automatic level is very easy to operate. The circular level is brought approximately to the centre of its run using the foot screws. Once the telescope is approximately level, the compensator is active and one has only to take readings on the staff kept at different points.

Electronic or digital levels

This is the latest form of the level, which eliminates the need to read the staff and record readings—two operations that are inherently error-prone.

The electronic level also uses a compensator mechanism with prisms as in the automatic level. A circular level is approximately centred to fix the telescope axis within the range of the compensator.

Electronic levels are used in conjunction with a special bar-coded staff. Software is built into the level to read the staff and send information to the digital display unit. Once the bar-coded staff is sighted, a button is pressed to take the reading. The reading is taken, stored, and displayed on a console screen in a matter of seconds.

The built-in software programs enable distances and elevations to be calculated and displayed. Many other facilities may also be provided depending upon the manufacturer. For example, a whole day's operations can be stored in memory modules, to be retrieved later. The data can also be fed to a computer through an RS-232 cable for further processing.

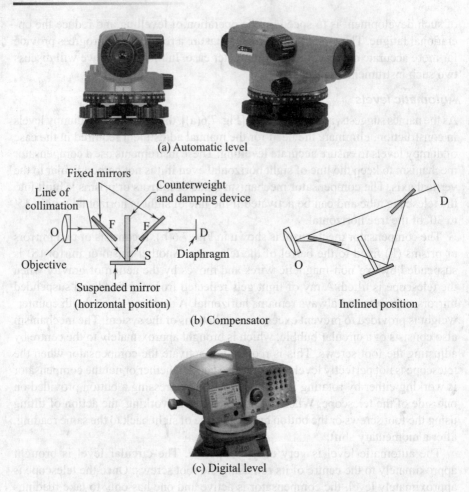

(a) Automatic level

(b) Compensator

(c) Digital level

Fig. 7.6 Modern levelling instruments, also see Plate 1

Electronic levels are used for all types of levelling operations—in precise levelling and in tacheometry. They speed up operations and eliminate most operator and manipulative errors. An electronic level is shown in Fig. 7.6(c)

DL100 Digital level Topcon's DL100 series levels are very powerful and accurate digital levels for many uses. These levels permit fully electronic reading of levels and calculation of elevation and heights with the press of a button. The levels are provided with a circular vial for levelling. The levels have compensator mechanisms to take care of small tilt in the axis.

Both electronic and optical readings are possible. For electronic reading, special bar-coded staff is used. The levelling operation consists of sight, focus, and measure by pressing a button.

The operating functions and software in the instruments include the following:
- n-times measurement giving average value and standard deviation
- Horizontal distance measurement to staff for contouring
- Height determination of intermediate points

- Calculation of difference in elevation
- Selectable minimum unit for reading
- Manual input of data or reading from memory cards
- Alphanumeric input function
- Swing correct function to reduce the effect of vibration
- Inverse staff mode, when the staff is held upside down

The level is suitable for precise levelling work as in first to fourth order levelling. The level is also very useful in topographic surveys, tunnel surveys, and mine surveying.

Salient features

The salient feature of the digital instruments are as follows:
- Compensator ± 12′; setting accuracy is ± 3″
- Height measurement accuracy – Electronic reading 0.4 mm with invar staff and 1.0 mm with fibre glass staff.
- Optical reading accuracy is 1.0 mm.
- Least count ranges between 0.1 mm and 0.01 mm.
- Measuring range is 100 m with fiberglass or aluminium staff and 60 m with invar staff.
- Circular vial sensitivity is 8″/2 mm division.
- The instrument can be powered by normal batteries or rechargeable battery pack with a working time of 10 hours.
- Staves are available in 5 m lengths made of fiberglass and aluminium and 3 m of invar.
- Sokkia's SDL30 digital level is similar in performance and specifications.

Topcon DL500 series level

This new series has a wave and read technology which allows the rodman to wave the rod. The instrument automatically senses the minimum reading when the rod is plump. The instrument features a pendulum compensator with magnetic damping. It has a 2000 data memory and can read electronically (with a RAB coded staff or optically with fibreglass or other types of staff.)

7.4.2   Levelling Staff

The levelling staff is the second instrument required for direct levelling. Levelling staves are of many types: single-piece staff, telescopic staff, folding staff, target staff, and so on. Levelling staves have to be prominent to be visible from a distance and as such are marked alternatively black and white. IS1799:1961 covers the specifications for levelling staves.

Solid staff   The metric solid staff is generally 3 m long and marked in centimetres or in 5 mm lengths. Each rectangular marking is either 10 mm or 5 mm long. Staves marked to smaller divisions are also available. They are generally made of good-quality wood of pine, deodar, etc. The markings are as shown in Fig. 7.7(e).

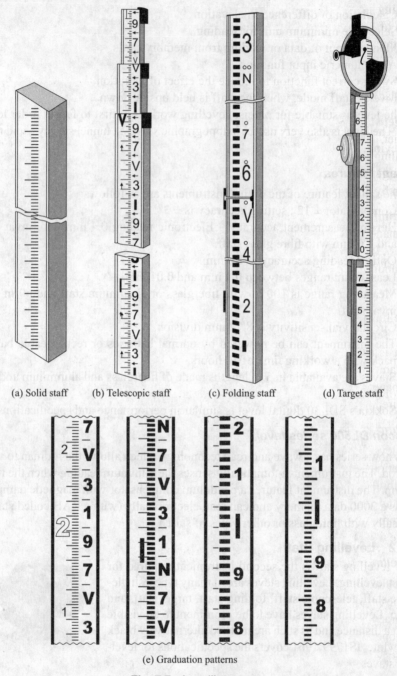

(a) Solid staff   (b) Telescopic staff   (c) Folding staff   (d) Target staff

(e) Graduation patterns

Fig. 7.7 Levelling staves

Telescopic staff A telescopic staff is made generally in three lengths of 1.5 m, 1.5 m, and 2 m. It is thus longer than a solid staff. Each length slides into the other so that the section of the shaft decreases from length to length [Fig. 7.76(b)]. The markings are done as in the case of a solid staff.

Folding staff A folding staff is generally 4 m long and is divided into two equal parts of 2 m each. It is made of well-seasoned wood about 75 mm wide and 18 mm thick. The staff is easy to carry as the graduated portion folds into the other [Fig. 7.7(c)]. A lock is provided so that the two portions do not fall apart automatically. Sometimes, provision is also made to separate them into single pieces. A circular bubble may be provided to check the verticality of the staff.

Target staff A target staff is generally 3 m long having a movable target with a vernier. The vernier helps to read up to one-tenths of the staff graduations. During levelling, the target is moved by the staff man till the level bisects the centre line of the target. Upon getting directions from the level man, the staff is fixed and readings taken by the staff man from the staff reading and the vernier [Fig. 7.7(d)].

Aluminium staff Levelling staves made of aluminium are also available. They are made of aluminium alloy and are much lighter in weight. The graduations [Fig. 7.7(e)] are etched with special weather-resistant and corrosion-resistant inks on anodized sections. They are available in solid and telescopic types.

7.5 Temporary Adjustments

Levelling instruments require two types of adjustments—temporary and permanent. Temporary adjustments are those that need to be done whenever the instrument is set up to take readings. Permanent adjustments are done to adjust the fundamental lines once in a while. Temporary adjustments are discussed here. Permanent adjustments are discussed in a later chapter. The temporary adjustments required for a levelling instrument are the following.

Setting up

The level is set up at a convenient point for ease of observation of stations whose elevations are to be measured. The instrument station has no other significance in levelling. The instrument is set up on a tripod. There is no need for centring the instrument in levelling. The tripod legs should be set well apart for stability of the instrument. They should be firmly fixed to the ground so that the tripod does not move when the telescope is turned.

Levelling the bubble tube

Levelling is a basic operation and should be carefully performed at every setting of the instrument. The whole operation of finding elevations depends upon the accuracy of levelling the spirit level. Levelling has been discussed in Chapter 4. The operation is the same in the case of levels also. Instruments may have either a three-screw or a four-screw levelling head.

Three-screw levelling head

After setting up and centring the instrument, levelling is done. The following procedure is adopted.

1. Swing the instrument and bring the telescope bubble parallel to any two of the foot screws. Centre the bubble by rotating the foot screws. For this, hold the

foot screws by the thumb and forefinger of each hand and *rotate both either inwards or outwards* [see Fig. 7.8(a)]. Also note that the bubble moves in the direction of movement of the left thumb during this operation.

2. Once the bubble traverses (or comes to the central position as seen from the graduation of the tube), swing the instrument and bring the bubble over the third foot screw. In this position, the bubble tube is at a right angle to the earlier position. Centre the bubble by rotating the third foot screw alone.

3. Bring the plate bubble to its previous position by swinging the instrument back. Check whether the bubble traverses. If not, centre the bubble using the two foot screws as before.

4. Repeat the procedure till the bubble traverses in both these positions.

5. Swing the instrument through 180° and check whether the bubble traverses. The bubble should traverse in all positions if the instrument has been properly adjusted.

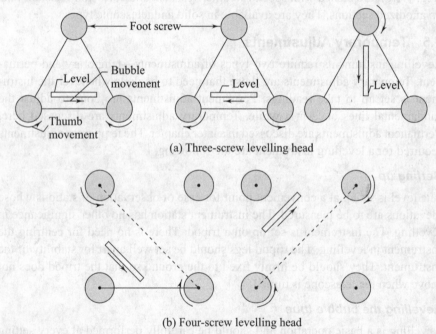

(a) Three-screw levelling head

(b) Four-screw levelling head

Fig. 7.8 Levelling operation

Four-screw levelling head

The following procedure is adopted with a four-screw levelling head.

1. After setting up and centring the instrument, bring the plate level parallel to any one pair of diagonally opposite foot screws. Operate these foot screws to centre the bubble [Fig. 7.8(b)].

2. Swing the instrument to bring the plate level parallel to the other pair of foot screws. Centre the bubble.

3. Swing the instrument back to the previous position. Check whether the bubble traverses, if not, centre it with the foot screws to which the level is parallel.

4. Swing it back, check the position of the bubble, and repeat the procedure.

5. Once the bubble traverses in the two orthogonal positions, swing it through 180°. The bubble should traverse in this position or in any other position.

Focusing the eyepiece

Focusing the eyepiece is the operation of bringing the cross hairs into focus. The focusing position varies with the eyesight of the observer. If the same observer takes the readings, this has to be done only once. To focus the eyepiece, do the following.

1. Keep a piece of white paper in front of the telescope, or direct the telescope towards a clear portion of the sky.

2. Looking through the telescope, adjust the vision by rotating the eyepiece till the cross hairs come into sharp and clear view.

3. If the eyepiece has graduations, note the graduation at which you get a clear view of the cross hairs. This can help in later adjustment if required.

Focusing the objective

The objective lens has to be focused whenever an object is sighted, as this depends upon the distance between the instrument and the object. There is a focusing screw on one side of the telescope, which is operated to focus the objective. This operation brings the image of the object in the plane of the cross hairs. This helps to exactly bisect the object, a levelling rod. To focus the objective, swing the instrument to bring the object into view by looking over the telescope. Rotate the focusing knob till the object is in sharp view along with the cross hairs.

7.6 Basic Levelling Operation and Terminology

We have briefly discussed the levelling operation. The instrument is set up at a convenient place for making observations to points whose elevations are to be determined. The basic operation remaining the same, the following points should be noted.

(a) The instrument is set up and levelled at every point it is kept. There is no centring of the level at these points.

(b) A number of observations can be taken from one instrument station. However, one instrument station may not be enough to cover all the points.

(c) Whenever the instrument is shifted from one instrument station to another, the elevation of the line of sight changes. To relate the elevations of the two consecutive lines of sight, readings are taken with the levelling staff kept at the same point from both instrument stations (see Fig. 7.9).

(d) Generally, the first sight is taken to a benchmark of known elevation. However, this is not necessary. The benchmark can be taken midway; any of the points where the staff is kept can be a benchmark or assigned an assumed elevation.

The following terminology is used in levelling operations and subsequent calculations for reduced levels (Fig. 7.9).

Staff station The staff station or simply station is a point where the levelling staff is kept for the purpose of determining its elevation.

Instrument station It is a point where the level is kept. This point has no special significance in the levelling operation. It is conveniently chosen for observing staff stations.

Backsight The backsight (BS) is a sight taken to a benchmark or a point of known elevation. Generally, the first sight taken after setting up the instrument at the start of the survey is always taken as a backsight. The first sight taken after shifting the instrument and setting it up at a new station is also recorded as a backsight.

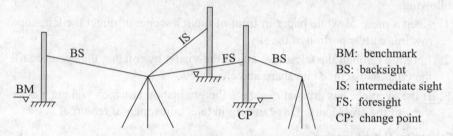

BM: benchmark
BS: backsight
IS: intermediate sight
FS: foresight
CP: change point

Fig. 7.9 Basic terms in recording and reducing levels

Intermediate sight Intermediate sights (IS) are those readings taken after a backsight but before the last sight with the same instrument station. Therefore, with the same instrument set-up, sights other than the first and last sights are classified as intermediate sights.

Foresight The forsight (FS) is the last sight taken before the instrument is shifted. The last reading taken in a levelling operation is always classified as a foresight. The foresight (other than the last sight taken during the operation) is taken to a staff station, to which a backsight is also taken immediately after shifting the level.

Change point (CP) or turning point (TP) The change point or turning point is a staff station to which a foresight and backsight are taken. The foresight is taken before shifting the instrument and the backsight is taken after shifting the instrument. This helps to relate the elevations of the lines of sight of two consecutive instrument stations.

Height of instrument or height of line of collimation This is the reduced level of the line of collimation with reference to the datum. It can be calculated from the staff reading taken to a staff station of known or assumed elevation. It has no relation to the height of the instrument from the ground on which it is set up [Fig. 7.10(a)].

Rise When point A is above point B, then A is said to have a rise over B [Fig. 7.10(b)]. The rise can be determined from the readings taken to A and B from the same height of collimation. If A has a rise over B, then the reading at A will be less than that at B with the same height of collimation.

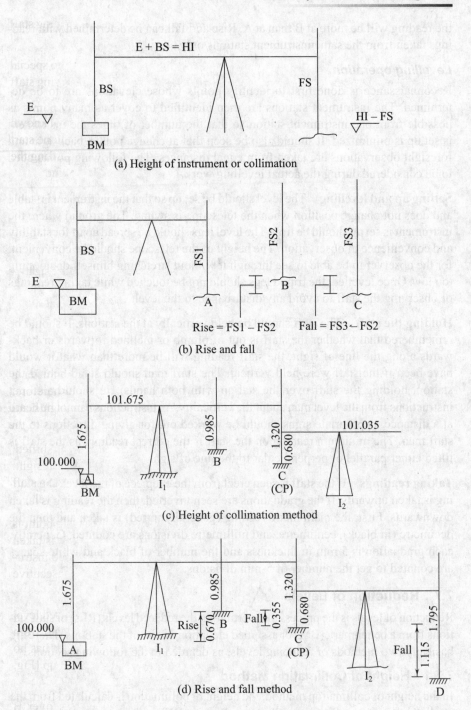

(a) Height of instrument or collimation

$E + BS = HI$

(b) Rise and fall

Rise = FS1 – FS2   Fall = FS3 – FS2

(c) Height of collimation method

(d) Rise and fall method

Fig. 7.10 Basic principles of reducing levels

Fall   It is the reverse of rise. When point B has a fall over A, point B is below the level of point A on the ground [Fig. 7.10(b)]. With the same height of collimation, the reading will be more at B than at A. Rise and fall can be determined with readings taken from the same instrument stations only.

Levelling operation

Reconnaissance is done first to identify points whose elevations are to be determined. The instrument stations are then identified to cover as many points as possible from one instrument station so that the number of times the instrument is set up is minimized. It should also be seen that at change points, backsight and foresight observations are taken from equal distances. The following points need to be considered during the actual levelling work.

Setting up and levelling   The level should be set up so that the instrument is stable and does not change position when the telescope is swung. The ground where the instrument is set up should be firm. The level legs should be spread apart for stability and convenience of observation. The height of the telescope should be convenient, for the observer to be able to see through it without stretching himself or stooping too low. Once levelled, the tripod legs should not be touched while taking readings or observing the staff to avoid any disturbance to the level.

Holding the staff   The staff should be held vertically at the stations. It should be remembered that whether the staff is out of plumb or inclined forwards or backwards along the line of sight, the staff reading will be more than what it would have been if the staff were held vertical. The staff man should stand behind the station, holding the staff over the station with both hands. He should also get instructions from the level man about the verticality. As instructions cannot be heard at a distance, appropriate signs should be worked out for giving directions to the staff man. The minimum reading on the staff is the correct reading, as the staff is tilted either parallel or perpendicular to the line of sight.

Taking readings   If the staff is seen erect from the eyepiece of the level, the reading is taken upwards. If the graduations are seen inverted, then the reading is taken downwards. First, the main metre reading (generally in red) is taken and then the decimetre (in black), centimetre, and millimetre divisions are counted. Generally, each graduation is 5 mm in thickness and the number of black and white spaces are counted to get the number of 5-mm divisions.

7.7   Reduction of Levels

Reduction of levels is the process of determining the reduced levels (RLs) of staff stations from a benchmark (BM) or assumed elevation of one of the stations. There are basically two methods of reducing levels, as detailed in the following sections.

7.7.1   Height of Collimation Method

In the height of collimation method, the height of collimation is calculated from the reading taken to a staff station of known elevation. When the instrument is shifted, the height of collimation changes. The two heights of collimation, one of which is known, are related by the foresight and backsight taken to a change point. The new height of collimation is calculated from the backsight and foresight. The following steps are involved in reduction of levels by this method.

1. Arrange the readings in tabular form, as shown in Table 7.1.

2. Table 7.1 shows only sample readings. There may be a large number of readings and a number of change points.

3. Columns 1 to 4 show data obtained from fieldwork and filled in from the field book; the form might be filled in during fieldwork itself. The first reading is always a backsight; enter the last reading as a foresight. There is one intermediate sight and one change point in the case shown.

Table 7.1 Sample format for height of collimation method of reducing levels

Staff station	Backsight	Intermediate sight	Foresight	RL of line of sight	Reduced levels	Remarks
A	1.675			101.675	100.00	BM
B		0.985			100.69	
C	0.680		1.320	101.035	100.355	CP
D			1.795		99.240	

4. A is given as an arbitrary benchmark of assumed elevation 100.000. Knowing the staff reading on a benchmark (a point of known elevation), calculate the RL of the line of sight or the height of collimation as $100.000 + 1.675 = 101.675$m [see Fig. 7.10(c)].

5. Consider the staff readings at B (an intermediate sight) and the foresight to C (a change point) from this line of collimation. To find the RLs of B and C, subtract these two readings from this height of collimation. Thus,

   Reduced level of B = $101.675 - 0.985 = 100.690$ m

   Reduced level of C = $101.675 - 1.320 = 100.355$ m

It may be seen that intermediate sights and foresights are deducted from the RL of the line of collimation.

6. C is a change point. After taking the foresight on C, shift the instrument and take a backsight on the same point C. Enter the readings to the same point in the same row, as shown in Table 7.1.

7. C is a point of known elevation now (it is like a benchmark). Repeat the same process as with the first line of collimation. Find the height of collimation by adding the RL of C and the backsight reading. RL of the height of collimation = $100.355 + 0.680 = 101.035$ m. Find the RL of D by subtracting 1.795 (reading on D) from 101.035 (height of collimation). RL of D = $101.035 - 1.795 = 99.240$ m.

This method is very simple to use. A set of readings, one backsight, intermediate sights, and one foresight are taken with one set-up (line of collimation). Similar sets are taken with other lines of collimations (other instrument stations) and are treated in the same way. A check on the calculation by this method is the following:

   Sum of backsights – sum of foresights = last RL – first RL

From the sample readings in Table 7.1,

Sum of backsights = 2.355,   sum of foresights = 3.155 m

Difference = 2.355 – 3.115 = – 0.76 m

Last RL – first RL = 99.240 – 100.000 = – 0.76 m

In this check, the intermediate sights are not involved and the calculations involving their values are not checked.

7.7.2 Rise and Fall Method

The second method for reducing levels is the rise and fall method [Fig. 7.10(d)]. The basic principle has already been discussed. *One staff station is above or below the other depending upon whether the reading is less or more, respectively. A higher point is said to have a rise and a lower point is said to have a fall.*

The following procedure is adopted to reduce levels by this method.

1. Arrange the readings in tabular form, as shown in Table 7.2.

Table 7.2   Rise and fall method of reducing levels

Staff station	Backsight	Intermediate sight	Foresight	Rise	Fall	Reduced level	Remarks
A	1.675					100.000	BM
B		0.985		0.69		100.69	
C	0.680		1.320		0.335	100.355	CP
D			1.795		1.115	99.240	

2. Columns 1 to 4 contain data collected from fieldwork. In this method, the height of collimation is not calculated. The differences in level between successive points are calculated and classified as rises or falls.

3. Rises and falls can be calculated with readings *from the same instrument station only. A reading is subtracted from the previous reading. If it is negative, it is a fall, and if positive, it shows a rise.*

4. The reading on A is 1.675 m and that on B is 0.985 m. As the reading on B is less, B is a higher point than A. B has a rise over A. The rise is 1.675 – 0.985 = 0.69 m.

5. Compare B and C. The reading on C, taken from the same instrument station from which the reading is taken on B, is 1.320. Therefore, compare 0.985 and 1.320 (and not 0.680, which has been taken from another instrument station). Point C is lower than B as the reading on C is higher. The fall is 1.320 – 0.985 = 0.335 (Or, subtracting the reading from the previous reading, 0.985 – 1.320 = – 0.335. The difference being negative, it is a fall).

6. Now compare C and D. Note that with the above principle, the readings to be compared are 0.680 and 1.795, which are taken with the same line of sight. The fall of D from C is 1.795 – 0.680 = 1.115.

7. Having calculated the rise or fall of different stations, the calculation of reduced levels is very simple. To the known reduced level of 100.000 of point A, add the rise of point B; from the reduced level of B, deduct the fall of C to get the RL of C; and so on. Thus,

RL of B = 100.000 + 0.69 = 100.690 m

RL of C = 100.69 – 0.335 = 100.335 m

RL of D = 100.335 – 1.115 = 99.240 m

This method also has an arithmetic check as follows:

Σ rise – Σfall = Σbacksight – Σforesight = last RL – first RL

Σ rise – Σfall = 0.69 – 1.450 = – 0.76

Σ backsight – Σforesight = 2.355 – 3.115 = – 0.76

Last RL – first RL = 99.240 – 100.000 = – 0.76

The rise and fall method has a better check on calculations, including all readings. The following examples illustrate both the methods described above.

Example 7.1 Readings taken in order during a levelling work are given below. Tabulate them suitably and find the reduced levels of all the points if the RL of the first point A is assumed to be 100.000.

Staff station	A	B	C	D	E	F	G
BS	0.684		0.864			2.845	
IS		1.246		1.684	0.964		
FS			1.105			1.368	0.748

Solution *Height of collimation method* Table 7.3 shows the solution by the height of collimation method. Arrange the staff stations and readings as shown in columns 1 to 4. The procedure for the solution is briefly outlined below. Note that there are two change points at C and F. Thus there are three heights of collimation.

1. Find the height of collimation by adding 100 + backsight to A = 100.684. RL of B = 100.684 – 1.246 = 99.438; RL of C = 100.684 – 1.105 = 99.579. This completes the readings with the first instrument station.

2. Shift the instrument after taking a foresight to C. We find the second height of collimation by adding the backsight to C to the RL of C. 99.579 + 0.864 = 100.443. Now we find the RLs of the remaining observations with this line of sight. RL of D = 100.443 – 1.684 = 98.759; RL of E = 100.443 – 0.964 = 99.479; RL of F = 100.443 – 1.368 = 99.075.

3. Shift the instrument and take a backsight to F. Find the new line of collimation from the RL of F and the backsight to F. Height of collimation = 99.075 + 2.845 = 101.920. RL of G = 101.920 – 0.748 = 101.172.

4. Arithmetic check: Sum of backsights = 4.393, sum of foresights = 3.221, difference = 4.393 – 3.221 = 1.172.

Last RL – first RL = 101.172 – 100.000 = 1.172

Rise and fall method The solution by the rise and fall method is shown in Table 7.4. The procedure is as follows.

Table 7.3 Solution of Example 7.1 by the height of collimation method

Staff station	Backsight	Intermediate sight	Foresight	Height of collimation	Reduced levels	Remarks
A	0.684			100.684	100.000	Assumed BM
B		1.246			99.438	
C	0.864		1.105	100.443	99.579	CP
D		1.684			98.759	
E		0.964			99.479	
F	2.845		1.368	101.920	99.075	CP
G			0.748		101.172	

Table 7.4 Solution of Example 7.1 by the rise and fall method

Staff station	Backsight	Intermediate sight	Foresight	Rise	Fall	Reduced level	Remarks
A	0.684					100.000	BM
B		1.246			0.562	99.438	
C	0.864		1.105	0.141		99.579	
D		1.684			0.820	98.759	
E		0.964		0.720		99.479	
F	2.845		1.368		0.404	99.075	
G			0.748	2.097		101.172	

1. From the staff readings, find the rise and fall of points. The rise and fall can be found from observations from the same instrument station. Thus, point B has a fall of 0.562 from A; point C has a rise of 0.141 with respect to B.
2. Shift the instrument and take a backsight to C. With the new line of sight, D has a fall of $1.684 - 0.864 = 0.820$. E has a rise of 0.720 above D. F has a fall of 0.404 from E.
3. Shift the instrument and take a backsight to F. The line of collimation changes. Thus, G has a rise of 2.097 from F.
4. Having calculated the rises and falls, it is a simple calculation to find the RLs by subtracting the falls and adding the rises. The result is shown in column 7 of Table 7.4. Thus, RL of $B = 100 - 0.562 = 99.438$, RL of $C = 99.438 + 0.141 = 99.579$, and so on.
5. Arithmetic check:

$$\Sigma \text{ rise} - \Sigma \text{ fall} = \Sigma \text{ BS} - \Sigma \text{FS} = \text{last RL} - \text{first RL}$$
$$\Sigma \text{ rise} = 2.958, \quad \Sigma \text{ fall} = 1.786, \quad \Sigma \text{ BS} = 4.393, \quad \Sigma \text{ FS} = 3.221$$
$$2.958 - 1.786 = 1.172, \quad 4.393 - 3.221 = 1.172$$
$$\text{Last RL} - \text{first RL} = 101.172 - 100 = 1.172$$

Hence the calculations are correct.

Example 7.2 The following staff readings were recorded in a levelling operation: 1.185, 2.604, 1.925, 2.305, 1.155, 0.864, 1.105, 1.685, 1.215, 1.545, and 0.605. A is the benchmark of reduced level 185.685 m. Find the RLs of all the other points by both the methods. The first reading was to point A and the instrument was shifted after the readings 2.604, 0.864, and 1.215.

Solution *Height of collimation method* The solution is given in Table 7.5.

Table 7.5 Solution to Example 7.2 by the height of collimation method

Staff station	Backsight	Intermediate sight	Foresight	Height of collimation	Reduced levels	Remarks
A	1.185			186.870	185.685	BM
B	1.925		2.604	186.191	184.266	CP
C		2.305			183.886	
D		1.155			185.036	
E	1.105		0.864	186.432	185.327	CP
F		1.685			184.747	
G	1.545		1.215	186.762	185.217	CP
H			0.605		186.157	

1. The instrument was shifted after the readings 2.604, 0.864, and 1.215. All these are foresight readings. The readings following these, 1.925, 1.105, and 1.545, are backsights. The first reading is a backsight and the last reading is a foresight. The remaining are intermediate sights.
2. Enter the readings as shown in Table 7.5 (columns 1 to 4).
3. The height of collimation for the first set-up is 185.685 + 1.185 = 186.870. RL of B = 186.870 − 2.604 = 184.266.
4. Now the instrument has been shifted; the new height of collimation = 184.266 + 1.925 = 186.191. Calculate the reduced levels of C, D, and E by subtracting 2.305, 1.155, and 0.864 from this height of collimation.
5. The instrument was shifted after the foresight at E. The new height of collimation = 185.327 + 1.105 = 186.432. Find the RLs of F and G by subtracting 1.685 and 1.215 from this height of collimation.
6. Now the instrument has been shifted; the new height of collimation = 185.217 + 1.545 = 186.762. RL of H = 186.762 − 0.605 = 186.157.
7. Arithmetic check: Sum of backsights = 5.760, sum of foresights = 5.288, difference = 0.472. Last RL − first RL = 186.157 − 185.685 = 0.472.

Rise and fall method   The solution by the rise and fall method is shown in Table 7.6. Columns 1 to 4 are filled as before.

1. Calculate the rises and falls from the readings. The rises are 2.305 − 1.155 = 1.150, 1.155 − 0.864 = 0.291, 1.685 − 1.215 = 0.470, and 1.545 − 0.605 = 0.940. The falls are 2.604 − 1.185 = 1.419, 2.305 − 1.925 = 0.380, and 1.685 − 1.105 = 0.580.
2. The reduced levels are obtained by adding the rises to and deducting the falls from the RL of the preceding point: 185.685 − 1.419 = 184.266, 184.266 − 0.380 = 183.886, 183.886 + 1.150 = 185.036, and so on.

Table 7.6 Solution to Example 7.2 by the rise and fall method

Staff station	Backsight	Intermediate sight	Foresight	Rise	Fall	Reduced level	Remarks
A	1.185					185.685	BM
B	1.925		2.604		1.419	184.266	CP
C		2.305			0.380	183.886	
D		1.155		1.150		185.036	
E	1.105		0.864	0.291		185.327	CP
F		1.685			0.580	184.747	
G	1.545		1.215	0.470		185.217	CP
H			0.605	0.940		186.157	

3. Arithmetic check: Sum of backsights = 5.760, sum of foresights = 5.288, difference = 0.472. Sum of rises = 2.851, sum of falls = 2.379, difference = 0.472. Last RL – first RL = 186.157 – 185.685 = 0.472.

Example 7.3 The staff readings taken during a survey in order are as follows: 2.365, 1.655, 0.695, 1.280, 2.355, 2.065, 1.755, 1.655, and 0.855. Tabulate the readings and find the reduced levels of all the stations. The instrument was shifted after the readings 0.695 and 2.065. The reading 0.695 was to a benchmark of elevation 201.655 m.

Solution   *Height of collimation method*   The readings are tabulated as shown in Table 7.7. The readings taken before shifting the instruments are foresights and the following readings are backsights.

Table 7.7   Solution to Example 7.3 by the height of collimation method

Staff station	Backsight	Intermediate sight	Foresight	Height of collimation	Reduced levels	Remarks
P	2.365			202.350	199.985	
Q		1.655			200.695	
R	1.280		0.695	202.935	201.655	BM RL 201.655
S		2.355			200.580	
T	1.755		2.065	202.625	200.870	CP
U		1.655			200.970	
V			0.855		201.770	

Height of collimation for the first set-up = 201.655 + 0.695 = 202.350
RL of P = 202.350 – 2.365 = 199.985
RL of Q = 202.350 – 1.655 = 200.695
Height of collimation for the second set-up = 201.655 + 1.280 = 202.935
RL of S = 202.935 – 2.355 = 200.580
RL of T = 202.935 – 2.065 = 200.870
Height of collimation for the third set-up = 200.870 + 1.755 = 202.625
RL of U = 202.625 – 1.655 = 200.970
RL of V = 202.625 – 0.855 = 201.770

Arithmetic check: Sum of backsights = 5.400, sum of foresights = 3.615; difference = 5.400 – 3.615 = 1.785. Last RL – first RL = 201.770 – 199.985 = 1.785.

Rise and fall method   The solution is shown in Table 7.8.

Table 7.8   Solution to Example 7.3 by the rise and fall method

Staff station	Backsight	Intermediate sight	Foresight	Rise	Fall	Reduced level	Remarks
P	2.365					199.985	
Q		1.655		0.71		200.695	
R	1.280		0.695	0.96		201.655	BM RL 201.655
S		2.355			1.075	200.580	
T	1.755		2.065	0.29		200.870	
U		1.655		0.10		200.970	
V			0.855	0.80		201.770	

The rises and falls are calculated as follows:

$$Q - 2.365 - 1.655 = 0.71 \text{ (rise)}, \quad R - 1.655 - 0.695 = 0.960 \text{ (rise)}$$
$$S - 2.355 - 1.280 = 1.075 \text{ (fall)}, \quad T - 2.355 - 2.065 = 0.29 \text{ (rise)}$$
$$U - 1.755 - 1.655 = 0.100 \text{ (rise)}, \quad V - 1.655 - 0.855 = 0.800 \text{ (fall)}$$

The staff station R is the benchmark of RL 201.655. The only thing to remember is that as you go up the table, the rise is subtracted and the fall is added. Thus, RL of Q = 201.655 − 0.96 = 200.695. Similarly, RL of P = 200.695 − 0.71 = 199.985.

As you go down the table, the rise is added and the fall is subtracted. Thus,

RL of S = 201.655 − 1.075 = 200.580

RL of T = 200.580 + 0.29 = 200.870

RL of U = 200.870 + 0.10 = 200.970

RL of V = 200.970 + 0.800 = 201.770

Arithmetic check: Sum of backsights − sum of foresights = 1.785, as calculated earlier. Sum of rise − sum of falls = 2.860 − 1.075 = 1.785. Last RL − first RL = 201.770 − 199.985 = 1.785.

Example 7.4 The following readings were taken during a levelling operation: 2.312, 1.185, − 2.365, 0.195, 0.885, 1.345, 0.905. The instrument was shifted after the fourth reading. The negative reading was taken with the staff inverted from a point below a sunshade. Find the reduced levels of all the points if the first reading was to a BM of assumed RL 100.000.

Solution *Height of collimation method* The solution by the height of collimation method is shown in Table 7.9.

The only special feature in this case is the inverted reading. The staff is held inverted for finding the reduced level of the underside of a beam or slab. The reading is taken with the inverted staff and should be given a negative sign. The rest of the procedure is the same as before.

Table 7.9 Solution to Example 7.4 by the height of collimation method

Staff station	Backsight	Intermediate sight	Foresight	Height of collimation	Reduced levels	Remarks
A	2.312			102.312	100.000	BM RL 100.000
B		1.185			101.127	
C		− 2.365			104.677	Staff held inverted
D	0.885		0.195	103.002	102.117	
E		1.345			101.657	
F			0.905		102.097	

The instrument is set up at two points only. The height of collimation for the first set-up is 100.000 + 2.312 = 102.312. With this, RL of B = 102.312 − 1.185 = 101.127, RL of C = 102.312 − (− 2.365) = 104.677, RL of D = 102.312 − 0.195 = 102.117. The height of collimation for the second set-up of the instrument is 102.117 + 0.885 = 103.002. With this height of collimation, RL of E = 103.002 − 1.345 = 101.657, RL of F = 103.002 − 0.905 = 102.097.

Arithmetic check: Last RL − first RL = 102.097 − 100.000 = 2.097. Sum of backsights − sum of foresights = 3.197 − 1.100 = 2.097.

Rise and fall method   The rises and falls are calculated as before. The solution is shown in Table 7.10.

Table 7.10   Solution to Example 7.4 by the rise and fall method

Staff station	Backsight	Intermediate sight	Foresight	Rise	Fall	Reduced level	Remarks
A	2.312					100.000	BM RL 100.000
B		1.185		1.127		101.127	
C		– 2.365		3.55		104.677	Staff held inverted
D	0.885		0.195		2.56	102.117	
E		1.345			0.46	101.657	
F			0.905	0.44		102.097	

The rises and falls are calculated as given below:

B: $2.312 - 1.185 = 1.127$ (rise),   C: $1.185 - (-2.365) = 3.550$ (rise)

D: $-2.365 - 0.195 = 2.560$ (fall),   E: $0.885 - 1.345 = 0.46$ (fall)

F: $1.345 - 0.905 = 0.44$ (rise)

The calculation of RLs is done as follows:

A: 100.000,   B: $100.000 + 1.127 = 101.127$

C: $101.127 + 3.55 = 104.677$,   D: $104.667 - 2.56 = 102.117$

E: $102.117 - 0.46 = 101.657$,   F: $101.657 + 0.440 = 102.097$

Example 7.5   The following staff readings were observed in sequence: 1.324, 2.605, 1.385, 0.638, 1.655, 1.085, 2.125, and 1.555. The instrument was shifted after the third and sixth readings. The third reading was taken to an arbitrary benchmark of elevation 0.000. Find the reduced levels of all the other points.

Solution   The readings are arranged as shown in Table 7.11. Note that the third reading is a foresight and the fourth reading is a backsight to the same staff station. Similarly, the sixth reading is a foresight and the seventh reading is a backsight.

Table 7.11   Solution of Example 7.5 by the height of collimation method

Staff station	Backsight	Intermediate sight	Foresight	Height of collimation	Reduced levels	Remarks
1	1.324			1.385	0.061	
2		2.605			– 1.220	
3	0.638		1.385	0.638	0.000	BM 0.000
4		1.655			– 1.017	
5	2.125		1.085	1.678	– 0.447	
6			1.555		0.123	

Height of collimation method   A brief explanation of the procedure for the solution is as follows.

1. Point 3 being a benchmark, the height of collimation for the first set-up is $0.00 + 1.385 = 1.385$. 0.638 is a reading taken to this station from the second set-up of the instrument and hence the height of collimation is $0.00 + 0.638 = 0.638$.
2. RL of station $1 = 1.385 - 1.324 = 0.061$, RL of station $2 = 1.385 - 2.605 = -1.220$.
3. From the second set-up of the instrument, RL of station $4 = 0.638 - 1.655 = -1.017$, RL of station $5 = 0.638 - 1.085 = -0.447$.
4. From the RL of station 5 and the backsight reading of 2.125, the height of collimation for the third set-up $= -0.447 + 2.125 = 1.678$. With this, RL of station $6 = 1.678 - 1.555 = 0.123$.
5. Arithmetic check: Sum of backsights $= 4.087$, sum of foresights $= 4.025$, difference $= 0.062$. Last RL $-$ first RL $= 0.123 - 0.061 = 0.062$.

Rise and fall method The solution by the rise and fall method is shown in Table 7.12.

1. The rises and falls are calculated as before:

   Point 2: $1.324 - 2.605 = -1.281$ (fall)

   Point 3: $2.605 - 1.385 = 1.220$ (rise)

   Point 4: $0.638 - 1.655 = -1.017$ (fall)

   Point 5: $1.655 - 1.085 = 0.57$ (rise)

   Point 6: $2.125 - 1.555 = 0.57$ (rise)

Table 7.12 Solution to Example 7.5 by the rise and fall method

Staff station	Backsight	Intermediate sight	Foresight	Rise	Fall	Reduced levels	Remarks
1	1.324					0.061	
2		2.605			1.281	−1.220	
3	0.638		1.385	1.220		0.000	BM
							0.000
4		1.655			1.017	−1.017	
5	2.125		1.085	0.57		−0.447	
6			1.555	0.57		0.123	

2. The reduced levels are calculated as follows: Point 3 is a benchmark of elevation 0.000. As mentioned earlier, as we go up the table, the rises are subtracted and the falls are added. Thus, RL of point $2 = 0.000 - 1.220 = -1.220$, RL of point $1 = -1.220 + 1.281 = 0.061$, RL of point $4 = 0.000 - 1.017 = -1.017$, RL of point $5 = -1.017 + 0.57 = -0.447$, RL of point $6 = -0.447 + 0.57 = 0.123$.
3. Arithmetic check: Difference between the sum of backsights and the sum of foresights $= 0.062$, difference between the last RL and the first RL $= 0.062$, sum of rises $= 2.36$, sum of falls $= 2.298$, difference $= 0.062$.

Example 7.6 The consecutive readings taken during a levelling operation are as follows: 0.685, 1.315, -1.825, -0.635, 1.205, 1.235, 2.631, 1.355, -2.015. The instrument was shifted after the third and sixth readings. The third reading was taken to a benchmark of assumed elevation 100.000. Find the reduced levels of other points.

Solution *Height of collimation method* The readings are entered as shown in Table 7.13. The third and sixth readings are foresights and the fourth and seventh readings are backsights.

Table 7.13 Solution to Example 7.6 by the height of collimation method

Staff station	Backsight	Intermediate sight	Foresight	Height of collimation	Reduced levels	Remarks
A	0.685			98.175	97.490	
B		1.315			96.860	
C	− 0.635		− 1.825	99.365	100.000	BM 100.000
D		1.205			98.160	
E	2.631		1.235	100.761	98.130	
F		1.355			99.406	
G			− 2.015		102.776	

1. The benchmark of 100.000 is taken to a point where the staff is held inverted. The point is, therefore, above the line of collimation. The height of collimation = 100.000 + (− 1.825) = 98.175.
2. RLs: point A: 98.175 − 0.685 = 97.49, point B: 98.175 − 1.315 = 96.860.
3. Obtain the height of the second line of collimation from the benchmark and the reading − 0.635 as 100 + (− 0.635) = 99.365.
4. RLs: point D: 99.365 − 1.205 = 98.160, point E: 99.365 − 1.235 = 98.130.
5. The height of collimation for the third set-up can be calculated from the RL of E and the backsight of 2.631 as 98.130 + 2.631 = 100.761.
6. RLs: point F: 100.761 − 1.355 = 99.406, point G: 100.761 − (− 2.015) = 102.776.
7. Arithmetic check: Sum of backsights = 2.681, sum of foresights = − 2.605, difference = 5.286, last RL − first RL = 102.776 − 97.490 = 5.286.

Rise and fall method   The solution by the rise and fall method is shown in Table 7.14.

Table 7.14 Solution to Example 7.6 by the rise and fall method

Staff station	Backsight	Intermediate sight	Foresight	Rise	Fall	Reduced levels	Remarks
A	0.685					97.490	
B		1.315			0.630	96.860	
C	− 0.635		− 1.825	3.140		100.000	BM 100.000
D		1.205			1.840	98.160	
E	2.631		1.235		0.030	98.130	
F		1.355		1.276		99.406	
G			− 2.015	3.370		102.776	

1. The rise or fall is calculated as the difference between the previous reading and the reading being considered.
2. Station B: 1.315 − 0.685 = 0.630 (fall), station C: 1.315 − (−1.825) = 3.140 (rise), station D: − 0.635 − 1.205 = − 1.840 (fall), station E: 1.205 − 1.235 = − 0.030 (fall), station F: 2.631 − 1.355 = 1.276 (rise), station G: 1.355 − (−2.015) = 3.370 (rise).
3. The BM is an intermediate point. Therefore, one has to go up the table for some stations and down the table for some other stations. As one goes up the table, the rises are subtracted and the falls are added; the reverse is true when one goes down the table.
4. RL of B = 100.000 − 3.140 = 96.860, RL of A = 96.860 + 0.630 = 97.490.

5. RL of D = 100.000 – 1.840 = 98.160, RL of E = 98.160 – 0.030 = 98.130, RL of F = 98.130 + 1.276 = 99.406, RL of G = 99.406 + 3.370 = 102.776.
6. Arithmetic check: Sum of rises = 7.786, sum of falls = 2.5, difference = 5.286, which is the same as the difference between the sum of backsights and foresights and the difference between the last RL and the first RL calculated earlier.

Example 7.7 The staff readings taken during a levelling operation are given below: 1.355, 1.605, 2.125, 0.685, 1.365, 2.015, 1.355, – 1.385, 0.685, 2.105, 1.685, 1.155, 1.105, 2.015, 1.085, 1.345, 1.355, – 2.015, 1.305, 1.655, 1.685, 1.455. The instrument was shifted after the 5th, 10th, 14th, and 19th readings. Arrange the data in tabular form and find the reduced levels of the points if the 12th reading was taken to a benchmark of RL 185.635.

Solution Height of collimation method The data is arranged in tabular form as given in Table 7.15. The 5th, 10th, 14th, and 19th readings are foresights and the readings next to these are backsights.

1. First calculate the heights of collimation. Station K is the benchmark of 185.635. The reading on the staff held at K is 1.155. The height of collimation is, therefore, 185.635 + 1.155 = 186.79.
2. RL of point J = 186.79 – 1.685 = 185.105. (Note that the backsight is to be subtracted, which is taken with this line of collimation.) From this RL and the foresight to point J, the height of collimation = 185.105 + 2.105 = 187.21.

Table 7.15 Solution to Example 7.7 by the height of collimation method

Staff station	Backsight	Intermediate sight	Foresight	Height of collimation	Reduced levels	Remarks
A	1.355			186.56	185.205	
B		1.605			184.955	
C		2.125			184.435	
D		0.685			185.875	
E	2.015		1.365	187.21	185.195	
F		1.355			185.855	
G		– 1.385			188.595	Staff inverted
H		0.685			186.525	
J	1.685		2.105	186.79	185.105	
K		1.155			185.635	BM RL 185.635
L		1.105			185.685	
M	1.085		2.105	185.86	184.775	
N		1.345			184.515	
P		1.355			184.505	
Q		– 2.015			187.875	Staff inverted
R	1.655		1.305	186.21	184.555	
S		1.685			184.525	
T			1.455		184.755	

3. 187.21 – 2.015 = 185.195 is the RL of station E and 185.195 + 1.365 = 186.56 is the height of the first line of collimation.

4. There is a change point at M. RL of M = 186.79 – 2.015 = 184.775. With this RL, the fourth height of collimation = 184.775 + 1.085 = 185.860.
5. There is yet another change point at R. In the same way as in the case of M, RL of R = 185.86 – 1.305 = 184.555 and the fifth height of collimation = 184.555 + 1.655 = 186.21.
6. With the height of collimations calculated, the calculation of the RLS of the other points is a simple matter.

Height of collimation = 186.56,   A: 186.56 – 1.355 = 185.205

B: 186.56 – 1.605 = 184.955,   C: 186.56 – 2.125 = 184.435

D: 186.56 – 0.685 = 185.875

Height of collimation = 187.21,   F: 187.21 – 1.355 = 185.855

G: 187.21 – (–1.385) = 188.595,   H: 187.21 – 0.685 = 186.525

Height of collimation = 186.79,   L: 186.79 – 1.105 = 185.685

Height of collimation = 185.86,   N: 185.86 – 1.345 = 184.515

P: 185.86 – 1.355 = 184.505,   Q: 185.86 – (–2.015) = 187.875

Height of collimation = 186.21,   S: 186.21–1.685 = 184.525

T: 186.21 – 1.455 = 184.755

7. Arithmetic check: Sum of backsights = 7.795, sum of foresights = 8.245, difference = – 0.450, last RL – first RL = 184.755 – 185.205 = – 0.450.

Rise and fall method   The solution is shown in Table 7.16.

Table 7.16   Solution to Example 7.7 by the rise and fall method

Staff station	Backsight	Intermediate sight	Foresight	Rise	Fall	Reduced level	Re-marks
A	1.355					185.205	
B		1.605		e	0.25	184.955	
C		2.125			0.52	184.435	
D		0.685		1.44		185.875	
E	2.015		1.365		0.68	185.195	
F		1.355		0.66		185.855	
G		– 1.385		2.74		188.595	Staff inverted
H		0.685			2.07	186.525	
J	1.685		2.105		1.42	185.105	
K		1.155		0.53		185.635	BM RL 185.635
L		1.105		0.05		185.685	
M	1.085		2.015		0.91	184.775	
N		1.345			0.26	184.515	
P		1.355			0.01	184.505	
Q		– 2.015		3.37		187.875	Staff inverted
R	1.655		1.305		3.32	184.555	
S		1.685			0.03	184.525	
T			1.455	0.23		184.755	

1. With the available data in columns 1 to 4, the rises and falls can be calculated. Starting from station B: $1.355 - 1.605 = -0.25$ (f), C: $1.605 - 2.125 = -0.52$ (f), D: $2.125 - 0.685 = 1.44$ (r), E: $0.685 - 1.365 = -0.68$ (f), F: $2.015 - 1.355 = 0.66$ (r), G: $1.355 - (-1.385) = 2.74$ (r), H: $-1.385 - 0.685 = -2.07$ (f), J: $0.685 - 2.105 = -1.42$ (f), K: $1.685 - 1.155 = 0.53$ (r), L: $1.155 - 1.105 = 0.05$ (r), M: $1.105 - 2.015 = -0.91$ (f), N: $1.085 - 1.345 = -0.26$ (f), P: $1.345 - 1.355 = -0.01$ (f), Q: $1.355 - (-2.015) = 3.37$ (r), R: $-2.015 - 1.305 = -3.32$ (f), S: $1.655 - 1.685 = -0.03$ (f), and T: $1.685 - 1.455 = 0.23$ (r).

2. K is the benchmark of RL 185.635. We have to go up the table for stations before K and down the table for stations after K. Going up the table, subtract the rises and add the falls and do the reverse when going down the table.

   RLs of stations before K: J: $185.635 - 0.53 = 185.105$, H: $185.105 + 1.42 = 186.525$, G: $186.525 + 2.07 = 188.595$, F: $188.595 - 2.74 = 185.855$, E: $185.855 - 0.66 = 185.195$, D: $185.195 + 0.68 = 185.875$, C: $185.875 - 1.44 = 184.435$, B: $184.435 + 0.52 = 184.955$, A: $184.955 + 0.25 = 185.205$.

   RLs of stations after K: L: $185.635 + 0.05 = 185.685$, M: $185.685 - 0.91 = 184.755$, N: $184.755 - 0.26 = 184.515$, P: $184.515 - 0.01 = 184.505$, Q: $184.505 + 3.37 = 187.875$, R: $187.875 - 3.32 = 184.555$, S: $184.555 - 0.03 = 184.525$, T: $184.525 + 0.23 = 184.755$.

3. Arithmetic check: Sum of rises $= 9.02$, sum of falls $= 9.47$, difference $= -0.45$, same as the difference between the sum of backsights and foresights and the difference between the last RL and the first RL.

Example 7.8 A page from an old levelling book is shown in Table 7.17. Some readings are not clearly decipherable. Compute the missing readings from the available data.

Table 7.17 Page from a field book (Example 7.8)

Staff station	Backsight	Intermediate sight	Foresight	Height of collimation	Reduced levels	Remarks
A				101.605	100.000	BM
B		1.285				
C	1.305				100.620	
D					99.060	
E	2.135				99.940	
F			1.045			
G						

Solution Against station A, there must be a backsight. Reduced level + backsight = height of collimation. Hence the backsight reading on A = $101.605 - 100.000 = 1.605$. Reduced level of B = $101.605 - 1.285 = 100.320$.

Point C is a change point, as there is a backsight entered against it. The corresponding foresight = $101.605 - 100.620 = 0.985$. There is a new instrument height for the backsight on C. This can be found as $100.620 + 1.305 = 101.925$.

There is an intermediate sight against D. This can be found from the height of collimation and RL as $101.925 - 99.060 = 2.865$.

Table 7.18 Solution to Example 7.8

Staff station	Backsight	Intermediate sight	Foresight	Height of collimation	Reduced levels	Remarks
A	1.605			101.605	100.000	BM
B		1.285			100.320	
C	1.305		0.985	101.925	100.620	
D		2.865			99.060	
E	2.135		1.985	102.075	99.940	
F			1.045		101.030	
G						

E is again a change point; foresight = 101.925 – 99.940 = 1.985 against E.

The height of collimation for the backsight on E = 99.940 + 2.135 = 102.075. RL of F = 102.075 – 1.045 = 101.030. The complete solution is shown in Table 7.18.

Example 7.9 The page of an old levelling book is shown in Table 7.19. Many readings are missing or cannot be read clearly. Find the missing readings and complete the page.

Table 7.19 Data for Example 7.9

Staff station	Backsight	Intermediate sight	Foresight	Rise	Fall	Reduced level	Remarks
P	1.785					100.000	BM
Q						99.72	
R				0.75			CP
S		1.635				100.70	
T				0.76			CP
U			1.315			101.17	

Solution The missing readings are found from the available data and the principle of the rise and fall method. RL of Q = 99.72, showing that there is a fall from P to Q; fall = 100.000 – 99.72 = 0.28. The intermediate sight at Q = 1.785 + 0.28 = 2.065. From Q to R, there is a rise of 0.75. Since R is a change point, the foresight reading on R = 2.065 – 0.75 = 1.315. The RL of R can be found as 99.72 + 0.75 = 100.47 (by adding the rise to the previous RL). From R to S, there is a rise, as can be seen from the RL of the two points; rise = 100.70 – 100.47 = 0.23. This rise can be used to calculate the backsight reading on R as 1.635 + 0.23 = 1.865. Similarly, T has a rise of 0.76 over S and hence the foresight reading on T = 1.635 – 0.76 = 0.875. The RL of T = 100.70 + 0.76 = 101.46.

The RLs of T and U show that there is a fall from T to U. The fall is 101.46 – 101.17 = 0.29. This will give the backsight reading on T as 1.315 – 0.29 = 1.025. The complete solution is shown in Table 7.20.

Table 7.20 Solution to Example 7.9

Staff station	Backsight	Intermediate sight	Foresight	Rise	Fall	Reduced level	Remarks
P	1.785					100.000	BM
Q		2.065			0.28	99.72	
R	1.865		1.315	0.75		100.47	CP
S		1.635		0.23		100.70	
T	1.025		0.875	0.76		101.46	CP
U			1.315		0.29	101.17	

7.8 Other Methods in Levelling

There are many types of levelling operations each with specific objectives. Some of these are differential or fly levelling, check levelling, profile levelling or sectioning, cross-sectioning, and reciprocal levelling.

Differential or fly levelling

Differential or fly levelling is resorted to when the difference in elevation between two points that are far apart is to be determined. Here, there is no reason to determine the reduced levels of the intermediate points. This method is adopted when one wants to establish a benchmark near the site of levelling work. Essentially, backsights and foresights are taken to reach a point B starting from point A. The essential principle remains the same. This relates the reduced level of point B with the known RL of point A (Fig. 7.11).

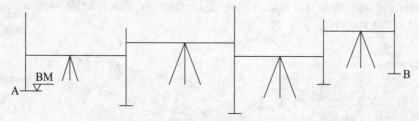

Fig. 7.11 Differential or fly levelling

Check levelling

Check levelling, as the name implies, is an operation done to check the levelling work done. A simple method is to run a series of levels as in fly levelling up to the starting point of the survey at the end of the day. This will check the accuracy of the levelling work.

Profile levelling

Profile levelling or sectioning is levelling done across a line, e.g., along the centre line of the road, to get a sketch of the profile. Essentially, the staff stations have to be ranged to be along a line. Readings are taken on the staff held at points along this line. When the levels are plotted along the distances in a line, we get a profile of the ground showing the elevations of the points. Profile levelling is discussed in detail in Chapter 20 on engineering surveys (Fig. 7.12)

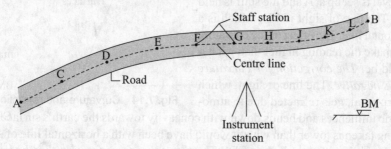

Fig. 7.12 Profile levellig

Cross-sectioning

Cross-sectioning involves taking levels across a line, e.g., the centre line of a road or a railway line. Depending upon the width of the road, a number of staff stations are fixed across the centre line. A series of such sections is taken at intervals along the centre line. This again gives a profile perpendicular to the longitudinal line and is useful in calculating the volume, etc. Cross-sectioning is dealt with in detail in Chapter 20 on engineering surveys (Fig. 7.13).

Reciprocal levelling

Reciprocal levelling is done when the distances are long and it is not possible to balance the lengths of the lines of sights. This happens, e.g., in the case of points lying on either side of a river, when it is not possible to maintain the level between the two points. This process eliminates many errors due to maladjustment of the instrument and those due to curvature and refraction. This is dealt with in detail later in this chapter.

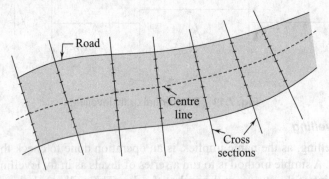

Fig. 7.13   Cross-sectioning

7.9   Curvature and Refraction

As explained earlier with respect to Fig. 7.1, a level surface is parallel to the mean spheroidal surface of the earth. The line of sight is horizontal and deviates from the level surface due to the curvature of the level surface. As shown in Fig. 7.14, the level is set up at A and the staff is held at B. The line of sight is CD, which is a horizontal line. The effect of the curvature is to make the reading higher than what it should be. *The correction for curvature is thus negative.* The line of sight, which is horizontal, gets refracted due to atmo-

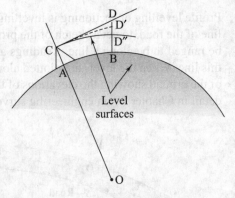

Fig. 7.14   Curvature and refraction

spheric influences and bends down with concavity towards the earth's surface. The reading taken is lower than what it would have been with a horizontal line of sight. The actual line of sight is thus CD′. *The correction for refraction thus has a sign opposite to that of curvature and is postive.*

7.9.1 Correction for Curvature

The correction for curvature can be derived as follows. Consider Fig. 7.15. A is the instrument station and B is the staff station. The horizontal line of sight intersects the staff at C. AB is the level surface through the line of sight. Taking the curvature into account, the reading should have been taken at B. However, due to the horizontal line of sight, the reading is taken at C. The reading corresponding to C is higher and hence the curvature causes the point to appear

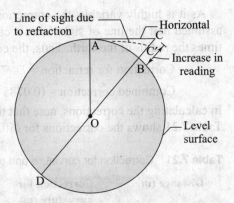

Fig. 7.15 Correction for curvature

lower than where it should be. The correction for curvature is to be subtracted from the reading. CB is the correction for curvature.

Let O be the centre of the earth. OA is a vertical line and the radius of the sphere. AC is horizontal and at right angles to AO. OB is also a radial line. CB when extended intersects the surface of the earth at D. From the right-angled triangle ACO,

$$AC^2 = OC^2 - OA^2 = (OB + CB)^2 - OA^2$$
$$= OB^2 + CB^2 + 2 \times OB \times CB - OA^2$$

$$OB = OA = radius, \quad 2 \times OB = diameter \text{ of the sphere}$$

CB is too small a quantity compared to the radius of the earth and hence its square can be neglected. Thus,

$$AC^2 = BD \times CB, \quad CB = AC^2/BD$$

Let AC = D, the distance between the staff and the level. BD = mean diameter of the earth. Thus the correction for curvature is given as follows.

Correction for curvature BC = D^2/diameter of the earth

If the mean diameter of the earth is taken as 12,740 km, then

Correction for curvature = $D^2 \times 1000/12,740$ m

$$= 0.07848D^2 \text{ m} \cong 0.0785D^2 \text{ m}$$

7.9.2 Correction for Refraction

Refraction is the phenomenon of light rays deviating from a straight line as they pass from a denser medium to a rarer medium or vice versa. As light rays pass through different layers of different densities, they get refracted. It is evident that refraction is a highly variable phenomenon in the case of the atmosphere. It does not remain constant and varies with factors such as temperature. However, as is shown in Fig. 7.15, as the rays of light go from the staff to the instrument, they take a curved path, with concavity towards the earth's surface. The horizontal line of sight does not remain horizontal because of refraction. Thus, the reading as observed from the instrument is at point C′ and not at C. Thus refraction makes the reading lower than what it should have been with a horizontal line of sight. The correction for refraction is thus of a sign opposite to that of curvature and is positive.

As it is highly variable, the correction for refraction is taken empirically. It is assumed that the line of sight takes a curved shape with a radius equal to seven times the radius of the earth. Thus, the correction for refraction is

$$\text{Correction for refraction} = D^2/7 \times \text{diameter of the earth} = 0.0112D^2$$

$$\text{Combined correction} = (0.0785 - 0.0112)D^2 = 0.0673D^2$$

In calculating the corrections, note that the distance D should be in kilometres. Table 7.21 shows the corrections for different distances.

Table 7.21   Correction for curvature and refraction

Distance (m)	Correction for curvature (m)	Correction for refraction (m)	Combined correction (m)
100	0.000785	0.000112	0.00067
250	0.0049	0.0007	0.0042
500	0.0196	0.0028	0.0168
1000	0.0785	0.0112	0.0673
2000	0.314	0.0448	0.2692

7.9.3   Distance to the Visible Horizon

Referring to Fig. 7.16, AB is a level surface passing through point A. AC is the height of the observer. The observer's line of sight becomes tangential at B. The distance AB is known as the *distance to the visible horizon*. If the observer has a higher elevation, say AC′, the distance to the visible horizon increases to AB′. This is the reverse process of finding the curvature correction: finding the distance when the height of the horizontal line of sight is known. If curvature alone is taken into account, then

$$h = 0.078D^2, \quad D = \sqrt{h/0.0785} = \sqrt{12.7388h} = 3.569\sqrt{h}$$

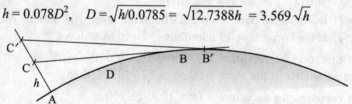

Fig. 7.16   Distance to the visible horizon

However, as we have seen, the line of sight does not remain horizontal as the observer views the horizon. The line of sight has a curvature concave to the earth's surface. Taking the curvature and refraction into account,

$$h = 0.0673D^2, \quad D = \sqrt{h/0.0673} = \sqrt{14.8588h} = 3.8547\sqrt{h}$$

In these formulae, D should be input in kilometres and h in metres. These formulae have been derived considering the fact that the height h is small compared to the radius of the earth. If h is not negligible, then AC = h and R is the radius of the mean spheroidal surface of the earth:

$$AC(AC + 2R) = D^2, \quad D = \sqrt{h(h + 2R)}$$

7.9.4 Dip of the Horizon

In Fig. 7.17, AC is the observer's height, B is the point of the visible horizon, and AB is the level surface through point A. If CD is the tangent to the level surface through C, then angle DCB is known as the dip of the horizon. From the properties of a circle, it is clear that the dip of the horizon is equal to the angle subtended by arc AB at the centre of the circle. Arc

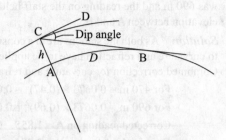

Fig. 7.17 Dip of the horizon

$AB = R\Theta$, where R is the mean radius of the spheroidal surface of the earth and Θ is the dip angle in radians. If arc AB is taken approximately equal to the distance to the visible horizon D, then dip of the horizon = distance to the visible horizon/R = D/R, in radians. D and R should be expressed in the same units. The following examples illustrate these concepts.

Example 7.10 Find the corrections for curvature and refraction and the combined correction if the distance is (a) 900 m and (b) 1500 m.

Solution

(a) For a sight distance of 900 m,

Correction for curvature = $0.0785 \times (0.9)^2 = 0.063$ m
Correction for refraction = $0.0112 \times (0.9)^2 = 0.009$ m
Combined correction = $0.0673 \times (0.9)^2 = 0.054$ m   or 0.063×0.009
= 0.054 m

(b) For a sight distance of 1500 m,

Correction for curvature = $0.0785 \times (1.5)^2 = 0.176$ m
Correction for refraction = $0.0112 \times (1.5)^2 = 0.025$
Combined correction = $0.0673 \times (1.5)^2 = 0.151$ or $0.176 - 0.025$
= 0.151 m

Example 7.11 Stations P and Q are 1600 m apart. A level was set up between P and Q such that the distance from P is 80 m. The readings taken on P and Q were 0.785 m and 2.735 m, respectively. Find the true difference in elevation between P and Q.

Solution As the distance to the staff at P is only 80 m, the correction for curvature and refraction is very small for the reading on P. However, the reading on the staff at Q has to be corrected, as the distance is large.

The combined correction for curvature and refraction = $0.0673D^2$, where D is the distance is in kilometres and the correction obtained will be in metres. For 1520 m,

Combined correction = $0.0673(1.52)^2 = 0.151$ m
Correct reading on staff at Q = $2.735 - 0.151 = 2.584$ m
True difference in elevation = $2.584 - 0.785$
= 1.799 m   (fall from P to Q)

Example 7.12 A level was set up between two stations A and B. The distance to station A was 470 m and the reading on the staff held at A was 1.855 m. The distance to station B was 690 m and the reading on the staff held at B was 2.385 m. Find the true difference in elevation between A and B.

Solution As both the stations are at a considerable distance from the level, the corrections to curvature and refraction have to be applied to the readings taken from both the stations. Combined correction for curvature and refraction:

For 470 m = $0.0673 \times (0.47)^2 = 0.015$ m

For 690 m = $0.0673 \times (0.69)^2 = 0.032$ m

Corrected reading on A = $1.855 - 0.015 = 1.840$ m

Corrected reading on B = $2.385 - 0.032 = 2.353$ m

True difference in elevation = $2.353 - 1.840$

$= 0.513$ m (fall from A to B)

Example 7.13 Find the distance to the visible horizon and the dip of the horizon if the height at which the observer stands on top of a tower is 80 m. Take the radius of the earth as 6370 km.

Solution Distance to the visible horizon $D = \sqrt{h/0.0673}$, where h is in metres and D is in kilometres. Distance to the visible horizon = $\sqrt{(80/0.0673)} = 34.5$ km. The dip of the horizon is approximately equal to D/R, where D is the distance to the visible horizon and R is the radius of the earth. Both D and R should be in the same units. Dip of the horizon = $34.5/6370 = 18' 37''$.

Example 7.14 The top of a lighthouse is just visible from a station at sea 40 km away. Find the height of the lighthouse.

Solution The height of the lighthouse is given by

$H = 0.0673D^2 = 0.0673(40)^2 = 107.68$ m

Example 7.15 The eye of an observer is 7.5 m above sea level and he was able to see a lighthouse 50 m high just above the horizon. Find the distance between the observer and the lighthouse.

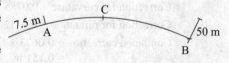

Fig. 7.18

Solution As shown in Fig. 7.18, the observer is at A and the lighthouse is at B. The horizontal sight meets the horizon at C. The distance between the observer and the lighthouse = AC + CB.

Distance AC = $3.8547 \sqrt{h} = 3.8547 \sqrt{7.5} = 10.556$ km

Distance CB = $3.8547 \sqrt{50} = 27.256$ km

Distance of observer from the lighthouse = $10.556 + 27.256$

$= 37.813$ km

Example 7.16 Two stations A and B are 12 km apart at the level of B. The reduced levels of A and B are 1284.675 m and 406.345 m, respectively. Find the distance between A and B at the level of station A.

Solution The perpendicular distance between the level surfaces through A and B = $1284.675 - 406.345 = 878.330$ m. Radius of the level surface through A = $6370 + 0.00878$ = 6370.00878 km. The length of the arc between radial lines subtending an angle at the centre is proportional to the radii. Therefore,

Arc length through A/arc length through B = 6370.00878/6370
Arc length through A = (6370.00878/6370) × arc length through B
$$= (6370.00878/6370) \times 12$$
$$= 12.0015 \text{ km (about 1.5 m more)}$$

Example 7.17 Two stations A and B are 200 m apart. The reduced levels of A and B are known to be 270.685 m and 269.345 m, respectively. A level was kept on the line AB. Readings were taken on staves held at A and B, the instrument being close to B. The readings observed were 1.565 at A and 2.985 at B. Find the angular error in the collimation of the instrument.

Solution   True difference in elevation = 270.685 – 269.345 = 1.34 m

Apparent difference in elevation = 2.985 – 1.565 = 1.42 m

This shows that the line of collimation is inclined downwards, as the reading on A has to be more to get the true difference in elevation. The difference between the two values = 1.42 – 1.34 = 0.08 m. This difference is over a distance of 200 m. Therefore,

$$\tan\alpha = 0.08/200, \quad \alpha = 1' \, 23''$$

The error in the line of collimation is therefore $1' \, 23''$.

Example 7.18   Two stations P and Q have a true difference in elevation (fall from P to Q) of 0.500 m. The level was kept in line with PQ and close to Q. The readings taken on P and Q 180 m apart were 1.865 and 1.305 m, respectively. Find the angular error in collimation. At what length of sight will the error due to collimation be the same as that due to curvature and refraction?

Solution   Apparent difference of elevation = 1.865 – 1.305 = 0.56 m

True difference in elevation = 0.500 m

Collimation error, therefore, is 0.56 – 0.50 = 0.06 m in a length of 200 m. The error in collimation or inclination of the line of sight is given by

$$\tan\alpha = 0.06/200 = 0.0003, \quad \alpha = 1' \, 02''$$

Error in the line of collimation = $1' \, 02''$

Combined correction due to curvature and refraction = $(6/7)(D^2/2R)$

Collimation error in length $D = D \tan (1' \, 02'') = 0.0003D$

These two are equal. Therefore

$$(6/7)(D^2/2R) = 0.0003D \quad \text{or} \quad D = 0.0003 \times 2R \times (7/6) = 4.46 \text{ km}$$

The distance for which the collimation error equals the combined error due to curvature and refraction is 4.46 km.

7.10   Balancing Backsight and Foresight Lengths

The readings from a level are accurate if the line of collimation is parallel to the axis of the bubble tube. If the line of sight is inclined when the bubble traverses, the error in the reading is proportional to the length of sight. To eliminate this error, it is necessary to balance the lengths of sights. Balancing the lengths of sights means making them nearly equal. This can be done in the case of backsights and foresights by selecting the instrument stations suitably. In ordinary work, the distances can be approximately estimated by any means. In more accurate work, the distances can be measured with a tape or by stadia methods.

In Fig. 7.19, let A and B be the two staff stations and C the instrument station. C is midway between A and B, not necessarily in line with them. The distances AC and CB are equal. When the bubble is in the centre of its run, let the line of sight be inclined at angle α.

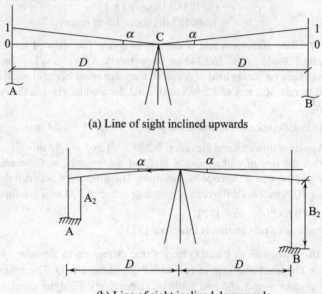

(a) Line of sight inclined upwards

(b) Line of sight inclined downwards

Fig. 7.19 Balancing lengths of foresights and backsights

From Fig. 7.19(a), A0 is the true staff reading if the line of sight is horizontal. A1 is the reading taken if the line of sight is inclined upwards. Similarly, B0 is the true staff reading at B if the staff were horizontal and B1 is the reading taken as the line of collimation is inclined upwards. It is clear from the figure that

$$A0 = A1 - D \tan\alpha \quad \text{and} \quad B0 = B1 - D \tan\alpha$$

If A is higher than B on the ground, then the true difference in level = A0 − B0 = A1 − D tanα − (B1 − D tanα) = A1 − B1. This shows that the true difference in level is obtained even with an inclined line of collimation if the sight distances are equal.

If the line of collimation was inclined downwards by an angle α when the bubble traverses, as shown in Fig. 7.19(b), let A0 and B0 be the true staff readings if the line of sight were horizontal and A2 and B2 be the readings taken. Again,

$$A0 = A2 + D \tan\alpha \quad \text{and} \quad B0 = B2 + D \tan\alpha$$

The difference in level is given by, assuming B to be higher than A,

$$A0 - B0 = A2 + D \tan\alpha - (B2 + D \tan\alpha) = A2 - B2$$

The true difference level is obtained even with an inclined line of sight if the sight distances are equal.

7.11 Reciprocal Levelling

Reciprocal levelling is a method of levelling to accurately determine the difference in elevation between two points that are a considerable distance apart with an intervening obstacle such as a river. As the level cannot be placed between them to balance the lengths of the lines of sight, reciprocal levelling is adopted.

Figure 7.20 shows the basic principle of reciprocal levelling. A and B are two stations with an intervening obstacle, a river. The objective is to find the difference in elevation between points A and B accurately. As shown in Fig. 7.20(a) the level

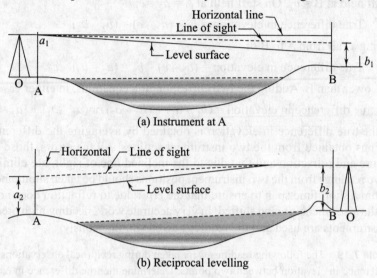

(a) Instrument at A

(b) Reciprocal levelling

Fig. 7.20 Reciprocal levelling

is set up near A on the line BA produced. The instrument is levelled and readings taken on the staff held at A and B. As the level is very close to A, the reading may have to be taken by looking through the object glass, with the eyepiece pointing to the staff held at B. The error in the reading on A is practically zero due to the inclined line of collimation and curvature and refraction. The reading on the staff held at B is considerable, as the distance is large. Let a_1 and b_1 be the readings taken. The reading at B is in error by an amount e due to the effect of curvature and refraction and the inclined line of collimation.

In the second step, the level is shifted to a point close to B [Fig. 7.20(b)] on the line AB produced. It is set up and levelled. Readings are taken on the staff held at A and B. The reading on B may have to be taken through the object glass as the distance to the staff is very small. Let a_2 and b_2 be the readings on the staff held at A and B. The reading on the staff B is practically without error as its distance from the level is very small. The reading a_2 is in error due to the large distance and the effect of curvature and refraction and the inclined line of collimation. Let e again be the error in the staff reading on A. The two errors (with the two positions of the level) are equal as the distance involved is the same. The corrected readings are the following.

Instrument at A

On staff held at A $= a_1$. On staff held at B $= b_1 - e$, where e = curvature correction – refraction correction ± correction due to inclined line of collimation.

True difference in elevation $= (b_1 - e) - a_1 = (b_1 - a_1) - e$

assuming a fall from A to B or

True difference in elevation $= a_1 - (b_1 - e) = (a_1 - b_1) + e$

if A is lower than B.

Instrument at B

On staff held at B = b_2. On staff held at A = $a_2 - e$.

True difference in elevation = $b_2 - (a_2 - e) = (b_2 - a_2) + e$

assuming a fall from A to B or

True difference in elevation = $(a_2 - e) - b_2 = (a_2 - b_2) - e$

if A is lower than B. Adding the two corresponding equations, in either case,

$2 \times$ true difference in elevation = $(b_1 - a_1) + (b_2 - a_2)$ or $(a_1 - b_1) + (a_2 - b_2)$

Thus the true difference in elevation is obtained by averaging the differences in elevations obtained from the two instrument settings. All the errors, those due to curvature and refraction and that due to the inclined line of sight, are eliminated. The two readings from the two instrument positions should be taken simultaneously or within a short time span to ensure that the error due to refraction does not vary while the instrument is being shifted. In very accurate work, needing high precision, two instruments are used and the readings taken simultaneously.

Example 7.19 The following readings were taken during reciprocal observations to find the difference in elevation between two points. Determine the true difference in elevation between them.

	Instrument at	A	B
	Staff at A	1.345	1.965
	Staff at B	1.525	2.015

If the reduced level of α is 128.945, find the reduced level of B.

Solution With the instrument at A, level difference = 1.965 – 1.345 = 0.620 m. With the instrument at B, level difference = 2.015 – 1.525 = 0.49 m.

True difference in elevation = (0.620 + 0.490)/2 = 0.555 m
This is a fall from A to B. Therefore,

Reduced level of B = 128.945 – 0.555 = 128.390 m

Example 7.20 The following readings refer to a reciprocal levelling operation between two points A and B. If the RL of A = 378.650, find the RL of B. If the distance between the stations is 950 m, find the collimation error, if any, of the instrument.

Instrument at	Staff reading at	
	A	B
A	0.656	2.097
B	0.867	2.298

Solution With the instrument at A, the difference in level between A and B is the following:

Fall from A to B = 2.097 – 0.656 = 1.441
With the instrument at B,

Fall from A to B = 2.298 – 0.867 = 1.431
True difference in elevation = (1.441 + 1.431)/2 = 1.436 m
RL of B = 378.650 – 1.436 = 377.214 m

To find the collimation error, we proceed as follows. With the instrument at A, the reading on B is affected by curvature and refraction and the collimation error. The correct reading on B is the following:

Correct reading on B = 2.097 – (curvature – refraction) – CE

where CE is the collimation error.

Combined error due to curvature and refraction = $0.0673 \times (0.95)^2$

$$= 0.0607 \text{ m}$$

Correct reading on B = 2.097 – 0.0607 – CE = 2.0363 – CE

Knowing the true difference in level, we can write

2.0363 – CE – 0.656 = 1.436   or   CE = – 0.0557 m

This shows that the line of collimation is inclined downwards.

Example 7.21   The following readings refer to the reciprocal level taken between two stations P and Q. Find the true difference in elevation between P and Q. If the instrument had a collimation error of 0.003/150 m and the distance between the stations was 1150 m, find the error due to refraction.

Instrument at	Staff reading at	
	P	Q
P	1.425	2.724
Q	1.429	2.504

Solution   The apparent differences in elevation with the two settings are the following. With the instrument at P,

Difference of elevation = 2.724 – 1.425 = 1.299 m

With the instrument at Q,

Difference of elevation = 2.504 – 1.429 = 1.075 m

True difference of elevation = (1.299 + 1.075)/2 = 1.187 m

Collimation error = 0.003/150 m for 1150 m

$$= 0.003 \times 1150/150 = 0.023 \text{ m}$$

Error due to curvature = $0.0785D^2$ for 1150 m or 1.15 km

$$= 0.0785 \times (1.15)^2 = 0.104 \text{ m}$$

With the instrument at P,

Correct reading at Q = 2.724 – (0.104 – R) – 0.023

where R is the error due to refraction.

True difference in elevation = 2.724 – 0.104 + R – 0.023 – 1.425

$$= 1.172 + R$$

With the instrument at Q,

Correct reading on P = 1.429 – (0.104 – R) – 0.023

True difference in elevation = 2.504 – [1.429 – (0.104 – R) – 0.023]

$$= 1.202 - R$$

Equating the two,

1.172 + R = 1.202 – R   or   R = 0.015

Correction due to refraction = 0.015 m

(Alternately, equate any of the above expressions to the true difference of elevation calculated: 1.172 + R = 1.187; R = 0.015 or 1.202 – R = 1.187; R = 0.015.)

Example 7.22   The following readings refer to the reciprocal observations from two points on either side of a river. Determine the true difference of elevation and the collimation error, if any, of the instrument. The distance between the stations = 1200 m.

Instrument at	Reading on staff at	
	A	B
A	1.115	1.765
B	1.750	2.315

Solution The apparent difference in elevation (DE) is calculated as follows. With instrument at A, DE = 1.765 − 1.115 = 0.650 m. With instrument at B, DE = 2.315 − 1.750 = 0.565 m.

True difference of elevation = (0.650 + 0.565)/2 = 0.6075 m

Let the collimation error be CE. Then, with the instrument at A,

Correct reading on B = 1.765 − combined error due to curvature and refraction − CE

Combined error due to curvature and refraction = 0.0785 × (1.2)^2
= 0.097

True difference of elevation = 1.765 − 0.097 − CE − 1.115 = 0.6075

CE = − 0.0545 in a length of 1200 m. The line of collimation is inclined downwards.

7.12   Field Problems in Levelling

During fieldwork, one may come across many situations in which to continue levelling work accurately some innovative measures have to be taken. The following are some of the common difficulties faced during fieldwork.

Levelling across steep slope

In levelling along a steep slope, there are two requirements for taking readings conveniently. The first is to equalize the backsight and foresight distances; one will be uphill and the other will be downhill. One reading will be to the bottom of the staff while the other will be to the top of the staff. The second is to ensure that readings can be taken to both positions from the level station, depending upon the slope. Equalizing the backsight and foresight distances, as we have discussed earlier, eliminates the error due to non-adjustment of the line of collimation. Selecting the instrument station suitably, one should attempt to equate the backsight and foresight distances. The instrument station can be taken anywhere and need not be in line with the chnge points.

As shown in Fig. 7.21, one reading will be very near the bottom of the staff while the other reading will be nearly at the top of the staff. Great care has to be taken while reading the staff at its top. Special care should be taken to keep the staff vertical, as any deviation from verticality will cause large errors in reading.

Determining the elevation of a high point like the soffit of a beam

Where the underside of a beam is at a reasonable height, inverted readings are taken as shown in Fig. 7.22. Since the staff is held inverted, the reading has to be given a negative sign. This takes care of the reduction of levels by any of the two methods. Care should be taken in taking inverted staff readings—in taking the main meter readings and the subunits.

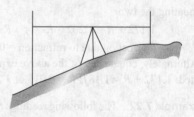

Fig. 7.21 Levelling across steep slope

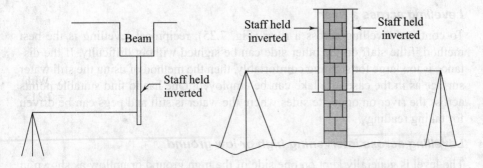

Fig. 7.22 Inverted staff reading   **Fig. 7.23** Levelling aross a wall

Levelling across a wall

To continue levelling work across a wall (Fig. 7.23), two methods can be employed. (i) When the wall is not very high, an inverted reading with the base of the staff held flush with the top of the wall is taken first from one side. The instrument is then transferred to the other side of the wall, set up, and levelled. An inverted reading, with the base of the staff held flush with the top of the wall, is now taken. This can relate the heights of the two lines of sight for continuing the levelling job across the wall. (ii) If the wall is very high and an inverted reading cannot be taken, one method may be to make a mark on one side of the wall at the height of the line of sight. Measure the distance from the top of the wall to the mark. The instrument is now transferred to the other side, set up, and levelled. A mark is made on the wall on the other side at the height of the line of sight and the distance measured. Knowing the RL of the top of the wall from the other side, the height of collimation can now be found.

Alternately, a mark can be made at a convenient distance from the top of the wall where marks are made on both sides. Inverted readings can be taken with the base of the staff held at this mark as in the first method.

Levelling across a lake

It is possible to go around a lake or pond and hence levelling across it is not a big problem (Fig. 7.24). If going around the obstacle is cumbersome, levelling across can be done by taking advantage of the water surface, which is a level surface when still. On either side of the lake, stout pegs are driven with their top surfaces flush with the water surface. Readings can be taken to the staff held on top of the pegs from either side. The two pegs being at the same level, the heights of collimation of the instrument stations on either side can be related.

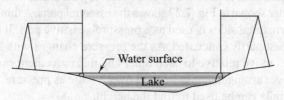

Fig. 7.24 Levelling across a lake

Levelling across a river

To continue levelling across a river (Fig. 7.25), reciprocal levelling is the best method if the staff on the other side can be sighted without difficulty. If the distance is too large for sighting comfortably, then the method of using the still water surface as in the case of a lake can be employed. One has to find suitable points across the river on opposite sides where the water is still and pegs can be driven for taking readings.

Levelling across intervening high or low ground

The level is generally kept on one side of the high ground or hollow as shown in Fig. 7.26. The staff is kept on the other side at a suitable location so that it can be bisected by the line of sight.

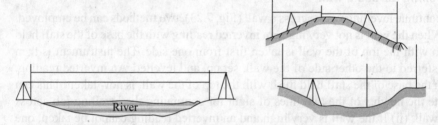

Fig. 7.25 Levelling across a river   **Fig. 7.26** Levelling across intervening high or low ground

7.13 Barometric Levelling

Barometric levelling is a method of rough levelling based on atmospheric pressure changes with height. It is well known that atmospheric pressure decreases with height. This can be used to estimate the elevation of a place. Two types of barometers are commonly used.

Mercury barometer

A mercury barometer uses a long glass tube dipped in a vessel containing mercury. The rise of the mercury column can be accurately read with a vernier. The instrument is not portable and hence finds little use in survey work, though it can be used in fixed stations at higher elevations.

Aneroid barometer

An aneroid barometer is a highly portable equipment and thus more commonly used even though it is less accurate than the mercury barometer. The section of the aneroid barometer shown in Fig. 7.27 shows the essential parts. A thin circular metallic box with corrugated sides is used as a pressure-sensitive part. It is hermetically sealed and at least partly evacuated. As the pressure changes with height, the two corrugated sides move relative to each other and this movement can be magnified by a suitable lever arrangement to be read over a scale as pressure. The following barometric formula can be used to find the height.

$$H = 18336.6 \log(h_1 - h_2) [1 + (T_1 + T_2)/500]$$

where H (in metres) is the difference in elevation between two stations A and B, h_1 is the barometer reading in centimetres at station A, h_2 is the barometer reading at station B, T_1 is the air temperature at station A, and T_2 is the air temperature at station B.

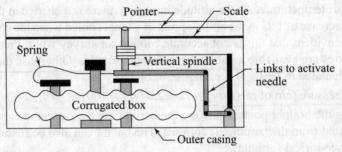

Fig. 7.27 Aneroid barometer

In the case of a mercury barometer, the same formula may be applied. Generally, a correction to the measured reading is applied as follows to allow for the temperature of mercury:

$$h = h'[1 + a(t_1 - t_2)]$$

where h is the corrected reading, h' is the observed reading, a is the difference in the coefficient of thermal expansion of glass and that of mercury, t_1 is the temperature of mercury at station A, and t_2 is the temperature of mercury at station B.

7.13.1 Barometric Methods

There are two methods of levelling or determining elevations with a barometer.

Single observation method

In this method, a single instrument is taken to the different stations, and observations on pressure and temperature are made. This results in errors due to variation in atmospheric parameters such as pressure and temperature while the instrument is being shifted.

Method of simultaneous observations

In this case a minimum of two sets of instruments, the barometer and a detached thermometer, are used for the purpose of eliminating the effects of changes in the atmosphere. One set of barometer and thermometer can be kept at a base station. The other set is taken to different stations whose elevations are to be determined. The observations of both sets of instruments are taken simultaneously at prefixed times so that the observations become simultaneous. This gives better results than with a single instrument.

7.14 Hypsometry

Hypsometry uses a hypsometer based upon the boiling point of water but can only give a rough indication of heights. This works on the principle that the boiling point of a liquid depends upon the atmospheric pressure, or the pressure of a boiling liquid is the same as that of the atmosphere. The liquid used is water, which is kept in a boiler and heated by a spirit lamp or some other means. A sensitive thermometer is used to measure the temperature of the boiling liquid. The thermometer is kept just above the water level so that the thermometer measures the temperature of the steam. A separate thermometer is used to measure the temperature of the air.

The boiling point of water is 100°C at sea level at a pressure of 76 cm of mercury and zero air temperature. As the altitude increases, there is a change in the boiling point. A decrease of 1°C in the boiling point of water is noted when the rise is about 30 m. It is evident that this is not accurate enough for survey work. Pressures and temperatures are available in tables for different altitudes. Otherwise the pressure can be calculated in terms of centimetres of mercury as

$$\text{Pressure (cm of mercury)} = 72 \pm 2.679(T - 100)$$

where T is the boiling point. Knowing the barometric pressure, the elevation can be calculated from the barometric formula. The height can also be found from the following empirical formula:

$$H = a(285.9t - 0.74t^2)$$

where H is the height above the datum (generally sea level, where water boils at 100°C), $t = 100 - T$, T is the boiling point, $A = [1 + (T_1 + T_2)/500]$ is a correction for air temperature, T_1 and T_2 are the air temperatures at the datum and at height H, respectively.

The following examples illustrate the above principles.

Example 7.23 Find the elevation of station Q from the following barometric data, given that the elevation of P is 198.735 m.

Station P: At 10 a.m, barometric pressure = 77.98 cm, air temperature = 20.45°C; at 2 p.m., barometric pressure = 78.08 cm, air temperature = 22.25°C.

Station Q: At 12 noon, barometric pressure = 76.15 cm, air temperature = 10.15°C.

Solution   The readings are taken at different times at the two stations. The probable value of the reading at 12 noon at station P is taken as the average of the two readings at 10 a.m. and 2 p.m. Thus the probable values at 12 a.m. at P are

Pressure = (77.98 + 78.08)/2 = 78.03 cm

Temperature = (20.45 + 22.25)/2 = 21.35°C

The values can now be substituted in the formula for height,

$$H = 18336.6 \, (\log h_1 - \log h_2) \, [1 + (T_1 + T_2)/500]$$

In this example,

$$h_1 = 78.03 \text{ cm}, \quad h_2 = 76.15 \text{ cm}, \quad T_1 = 21.35°C, \quad \text{and } T_2 = 10.15°C$$

Substituting,

$$H = 18336.6(\log 78.03 - \log 76.15) \, [1 + (21.35 + 10.15)/500]$$

$$= 206.45 \text{ m}$$

Elevation of station Q = 198.735 + 206.45 = 405.185 m

Example 7.24 Determine the difference in elevation between two points A and B from the following hypsometry data:

Station A: Boiling point—99.56°C, air temperature—15.85°C

Station B: Boiling point—97.05°C, air temperature—14.05°C

Solution There can be two methods of solution. We can find the height above sea level of points A and B and then find the difference in elevation between them, or find the barometric pressure from the boiling point readings and then use the barometric formula.

(a) *Height from sea level* The height is given by the formula $H = (285.9t + 0.74t^2)a$.

Station A: $t = 100 - 99.56 = 0.44$ and $a = [1 + (15.85 + 14.05)/500] = 1.0598$

$$H = (285.9 \times 0.44 + 0.74 \times 0.44^2) \times 1.0598 = 133.47 \text{ m}$$

Station B: $t = 100 - 97.05 = 2.95$

$$H = (285.9 \times 2.95 + 0.74 \times 2.95^2) \times 1.0598 = 900.66 \text{ m}$$

Difference in elevation $= 900.66 - 133.47 = 767.19$ m

(b) *Barometric pressure* The barometric height at a station is given by the formula

$$h = 76.0 - 2.679t$$

If h_1 and h_2 are the barometric heights at stations A and B, respectively,

$$h_1 = 76 - 2.679 \times 0.44 = 74.821 \text{ cm}$$

$$h_2 = 76 - 2.679 \times 2.95 = 68.097 \text{ cm}$$

Substituting these values in the barometer formula,

$$H = 18336.6 \,(\log h_1 - \log h_2)\, [1 + (T_1 + T_2)/500]$$

$$= 18336.6 \,(\log 74.821 - \log 68.097)\, [1 + (15.85 + 14.05)/500]$$

$$= 794.72 \text{ m}$$

7.15 Precision of Levelling

The precision required, or the degree of precision that can be achieved, of a levelling work depends upon the nature of the survey, the terrain, environmental conditions, skill of personnel, importance of the survey, and the instruments used. The precision required of a levelling job, more accurately the permissible error, is given by the general formula $E = C\sqrt{K}$. Here, K is the distance in kilometres and C is a constant. The values generally recommended are as follows.

   (a) Rough levelling work: $\pm 100\sqrt{K}$
   (b) Ordinary levelling, construction surveys: $\pm 24\sqrt{K}$
   (c) Accurate levelling for establishing benchmarks, etc: $\pm 12\sqrt{K}$
   (d) Precise levelling: $\pm\sqrt{K}$ to $\pm 4\sqrt{K}$

7.16 Errors in Levelling

Errors in levelling can be classified into instrumental errors, personal errors, and errors due to natural causes.

7.16.1 Instrumental Errors

Instrumental errors arise from maladjustments of the level. These can be of the following types.

Line of collimation not parallel to the bubble axis When the bubble is in the centre, the bubble tube axis is horizontal. The line of collimation, if parallel to the bubble tube axis, traverses a horizontal plane when moved around to sight objects. The basic principle of levelling is based on this parallelism of the two lines. This can be adjusted to be so by permanent adjustment. The line of collimation if inclined upwards or downwards introduces a collimation error which depends upon the length of sight. Adjusting the backsight and foresight lengths is one way to minimize the error.

Sensitivity of bubble   The bubble is sometimes sluggish and may be apparently centred even though the bubble axis is not horizontal. This will cause errors in the readings. If the bubble is oversensitive, levelling the bubble tube may take a long time.

Faulty levelling staff   The graduations on the levelling may be faulty but this is not a serious problem in ordinary surveys. In the case of a folding or telescopic staff, the moving parts may wear out and cause errors in readings.

7.16.2 Operational Errors

Carelessness in operating the equipment can cause errors in levelling work.

Careless levelling of bubble tube   Levelling of the bubble tube is a basic operation to make the bubble traverse and is a fundamental requirement of levelling. This should be done carefully at every setting of the level. Except in the case of tilting level, all other levels need careful levelling of the bubble tube.

Bubble not at centre while sighting   After the bubble tube has been levelled, it must be ensured that the bubble is in the centre while sighting. If not, the bubble can be centred before taking the reading.

Staff not held plumb   The levelling staff has to be vertical while taking readings. Whether the staff is out of plumb sideways or along the line of sight, the effect is to make the reading larger. One method is to swing the staff to and fro and note the minimum reading which will be in the plumb position.

Parallax   This is a temporary adjustment and if not done properly will result in a hazy view of the staff. Eye lens adjustment has to be done once for the eye of the observer. The object lens has to be adjusted to the length of sight. The staff must be in sharp and clear view before taking the reading.

Settlement of level and staff   The level should be set up on firm ground so that the legs are stable. The tripod legs should not be touched while taking readings. The staff should be held on firm ground or can be provided with some support for taking readings. The staff is likely to settle between backsight and foresight readings or its position may change.

7.16.3 Errors Due to Natural Causes

Curvature   As explained earlier, a level surface is parallel to the mean spheroidal surface of the earth. The line of sight is horizontal and hence the effect of curvature is to make the readings appear large. This is apparent, or needs to be taken into account, only in the case of long sights. This can also be eliminated by reciprocal levelling.

Refraction   It is a phenomenon whereby light rays bend while travelling through different media or the same medium with different densities. It is highly variable, as the atmospheric conditions change with time. The effect of refraction is opposite to that of curvature and is generally taken to be one-seventh of the effect of curvature.

Effect of temperature   High temperature does not have any perceptible effect on levelling in ordinary work. In precise levelling work, these effects are reduced by shielding the level and staff. An invar staff with a low coefficient of thermal expansion is used in precise levelling. The level may get heated up and the bubble may go off-centre due to differential heating. Most of these effects are random and tend to cancel over a number of readings.

Effect of wind   In windy conditions the stability of the instrument may be affected. It also becomes difficult to keep the levelling staff plumb and stable and to sight and read.

Other mistakes

Many other small errors or mistakes can take place due to carelessness. These include (i) reading the staff in the wrong direction, upwards instead of downwards or vice versa, (ii) reading the stadia hair instead of the central horizontal hair if there are three cross hairs, (iii) reading the staff wrongly, (iv) not taking the foresight and backsight at the same point, (v) entering the readings wrongly, (vi) not extending a telescopic staff fully before taking the reading, (vii) missing some readings or recording digits wrongly, etc.

Summary

Levelling is a method of determining the relative heights of different points with reference to a datum surface. A level surface is parallel to the mean spheroidal surface of the earth. A horizontal line at a point on the surface is a line tangential to the level surface through the point. The essential requirement to find the relative elevations of points is to have a line of sight that remains in a horizontal plane. We also need a vertical measuring rod with which heights can be measured.

A level is essentially an instrument that provides the horizontal line of sight. A level consists of a levelling head, a spirit level, and a telescope. The levelling head is used to bring the bubble of the spirit level to the centre of its run in all directions. The line of collimation then provides a horizontal line of sight. The dumpy level, Wye level, and reversible level were common in earlier days. A tilting level is one in which a tilting screw is provided to tilt the telescope at the time of taking a reading. The vertical axis need not be perpendicular in such a case, but the horizontal bubble tube axis or the line of collimation is made horizontal whenever a reading is taken. This speeds up the work. Automatic and electronic levels are modern instruments that are very precise and easy to handle and operate. The temporary adjustments of a level include setting up, levelling, focusing the eyepiece, and focusing the objective. No centring is required for a level over a station. The station where the instrument is kept is chosen considering the convenience for making observations and has no other significance.

The levelling staff is a graduated rod, generally made of wood, but aluminium rods are also available. The staff can be a solid single piece, telescopic, or folding. The target staff has a target which is moved so that it bisects the line of sight and a reading is taken by the staff man with a vernier on the target staff. The staff is generally 3 m to 4 m in length.

During levelling, the levelling staff is kept at points whose elevations are to be determined. One of the points must be a benchmark of known elevation or a point whose elevation is assumed. Once the readings are taken, the reduced levels of all the points are found by the method of reduction of levels. If the instrument station is changed during levelling, the height of the instrument at different stations will be different. To relate the heights of collimation at two consecutive stations, readings are taken to the staff held at the same point from the two instrument stations.

In the height of collimation method of reduction of levels, the height of the line of collimation, meaning the elevation of the line of collimation, is found at each setting of the level. The first reading with the level is called *backsight* and the last reading before shifting the level is known as *foresight*. The rest of the readings are classified as intermediate sights. The last reading during a levelling job is always entered as the foresight. A station to which a foresight and a backsight are taken is known as a *change point*. Heights of collimation are related by two readings taken to the same station. Reduced levels (RLs) of stations are found by subtracting the staff readings from the height of collimation. The arithmetic check in this method is that the sum of backsights minus the sum of foresights is equal to the last RL minus the first RL.

In the rise and fall method, the rise and fall of a point from the previous point is determined from the readings taken (from the same instrument station). A point has a rise when the reading at this point is less than that on the previous point and a fall when the reading is greater. Knowing the rise and fall of points, their reduced levels can be obtained by adding the rise to or subtracting the fall from the elevation of the previous point. An arithmetic check in this method that the sum of backsights minus the sum of foresights is equal to the sum of rise minus the sum of falls and is equal to the last RL minus the first RL.

The line of sight being horizontal, the measured readings do not refer to a level surface. This difference is insignificant if the length of sight is small. If the distances are large, it is necessary to make a correction for the curvature of the level surface. As the horizontal line of sight makes a reading larger, the correction for curvature is always negative. The curvature correction is $0.0785D^2$ m, where D is the distance in kilometres. The line of sight is also affected by the refraction of light rays through the atmosphere. The correction for refraction is one-seventh the curvature correction but is positive. The combined correction is $0.0673D^2$.

Reciprocal levelling is a method of levelling by which errors due to collimation, curvature, and refraction are eliminated. The instrument has to be kept near both the stations whose difference in elevation is to be determined. The true difference in elevation is the average of the apparent differences in elevation.

Errors can creep into levelling work due to faulty instruments as well as due to faulty manipulation, reading, and recording. These errors can be eliminated by careful adjustments of the instrument and careful reading and recording.

Though barometric levelling (based on atmospheric pressure) and hypsometry (based on the boiling point of a liquid changing with pressure) can be used for determining elevations, they are rough estimates and are used only in rough estimations of elevations.

Exercises

Multiple-Choice Questions

1. An example of a level surface is
   - (a) the top surface of a dining table
   - (b) the floor surface of a building
   - (c) the still water surface of a lake
   - (d) the surface of a piece of plywood

2. A plumb line is a line
   (a) lying on a level surface
   (b) lying on a horizontal plane
   (c) perpendicular to a level surface
   (d) that joins two points on ground
3. In levelling centring of the level over the station point
   (a) is done using the tripod legs
   (b) is not required to be done
   (c) is done using the foot screws
   (d) is an important step
4. In reduction of levels using the height of instrument method, height of instrument refers to
   (a) height of the line of sight over the instrument station
   (b) height of the centre of telescope from the plane of foot screws
   (c) the reduced level of the line of sight
   (d) the reading on the staff from the instrument
5. When the reading is taken on a staff held at a point of known elevation of 123.45 m, the staff reading is recorded as 1.875 m. The height of instrument is
   (a) 1.875 m
   (b) 121.575 m
   (c) 123.45 m
   (d) 125.325 m
6. The three consecutive readings taken from a level are 1.325 m, 0.985, and 2.546 m. If the instrument was shifted after the first reading, the rise or fall of the last point is
   (a) 1.221 m, rise
   (b) 1.561 m, rise
   (c) −1.221 m, fall
   (d) −1.561 m, fall
7. Three consecutive readings from a level are 1.455 m, 0.875 m, and − 2.345 m. The rise or fall of the last point is
   (a) 3.79 m, rise
   (b) −0.89 m, fall
   (c) 3.22 m, rise
   (d) 1.455 m, rise
8. Seven readings were taken with a level as 1.345, 1.525, 0.785, 1.105, 1.855, 2.005, 1.985. If the instrument was shifted after the third and fifth readings, the readings recorded under the backsight column would be
   (a) 1.345, 0.785, 1.855
   (b) 1.105, 2.005
   (c) 1.345, 2.005
   (d) 1.345, 1.105, 2.005
9. A negative reading in the levelling data means the staff is
   (a) read upside down
   (b) kept upside down
   (c) read through the objective lens
   (d) read with the lower cross hair
10. Two consecutive readings in the levelling data are 1.345 m and 1.865 m. The first is a foresight and the second is a backsight. Then,
    (a) the fall from the first point to second point is 0.52 m
    (b) the rise from the first point to second point is 0.52 m
    (c) the two readings are taken from the same instrument station
    (d) the two readings are taken to the same point from two instrument stations
11. The folowing is a set of readings taken with a level: 1.565, 0.985, 1.235, 2.545, 3.455, 1.875, 1.985, 0.865, and 1.285. If the instrument was shifted after the second and fifth readings, then the entries in the foresight column would be
    (a) 1.565, 1.235, 1.875
    (b) 0.985, 3.455, 1.285
    (c) 1.235, 3.455, 1.985
    (d) 1.235, 1.875, 1.285
12. An arithmetic check for checking the reduction of level data in the height of collimation method is that last RL − first RL is equal to
    (a) sum of back sights − sum of intermediate sights
    (b) sum of intermediate sights − sum of foresights
    (c) sum of back sights − sum of foresights
    (d) sum of back sights and foresights − sum of intermediate sights.
13. An arithmetic check in the reduction of levels by the rise and fall method is stated as last RL − first RL is equal to
    (a) sum of back sights − sum of intermediate sights

(b) sum of foresights – sum of intermediate sights

(c) sum of rises – sum of falls

(d) sum of back sights – sum of foresights

14. Correction for curvature is done due to

(a) curved surface of the lens in the level

(b) chromatic aberration

(c) spherical aberration

(d) the curved nature of a level surface

15. Correction for refraction is done because

(a) of spherical aberration of the objective lens

(b) the line of sight is horizontal and deviates from a level surface

(c) of chromatic aberration

(d) the line of sight is refracted and deviates from its straight path

16. The correction due to refraction and curvature have

(a) the same sign   (c) opposite signs sometimes

(b) opposite signs   (d) the same sign sometimes

17. Distance to the visible horizon is given by

(a) $0.0673 \, D^2$ (b) $0.0112 \, D^2$ (c) $0.0785 \, D^2$ (d) $3.8547 \, \sqrt{h}$

18. Dip of the horizon is the angle

(a) made by the line of sight with the horizontal

(b) between the line of sight and the level surface

(c) between the line of sight and the tangent to the level surface

(d) made by the line of sight with the plumb line

19. A staff is held at a distance of 500 m from a level. If the reading on the staff is 1.875 m, the reading corrected for curvature is

(a) 1.858 m   (b) 1.872 m   (c) 1.870 m   (d) 1.855 m

20. A staff is held at a distance of 300 m from a level. If the reading on the staff is recorded as 1.345 m, the corrected reading after accounting for refraction is

(a) 1.344 m   (b) 1.338 m   (c) 1.339 m   (d) 1.340 m

21. If the reading on a staff held at a distance of 600 m is 2.365 m, then the reading corrected for curvature and refraction is

(a) 2.337 m   (b) 2.341 m   (c) 2.361 m   (d) 2.364 m

22. The distance to the visible horizon for an observer standing on a tower of height 40 m is

(a) 23.91 km   (b) 24.12 km   (c) 24.38 km   (d) 25.12 km

23. The dip angle for the observer standing at a height of 75 m is (take the radius of the earth as 6370 km)

(a) 18 minutes   (c) 42 seconds

(b) 19 seconds   (d) 56 seconds

24. The eye of an observer is 10 m above the ground. He was able to see the top of a lighthouse 60 m high just at the level of the horizon. The distance between the observer and the light house is

(a) 12.18 km   (b) 27.25 km   (c) 28.85 km   (d) 42.05 km

25. Balancing the sight lengths for backsights and foresights is done to eliminate the error due to

(a) faulty staff   (c) curvature and refraction

(b) faulty centring of level   (d) small inclination of line of sight

26. Reciprocal levelling is a method of levelling that eliminates the error due to

(a) curvature

(b) inclination of line of collimation

(c) curvature and refraction

(d) curvature and refraction and inclination of line of collimation

27. If in a reciprocal levelling operation, the readings are as follows: From near A: on staff held at A – 1.345 m; on staff at B at 800 m – 1.675 m; From near B: on staff held at B – 1.485 m and on staff at A – 1.875 m. The true difference in elevation between A and B is

    (a) 0.33 m fall from A to B   (c) 0.36 m fall from A to B

    (b) 0.39 m rise from A to B   (d) 0.36 m rise from A to B

28. When you have to cross a water body while performing levelling operations, then the best method is

    (a) fly levelling   (c) reciprocal levelling

    (b) profile levelling   (d) differential levelling.

29. Barometric levelling is based on the principle that

    (a) temperature decreases with height

    (b) boiling point of a liquid varies with pressure

    (c) atmospheric pressure decreases with height

    (d) melting point of a solid varies with pressure

30. Hypsometry is a method of

    (a) surveying of water bodies

    (b) determining elevations based on the boiling point of liquids

    (c) measuring distances

    (d) finding temperatures at different heights

Review Questions

1. Draw a neat sketch to indicate the following: level surface, horizontal surface, vertical line, level line, and horizontal line.
2. Bring out the differences between a dumpy level and a tilting level.
3. Explain the principle of direct levelling.
4. Explain the height of collimation method of reduction of levels.
5. Explain the rise and fall method of reduction of levels.
6. Explain the differences between the height of collimation method and the rise and fall method of reduction of levels.
7. Explain briefly the following: fly levelling, check levelling, and reciprocal levelling.
8. Explain the terms profile levelling and cross-sectioning.
9. Explain the principle, instrument, and method of barometric levelling.
10. What is a hypsometer? On what principle does it work?
11. Describe the method used to carry on levelling work across (a) a river, (b) a wall, and (c) high ground.
12. Explain any three important errors caused by a faulty levelling instrument.
13. Explain the operational errors possible in levelling and the precautions that should be taken to prevent them.
14. Show that errors due to curvature and refraction as well as any collimation errors are eliminated by reciprocal levelling.

Problems

1. The following readings were taken in sequence during a levelling work: 1.605, 2.150, 1.385, 1.895, 1.365, 2.105, 1.950, 0.985, 1.305, 1.185, 1.305, – 2.105, 1.385, 1.005, 1.155 and 1.145. The first reading was to a benchmark of assumed elevation 100.000. Find the RLs of the remaining stations if the instrument was shifted after the 3rd, 7th, 10th and 13th readings. Use the height of collimation method.

2. Eight readings were taken with a level in sequence as follows: 1.585, 1.315, 2.305, 1.225, 1.325, 1.065, 1.815, and 2.325. The level was shifted after the third and sixth readings. The second change point was a benchmark of elevation 186.975. Find the reduced levels of the remaining stations. Use the rise and fall method.

3. The following readings were taken with a level in sequence: 2.315, 1.615, 1.805, 1.115, − 2.345, 1.345, 2.105, 1.305, and 1.025. The level was shifted after the third and sixth readings. The fifth reading was to a station whose elevation is assumed to be 0.00. Find the reduced levels of the remaining points. Use the height of collimation method and check with the rise and fall method.

4. Sixteen readings were taken with a level, with the instrument being shifted after the 3rd, 7th, 11th, and 14th readings. The last reading was to a benchmark of elevation 203.565. Find the reduced levels of the remaining points. The readings are as follows: 1.305, 1.815, 1.015, 1.245, 1.355, − 2.015, 0.805, 2.015, 1.085, 1.315, 0.985, 1.305, 1.415, 1.565, and 1.685. Use the rise and fall method.

5. The following readings were taken with a level: 1.005, 1.315, 1.865, 0.965, 1.405, 1.555, 0.865, 1.345, 1.110, 0.965, and 1.715. The instrument was shifted after the third, sixth, and eighth readings. Tabulate the readings and find the reduced levels of all the points by the height of collimation method. The fifth reading was to a benchmark of elevation 302.540.

6. The following consecutive readings were taken with a level: 1.115, 0.745, 1.245, 1.065, 0.785, 1.315, 2.150, 0.845, 1.150, − 2.365, 1.360, 1.575, and 1.840. The instrument was shifted after the 4th, 8th, and 11th readings. Tabulate the readings in a proper format and find the reduced levels using the rise and fall method if the eighth reading was taken to a benchmark of reduced level 325.675.

7. The following staff readings were obtained during a levelling work with the instrument being shifted after the 4th, 7th, and 10th readings: 2.305, 0.940, 0.865, 1.325, 2.905, 1.185, 1.205, 2.015, 1.365, 0.985, and 1.785. Find the reduced levels of the remaining points if the RL of the second turning point is 0.000.

8. The following ten readings were taken with a level, the instrument being shifted after the fifth and eighth readings: 1.315, 0.965, − 2.345, 1.105, 0.875, 1.155, 1.305, 1.675, 1.345, and 1.875. The RL of the first turning point is 100.000. Find the reduced levels of the remaining points by the height of collimation method.

9. Some observations are missing from the page of a field book shown below. Find the missing readings from the available data.

Staff station	Backsight	Intermediate sight	Foresight	Height of collimation	Reduced level	Remarks
A					100.91	
B		1.085				
C		2.125				BM RL 100
D	1.315				101.26	
E			1.325	102.235		
F					101.61	

10. The page of an old field book is shown below. Some readings are not clear. Determine these readings from the available data.

Staff station	Backsight	Intermediate sight	Foresight	Rise	Fall	RL	Remarks
P	0.635					215.915	
Q					0.680		
R			0.865				BM RL 215.685
S		0.785		0.43			
T	0.935				0.32		
U						215.715	

11. Determine the corrections due to (a) curvature and (b) refraction if the length of sight is (i) 1200 m and (b) 1800 m.

12. Two stations A and B are 1200 m apart. A level was set up between the two stations 100 m away from A. The readings observed were 1.375 m on A and 2.465 on B. Find the true difference in elevation between A and B.

13. Two stations P and Q were 1250 m apart and a level was set up between them 500 m away from P. The staff readings on P and Q were 0.985 and 1.205, respectively. Find the true difference in elevation between P and Q.

14. Find the distance to the visible horizon and the dip of the horizon if the height of the observer at eye level is 75 m above sea level.

15. An observer at sea was just able to see the top of a lighthouse when 60 km away from it. Find the height of the lighthouse.

16. An observer standing on the deck of a ship was able to just see the top of a lighthouse at the horizon. If the observer's eye is 8 m above sea level and the lighthouse is 60 m high, find the distance of the observer from the lighthouse.

17. An observer standing on the deck of a ship was able to see the top of a lighthouse 60 km away at the horizon. If the observer's eye is 6 m above sea level, find the height of the lighthouse.

18. P and Q are two stations on the surface, P having an elevation 750 m higher than that of Q. If the horizontal distance between P and Q is 10 km at the level of Q, find the horizontal distance between the two stations at the level of P.

19. Two stations P and Q are 200 m apart. The RLs of the two stations are 221.384 (P) and 220.135 (Q). A level was set up in line with P and Q, near Q, and readings were taken on a staff held at P and Q, centring the bubble before taking the reading. If the readings are 0.865 and 2.184 on the staff held at P and Q, respectively, determine whether the line of collimation is in adjustment and, if not, the angular error of collimation.

20. The true difference of elevation between two points A and B, 150 m apart, is 0.984 m. A level was kept along the line AB produced, but close to B, and readings were taken on the staff held at A and B. If the readings are 2.055 and 1.051 on A and B, respectively, find the angular error in the line of collimation and also the length of sight for which the error due to the angular error in collimation equals that due to curvature and refraction.

21. The following readings are reciprocal levelling observations across a river between two points A and B. Find the true difference in elevation between the two points.

Instrument at	Staff at A		Staff at B
A	1.441	2.613	
B	1.772	2.950	

22. The following readings refer to reciprocal levelling observations between two points A and B 1000 m apart. The reduced level of A is 193.835 m. Find the reduced level of B and the collimation error, if any, of the instrument.

Instrument at	Staff at A		Staff at B
A	1.279		2.918
B	1.110	2.739	

23. The distance between two stations P and Q is 900 m. The observations from reciprocal levelling were 1.686 on staff at P, 3.107 on staff at Q with the instrument at P and 0.560 on staff at P, 2.015 on staff at Q with the instrument at Q. Find the true difference in elevation and the collimation error in the instrument.

24. Two stations P and Q were on either side of a river 1200 m apart. The instrument was kept near P and the readings on the staff at P and Q were 1.701 and 2.427, respectively. The instrument was then shifted to Q and the readings on the staff held at P and Q were 0.805 and 1.285, respectively. If the reduced level of P was 203.135, find the RL of Q. Also find the error due to refraction if the collimation error of the instrument is 0.002 m in 100 m.

25. A and B are two stations, A being lower in altitude. The barometric pressure and temperature at the two stations are as follows:

> Station A: Barometric pressure = 76.5 cm of mercury
> Air temperature = 17.2°C
> Station B: Barometric pressure = 74.8 cm of mercury
> Air temperature = 6.8°C

Find the difference in elevation between the two points.

26. Find the difference in elevation between two points P and Q with the following observations with a hypsometer:
Station P: Boiling point = 99.4°C,   air temperature = 16.15°C
Station Q: Boiling point = 97.1°C,   air temperature = 13.74°C

8 Computation of Areas

Introduction

The primary objective of surveying is to prepare a plan or a map. Once a plan is prepared using survey data, one is often required to find the area of the land or the volume of earthwork required for roads, railway embankments, etc. and the volume of reservoirs. In this chapter, we will deal with the computation of areas. Chapter 10 deals with the computation of volumes.

To find the area of a demarcated land, we can use a plan if available, different computation techniques, or a planimeter. The area can also be computed from survey data if a plan is not available or even otherwise. The areas of regular geometrical figures can be determined using standard formulae listed in Appendix 3.

In general, if an area has irregular boundaries, the common practice is to run a traverse within the area as close to the boundary as practicable, as shown in Fig. 8.1. From the traverse lines, offsets are taken to the irregular boundary. In such a case, it will be necessary to find first the area of the traverse or the skeletal frame and then the area of the tract of land between the traverse lines and the boundary.

Many methods will be outlined in this chapter to find the area of a traverse and the area between the traverse line and an irregular boundary. It must be clearly understood that the area calculated is that of the projected plan on a horizontal plane and the area of the actual surface is not obtained.

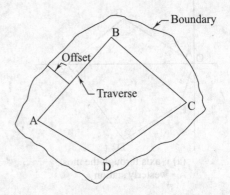

Fig. 8.1 Skeletal area and curved boundary

The units of area are square metre (sq. m), are (100 sq. m), and hectare (100 are or 10,000 sq. m). Areas measured in other unit systems can be converted into SI units using the conversion factors given in Appendix 1.

We start with the methods to find the area of a traverse.

8.1   Area of a Traverse

A traverse is generally a closed polygon and there are many methods to find its area. The following methods will be discussed here:

1. From coordinates of vertices
2. From latitudes and double meridian distances
3. From departures and total latitudes
4. By dividing into triangles

8.1.1   Area from Coordinates of Vertices

Consider a closed polygon ABCDEA representing a traverse, as shown in Fig. 8.2(a). The coordinates are calculated by placing the traverse in an x-y coordinate system. The most westerly station could be placed at the origin of the x-y coordinate system, which will make all the x-coordinates positive, as in Fig. 8.2(a). Alternately, the origin of the coordinate system could be chosen in such a way that all the x- and y-coordinates are positive, as in Fig. 8.2(b). The formulae to be derived will be applicable in all cases if proper signs are given to the coordinates as per the origin selected.

Calculation of coordinates

The coordinates of a traverse can be calculated from the latitudes and departures of lines, which are obtained from the lengths and bearings or other angles measured in the traverse. If the whole circle bearings are available, they can be converted into reduced bearings. The latitude of a line is parallel to the y-axis and the departure of a line is parallel to the x-axis. If l is the length of a line and θ is its reduced bearing, then the latitude of the line is $l \cos \theta$ and the departure of the line is $l \sin \theta$. In the traverse shown in Fig. 8.2(a), we calculate the coordinates with the data given in Table 8.1.

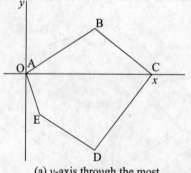

(a) y-axis through the most
westerly station

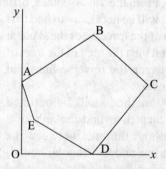

(b) y-axis through westerly station
and x-axis through southerly station

Fig. 8.2   Selection of coordinate axes

Table 8.1 Calculation of coordinates (A is the origin)

Line	Length (m)	Reduced bearing	Latitude	Departure	Station	x-coordinate	y-coordinate
AB	150.65	N62° 30′E	69.56	133.63	A	0	0
BC	142.85	S59° 45′E	− 71.96	123.4	B	133.63	69.56
CD	168.9	S30° 30′W	− 145.53	− 85.72	C	257.03	− 2.4
DE	121.55	N72° 15′W	37.05	− 115.76	D	171.31	− 147.93
EA	124.02	N26° 37′W	110.88	− 55.55	E	55.55	− 110.88

In Table 8.1, columns 1, 2, and 3 are given data. The latitudes and departures of the lines are calculated as explained in Chapter 7. The coordinates are calculated for the case shown in Fig. 8.2(a), i.e., with the origin at A. Knowing that the y-axis is parallel to the latitudes and the x-axis is parallel to the departures, the coordinates are calculated from known coordinates by algebraically adding the departures to the x-coordinates and the latitudes to the y-coordinates moving in the clockwise direction. The calculation of coordinates is done as follows. The coordinates of A are (0, 0) as A is the origin.

Station B: x-coordinate = 0 + 133.63 = 133.63

y-coordinate = 0 + 69.56 = 69.56

Station C: x-coordinate = 133.63 + 123.4 = 257.03

y-coordinate = 69.56 − 71.96 = − 2.4

Station D: x-coordinate = 257.03 − 85.72 = 171.31

y-coordinate = − 2.4 − 145.53 = − 147.93

Station E: x-coordinate = 171.31 − 115.76 = 55.55

y-coordinate = − 147.93 + 37.05 = − 110.88

As a check, if we add the departure and latitude of EA to the coordinates of E, we get the coordinates of A (0, 0).

Table 8.2 Calculation of coordinates (O is the origin)

Line	Length (m)	Reduced bearing	Latitude	Departure	Station	x-coordinate	y-coordinate
AB	150.65	N62° 30′E	69.56	133.63	A	0	147.93
BC	142.85	S59° 45′E	− 71.96	123.4	B	133.63	217.49
CD	168.9	S30° 30′W	− 145.53	− 85.72	C	257.03	145.53
DE	121.55	N72° 15′W	37.05	− 115.76	D	171.31	0
EA	124.02	N26° 37′W	110.88	− 55.55	E	55.55	37.05

For the case shown in Fig. 8.2(b), the origin is taken such that the y-axis passes through the most westerly station and the x-axis passes through the most southerly station. This makes all the coordinates positive, leading to fewer mistakes in calculations. The coordinates are calculated in the same way as in the previous case. This is shown in Table 8.2.

You will notice from Table 8.2 that the coordinates of all the stations are positive. From Fig. 8.2(b), it is clear that the origin is 147.93 (37.05 + 110.88) m below A and 171.31 (115.76 + 55.55) m to the left of D. This makes the coordinates of A (0, 147.93) and those of D (171.31,0). Starting from A, the coordinates can be calculated as in the previous case, as follows.

The coordinates of A are (0,147.93)

Station B: x-coordinate = 0 + 133.63 = 133.63

y-coordinate = 147.93 + 69.56 = 217.49

Station C: x-coordinate = 133.63 + 123.4 = 257.03

y-coordinate = 217.49 − 71.96 = 145.53

Station D: x-coordinate = 257.03 − 85.72 = 171.31

y-coordinate = 145.53 − 145.53 = 0

Station E: x-coordinate = 171.31 − 115.76 = 55.55

y-coordinate = 0 + 37.05 = 37.05

Thus, given the length and bearings of a line, the coordinates of the station points can be calculated with reference to the origin selected. These coordinates can be used to calculate the area of the traverse.

To calculate the area of a traverse, refer to the general traverse shown in Fig. 8.3. Let the coordinates of A be (x_a, y_a), B be (x_b, y_b), C be (x_c, y_c), D be (x_d, y_d), and E be (x_e, y_e). Perpendiculars are drawn from the vertices to the x-axis. The feet of these perpendiculars are named as A′, B′, C′, D′, and E′ in Fig. 8.3. ABB′A′, BCC′B′, etc. are trapeziums whose areas can be found from formulae. It is clear from the figure that

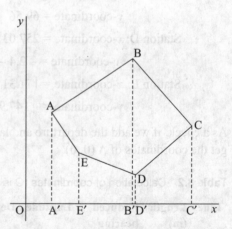

Fig. 8.3 Area from coordinates

$$\text{Area of ABCDE} = \text{area of ABB′A′} + \text{area of BCC′B′} − \text{area of AEE′A′}$$
$$− \text{area of EDD′E′} − \text{area of DCC′D′}$$

$$\text{Area of ABB′A′} = (1/2)(\text{AA′} + \text{BB′})(\text{OB′} − \text{OA′}) = (1/2)(y_a + y_b)(x_b − x_a)$$

$$\text{Area of BCC′B′} = (1/2)(\text{BB′} + \text{CC′})(\text{OC′} − \text{OB′}) = (1/2)(y_b + y_c)(x_b − x_c)$$

Area of AEE'A' = $(1/2)(AA' + EE')(OE' - OA') = (1/2)(y_a + y_e)(x_a - x_e)$

Area of EDD'E' = $(1/2)(EE' + DD')(OD' - OE') = (1/2)(y_d + x_e)(x_d - x_e)$

Area of DCC'D' = $(1/2)(DD' + CC')(OC' - OD') = (1/2)(y_c + y_d)(x_c - x_d)$

Area of ABCDE = $(1/2)(y_a + y_b)(x_b - x_a) + (1/2)(y_b + y_c)(x_b - x_c)$

$$- (1/2)(y_a + y_e)(x_a - x_e) - (1/2)(y_d + x_e)(x_d - x_e)$$

$$- (1/2)(y_c + y_d)(x_c - x_d)$$

Simplifying,

Area of ABCDE = $(1/2)[(y_a + y_b)(x_b - x_a) + (y_b + y_c)(x_b - x_c)$

$$- (y_a + y_e)(x_a - x_e) - (y_d + x_e)(x_d - x_e)$$

$$- (y_c + y_d)(x_c - x_d)]$$

$$= (1/2)(y_a x_b - y_a x_a + y_b x_b - y_b x_a + y_b x_c - y_b x_b + y_c x_c$$

$$- y_c x_b - y_a x_e + y_a x_a - y_e x_e + y_e x_a - y_e x_d + y_e x_e - y_d x_d$$

$$+ y_d x_e - y_d x_c + y_d x_d - y_c x_c + y x_d)$$

This reduces to

Area of ABCDE = $(1/2)[x_a(y_e - y_b) + x_b(y_a - y_c) + x_c(y_b - y_d)$

$$+ x_d(y_c - y_e) + x_e(y_d - y_e)]$$

This formula can be used to find the area of a traverse from coordinates. This formula need not be remembered and can be easily recalled by writing the coordinates as given below:

y	y_a	y_b	y_c	y_d	y_e	y_a
x	x_a	x_b	x_c	x_d	x_e	y_a

Note that since the traverse closes as ABCDEA, the coordinates of A appear in the beginning and at the end. In this determinant form, we find the algebraic sum of products of the terms in the first row with the terms in the second row of the next column, e.g., $y_a x_b$, $y_b x_c$, $y_c x_d$, and so on, and subtract the products of terms in the second row with terms of the first row of the next column, e.g., $x_a y_b$, $x_b y_c$, $x_c y_d$, and so on. Figure 8.4 shows how the products are to be found. The products of the terms connected with the solid line are added algebraically and the products of the terms connected by dotted lines are to be subtracted.

The general procedure for finding the area of a traverse using coordinates is the following.

1. Select the x-y coordinate axis system. The origin must preferably be through the most westerly station.

2. Calculate the latitudes and departures of the lines from the lengths and bearings.

y $\quad y_A \quad y_B \quad y_C \quad y_D \quad y_E \quad y_A$

x $\quad x_A \quad x_B \quad x_C \quad x_D \quad x_E \quad x_A$

(a) General form

y $\quad 0 \quad 69.56 \quad 2.4 \quad 147.93 \quad -110.88 \quad 0$

x $\quad 0 \quad 133.63 \quad 257.03 \quad 171.31 \quad 55.55 \quad 0$

(b) y-axis through A

y $\quad 147.93 \quad 217.49 \quad 145.53 \quad 0 \quad 37.05 \quad 147.93$

x $\quad 0 \quad 133.63 \quad 257.03 \quad 171.31 \quad 55.55 \quad 0$

(c) y-axis through A and x-axis though D

Fig. 8.4 Writing coordinates in determinant form

3. Calculate the (x, y) coordinates of the stations of the traverse from the known coordinates of a station. If the most westerly station is selected as the origin, its coordinates are $(0, 0)$. Add the departures and latitudes algebraically to the coordinates of the previous station to get the (x, y) coordinates of a station.
4. Use the formula or arrange the coordinates in determinant form in two rows. Find the algebraic sum of the products and divide the sum by 2 to obtain the area.

In the example given in Table 8.1, the area can be determined [from Fig. 8.4(b)] as follows.

$$2 \times \text{area} = 69.56 \times 257.03 + (-2.4) \times 171.31 + (-147.93) \times 55.55$$
$$- [133.63 \times (-2.4) + 257.03 \times (-147.93) + 171.31 \times (-110.88)]$$
$$= 66,588.4 \text{ sq. m}$$

Thus

$$\text{Area} = 33,294.2 \text{ sq. m}$$

In the same example, with the origin at O, from the coordinates given in Table 8.2, and referring to Fig. 8.4(c),

$$2 \times \text{area} = 147.93 \times 133.63 + 217.49 \times 257.03 + 145.53 \times 171.31$$
$$- (133.63 \times 145.53 + 171.31 \times 37.05 + 55.55 \times 147.93)$$
$$= 66,588.4 \text{ sq. m} \quad \text{(as before)}$$

and

$$\text{Area} = 33,294.2 \text{ sq. m}$$

8.1.2 Area from Latitudes and Double Meridian Distances

The second method to calculate the area of a traverse uses the concept of double meridian distances. The double meridian distance (DMD) of a line of a traverse is the sum of the meridian distances of the two ends of a line. The meridian distance is parallel to the departures or the east–west direction.

Calculating the double meridian distance

For calculating the meridian distances, it is necessary to select a reference meridian. The reference meridian (parallel to the north–south direction) can be selected anywhere, but it is convenient to select the most westerly station of the traverse. The meridian distance of a station point is the perpendicular distance of the station from the reference meridian. The double meridian distance of a line is the sum of the meridian distances of the end points of the line.

Figures 8.5(a) and (b) show a traverse with a reference meridian selected arbitrarily and at the most westerly station of the traverse, respectively. In case (a), the reference meridian is at a distance x from A. In case (b), the reference meridian is through A, the most westerly station.

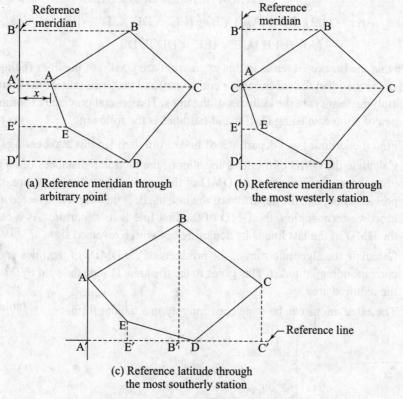

(a) Reference meridian through arbitrary point

(b) Reference meridian through the most westerly station

(c) Reference latitude through the most southerly station

Fig. 8.5   Double meridian distances

In case (a), the DMD of AB = x + BB′ = x + x + departure of AB. DMD of BC = BB′ + CC′ = x + departure of AB + x + departure of AB + departure of BC. (Note that BB′ = x + departure of AB and CC′ = x + departure of AB + departure of BC.) DMD of BC = DMD of AB + departure of AB + departure of BC. This gives us a general rule for calculating the DMD of a line: The DMD of a line is equal to the DMD of the previous line + departure of the previous line + departure of the line itself.

In case (b), the DMD of AB = 0 + departure of AB. DMD of BC = BB′ + CC′ = DMD of AB + departure of AB + departure of BC, noting that CC′ = departure of AB + departure of BC. The general rule derived in case (a) applies to this case as well.

Knowing the double meridian distances and latitudes of the lines, the area of the traverse can be obtained as half the sum of the products of the latitudes and the corresponding double meridian distances. For example, in Fig. 8.5(a), the reference meridian is through a point x m west of A. AA′, BB′, CC′, DD′, and EE′ are the perpendiculars drawn to the reference meridian. B′BCC′, C′CDD′, etc. are trapeziums and their area can be found using formulae. From the figure, it is clear that

$$\text{Area of ABCDE} = \text{ area of B′BCC′} + \text{area of C′CDD′} - \text{area of B′BAA′}$$
$$- \text{area of A′AEE′} - \text{area of E′EDD′}$$
$$\text{Area} = (1/2)[(BB′ + AA′)A′B′ + (CC′ + DD′)C′D′ - (BB′ + AA′)A′B′$$
$$- (AA′ + EE′)A′E′ - (EE′ + DD′)E′D′]$$

Note that the bracketed terms in this expression are DMDs of the lines (being the sum of the meridian distances of their end points) and the terms outside the brackets (multiplying terms) are the latitudes of the lines. The general procedure for finding the area of a traverse using DMDs and latitudes is the following.

1. Find the latitudes and departures of lines from their lengths and bearings.
2. Calculate the DMD of a line as the sum of the meridian distances of its ends or as the algebraic sum of the DMD of the previous line, the departure of the previous line, and the departure of the line itself. If the origin is selected at the most westerly station, the DMD of the first line is its departure. As a check, the DMD of the last line is its departure but with a reversed sign.
3. Calculate the algebraic sum of the products of the DMDs of the lines and the corresponding latitudes. This gives twice the area. Divide the sum by 2 to get the required area.
4. The calculations can be done conveniently in a tabular form.

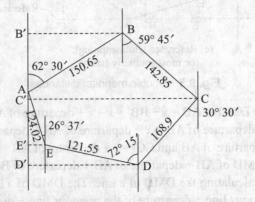

Fig. 8.6 Traverse with sample data (Table 8.1)

Note: It must be clear from the above discussion that the area of a traverse can be found using many methods. We used double meridian distances and latitudes. Similarly, it will be possible to use double latitudes and departures as well. In Fig. 8.5(c), for the traverse ABCDEA, we take the most southerly station D as the reference station and the east–west line through D as the reference line. The double latitude of a line is the algebraic sum of the latitudes of its end points. Therefore,

$$\text{Area of ABCDE} = \text{area of } (A'ABB'A' + B'BCC' - A'AEE'$$
$$- E'ED - DCC')$$
$$= (1/2) [(A'A + BB')A'B' + (BB' + CC')B'C'$$
$$- (AA' + EE')A'E' - (EE')E'D - (DC')CC']$$

The bracketed terms are double latitudes of the lines and the multiplying terms are the departures.

In Fig. 8.6, the traverse discussed earlier is shown with the reference meridian through the most westerly station A. The calculation of DMDs and areas is shown in Table 8.3.

Table 8.3 Area from double meridian distances and latitudes

Line	Length (m)	Bearing	Latitude (m)	Departure (m)	Double meridian distance	2 × area (latitude ×DMD; sq. m)
AB	150.65	N62° 30′E	69.56	133.63	133.63	9295.3
BC	142.85	S59° 45′E	− 71.96	123.4	390.66	− 28,111.9
CD	168.9	S30° 30′W	−145.53	− 85.72	428.34	− 62,336.3
DE	121.55	N72° 15′W	37.05	− 115.76	226.86	8405.2
EA	124.02	N26° 37′W	110.88	− 55.55	55.55	6159.4

In Table 8.3, the DMDs are calculated as follows:

DMD of AB = departure of AB = 133.63 m

DMD of BC = DMD of AB + departure of AB + departure of BC
$\qquad$ = 133.63 + 133.63 + 123.4 = 390.66.

DMD of CD = 390.66 + 123.4 + (− 85.72) = 428.34 m

DMD of DE = 428.34 + (− 85.72) + (− 115.76) = 226.86 m

DMD of EA = 226.86 + (− 115.76) + (− 55.55) = 55.55 m

Check: DMD of the last line is the departure of the line with a reversed sign.

The total of the last column of the table = 66,588.4 sq. m. Therefore, area of AB-CDE = 33,294.2 sq. m.

8.1.3 Area from Total Latitudes and Departures

A third method to calculate the area of a traverse is to use the total latitude and departure of a line. Any one of the stations can be taken as a reference station. The latitudes and departures can be calculated from the lengths and bearings of the lines of the traverse.

Calculating the total latitude

The total latitude of a station point (vertex of the traverse) is the algebraic sum of the latitudes of all the lines preceding that station from the reference station. Thus referring to Fig. 8.7(a), taking D, the most southerly station, as the reference point, total latitude of A = latitude of DA, total latitude of B = latitude of DA + latitude of AB, total latitude of C = algebraic sum of the latitudes of DA, AB, and BC, and so on. Taking the most southerly station as the reference station only makes the total latitudes of all the stations positive. The results will be the same whichever station is taken as the reference station.

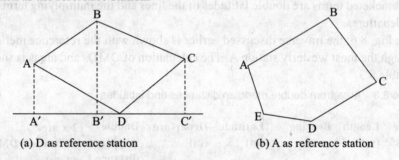

(a) D as reference station   (b) A as reference station

Fig. 8.7   Area from total latitudes and departures

In Fig. 8.7(a), AA', BB', and CC' are the total latitudes of the stations A, B, and C with respect to the reference station D. Total latitude of D = 0.

$$\text{Area of ABCD} = (1/2)[(AA' + BB')A'B' + (BB' + CC')B'C' - AA'$$
$$\times DA' - DC' \times C'C]$$
$$= (1/2)[AA' \times A'B' + BB' \times A'B' + B'B \times B'C' + C'C$$
$$\times B'C' - AA' \times DA' - DC' \times C'C]$$
$$= (1/2)[AA' (A'B' - DA') + BB'(B'A' + B'C')$$
$$+ CC' (B'C' - DC']$$

AA', BB', and CC' are the total latitudes of stations A, B, and C with respect to the reference station D. The term within brackets is the algebraic sum of the departures of the lines meeting at the respective stations. Thus the total latitudes multiplied by the algebraic sum of the departures of lines meeting at that station give twice the area. To find the area, we multiply the total latitude of a station by the algebraic sum of the departures of the lines meeting at that point. The total latitude of A is multiplied by the sum of the departures of DA and AB, the total latitude of B is multiplied by the algebraic sum of the departures of AB and BC, and so on. The algebraic sum of such products gives twice the area of the traverse. The general procedure for calculating the area using total latitudes is the following.

1. Calculate the latitudes and departures of the lines from their lengths and bearings.

2. Select one of the stations as the reference station. The total latitude of the reference station is zero.

3. Calculate the total latitude of the stations as the algebraic sum of the latitudes of the lines between the reference station and the station being considered.

4. For each station, find the algebraic sum of the departures of the adjacent lines (or the lines meeting at the station).

5. Find the algebraic sum of the products of the total latitudes and the sum of departures. This gives twice the area. Find the area by dividing the sum by 2.

6. Do the calculations conveniently in a tabular form.

Referring to Fig. 8.7(b), the most westerly station A is selected as the reference station. The total latitudes of the stations are A = 0, B = BB′, C = CC′, D = DD′, E = EE′. In the cases just discussed, the area can be calculated as follows using the data of Table 8.1. Table 8.4 shows the method of calculation.

Table 8.4 Area by total latitudes and sum of adjacent departures

Line	Length (m)	Bearing	Latitude (m)	Departure (m)	Station	Total latitude	Sum of departure of adjacent lines	2 × area (total lat. × sum of dep.; sq. m)
AB	150.65	N62° 30′E	69.56	133.63	B	69.56	257.03	17,879
BC	142.85	S59° 45′E	−71.96	123.4	C	−2.4	37.68	−90.43
CD	168.9	S30° 30′W	−145.53	−85.72	D	−147.93	−201.48	29,805
DE	121.55	N72° 15′W	37.05	−115.76	E	−110.88	−171.31	18,994.8
EA	124.02	N26° 37′W	110.88	−55.55	A	0.0	78.08	0

The total latitudes in Table 8.4 are calculated as follows:

Total latitude of A = 0,   total latitude of B = 0 + 69.56 = 69.56

Total latitude of C = 69.56 + (− 71.96) = − 2.4

Total latitude of D = − 2.4 − 145.53 = − 147.93

Total latitude of E = − 147.93 + 37.05 = − 110.88

Similarly, in the sum of departure of adjacent lines column, the sum for B = 133.63 + 123.4 = 257.03, for C = 123.4 − 85.72 = 37.68, for D = − 85.72 − 115.76 = −201.48, for E = − 115.76 − 55.55 = − 171.31, and for A = − 55.55 + 133.63 = 78.08.

The algebraic sum of the products in the last column = 62,588.4. Therefore, area = (1/2)(62,588.4) = 31,294.2 sq. m. Note: A negative sign to the calculated area has no significance and should be ignored.

8.1.4 Area by Dividing into Triangles

If appropriate measurements are taken during the fieldwork, it will be possible to find the area of a traverse from the area of the triangles into which it can be divided. In the traverse shown in Fig. 8.8, if a point inside the traverse is taken appropriately, the traverse can be divided into a number of triangles. In Fig. 8.8, the field measurements to be taken include the angles at O and the distances OA, OB, OC, OD, and OE. The lengths of the traverse lines can also be taken as a check. For example, knowing ∠AOB and the lengths OA and OB,

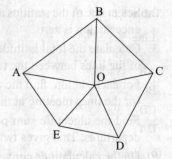

Area of triangle OAB = (1/2) OA × OB sin (∠OAB)

Similarly the areas of the other triangles can be calculated. The sum of these areas gives the area of the traverse.

Fig. 8.8 Area of traverse with central station

If the three sides of a triangle are known, the area can also be calculated using Hero's formula. If OA, AB, and OB are known,

$$\text{Area of AOB} = \sqrt{[S(S - OA)(S - AB)(S - OB)]}$$

where S is the half perimeter $= (OA + AB + OB)/2$. Similarly, the areas of other triangles can be calculated. The following examples further illustrate the principles involved.

Example 8.1 The lengths and bearings of a four-sided traverse are as follows. The lengths are AB = 320 m, BC = 440 m, CD = 390 m, and DA = 513.8 m. The reduced bearings of the lines are AB = N45° 15′E, BC = S71° 30′E, CD = S30° 15′W, and DA = N60° 44′W. Find the area of the traverse from (a) coordinates and (b) latitudes and double meridian distances.

Solution From the bearings and lengths, an approximation of the traverse is shown in Fig. 8.9(a). The calculations are conveniently done in tabular form as given in Table 8.5.

(a) *Area from coordinates* First the latitudes and departures are calculated. Taking A as the origin, the coordinates are calculated by algebraically adding the departures for the x-coordinates and the latitudes for the y-coordinates. The x- and y-coordinates are shown in Table 8.5.

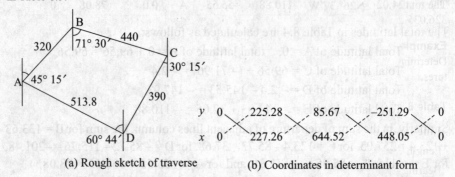

(a) Rough sketch of traverse

(b) Coordinates in determinant form

Fig. 8.9 Traverse for Example 8.1

The coordinates in determinant form are shown in Fig. 8.9(b). The products with the terms of the first row = 225.28 × 644.52 + 85.67 × 448.05 = 183,581.8. The products with the terms of the second row = 227.26 × 85.67 + 644.53 × (− 251.23) = − 142,453.4. Therefore, 2 × area = 1,835,818.8 + 142,453.4 = 326,035.2. Area = 163,017.6 sq. m.

Table 8.5 Area of traverse by coordinates

Line	Length (m)	Bearing	Latitude	Departure	Station	x-coordinate	y-coordinate
AB	320	N45° 15′E	225.28	227.26	B	227.26	225.28
BC	440	S71° 30′E	−139.61	417.26	C	644.52	85.67
CD	390	S30° 15′W	−336.9	−196.47	D	448.05	−251.23
DA	513.8	N60° 44′W	251.23	−448.05	A	0	0

(b) *Area from double meridian distances* The calculations are conveniently done in tabular form as shown in Table 8.6. The latitudes and departures are calculated from the lengths and bearings. Taking A, the most westerly station, as a point through the reference meridian, the double meridian distances of all the lines are calculated.

Table 8.6 Area of traverse from DMDs and latitudes

Line	Length (m)	Bearing	Latitude	Departure	DMDs	DMD × latitude
AB	320	N45° 15′E	225.28	227.26	227.26	51,197.1
BC	440	S71° 30′E	− 139.61	417.26	871.78	− 368,086.8
CD	390	S30° 15′W	− 336.9	− 196.47	1092.57	− 121,709.2
DA	513.8	N60° 44′W	251.23	− 448.05	448.05	112,563.6

The DMDs of the lines are calculated as follows:

DMD of AB = departure of AB = 2227.26

DMD of BC = DMD of AB + departure of AB + departure of BC

= 227.26 + 227.26 + 417.26 = 871.78

DMD of CD = 871.78 + 417.26 + (− 196.47) = 1092.57

DMD of DA = 10,922.57 + (− 196.47) + (− 448.05) = 448.05

Check: The DMD of the last line is equal to the departure of the line with a reversed sign. The total of the last column in Table 8.6 is equal to twice the area. This algebraic sum is 326,035.2. This gives the area = (1/2)(326,035.2) = 163,017.6 sq. m.

Example 8.2 The lengths and bearings of the lines of a traverse are given in Table 8.7. Determine the area of the traverse from the coordinates and the total latitudes and departures.

Table 8.7 Traverse data

Line	AB	BC	CD	DE	EF	FA
Length	350	330	310	370	300	288
Bearing	60° 25′	94° 30′	154° 15′	244° 30′	277° 15′	331° 46′

Solution An approximate sketch of the traverse is shown in Fig. 8.10.

From coordinates The most westerly station, A, is selected as the origin. The bearings given are whole circle bearings and will have to be changed to reduced bearings. The latitudes and departures of the lines are then calculated. The coordinates of station A are (0, 0). The other coordinates are calculated from the latitudes and departures of the lines as shown in Table 8.8.

Table 8.8 Calculation of coordinates

Line	Length	Bearing	Latitude	Departure	Station	x-coordinate	y-coordinate
AB	350	N62° 25′E	172.79	304.37	A	0	0
BC	330	S85° 30′E	− 25.89	329	B	304.37	172.79
CD	310	S25° 45′E	− 279.21	134.68	C	633.37	146.9
DE	370	S64° 30′W	− 159.29	− 333.96	D	768.05	− 132.31
FA	300	N82° 45′W	37.86	− 297.6	E	434.09	− 291.6
FA	288	N28° 14′W	253.74	− 136.49	F	136.49	− 253.4

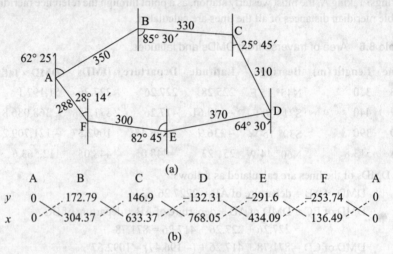

Fig. 8.10 Traverse for Example 8.2

Taking A as the origin, the coordinates are calculated as follows:

Coordinates of A = (0, 0)

B: x-coord. = 0 + 304.37 = 304.37

y-coord. = 0 + 172.79 = 172.79

C: x-coord. = 304.37 + 329 = 633.37

y-coord. = 172.79 − 25.89 = 146.9

D: x-coord. = 633.37 + 134.68 = 768.05

y-coord. = 146.9 − 279.21 = − 132.31

E: x-coord. = 768.05 − 333.96 = 434.09

y-coord. = 132.31 − 159.29 = − 291.6

F: x-coord. = 434.09 − 297.6 = 136.49

y-coord. = − 291.6 + 37.86 = 253.74

Check: A: x-coord. = 136.49 − 136.49 = 0, y-coord. = − 253.74 + 253.74 = 0. The x-y co-ordinates are written in determinant form in Fig. 8.10(b). The area can then be computed as follows:

$$\text{Area} = (1/2)[(172.79 \times 633.37 + 146.9 \times 768.05 + (- 132.31) \times 434.09$$
$$+ (- 2291.6) \times 136.49 - 304.37 \times 146.9 + 633.37 \times (- 132.31)$$

$$+ 768.05 \times (-291.6) + 434.09 \times (-253.04)]$$
$$= (1/2)(498,230.4) = 249,115.2 \text{ sq. m}$$

From total latitudes and sum of departures The total latitudes are calculated by taking A, the most westerly station as the reference station. Thus, the total latitude of A = 0, of B = 0 + 172.79 = 172.79, of C = 172.79 + (– 25.89) = 146.9, of D = 146.9 + (– 279.21) = – 132.31, of E = – 132.31 – 159.29 = – 291.60, of F = – 291.6 + 37.86 = – 253.74. The sum of adjacent departures for each station is calculated as

B: 304.37 + 329 = 633.37,   C: 329 + 134.68 = 463.68

D: 134.68 – 333.96 = – 199.28,   E: – 333.96 – 297.6 = – 631.56

F: – 136.4 – 2297.6 = – 434.09,   A: – 136.49 + 304.37 = 167.88

These are tabulated in Table 8.9.

Table 8.9 Calculation of total latitudes and sum of departures

Line	Length	Bearing	Latitude	Departure	Station	Total latitude	Sum of adjacent departures
AB	350	N62° 25′E	172.79	304.37	B	172.79	633.37
BC	330	S85° 30′E	– 25.89	329	C	146.9	463.68
CD	310	S25° 45′E	– 279.21	134.68	D	–132.31	– 199.28
DE	370	S64° 30′W	– 159.29	– 333.96	E	–291.6	– 631.56
FA	300	N82° 45′W	37.86	– 297.6	F	–253.74	– 434.09
FA	288	N28° 14′W	253.74	– 136.49	A	0	167.88

2 × area = algebraic sum of products of total latitude and sum of adj. departures

$$= 172.79 \times 633.37 + 146.9 \times 463.68 + (-132.31) \times (-199.28)$$
$$+ (-291.6) \times (-631.56) + (-253.74) \times (-434.09)$$
$$= 109,440 + 68,114.6 + 26,366.7 + 184,162.9 + 110,146$$
$$= 498,230.2$$

Area = 249,115.1 sq. m

Example 8.3 For settling a property dispute, a surveyor was asked to demarcate an area of nearly 18,500 sq. m out of a larger area. After surveying the area and consulting the clients, he found that the area will have the shape of a parallelogram with sides ratio 1:2.8; the shorter side will be along the bearing 149° 30′ and the longer side will be along the east–west direction. Determine the lengths of the sides of the area.

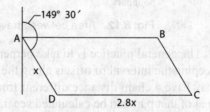

Fig. 8.11 Traverse for Example 8.3

Solution The situation can be represented as shown in Fig. 8.11.

We can use any of the methods outlined above. Assume the shorter side is of length *x* and the longer side of length 2.8*x*. We use the DMD method here. Table 8.10 shows the calculation of DMDs and the products of the latitudes and DMDs.

Table 8.10   Calculation of DMDs

Line	Length (m)	Bearing	Latitude	Departure	DMD	DMD×latitude
AB	2.8x	N90°E	0	2.8x	2.8x	0
BC	x	S30° 30′E	−0.8616x	0.5075x	5.6x	−4.825x²
					+0.5075x	−0.4372x²
CD	2.8x	N90°W	0	−2.8x	2.8x	0
					+0.5075x	
DA	x	N30° 30′W	0.8616x	−0.5075x	0.5075x	0.4372x²

Neglecting the negative sign, $2 \times \text{area} = 4.825x^2 = 2 \times 18,500$; $x = 87.569$ m. Shorter side = 87.569 m and longer side = $2.8 \times 87.569 = 245.193$ m.

8.2   Area Between a Traverse Line and an Irregular Boundary

In surveying the area of a piece of land with an irregular boundary, as mentioned earlier, a traverse is run within the area as close to the boundary as possible. Offsets are then taken from the survey line to the irregular boundary. As shown in Fig. 8.12(a), the survey consists of a traverse and offsets to the irregular boundary. In this section, we will look at methods to find the area of the land between the survey line and the boundry.

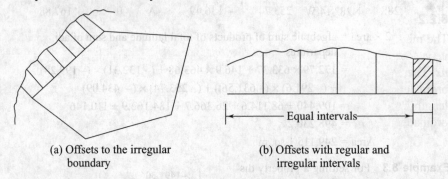

(a) Offsets to the irregular   (b) Offsets with regular and
    boundary   irregular intervals

Fig. 8.12   Area between traverse line and irregular boundary

The general practice is to take perpendicular offsets to points on the boundary, keeping the interval of offsets along the chain line constant. The first or last offset may have a chain distance different from the constant interval, in which case the area of that part can be calculated separately and added. As shown in Fig. 8.12(b), the shaded area, which has an offset interval different from the constant interval used for all other offsets, can be calculated separately. In case the offset interval cannot be kept constant due to obstacles or any other reason, the area can be calculated by the method of coordinates outlined earlier. In general, the larger the number of sections at which offset is taken (or smaller the offset interval), the more accurate the area calculated. Some of the methods to calculate the area are as follows.

8.2.1 Method of Mid-ordinates

There are many ways to find the area of such a narrow tract of land. In the method of mid-ordinates, the offsets are taken at the midpoints of the sections of constant interval into which the chain line is divided (see Fig. 8.13). y_1, y_2, y_3, etc. are the

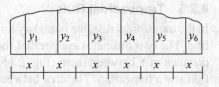

Fig. 8.13 Area from mid-ordinates

ordinates at the midpoints of the sections of length x, which is the constant offset interval. The offsets are taken at $x/2, 3x/2, 5x/2$, etc. chainages along the length of the line. The area is calculated as the area of rectangle of width x and height y_n. Thus, in Fig. 8.13, the area of the land between the survey line, the first and last offsets, and the boundary line can be calculated as follows:

$$\text{Area} = xy_1 + xy_2 + xy_3 + \cdots + xy_{n-1} + xy_n = x(y_1 + y_2 + y_3 + \cdots + y_{n-1} + y_n)$$
$$= (L/n)(y_1 + y_2 + y_3 + \cdots + y_{n-1} + y_n) \quad \text{(as } x = L/n\text{)}$$

In the above formula, x is the constant distance between the offsets, n is the number of sections, L is the length of the chain line, and $y_1, y_2, y_3, \cdots y_{n-1}, y_n$ are the perpendicular offsets. The mid-ordinate formula can be stated as follows: *Multiply the sum of all the mid-ordinates by the distance between the sections or by the length of the line divided by the number of sections*. This method requires the offsets to be taken at the midpoint of the section. In all the other methods that follow, the offsets are taken at the section points.

8.2.2 Method of Average Ordinates

The method of average ordinates is explained with reference to Fig. 8.14. $y_0, y_1, y_2, y_3, \cdots, y_{n-1}, y_n$ are the offsets taken at chainages $0, x, 2x, 3x, 4x$, and so on up to nx. There are n divisions and $n + 1$ ordinates. The method involves averaging the ordinates by dividing their sum by $n + 1$. The average ordinate is multiplied by the length L ($= nx$) to give the area of the piece of land lying between the first and last ordinates, the chain line, and the boundary. Thus,

Average ordinate $= (y_0 + y_1 + y_2 + \cdots + y_n)/(n + 1)$

Length of base line $= nx = L$

Area $= (y_0 + y_1 + y_2 + \cdots + y_n)[L/(n + 1)]$

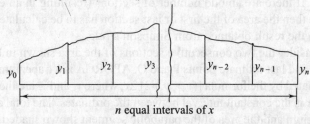

n equal intervals of *x*

Fig. 8.14 Area from average ordinate

If the ordinates have a reasonably constant length with only a little variation, this will give a result close to the true value. The average ordinate rule can be stated as follows: *Find the average of the ordinates and multiply by the length of the line.* Note that there are $n + 1$ ordinates.

8.2.3 Trapezoidal Rule

In the trapezoidal rule, the boundary line between two ordinates is taken as a straight line. The two ordinates are parallel (being perpendicular to the chain line) with a distance of x between them (Fig. 8.15). Thus the area corresponding to each section x is a trapezium. The area between ordinates y_0 and y_1 is $= (x/2)(y_0 + y_1)$. The area between ordinates y_1 and y_2 is $= (x/2)(y_1 + y_2)$, and so on. The area of the last section is $= (x/2)(y_{n-1} \ y_n)$.

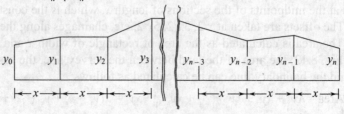

Fig. 8.15 Trapezoidal rule

Adding all the areas,

$$\text{Total area} = (x/2)[y_0 + 2y_1 + 2y_2 + \cdots + 2y_{n-1} + y_n)$$

This can also be expressed as

$$\text{Area} = x[(y_0 + y_n)/2 + y_1 + y_2 + y_3 + \cdots + y_{n-1}]$$

In this formula, note that $x = L/n$ is the constant interval between offsets and y_0, $y_1, y_2, y_3, \ldots, y_{n-1}, y_n$ are the perpendicular offsets. The trapezoidal formula can be stated as follows: *To the sum of the first and last ordinates add twice the sum of the other ordinates and multiply the result by half the interval between the offsets* or *to the average of the first and last ordinates add the sum of the other ordinates and multiply this sum by the distance between the ordinates.*

8.2.4 Simpson's Rule or Parabolic Rule

In Simpson's rule, the boundary line segment between the ordinates is assumed to be a parabola hence the name parabolic rule. The area is shown in Fig. 8.16(a). The formula finds the area using three consecutive ordinates and two sections at a time. Hence, *the formula requires an even number of sections or an odd number of ordinates.* If there are an odd number of sections (resulting in an even number of ordinates) then the area of the first or last section has to be calculated separately and added to the result obtained from Simpson's rule.

Let us consider the two consecutive sections of the area shown in Fig. 8.16(b). Join the first and last ordinate points P and Q. APQRCA is a trapezium and its area can be calculated by the formula $[(y_0 + y_2)/2] \ 2x$, where y_0 and y_2 are the ordinates at the end and x is the constant interval between the ordinates. The total area consists of this trapezium and the area of the parabolic segment shown shaded. This area is two-thirds of the area of the enclosing parallelogram. The base of this parallelogram can be taken as $2x$ and the height as MQ. The height $MQ = BM - BQ$. $BM = y_1$ and $BQ = (y_0 + y_2)/2$. Thus the shaded area is the following:

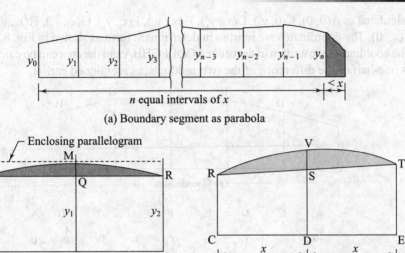

(a) Boundary segment as parabola

(b) Area between three ordinates

(c) Trapezium and parabolic areas

Fig. 8.16 Simpson's rule

Shaded area = $(2/3)[y_1 - (y_0 + y_2)/2](2x)$

The total area between the first and last ordinates, the two sections of the chain line, and the boundary is the sum of these two areas:

Area APMRC = $(y_0 + y_2)(2x/2) + (2/3)[y_1 - (y_0 + y_2)/2](2x)$

$= (x/3)[y_0 + 4y_1 + y_2]$

The next two sections will be as shown in Fig. 8.16(c) and the area will be given as follows:

Area CETVR = $(x/3)[y_2 + 4y_3 + y_4]$

When we sum up these partial areas,

Total area = $(x/3)[y_0 + 4y_1 + 2y_2 + 4y_3 + 2y_4 + \ldots + 2y_{n-2} + 4y_{n-1} + y_n]$

Simpson's rule can be stated as follows: *To the sum of the first and last ordinates, add twice the sum of the third, fifth, seventh, etc. ordinates and four times the sum of the second fourth, sixth, etc. ordinates and multiply this sum by x/3, where x is the distance between the ordinates.*

To apply Simpson's rule, we need to have *even number of sections* (or *odd number of ordinates*).

8.2.5 Areas from Coordinates

We have outlined the method of coordinates earlier. While the rules or methods stated so far all require ordinates at equal intervals to apply the formula fully or partially, the method of coordinates is particularly suitable when the ordinates are not taken at equal intervals due to any reason.

In Fig. 8.17(a), an area is shown with ordinates named as before. The distances from A to the ordinates (offset points) are x_1, x_2, x_3, x_4, x_5 and C, D, E, F, G, H are points on the boundary. With A as the origin, the coordinates of all the points can be calculated as A(0, 0), C(0, y_0), D(x_1, y_1), E (x_2, y_2), F(x_3, y_3), G(x_4, y_4), H(x_5, y_5), and B(x_5, 0). The coordinates are written in determinant form as shown in Fig. 8.17(b). The coordinates are written in the order ACDEFGHBA and the area can be calculated as one-half of the difference of the two products, as mentioned earlier.

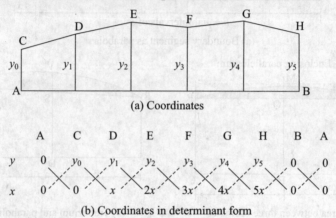

(a) Coordinates

(b) Coordinates in determinant form

Fig. 8.17 Area of an irregular boundary by coordinates

$$\text{Area} = (1/2)[y_0x_1 + y_1x_2 + y_2x_3 + y_3x_4 + y_4x_5 + y_5x_5]$$
$$- [x_1y_2 + x_2y_3 + x_3y_4 + x_4y_5]$$

Comparison of rules

The average ordinate rule will not give correct results if the range of values of the ordinates is large. The trapezoidal rule and Simpson's rule are the most commonly used rules. Simpson's rule is considered generally more accurate, as the assumption of a curved boundary is more realistic compared to the assumption of a straight-line boundary between ordinates made in the trapezoidal rule. The method of coordinates and the trapezoidal rule give the same results, as both are based on the same assumption. If the boundary is concave towards the survey line, the trapezoidal rule will calculate less area than the Simpson's rule, while if the boundary is convex towards the survey line, Simpson's rule will calculate a lower value. The following examples illustrate the use of these formulae.

Example 8.4 Offsets were taken from a chain line to a curved boundary. The chain line was 50 m long and was divided into 10 sections and offsets were taken to the middle of each section at 5 m, 15 m, 25 m, 35 m, and 45 m. The lengths of the offsets were 5.4 m, 6.8 m, 8.4 m, 7.5 m, and 7.2 m. Calculate the area between the chain line and the boundary using the mid-ordinate rule.

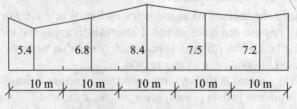

Fig. 8.18 Area for Example 8.4

Solution The situation is shown in Fig. 8.18. According to the mid-ordinate rule,

Area = $(5.4 + 6.8 + 8.4 + 7.5 + 7.2) \times 10 = 353$ sq. m

Example 8.5 Eleven offsets were taken from a chain line to a curved boundary at 10-m intervals and the lengths of the offsets from the left end are (in metres) 3.8, 5.1, 6.5, 6.8, 5.9, 6.2, 7.0, 6.6, 5.8, and 4.2. Determine the area between the chain line, the curved boundary, and the first and last offsets by (a) the average ordinate rule, (b) the trapezoidal rule, and (c) the parabolic rule.

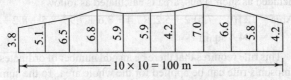

Fig. 8.19 Area for Example 8.5

Solution The area is shown in Fig. 8.19.

(a) *Average ordinate rule* The average of the offsets is determined and multiplied by the total length of the line to get the area. The sum of the offsets is the following:

Sum of offsets = 3.8 + 5.1 + 6.5 + 6.8 + 5.9 + 5.9 + 6.2 + 7.0 + 6.6 + 5.8

+ 4.2 = 63.8 m

There are 11 ordinates. Thus,

Area = $(63.8/11) \times 100 = 580$ sq. m

(b) *Trapezoidal rule* By this method, the area is given by multiplying the offset interval by the sum of the average of the first and last offsets and the sum of the remaining offsets. The offset interval is 10 m. The first offset is 3.8 m and the last offset is 4.2 m.

Area = 10[(3.8 + 4.2)/2 + 5.1 + 6.5 + 6.8 + 5.9 + 5.9 + 6.2 + 7.0 + 6.6

+ 5.8] = 598 sq. m

(c) *Parabolic rule* The parabolic rule requires even number of sections (or odd number of ordinates). There are 11 ordinates, which means that the parabolic rule can be applied for the whole area. The rule states that one-third the offset interval is multiplied by the sum of the first and last offsets, four times the sum of the second, fourth, sixth, etc. ordinates and two times the sum of the third, fifth, seventh, etc. ordinates. Thus,

Area = (10/3)[3.8 + 4.2 + 4(5.1 + 6.8 + 5.9 + 7.0 + 5.8)

+ 2(6.5 + 5.9 + 6.2 + 6.6)]

= 602.67 sq. m

Example 8.6 A chain line was divided into eight sections of 12 m each and offsets were taken from the chain line to a hedge. The lengths of the offsets were (in metres): 0, 5.2, 7.4, 8.6, 7.9, 8.5, 8.2, 9.1, and 7.6. Find the area between the chain line, the first and last offsets, and the boundary by (a) the average ordinate rule, (b) the trapezoidal rule, and Simpson's rule.

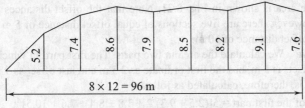

Fig. 8.20 Area for Example 8.6

Solution   The area is shown in Fig. 8.20.

Average ordinate rule   The sum of the ordinates is divided by the number of ordinates and multiplied by the length of the line. The sum of the ordinates is the following:

$$Sum = 0 + 5.2 + 7.4 + 8.6 + 7.9 + 8.5 + 8.2 + 9.1 + 7.6 = 62.5$$
$$Area = (62.5/9) \times 12 \times 8 = 666.67 \text{ sq. m}$$

Trapezoidal rule   To the average of the first and last offsets, the remaining offsets are added and this sum is multiplied by the equal distance between offsets. The first ordinate is zero and should be included as such. Thus, area is calculated as follows:

$$Area = 12 [(0 + 7.6)/2 + 5.2 + 7.4 + 8.6 + 7.9 + 8.5 + 8.2 + 9.1]$$
$$= 704.4 \text{ sq. m}$$

Simpson's rule   This rule requires that there be an odd number of ordinates. There are nine ordinates and Simpson's rule can be applied for the whole area. To the sum of the first and last ordinates, we add four times the even ordinates and two times the odd ordinates. This sum multiplied by one-third the equal distance between the offsets gives the area. The area can thus be calculated as follows:

$$Area = (12/3)[0 + 7.6 + 4(5.2 + 8.6 + 8.5 + 9.1) + 2(7.4 + 7.9 + 8.2)]$$
$$= 720.8 \text{ sq. m}$$

Note that zero should be included as the first ordinate.

Example 8.7   The offsets (in metres) taken from a chain line to a curved boundary are given below:

Chainage	0	5	10	15	20	25	35	45	55	65
Offset	2.5	3.8	8.4	7.6	10.5	9.3	5.8	7.8	6.9	8.4

Find the area between the chain line, the first and last offsets, and the boundary by (a) the trapezoidal rule, (b) Simpson's rule, and (c) coordinates.

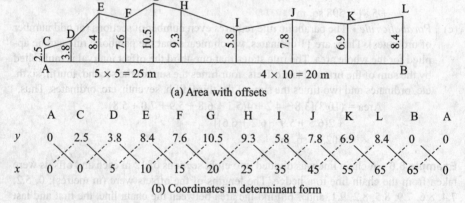

(a) Area with offsets

(b) Coordinates in determinant form

Fig. 8.21   Area for Example 8.7

Solution   The area is shown in Fig. 8.21. Note that the offset distances are not equal throughout. However, there are five sections of equal offset distance of 5 m and four sections of equal offset distance of 10 m.

Trapezoidal rule   We calculate the area in two parts. The first part, in which the distance between the offsets is 5 m, has the first ordinate as 2.5 m and last offset as 9.3 m. The area of the first part is, therefore, calculated as follows:

$$Area of the first part = 5[(2.5 + 9.3)/2 + 3.8 + 8.4 + 7.6 + 10.5]$$
$$= 181 \text{ sq. m}$$

For the seond part, for which the offset distance is 10 m, the first offset is 9.3 m and the last offset is 8.4 m. The area of this part is calculated as follows:

Area of the second part = 10[(9.3 + 8.4)/2 + 5.8 + 7.8 + 6.9] = 293.5 sq. m

Total area = 474.5 sq. m

Simpson's rule   Simpson's rule requires an odd number of ordinates. The 5-m offset interval section has six ordinates and hence one section has to be calculated separately. We calculate the area between the first and second ordinates separately as the area of a trapezium. The remaining part of the 5-m sections will be calculated by applying Simpson's rule. The 10-m section has five ordinates and hence this area can be calculated by Simpson's rule directly.

Area between first and second ordinates = 5(2.5 + 3.8)/2 = 15.75 sq. m
Area between second and sixth ordinates = (5/3)[3.8 + 9.3 + 4(8.4 + 10.5) +
2 × 7.6]
= 173.17 sq. m
Area between sixth and tenth ordinates = (10/3)[9.3 + 8.4 + 4(5.8 + 6.9)
+ 2 × 7.8] = 280.33
Total area = 469.25 sq. m

Coordinates method   The method of coordinates can be applied irrespective of the offset distances. AB is the chain line and C, D, E, F, G, H, I, J, K, and L are the points on the boundary. The coordinates of these points, taking A as the origin, are A(0, 0), B(65, 0), C(0, 2.5), D(5, 3.8), E(10, 8.4), F(15, 7.6), G(20, 10.5), H(25, 9.3), I(35, 5.8), J(45, 7.8), K(55, 6.9), and L(65, 8.4). These coordinates are arranged in the determinate form, as shown in Fig. 8.21(b). The area can be found as follows:

Area = (1/2)[(2.5 × 5 + 3.8 × 10 + 8.4 × 15 + 7.6 × 20 + 10.5 × 25 + 9.3
× 35 + 5.8 × 45 + 7.8 × 55 + 6.9 × 65 + 8.4 × 65) − (5 × 8.4 + 10
× 7.6 + 15 × 10.5 + 20 × 9.3 + 25 × 5.8 + 35 × 7.8 + 45 × 6.9 + 5
× 8.4)]
= (1/2)(2601 − 1652) = 474.5 sq. m

Note that the trapezoidal rule and the coordinates method give the same result because both consider the boundary to be straight line between ordinates.

Example 8.8   The offsets from a chain line to a curved fence are given below:

Chainage	0	7	12	18	25	32	42	48	55	65
Offsets	3.4	6.2	6.8	5.9	8.4	6.2	10.3	11.5	9.8	8.5

All distances are in metres. Find the area between the chain line, the boundary, and the first and last offsets.

Solution   A sketch of the area is shown in Fig. 8.22. In this case, we notice that the offsets are not taken at regular intervals. Assuming that the boundary between the offsets is a straight line, the area can be divided into nine trapeziums. Thus, the area has to be calculated by the coordinates method or by calculating the areas of the individual trapeziums and summing them up.

Area of a trapezium = (sum of adjacent ordinates/2) × distance between ordinates

Area = (3.4 + 6.2)/(2 × 7) + (6.2 + 6.8)/(2 × 5) + (6.8 + 5.9)/(2 × 6)
+ (5.9 + 8.4)/(2 × 7) + (8.4 + 6.2)/(2 × 7) + (6.2 + 10.3)/(2 × 10)
+ (10.3 + 11.5)/(2 × 6) + (11.5 + 9.8)/(2 × 7) + (9.8 + 8.5)/(2 × 10)
= 33.6 + 32.5 + 38.1 + 50.05 + 51.1 + 82.5 + 65.4 + 74.55 + 91.50
= 519.3 sq. m

Coordinates method   The base line is named AB and the points on the boundary are named C, D, E, etc. The coordinates of these points are A(0, 0), C(0, 3.4), D(7, 6.2), E(12, 6.8), F(18, 5.9), G(25, 8.4), H(32, 6.2), I(422, 10.3), J(48, 11.5), K(55, 9.8), L(65, 8.5), B(65, 0). These coordinates can be arranged in determinant form as shown in Fig. 8.22(b). The area can then be found as follows:

$$\text{Area} = (1/2)[(3.4 \times 7 + 6.2 \times 12 + 6.8 \times 18 + 5.9 \times 15 + 8.4 \times 32 + 6.2 \times 42 + 10.3 \times 48 + 11.5 \times 55 + 9.8 \times 65 + 8.5 \times 65) - (7 \times 6.8 + 12 \times 5.9 + 18 \times 8.4 + 25 \times 6.2 + 32 \times 10.3 + 42 \times 11.5 + 48 \times 9.8 + 55 \times 8.5)]$$

$$= 519.3 \text{ sq. m}$$

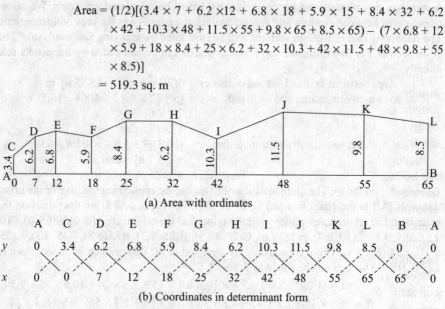

(a) Area with ordinates

(b) Coordinates in determinant form

Fig. 8.22   Area for Example 8.8

8.3 Determining Areas from Plans

Often one is required to find areas from plans drawn to scale available instead of survey data. In such cases, it is possible to determine the area by the following methods.

8.3.1 Determining Area Using a Tracing Paper

A tracing paper, with either squares, triangles, or straight lines drawn on it, can be used for determining areas from plans. Squares are more commonly used. The plan would have been drawn to a scale that will be indicated in it. Knowing the scale, it will be possible to calculate the actual area that each square in the tracing paper represents. Knowing this, it will be possible to find the area of the plan as follows. The tracing paper is laid over the plan. Within the boundary of the area, the number of full squares is counted. Along the boundary of the plan it will be necessary to use judgement to find the area. Generally, more than half squares are counted as full squares and less than half squares are discarded. This is called *balancing* or *equalizing*. Exactly half squares may be counted as half-squares themselves. With the number of squares counted, it will be possible to find the area from the scale.

As an example, consider the area shown in Fig. 8.23. A tracing paper with squares drawn on it is used to find the area. The plan of the area is drawn to a scale of 1 cm = 200 m. With the tracing paper placed on the area, the number of full squares is counted first. Then, along the boundary, squares covered by more than half by the boundary are counted. Then, the exactly half squares are counted. Let the number of full squares be 54, more than half squares counted full be 18, and exactly half squares be 8. Each square is 1 sq. mm in area. The area represented by it as per scale will be = $200 \times 200/100$ sq. m or 400 sq. m.

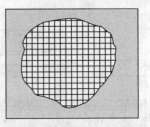

No. of sq. mm counted = $54 + 18 + 8/2 = 76$

Area = $76 \times 400 = 30{,}400$ sq. m

Fig. 8.23   Area using tracing paper with squares

In using a tracing paper with horizontal lines, the area can be divided into a number of strips. In order to fit the area exactly within the boundary of the area, the longest dimension can be measured and this can be divided into an equal number of divisions. A tracing paper ruled with this strip width can be used for measuring the area as shown in Fig. 8.24.

Equalizing or balancing is again done at the boundary as shown in Fig. 8.24. In the tracing paper, the middle lines of the strips are also marked in a different colour or as dotted lines. The idea is to measure the middle line lengths of each strip after balancing. Thus the lengths of A′B′, C′D′, E′F′, etc. are measured. The sum of these lengths multiplied by the constant distance between the lines gives the area of the plan. This multiplied by the scale factor for the drawing will give the actual area on the ground.

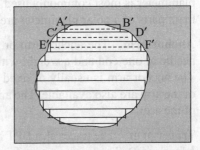

Fig. 8.24   Area using tracing paper with parallel lines

To aid the balancing at the boundary and the measurement of length, one can use the computing scale shown in Fig. 8.25. The computing scale is a scale with an attachment with a cursor line that can be moved along the scale. The cursor helps to finely equalize at the boundary and the movement of the cursor measures the length. To use the computing scale, the cursor is set at the zero of the scale and kept at the left end of the strip. The scale is physically moved to position it at a balancing position. The cursor can now be moved to the right end and placed at a balancing position. The movement of the cursor measures the length of the line. The cursor can now be moved to the left end of the second line and placed to balance the boundary area with the same reading obtained at the right end of the line. The process can be repeated for all the lines. The final reading on the scale gives the total length of the line.

As an example, consider the area shown in Fig. 8.25(b). The plan is drawn to a scale of 1:250. The distance between the lines is 10 mm. The total length measured, either as the middle lengths of the strips or using a computing scale, is 624 mm. The area as measured on the plan is, thus, 6240 sq. mm. Each square millimetre of the plan measures an area of 250×250 sq. mm or 62,500 sq. mm. The total area is thus $62{,}500 \times 6240$ sq. mm, which is = 390,000,000 sq. mm or 390 sq. m.

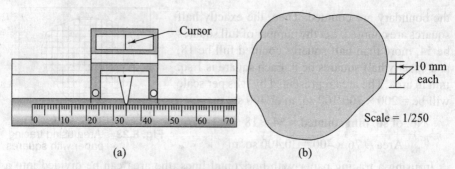

(a)   (b)

Fig. 8.25 Computing scale

8.3.2 Determining Area Using a Planimeter

The planimeter is an instrument used to measure areas. There are two types of planimeters—the polar planimeter and the roller planimeter. Of these, the polar planimeter is most commonly used. A polar planimeter is shown in Fig. 8.26. The main parts of a polar planimeter are the following.

Anchor arm The anchor arm or polar arm AP (refer to Fig. 8.26) has a pivot needle at one end and a weight with a pin below to anchor the arm to the paper. The anchor arm is usually of fixed length, but instruments with variable lengths of the arm are also available. The anchor arm is also graduated to know the length of the arm.

Tracer arm The tracer arm BPT has a pivot point at P, a tracing point at T, and a free end B. The tracing point T is a fine needle point which is used to trace the boundary of the area. The point may also have a magnifying attachment to enable accurate tracing of the boundary. The tracing arm is also graduated.

Measuring unit The measuring unit has a graduated disc with an index which is attached with a worm gear system to a wheel which has a graduated drum. The drum also has an attached vernier for finer reading. The wheel of the measuring unit may be attached to the tracing arm, either between the pivot point and the tracing point or between the pivot point and the free end of the tracing arm. When attached to the tracing arm, the wheel axis is parallel to the tracing arm (or the plane of the wheel is at right angles to the tracing arm). The measuring unit is fixed to the tracing arm with a fixing screw, which also has a slow motion screw for accurate placement of the bevelled edge of the measuring unit on the tracing arm.

When the tracing point is moved over the boundary of the area, the wheel rolls due to movement perpendicular to the tracing arm and slides due to movement parallel to the tracing arm. The sliding does not cause any change in the reading of the graduated drum but the movement perpendicular to the tracing arm causes a change in the reading. The difference in the reading can be used to calculate the area as will be explained later.

The reading from the planimeter consists of four digits—the reading of the index on the graduated disc, where each division corresponds to one full rotation of the wheel. The graduated drum of the wheel has 100 divisions enabling the reading of the 1/10th and 1/100th divisions of the index reading. The vernier attached enables the reading of one-tenth of a division of the wheel. The reading thus consists of four digits, index reading, 1/10th and 1/100th from the drum, and 1/1000th from the vernier.

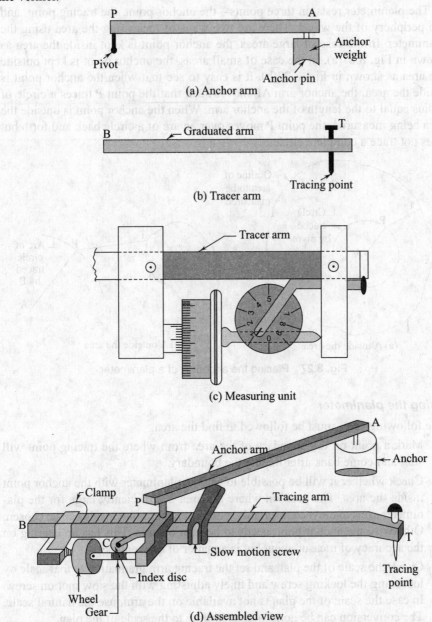

(a) Anchor arm

(b) Tracer arm

(c) Measuring unit

(d) Assembled view

Fig. 8.26 Planimeter

The graduated or index disc has 10 divisions, each division corresponding to one revolution of the wheel. One full revolution over the index on the disc corresponds to 10 revolutions of the wheel. It will be necessary to note down the number of times the index has been crossed while tracing the direction in which the zero has moved over the index. The graduations on the tracing arm consist of scales of the drawing and the area covered due to unit rotation of the wheel for that scale. Also, the constant to be used for finding the area is marked near the graduation.

The planimeter rests on three points—the anchor point, the tracing point, and the periphery of the wheel. There are two ways of measuring the area using the planimeter. In the case of large areas, the anchor point is kept inside the area as shown in Fig. 8.27(a). In the case of small areas, the anchor point is kept outside the area as shown in Fig. 8.27(b). It is easy to see that when the anchor point is inside the area, the anchor arm AP moves such that the point P traces a circle of radius equal to the length of the anchor arm. When the anchor point is outside the area being measured, the point P moves over an arc of a circle back and forth but does not trace a complete circle.

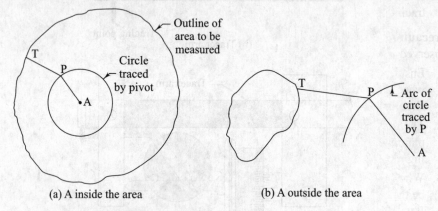

(a) A inside the area   (b) A outside the area

Fig. 8.27 Placing the anchor A of a planimeter

Using the planimeter

The following steps must be followed to find the area.

1. Mark a point on the boundary of the area from where the tracing point will start and come back after tracing the boundary.

2. Check whether it will be possible to use the planimeter with the anchor point inside the area. The position where the area is sufficiently large for the planimeter arms to move comfortably with the anchor point inside can be chosen. Otherwise the anchor point needs to be kept outside. This has no bearing on the accuracy of measurement but is a matter of convenience.

3. Check the scale of the plan and set the tracing arm graduation to that scale by loosening the locking screw and finely adjusting with the slow motion screw. In case the scale of the plan is not available on the arm, use the natural scale. The conversion can be done later according to the scale of the plan.

4. Keep the tracing point on the boundary of the area at the point marked. Note down the initial reading on the disc, the graduated drum, and the vernier. Generally, it is not necessary to set the reading to zero.

5. Trace the boundary carefully in a *clockwise* direction all around with the tracing point. Note the final reading once the tracing point reaches the starting point. As the tracing proceeds, note down the number of times the zero crosses the index

mark on the graduated disc and the direction in which it crosses, clockwise or counterclockwise.

6. Repeat the measurements at least once more as a check.
7. Calculate the area using the formula

$$\text{Area} = M(\text{FR} - \text{IR} \pm 10N + C)$$

where M is the multiplicative constant, FR is the final reading, IR is the initial reading, N is the number of times the zero of the index disc crosses the fixed index, and C is the additive constant. The formula is explained in the sections to follow.

8. If the anchor point is inside the area, include the additive constant given on the tracing arm. For anchor points outside the area, the additive constant is zero.

Precautions in using the planimeter The following precautions need to be observed while using the planimeter.

1. Ensure that the instrument is clean and dust has not accumulated on it. The wheel should move freely without any jerk. The plan on which the wheel moves should similarly be clean.
2. The plan should be kept horizontal on a firm table for free movement of the planimeter. The plan should not have been folded, which will cause jerks during movement.
3. When the index mark is on zero, all the readings should be zero. This may not be the case in most instruments due to imperfections in the worm gear system. It will be preferable to read the initial reading rather than setting the instrument to zero.
4. Always trace the boundary in the clockwise direction.
5. The tracing point should be kept exactly on the outline of the plan at all times. French curves or a scale may be used to ensure this.
6. Measure the area at least twice and preferably with a changed position of the anchor point to average out the adjustment errors.
7. The area can be measured with the anchor point inside the area in the case of large drawings. The area can also be divided into parts and each part measured with the anchor point outside.
8. Take accurate readings at the index disc, graduated drum, and the vernier.
9. One should carefully note down the number of times the zero crosses the index during the tracing and the direction in which it does so.

Derivation of planimeter formula

The theory of the planimeter can be explained with respect to the movement of the tracing arm, and the formula mentioned above can be derived as follows.

Case 1: Anchor point outside the area Referring to Fig. 8.28, the plan of the area A_0 is to be measured. The anchor point A is outside the area and AP is the length of the anchor arm. PT is the effective length of the tracing arm, i.e., the length between the pivot and the tracing point. The tracing arm is moved clockwise along the boundary of the area A_0. As the tracing arm moves, the arm AP is constrained to move along a circular arc with centre at A, while the tracing point has to translate and rotate. This movement will cause sliding (movement parallel to the tracing arm) and movement perpendicular to the tracing arm. The wheel will record only the movement perpendicular to the arm.

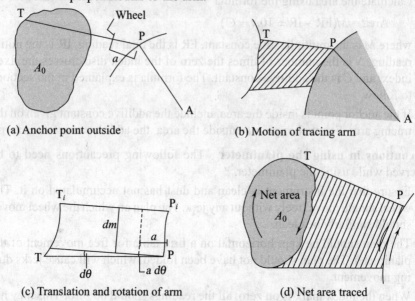

(a) Anchor point outside

(b) Motion of tracing arm

(c) Translation and rotation of arm

(d) Net area traced

Fig. 8.28 Theory of the planimeter

Considering the motion shown in Figs 8.28(b) and (c), due to a small movement of the tracing point along the boundary, the tracing arm covers the area shown hatched in Fig. 8.28(b). This is shown enlarged in Fig. 8.28(c), where the movement of the tracing arm can be considered equivalent to a translation dp and a rotation $d\theta$. In the case shown in Fig. 8.28, the wheel is shown to be between the pivot and the tracing point. The movement of the wheel is equal to $dm = dp$ + the movement due to rotation $d\theta$. If the distance of the plane of the wheel and the pivot is a, then the total movement dm as recorded by the wheel is equal to $dm = dp + a\,d\theta$ from Fig. 8.28(c). $dp = dm - a\,d\theta$. The area traversed by the tracing arm = PT × dp + (1/2) PT × PT × $d\theta$, considering the elementary area, dA_0, as the combination of a rectangle and a triangle.

If we take PT = l, then $dA_0 = l\,dp + (1/2)l^2 d\theta$. Substituting for dp,

$$DA_0 = l(dm - ad\theta) + (1/2)l^2 d\theta$$

It is clear from Fig. 8.28 that as the tracing point goes along the boundary and returns to the starting point, the tracing arm moves downwards and upwards and the net area traced by the arm is the area A_0. This is illustrated in Fig. 8.28(d). The area traced outside during motion in one direction gets cancelled by the motion in the reverse direction.

$$dA_0 = l\,dm - la\,d\theta + (1/2)l^2\,d\theta$$

The area A_0 can be obtained by integrating the expression

$$\text{Area } A_0 = l \int dm - la \int d\theta + (1/2)l^2 \int d\theta$$

as l and a are constant. When the anchor arm is outside the area, the net angle swept by the arm is zero. So $\int d\theta = 0$. Thus, area $A_0 = l \int dm$. This integral gives the total movement of the wheel and is the difference between the final and initial readings and 10 times the number of times the zero crosses the index mark. If d is the diameter of the wheel, one revolution = πd. If we set $l\pi d = M$, a multiplying constant for the instrument, then

$$\text{Area } A_0 = M(\text{total movement of the wheel}) = M(\text{FR} - \text{IR} \pm 10N)$$

Here M is the multiplying constant = $l\pi d$, πd being a constant; M depends on l. By adjusting the length of the tracing arm, the value of M can be changed and kept at convenient values and for different, commonly used scales of the plan. In the formula, for $\pm 10N$, we use the plus sign when the zero crosses the index in the clockwise direction and the minus sign when the zero crosses the index in the counterclockwise direction.

Case 2: Anchor point inside the area When the anchor point is kept inside the area, from Fig. 8.29, it is clear that the pivot P traces the outline of a circle or the anchor arm traces the area of a circle as the tracing point moves over the boundary of the area. Thus, the measurement taken from the movement of the tracing arm and the wheel is only the area outside the circle traced by the anchor arm. The area of the circle traced by the anchor arm has to be added to the area obtained from the tracing arm.

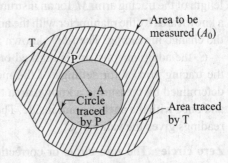

Fig. 8.29 Anchor point outside the area

This is the additive constant for a certain length of the anchor arm.

If A_0 is the area of the plan, then

A_0 = area traced by the tracing arm + area traced by the anchor arm

$$= l \int dm - la \int d\theta + (1/2)l^2 \int d\theta + \pi r^2$$

where r is the length of the anchor arm. In this case $\int d\theta = 2\pi$, as the arm traces a full circle. Thus,

$$A_0 = M[\text{FR} - \text{IR} \pm 10N] - la\,2\pi + (1/2)l^2 2\pi + \pi r^2$$

$$= M[\text{FR} - \text{IR} \pm 10N] + \pi(l^2 - 2al + r^2)$$

This equation can be written as

$$A_0 = M[\text{FR} - \text{IR} \pm 10N + C]$$

where C is the additive constant and is given by

$$C = [\pi(l^2 - 2al + r^2)]/M$$

Considering the two cases discussed above, the planimeter formula can be written as a single formula covering both the cases as $A_0 = M[\text{FR} - \text{IR} \pm 10N + C]$. The constant C is considered only when the anchor point is inside the area.

We must be clear as to what each term in the formula represents. M is a multiplying constant and is equal to $l\pi d$, where l is the length of the tracing arm (length between the pivot and the tracing point) and πd is the perimeter of the wheel rim which actually rolls. Since the perimeter is constant, M depends on the length of the tracing arm. The length of the tracing arm is marked on this basis to get convenient values of the constant M. C is the additive constant used only when the anchor point is inside the area. This constant depends upon the tracing arm length l, the distance a between the wheel and the pivot, and the length of the anchor arm between the pivot and the anchor pin. IR and FR are readings of four digits taken from the index disc, wheel, and vernier. N is the number of times the zero of the disc has passed the index during tracing. This is taken to be positive when the zero passes the index in a clockwise direction and negative when the zero passes the index in a counterclockwise direction.

M, the multiplier constant, is the area covered for one revolution of the wheel by the tracing arm. Manufacturers provide the value of M for different values of the length of the tracing arm. M, for an instrument, can also be determined by measuring a known area by the planimeter with the anchor point outside the area, and recording the change in reading. Then $M = \text{known area}/(\text{FR} - \text{IR} \pm 10N)$.

C, the additive constant, is marked on the tracing arm for different lengths of the tracing arm or for settings of the measuring unit on the arm. It can also be determined by measuring a known area twice, once with the anchor point outside and once with the anchor point inside. The difference in the two areas based on the readings gives the value of C.

Zero circle   The zero circle or correction circle, as it is sometimes called, is a circle of area $= \pi(l^2 \pm 2al + r^2)$. It can be shown that when the tracing point moves over the zero circle, the wheel slides and does not cause any change in the reading. There are two cases depending upon whether the wheel is kept between the pivot and the tracing point or beyond the pivot, away from the tracing point. These are shown in Fig. 8.30.

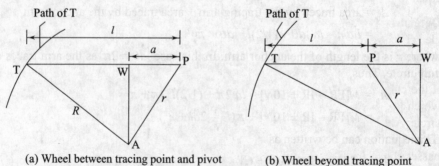

(a) Wheel between tracing point and pivot (b) Wheel beyond tracing point

Fig. 8.30   Concept of zero circle

In Fig. 8.30(a), the case when the wheel lies between the tracing point and the pivot is shown. The case of pure sliding occurs when the plane of the wheel passes through the anchor point A. AP is the anchor arm of length r, P is the pivot, PT is the length l of the tracing arm, a is the distance of the wheel from P, W is the wheel, and AW is perpendicular to PT.

The zero circle is a circle with A as the centre and AT as its radius. Then from the right-angled triangle AWT,

$$AT^2 = AW^2 + WT^2$$

But, $AW^2 = AP^2 - PW^2$; therefore,

$$AT^2 = WT^2 + (AP^2 - PW^2)$$

Setting R as the radius AT of the zero circle, $WT = l - a$, $AP = r$, and $PW = a$, and substituting,

$$R^2 = (l - a)^2 + (r^2 - a^2)$$
$$= l^2 - 2al + r^2$$

In Fig. 8.30(b), the wheel is placed beyond the pivot point, away from the tracing point. In this case,

$$AT^2 = AW^2 + WT^2$$

But,

$$AW^2 = AP^2 - WP^2$$
$$= AP^2 + PW^2$$

and

$$WT^2 = WP^2 + PT^2$$

Substituting, $AT = R$, $AP = r$, $PW = a$, and $PT = l$

$$R^2 = r^2 - a^2 + (l + a)^2$$
$$= l^2 + 2al + r^2$$

In either of the above cases, $R^2 = l^2 \pm 2al + r^2$; the plus sign is used when the wheel is placed beyond the pivot away from the tracing point and the minus sign is used when the wheel is placed between the pivot and the tracing point.

The area of the zero circle $= \pi R^2 = \pi(l^2 \pm 2al + a^2)$, which we have seen earlier is equal to MC, where M is the multiplying constant and C is the additive constant.

Determining the area of zero circle

The area of the zero circle can be determined in many ways. An obvious way is to determine the product MC, if the constants M and C are known, which is equal to the area of the zero circle. Another way is to replicate the condition that the plane of the wheel passes through the anchor point and PT is perpendicular to the line joining the wheel and the anchor point. On a paper, mark two lines WT and AW at right angles to each other as shown in

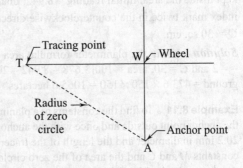

Fig. 8.31 Determining the area of a zero circle

Fig. 8.31. Set the anchor point at A, the tracing point at T, and the wheel at W. Measure the distance AT, which gives the radius of the zero circle. Area of zero circle = $\pi \times$ (AT)^2. Another method is to measure a plan whose area is known using the planimeter with the anchor point inside. The difference between the known area and the measured area gives the area of the zero circle.

Another way is to measure an unknown area in two ways—once with the anchor point inside and once with the anchor point outside. With the anchor point outside, the area can be calculated. This known value of the area can be used to calculate the area of the zero circle.

Care of the planimeter

To get accurate results, the planimeter should be kept inside its box when not in use to prevent dust accumulating on the exposed parts. In some instruments, the wheel and the assembly of the measuring unit come protected. The gear system is very sensitive and requires careful handling. The plastic coating on which the graduations are marked should be kept clean. The tracer arm must be dust-free before adjusting its length and locking it using a slow motion screw.

Example 8.9 A planimeter was used to measure the area from a plan drawn to a scale of 1 cm = 100 m. The tracer arm was set to natural scale and the anchor arm was kept outside the figure. Initial reading = 6.973; final reading = 2.921. For the natural scale, M = 100 sq. cm and the zero of the disc passed the index mark once in a clockwise direction. Find the area of the ground represented by the plan.

Solution From planimeter readings, the area is given by

$$\text{Area} = M[\text{FR} - \text{IR} \pm 10N + C]$$

FR = 2.921, IR = 6.973, N = +1, and C = 0 (for the anchor point inside)

Therefore

$$\text{Area} = 100[2.921 - 6.973 + 10] = 594.8 \text{ sq. cm}$$

Scale is 1 cm = 100 m; 1 sq. cm of area represents 100×100 sq. m of land area. Therefore,

$$\text{Area of the ground} = 594.8 \times 100 \times 100 = 5,948,000 \text{ sq. m}$$
$$= 594.8 \text{ hectares}$$

Example 8.10 A planimeter set to natural scale (with M = 100 sq. cm) was used to measure the area of land drawn on a plan to a scale of 1 cm = 150 m. The anchor point was kept inside the area. Initial reading = 8.942, final reading = 3.678, and the zero crossed the index mark twice in the counterclockwise direction. Find the area of the land in hectares. C = 30 sq. cm.

Solution Using the planimeter formula, area = $M[\text{FR} - \text{IR} \pm 10N + C]$. Noting that N = – 2 and C = 30, area = 100[3.678 – 8.942 – 20 + 30] = 473.6 sq. cm. Therefore, area of ground = 473.6 $\times$ 150 $\times$ 150 = 1065.6 hectares.

Example 8.11 To find the constants of a planimeter, a given area was measured once with the anchor point outside and once with the anchor point inside. The planimeter's wheel was 20.2 mm in diameter and the length of the tracer arm was 157.6 mm. Find the value of the constants M and C and the area of the zero circle from the following data.

Anchor point inside: Initial reading = 0.012, final reading = 9.884, the zero crossed the index twice in the counterclockwise direction during tracing.

Anchor point outside: Initial reading = 2.192, final reading = 4.352, the zero crossed the index once in the clockwise direction.

Find the constants M and C and the area of the zero circle.

Solution The multiplying constant $M = l\pi d = 15.7 \times \pi \times 2.02 = 100.01$ cm. With the anchor point outside, the area is given by

$$\text{Area} = 100.01(4.352 - 2.192 + 10 \times 1) = 1216.12 \text{ sq. cm}$$

With the anchor point inside, area = $100.01(9.884 - 0.012 - 10 \times 2 + C)$. Equating this to the area calculated above, $C = 22.29$. Therefore,

$$\text{Area of zero circle} = MC = 100.01 \times 22.29 = 2229.22 \text{ sq. cm}$$

Example 8.12 A planimeter is to be set to have the multiplying constant $M = 100$ and additive constant $C = 30$. The wheel diameter is 2 cm. Find the lengths of the tracing arm and anchor arm required to obtain these values of M and C when the wheel is set (a) beyond the pivot and (b) between the pivot and tracing point by 2.93 cm.

Solution Multiplying constant $M = l\pi d$, $100 = l\pi \times 2$, $l = 15.92$ cm

$$\text{Area of zero circle} = 30 \times 100 = 3000 \text{ sq. cm}$$

When the wheel is set beyond the pivot, $\pi(l^2 + 2al + r^2) = 3000$, from which $r = 24.67$ cm. When the wheel is between the pivot and tracing point, $\pi(l^2 - 2al + r^2) = 3000$, from which $r = 28.11$ cm.

8.3.3 Digital Planimeter

The digital planimeter is an improvement over the conventional planimeter. Figure 8.32 shows the photograph of a digital planimeter.

Fig. 8.32 Digital planimeter

Digital planimeter is a microprocessor-based instrument that has sensors to determine the length of a line or the area enclosed by lines. A typical instrument may have the following features.

Control panel The control panel has a display panel which displays the length or area. It has a number of keys—digits 0 to 9, decimal point, arithmetic operation (for addition, subtraction, multiplication, and division), and equal-to. A set key, together with the digital keys, allows to set units and scale values. Units, vertical scales, and horizontal scales can be set independently. Apart from these, clear entry and clear keys are provided. Finally, a summation key and memory-related keys, such as memory store and memory recall, are available. As in a calculator, the display may have two to three lines of diplay.

Roller   A high friction roller with a diamond-inlaid wheel is used to provide stable and precise measurements.

Tracer arm   It is the arm that is moved while measuring lengths and areas. The power switch may be fixed to this arm, which will also release the arm for movement. A magnifying lens may be provided over the tracing point in the arm, for a clearer view of lines and points. The tracing point is kept over points or moved over curves.

Mode switch   It facilitates the shift from the point mode to the continuous mode.

Operation

The digital planimeter is simple to use. The mode has to be selected first. For example, in the case of a straight line boundary, the point mode is used. In this mode, the tracer point is kept over the vertices or the end points of the lines one after the other and the lengths of the lines are measured. In the case of curved boundaries, the continuous mode is used. The tracer point has to be moved along the boundary curve.

The procedure for measuring lengths and areas is as follows.

1. Lift the fixing lever of the tracer arm (or any other switch provided) to switch on the supply.
2. With the help of the set key and the number keys, set the units, vertical scales, and horizontal scales. Depress the set key repeatedly to set the units and the two scales.
3. Set the tracer point on the starting point and press the start switch to start the operation.
4. In the case of a straight line, move the tracer point to the other point of the line. In the case of curved lines, move the tracer point carefully over the boundary.
5. Press the end key after the measurement to display the lengths and areas in terms of the set units and scales.
6. Switch the instrument off after measurement.

The range and resolution of digital planimeters are different in different models. These instruments are lightweight and powered by rechargeable batteries with a certain number of hours (40 to 60 hours) of operation. Typical specifications of a planimeter are as follows.

Measurement functions: Coordinate, area, line length, side length, radius, angle

Measurement modes: Line (Point), curve (Stream), circular arc (Arc)

Measuring units: Metric (mm, cm, m, km), foot (in., ft, acre/yd, mi), user-defined units

Display: 140×30 dot graphic liquid crystal (17 digits $\times$ 3 lines)

Measuring range: 380 mm $\times$ 10 m

Linear resolution: 0.05 mm

Accuracy: $\pm 0.1\%$

Power: Rechargeable Ni-Cd cell (chargeable with the ac adapter supplied)

Charging time: About 10 hr in normal use/about 15 hr after complete discharge

Operating time: About 40 hr of continuous operation after full charge

Dimensions: $350 \times 43 \times 165$ mm (main body), $365 \times 65 \times 195$ mm (case)

Weight approx: 1 kg (main body only)

8.4 Partitioning of Land

Often, one is required to divide an area or land into parts with given specifications. The methods of traverse computations are used in such cases. Some typical cases are discussed here.

8.4.1 Partitioning Through a Given Point

Referring to Fig. 8.33(a), run a line through a point P so that the area is divided in a given ratio or a certain area is marked out. ABCDE is a traverse which has been corrected for closing errors. The procedure for partitioning through a given point is the following.

1. Connect point P to one of the vertices of the traverse so that the approximate area will be given by moving this line through some angle.

2. Calculate the latitudes and departures from the lengths and bearings of the lines.

3. Determine the position of line PQ such that area PBCDQ is the required area.

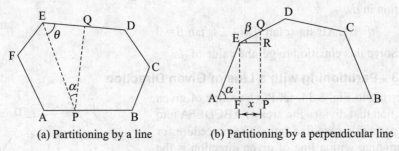

| (a) Partitioning by a line | (b) Partitioning by a perpendicular line |

Fig. 8.33 Partitioning of land

4. Calculate the area of ABCDEF using any of the methods outlined in the preceding sections.

5. Join P to E. Find the length and bearing of the closing line PE from the closed polygon PBCDEP.

6. Compute the area of the polygon PBCDEP using any of the methods outlined in the preceding sections.

7. Let A be the area of polygon ABCDEF, A' the area of PBCDEP, and A_0 the required area. $A' - A_0$ is the area of triangle EPQ.

8. In triangle EPQ, calculate the angle θ from the bearings of ED and EP. With the length of EP, the angle θ, and the known area, calculate the length PQ as

$$\text{Area of EPQ} = (1/2)\text{EP} \times \text{EQ} \times \sin \theta$$

$$\text{EQ} = 2 \times \text{area of EPQ}/(\text{EP} \sin \theta)$$

9. Calculate the angle α from

$$\tan \alpha = EQ \sin \theta/(EP - EQ \cos \theta)$$

Calculate the length PQ, using the sine rule, from triangle EPQ as

$$PQ = EQ \sin \theta/\sin \alpha$$

8.4.2 Partitioning with a Perpendicular Line

Referring to Fig. 8.33(b), ABCDEF is a closed traverse. PQ is a perpendicular line which divides the area in a given ratio. The procedure for partitioning with a perpendicular line is as follows.

1. Draw EF perpendicular to AB (parallel to PQ) and ER perpendicular to PQ. Determine angles α and β from the bearings of the lines.
2. APQEA is a specified area $= A_1$. Let d be the distance between EF and PQ.
3. From triangle EAP,

   $$AF = AE \cos \alpha$$
   $$EF = AE = \sin \alpha$$
   $$AP = AE \cos \alpha + d$$

4. Area of AEF $= (1/2)AE^2 \sin \alpha \cos \alpha$. Area of EFPQ $= A_0 = A_1 - (1/2)AE^2 \sin \alpha \cos \alpha$.
5. Area of EFPQ = area of rectangle EFPR and area of triangle ERQ $= d \times AE \sin \alpha + (d^2 \tan \beta)/2$ (because QR $= d \tan \beta$).
6. Therefore, $A_0 = d \times AE \sin \alpha + (d^2 \tan \beta)/2$. This gives rise to a quadratic equation in d:

   $$d^2 + (2AE \sin \alpha/\tan \beta)d - 2A_0/\tan \beta = 0$$

   Solve this equation to get the value of d.

8.4.3 Partitioning with a Line of Given Direction

Referring to Fig. 8.34, let PQ be a line of given direction that divides the area of ABCDEFA into two parts in a specified ratio. The procedure for partitioning with a line of given direction is the following.

1. Let PQ be the line of specified direction. Calculate the area of ABCDEFA from the lengths and bearings of the lines using any of the methods outlined in the preceding sections.
2. Draw a line through CG parallel to PQ. Area of APQCR = area of traverse ABCG + area of trapezium CGPQ. The distance x between CG and PQ has to be calculated.

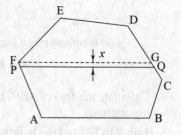

Fig. 8.34 Partitioning a land with a line of given direction

3. In the closed traverse ABCG, find the missing lengths AG and GC using the methods outlined in Chapter 6. Find the area of the traverse ABCG using any method.
4. Calculate the angles marked α and β in Fig. 8.34 from the known bearings of the lines.

5. If A is the area of the given traverse, A_1 the specified area, and A_2 the area of traverse ABCG, then $A_1 = A_2 +$ area of trapezium CGPQ. Calculate the area of the trapezium as $(1/2)[CG + PQ]x$, where x is the height.

$$PQ = CG - x \cot \alpha + x \cot \beta$$

Therefore,

$$\text{Area of trapezium} = (1/2)(CG + CG - x \cot \alpha + x \cot \beta)x$$

Determine the value of x from this equation.

8.4.4 Partitioning a Trapezium into Two Parts with a Parallel Line

Referring to Fig. 8.35, let ABCD be the trapezium. This trapezium is to be divided into two parts by a line PQ parallel to the parallel sides of the trapezium such that the areas are in a given ratio. We use the following procedure to locate the position of line h1 and the length of line PQ.

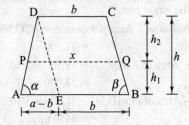

Fig. 8.35 Partitioning a trapezium

1. Let area of ABCD be A, that of ABQP be A_1, and that of PQCD be A_2. Let H be the height of the trapezium and h_1 and h_2 be the heights into which PQ divides H. PQ = x ratio of areas, $A_1/A_2 = r_1/r_2$. Also, let AB = a and CD = b.

2. $A_1/A_2 = r_1/r_2$ gives $(A_1 + A_2)/A_2 = (r_1 + r_2)/r_2$ and $A_2 = Ar_1/(r_1 + r_2)$. Similarly, $A_1 = Ar_2/(r_1 + r_2)$.

3. Draw line DE parallel to BC. From similar triangles DPF and DAE,

$$h/h_2 = (a - b)/(x - b)$$

(Note that PF = $x - b$ and AE = $a - b$.)

$$h_2 = h(x - b)/(a - b), \qquad h_1 = h - h_2 = h(a - x)/(a - b)$$

4. If α and β are the angles shown, then

$$(a - b) = h(\cot \alpha + \cot \beta)$$

$$h = (a - b)/(\cot \alpha + \cot \beta)$$

From this,

$$h_1 = (a - x)/(\cot \alpha + \cot \beta)$$

$$h_2 = (x - b)/(\cot \alpha + \cot \beta)$$

5. $A = (1/2)(a + b)h$

$$= (1/2)(a + b)(a - b)/(\cot \alpha + \cot \beta)$$

$$= (1/2)(a^2 - b^2)/(\cot \alpha + \cot \beta)$$

Similarly,

$$A_1 = (1/2)(a^2 - x^2)/(\cot \alpha + \cot \beta)$$

$$A_2 = (1/2)(x^2 - b^2)/(\cot \alpha + \cot \beta)$$

6. As $A_1 = Ar_2/(r_1 + r_2)$, we have

$$\frac{(1/2)(a^2 - x^2)}{\cot\alpha + \cot\beta} = \frac{(1/2)(a^2 - b^2)r_2}{-(\cot\alpha + \cot\beta)(r_1 + r_2)}$$

$$x = \sqrt{[(r_2 b^2 + r_1 a^2)/(r_1 + r_2)]}$$

Knowing x, h_1 and h_2 can be calculated. The following examples illustrate these principles.

Example 8.13 In Fig. 8.36, find the position of line PQ, parallel to AB, such that it divides the area into two parts in the ratio 7:3, PQBA being the larger part.

Solution Area of trapezium $= (1/2)(170 + 250) \times 150$

$\qquad = 31,500 \text{ m}^2$

Area of ABQP, $A_1 = 31,500 \times 7/10$

$\qquad = 22,050 \text{ m}^2$

Area of PQCD, $A_2 = 31,500 \times 3/10$

$\qquad = 9450 \text{ m}^2$

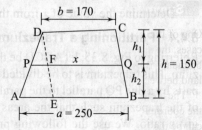

Fig. 8.36 Trapezium of Example 8.13

Length of PQ, $x = \sqrt{[(7 \times 170^2 + 3 \times 250^2)/(7+3)]} = 197.43$ m

$\qquad h_1 = 150(250 - 197.43)/(250 - 170) = 98.56$ m

$\qquad h_2 = 150 - 98.56 = 51.44$ m

Example 8.14 In the traverse shown in Fig. 8.37, find the position of the perpendicular line PQ such that the area of APQE = 25,000 m². The traverse data is as follows:

Line	AB	BC	CD	DE	EA
Length	405.9	182	175	171	212
Bearing	N90° 00′E	N00° 00′E	N60° 00′W	S60° 00W	S30° 00′W

Solution The latitudes and departures can be calculated as shown below.

Line	Length	Bearing	Latitude	Departure
AB	405.85	N90°E	0	405.85
BC	182	N00°E	182	0
CD	175	N60° 00′W	87.5	− 151.55
DE	171	S60° 00′W	− 85.5	− 148.1
EA	212.4	S30° 00′W	− 184.00	− 106.2

Draw EF perpendicular to AB and ER perpendicular to PQ. The area of triangle EAF can be found as follows:

$\qquad$ AF = AE cos 60° = 184 m

$\qquad$ EF = AE sin 60° = 106.2 m

$\qquad$ Area of triangle EAF = $(1/2) \times 184 \times$ 106.2 = 9770.4 m²

As the area required is 25,000 m²,

$\qquad$ Area of EFPQ = 25,000 − 9770.4 = 15,229.6 m²

The area of EFPQ consists of a rectangle and a triangle.

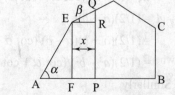

Fig. 8.37 Traverse of Example 8.14

$\qquad$ Area of rectangle EFPR = 106.2x

where x is the distance FP.

Area of triangle ERQ: $ER = x$, $RQ = x \tan 30°$, area $= (1/2)x^2 \tan 30°$

Therefore,

$106.2x + (1/2)x^2 \tan 30° = 15,229.6$ m^2

The solution of this quadratic equation is $x = 110.32$ m.

Summary

One of the results expected from survey data is a plan of the land surveyed, and in many cases, the area of the land has to be calculated. The area of a piece of land with an irregular boundary is calculated as the sum of the area of a traverse within the area and the tract of land between the survey line and the irregular boundary.

The area of a traverse can be calculated by many methods. In the method of coordinates, the coordinates of the stations of a traverse are determined with respect to an origin, which can be the most westerly station or any other arbitrarily chosen point. The area can then be calculated as half of the difference in the products of terms from the two rows by writing them in determinant form.

The area of a traverse can also be calculated as half the algebraic sum of the products of the double meridian distances (DMD) of lines and their respective latitudes. The reference meridian for calculating the DMD is the north–south line through the most westerly station or a parallel meridian through any point. If the reference meridian passes through the most westerly station, then the DMD of any line = DMD of the previous line + departure of previous line + departure of line itself.

A third method to find the area of a traverse is to use the total latitudes and departures. Starting from a reference station (generally the most westerly station is selected) the total latitude of a station is the algebraic sum of the latitudes of the lines prior to that station from the reference point. The algebraic sum of the products of the total latitudes and departures gives twice the area.

When offsets are taken between a survey line and the boundary, the area can be found by the following methods.

- *Mid-ordinate rule* In this method, the line is divided into n equal sections, and ordinates are taken to the middle of the sections. Then

    Area $= (L/n)(y_1 + y_2 + y_3 + \cdots + y_{n-1} + y_n)$

    where L is the length of the line, n is the number of ordinates taken, and $y_1, y_2, y_3, \cdots, y_{n-1}, y_n$ are the ordinates.

- *Average ordinate rule* In this method, the line is divided into equal sections and ordinates are taken at the section points. Then

    Area $= (y_0 + y_1 + y_2 + \cdots + y_n)[L/(n + 1)]$

- *Trapezoidal rule* With the usual notations, this rule gives the area as follows:

    Area $= x[(y_0 + y_n)/2 + y_1 + y_2 + y_3 + \cdots + y_{n-1}]$

    where $x = L/n$.

- *Simpson's rule or parabolic rule* This rule gives the area as

    Area $= (x/3)[y_0 + 4y_1 + 2y_2 + 4y_3 + 2y_4 + \cdots + 2y_{n-2} + 4y_{n-1} + y_n]$

    where $x = L/n$.

The area of a given plan can be calculated by the tracing paper method, whereby the area is measured as the total area of triangles, squares, rectangles, or trapeziums which are marked onto a tracing paper by keeping it over the area for counting the number of geometrical figures into which the area can be divided.

The area of a given plan can also be found using a planimeter. The planimeter consists of a tracer arm to which a measuring unit having a wheel is attached and an anchor arm with a weight and a pin at one end by which it is fixed to the plan. Both the arms may be graduated and can be set according to requirement. The wheel measures the area by rolling, which is recorded through a graduated drum, vernier, and a graduated disc. The planimeter formula is area = $M[FR - IR \pm 10N + C]$. M and C are constants marked on the tracing arm. The constant C is added only when the anchor point is kept inside the area for measurement. The zero circle is a circle of area = $MC = \pi (l^2 \pm 2al + r^2)$; the plus sign is used when the wheel is placed beyond the pivot away from the tracing point.

Exercises

Multiple-Choice Questions

1. One hectare is equal to
   (a) 100 m²   (b) 1000 m²   (c) 10000 m²   (d) 100000 m²
2. Area of a traverse by coordinates can be obtained by taking
   (a) any point as origin and finding the coordinates of the nodes of the traverse
   (b) the origin such that at least one of the two coordinates is positive
   (c) the origin such that all coordinates are positive
   (d) only one of the points of the traverse as origin
3. If the length of line AB is 135.8 m and its bearing is N 38° 45' W, the coordinates of B are, taking the coordinates of A (–100, 100),
   (a) (–15, 205.91)   (c) (205.91, 185)
   (b) (15, –5.91)   (d) (–185, 205.91)
4. If the coordinates of two consecutive points of a traverse are A (23.5, 134.5) and B (176.5, –23.5), then the length of the line is
   (a) 189.1 m   (b) 219.94 m   (c) 234.8 m   (d) 314.53 m
5. If the coordinates of point A are (35.6, 121.8) and that of B are (–65.8, 182.8) then the bearing of line AB is
   (a) N 58 58'W   (b) S 58 58' E   (c) N31 02'E   (d) S 31 02'W
6. If the length and bearing of a line AB are 134.5 m and S 30° 45'E, then the coordinates of B, taking the coordinates of A as (0,0), are
   (a) (115.59, 68.76)   (c) (–68.76, 115.59)
   (b) (115.59, –68.76)   (d) (68.76, –115.79)
7. Meridian distance of a station point of a traverse is the
   (a) projection of a line from the point on the north–south direction
   (b) mean distance of a line from the point to one of the coordinate axes
   (c) perpendicular distance of the point from a reference meridian
   (d) length of the line itself
8. Double meridian distance of a line in a closed traverse is
   (a) the sum of the latitudes of the line and the previous line
   (b) the sum of the meridian distances of the end points of a line
   (c) the sum of the latitude and departure of the line
   (d) the product of latitude and departure of the line
9. Traverse data from a traverse is as follows: AB: length – 98.6 m; bearing 45° 30'; BC: length –110.85 m, bearing –155° 30'. If the double meridian distance of AB is 70.33 m, the double meridian distance of BC is
   (a) –31.75 m   (b) 116.3 m   (c) 139.43   (d) 186.62

10. If the length and bearing of line AB are 132.5 m and N 45 45′E and the reference meridian direction passes through A, then the double meridian distance of AB is
    (a) 92.45 m   (b) 94.91 m   (c) 132.5 m   (d) 136 m

11. In the double meridian distance (DMD) of finding areas, the area is given by
    (a) ½ [algebraic sum of products of DMDs and corresponding departures]
    (b) ½ [algebraic sum of products of DMDs and corresponding latitudes]
    (c) ½ [algebraic sum of products of DMDs and lengths]
    (d) ½ [algebraic sum of products of DMDs of adjacent lines]

12. The total latitude of the vertex (station point) of a traverse is the algebraic sum of
    (a) latitudes of the previous line and the next line
    (b) latitude and departure of the previous line
    (c) latitude and departure of the next line
    (d) latitudes of all lines preceding that station from the reference station

13. The area of a traverse from total latitudes is given by
    (a) ½ [Σ total latitude × lengths of the line meeting at that point]
    (b) ½ [Σ total latitude × the departures of the previous line from that point]
    (c) ½ [Σ total latitude of a point × algebraic sum of departures of lines meeting at the point]
    (d) ½ [Σ total latitude x departure of the next line from that point]

14. The data from a traverse is as follows: AB: length 132.45 m, bearing –N 65° 45′W; BC: length –109.65 m; bearing –S 65° 35′ W. If A is the reference station, the total latitude of B is
    (a) –54.4 m   (b) –45.72 m   (c) 54.4 m   (d) –99.72 m

15. The total latitude of the reference station is
    (a) zero
    (b) latitude of the first adjacent line
    (c) average latitude of the lines meeting at that station
    (d) sum of the latitudes of the lines meeting at the reference station

16. Trapezoidal rule can be applied, in the case of offsets at equal intervals
    (a) only when the number of offsets are even
    (b) only when the number of offsets are odd
    (c) in the case of even or odd offsets
    (d) only when the offsets do not differ very much in value

17. Trapezoidal rule can be stated as
    (a) determine the average of the ordinates and multiply this sum by the interval between them
    (b) find the sum of ordinates and multiply this sum by the interval between the ordinates
    (c) determine the sum of ordinates and multiply by half the interval between the ordinates
    (d) find the average of first and last ordinates and to this add the sum of other ordinates and multiply by the interval between ordinates

18. Simpson's rule or parabolic rule can be stated as
    (a) take the average of three consecutive ordinates and multiply by twice the offset interval between them
    (b) taking three consecutive ordinates, find the sum of 2 × central ordinate + the sum of other ordinates and then multiply this sum by twice the offset interval
    (c) to the sum of first and last ordinates, add twice the sum of odd ordinates and three times the sum of even ordinates and multiply by the offset interval

    (d) to the sum of first and last ordinates, add twice the sum of odd ordinates and four times the sum of even ordinates and multiply this sum by one-third the offset interval

19. Simpson's rule or parabolic rule with common offset interval can be applied to the whole area directly
    (a) only if the number of sections are even
    (b) only if the number of sections are odd
    (c) for even or odd number of sections
    (d) only if the number of ordinates are even

20. Area obtained by the following two methods will give the same value
    (a) average ordinate method and trapezoidal rule
    (b) trapezoidal rule and parabolic rule
    (c) trapezoidal rule and coordinates method
    (d) parabolic rule and method of coordinates

21. If the interval between ordinates is not constant, then the most appropriate rule to find the area is
    (a) average ordinate method   (c) trapezoidal rule
    (b) coordinates method   (d) parabolic rule

22. If three consecutive ordinates are taken at 2 m intervals from a traverse line and measured as 1.8 m, 2.5 m and 2.0 m, then the area between the traverse line, the first and last ordinates and the boundary, by trapezoidal rule, is
    (a) $21.6\,m^2$ (b) $17.6\,m^2$ (c) $8.8\,m^2$ (d) $7.6\,m^2$

23. If three consecutive ordinates taken from a survey line at 3 m intervals are 1.3 m, 2.8 m and 1.85 m, then by parabolic rule, the area between the line, the first and last ordinates and the boundary is
    (a) $16.84\,m^2$ (b) $14.35\,m^2$ (c) $8.75\,m^2$ (d) $5.95\,m^2$

24. If three consecutive ordinates are taken at intervals of 1m, and 1.5 m and are recorded as 1.8 m, 2.35 m and 1.75 m, then the area between the traverse line, the first and last ordinates and boundary is
    (a) $5.15\,m^2$ (b) $6.025\,m^2$ (c) $7.875\,m^2$ (d) $12.05\,m^2$

25. A planimeter is an instrument used for
    (a) checking whether a given surface is plane
    (b) checking whether the plane table surface is level
    (c) finding areas from plans and maps
    (d) finding the slope of a given terrain

26. In the planimeter formula, $A = M(\text{FR} - \text{IR} \pm 10\,N) + C$, the additive constant is considered when
    (a) the anchor point is inside the area being measured
    (b) the anchor point is outside the area being measured
    (c) the anchor point is inside or outside the area
    (d) the area is traced in the counter-clockwise direction

27. Zero circle with reference to planimeter is (M is the multiplying constant and C is the additive constant)
    (a) the additive constant C (c) the multiplying constant M
    (b) the constant given by MC (d) the constant given by M/C

Review Questions

1. Explain the method of calculating the area of a traverse from coordinates. What is the effect of changing the origin on the coordinates and the area?
2. Explain the concept of double meridian distance. Show that the algebraic sum of the product of DMDs and the corresponding latitudes give twice the area of a traverse.

3. Will the area of a traverse calculated by different methods give identical results? Explain.
4. Derive the trapezoidal rule. What is the assumption made in deriving this rule?
5. Derive the parabolic or Simpson's rule. On what assumption is it derived?
6. The parabolic rule always gives greater values of area than the trapezoidal rule. Is this true? Explain.
7. Explain the construction and working principle of a planimeter.
8. What is the zero circle? Under what condition does the zero circle get traced by the tracing point? Describe the methods used to find the area of a zero circle.
9. What are the constants of the planimeter? Explain how they can be determined for a given instrument.

Problems

1. From the data given below, calculate the area of the traverse by (a) the coordinates method, (b) double meridian distances, and (c) total latitudes.

Line	AB	BC	CD	DE	EA
Length	402	398	430	490	274.5
Bearing	N60° 45′E	S70° 30′E	S22° 30′W	N81° 45′W	N16° 11′W

2. The following table gives the details of a traverse conducted:

Line	AB	BC	CD	DE	EA
Length	330	235	250	348	435.3
Bearing	N0° 0′E	S48° 30′E	S0° 0′E	N0° 0′W	N21° 15′W

   Find the area of the traverse by (a) double meridian distances, (b) total latitudes, and (c) coordinates.

3. Find the area of the following traverse by (a) the coordinates method and (b) the double meridian distances method.

Line	AB	BC	CD	DE	EF	FA
Length	310	260	400	350	160	264.7
Bearing	75° 15′	90° 0′	180° 0′	270° 0′	0° 0′	N52° 31′W

4. Find the area of the following traverse by (a) coordinates, (b) double meridian distances, and (c) total latitudes with the following data.

Line	AB	BC	CD	DA
Length	310.5	340.8	405.2	279.3
Bearing	225°	50° 30′	305° 45′	213° 12′

5. A chain line was divided into eight sections of 12 m each. Offsets were taken from the middle of each section to a boundary. The offsets were measured (in metres) as follows from the left end: 5.63, 6.84, 7.23, 6.95, 7.58, 5.97, 5.84, and 4.95. Find the area between the chain line and the boundary.

6. The offsets taken from a chain line are given below. Find the area between the first and last offsets, the boundary, and the chain line using (a) the average ordinate rule, (b) the trapezoidal rule, and (c) Simpson's rule.

Chainage (m)	0	10	20	30	40	50	60	70	80
Offset (m)	4.25	5.38	6.94	6.84	6.25	6.34	6.14	7.23	5.9

7. The following offsets were taken from a chain line to a boundary. Find the area between the first and last offsets, the chain line, and the boundary using (a) the average ordinate rule, (b) the trapezoidal rule, and (c) the prismoidal rule.

Chainage (m)	0	15	30	45	60	75
Offset (m)	3.4	4.8	5.6	6.9	6.98	7.2

8. The offsets taken from a survey line to a boundary are given below. Find the area between the survey line, the first and last offsets, and the boundary by the trapezoidal rule and Simpson's rule.

Chainage (m)	0	8	16	24	32	40	48
Offset (m)	0	4.5	4.8	5.2	4.8	5.4	4.8

9. Offsets were taken from a survey line at 12-m intervals and the lengths of the offsets were (starting from the left): 0, 3.8, 4.4, 5.2, 4.8, 6.4, 5.9, and 0 m. Find the area between the survey line, the first and last offsets, and the boundary by the trapezoidal rule and Simpson's rule.

10. Offsets were taken from a survey line to a boundary starting from the left station. The line was divided into five sections of 10 m each first and later into four sections of 15 m each. The lengths of the offsets in metres were 0, 5.4, 5.6, 5.8, 6.2, 6.4, 6.8, 7.2, 7.5, and 4.8. Find the area between the survey line, the first and the last offsets, and the boundary by the trapezoidal rule and Simpson's rule.

11. From the chainages and offsets given below, find the area between the boundary, the first and last offsets, and the survey line (a) by the coordinates method and (b) by considering individual areas as trapeziums (all measurements in metres).

Chainage	0	6	11	16	22	29	35
Offset	1.6	2.4	3.8	4.9	7.6	6.8	5.8

12. From the following measurements, find the area between the first and last offsets, chain line, and the boundary.

Chainage (m)	0	12	20	25	34	42	52
Offset (m)	0	6.9	7.6	9.8	10.2	9.9	6.8

13. The readings from a planimeter were the following: initial reading = 5.673, final reading = 3.638, the zero crossed the index twice in the clockwise direction. The anchor point was kept outside, the instrument set to natural scale with $M = 100$ sq. cm. Find the area of the land if the plan was made to a scale of 1 cm = 200 m.

14. The readings from a planimeter were the following: initial reading = 9.684, final reading = 3.876, the zero crossed the index twice in the counterclockwise direction. The anchor point was kept inside the area and the constants were $M = 101.8$ sq. cm and $C = 30$. Find the area of the plan.

15. Find the area of the zero circle from the following data: diameter of wheel = 20.5 mm, length of tracing arm = 15.82 cm, length of pole arm = 25 cm. The wheel is set beyond the pivot by 30 mm.

16. An area was measured twice, once with the anchor point inside and once with the anchor point outside. The readings were the following:

Anchor point outside: initial reading = 9.868, final reading = 3.493

Anchor point inside: initial reading = 4.743, final reading = 8.368

The diameter of the wheel is 2.02 cm and the length of the tracing arm is 15.76 cm. Find the area of the zero circle.

17. A rectangular area of sides 250 × 750 mm was measured by a planimeter twice, once with the anchor point inside and once with the anchor point outside. The readings were as follows:

Anchor point outside: initial reading = 3.398, final reading = 9.648, zero crossed the index mark twice in the clockwise direction

Anchor point inside: initial reading = 1.638, final reading = 7.888, zero crossed the index mark once in the counterclockwise direction

Find the constants and the area of the zero circle.

18. A plan was drawn to a scale of 1 cm = 250 m. A planimeter set to natural scale was used to measure the area, and the readings obtained were as follows from two different positions of the anchor point, both outside the area. Find the average area in hectares from the following data.

Position 1. Initial reading = 0.846, final reading = 9.634, zero crossed the index twice in the clockwise direction

Position 2: Initial reading = 0.134, final reading = 1.350, zero crossed the index once in the clockwise direction

The multiplying constant = 120 sq. cm.

9

Measurement of Elevations—Contouring

Introduction

In Chapter 7, we discussed levelling. Levelling is the operation that is used to determine the relative heights of points. Contouring is basically a levelling operation. The basic operation, instruments, and methods remaining the same, the objective is different. In levelling, the objective was to find the elevations or reduced levels of selected points. In contouring, on the other hand, we find the reduced levels of selected points as in levelling, but with the objective of determining points with the same elevation and plotting them on a plan. Since such points cannot be determined or selected easily in advance, two methods are employed—(a) to determine these points by a laborious process of trial and error and (b) to determine these points using some interpolation technique. It is necessary to get the levelling data first to interpolate and locate such points. Both horizontal and vertical control are necessary to prepare contour maps. Contour maps give a wealth of information about the topography of the terrain, based on which many calculations for engineering projects are done.

9.1 Contours and Contouring

A contour is a line joining points having the same elevation. More precisely, a contour line is the intersection of a level surface with the ground. An example of contour lines is shown in Fig. 9.1(a). To be more illustrative, consider the contour plan of a pond shown in Fig. 9.1(b). Each line in the contour map joins points of the same elevation. In this figure, each contour line is the intersection of the water surface at that level with the ground. There are an infinite number of points on the line. Obviously, the elevation of all the points cannot be actually measured. Some points may be obtained by direct measurement and some may be obtained by interpolation.

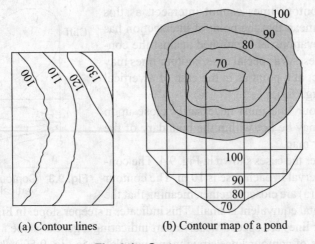

(a) Contour lines   (b) Contour map of a pond

Fig. 9.1 Contour maps

Contouring is the method, essentially of levelling, of locating contour lines. In order to plot contour lines, we need both vertical and horizontal distances. The method, thus involves locating points on the ground by distance measurement and finding their elevations using a level.

9.1.1 Terminology

The following basic terminology needs to be understood (see Fig. 9.2).

A *contour line or contour* is a line joining points having the same elevation. Alternately it is a line of intersection of a level surface with the ground. A contour line should always be marked with its elevation from the datum.

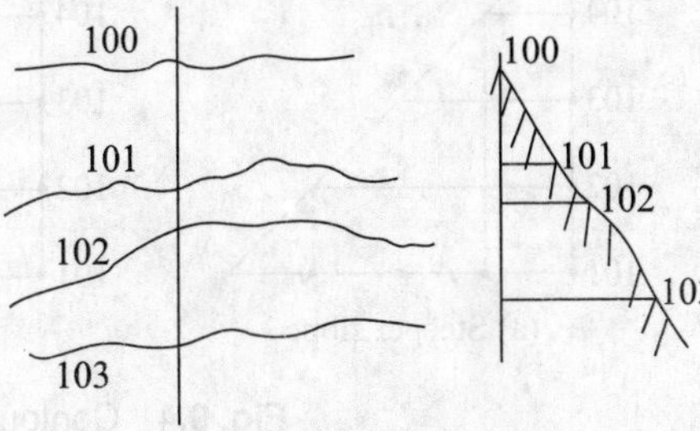

Fig. 9.2 Contour interval

The contour interval is the difference in elevation between successive contour lines. It is the vertical distance between two successive contour lines. In Fig. 9.2, the contour interval is 1 m. The contour interval is selected based upon many factors such as the nature of the ground and the purpose of the contour plan.

The horizontal equivalent is the horizontal distance between two successive contour lines. If this distance is small, it indicates steeper a slope. This will not be the same at every point on the lines.

A contour map is a map showing contour lines of different elevations at some selected contour interval. It provides considerable information about the topography of the terrain, as it indicates both horizontal and vertical distances.

9.2 Characteristics of Contours

From the very definition of a contour line, the following characteristic features are evident.

(a) Two contour lines cannot intersect, as this would mean that the point of intersection has two elevations as mentioned against the contour lines. As a special case, contour lines may intersect at a point as in the case of a vertical cliff (Fig. 9.3).

(b) A contour line must necessarily close upon itself, may be not within the boundary of the contour map.

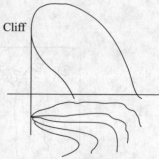

Fig. 9.3 Contour lines in a vertical cliff

(c) Consider the cases shown in Fig. 9.4. The contour interval in each case is 10 m. The contour lines in (a) are closer together, meaning that the horizontal equivalent is small. This indicates a steeper slope. In Fig. 9.4(b), the contour lines are spaced widely apart indicating a gentler slope.

(d) A series of contour lines straight and parallel as in Fig. 9.5(a) shows a plane surface. The series of contour lines in Fig. 9.5(b) shows a steep slope. Figure 9.5(c) depicts a gentler slope of the terrain. A uniform slope is indicated by the contour lines in Fig. 9.5(d), where they are equally spaced.

(e) The contour map of a small hillock will be as shown in Fig. 9.6(a). Such a map will have contours of higher elevation inside. Figure 9.6(b) shows the contour

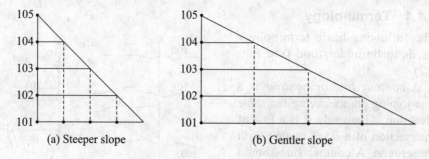

Fig. 9.4 Contour lines and slopes

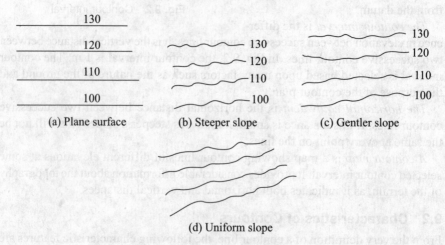

Fig. 9.5 Contour lines and slopes

map of a pond. Here, the contours of higher elevation will be outside. These lines, of course, indicate the meeting points of the water surface with the ground when the water level reaches the respective elevations.

(f) In the case of a ridge, the contour lines will be as shown in Fig. 9.7(a). Here, the contour lines cross the ridge line at right angles and will be so curved to have the concave side towards higher ground. Similarly, in the case of a valley, the contour lines cross the valley line at right angles. The V-shaped curves across the valley line will have their convex side towards higher ground as shown in Fig. 9.7(b).

(g) A contour line passing through a point is at right angles to the line of maximum slope at that point (Fig. 9.8).

(h) Two contour lines of the same elevation cannot unite and continue as one line, nor can a contour line split and continue in different directions, as such a knife-edge ridge or depression does not occur in nature.

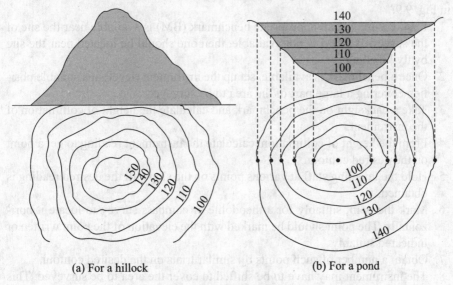

| (a) For a hillock | (b) For a pond |

Fig. 9.6 Contour lines

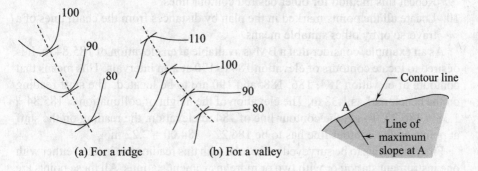

| (a) For a ridge | (b) For a valley | Line of maximum slope at A |

Fig. 9.7 Contour lines   **Fig. 9.8** Line of maximum slope

9.3 Contouring Methods

There are several methods of contouring. The final objective being the preparation of a contour map, the equipment required includes those for distance measurement and those for elevation measurement. A chain or tape is used in ordinary contouring jobs. A level with a levelling staff is required for finding the elevations of the points. The following methods are commonly used.

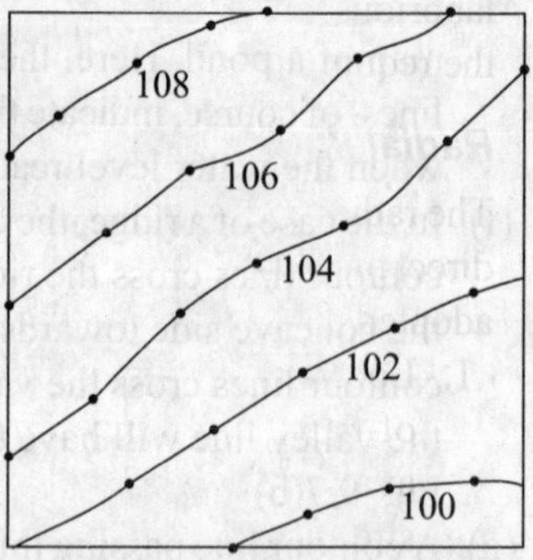

Fig. 9.9 Direct method of contouring

9.3.1 Direct Method

In the direct method, the desired contours are obtained by trial and error. It is a very laborious but accurate method. It can be used when the extent of survey is small and greater accuracy is desired in locating contours. The procedure is as follows and illustrated in Fig. 9.9.

1. First, ensure that an appropriate benchmark (BM) is available near the site of the survey. If a BM is not available, then one should be located near the site by fly levelling.
2. Once a benchmark is available, set up the instrument (level) at a suitable position covering a large part of the area to be surveyed.
3. Take a backsight on the benchmark and calculate the height of collimation of the instrument.
4. From the height of collimation, calculate the instrument reading to get a point on the desired contour.
5. Hold the levelling staff at various points on the slope till the desired reading is obtained.
6. Mark the point suitably for a theodolite or compass survey to locate it horizontally. The point should be marked with the elevation of the point written or indicated suitably.
7. Obtain a number of such points by similar trials on the desired contour.
8. The instrument may have to be shifted to cover the area to be surveyed. This completes contouring of one contour line.
9. Repeat this method for other desired contour lines.
10. Locate all the points marked in the plan by distances from the chain lines of a traverse or by other suitable means.

As an example, consider that a BM is available at an elevation of 185.84 m. It is desired to locate contours of elevation 184 to 190 at 2-m intervals. This means that contours of elevation 184, 186, 188, and 190 are to be located. The staff reading on the benchmark is 1.38 m. The elevation of the height of collimation is 185.84 + 1.38 = 186.22. To get the contour line of 184 m elevation, the reading on the staff at points on the contour line has to be 186.22 − 184.00 = 2.22 m.

First, the points to be surveyed in the area with this reading are located either with one instrument station or with two or more instrument stations. All these points are marked suitably for locating by distance. A similar effort is required to locate all the desired contour lines. One contour is located separately to avoid confusion. Though laborious, this method gives points having the required elevation.

Radial line method

The radial line method is a variation of the direct method. The following procedure is adopted as illustrated in Fig. 9.10.

1. Locate a benchmark near the centre of the area by fly levelling from an available benchmark from a nearby site.
2. Lay out radial lines from the centre point conveniently to cover the area suitably.
3. Benchmarks can also be located at the end of radial lines if required.

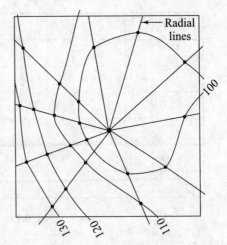

Fig. 9.10 Radial line method

4. Locate points on these radial lines having the required elevation after calculating the staff reading required for that elevation. This can be done conveniently by moving inwards or outwards with the staff.
5. Once a point is located, mark it suitably with an arrow or other marks, with a tag indicating the elevation of the contour.
6. Repeat the process with all the radial lines and for all the desired contours.
7. Then locate the points on the horizontal plane by a chain or tape by measuring distances from the centre point already marked.

Direct methods are laborious and time-consuming. The only advantage is that the points located are correct and no interpolation errors occur. Two teams are often employed to implement these methods. A levelling team marks points with the required elevations. A second team locates these points from the established lines with a chain or a tape and a compass or a theodolite to get their relative horizontal positions.

9.3.2 Indirect Methods

In the indirect method, instead of locating points of required elevations, a general survey is conducted to obtain the vertical and horizontal distances of the points. When a large number of such points are available, the points of required elevation are obtained by interpolation. This method is fast compared to the direct methods. The following indirect methods are generally used.

By spot levels

This method is generally employed in the case of areas which are not very extensive and are reasonably undulating. The following procedure is adopted as illustrated in Fig. 9.11.

1. Divide the area into squares of reasonable dimensions. 5-m to 20-m squares are generally employed. The dimensions of the squares are governed by the nature of the ground and the number of contour lines required.
2. Take levels at the corners of the squares only. This is known as *spot levelling*.

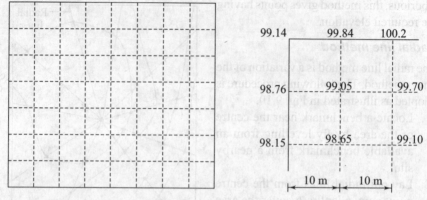

Fig. 9.11 Spot levels

3. Depending upon the undulations of the ground, levels may be taken at intermediate points on the sides of the squares.
4. Once these levels are available, determine the contours of required elevation
   by interpolation.
5. The dimensions of the squares should be large enough to have the survey done
   rapidly, but small enough to locate points of desired elevations and take care
   of the undulations of the ground.
   The method of interpolation will be discussed later.

By cross sections

This method is most commonly employed in route surveys for roads. The following
procedure is adopted (see Fig. 9.12).

1. Take sections at right angles to the centre line of the road. They need not be at
   right angles and can be inclined to the centre line if required.
2. The distance between sections depends upon the terrain. They should not be
   less than 20 m in hilly terrain, but can be 50 to 100 m in plane country.
3. Take levels at a number of points along the section line.
4. Locate the points horizontally by a chain or tape.
5. Obtain the contours of required elevation by interpolation.
6. Join such points to get the contour lines.

By Tacheometry

Tacheometry is a method of finding the distances and elevations simultaneously.
This is suitable in the case of hilly terrain, where measuring distances with a chain
or tape is laborious and time-consuming. The method is discussed in detail in
Chapter 14.

Tacheometry involves a tacheometer or a theodolite fitted with stadia wires.
The basic principle of the method is shown in Fig. 9.13. θ is the vertical angle
as measured by the central cross hair. The intercept between the two outer cross
hairs is S. The horizontal and vertical distances (with respect to the line of collimation) from the instrument station to the staff station are given, respectively, by

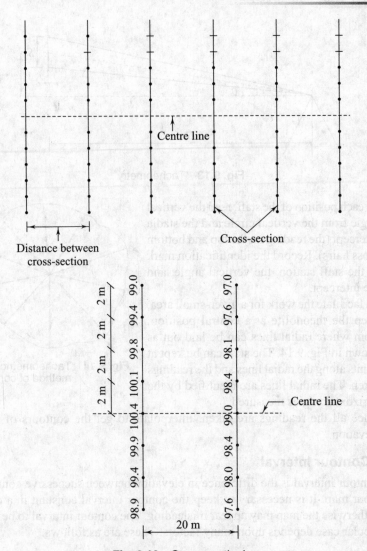

Fig. 9.12 Cross-sectioning

$$H = K_1 S \cos^2 \alpha + K_2 \cos \alpha$$

$$V = H \tan \alpha$$

where S is the intercept on the staff with the top and bottom cross hairs, α is the vertical angle, and K_1 and K_2 are constants of the instrument. The horizontal distance from the instrument and the vertical height from the line of collimation are thus calculated with a tachometer or a theodolite.

The following procedure is generally adopted.

1. Set up the instrument at a suitable place having a commanding view of the area. Carry out the temporary adjustments as required.
2. Hold the staff on a benchmark of known elevation to get the elevation of the line of sight.
3. Then hold the staff in positions marked either on the vertices of squares or otherwise.

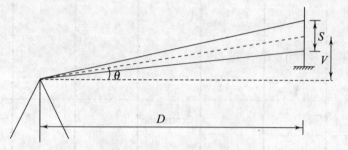

Fig. 9.13 Tacheometry

4. At each position of the staff, read the vertical angle from the vertical circle and the stadia intercept (the reading on the top and bottom cross hairs). Record the identification mark of the staff station, the vertical angle, and the intercept.

5. To facilitate the work for a given small area, keep the theodolite at a central position, from where radial lines can be laid out as shown in Fig. 9.14. The staff can be kept at points along the radial lines and the readings taken. The radial lines are identified by the horizontal angles measured.

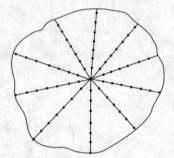

Fig. 9.14 Tacheometric method of contouring

6. Once all the readings are taken, interpolate to get the contours of desired elevation.

9.4 Contour Interval

The contour interval is the difference in elevation between successive contours in a contour map. It is necessary to keep the contour interval constant in a contour map, otherwise the map may appear misleading. The contour interval to be used in a particular case depends upon many factors. These are as follows.

Nature of terrain   The contour interval depends upon the nature of the ground—whether it is undulating or flat. On a very flat terrain, smaller intervals can be adopted, as the contours will be spaced sufficiently apart. In a hilly terrain or undulating ground, a larger interval is adopted, otherwise the contours will come too close for plotting due to the steep slope.

Objective and extent of survey   To a large extent, the objective of the survey decides the contour interval. If the contour map is to be used for design or for earthwork computations when accuracy is required, the contour interval can be small. On the other hand, when the survey area is very extensive as in the case of reservoirs, communication lines, or drainage purposes, the contour interval can be large.

Scale of map   The contour interval should be such that it can be plotted to scale and is distinguishable clearly in the map. If the scale is small, the contour interval has to be large. In the case of large-scale drawings, the contour interval can be small. In general, the contour interval should be inversely proportional to the scale to which it is proposed to draw the map.

Time and expense A small contour interval will involve more fieldwork and paperwork for interpolation and plotting. A large contour interval will take less time and money to conduct the field job and subsequently less paperwork. The time and money available is also thus a factor in deciding the contour interval. The contour interval is fixed after considering all the above-mentioned factors. For general topographic work, the contour interval in metres may be taken as 25/100,000. Tables 9.1 and 9.2 give some guidelines for adopting contour intervals based on the purpose and scale of the map.

Table 9.1 Contour interval based upon the objective of survey

Purpose	Scale	Contour interval (m)
Location surveys	1:5000	2
	1:20,000	
Town planning schemes reservoirs	1:5000	0.5–1
	1:10,000	
Building sites	1:1000	0.2–0.5

Table 9.2 Contour interval based upon scale and terrain

Scale of map	Terrain	Contour interval (m)
Small (1:10,000 or more)	• Flat	0.2–0.5
	• Hilly	1–2
Medium (1:1000 to 1:10,000)	• Flat	0.5–1
	• Hilly	2–3
Large (1:1000 and less)	• Flat	0.2–0.5
	• Hilly	1–1.5

9.5 Interpolation of Contours

Once the survey work, by indirect methods, for the determination of contour lines is completed, contours of desired elevation have to be obtained by interpolation. The contour interval is decided depending upon the use of contour maps. The scale is so chosen so that the contour lines can be clearly drawn on paper. Interpolation can be done by any of the following methods.

By estimation

The positions of the required contour points between the known elevations are estimated and marked. The method is rough and not suitable for general survey work. The points are then joined by smooth curves to obtain the required contour lines.

By calculation

In this case, the required points are obtained by linear interpolation from the elevations of the guide points. As an example, see Fig. 9.15. The four points of the square have the elevations shown. It is required to find contours at 0.2-m intervals between these points. The following method of calculation is employed. On the line AB, the elevation of A is 100.15 m and that of B is 101.05 m. We locate contours of elevations 100.2, 100.4, 100.6, 100.8, and 101 between A and B. The distance AB is 20 m. The difference in elevation between A and B is 101.05 – 100.15 = 0.9 m in a length of 20 m. From this, the distances of the required contours can be calculated as follows:

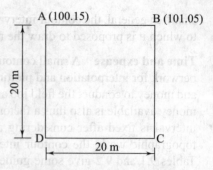

Fig. 9.15 Linear interpolation

100.2-m contour: Difference of elevation from A

$$= 100.2 - 100.15$$
$$= 0.05 \text{ m}$$

Distance from A = 20 × 0.05/0.9 = 1.11 m

100.4-m contour: Difference from elevation of A = 100.4 – 100.15
$$= 0.25 \text{ m}$$

Distance from A = 20 × 0.25/0.9 = 5.55 m

100.6-m contour: Difference in elevation from A = 100.6 – 100.15
$$= 0.45 \text{ m}$$

Distance from A = 20 × 0.45/0.9 = 10 m

100.8-m contour: Difference in elevation from A = 100.8 – 100.15
$$= 0.65 \text{ m}$$

Distance from A = 20 × 0.65/0.9 = 14.44 m

101-m contour: Difference in elevation from A = 101 – 100.15 = 0.85 m

Distance from A = 20 × 0.85/0.9 = 18.89 m

Similar computations can be done on the other sides of the squares. This is illustrated with examples later.

By graphical methods

In place of calculation, linear interpolation can be done by graphical methods. Graphical methods use tracing paper or cloth. There are two methods for interpolating contours.

1. In the first method, a tracing paper marked as shown in Fig. 9.16 is used. The vertical axis gives the contour intervals at 0.2 m or as per requirement. A series of horizontal lines is marked against each contour interval. If A and B are points surveyed with known elevations, the tracing paper is kept such that A and B fall on the respective elevation lines simultaneously. This is done by turning and adjusting the tracing paper. Once the two points are correctly placed on their elevation lines, the required contour can be marked by making a pinprick on the line AB against the point of intersection with the required contour elevation horizontal line. Generally the following procedure is adopted.

    (i) Prepare the tracing paper as per requirement. On the vertical axis, mark the lines suitably with a constant interval depending upon the contour interval desired. Draw horizontal lines to the required length.

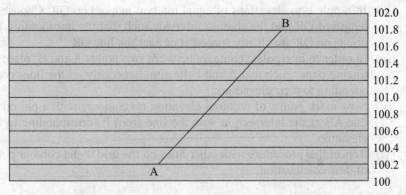

Fig. 9.16 Graphical method of interpolation

(ii) Once the tracing paper is ready, keep it on the grid points of known elevations.

(iii) On any line AB, the elevations of the end points A and B are known. Keep the tracing paper such that the points A and B lie on lines marking their elevations. To get such a placement, keep line A on the horizontal line of its elevation. Turn the tracing paper such that point B also lies on the horizontal line showing its elevation.

(iv) Mark the intersection points of AB and the horizontal lines of required elevation on paper with a pin.

(v) Designate the points so marked with their elevations.

(vi) Repeat this procedure with all the lines of the grid.

(vii) Draw contour lines that join lines of equal elevation on the map.

2. A second graphical method is to use a tracing paper with lines as shown in Fig. 9.17. The principle remains essentially the same. The procedure is the following.

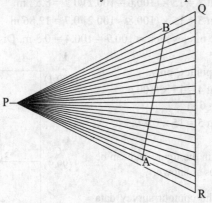

Fig. 9.17 A second graphical method of interpolation

(i) Draw a line QR conveniently on the tracing paper. Choose a pole P such that it lies on the perpendicular bisector of line QR. Choose the distance of P from QR conveniently.

(ii) Mark points on the line QR as per the required elevations and contour interval.

(iii) Join pole P to elevations of equal interval marked on QR. Choose the length of QR and the contour intervals such that the lines joining P to points on QR are clearly spaced at or near the line QR.

(iv) In order to interpolate contours between two points A and B, place the tracing paper such that A and B lie simultaneously on the lines corresponding to their elevations.

(v) Now mark points of required elevation on contours with a pin on the line AB at the intersection with the line from P corresponding to that elevation.

(vi) Repeat this procedure with other lines on the grid to get contour points of desired elevation.

Example 9.1 In the square shown in Fig. 9.18, the elevation of the points at the corners are given. Determine the positions of the points of 100.4-, 100.6-, and 100.8-m contours on the grid.

Solution We interpolate on all the lines of the square grid where the required points can be. It can be seen from the elevations of the four points that points of the required elevation can be found only along lines AB and BC. CD will have contours of elevation less than 100.4 m only and AD will have contours of elevation less than 100.2 m only.

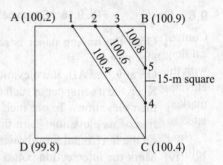

Fig. 9.18

On line AB: Difference in elevation = 100.9 – 100.2 = 0.7 m in 15 m. Distances from A of points of elevation:

100.4 m: point 1 = 15 × (100.4 – 100.2)/0.7= 4.28 m

100.6 m: point 2 = 15 × (100.6 – 100.2)/0.7 = 8.57 m

100.8 m: point 3 = 15 × (100.8 – 100.2)/0.7 = 12.86 m

On line BC: Difference of elevation = 100.9 – 100.4 = 0.5 m. Distances from C of points of elevation:

100.4 m is at point C only.

100.6 m: point 4 = 15 × (100.6 – 100.4)/0.5 = 6 m

100.8 m: point 5 = 15 × 0.4/0.5 = 12 m

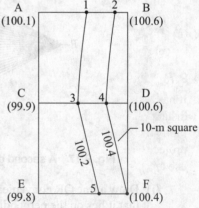

Fig. 9.19

The points are marked in the figure and can be joined by smooth curves as shown.

Example 9.2 With the contour survey data shown in Fig. 9.19, find the points of the 100.2 and 100.4 contours.

Solution From the given elevations of the corner points, it is clear that the points of elevation 100.2 and 100.4 will only be along the horizontal lines AB, CD, and EF.

On line AB: Difference in elevation = 100.6 – 100.1 = 0.5 m over 10 m. Distances from A of points of elevation:

100.2 m: point 1   = $10 \times 0.2/0.5$

                          = 4 m

100.6 m: point 2   = $10 \times 0.4/0.5 = 8$ m

On line CD: Difference in elevation = $100.6 - 99.9 = 0.7$ m over 10 m. Distance from C of points of elevation:

100.2 m: point 3   = $10 \times 0.3/0.7 = 4.3$ m

100.4 m: point 4   = $10 \times 0.5/0.7 = 7.1$ m

On line EF: The 100.4-m contour is at point F. The 100.2-m contour is at point 5 at a distance of $10 \times 0.4/0.66.67$ m.

These points can be marked and contour lines drawn as shown in Fig. 9.19.

9.6 Preparing Contour Maps

Contour maps are prepared with the survey data for horizontal distances and vertical heights of points. In the direct method of contouring, points have to be plotted with the horizontal distances measured in the field. These points are of the desired elevations. In the indirect method, the grid or radial lines are plotted and distances marked for points whose elevations have been obtained. This data is then interpolated to obtain points of desired elevation.

Smooth curves are to be drawn through points having the same elevation. The following points should be kept in mind while plotting the contour lines.

(a) The contour interval in a map should be kept constant to avoid apparent distortion of the drawing.

(b) The contour lines should be plotted with thick lines. Broken lines may also be used if there are other lines also to be included in the same drawing.

(c) It should be remembered that a contour map is plotted to a horizontal scale only. There is no height scale used for elevations.

(d) The elevations of all contour lines must be mentioned against the lines. Different practices are followed in writing the contour elevation. It can be written above the line at a readable point, the line may be broken to write the elevation, or some other convenient way can be used. In a large map with long contour lines, the elevations may be written at more than one point for ease of reading.

(e) If a contour line does not close within the limits of the drawing, the contour elevation must be written at both the ends of the line for ease of reading.

9.7 Use of Contour Maps

It has been mentioned earlier that contour maps, with both horizontal and vertical distances included in the drawing, provide a wealth of information about the topography of the land surveyed. Such maps are important for undertaking engineering projects. Some of the uses of contour maps are outlined here. More details of their use are discussed in Chapter 13 on setting out works.

9.7.1 Developing Cross Sections

One of the uses of contour maps is to develop cross sections, from which the general shape of the terrain can be determined. This can be used to determine the earthwork required to lay engineering works such as communication lines. Figure 9.20 shows how this is done. The following procedure is adopted.

1. The contour map is given in Fig. 9.20(a). If it is required to lay any work along the direction AB, draw the line AB on the contour plan.
2. Draw the section by setting the *x-y* axes as shown in Fig. 9.20(b). Project the intersection points of line AB with the contour lines down to the *x*-axis first.
3. Mark the elevations of the intersection points in the *y*-direction to some suitable scale.
4. Join the points so obtained together by a smooth curve. This curve shows the general shape of the ground.
5. Knowing the depth at which any work is to be laid, carry out the computation of the earthwork from the cross section so obtained.

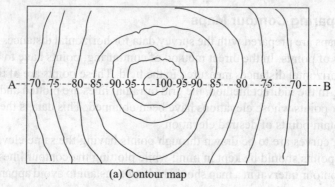

(a) Contour map

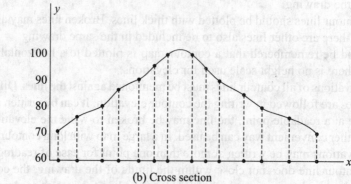

(b) Cross section

Fig. 9.20   Cross sections from contours

9.7.2   Setting Contour Gradients

A contour gradient is a line in a plan having a constant slope. This line will pass through the different contour lines at different points but is assumed to be a straight line between contour lines. Knowing the contour interval and the horizontal equivalent, the contour gradient can be worked out on the plan. The following procedure is used.

Figure 9.21 shows a contour plan. The contour interval is 0.5 m. The contour plan is drawn to a scale of 1:10. Suppose we have to find a contour gradient of 1 in 40. This gradient is the same as 0.5:20. The horizontal equivalent is thus 20 m. The gradient of 1 in 40 will thus have a horizontal distance of 20 m. This works out to 2 cm according to the scale in which the plan is drawn.

1. The starting point of the contour gradient is A. With A as the centre and a 2 cm radius, draw an arc to cut the next contour at point 1. A-1 is thus a line having a gradient of 1 in 40.
2. With 1 as the centre and a radius of 2 cm, draw an arc to cut the next contour at point 2. 1-2 is thus a line having a gradient of 1 in 40.
3. Proceed similarly to get other points 3, 4, etc.
4. A-1-2-3-4, etc. is the contour gradient. These points can be located on the ground by measurement.

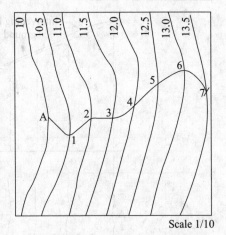

Scale 1/10

Fig. 9.21 Contour gradient

The contour gradient can also be obtained directly in the field. A theodolite or level can be used conveniently for this. The following procedure can be adopted.

1. Set up and level the instrument at a commanding position in the field from which a number of contour lines can be tackled.
2. Take a reading on the staff held at the starting point A.
3. The horizontal equivalent has been calculated as 20 m. Therefore, use a 20-m chain or the 20 m length of a longer tape.
4. With the zero end of the tape at A and the staff held at the 20-m end, a person moves till one gets the required reading for the next contour line. For example, if the reading on staff at A is 1.25 m, then the reading on the next staff will be $1.25 + 0.5 = 1.75$ m. This point can be located and marked on the ground.
5. A number of such points can be located with one instrument station. If required, the instrument can be shifted and the process repeated to locate more points.

9.7.3 Determining Intervisibility Between Stations

A contour map can be used to ascertain the intervisibility between stations in the case of undulating ground. This is particularly important in the case of a triangulation survey, where the distance between the stations can be considerably large. The procedure is as follows.

1. In the contour plan, mark the two stations between which intervisibility has to be checked. In Fig. 9.22, A and B are the two stations whose intervisibility has to be checked.
2. Join the points A and B by a straight line.
3. Draw the cross section along the line AB following the procedure given earlier.
4. We draw the *x*- and *y*-axis and project the points A and B and the intervening points on the contour lines at which the line AB intersects them. The elevations are marked to a suitable scale for each of these points. Join the points by a smooth curve to get the shape of the ground.
5. On the cross section, join the line A and B.
6. If the line AB cuts the ground at some points, then the ground is too high there for the stations A and to be intervisible.
7. If the line is clear of all the ground points, then the stations are intervisible.
8. Both these cases are illustrated in Fig. 9.22.

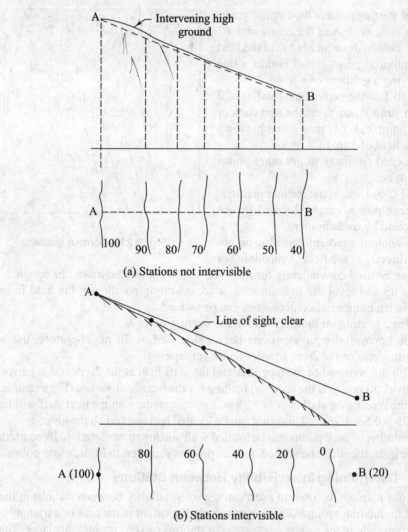

(a) Stations not intervisible

(b) Stations intervisible

Fig. 9.22 Intervisibility of stations

9.7.4 Locating the Watershed Line at a Point

The watershed line for a point encloses the area from which water will flow through that point. This is important in all watershed management problems. A watershed line has the following characteristics.

(a) It passes through the ridge lines (points of high elevation with respect to the ground on either side).

(b) It is perpendicular to the contour lines.

(c) In general, it follows the ridge lines.

The watershed line can be located from the contour plan. Figure 9.23 shows the watershed line for a dam. Depending upon the height of the dam and the contour lines surrounding it, the watershed line is drawn based on the features outlined above. The extent of the catchment area for the dam can be decided from such a drawing. The area and volume of water stored can also be calculated.

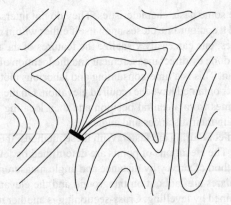

Fig. 9.23 Determining the watershed line

9.7.5 Determining Reservoir Capacity

The calculation of the capacity of a reservoir is possible using a contour plan. Referring to Fig. 9.24, the following procedure is adopted.

1. From the contour plan, decide the capacity up to the elevation of the dam or any other obstacle.

2. If the elevation of 150 m is the maximum water level, contour lines below that level.

3. Depending upon the contour interval decided earlier, contour lines of 130 m, 110 m, etc. (at 20-m intervals) are available.

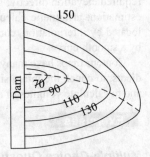

Fig. 9.24 Capacity of reservoir

4. Measure the area enclosed by the contour lines at 150 m, 130 m, 110 m, etc. by using a planimeter.

5. Knowing these areas and the vertical interval of 20 m between the lines, calculate the capacity of the reservoir using the formulae.

A detailed discussion on the calculation of reservoir capacities has been included in Chapter 10.

Summary

Contouring is a very important aspect of surveying. A contour map provides a wealth of information on many aspects concerning the design and execution of engineering projects.

A contour is the intersection of a level surface with the surface of the earth. In other words, a contour is a line joining points having the same elevation. The contour interval is the difference in elevation between successive contour lines. In practice, the same contour interval is used in a map to avoid distortion of the map. The contour interval is selected based on the nature of the ground, the purpose of the map, the scale of the map, and the time and money available.

Contouring is essentially a levelling operation, but a contour map is plotted on a horizontal scale only, whereby distance measurement becomes a part and parcel of the contouring operation. A level and tape are essential equipment for the job.

Contour lines have some basic characteristics. They do not intersect (except in the case of a vertical cliff), and a contour line closes upon itself either within or outside the map. By looking at contour lines, one can easily visualize the features of the terrain.

Contouring can be done by direct measurement and the location of points of the required elevation on the ground. This is a time-consuming and laborious method. It is done by trial and error, but the points obtained have the required elevation. On the ground, it can be done by setting out radial lines from a central position and then locating points of the required elevation and measuring their distances and angles.

Indirect methods of contouring are faster. With these methods, the contouring operation is similar to levelling. The elevations of points are determined by levelling, and the points are located by any method of survey by distance and angle measurement. Spot levelling is one method where squares are laid out on the ground and the elevations of the corners of the squares are determined by levelling. Cross-sectioning is another method generally used in the case of roads, railways, etc. Tacheometry can be used when elevations and distances are measured simultaneously. In the indirect methods we do not obtain the points of the required elevation directly; we use interpolation techniques. Interpolation can be done by estimation, arithmetic calculations, or graphical methods. Once such points have been located by interpolation, contour lines are drawn by joining points of the same elevation by a smooth curve.

Contour maps have many uses—to locate lines of uniform slope known as contour gradients, to check the intervisibility of stations, to draw sections, to locate watershed lines, and to calculate volumes of reservoirs.

Exercises

Multiple-Choice Questions

1. The difference between contouring and levelling is that
   (a) contouring focuses on distance measurements while levelling focuses on elevation
   (b) contouring is an angle measuring operation while levelling is for measuring heights
   (c) contouring focuses finding points having a given elevation while levelling is to find the elevation of given points
   (d) contouring and leveling are different names for the same operation
2. A contour or contour line is a line joining
   (a) points having the same elevation
   (b) a set of points having different elevations
   (c) a set of equidistant points having different elevations
   (d) joining points of a traverse surveyed
3. Contour interval is
   (a) the distance between contours in a contour map
   (b) the distance between points on a contour line whose elevations are found
   (c) the distance between the level and a point selected for contouring
   (d) the difference in elevation between successive contour lines
4. Two contour lines
   (a) will always intersect
   (b) may sometimes intersect
   (c) will not intersect except in special cases
   (d) may intersect in case of a ridge or valley

5. If the horizontal distance between contour lines having the same contour interval is small in a contour map, it indicates
   (a) a plane surface　　　　　　　　(c) surface with gentler slope
   (b) a surface with very steep slope　　(d) the ridge of a mountain
6. If we draw contour lines of a plane surface, they will appear as
   (a) curved lines spaced far apart
   (b) curved closed lines
   (c) straight lines uniformly spaced
   (d) straight lines inclined at an angle to the edges of the surface
7. A contour line
   (a) necessarily closes upon itself, even if outside the map
   (b) is always an open curve, not closing upon itself
   (c) may sometimes close upon itself
   (d) may sometimes be an open curve not closing upon itself
8. When the contour lines having the same contour interval are farther apart, it shows a
   (a) gentle slope　　　　　　　　　(c) plane surface
   (b) very steep slope　　　　　　　(d) ridge or a valley
9. The series of contour lines shown represents a

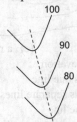

   (a) ridge of a mountain　　　　　　(c) hillock
   (b) valley　　　　　　　　　　　　(d) pond
10. The series of contour lines shown represents a

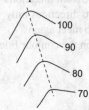

   (a) ridge of a mountain　　　　　　(c) hillock
   (b) valley　　　　　　　　　　　　(d) pond
11. The contour interval in the figure shown is

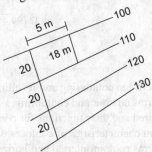

   (a) 20 m　　　(b) 18 m　　　(c) 10 m　　　(d) 5 m

12. The contour map shown represents

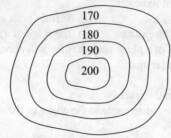

    (a) a hillock   (b) a pond   (c) a ridge   (d) a valley

13. The contour map shown represents

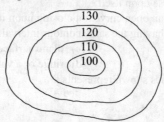

    (a) a hillock   (b) a pond   (c) a ridge   (d) a valley

14. Two contour lines of the same elevation
    (a) cannot unite and continue as one line
    (b) can unite but will continue as separate lines
    (c) can unite and continue as one line
    (d) cannot unite but can continue as separate lines

15. A contour line
    (a) cannot split and continue in different directions
    (b) can split and continue in different directions
    (c) can split but continue in two directions only
    (d) can split but has to return to meet again

16. A line perpendicular to a point on the contour line in plan
    (a) is along a plane surface
    (b) is along the least slope line
    (c) is along the steepest slope
    (d) shows an upward slope

17. In the figure shown, the contour line of 100
m elevation will intersect AB at
    (a) 5m from A
    (b) 5m from B
    (c) 10 m from A
    (d) 10 m from B

A (98.8)   B (100.6)

—10-m square

D (99.4)   C (100.6)

Review Questions

1. Explain the differences between contouring and levelling.
2. Define and explain the terms contour and contouring.
3. List the instruments required and their functions for contouring.
4. Explain any five important characteristics of contours with neat sketches.
5. Define and explain the terms contour interval and horizontal equivalent.
6. Explain the factors to be considered while deciding the contour interval for a survey.

7. Explain the direct methods of contouring. Explain the advantages and disadvantages of these methods.
8. Explain the indirect methods of contouring. What are the advantages and disadvantages of these methods?
9. Explain the method of preparing a contour plan.
10. Explain briefly, with the help of neat sketches, the uses of contour maps.
11. Define the term contour gradient. Explain the method to locate a given contour gradient in a contour plan.
12. Explain the method to locate a contour gradient in the field with a level.

Problems

1. In the grid shown in Fig. 9.25, find the positions of the contour lines of elevation 10.2, 100.4, and 100.6 m.
2. In the grid shown in Fig. 9.26, find the positions of the contours of elevation 100.2, 10.4, 100.6, and 100.8 m.

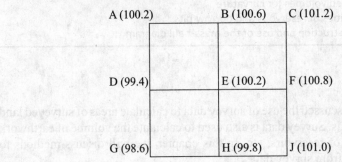

A (100.2)   B (100.6)   C (101.2)

D (99.4)   E (100.2)   F (100.8)

G (98.6)   H (99.8)   J (101.0)

Fig. 9.26

Computation of Volumes

Learning Objectives

After going through this chapter, the reader will be able to
- compute the area of different types of cross sections
- calculate volumes using trapezoidal and prismoidal formulae
- make prismoidal correction to calculated volumes
- correct computed volumes for curvature
- calculate capacities of reservoirs and borrow pits
- explain the construction and use of the mass-haul diagram

Introduction

In Chapter 8, we discussed the use of survey data to calculate areas of surveyed land by different methods. Survey data is also used to calculate the volume of earthwork required, capacity of reservoirs, etc. In this chapter, we will discuss methods to calculate volumes from survey data.

Levelling of an area provides us with the heights of points on the surface of the earth. Contouring combines the data on heights with distances along the ground. Such data is necessary to calculate the volumes.

The generally used unit of volumes is cubic metre. Today, it has become mandatory to use the metric or SI units. However, in many countries, FPS units such as cubic feet and cubic yards are used. To convert from one set of units to another, conversion factors can be used. These are given in Appendix 1.

The volumes of simple solids are calculated using the formulae given in Appendix 3. However, in practice, the areas and volumes encountered are complex in shape. The final form of the structure may be arrived at by cutting (where you remove earth), filling (where you add earth from outside), or a combination of filling and cutting. The areas of such sections are calculated from the general formulae to be derived in the next section. Once the area is known, the length factor can be added. The length may be a straight line or a curve and hence a curvature correction may be necessary.

Pappus' Theorems

Theorem 1 (for areas): *When a line or an arc rotates about another fixed line, the area swept is given by the length of the line or arc multiplied by the length traced by the centroid of the line or arc.*

Theorem 2 (for volumes): *When an area revolves about an axis, the volume generated is equal to the area multiplied by the length traced by the centroid of the area.*

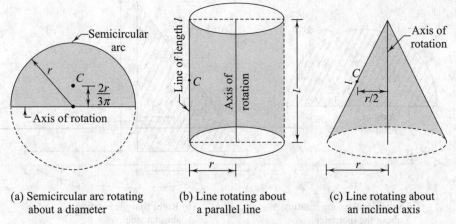

(a) Semicircular arc rotating
    about a diameter

(b) Line rotating about
    a parallel line

(c) Line rotating about
    an inclined axis

Fig. 10.1 Pappus' theorem for areas

For the first theorem, consider a semicircular arc rotating about its diameter as shown in Fig. 10.1(a). The length of the arc is πr. As it revolves about its diameter, it traces the surface of a sphere in one revolution. The centroid of the arc is at $2r/3\pi$ from the base diameter. The path traced by the centroid is the perimeter of a circle of radius $2r/3\pi$. The perimeter is $2\pi \times 2r/3\pi$. Thus the area traced is given by the length of the arc × length traced by the centroid = $\pi r \times 4\pi r/3\pi = (4/3)\pi r^2$, which is the surface area of the sphere.

As a second example, consider a line rotating about another line as shown in Figs 10.1(b) and (c). In Fig. 10.1(b), the area obtained is the curved surface area of a cylinder, and in Fig. 10.1(c), the area obtained is the curved surface area of a cone.

In Fig. 10.1(b), the length of the line is l and the path traced by the centroid of the line is $2\pi r$. The curved surface area is $2\pi rl$, which is the curved surface area of a cylinder. In Fig. 10.1(c), the length of the line is l and the path traced by the centroid is $2\pi(r/2)$. Thus the curved surface area obtained is πrl, which is the curved surface area of a cone.

For the second theorem, consider the semicircular area rotating about its diameter shown in Fig. 10.2(a). The volume obtained in one revolution is that of a sphere. The area is $\pi r^2/2$. The centroid of a semicircular area is at $4r/3\pi$ from the base and the length traced by the centroid during rotation is $2\pi(4r/3\pi)$. The volume is thus equal to $\pi r^2 \times 2\pi(4r/3\pi) = (4/3)\pi r^3$, which is the volume of a sphere.

As a second example, consider the case of a right-angled triangle rotating about one of its sides [Fig. 10.2(b)]. The volume generated is that of a cone. If the dimensions are the side h about which rotation takes place and is the other side r, then the area is $rh/2$ and the path traced by the centroid is $2\pi r/3$ (the centroid is at $r/3$ from the side). The volume obtained is therefore $rh/2 \times 2\pi r/3 = \pi r^2 h/3$, which is the volume of a cone.

These theorems will be useful when the length part of the volume is curved. We start with the calculation of cross-sectional areas of sections normally encountered in practice.

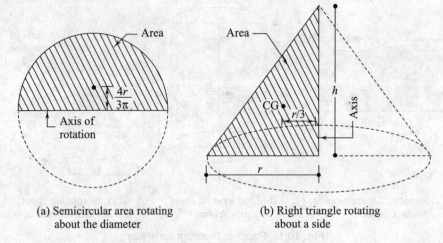

(a) Semicircular area rotating   (b) Right triangle rotating
    about the diameter   about a side

Fig. 10.2 Pappus' theorem for volume

10.1 Areas of Cross Sections

The common cross sections one comes across in practice are shown in Fig. 10.3. The following symbols will be used to indicate the various parameters of these sections.

Formation width It is the width of the subgrade (b).

Depth It is the depth of cutting or filling at the centre line (h).

Half-breadth This is the horizontal distance from the centre to the intersection of the original ground with the side slopes (w_1 and w_2).

Side slope s to 1 is the side slope, s horizontal to 1 vertical [i.e., the ground rises (or falls) by 1 m over a horizontal length of s metres].

Transverse slope r to 1 is the transverse slope of the ground, r horizontal to 1 vertical [i.e., the ground rises (or falls) by 1 m over a horizontal length of r metres].

Side heights h_1 and h_2 are the heights from the formation level to the intersection of the side slope with the ground. These parameters have been indicated in the sections shown in Fig. 10.3.

The following points must be clearly understood in dealing with slopes, as illustrated in Figs 10.3(f) and (g).
(a) If s:1 is the slope given, then for a given height h, the horizontal distance $= hs$. This is clear from the similar triangles shown [Fig. 10.3(f)].
(b) If s:1 is the slope given and the horizontal distance is d, then the vertical distance is given by d/s. This is again clear from the similar triangles shown [Fig. 10.3(g)].
The area of each of these cross sections can be calculated as follows.

10.1.1 Single-level Section

A single-level section (Fig. 10.4) may be formed by cutting or filling. It is just a trapezium and formulae are readily available for calculating the area of such a section. In this section, AB is the subgrade of width b, w_1 and w_2 are half-widths—

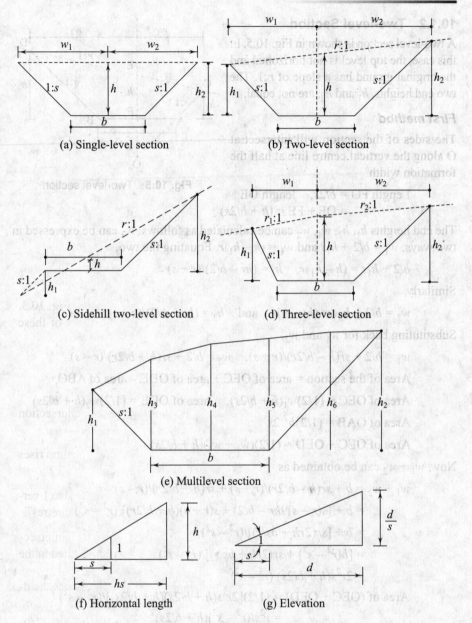

Fig. 10.3 Types of sections

equal in this section, and h_1 and h_2 are end heights—equal to h, the height at midsection. Therefore, $w_1 = w_2$ and $h_1 = h_2$, $w_1 = w_2 = b/2 + hs$; $w_1 + w_2$ can be calculated as $b + 2hs$, where $h = h_1 = h_2$.

Area of the section = $(1/2)[2 \times$ half-breadth $+ b] \times h = h(b + hs)$

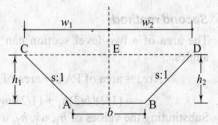

Fig. 10.4 Single-level section

10.1.2 Two-level Section

A two-level section is shown in Fig. 10.5. In this case, the top level is not horizontal and the original ground has a slope of $r{:}1$. The two end heights, h_1 and h_2, are not equal.

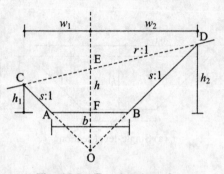

Fig. 10.5 Two-level section

First method

The sides of the section will intersect at O along the vertical centre line at half the formation width.

$$\text{Length FO} = b/2s, \quad \text{length OE}$$
$$= \text{OF} + \text{FE} = (h + b/2s)$$

The end heights h_1, h_2, w_1, w_2 can be calculated as follows. w_1 can be expressed in two ways: $w_1 = b/2 + h_1 s$ and $w_1 = (h - h_1)r$. Equating the two,

$$b/2 + h_1 s = (h - h_1)r, \quad h_1 = (hr - b/2)/(r + s)$$

Similarly,

$$w_2 = b/2 + h_2 s = (h_2 - h)r \quad \text{and} \quad h_2 = (hr + b/2)/(r - s)$$

Substituting back for w_1 and w_2,

$$w_1 = b/2 + sr(h - b/2r)/(r + s), \quad w_2 = b/2 + sr(h + b/2r)/(r - s)$$

Area of the section = area of OEC + area of ODE – area of ABO

Area of OEC = $(1/2)w_1(h + b/2s)$, area of ODE = $(1/2)w_2(h + b/2s)$

Area of OAB = $(1/2)b^2/2s$

Area of OEC + OED = $(1/2)(w_1 + w_2)(h + b/2s)$

Now, $w_1 + w_2$ can be obtained as

$$
\begin{aligned}
w_1 + w_2 &= b + sr(h - b/2r)/(r + s) + sr(h + b/2r)/(r - s)\\
&= b + [s(r - s)9hr - b/2) + s(r + s)(h + b/2r)]/(r^2 - s^2)\\
&= b + [sr(2rh + bs/r)]/(r^2 - s^2)\\
&= [b(r^2 - s^2) + sr(2rh + bs/r)]/(r^2 - s^2)\\
&= 2r^2s(h + b/2s)/(r^2 - s^2)
\end{aligned}
$$

$$
\begin{aligned}
\text{Area of (OEC + OED)} &= (1/2)[2r^2s(h + b/2s)(h + b/2s)]/(r^2 - s^2)\\
&= r^2s/(r^2 - s^2)(h + b/2s)^2
\end{aligned}
$$

Area of section = $[r^2s/(r^2 - s^2)](h + b/2s)^2 - b^2/4s$

Second method

The area of a two-level section can also be calculated as the area of four triangles.

$$
\begin{aligned}
\text{Area} &= \text{area of FAC} + \text{area of FCE} + \text{area of FDE} + \text{area of FBD}\\
&= (1/2)(b/2)h_1 + (1/2)hw_1 + (1/2)hw_2 + (1/2)(b/2)\,h_2
\end{aligned}
$$

Substituting the values of h_1, w_1, h_2, w_2 as calculated earlier,

$$\text{Area} = (1/2)(b/2)(h_1 + h_2) + (1/2)h(w_1 + w_2)$$

$$h_1 + h_2 = (hr - b/2)/(r + s) + (hr + b/2)/(r - s) = (2hr^2 + bs)/(r^2 - s^2)$$

$$w_1 + w_2 = b/2 + s(hr - b/2)/(r + s) + b/2 + s(hr + b/2)/(r - s)$$

$$= b + [s(r - s)(hr - b/2) + s(r + s)(hr + b/2)]/(r^2 - s^2)$$

$$= (br^2 + 2shr^2)/(r^2 - s^2)$$

$$\text{Area} = (1/2)(b/2)[(2hr^2 + bs)]/(r^2 - s^2) + (1/2)h[br^2 + 2shr^2]/(r^2 - s^2)$$

$$= (r^2sh^2 + br^2h + \frac{b^2s}{4})/(r^2 - s^2)$$

Area using coordinates

A simple way to calculate area is to use coordinates. The calculation of coordinates will involve the same calculations as for h_1, h_2, w_1, w_2, etc. From Fig. 10.5, the coordinates are calculated as A $(0, - b/2)$, B $(0, b/2)$, D (w_2, h_2), and C $(- w_1, h_1)$, with the origin at the midpoint of AB. The coordinates are written in determinant form as follows:

Point	A	B	D	C
y	$- b/2$	$b/2$	h_2	h_1
x	0	0	w_2	$- w_1$

Then,

$$2 \times \text{area} = [(b/2)w_2 - w_1h_2] - (h_1w_2)$$

10.1.3 Sidehill Two-level Section

A sidehill two-level section is shown in Fig. 10.6. In this section, the ground cuts the embankment width at some point, and hence a part of the section is obtained by cutting and a part of the section is in filling. In the section shown, the right part is obtained by cutting while the left part is obtained by filling. With the same symbols as before (shown in Fig. 10.6),

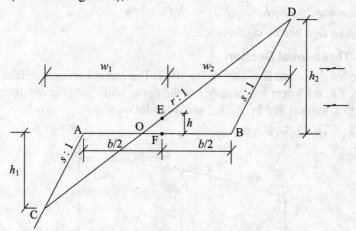

Fig. 10.6 Sidehill two-level section

$$w_1 = b/2 + sh_1 = (h + h_1)r, \quad h_1 = (b/2 - hr)/(r - s)$$

$$w_2 = sh_2 + b/2 = (h_2 - h)r, \quad h_2 = (hr + b/2)/(r - s)$$

$$w_1 = b/2 + s(b/2 - hr)/(r - s) = rs(b/2r - h)/(r - s)$$
$$w_2 = b/2 + sh_2 = rs(h + b/2r)/(r - s)$$
$$\text{Area in filling} = (1/2)h_1 \times \text{AO} = (1/2)h_1 \times (b/2 - rh)$$
$$= (1/2)(b/2 - hr)r(r - s)(b/2 - rh)$$
$$= (b/2 - rh)^2/2(r - s)$$
$$\text{Area in cutting} = (1/2)h_2 \times \text{BO} = (1/2)(b/2 + rh)(b/2 + rh)$$
$$= (1/2)(b/2 + rh)^2/(r - s)$$

Area using coordinates

The coordinates method can be used in the case of this type of section as well. The area in cutting and filling is calculated separately:

Area in filling = area of OAE,   area in cutting = area of OBE

Area in filling

The coordinates are $A(-[b/2], 0)$, $O(-rh, 0)$, and $E([-w_1 + rh], -h_1)$ with F as the origin. Writing these in determinant form,

Point	A	O	E	A
y	0	0	$-h_1$	0
x	$-(b/2)$	$-rh$	$-(w_1)$	$-(b/2)$

$$2 \times \text{area} = h_1 b/2 - (h_1 rh) = h_1(b/2 - rh)$$

Area in cutting

The area in cutting is OBC. The coordinates with F as the origin are

Point	O	B	C	O
y	0	0	h_2	0
x	$-rh$	$b/2$	w_2	$-rh$

$$2 \times \text{area} = -h_2 rh - h_2(b/2) = -h_2(b/2 + rh)$$

The negative sign has no significance.

10.1.4   Three-level Section

The three-level section is shown in Fig. 10.7. The natural ground surface has two slopes, $r_1{:}1$ in the lower half and $r_2{:}1$ in the upper half. With the notations shown in the figure, we can find h_1, h_2, w_1, and w_2 as follows:

$$w_1 = (h - h_1)r_1 = b/2 + h_1 s$$

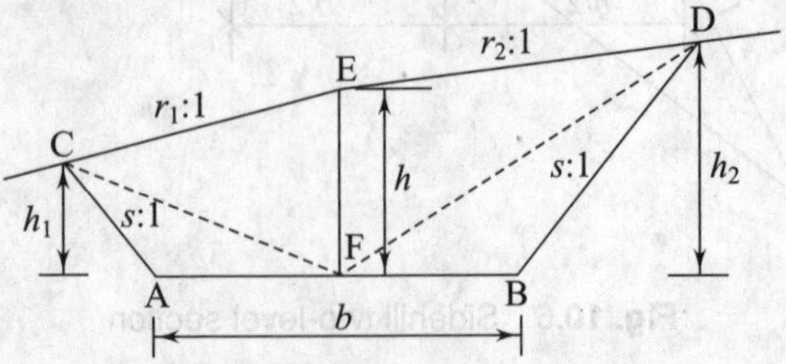

Fig. 10.7   Three-level section

From this,

$$h_1 = (hr_1 - b/2)/(r_1 + s)$$

Substituting back,

$$w_1 = b/2 + s(hr_1 - b/2)/(r_1 + s) = r_1 s(h + b/2s)/(r_2 - s)$$

Similarly,

$$w_2 = (h_2 - h)r_2 = b/2 + h_2 s, \quad h_2 = (hr_2 + b/2)/(r_2 - s)$$

from which

$$w_2 = b/2 + s(hr_2 + b/2)/(r_2 - s)$$

$$= r_2 s(h + b/2s)/(r_2 - s)$$

Knowing these values, the area of the section can be calculated as the sum of the areas of four triangles, namely ACF, FCE, EFD, and DFB. Area of triangle ACF = $(1/2)\, h_1 b/2$, area of triangle CFE = $(1/2)hw_1$, area of triangle EFD = $(1/2)w_2 h$, and area of triangle DFE = $(1/2)\, h_2 b/2$.

$$\text{Total area} = (b/4)(h_1 + h_2) + (1/2)h(w_1 + w_2)$$

The values calculated can be substituted in this formula to get the area.

Area from coordinates

The coordinate method can be used to calculate the area as well. The coordinates with F as the origin are A$(- b/2, 0)$, B$(b/2, 0)$, D(w_2, h_2), E$(0, h)$, and C$(- w_1, h_1)$. The order of writing is ABDECA. The coordinates can be put into determinant form as follows:

Point	A	B	D	E	C	A
y	0	0	h_2	h	h_1	0
x	$- b/2$	$b/2$	w_2	0	$- w_1$	$- b/2$

$$2 \times \text{area} = [- w_1 h - h_1 b/2] - [h_2 b/2 + hw_2]$$

10.1.5 Multilevel Section

A multilevel section is shown in Fig. 10.8. The area of such a section is calculated by the coordinate method only, as the slopes are many and it is easier to work with coordinates. For the section shown in Fig. 10.8, the coordinates with the centre point of formation as the origin are written as follows:

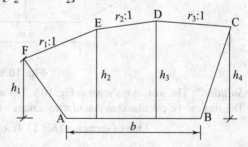

Fig. 10.8 Multilevel section

Point	A	B	C	D	E	F	G	A
y	0	0	h_1	h_2	h_3	h_4	h_5	0
x	$- b/2$	$b/2$	x_c	x_d	x_e	x_f	x_g	$- b/2$

Then, the area can be calculated as

$$2 \times \text{area} = [h_1 x_d + h_2 x_e + h_3 x_f + h_4 x_g - h_5 b/2]$$
$$- [h_1 b/2 + h_2 x_c + h_3 x_d + h_4 x_e + h_5 x_f]$$

The areas of different sections are summarized in Table 10.1.

Table 10.1 Areas of cross sections

Type of section	Area	Remarks
Single level	$h(b + hs)$	h—height at centre, b—formation width, s—side slope
Two-level	$[r^2s(h + b/2s)^2]/(r^2 - s^2)$ $- b^2/4s$ or $(r^2sh^2 + r^2bh + bs^2/4)/(r^2 - s^2)$	r—transverse slope
Sidehill two-level	Filling: $(b/2 - sh)^2/2(r - s)$ Cutting: $(b/2 + sh)^2/2(r - s)$	–
Three-level	$(h_1 + h_2)b/4 + (w_1 + w_2)h/2$	$h_1 = (hr_1 - b/2)/(r_1 + s)$ $h_2 = (hr_2 + b/2)/(r - s)$ $w_1 = [r_1s(h + b/2s)]/(r_1 + s)$ $w_2 = [r_2s(h + b/2s)]/(r_2 - s)$ r_1, r_2 are transverse slopes; h_1, h_2 end heights; and w_1 and w_2 half-breadths
Multilevel	–	Determined by coordinate method

The following examples illustrate the application of these formulae.

Example 10.1 Determine the cross-sectional area of a single-level section with the following data: width of subgrade = 6 m, height at the centre = 1.85 m, and side slopes = 2:1.

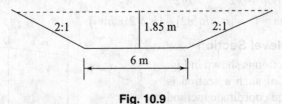

Fig. 10.9

Solution The section is shown in Fig. 10.9. Top width = $b + 2hs = 6 + 2 \times 2 \times 1.85 = 13.4$ m. The area can be calculated as that of a trapezium = $(a + b)(h/2)$.

$$\text{Area of section} = (6 + 13.4) \times 1.85/2 = 17.95 \text{ sq. m}$$

Example 10.2 A two-level section is shown in Fig. 10.10. Find the area of the section.

Solution The area can be obtained by the formula derived for the section:

$$\text{Area} = [s(b/2)^2 + (b/2)^2(bh + sh^2)]/(r^2 - s^2)$$

where h is the height at midsection, s is the side slope, r is the transverse slope, and b is the formation width. Substituting,

$$\text{Area } A = [1.5 \times 4^2 + 4^2(8 \times 2.8 + 1.5 \times 2.8^2)]/(4^2 - 1.5^2) = 41.5 \text{ m}^2$$

One need not remember the formula for the sections. One can derive the area from the basic principles or use the method of coordinates. Both are illustrated below.

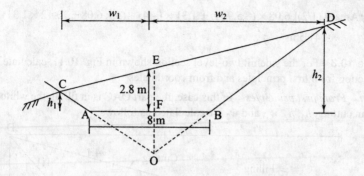

Fig. 10.10

From basic principles   Let the sides of the section meet at point O as shown. Let h_1 and h_2 be the heights of the top ends of the section and w_1 and w_2 be the half-widths. These parameters can be derived as follows:

$$w_1 = (h - h_1)r = (b/2) + h_1 s$$

or

$$(h - h_1)r = (b/2) + h_1 s$$

from which

$$h_1 = [h \times r - (b/2)]/(r + s) = (2.8 \times 4 - 4)/(4 + 1.5) = 1.31 \text{ m}$$

Similarly,

$$w_2 = (h_2 - h)r = b/2 + h_2 s$$

and

$$h_2 = (hr + b/2)/(r - s)$$
$$= (2.8 \times 4 + 4)/(4 - 1.5) = 6.08 \text{ m}$$

From these values,

$$w_1 = (2.8 - 1.31) \times 4 = 5.96 \text{ m}, \quad w_2 = (6.08 - 2.8) \times 4 = 13.12 \text{ m}$$

Also,

$$OE = OF + FE, \quad OF = b/2s = 8/(2 \times 1.5) = 2.667 \text{ m}, \quad OE = 5.467 \text{ m}$$

Area = area of triangle CEO + area of triangle DEO – area of triangle ABO

Area of triangle CEO = $(1/2)w_1 \times OE = (1/2) \times 5.964 \times (2.8 + 2.667)$
$$= 16.30 \text{ m}^2$$

Area of triangle DEO = $(1/2) \times w_2 \times OE$
$$= (1/2) \times 13.12 \times (2.8 + 2.667) = 35.86 \text{ m}^2$$

Area of triangle ABO = $(1/2) \times 8 \times 2.667 = 10.668 \text{ m}^2$

Area of section = $16.30 + 35.86 - 10.668 = 41.5 \text{ m}^2$

By coordinates   Considering the origin to be at the midpoint of AB, by the coordinates method, the values of $h_1, h_2, w_1,$ and w_2 are to be calculated as shown before. The coordinates are arranged in determinant form as follows:

Point	A	B	D	C	A
y-coord.	0	0	6.08	1.31	0
x-coord.	– 4	4	13.12	– 5.964	– 4

Area $= (1/2)[\,6.08 \times (-5.964) + 1.31 \times (-4) - (4 \times 6.08 + 13.12 \times 1.31)$

$= 41.5 \text{ m}^2$

Example 10.3 For the sidehill two-level section shown in Fig. 10.11, calculate the area of the section from first principles and from coordinates.

Solution *From first principles* In this case, the part OAC is in filling the while the part OBD is in cutting. h_1, h_2, w_1, and w_2 are calculated as follows:

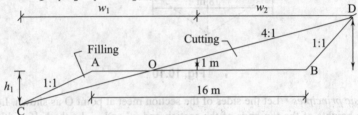

Fig. 10.11

$w_1 = b/2 + h_1 s = (h_1 + h)r$

Here, $b/2 = 8$ m, $s = 1$, and $r = 4$. From this,

$h_1 = (b/2 - hr)/(r - s) = (8 - 1 \times 4)/(4 - 1) = 1.33$ m

Similarly,

$w_2 = b/2 + h_2 s = (h_2 - h)r, \quad h_2 = (b/2 + hr)/(r - s)$

$= (8 + 4 \times 1)/(4 - 1) = 4$ m

$w_1 = 8 + 4/3 = 9.33$ m, $\quad w_2 = 8 + 4 \times 1 = 12$ m

Area of the section = area of triangle OAC + area of triangle OBD

The bases of triangles OAC and OBC are AO and OB, the lengths of which can be determined from the principle of similarity of triangles:

$4/(4/3) = (8 + x)/(8 - x)$

where x is the distance by which O is shifted from centre. $x = 4$ m.

AO $= 4$ m and OB $= 12$ m

Area in filling $= (1/2) \times 4 \times 4/3 = 2.67 \text{ m}^2$

Area in cutting $= (1/2) \times 4 \times 12 = 24 \text{ m}^2$

From coordinates The coordinates of the different points are A $(-4, 0)$, C $(-5.33, -1.33)$, O $(0, 0)$, D $(16, 4)$, B $(12, 0)$. The coordinates are written with O as the origin. The order is ACOA and ODBO. The coordinates are written in this order as

y	0	-1.33	0	0	y	0	4	0	0
x	-4	-5.33	0	-4	x	0	16	12	0

From these coordinates,

Twice the area of filling $= -4 \times 1.33 = 5.33 \text{ m}^2$

Area in filling $= (1/2) \times 5.33 = 2.67 \text{ m}^2$

Twice the area in cutting $= (1/2) \times 4 \times 12 = 48 \text{ m}^2$

Area in cutting $= (1/2) \times 48 = 24 \text{ m}^2$

Example 10.4 Find the area of a three-level section with the following data: formation width = 9.2 m, side slopes = 2:1, natural ground slope = 15:1 in the higher half and 8:1 in the lower half. Height at midsection = 2.15 m.

Solution The sketch of the section is shown in Fig. 10.12.

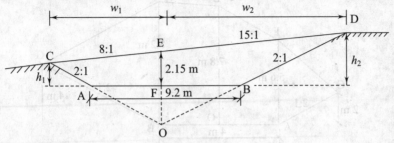

Fig. 10.12

Area of section by formula

$$\text{Area} = (h_1 + h_2) \times b/4 + (w_1 + w_2)h/2$$
$$= (1.26 + 2.835) \times 9.2/4 + (7.12 + 10.275) \times 2.15/2$$
$$= 18.70 + 9.42 = 2812 \text{ m}^2$$

Area of section from first principles

$$w_1 = (h - h_1)r_1 = b/2 + sh_1$$
$$h_1 = (hr_1 - b/2)/(r_1 + s)$$
$$= (2.15 \times 8 - 4.6)/(8 + 2) = 1.26 \text{ m}$$
$$w_2 = (h_2 - h)r_2 = b/2 + h_2 s$$
$$h_2 = (b/2 + hr_2)/(r_2 - s)$$
$$= (4.6 + 2.15 \times 15)/(15 - 2) = 2.835 \text{ m}$$
$$w_1 = (2.15 - 1.26) \times 8 = 7.12 \text{ m}, \quad w_2 = (2.835 - 2.15) \times 15 = 10.275 \text{ m}$$

OF = 4.6/2 = 2.3 m   OE = 2.3 + 2.15 = 4.45 m

Area = area of CEO + area of DEO − area of ABO

Area of CEO = $(1/2)$OE $\times w_1 = (1/2) \times 4.45 \times 7.12 = 15.84 \text{ m}^2$

Area of DEO = $(1/2)$OE $\times w_2 = (1/2) \times 4.45 \times 10.275 = 22.86 \text{ m}^2$

Area of ABO = $(1/2)$AB $\times$ OF = $(1/2) \times 9.2 \times 2.3 = 10.58 \text{ m}^2$

Area of section = 15.84 + 22.86 − 10.58 = 28.12 m²

Area from coordinates The section is ABCDEA. With the origin at the midpoint of AB, the coordinates are A(− 4.6, 0), B(4.6, 0), D(10.275, 2.895), E(0, 2.15), C(− 7.12, 1.26). These are written in determinant form as follows:

y	0	0	2.895	2.15	1.26	0
x	− 4.6	4.6	10.275	0	− 7.12	− 4.6

$$2 \times \text{area} = - 7.12 \times 2.15 - 1.26 \times 4.6 - 4.6 \times 2.875 - 10.275 \times 2.15$$
$$= 28.12 \text{ m}^2$$

Example 10.5 Determine the area of the multilevel section shown in Fig. 10.13.

Solution The area is determined by the coordinates method. With the origin at the centre of the base, the coordinates are A(− 4, 0), B(4, 0), C(8, 4), D(4, 8.2), E(0, 7.8), F(− 4, 5.6), and G(− 4, 2). These are written in determinant form for ABCDEFGA as

y	0	0	4	8.2	7.8	5.6	2	0
x	− 4	4	8	4	0	− 4	− 8	− 4

Twice the area = $[4 \times 4 + 7.8 \times (- 4) + 5.6 \times (- 8) + 2 \times (- 4)]$

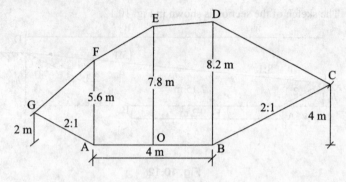

Fig. 10.13

$$- [4 \times 4 + 8 \times 8.2 + 7.8 \times 4 + 2 \times (- 4)]$$
$$= -172.4$$
Area = 86.4 m^2

Note: In this particular case, the area can be found as the area of triangles and trapeziums.

Area of AFD = $(1/2) \times 5.6 \times 4 = 11.2$ m^2

Area of AFEO = $(5.6 + 7.8) \times 4/2 = 26.8$ m^2

Area of OEDB = $(1/2) \times (7.8 + 8.2) \times 4 = 32$ m^2

Area of BDG = $(1/2) \times 8.2 \times 4 = 16.4$ m^2

Total area = 86.4 m^2

10.2 Volume from Cross Sections

Figure 10.14 illustrates a volume with three sections at equal intervals. A_1, A_m, and A_2 are the areas of the three sections shown. Such a volume is known as a *prismoid*. A prismoid is a volume made up of end areas which lie in parallel planes. The two ends need not be made of the same number of sides. The volume can be calculated by two methods—prismoidal formula and trapezoidal formula. This is similar to the case of areas, where we had two main formulae—parabolic formula and trapezoidal formula.

The trapezoidal formula, also known as the end-area formula, uses the average area between two sections. The volume is calculated as a sum of volumes due to wedges and prisms. The prismoidal formula, also known as Simpson's formula for volumes, calculates volume by considering it to be composed of pyramids with bases at the sides and end faces.

The trapezoidal formula overestimates the volume by considering only prisms and wedges. A negative correction applied to this volume is known as the *prismoidal correction*. The actual volume in practice need not be due to a prismoid. The end areas are assumed to be lying in parallel planes in these two formulae. However, as in the case of roads having curved profiles, the plan being curved, a correction known as *curvature correction* has to be applied. These corrections are dealt with in the following sections.

10.2.1 Prismoidal Formula

The prismoidal formula can be derived as follows. Referring to Fig. 10.14, the prismoid can be considered to be composed of prisms and pyramids [Fig. 10.14(b)] or of pyramids alone [Fig. 10.14(c)]. In either case, the volume can be calculated as explained in the following.

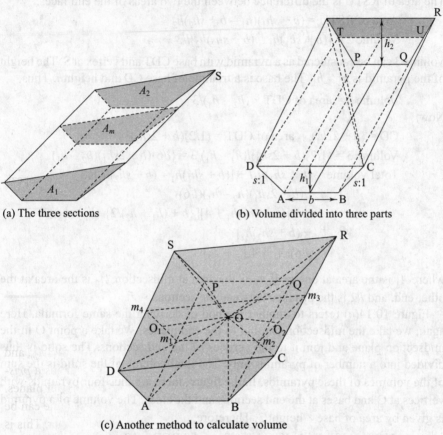

(a) The three sections   (b) Volume divided into three parts

(c) Another method to calculate volume

Fig. 10.14 Volume of cross sections with length

In Fig. 10.14(b), we consider a plane through the side CD and parallel to the base ABQP. We thus have a prism with end faces ABCD and PQUT. The remaining volume consists of two pyramids—one with base RSTU and vertex at C and the other with base DCT and vertex at S. Let the subscripts 1 and 2 refer to values at sections ABCD and PQRS.

Area of section ABCD $= (b + sh_1)h_1$

Area of section PQRS $= (b + sh_2)h_2$

Area of section midway between the two $= (1/2)[(b + sh_1)h_1$
$$+ (b + sh_2)h_2]$$
$$= [b + s(h_1 + h_2)/2](h_1 + h_2)/2$$

The total volume can be computed as the sum of the volumes of the three solids.

Volume of prism = area of ABCD × length

Volume 1 = $(b + sh_1)h_1 l$

Volume of pyramid with vertex at C = area of RSTU × $l/3$

The area of RSTU is the difference between the two areas of the end faces.

Area of RSTU = $(b + sh_1)(h_1 - b + sh_2)h_2$

Volume 2 = $[(b + sh_1)h_1 - (b + sh_2)h_2]l/3$

Volume 3 can be considered as a pyramid with base CDT and vertex at S. The height of the pyramid is $h_2 - h_1$. The base is a triangle of base CD and height l. Thus,

Volume 3 = area of CDT × $(h_2 - h_1)/3$

Now,

$$CD = (b + 2sh_1), \quad \text{area of CDT} = (1/2)(b + 2sh_1)l$$

Volume 3 = $(1/2)(b + 2sh_1)l(h_2 - h_1)/3 = (l/6)(b + 2sh_1)(h_2 - h_1)$

Total volume = $(b + sh_1)h_1 l + [(b + sh_1)h_1 - (b + sh_2)h_2]l/3$

$$+ (b + 2sh_1)(h_2 - h_1)(l/6)$$

$$= (l/6)[(b + sh_1)h_1 + 4][\{b + (h_1 + h_2)/2\} (h_1 + h_2)/2]$$

$$+ (b + sh_2)h_2]$$

$$= (l/6)[A_1 + 4A_m + A_2]$$

where A_1 is the area at one end, A_m is the area at midsection, A_2 is the area at the other end, and $l/2$ is the distance between the sections.

Figure 10.14(c) refers to another method of deriving the same formula. Here again we take the midsection parallel to the end planes. We take a point O in the midsection plane and join it to the vertices of the end sections. The solid is thus divided into a number of pyramids only. The total volume of the solid is the sum of the volumes of these pyramids. In the figure, there are thus four pyramids with vertices at O and bases at the end sections and the sides. The volume of a pyramid is given by area of base × height/3. Therefore,

Volume of solid 1, $V_1 = 1/3 \times$ area of ABCD × $l/2$

$$= 1/3 \times (b + sh_1)h_1 l/2 = A_1 \times l/6$$

Volume of solid 2, $V_2 = 1/3 \times$ area of PQRS × $l/2$

$$= 1/3 \times (b + sh_2)h_2 l/2 = A_2 \times l/6$$

Volume of solid 3, $V_3 =$ area of APSD × OO_1

$$= (1/3) m_1 m_4 l \times OO_1$$

Volume of solid 4, $V_4 =$ area of BQRC × OO_2

$$= (1/3) m_2 m_3 l \times OO_2$$

where OO_1 and OO_2 are perpendiculars from O to the sides of the midsection planes.

$$V_3 = 1/3 \times l \times \text{twice the area of triangle } m_1 m_4 O$$

Similarly,

$V_4 = 1/3 \times l \times$ twice the area of triangle m_2m_3O

Therefore,

$$V_3 + V_4 = 2/3 \times l \times \text{area of midsection} = 2/3 \times l \times A_m$$

Adding the volumes,

$$V = A_1 l/6 + A_2 l/6 + (2/3)l \times A_m$$
$$= (l/6)(A_1 + A_2 + 4A_m)$$

where l is the length of the prismoid, A_1 and A_2 are the end areas, and A_m is the middle area.

To apply the prismoidal formula, two equal lengths with three cross sections are taken. Thus it is necessary to have an *odd number* of cross sections to apply the prismoidal formula. The formula can be generalized as follows: If $A_1, A_2, A_3, \ldots,$ A_n are the areas of cross sections at equal intervals l, then

Volume between A_1 and $A_3 = (l/6)(A_1 + A_3 + 4A_2)$

Volume between A_3 and $A_5 = (l/6)(A_3 + A_5 + 4A_4)$

Volume between A_5 and $A_7 = (l/6)(A_5 + A_7 + 4A_6)$, and so on.

Volume between A_1 and $A_n = (l/6)[A_1 + A_n + 4(A_2 + A_4 + A_6 + \ldots)$
$$+ 2(A_3 + A_5 + A_7 + \ldots)]$$

assuming n is odd. If n is even, then the formula can be applied to $n - 1$ sections and the volume of the last section calculated separately.

10.2.2 Trapezoidal Formula

The trapezoidal or average end-area formula is much simpler and calculates the volume as average area ¥ length. In Fig. 10.15, the same notations are used and volume is derived as

Volume $= (A_1 + A_2) \times l/2$ (volume between sections 1 and 2)

Volume $= (A_2 + A_3) \times l/2$ (volume between sections 2 and 3)

and so on. Thus, If $A_1, A_2, A_3, \ldots, A_n$ are the areas of cross sections at equal intervals l, then

Volume $= (l/2)[A_1 + A_n + 2(A_2 + A_3 + A_4 + \ldots + A_{n-1})]$

The trapezoidal formula can be applied to odd or even number of sections directly.

10.3 Prismoidal Correction

The volume calculated by the trapezoidal formula is generally more than that by the prismoidal formula. If the volumes are calculated by the trapezoidal formula, then a correction is applied on the assumption that the prismoidal formula gives a more accurate value of the volume. This correction is always subtractive and is known as the prismoidal correction. The trapezoidal

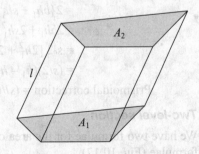

Fig. 10.15 Trapezoidal formula

formula is easier to apply, and often volumes are calculated using it and a correction applied later. The correction depends upon the cross section, and hence different values of the correction have to be used depending upon the type of cross section. The prismoidal correction can be derived for different cross sections as follows.

Single-level section

Considering the length l of the volume (Fig. 10.16), let A_1 and A_2 be the end areas and A_m the middle area. Using the standard notation adopted earlier, $A_1 = h_1(b + sh_1)$ and $A_2 = h_2(b + sh_2)$. To calculate the midsection area A_m, the central height at midsection is taken as the average height $= (h_1 + h_2)/2$. Then,

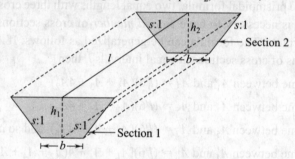

Fig. 10.16 Prismoidal corrections

$A_m = [b + s(h_1 + h_2)/2](h_1 + h_2)/2$

Volume by trapezoidal formula $= (A_1 + A_2)(l/2)$

Volume by prismoidal formula $= (A_1 + 4A_m + A_2)(l/6)$

Difference $= (A_1 + A_2)(l/2) - (A_1 + 4A_m + A_2)(l/6)$

$= (l/6)(3A_1 + 3A_2 - A_1 - 4A_m - A_2)$

$= (l/6)(2A_1 + 2A_2 - 4A_m)$

$= (l/3)(A_1 + A_2 - 2A_m)$

This formula can be used in all cases by taking appropriate formulae for sectional areas. Now,

$$A_1 + A_2 - 2A_m = 2\{bh_1 + sh_1^2 + bh_2 + sh_2^2$$
$$- [b + s(h_1 + h_2)/2](h_1 + h_2)/2\}$$
$$= 2\{bh_1 + sh_1^2 + bh_2 + sh_2^2 - [b(h_1 + h_2)/2 + s(h_1 + h_2)2/4]\}$$
$$= 2bh_1 + 2sh_1^2 + 2bh_2 + 2sh_2^2 - [2bh_1 + 2bh_2 + s(h_1 + h_2)2]/2$$
$$= s/2 [2h_1^2 + 2h_2^2 - h_1^2 - h_2^2 - 2h_1h_2]$$
$$= (s/2)(h_1 - h_2)^2$$

Prismoidal correction $= (sl/6)(h_1 - h_2)^2$

Two-level section

We have two formulae for the area of a two-level section. Let us use one of these formulae (Fig. 10.17).

Area = $[r^2s/(r^2 - s^2)(h + b/2s)^2] - b^2/4s$

Area at section 1 (height h_1), $A_1 = [r^2s/(r^2 - s^2)(h_1 + b/2s)^2] - b^2/4s$

Area at section 2 (height h_2) $A_2 = [r^2s/(r^2 - s^2)(h_2 + b/2s)^2] - b^2/4s$

Area at midsection $A_m = \{r^2s/(r^2 - s^2)[(h_1 + h_2)/2 + b/2s]^2\} - b^2/4s$

Prismoidal correction = $(l/3)(A_1 + A_2 - 2A_m)$

$$A_1 + A_2 - 2A_m = [r^2s/(r^2 - s^2)]\{(h_1 + b/2s)^2 + (h_2 + b/2s)^2 - 2[(h_1 + h_2)/2 - b/2s]^2\} - b^2/4s - b^2/4s + 2b^2/4s$$

$$A_1 + A_2 - 2A_m = [r^2s/(r^2 - s^2)][h_1^2 + b^2/4s^2 + h_1b/s + h_2{}^2 + b^2/4s^2 + h_2b/s - (1/2)(h_1^2 + h_2^2 + 2h_1h_2 - b^2/s^2)]$$

$$= [r^2s/(r^2 - s^2)](1/2)[h_1^2 + h_2^2 - 2h_1h_2]$$

Prismoidal correction $C_p = (l/6)[r^2s/(r^2 - s^2)](h_1 - h_2)^2$

(a) Section at one end (b) Section at other end

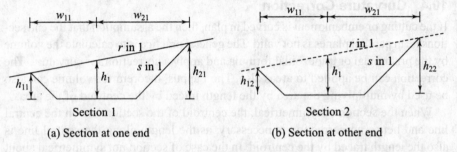

Fig. 10.17 Two-level section

Note that h_1 and h_2 are the heights at the centre at sections 1 and 2. Another formula for prismoidal correction for a two-level section is

$$C_p = (l/6s)(w_{11} - w_{12})(w_{21} - w_{22})$$

Sidehill two-level section

Two sidehill two-level sections are shown in Fig. 10.18. The following formulae can be used:

$$C_p \text{ (filling)} = (l/12s)(w_{11} - w_{12})[(b/2 + rh_1) - (b/2 + rh_2)]$$
$$= [l/12(r - s)]r^2(h_1 - h_2)^2$$

$$C_p \text{ (cutting)} = (l/12s)(w_{21} - w_{22})[(b/2 - rh_1) - (b/2 - rh_2)$$
$$= [l/12(r - s)]\, r^2(h_1 - h_2)^2$$

(a) Section at one end (b) Section at the other end

Fig. 10.18 Sidehill two-level section

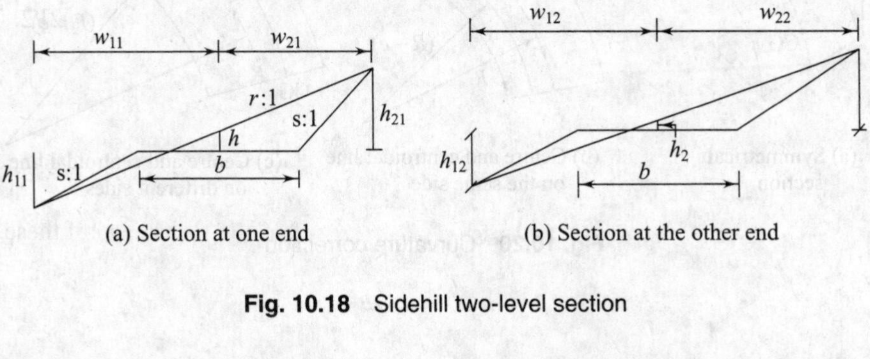

Three-level section

Two three-level sections are shown in Fig. 10.19.

$$C_p = (l/12)(h_1 - h_2)[(w_{11} + w_{21}) - (w_{12} + w_{22})]$$

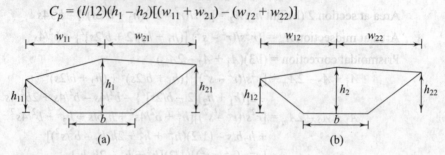

Fig. 10.19 Three-level section

10.4 Curvature Correction

If the cutting or embankment is curved in plan, then the assumption that the end sections lie in parallel planes is not valid. The general practice is to calculate the volume by the prismoidal or trapezoidal formula and apply a correction for curvature. The correction can be applied to areas also. The Pappus' theorem for volume can also be used by multiplying the area by the length traced by the centroid of the area.

When the section is symmetrical, the centroid of the section lies on the central line and hence no correction is necessary, as the length along the central line is also the length traced by the centroid. In the case of section not symmetrical about the central line, the length along the central line will be different from the length traced by the centroid. As shown in Fig. 10.20, two cases can arise—one, when the centroid and the centre of curvature are on the same side [Fig. 10.20(b)] and two, when the centroid and the centre of curvature are on different sides of the central line [Fig. 10.20(c)]. In the former case, the path traced by the centroid is less than the central line length; in the latter case, the path traced by the centroid is more than the length of the central line.

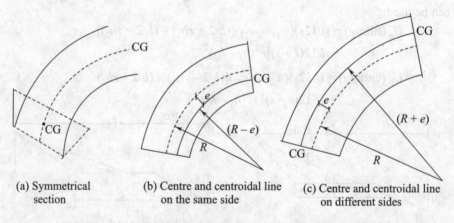

(a) Symmetrical section

(b) Centre and centroidal line on the same side

(c) Centre and centroidal line on different sides

Fig. 10.20 Curvature correction

In Fig. 10.20(b), the path traced by the centroid is due to an arc of radius $(R - e)$, where R is the radius of curvature and e is the eccentricity of the centroid from the centre line. The central angle α is l/R radians. The path traced by the centroid $= (R - e)l/R = l(1 - e/R)$. In Fig. 10.20(c), similarly, the path traced by the centroid $= l(1 + e/R)$. In general, the path traced by the centroid $= l(1 \pm e/R)$.

If the section is not uniform at different sections, the average of the eccentricity at the two end sections can be taken. If the eccentricity at section 1 is e_1 and the eccentricity at section 2 is e_2, then the eccentricity for applying the correction can be taken as $e = (e_1 + e_2)/2$.

The curvature correction for different sections is as follows.

Single-level section   No correction is required, as the area is symmetrical about the centre line.

Two-level section   In the case of a two-level section, with the usual notations,

$$\text{Eccentricity } e = w_1 w_2 (w_1 + w_2)/3AR$$

where A is the area in cutting and filling.

$$\text{Curvature correction } C_c = (l/6R)(w_1^2 - w_2^2)(h + b/2s)$$

Sidehill two-level section   In the case of a sidehill two-level section, there are two areas, one in filling and one in cutting. With the notations used so far,

$$\text{Eccentricity } e \text{ (cutting)} = (1/3)(w_2 + b/2 - rh)$$
$$\text{(assuming this to be the larger area)}$$
$$\text{Eccentricity } e \text{ (filling)} = (1/3)(w_1 + b/2 + rh)$$
$$\text{Curvature correction } C_c = Ael/R$$

Three-level section   The two-level section formula is used for calculating the curvature correction for this section also.

Table 10.2 summarizes the prismoidal and curvature corrections for different sections.

Table 10.2   Summary of corrections to volumes

Type of section	Prismoidal correction	Curvature correction
Single-level section	$(ls/6)(h_1 - h_2)^2$	No correction required
Two-level section	$(l/6)[r^2 s/(r^2 - s^2)](h_1 - h_2)^2$	$\pm (l/6R)(w_1^2 - w_2^2)$
	$(l/6s)(w_{11} - w_{21})(w_{12} - w_{22})$	$(h + b/2s)$
Sidehill two-level section	$[(l/12)(r - s)]r^2(h_1 - h_2)^2$	$r(A/3R)(w_1 + b/2 - rh)$
		$(A/3R)(w_1 + b/2 + rh)$
Three-level section	$(l/12)[(w_{11} + w_{21})$	$\pm (l/6R)(w_1^2 - w_2^2)$
	$- (w_{12} - w_{22})$	$(h + b/2s)$

Example 10.6   A level section has a formation width of 12 m, side slopes 1.5:1, and central height of 2.4 m. Find the volume of earthwork in a length of 200 m.

Solution   The area of the section has to be found first.

$$\text{Area} = h(b + sh) = 2.4\,(12 + 1.5 \times 2.4) = 37.44 \text{ m}^2$$

As the area does not change over the length, the volume can be calculated directly as area × length:

$$\text{Volume} = 37.44 \times 200 = 7488 \text{ m}^3$$

Example 10.7   An embankment has side slopes 1.5:1 and is level in the transverse direction. The depths at the centre at 20-m intervals are 1.8 m, 2.4 m, 3.0 m, and 3.6 m. Find the volume by the trapezoidal formula.

Solution   We will first calculate the areas at the different sections as shown in Table 10.3.

Table 10.3   Calculation of areas (Example 10.7)

Section at	Formation width (b)	Height at centre (h)	Side slope (s:1)	bh	sh²	Area (m²)
0	8.8	1.8	1.5:1	15.8	4.86	0.66
20	8.8	2.4	1.5:1	21.12	8.64	29.76
40	8.8	3.0	1.5:1	26.4	13.5	39.90
60	8.8	3.6	1.5:1	31.68	19.44	51.12

Volume = 20[(20.66 + 51.12)/2 + 29.76 + 39.90] = 2111 m³

Example 10.8   A single-level section has a formation width of 7.5 m and side slopes 2:1. The depth of cutting at the centre at 30-m intervals are 1.8 m, 2.175 m, 2.55 m, 2.925 m, and 3 m. Find the volume of earthwork in the length of 120 m.

Solution   The area of the sections has to be calculated first. This is done in Table 10.4.

Table 10.4   Sectional areas (Example 10.8)

Chainage	Height	bh	sh²	Area
0	1.8	13.5	8.48	19.98
30	2.175	16.31	9.46	25.77
60	2.55	19.12	13.00	32.12
90	2.925	21.94	17.11	39.05
120	3.0	22.5	18.00	40.50

By the prismoidal formula:

$$V = (30/3)[19.98 + 40.50 + 4(25.77 + 39.05) + 2 \times 32.12] = 3840 \text{ m}^3$$

By trapezoidal formula:

$$V = (30/2)[19.98 + 40.50 + 2(25.77 + 32.12 + 39.05)] = 3829.8 \text{ m}^3$$

Example 10.9   The two-level section shown in Fig. 10.21 is constant over a length of 120 m. Find the volume of earthwork in this length.

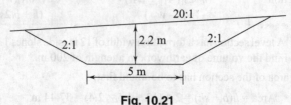

Fig. 10.21

Solution The area of the section is calculated using any of the formulae given:

$$\text{Area} = [s(b/2)^2 + r^2(bh + sh^2)]/(r^2 - s^2), \quad s = 2, \quad r = 20 \quad h = 2.2 \text{ m}$$

$$\text{Area} = [2 \times 2.5^2 + 20^2(5 \times 2.2 + 2 \times 2.2^2)]/(20^2 - 2^2) = 20.92 \text{ m}^2$$

$$\text{Volume} = 20.92 \times 120 = 2510.4 \text{ m}^3$$

Example 10.10 The ground levels (GLs) at 20-m intervals on a stretch of ground are as given below:

Chainage	0	20	40	60	80	100	120
GL	214.2	214.8	215.1	216.1	216.9	217.4	218.2

Table 10.5 Calculation of area at sections

Chainage	Ground level	Reduced level of formation	Depth of cutting	bh	sh²	area
0	214.2	212.66	1.54	9.24	4.74	13.98
20	214.8	213.16	1.64	9.84	5.38	15.22
40	215.1	213.66	1.44	8.64	4.15	12.79
60	216.1	214.16	1.94	11.64	7.53	19.17
80	216.9	214.66	2.24	13.44	10.03	23.47
100	217.4	215.16	2.24	13.44	10.03	23.47
120	218.2	215.66	2.54	15.24	12.90	28.14

The depth of cutting at chainage zero is 1.54 m and the formation goes at a longitudinal upward slope of 1 in 40. Find the depth of cutting at each section and the volume of earthwork in this stretch if the formation width is 6 m and the side slope is 2:1. The ground is level in the transverse direction.

Solution The calculation of areas is shown in Table 10.5

By the trapezoidal formula, the volume V_t is calculated as

$$\text{Volume } V_t = (20/2)[13.98 + 28.14 + 2(15.22 + 12.79 + 19.17 + 23.47 + 23.47)]$$

$$= 2303.6 \text{ m}^3$$

By the prismoidal formula, the volume V_p is calculated as

$$\text{Volume } V_p = (20/3)[13.98 + 28.14 + 4(15.22 + 19.17 + 23.47) + 2(12.79 + 23.47)]$$

$$= 2307.2 \text{ m}^3$$

Example 10.11 From the details shown in Fig. 10.22, find the volume of the tank. The reduced level (RL) of the base of the tank is 121.56 m and that of the ground at the top level is 125.96 m.

Solution The depth of the tank is 125.96 – 121.56 = 4.4 m. The dimensions at the top level of the tank can be found as

$$\text{Length} = 72 + 4.4 + 4.4 \times 2$$

$$= 85.2 \text{ m}$$

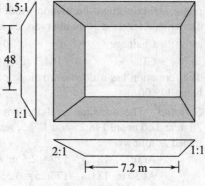

Fig. 10.22

Width of the tank = 48 + 4.4 + 4.4 × 1.5 = 59 m

The end areas are 72 × 48 = 3456 m² and 85.2 × 59 = 5026.8 m² by the trapezoidal formula,

$$V_t = (3456 + 5026.8) \times 4.4/2 = 18{,}662.16 \text{ m}^3$$

To find the volume by the prismoidal formula, we have to find the area at mid-depth. The dimensions at mid-depth are

Length = 72 + 2.2 + 4.4 = 78.6 m,   width = 48 + 2.2 + 3.3 = 53.5 m

Area at mid-depth = 78.6 × 53.5 = 4205.1 m²

By the prismoidal formula,

$$V_p = (2.2/3)[3456 + 5026.8 + 4 \times 4205.1] = 18{,}555.68 \text{ m}^3$$

Example 10.12   A two-level section has a formation width of 10 m and side slopes of 2:1. The transverse slope of the ground is 8:1. The central heights at 20-m intervals are 2 m, 2.4 m, and 3.0 m. Find the volume of earthwork in the length of 40 m.

Solution   The section is shown in Fig. 10.23. The areas at different sections are the following.

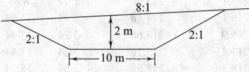

Fig. 10.23

Chainage 0: area = $[r^2 s/(r^2 - s^2)](h + b/2s)^2 - b^2/4s$

$$= 8^2 \times 2/(8^2 - 2^2)(2 + 5/2 \times 2)^2 - 10^2/4 \times 2 = 30.69 \text{ m}^2$$

Chainage 20: area= $8^2 \times 2(8^2 - 2^2)(2.4 + 10/2 \times 2)^2 - 10^2/4 \times 2 = 38.71 \text{ m}^2$

Chainage 40: area= $8^2 \times 2/(8^2 - 2^2)(3 + 10/2 \times 2)^2 - 10^2/4 \times 2 = 52.02 \text{ m}^2$

Volume by the trapezoidal formula:

$$V_t = (30.69 + 38.71) \times 20/2 + (38.71 + 52.02) \times 20/2 = 1601.3 \text{ m}^3$$

Volume by the prismoidal formula:

$$V_p = (20/3)[30.69 + 52.02 + 4 \times 38.71] = 1583.66 \text{ m}^3$$

Example 10.13   Find the volume of cutting in a length of 60 m with the following data for a two-level section using the prismoidal and trapezoidal formulae. Also calculate the prismoidal correction. Formation width = 9 m, side slopes = 2:1, transverse slope = 6:1. The ground levels at 30-m intervals are as given below:

Chainage	0	30	60
GL	181.5	181.8	182.4

The formation has a downward slope of 1 in 40 with the formation level at 0 chainage being 179.00.

Solution   The formation levels at 30 m change by 30/40 = 0.75 m. The formation levels at 30 m and 60 m are 178.25 and 177.5, respectively. The depth of cutting can be then determined as follows:

At 0:   181.5 – 179 = 2.5 m

At 30 m: 181.8 – 178.25 = 3.55 m

At 60 m: 182.4 – 177.5 = 4.9 m

The area of a two-level section can be calculated as

$$A = [r^2s/(r^2 - s^2)](h + b/2s)^2 - b^2/4s$$

The areas of the three sections can be calculated as in Table 10.6.

Table 10.6 Areas at different sections

Chainage	$r^2s/(r^2 - s^2)$	$(h + b/2s)^2$	$b^2/4s$	Area
0	2.25	$[2.5 + 9/(2 \times 2)]^2 = 22.56$	10.125	40.64
30	2.25	$(3.55 + 9/2 \times 2)^2 = 33.64$	10.125	65.56
60	2.25	$(4.9 + 9/2 \times 2)^2 = 51.12$	10.125	104.9

Volume by the trapezoidal formula:

$$V_t = (30/2)[40.64 + 104.9 + 2 \times 65.56] = 4150 \text{ m}^3$$

Volume by the prismoidal formula:

$$V_p = (30/3)[40.64 + 104.9 + 4 \times 65.56] = 4078 \text{ m}^3$$

$$\text{Prismoidal correction} = (ls/3)[r^2/(r^2 - s^2)][(h_1 - h_m)^2 + (h_m - h_2)^2]$$
$$= 30 \times (2/3)[6^2/(6^2 - 2^2)][(3.55 - 2.5)^2 + (4.9 - 3.55)^2]$$
$$= 65.81 \text{ m}^3$$

Example 10.14 Find the volume by the prismoidal and trapezoidal formulae and the prismoidal correction with the following data: Formation width = 9 m, side slope = 1.5:1 in cutting and 2:1 in filling, and transverse slope of the ground is 1 in 4. The depths of cutting at the centre are 0.5 m, 1.0 m, and 1.5 m at three sections 20 m apart. Find the volume in the 40 m length.

Solution The sections are shown in Fig. 10.24. The values of h_1, h_2, w_1, and w_2 can be calculated and are as given in Table 10.7.

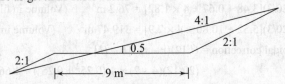

Fig. 10.24

Table 10.7 Parameters for sections and areas

Section at	h_1	h_2	w_1	w_2	x	Area in filling (m^2)	Area in cutting (m^2)
0	1.75	2.2	8	7.8	3.98	3.48	5.52
20	1.25	2.6	7	6.8	2.92	1.82	7.90
40	0.75	3.0	6	5.8	1.8	0.67	10.80

To calculate h_1, we use the equality

$$w_1 = (h_1 + h)r = b/2 + h_1s$$
$$h_1(r - s) = b/2 - hr$$
$$h_1 = (b/2 - hr)/(r - s)$$

$b/2 = 4.5$ m, $r = 4$, $s = 2$. For example, in the section at 0 m,

$$h_1 = (4.5 - 0.25 \times 4)/(4 - 2) = 1.75 \text{ m}$$

To calculate h_2,

$$w_2 = (h_2 - h)r = b/2 + h_2 s$$

$$h_2(r - s) = b/2 + hr$$

$$h_2 = (b/2 + hr)/(r - s)$$

$b/2 = 4.5$ m, $r = 4$, $s = 1.5$. For example, in the section at 0 m,

$$h_2 = (4.5 + 0.25 \times 4)/(4 - 1.5) = 2.2 \text{ m}$$

Once the values of h_1 and h_2 are known, w_1 and w_2 can be calculated. At the 0-m section w_1 = $4.5 + 1.75 \times 2 = 8$ m and $w_2 = 4.5 + 2.2 \times 1.5 = 7.8$ m.

x m is the distance from A of the point where the ground surface intersects the formation. In this section, x is the base of the triangle giving the area of filling and $(9 - x)$ is the base of the triangle giving the area of cutting. The value of x is obtained from similar triangles CA′E and COF. We can write

$$(b/2 - x)/h = (x + h_1 s)/ h_1$$

At the 0-m section,

$$(4.5 - x)/0.25 = (x + 1.75 \times 2)/1.75$$

which gives $x = 3.5$ m. Similarly, the value of x for other sections can be calculated. Area in filling = $xh_1/2$ and area in cutting = $(9 - x)h_2/2$. These are listed in the last two columns of the table.

Knowing the areas of the sections, the volumes can be calculated from the trapezoidal and prismoidal formulae.

By the trapezoidal formula:

$$V_t = (20/2)[3.48 + 0.67 + 2 \times 1.82] = 77.9 \text{ m}^2 \quad \text{(Volume in filling)}$$

$$V_t = (20/2)[5.52 + 10.80 + 2 \times 7.90] = 321.2 \text{ m}^3 \quad \text{(Volume in cutting)}$$

By the prismoidal formula:

$$V_p = (20/3)[3.48 + 0.67 + 4 \times 1.82] = 76.2 \text{ m}^3 \quad \text{(Volume in filling)}$$

$$V_p = (20/3)[5.52 + 10.80 + 4 \times 7.9] = 319.47 \text{ m}^3 \quad \text{(Volume in cutting)}$$

$$\text{Prismoidal correction} = (l/12)(r - s)r^2[h - H]^2$$

$$= (20/12)(4 - 2) \times 4^2(0.25^2 + 0.25^2) = 1.667 \text{ m}^3$$

Example 10.15 A two-level section has a formation width of 10 m, side slopes 1.5:1, and transverse slope 6:1. The height along the centre line is 2 m. Find the curvature correction for a length subtending an angle of 30° if the radius of the centre line is 180 m.

Solution The section is shown in Fig. 10.25. We calculate the values of h_1, h_2, w_1, and w_2 from the given data.

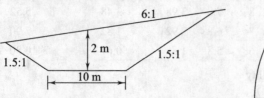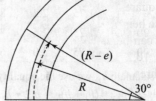

Fig. 10.25

$$b/2 + h_1 \times 1.5 = (2 - h_1)8 = w_1, \quad h_1 = 1.16 \text{ m}$$

$b/2 + h_2 \times 1.5 = (h_2 - 2)8 = w_2, \quad h_2 = 3.23$ m

$w_1 = 5 + 1.16 \times 1.5 = 6.74$ m, $\quad w_2 = 5 + 3.23 \times 1.5 = 9.845$ m

Area of section $= [s(b/2)^2 + r^2 bh + r^2 sh^2]/(r^2 - s^2)$

$\qquad\qquad\qquad = [1.5 \times 5^2 + 8^2 \times 10 \times 2 + 8^2 \times 1.5 \times 2^2]/(8^2 - 1.5)^2$

$\qquad\qquad\qquad = 27.55$ m^2

Eccentricity of centroid $= w_1 w_2 (w_1 + w_2)/3Ar$

$\qquad\qquad\qquad\qquad = 6.74 \times 9.845 \times 16.585/3 \times 27.55 \times 8$

$\qquad\qquad\qquad\qquad = 1.66$ m

Curvature correction $= Ael/R$,

where R is the radius of the curve.

$R = 180$ m and $l = 180 \times 30 \times \pi/180 = 94.24$ m.

$\qquad$ Curvature correction $= 27.55 \times 1.66 \times 94.24 /180 = 23.94$ m^3

10.5 Capacity of Reservoirs

The capacity of a reservoir is calculated from contour plans. As shown in Fig. 10.26, a number of contour lines are available at a selected contour interval. The area within a contour line is measured using a planimeter or other means. Such sectional areas are then used to calculate the volume of water that can be stored in the reservoir using either the trapezoidal or the prismoidal formula. The contour interval is the distance between the sections. If $A_1, A_2, A_3, A_4, \ldots$ are the areas of the sections and d is the contour interval, then the volume can be obtained as follows.

By the trapezoidal formula:

$$V_t = (d/2)[A_1 + A_n + 2(A_2 + A_3 + \ldots + A_{n-1})]$$

By the prismoidal formula:

$$V_p = (d/3)[A_1 + A_n + 4(A_2 + A_4 + \ldots) + 2(A_3 + A_5 + \ldots)]$$

Note that A is the area enclosed by a contour line and not the area between contour lines.

10.6 Volume of Borrow Pits

Volumes of borrow pits are obtained from spot levels. Spot levels are levels obtained at predetermined points, either at corners of squares, rectangles, or triangles. Using these height values, the volumes of borrow pits can be calculated as follows. Refer to Fig. 10.27. Consider the prism composed of three corners giving a triangular area in plan. If h_1, h_2, and h_3 are the corner heights and A is the area of the triangle in plan, then the volume of the prism is obtained by multiplying the area by the average height as:

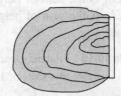

Fig. 10.26 Calculating the capacity of reservoirs

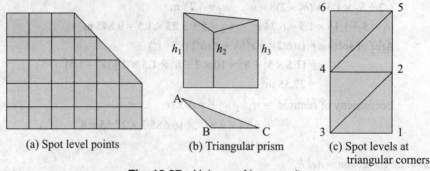

(a) Spot level points (b) Triangular prism (c) Spot levels at triangular corners

Fig. 10.27 Volume of borrow pits

Volume of triangular prism = $A(h_1 + h_2 + h_3)/3$

In the case of a rectangular or square prism, using the same principle, if h_1, h_2, h_3, and h_4 are the heights at the corners and A is the area of the square or rectangle, then

Volume of rectangular or square prism = $A(h_1 + h_2 + h_3 + h_4)/4$

If the volume consists of three-sided figures only, then it can be seen that some heights are used once, some heights twice, and some heights thrice. In Fig. 10.27 (c), 1 to 6 show the points at which heights h_1 to h_6 are taken. In calculating the volume, heights 1 and 6 are used only once, heights 3 and 5 are used twice, and heights 2 and 4 are used thrice. This depends upon the number of triangles for which they form vertices. The volume in such a case, assuming that all the triangles have the same base area, is

Volume = $A[1(h_1 + h_6) + 2(h_3 + h_5) + 3(h_2 + h_4)]/3$

In general, for volumes consisting of equal triangular prisms only, the volume is given by

Volume = $(A/3)[\Sigma h_1 + 2\Sigma h_2 + 3\Sigma h_3]$

where Σh_1 is the sum of heights used only once, Σh_2 is the sum of heights used twice, and Σh_3 is the sum of heights used thrice.

Similarly, if the base of the borrow pit consists of four-sided figures of equal area, then the volume can be computed on similar principles as

Volume = $(A/4)[\Sigma h_1 + 2\Sigma h_2 + 3\Sigma h_3 + 4\Sigma h_4]$

where Σh_1 is the sum of heights used only once, Σh_2 is the sum of heights used twice, Σh_3 is the sum of heights used thrice, and Σh_4 is the sum of heights used four times.

Example 10.16 Find the capacity of a reservoir from the contour data given in Table 10.8. The plan is drawn to a scale of 1:4000

Table 10.8 Data for Example 10.16

Contour	Area (cm²)	Contour	Area (cm²)
260	400	248	205
258	367.5	246	177.5
256	327.5	244	147.5
254	310	242	115
252	277.5	240	0
250	243.75		

Table 10.9 Areas (Example 10.16)

Contour	Area (m² × 1000)	Contour	Area (m² × 1000)
260	640	248	328
258	588	246	284
256	524	244	236
254	496	242	184
252	444	240	0
250	390		

Solution   As the plan is drawn to a scale of 1:4000 (1 cm = 40 m), the areas measured by the planimeter are to be multiplied by 1600 to get the area in square metres. These areas are given in Table 10.9.

Volume by the trapezoidal formula:

$$V_t = (2/2)[640 + 0.0 + 2(588 + 524 + 496 + 444 + 390 + 328 + 284 +$$
$$236 + 184)] \times 1000$$
$$= 7588 \times 1000 \text{ m}^3$$

Volume by the prismoidal formula:

$$V_p = (2 \times 1000/3)[640 + 0 + 4(588 + 496 + 390 + 284 + 184) + 2(524 + 444 +$$
$$328 + 236)]$$
$$= 7648 \times 1000 \text{ m}^3$$

Example 10.17 P, Q, R, and S are four points at the corners of an excavation. The reduced levels of the four points are 121.38 m, 119.64 m, 120.32 m, and 121.68 m, respectively. The reduced level of the excavation is uniform at 118.17 m. PS and QR are parallel sides of lengths 3 m and 7.5 m, respectively. The distance between PS and QR is 6 m. Find the volume of the excavation.

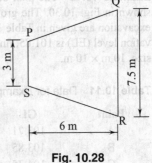

Fig. 10.28

Solution   The plan is shown in Fig. 10.28.

Area of excavation = $(3 + 7.5) \times 6/2 = 31.5 \text{ m}^2$

The heights of the excavation at the four points are the following:

At P, height = 121.38 − 118.17 = 3.21 m

At Q, height = 119.64 − 118.17 = 1.47 m

At R, height = 120.32 − 118.17 = 2.15 m

At S, height = 121.68 − 118.17 = 3.51 m

Volume = area × average height

= $(31.5/4)(3.21 + 1.47 + 2.15 + 3.51) = 81.42 \text{ m}^3$

Example 10.18 Four squares of 10 × 10 m² were made and excavation done on an area of 20 × 20 m² as shown in Fig. 10.29. The depth of the excavation at the nine points are given in Table 10.10. Find the volume of earth excavated.

Table 10.10 Data for Example 10.18

Point	Depth (m)	Point	Depth (m)
A	2.8	F	2.4
B	3.2	G	2.4
C	2.6	H	2.9
D	2.6	I	2.2
E	3.1		

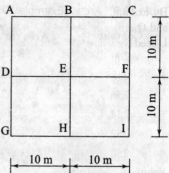

Fig. 10.29

Solution From the figure, it is clear that the heights of A, C, G, and I are used only once; those of B, D, F, and H are used twice; and those of E used four times in calculating the volume of the four prisms.

$$\text{Volume} = A\Sigma h = A(\Sigma h_1 + 2\Sigma h_2 + 3\Sigma h_3 + 4\Sigma h_4)/4$$

$$\Sigma h_1 = 2.8 + 2.6 + 2.4 + 2.2 = 10$$

$$\Sigma h_2 = 3.2 + 2.6 + 2.4 + 2.9 = 11.1$$

$$\Sigma h_3 = 0 \text{ and } \Sigma h_4 = 3.1$$

Area $A = 10 \times 10 = 200 \text{ m}^2$

Volume = $(100/4)(10 + 2 \times 11.1 + 4 \times 3.1) = 1115 \text{ m}^3$

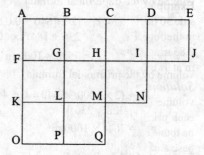

Fig. 10.30

Example 10.19 The plan of an excavation is shown in Fig. 10.30. The ground levels before excavation are given in Table 10.11. If the excavation level (EL) is 101.85, find the volume of the excavation. The squares are of uniform size, 10 m × 10 m.

Table 10.11 Data for Example 10.19

Point	GL	Point	GL	Point	GL
A	103.71	G	104.25	M	104.75
B	103.85	H	104.65	N	105
C	104.35	I	104.95	O	104.7
D	104.95	J	104.29	P	105.05
E	104.09	K	104.45	Q	104.85
F	104.05	L	104.95		

Solution The depth of excavation at each point (GL – EL) can be worked out as in Table 10.12.

Table 10.12 Solution of Example 10.19

Point	Depth	Point	Depth	Point	Depth
A	1.86	G	2.4	M	2.9
B	2.0	H	2.8	N	3.15
C	2.5	I	3.1	O	2.85
D	3.1	J	2.44	P	3.2
E	2.24	K	2.6	Q	3.00
F	2.2	L	3.1		

By examining Fig. 10.30, in calculating the volume, the heights of points A, E, J, N, Q, and O are used once; B, C, D, F, K, and P used twice; I and M used thrice, and G, H, and L used four times.

Sum of heights used only once = $1.86 + 2.24 + 2.44 + 3.15 + 3 + 2.85$

$$= 15.54 \text{ m}$$

Sum of heights used twice = $2 + 2.5 + 3.1 + 2.2 + 2.6 + 3.2 = 15.6 \text{ m}$

Sum of heights used thrice = $3.15 + 2.90 = 6.05 \text{ m}$

Sum of heights used four times = $2.4 + 2.8 + 3.1 = 8.3 \text{ m}$

Volume = $AH/4$, where A is the area = $10 \times 10 = 100 \text{ m}^2$ and H is the sum of heights multiplied by the number of times used.

Volume = $100(15.54 + 2 \times 15.6 + 3 \times 6.05 + 4 \times 8.3)/4 = 2452 \text{ m}^3$

Example 10.20 An excavation in the form a square of side 30 m, shown in Fig. 10.31, has the following heights at the four corners and the intersection of the diagonals: 2.85 m at P, 2.45 m at Q, 3.15 m at R, 3.55 m at S, and 2.95 m at T. Find the volume of earth excavated from the pit.

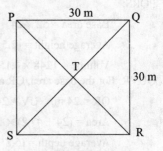

Fig. 10.31

Solution There are four triangular prisms, and the total volume is the sum of the volumes of these prisms. For each prism, the average depth of the excavation should be found by averaging the depths at the three corners. The base area for each triangular prism is the same.

Base area = $30 \times 15/2 = 225 \text{ m}^2$

Prism PQT: Average height = $(2.85 + 2.45 + 2.95)/3 = 2.75 \text{ m}$

Volume = $2.75 \times 225 = 618.75 \text{ m}^3$

Prism QTR: Average height = $(2.45 + 3.15 + 2.95)/3 = 2.85 \text{ m}$

Volume = $2.85 \times 225 = 641.25 \text{ m}^3$

Prism STR: Average height = $(3.55 + 2.95 + 3.15)/3 = 3.22 \text{ m}$

Volume = $3.12 \times 225 = 723.75 \text{ m}^3$

Prism PTS: Average height = $(2.85 + 2.95 + 3.55)/3 = 3.12 \text{ m}$

Volume = $3.12 \times 225 = 701.25 \text{ m}^3$

Total volume = $618.75 + 641.25 + 723.75 + 701.25 = 2685 \text{ m}^3$

Alternatively, since the base areas are the same, the formula $V = AH$, where

$$H = (\Sigma h_1 + 2\Sigma h_2 + 3\Sigma h_3)/3$$

can be used. In this case the four corner heights, P, Q, R, and S, are used twice and that of T is used four times.

Base area = $30 \times 15/2 = 225 \text{ m}^2$

Volume = $225 \times [2(2.85 + 2.45 + 3.15 + 3.55) + 4 \times 2.95]/3 = 2685 \text{ m}^3$

Example 10.21 The plan and elevation of an excavation is shown in Fig. 10.32. Find the volume excavated from the pit.

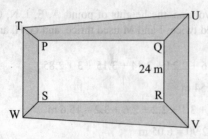

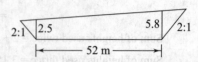

Fig. 10.32

Solution The volume of excavation can be considered as the sum of the volumes of five prisms—PQRS, QUVR, TPSW, TUQP, and SRVW. For each prism, the area is equal to the base area multiplied by the average depth of excavation at the four corners. These can be calculated as follows.

PQRS:

Base area = $52 \times 24 = 1248$ m^2

Average height = $(2.5 + 2.5 + 5.8 + 5.8)/4 = 4.15$ m

Volume = $1248 \times 4.15 = 5179.2$ m^3

QUVR: For the base area, QR and UV are parallel sides:

QR = 24 m,   UV = $24 + 2 \times 11.6 = 47.2$ m

Area = $(24 + 47.2) \times 11.6/2 = 412.96$ m^2

Average depth = $(5.8 + 5.8 + 0 + 0)/4 = 2.9$ m

Volume = $412.96 \times 2.9 = 1197.58$ m^3

TPSW: Sides PS and TW are parallel:

PS = 24 m,   TW = $24 + 5 + 5 = 34$ m

Area = $(24 + 34) \times 5/2 = 145$ m^2

Average depth = $(2.5 + 2.5 + 0 + 0)/4 = 1.25$ m

Volume = $145 \times 1.25 = 181.25$ m^3

TUQP: Length TU = $52 + 5 + 11.6 = 68.6$ m. We find the area of TUQ′P′ and subtract the area of two triangles TPP′ and UQQ′.

Area = $68.6(11.6 + 5)/2 - (5^2 + 11.6^2)/2 = 489.6$ m^2

Volume = $489.6 \times (5.8 + 11.6 + 0 + 0)/4 = 2129.76$ m^3

WSRV: Volume is the same as that of TUQP.

Volume = 2129.76 m^3

Total volume = $5179.2 + 1197.58 + 181.25 + 2129.76 + 2129.76$

= $10,835.55$ m^3

10.7 Mass-haul Diagram

Mass-haul diagrams are drawn to calculate the volume of cutting and filling in a job and to estimate the amount of fill to be imported and the haulage of earth from cutting to filling. It is necessary to understand the following terminology to study mass-haul diagrams.

Free haul The distance for which no extra charge is levied.

Haul distance The distance from the face of the excavation to the point where the earth is deposited.

Average haul distance The distance from the centre of gravity (CG) of the cutting to the CG of the filling.

Borrow The earth obtained from outside sources such as borrow pits.

Bulking The increase in volume when the earth is excavated over its original volume.

Shrinkage The reduction in volume when the earth is compacted from its original in situ volume.

A mass-haul diagram is used in conjunction with the longitudinal section of a road or railway line and can be used to find the following:
- The distance over which the cut volume and fill volume balance
- The quantity of earth to be moved and the direction of movement
- Areas requiring borrowing of materials
- Areas where earth has to be thrown away
- The most economical movement for the excavated earth

We illustrate this principle with an example.

Example 10.22 The data in Table 10.13 give the areas of cutting or filling between the chainages indicated. Draw a mass-haul diagram and determine the amount of earth to be borrowed for the maximum haul distance.

Table 10.13 Data for Example 10.22

Chainage (m)	Cutting (m²)	Filling (m²)
0	–	–
30	18	–
60	40	–
90	44	–
120	50	–
150	40	–
180	20	–
210	0	–
240	–	20
270	–	30
300	–	40
330	–	50
360	–	60
390	–	48
420	–	30
450	–	0

Solution The longitudinal section is shown in Fig. 10.33(a). This diagram shows the chainages along the x-axis and the areas of cutting and filling along the y-axis. Cutting is considered positive and filling is considered negative.

We calculate the volumes of cutting and filling from the average end-area method. These have been calculated and listed in Table 10.14.

Table 10.14 Computation of volumes and cumulative volumes (Example 10.22)

Chainage	Area of cutting (m²)	Area of filling (m²)	Volume of cutting (m³)	Volume of filling (m³)	Cumulative volume (m³)
0	0	–	–	–	–
30	18	–	270	–	270
60	40	–	870	–	1140
90	44	–	1260	–	2400
120	50	–	1410	–	3810
150	40	–	1350	–	5160
180	20	–	900	–	6060
210	0	–	300	–	6360
240	–	20	–	300	6060
270	–	30	–	750	5310
300	–	40	–	1050	4260
330	–	50	–	1350	2910
360	–	60	–	1650	1260
390	–	48	–	1620	– 360
420	–	30	–	1170	– 1530
450	–	–	–	450	– 1980

The mass-haul diagram is plotted in Fig. 10.33(b). This diagram is plotted with chainages along the x-axis and cumulative volumes along the y-axis.

The following characteristics of the mass-haul diagram are evident from the graph plotted in this example.

(a) A positive slope of the mass-haul diagram (as in part AD) indicates excavation and a negative slope (as in part DB) indicates filling or embankment.

(b) The y-coordinate at any point of the curve gives the net volume (volume of cutting – volume of filling).

(c) The maximum point on the curve (D) indicates the end of the excavation and the minimum point (C) indicates the end of the embankment.

(d) When the curve crosses the x-axis, as at point B, this shows that the excavation and filling are balanced up to this point. AB is thus known as the balancing line.

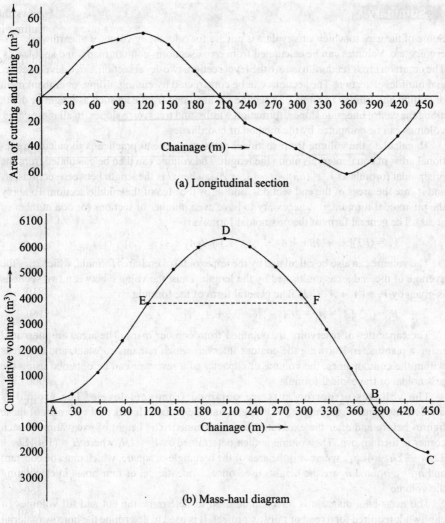

(a) Longitudinal section

(b) Mass-haul diagram

Fig. 10.33

(e) Any other horizontal line is also a balancing line, such as EF, when starting from E.

(f) The length of the balancing line is the maximum haul distance when starting from A. EF is the maximum haul distance when starting from E.

(g) The area bound by the curve and the balancing line is known as the *haul* and is equal to volume × distance.

(h) The minimum haul distance is obtained when the algebraic sum of the areas cut off by the balancing line is a minimum.

In this particular example, over the distance selected, there is an excess of filling, requiring earth to be borrowed. By computing the area under the curve, it is possible to calculate the haulage requirements (number of vehicles of a particular capacity required) and the haulage distances.

Summary

Some of the uses to which survey data is put are for calculating volumes of earthwork, reservoirs, etc. Volumes can be calculated from cross sections, contour maps, and spot levels. The common cross sections used are the level section, two-level section, three-level section, and multilevel section. The sections can be constructed by cutting, filling, or embankment or cutting and filling. The area of a section is calculated from data such as height or depth along the centre line, side slopes, formation widths, and transverse slopes. In all these cases, volumes can be computed by the method of coordinates.

To calculate the volume from sectional areas, the general practice is to calculate sectional areas at equal intervals along the length. The volume can then be calculated from the prismoidal formula as $V_p = (l/3)[A_1 + 4A_m + A_2]$, where l is the length between sections, A_1 and A_2 are the areas of the end sections, and A_m is the area of the middle section. To apply the prismoidal formula, it is necessary to have even number of sections (or odd number of areas). The general form of the prismoidal formula is

$$V_p = (l/3)[A_1 + A_n + 4(A_2 + A_4 + \ldots) + 2(A_3 + A_5 + \ldots)]$$

The volume can also be calculated by the trapezoidal or end-area formula, which uses the average of the end areas multiplied by the length. Thus, the volume between two sections is given by $V_t = (A_1 + A_2)(l/2)$. The general form of the formula is

$$V_t = (l/2)[A_1 + A_n + 2(A_2 + A_3 + \ldots + A_{n-1}].$$

The capacities of reservoirs are obtained from contour maps. The areas are measured using a planimeter. Knowing the contour interval, which remains constant, and the areas within the contour lines, the volume or capacity of a reservoir can be obtained using the prismoidal or trapezoidal formula.

The volumes of borrow pits are obtained from spot levels. The ground is divided into squares, rectangles, or triangles and the reduced levels of the corners of these figures before and after the excavation are determined. The height of excavation at each corner is then known. The volume is then determined as $V = AH$, where $H = (1/4)[\Sigma h_1 + 2\Sigma h_2 + 3\Sigma h_3 + 4\Sigma h_4]$, where A is the area of the rectangle or square, which must be constant and h_1, h_2, h_3, and h_4 are the heights used once, twice, thrice, or four times in calculating the volume.

The mass-haul diagram is a technique used to determine the cut and fill volumes of earthwork required for a road or railway project. It is used to determine the borrow material required, the haul distances, and the most economic method of hauling the earth.

Exercises

Multiple-Choice Questions

1. Pappus' theorem of areas states that a curve or line rotating about a given axis generates a surface having area equal to the length of the line multiplied by
   (a) distance from the axis
   (b) the average distance from the axis
   (c) the shortest distance from the axis
   (d) the distance through which the CG of the line moves
2. Pappus' theorem of volumes states that when a given area rotates about a given axis, it generates a solid whose volume is given by the area multiplied by the
   (a) distance from the axis   (c) distance moved by the CG of the area
   (b) average distance from the axis   (d) shortest distance from the axis

3. A straight line AB lies such that its end A is 2 cm and end B is 5 cm from an axis. If AB rotates about this axis, it will generate
   (a) a conical surface
   (b) a cylindrical surface
   (c) curved surface of a frustum of a cone
   (d) the volume a cone

4. When a right angled triangle rotates about the axis shown, the volume generated is

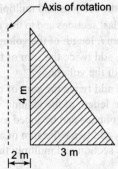

Axis of rotation

4 m

2 m   3 m

   (a) 18 cm³   (b) 56.54 cm³   (c) 113 cm³   (d) 226 cm³

5. The volume of the single-level section shown for a length of 10 m is

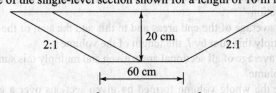

2:1   20 cm   2:1

60 cm

   (a) 2 m³   (b) 1.4 m³   (c) 1.2 m³   (d) 1 m³

6. In the two-level section shown, the width w¹ is

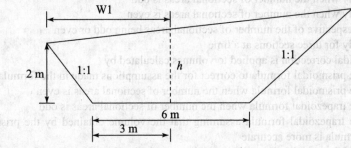

W1

2 m   1:1   h   1:1

6 m

3 m

   (a) 4 m   (b) 5 m   (c) 6 m   (d) 7 m

7. In the two-level section shown, the height h at the centre is
   (a) 3 m   (b) 4.5 m   (c) 5.5 m   (d) 7 m

8. In the three-level section shown, the width w1 is
   (a) 5 m   (b) 10 m   (c) 12 m   (d) 14 m

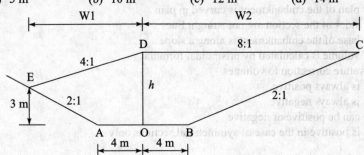

W1   W2

D   8:1   C

4:1

E   h

3 m   2:1   2:1

A   O   B
4 m   4 m

9. In the three level-section shown, the width w_2 is
   - (a) 6.67 m
   - (b) 8.67 m
   - (c) 12.67 m
   - (d) 14.67 m
10. In the three-level section shown, the height at the centre OD is
    - (a) 3.5 m
    - (b) 4 m
    - (c) 4.5 m
    - (d) 5.5 m
11. The trapezoidal formula for volumes can be stated as
    - (a) find the average area of the sections and multiply by the length of the volume
    - (b) add the areas of first and last sections and to this add the average area of remaining sections. multiply the sum l, length of the volume
    - (c) to the sum of end areas, add twice the sum of the remaining areas and multiply the sum by l, the length of the volume
    - (d) to the sum of end areas, add twice the sum of remaining areas and multiply the sum by $l/2$, where l is the length of the volume
12. The prismoidal formula for volumes can be stated as
    - (a) To the sum of end areas, add twice the sum of areas odd sections and four times the sum of areas of even section and multiply this sum by l, the length of the volume
    - (b) To the sum of end areas, add twice the sum of areas of odd sections and four times the sum of areas of even sections and multiply this sum by $l/6$, where l is the length of the section
    - (c) Find the average of the end areas and to this add the sum of the remaining areas and multiply this sum by l, the length of the volume
    - (d) Find the average of all sectional areas given and multiply this sum by l, the length of the volume
13. To determine the whole volume formed by given sections over a given length, the prismoidal formula can be applied
    - (a) only when the number of sectional areas is odd
    - (b) only when the number of sectional areas is even
    - (c) irrespective of the number of sectional areas being odd or even
    - (d) only for three sections at a time
14. Prismoidal correction is applied to volumes calculated by
    - (a) the prismoidal formula to correct for the assumptions made in the formula
    - (b) the prismoidal formula when the number of sectional areas is even
    - (c) the trapezoidal formula when the number of sectional areas is odd
    - (d) the trapezoidal formula assuming that the volume obtained by the prismoidal formula is more accurate
15. Prismoidal correction is applied to volumes calculated by the _____ formula and is always _____.
    - (a) prismoidal, subtractive
    - (c) trapezoidal, subtractive
    - (b) prismoidal, additive
    - (d) trapezoidal, additive
16. Curvature correction to volumes is applied when the
    - (a) plan of the embankment is curved in plan
    - (b) sides of the section are not straight lines
    - (c) base of the embankment is along a slope
    - (d) volume is calculated by prismoidal formula
17. Curvature correction to volumes
    - (a) is always positive
    - (b) is always negative
    - (c) can be positive or negative
    - (d) is positive in the case of symmetrical sections only

18. A mass-haul diagram is a diagram showing
    (a) the amount of cutting and filling along the length
    (b) the longitudinal section of the site
    (c) the cross section of the site
    (d) cumulative volume of cutting and filling along the length
19. A positive slope of mass-haul diagram indicates
    (a) that excavation has to be done at these places
    (b) that filling has to be done at these places
    (c) larger amount of excavation
    (d) larger amount of filling
20. A negative slope of the mass-haul diagram indicates
    (a) that excavation has to be done at these places
    (b) that filling has to be done at these places
    (c) larger amount of excavation
    (d) larger amount of filling
21. Where a mass-haul diagram crosses the X- or chainage axis, it shows
    (a) there is net volume of excavation
    (b) there is net volume of filling
    (c) that at that point excavation and filling are balanced
    (d) the point from where excavation has to start

Review Questions

1. What is a multilevel section? Draw a sketch and explain the method to find the area of such a section.
2. Derive the formula for the area of a two-level section.
3. Derive the formula for the area of a three-level section.
4. State the prismoidal formula for volume and explain the meaning of each term in the formula. What is the condition for applying this formula?
5. State the trapezoidal formula for volume and explain the meaning of each term in the formula.
6. Explain the term prismoidal correction. Derive the prismoidal correction for a two-level section.
7. Explain the basic principle of curvature correction for volumes. Derive the formula for the curvature correction of a two-level section.
8. Explain the method to find the capacity of a reservoir.
9. Explain the method to find the volume of borrow pits from spot levels.
10. What is a mass-haul diagram? Explain the method of preparing such a diagram and its use.

Problems

1. Find the area of a single-level section if the formation width is 8 m and the side slopes are 2:1. The height at the centre of the section is 2.5 m.
2. Find the area of a two-level section with the following data: formation width = 10 m, side slopes = 1.5:1, transverse slope of the ground = 4:1, and height at the centre = 2.8 m.
3. A sidehill two-level section has a formation width of 12 m, side slopes at 2:1, and a transverse slope of the ground of 6:1. The height at the centre of the section is 0.35 m. Find the areas of filling and cutting in the section.
4. Find the areas of cutting and filling in a sidehill two-level section if the formation width is 6 m, side slopes 2:1, and transverse slope of the ground 3:1. The height at the centre of the section is 0.5 m.

5. Find the area of a two-level section with the following data: formation width = 12 m, side slope = 2:1, formation level = 125.00 m. The levels of the points of intersection of the side slopes with the ground are 126.75 on the left and 129.25 on the right.

6. A two-level section has the following data: formation width = 8 m, side slope = 2:1 on the left and 1:1 on the right, transverse slope of the ground = 8:1. Find the area of the section if the height at the centre is 2.8 m.

7. Find the area of a three-level section with the following data: Formation width = 10 m, side slope = 2:1, transverse slope = 4:1 on the left half and 6:1 on the right half, height at the centre = 2 m.

8. Find the area of a three-level section with the following data: Formation width = 8 m, central height = 1.8 m, side slopes = 2:1 on the left and 1:1 on the right, transverse slope = 6:1 on the left half and 10:1 on the right half.

9. Find the area of the multilevel section shown in Fig. 10.34.

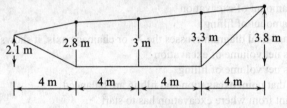

Fig. 10.34

10. Find the area of the multilevel section shown in Fig. 10.35.

11. A level section has a formation width of 6 m and side slope 2:1. The central height is 2.5 m. The cross section is constant over a length of 40 m. Find the volume of earthwork.

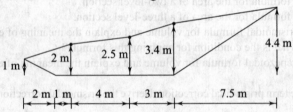

Fig. 10.35

12. A level section has a formation width of 8 m and side slopes 1.5:1. The formation level has a downward slope of 20:1. If the central height is 2.4 m at the start, find the height at 30-m intervals and find the volume in a length of 60 m by the trapezoidal and prismoidal formulae.

13. A two-level section has a formation width of 16 m, side slopes 2:1, and transverse slope 6:1. The formation level rises at a gradient of 1 in 100. Find the volume of earthwork in a length of 80 m by sections at 20-m intervals by the trapezoidal and prismoidal formulae. The depth of cutting at the first section is 3.4 m.

14. Find the volume of earthwork by prismoidal and trapezoidal formulae with the following data: formation level = 201.85 m, formation width = 6 m, downward slope of formation = 100:1, side slope = 2:1, and transverse slope = 6:1; the ground has an upward gradient of 50:1. The depth of cutting at 0 m is 1.65 m. Find the depth of cutting at 30 m and 60 m and the volume in the 60 m length using the trapezoidal and prismoidal formulae. Compute the prismoidal correction.

15. The ground levels at 20-m intervals are as follows:

Chainage	0	20	40	60	80
Ground level	101.5	101.55	101.68	101.79	102.24

The depth of cutting at the first section is 1.5 m. The formation has a downward slope of 1 in 40. The formation width is 8 m and side slope is 2:1. Find the formation level and depth of cutting at each section. Find the volume of earthwork in this length by the prismoidal formula.

16. Find the volume by the prismoidal and trapezoidal formulae with the following data:

Chainage	0	30	60
Central depth	1.85	2.15	2.45

Formation width = 12 m, side slopes = 2:1, transverse slope = 6:1.

17. The section of an embankment is shown in Fig. 10.36. The heights of the embankment at different sections are given below:

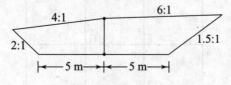

Fig. 10.36

Chainage	0	20	40	60	80	100
Height	2.5	2.85	3.15	3.40	3.80	4.2

Calculate the volume using the trapezoidal formula.

18. A two-level section has the following data. Formation width = 10 m; side slopes = 1.5:1; transverse slope = 6:1; depths of cutting at 20-m intervals at five sections: 2 m, 2.25 m, 2.85 m, 3.2 m, and 3.55 m. Find the volume by the prismoidal formula and the trapezoidal formula. Also calculate the prismoidal correction.

19. Find the volume by the prismoidal formula with the following data:

Chainage	0	30	60	90	120	150
Depth	1.1	1.3	1.5	1.8	1.5	1.35

Formation width = 6 m, side slope = 2:1, transverse slope = 4:1.

20. The following data refers to an embankment that is curved in plan. Formation width = 12 m, side slopes = 2:1, transverse slope = 6:1. The depth at the centre is 2.2 m. The radius of curvature is 180 m. Find the curvature correction for a length of arc that subtends an angle of 30° at the centre.

21. A three-level section has a formation width of 12 m. The side slope is 2:1. The ground slopes are 6:1 on the left half and 10:1 on the right half. The height along the centre line is 2.5 m. The centre line lies along a curve of radius 200 m and subtends an angle of 20° at the centre. Find the curvature correction for this length.

22. Find the capacity of a reservoir from the data given in Table 10.15 if the plan from which the area is measured is drawn to a scale of 1 cm = 30 m. Find the capacity between the levels 252 m and 266 m.

Table 10.15 Data for Problem 10.22

Contour	Area (cm²)	Contour	Area (cm²)
252	105	260	1055
254	254	262	1265
256	485	264	1395
258	735	266	1590

23. Find the volume of excavation from the plan given in Fig. 10.37 and the original ground levels given in the Table 10.16 if the excavation level is 110.55 m.

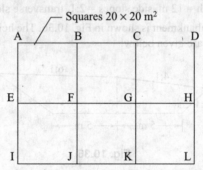

Fig. 10.37

Table 10.16 Data for Problem 10.23

Point	Ground level (m)	Point	Ground level (m)
A	111.85	G	116.85
B	112.35	H	117.55
C	114.65	I	112.45
D	116.75	J	114.55
E	112.55	K	115.85
F	115.25	L	116.75

24. Find the volume of excavation from the following data: base of excavation rectangle of sides 24 m × 48 m, side slope 2:1, and the original ground has a slope of 1 in 12 along the centre line along the longer edges. The depths of excavation at the two corners are 3.6 m at the lower end and 7.6 m at the higher end.
25. An excavation is in the form of a rectangle, 40 m × 18 m. The depths of the excavation at the corners are 2.85, 3.25, 2.55, and 3.05 m. The depth of excavation at the intersection of the diagonals is 2.9 m. Find the volume of the excavation.
26. Draw a mass-haul diagram from the data given in Table 10.17 and determine the borrow material or wastage and the maximum haul distance.

Table 10.17

Chainage (m)	Area of cutting (m²)	Area of filling (m²)
0	–	–
40	15	–
80	25	
120	35	–
160	45	
200	30	–
240		
280	–	25
320		45
360		60
400	–	75
440	–	55
480	–	45
520		25
560	–	0

CHAPTER

11

Minor Instruments

Learning Objectives

After going through this chapter, the reader will be able to
- explain the purpose of minor instruments
- describe the construction, use, and testing of the hand level
- describe the construction, use, and testing of different types of clinometers
- describe the construction, working, use, and testing of the sextant
- explain the parts and function of the pantograph and the eidograph

Introduction

In this chapter, we will discuss certain instruments classified as minor instruments. Such instruments are generally not very precise but are used commonly in reconnaissance operations and preliminary surveys. Most of these are held in the hand for use. They are generally very useful for a general idea of the terrain for the purpose of planning a detailed survey.

11.1 Hand Level

A hand level is a simple instrument to use. It is used in reconnaissance operations and preliminary surveys to obtain the levels of salient points before taking up a detailed survey for determining levels. As the name suggests, a hand level is a hand-held instrument. It is used when the line of sight is not long and when high-precision setting is not required.

The hand level [Fig. 11.1(a)] consists of a 150-mm-long metallic tube of circular or rectangular cross section. It has a spirit level attached to the top of the tube with screws. The tube has a rectangular opening just below the level through which an image of the bubble of the level can be seen on a mirror fixed inside the tube. The mirror occupies only half the width of the tube, the other half being open. E is the eyehole through which the observer can see the bubble's image as well as a staff held at a distance. There is a horizontal cross hair at the other end of the tube, which acts as the reference mark for reading the staff. The observer can hold the hand level in such a way that the bubble is centred (bisected by the cross hair) while viewing the staff.

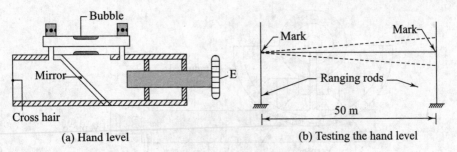

| (a) Hand level | (b) Testing the hand level |

Fig. 11.1 Hand level

Using the hand level

To use the hand level, proceed as follows.

1. Hold the instrument in hand or against a ranging rod at eye level.
2. Have a helper hold a levelling staff at the point whose elevation is required.
3. Looking through the eyehole, observe the image of the bubble reflected in the mirror and the staff through the empty part along with the cross hair at the other end of the tube.
4. Adjust the hand level so that the cross hair bisects the image of the bubble. Note the reading on the staff.

A hand level can be tested and adjusted, if there is a provision for adjustment. The cross wire is housed in a frame that may be adjustable. To test the hand level, proceed as follows [Fig. 11.1(b)].

1. Take two ranging rods, make a mark on them at about eye level.
2. Set these ranging rods about 50 m apart on fairly level ground.
3. Holding the hand level at the level of the mark on the first ranging rod, sight the mark on the other ranging rod.
4. Shift the hand level to the other ranging rod. Keep it at the mark on the ranging rod, centre the bubble, and sight the ranging rod.
5. If the line of sight bisects the mark made on the ranging rod, the instrument is working properly. If not, adjust the cross wire to bisect the mark on the ranging rod.
6. Repeat the test and adjustment till the instrument starts working properly.

A hand level is useful for rough work in contouring and sectioning.

11.2 Clinometers

A clinometer (one form was discussed in Chapter 2) is a hand-held instrument. Clinometers are used for rough measurement of vertical angles and the slope of the ground, and for setting a point along a grade. A few forms of the clinometer are discussed in the following sections.

11.2.1 Abney Level

The Abney level [Fig. 11.2(a)] is a form of clinometer used for measuring slopes and vertical angles. It is a very popular form. It finds immense use for speedy work that does not require very high precision.

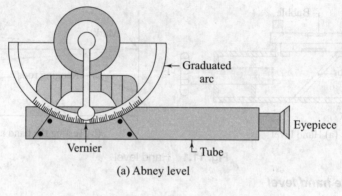

(a) Abney level

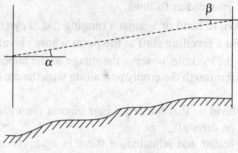

(b) Testing the Abney level

Fig. 11.2   Abney level (clinometer)

The Abney level consists of a square tube with an eyepiece or a small power telescope at one end. Inside the tube a mirror is kept at 45° to the horizontal and occupies half the width of the tube. A cross wire is fixed at the far end of the tube, with which a staff can be bisected. On top of the tube is a bubble tube with a vertical vernier arm (or the main scale in some instruments) attached to it. When the tube is held at an angle, the bubble will go off-centre. As it is brought to the centre (bisected by the cross wire) by operating a milled-headed screw, the vernier arm (or main scale) rotates against a main scale (or vernier arm). The main scale is in the form of a semicircular graduated arc. The graduations on the arc run in either direction from zero at the centre to 90° at the outer ends. The main scale is also graduated in % slopes. The bubble is seen reflected in the mirror along with the staff seen directly from the eyepiece end. The vernier is of the extended type with an index mark at the centre and graduations running in either direction.

Using the Abney level

The Abney level is used for various purposes as follows.

Reading vertical angles   The procedure is as follows.

1. Hold the instrument at eye level. Raise or lower the tube to sight the object with the cross hair.
2. The bubble will be off-centre in this position, as the tube is held in an inclined position. Bring the bubble to the centre, bisected by the cross wire, by using the milled-headed screw.

3. The vernier arm moves when the bubble is brought to the centre. Read the vertical angle from the main scale and the vernier. Angles can be read with the vernier to an accuracy of 10′.

Finding the slope of the ground The procedure is as follows.

1. To measure the slope, it will be necessary to have a line that can indicate the slope. To do this, measure the height of the observer's eye and place a mark on a ranging rod at the same height.
2. Holding the Abney level at eye level, sight the ranging rod held at a point. Bisect the mark on the ranging rod with the cross wire.
3. The bubble will be off-centre at this position. Operate the milled-headed screw to bring the bubble to the centre of its run, bisected by the cross wire.
4. Read the angle or slope from the main scale and vernier scale.

Locating points on a grade To locate points, or trace grade points of a given gradient or slope, the following procedure is adopted.

1. As in the previous case, obtain a line of sight by marking a line on a ranging rod at the same height as the observer's eye.
2. Set the instrument to the angle of slope required by using the vernier scale.
3. Stand at a point holding the Abney level at eye level.
4. By trial and error, direct an assistant holding the ranging rod to a point where he bisects the mark, the bubble being centred at the same time. The line joining the eye of the observer to the mark on the rod is a line parallel to the slope required.
5. Mark the point.
6. Move to this point with the instrument to locate another point along the same slope or gradient.

Testing the Abney level

The Abney level can be tested and adjusted using the following method [Fig. 11.2(b)].

1. Select two points along a sloping ground, A and B. On two ranging rods, make a mark at a convenient height, say 1.5 m.
2. Place the ranging rods at A and B.
3. Holding the level at the mark on the ranging rod at point A, observe the mark on the ranging rod at B. Bring the bubble to the centre. Note down the angle. Let this angle be α.
4. Move with the instrument to point B. Holding the level at the mark on the ranging rod, bisect the mark on the ranging rod at A. Bring the bubble to the centre. Note down the angle of the slope. Let this angle be β.
5. It is obvious that one of the angles will be an angle of elevation and the other will be an angle of depression. If the two angles are equal, the instrument is in adjustment. If not, adjustment is required.
6. Set the instrument to read the mean of the two angles, i.e., $(\alpha + \beta)/2$.

7. With this angle set, hold the instrument at one of the stations and sight the mark on the ranging rod. The bubble will be off the centre. Bring the bubble to the centre of its run by the capstan-headed nuts holding the bubble tube. Repeat the test and adjustment until the instrument reads correctly.

11.2.2 De Lisle's Clinometer

De Lisle's clinometer (Fig. 11.3) is used for measuring vertical angles and slopes. The instrument consists of a mirror held in a frame suspended from gimbals. The frame is connected to a ring from which it can be suspended, either held in the hand or hung from a nail on a rod or pole. The mirror occupies half the width of the frame, the other half being open. The edge of the mirror acts as a vertical reference line. A heavy semicircular arc is connected to the frame. The arc has graduations marked in gradients. The arc is a heavy metal arc jointed on a vertical axis so that it can be moved towards or away from the observer to measure angles of elevation or depression. A radial arm with a sliding weight is attached to the centre of the arc. When at the extreme outer end of its slide, the weight balances the weight of the metal arc in the horizontal position and the mirror lies in a vertical plane.

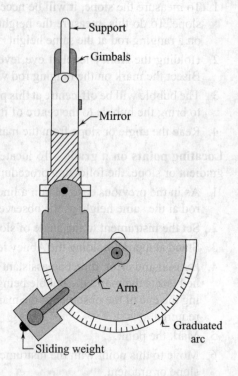

Fig. 11.3 De Lisle's clinometer

The arm has a bevelled edge which acts as an index. When the arc is moved, the mirror becomes inclined to the vertical. The inclination of the mirror to the vertical is the same as the inclination to the horizontal of the line joining the eye to the point at which it appears in the mirror. The line of sight is made horizontal by sliding the weight to the outer end and turning the radial arm back to its full extent.

Using De Lisle's clinometer

The clinometer can be used to measure vertical angles, determine slopes, and set gradients as follows.

Measuring vertical angles To measure vertical angles, proceed as follows.

1. Move the sliding weight to the inner point. Hold the instrument at arm's length by suspending it by the thumb.
2. Move the arc forward (towards the observer) for an angle of elevation and backward (away from the observer) for an angle of depression.
3. Sight a high or low object through the open space by the side of the mirror. Move the arc to coincide the reflection with the view of the object seen directly.

4. The reading of the graduated arc gives the tangent of the vertical angle to the object from a horizontal line through the eye.

Measuring slope   To measure the slope of the ground, proceed as follows.

1. In measuring a slope, hold the instrument at the low or high point of the slope as per convenience.
2. Mark a ranging rod with a line at the level of the observer's eye and hold it at the other end of the slope.
3. Move the arc till the reflection in the mirror coincides with the view of the mark seen directly.
4. Read the arc to obtain the gradient.

Locating points on a given grade   To locate points on a given grade, proceed as follows.

1. Depending upon whether an upgrade or downgrade is required, turn the arc away from or towards you.
2. Set the arm to the given grade. The mirror is now inclined at this angle to the vertical.
3. Mark a ranging rod at a point at the same height as the observer's eye.
4. Ask an assistant to move with this ranging rod to a convenient distance. Direct him/her to move till the reflection and the mark on the ranging rod seen directly are coincident. Mark this point. The line joining the instrument station to the point where the ranging rod is held is on the given gradient.
5. Locate further points similarly.

11.2.3   Foot Rule Clinometer

A foot rule clinometer (Fig. 11.4) is a simple device with two arms hinged at one end. The arms can be inclined by rotating them about the hinge. A brass arc (a quadrant) carrying graduations from 0 to 90° is fixed at the hinge. The angles are read with an index on the arm. Each arm carries a simple spirit level. The arm is horizontal when the level is at the centre of its run. A magnetic compass may also be

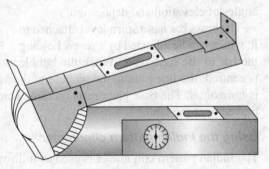

Fig. 11.4   Foot rule (line meter)

provided in the foot rule clinometer for determining the meridian. Two sights are fixed on the top arm to provide a line of sight. One of the brass plates has a pinhole and the other has a glass with a cross marked on it. The line joining the pinhole to the cross is the line of sight. The instrument may be provided with gradients and hypotenusal allowance in addition to angles.

Using the foot rule clinometer

The procedure for measuring a vertical angle or slope is as follows.

1. Hold the foot rule clinometer at eye level such that the bubble is horizontal in the lower arm.
2. Raise the upper arm to sight the object, which is a ranging rod with a mark at the eye level of the observer.
3. Once the object is bisected, read the angle or slope on the graduated arc.

If a slope is to be measured, place the clinometer along the slope. The lower bubble tube is off-centre. Turn the upper arm to bring the bubble to the centre of its run. The reading on the arc gives the slope in degrees or in the form of a gradient.

11.2.4 Indian Pattern Clinometer

The Indian Pattern clinometer (Fig. 11.5) is a simple instrument used extensively in plane tabling. This instrument helps in determining the elevations of points located on the plane table so that a topographic map can be prepared.

The instrument consists of a base plate on which a brass bar that holds a vane at either end is fixed. Both vanes can be folded onto the bar when the instrument is not in use. One of the vanes (the eye vane) has a pinhole while the other (the object vane), fixed at the opposite end, is longer and has a long vertical slit. The object or sight vane has a frame carrying a cross wire that can be moved along the base plate by a milled-headed screw. The sight vane is graduated, with zero against the pinhole in the horizontal position, in degrees and tangents towards the top and bottom for angles of elevation and depression.

The brass bar has a spirit level attached to it, which can be adjusted by a screw holding the bar to the base plate. When the bubble

Fig. 11.5 Indian pattern clinometer

is centred, the line joining the pinhole to the zero of the scale on the object vane is horizontal. The base plate rests on three smooth points so that it can be easily moved over the table.

Using the Indian pattern clinometer

The Indian pattern clinometer is used as follows.

1. Set up and level the plane table on which the clinometer is to be used.
2. Place the clinometer on the plane table, unfold the vanes, and level the clinometer using the spirit level on the bar.
3. Measure the height of the eyehole on the vane from the ground to obtain the height of the line of sight.
4. Looking through the eyehole, adjust the cross wire to bisect the object. Note the vertical angle and tangent of the vertical angle marked on either side.
5. RL of the point = RL of the plane table station + height of the eyehole + distance × tangent value.

Testing the Indian pattern clinometer

To test the clinometer, we use the fact that when the bubble is centred, the line of sight when horizontal must pass through the zero of the graduations. The following procedure is adopted.

1. Place the clinometer over a levelled plane table. Bring the bubble to the centre of its run.
2. Establish, on a distant point, a mark on a ranging rod at the same height as the height of the eyehole.
3. With the bubble centred, observe the cross wire bisecting the mark through the eyehole.
4. If the wire is against the zero graduation, the instrument is in adjustment. If not, bring the cross wire against the zero graduations using the levelling screw on the base plate. The bubble will go off the centre. Centre the bubble using the capstan-headed screw holding the bubble tube.
5. Repeat the test and adjustment until the instrument reads correctly.

11.2.5 Ceylon Ghat Tracer

The Ceylon ghat tracer (Fig. 11.6), popular in earlier years, is another form of clinometer. It can be used to find slopes or locate points on a gradient. The instrument consists of a hollow brass tube suspended from a triangular bracket with a hook that can be used to hang the instrument from a vertical rod. The brass tube has an eyehole at one end and a cross wire at the other. The line of sight is defined by the line joining the eyehole to the intersection of the cross hair.

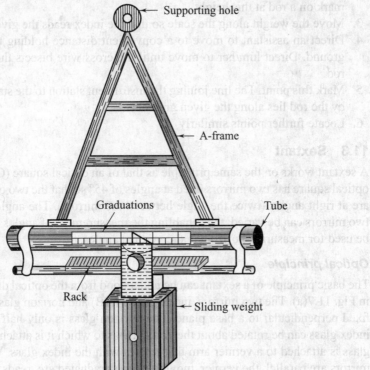

Fig. 11.6 Ceylon ghat tracer

The brass tube has a weight attached to it which can be moved along it by a rack and pinion arrangement by turning a milled-headed screw. The weight has an index that slides along the graduations on the flat part of the tube. The graduations are in the form of gradients marked from the centre outwards in either direction to measure rising or falling gradients.

Using the ghat tracer

The ghat tracer is used by hanging it freely from a vertical staff. The following procedure is adopted to measure a slope.

1. Hold the instrument at a station, suspending it from the vertical staff at one end of the slope.
2. Measure the height of the axis of the sighting tube from the base of the vertical staff.
3. On a ranging rod or another such rod, mark this height.
4. Have an assistant with the marked rod move along the slope to the end of the slope.
5. Now move the weight along the tube while looking through the eyehole until the cross hair bisects the mark on the rod.
6. Note the reading against the index mark on the weight. This gives the slope of the ground.

To locate points along a given gradient, the following procedure is adopted.

1. Hold the ghat tracer at the starting station, suspended from the vertical staff.
2. Measure the height of the axis of the tube from the base of the staff. Make a mark on a rod at this height.
3. Move the weight along the scale so that the index reads the given gradient.
4. Direct an assistant to move to a convenient distance holding the rod on the ground. Direct him/her to move until the cross wire bisects the mark on the rod.
5. Mark this point. The line joining the instrument station to the station occupied by the rod lies along the given gradient.
6. Locate further points similarly.

11.3   Sextant

A sextant works on the same principle as that of an optical square (Chapter 2). An optical square has two mirrors fixed at angles of 45° so that the two objects sighted are at right angles (twice the angle between the mirrors). The angle between the two mirrors can be varied, thus enabling the measurement of angles. A sextant can be used for measuring horizontal and vertical angles.

Optical principle

The basic principle of a sextant can be understood from the optical diagram shown in Fig. 11.7(a). The two mirrors, index glass I (AB) and horizon glass H (CD), are fixed perpendicular to a base plane. The horizon glass is only half silvered. The index glass can be rotated about the vertical axis to which it is attached. The index glass is attached to a vernier arm that rotates with the index glass. When the two mirrors are parallel, the vernier, moving over a graduated arc, reads zero.

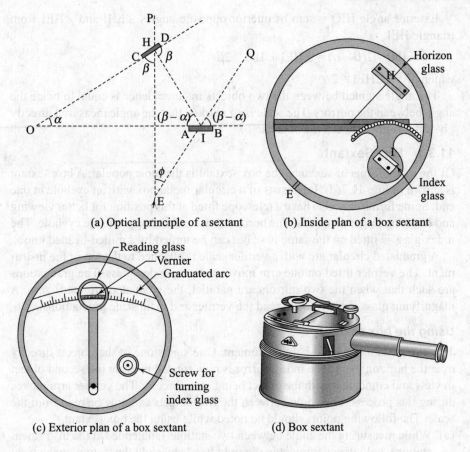

(a) Optical principle of a sextant

(b) Inside plan of a box sextant

(c) Exterior plan of a box sextant

(d) Box sextant

Fig. 11.7 Box sextant

Two objects P and Q are sighted through the eyehole E. The mirror is rotated to coincide them. One of the objects (P) is sighted directly over the mirror H (over the unsilvered portion) and the other object (Q) is sighted after reflection from the two mirrors.

The angle between the index glass I and the horizon glass H (angle between BA and CD) is α. Let the angle between P and Q be ϕ.

$$\angle IOH = \alpha \text{ and } \angle QEP = \phi$$

In triangle IHO, exterior angle DHI = sum of interior opposite angles:

$$\angle DHI = \angle HOI + \angle HIO, \angle DHI = \alpha + \angle HIO$$

Let angle DHI = β. We get

$$\beta = \alpha + HIO, \angle HIO = \beta - \alpha$$

$\angle BIQ$ is also equal to $(\beta - \alpha)$, as the angle of incidence is equal to the angle of reflection. $\angle DHI = \beta = \angle EHO$ for the same reason.

$$\angle HIQ = 180 - 2(\beta - \alpha) \text{ and } \angle EHI = 180 - 2\beta$$

from the straight lines AB and CD.

Exterior angle HIQ = sum of interior opposite angles ∠EHI and ∠HEI from triangle HEI.

$$180 - 2(\beta - \alpha) = \angle HEI + 180 - 2\beta$$

which gives ∠HEI = 2α.

The angle sighted between the two objects in coincidence is equal to twice the angle between the mirrors. The scale is so marked that the angle measured directly gives the angle between the objects.

11.3.1 Box Sextant

Of the many forms of sextants, the box sextant is the most popular. A box sextant is shown in Fig. 11.7(d). It consists of a circular metal box with an eyehole at one end. Some forms may also have a telescope fitted at this position for better viewing and covering larger distances. The horizon glass is fixed opposite the eyehole. The index glass is fitted on the same base but can be turned by a milled-headed knob.

A graduated circular arc with a vernier scale is attached to the top of the instrument. The vernier fitted on one arm moves with the index glass. The graduations are such that when the two mirrors are parallel, the graduated arc reads zero. A magnifying glass is provided to read the vernier and main scale graduations.

Using the box sextant

The box sextant is a hand-held instrument. One sights one of the objects directly over the horizon glass. The index mirror is now rotated to bring the second object in view and coincide it with the object being seen directly. The vernier arm moves during this process. The angle between the two objects can now be read from the scale. The following points should be noted while using the box sextant.

(a) While measuring the angle between two stations (subtended at the instrument station), hold the instrument in the right hand and sight the station on the right side through the unsilvered portion of the horizon glass.

(b) When two points are at unequal distances, sight the nearer point directly through the horizon glass.

(c) Sight the brighter object by reflection.

(d) The instrument can be turned upside down if necessary. If two stations are at different elevations, the instrument can be tilted.

(e) If the angle between the two stations is large (very near the range of the instrument), the angle can be measured in parts using an intermediate point.

Measuring horizontal angles   The following procedure is adopted for measuring horizontal angles.

1. Suppose the measurement of angle AOB is required. Place two ranging rods at A and B.

2. Hold the instrument over station O. With the graduated circle set to zero, sight the ranging rod at A. The ranging rod is seen directly over the horizon glass.

3. Rotate the knob to activate the vernier and the index mirror till the ranging rod at B is seen. Rotate the knob gradually to coincide the ranging rod at B with the ranging rod seen directly.

4. Read the vernier and record the angle between OA and OB.

Measuring vertical angles   The following procedure is adopted for measuring vertical angles.

1. While measuring vertical angles, the angle between the horizontal line at the eyehole level and the object (high or low) is measured. First establish a point at eye level. Mark a ranging rod with this height.
2. Hold the instrument vertically. (Turn the box to a vertical position.) Sight the point on the ranging rod directly over the horizon glass.
3. Rotate the knob to bring the other object into position and coincide it with the mark on the ranging rod.
4. Read the graduated arc and the vernier to record the vertical angle.

Testing and adjusting the box sextant   There are four parts in a box sextant requiring adjustments.

1. The two mirrors, the index glass, and the horizon glass must be placed with their planes perpendicular to the base or plane of the graduated arc.
2. When the two mirrors are parallel, the reading on the graduated arc should be zero.
3. The eyehole should be placed such that the horizontal line through it passes just over the horizon glass.

Generally, only two adjustments are possible in a box sextant.

Making the plane of the horizon glass perpendicular to the plane of the graduated arc   The following procedure is adopted [Fig. 11.8(a)].

1. Set the vernier to read zero on the main scale. The two mirrors are now parallel.
2. With the instrument held vertically, sight a high object, such as the spire of a temple.
3. Check for any vertical displacement of the object (between the image seen directly and the reflection). A displacement exists if the horizon glass is not perpendicular to the plane of the arc.
4. Turn the two screws provided at the top for adjustment of the horizon glass appropriately to eliminate any vertical displacement of the images. When the two images exactly coincide, the horizon glass will be at right angles to the base.

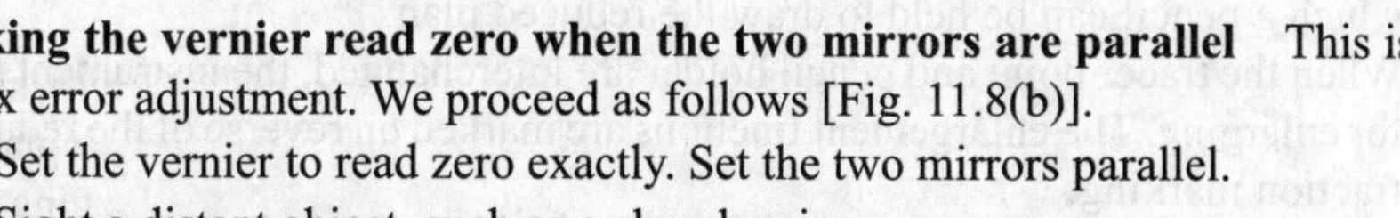

(a) Vertical displacement   (b) Lateral displacement

Fig. 11.8   Testing the box sextant

Making the vernier read zero when the two mirrors are parallel   This is the index error adjustment. We proceed as follows [Fig. 11.8(b)].

1. Set the vernier to read zero exactly. Set the two mirrors parallel.
2. Sight a distant object, such as a church spire.
3. Check for any horizontal displacement between the image seen directly and the image seen by reflection.

4. Adjust the horizon glass with the screw provided. By turning the screw, the horizon glass can be turned and made parallel to the index glass. This eliminates the lateral displacement of the images.

11.4 Pantograph

A pantograph (Fig. 11.9) is an instrument used to enlarge or reduce plans. The pantograph has the following parts.

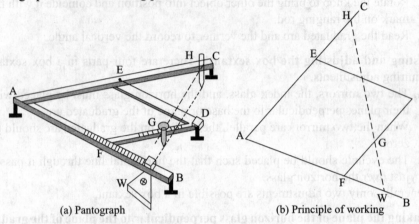

(a) Pantograph   (b) Principle of working

Fig. 11.9 Pantograph

(a) Four rods of tubular section—AB, AC, FD, and DE. The rods AB and AC are hinged at A and are longer than the other two rods. The rods DF and DE are shorter in length and are attached to the longer arms at F and E. The connections are such that the opposite sides are equal. AF = DE and AE = DF. AEDF remains a parallelogram for all positions of the instrument.

(b) The longer arm AB carries a weight W near B called a fulcrum. The instrument moves around this point when it is being used to trace a plan.

(c) The arm AB carries an outer tube that can be slid along the arm and serves as an index to the graduations on AB. AB has graduations indicating the reducing fractions—1/2, 1/4, etc. The index can be slid along the bar and then clamped in position.

(d) The arm DF also has graduations with a sliding index as in arm AB.

(e) The ends B and C are kept on castors or rollers for smooth movement of the instrument on paper.

(f) The arm AC carries a tracer point at H which moves along the boundary of the plan while reducing. The arm DF similarly carries a pencil holder at G, in which a pencil can be held to draw the reduced plan.

(e) When the tracer point and pencil holder are interchanged, the instrument is set for enlarging. The enlargement fractions are marked on reverse of the reducing fraction marking.

Principle of the pantograph

The working principle of the pantograph is that of similar triangles. It can be understood from Fig. 11.9(b). It is mainly used for reducing, as it does not give satisfactory results for enlargement.

The fulcrum, the pencil point, and the tracing point always remain in a straight line. Since AEDF is a parallelogram for all positions, FG is parallel to AE. FG||AE. The two triangles WGF and WHA are similar triangles. Hence,

$$\frac{WG}{WH} = \frac{FW}{AW}$$

FW/AW is the reduction factor. Any movement of the tracing point H will result in a corresponding reduced movement of *G*.

Using the pantograph

For reducing plans, the procedure is as follows.
1. Set the tracing point at H and the pencil point at G.
2. Set the instrument over a convenient portion of the paper so that the tracing point can be moved smoothly over the given plan.
3. Set the index marks for the desired reduction.
4. Set the tracing point at one point on the given plan. Trace smoothly the boundary of the given plan. The pencil point marks a reduced plan to the set fraction.

For enlargement, the same procedure is adopted by interchanging the tracing point and the pencil point.

11.5 Eidograph

The eidograph is an improved version of the pantograph used for the same purposes. The eidograph is more stable and steady and gives better results. It has the following parts (Fig. 11.10).

(a) Three brass tubular arms AB, CD, and EF are used and interconnected. Each tube is graduated to 200 divisions and 100 divisions on either side of the centre. There is also a vernier scale provided on each arm so that it can be set to one-tenth of the main division.

(b) The centre beam EF carries a lead weight W covered with brass and has three pins to fix it on paper. W acts as a support for the instrument; the eidograph moves around this weight when used. A pin, forming the fulcrum about which the instrument moves, projects from the upper part of the weight and fits into a socket attached to a sliding box G. The box can be moved over the beam and fixed in any position.

(c) The centre beam EF carries a wheel at either end, fitted into sockets attached to these ends. The wheels are connected by two steel bands attached to their circumferences. This ensures that the wheels move in unison. Using the screws provided, these steel bands can be adjusted so that the beams AB and CD are parallel and at the same time the bands have sufficient tension to steady the motion of the parallel arms AB and CD.

(d) The parallel arms AB and CD pass through sliding boxes at either end of the central arm EF. These boxes can be fitted at any position on the arms and clamped.

(e) At the end of one of the parallel arms is the tracing point T. The other arm carries at its end a pencil holder. There is also an arrangement to lift the pencil from the paper with a lever.

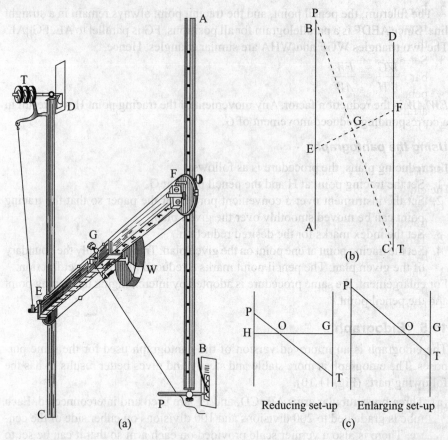

Fig. 11.10 Eidograph

Principle of operation

The working principle of the eidograph can be understood from Fig. 11.10(b). When the instrument is set to the correct values in each of the arms, the fulcrum, the tracing point, and the pencil point will lie in a straight line. If the value x is set in the central arm on the pencil side, and on each of the arms shown, the plan will be reduced. However, if the value is set in the central arm on the tracing point side, and on the other arms shown, the original plan will be enlarged.

The fact that the two arms AB and CD are parallel for any setting gives two similar right-angled triangles OGT and OHP. Any movement of T will result in the movement of P in the proportion set in the instrument.

Using the eidograph

To use the eidograph, the following procedure is adopted.

1. Set the instrument to the value of reduction or enlargement. To find this proportion, do the following calculation. If the ratio is 1:n, then the slide in the central beam is set to a value given by $100(n-1)/(n+1)$. For example, if the proportion is 1:4, then the value to be set is $100(4-1)/(4+1) = 60$.

2. For reduction of plans, set this value in the central beam on the pencil side from the centre. For enlargement, set the same value on the tracing point side; this will enlarge the plan to four times its original size.
3. Set the same values on the beams AB and CD. For reduction, lengthen the bar carrying the tracing point to the set value and reduce the bar carrying the pencil in length by setting it to the same value. For enlargement, the reverse arrangement is used [Fig. 11.10(c)].
4. Once the instrument is set like this, the tracing point is moved across the boundary of the plan and a reduced or enlarged drawing is produced by the pencil point.

Testing and adjusting the eidograph

The eidograph can be tested and adjusted as follows.
1. Set the instrument to read zero. Make a mark with the pencil point and the tracing point.
2. Turn the instrument around so that the tracing point comes over the mark made by the pencil point.
3. Check whether the pencil is over the mark made by the tracing point. If the pencil comes over this point, the instrument is in adjustment.
4. If not, make another mark with the pencil. Mark a point halfway between these two points.
5. Adjust the instrument by turning the screws on the steel band to bring the pencil point over the point marked midway.

Summary

Minor instruments are used for speedy work that does not require very high precision. These instruments are handy to use in reconnaissance operations and preliminary surveys to get a general idea about the terrain. They also find use in difficult terrain to measure angles and slopes, and locate points on a given gradient approximately. They are mostly held in the hand for use.

A hand level is a simple instrument used to find elevations of points. Clinometers are instruments used to measure vertical angles, determine slopes, and set points on a given grade. The various forms commonly used are the Abney level, Indian pattern clinometer, De Lisle's clinometer, foot rule clinometer, and Ceylon ghat tracer.

Sextants are used to measure horizontal angles. The sextant works on the same principle as that of an optical square. A box sextant is the most commonly used form in land surveys. It works on the principle that if rays of light are reflected from two mirrors set at an angle, the angle between the rays is double this angle after reflection.

A pantograph is a simple device used to enlarge or reduce plans. An eidograph is used for the same purpose as a pantograph but is more stable and reliable.

Exercises

Multiple-Choice Questions

1. A hand level essentially consists of
    (a) a telescope and a spirit level
    (b) a telescope and a graduated vertical circle
    (c) a spirit level and a sighting tube with a half-mirrored glass
    (d) a spirit level and a graduated vertical rod

2. An Abney level is mainly used for finding
   (a) horizontal angles
   (b) reduced levels of points
   (c) contour lines
   (d) vertical angles and gradients
3. De Lisle's clinometer is an instrument used for finding
   (a) horizontal angles
   (b) reduced level of points
   (c) contour lines
   (d) vertical angles and measuring slopes
4. The Indian pattern clinometer
   (a) is mainly used with a plane table to find elevations of points
   (b) is fitted with a telescope and spirit level to find elevations
   (c) is used to find angles and slopes
   (d) is used to find horizontal distances
5. The Ceylon Ghat tracer is used to
   (a) trace from maps
   (b) find slopes and locate points on a gradient
   (c) find horizontal distances
   (d) find horizontal angles
6. The optical principle of a sextant is that when two objects are in coincidence, the angle between the mirrors
   (a) and the two objects is the same
   (b) is twice the angle between the objects
   (c) is half the angle between the objects
   (d) and the two sighted objects is in a fixed ratio
7. The box sextant can be used to
   (a) enlarge or reduce plans
   (b) measure distances
   (c) determine reduced levels of given points
   (d) measure angles and set gradients
8. The basic principle of the pantograph is
   (a) that of similarity of triangles
   (b) that the opposite sides of a parallelogram are equal
   (c) the Pythogoras theorem
   (d) that the sum of the four internal angles of a parallelogram is 4 right angles
9. The eidograph is used for
   (a) drawing profiles of the ground
   (b) drawing cross sections
   (c) reducing or enlarging plans
   (d) drawing lines on the plane table
10. The working principle of the eidograph is
   (a) that the opposite sides of a parallelogram are equal
   (b) the Pythogoras theorem
   (c) similarity of two right angled triangles
   (d) that the sum of the two adjacent internal angles is 2 right angles

Review Questions

1. Give a classification of minor instruments and state their functions
2. Draw a neat sketch of a hand level and describe its construction, use, and testing.
3. Show with a neat sketch the construction and use of an Abney level. Explain the testing and adjustment of the Abney level
4. Describe the construction and use of De Lisle's clinometer

5. Draw a neat sketch of the Indian pattern clinometer and describe its construction, use, and testing.
6. Draw a neat sketch of the Ceylon ghat tracer. Describe its construction and use
7. Explain the principle of working of a sextant. How is it different from an optical square?
8. Draw a neat sketch of a box sextant and describe its construction. How is it used for measuring angles? Describe the procedure for testing and adjusting a box sextant.
9. Explain the functioning of a pantograph with a neat sketch.
10. With a neat sketch, show the parts of an eidograph and explain its construction. Explain how it is used to enlarge and reduce drawings.

12

Permanent Adjustments of the Level and Theodolite

Learning Objectives

After going through this chapter, the reader will be able to

- explain the principle of reversion
- state the fundamental lines of a level
- state the desired relationships between the fundamental lines of a level
- explain the procedure for testing and adjusting dumpy and tilting levels
- state the fundamental lines of a theodolite
- explain the desired relationships between the fundamental lines of a theodolite
- explain the procedure for testing and adjusting a theodolite
- explain the importance of adjustments of levels and theodolites

Introduction

We have already discussed the use of level and theodolite in surveying. A level is basically used to determine elevations. A theodolite, on the other hand, is a more versatile instrument. Though basically intended to measure horizontal and vertical angles, a theodolite can also be used as a level.

A surveying instrument needs to be set up at the required station and certain adjustments are required before it can be used. Such adjustments are known as temporary adjustments.

Over a period of time, with continuous use, it is seen that a surveying instrument starts to malfunction and does not give correct, real results. The instrument is then said to be out of adjustment and needs to be adjusted to give accurate results. Survey instruments have certain fundamental lines and there must exist a specified relationship between these lines for the instrument to be in correct adjustment. Permanent adjustments are intended to check and, if necessary, make corrections to ensure that the relationships between the fundamental lines are maintained. In this chapter we will study permanent adjustments of levels and theodolites. There are many forms of instruments developed by different manufacturers. In this chapter, we will discuss the most common forms and their adjustments.

12.1 Principle of Reversion

The principle of reversion states that if there is any error in any line (with respect to its relationship with another line) the error gets doubled when the position of the line is reversed with respect to the other line. This principle is the most commonly used method to detect and adjust errors. In Fig. 12.1, lines AB and CD are considered to be at right angles in an instrument in adjustment. If the error is α, i.e., the angle between the lines is $(90° - \alpha)$, then when the orientation of line AB is reversed, the error is 2α, i.e., the angle between the lines is $(90° - 2\alpha)$.

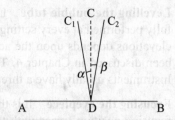

This principle also leads to methods of eliminating errors, wherein observations in normal and reversed positions are taken. Then the average of both the readings is taken to be the correct reading even though the instrument is not in adjustment.

Fig. 12.1 Principle of reversion

12.2 Adjustments of Level

In terms of function, a level is supposed to provide a horizontal line of sight in all directions. The line of sight remains in a horizontal plane when the instrument is set up and adjusted for use. As the telescope is rotated in azimuth, the line of sight traverses a horizontal plane.

There are different types of levels. We will study the adjustments of only two of these types—dumpy level and tilting level. Modern automatic levels are made in such a way that they are very stable and remain in adjustment for a much longer time.

The fundamental lines of a level (Fig. 12.2) are as follows.

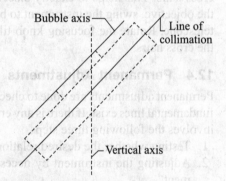

1. The line of sight or collimation is the line joining the optical centre of the objective to the intersection point of the cross hairs.
2. The axis of the bubble tube is the line tangential to the longitudinal curved profile of the bubble tube.
3. The vertical axis is a line perpendicular to the line of collimation and passes through the centre of the instrument.

Fig. 12.2 Fundamental lines of a level

The desired relationships among these lines are as follows.
1. The axis of the bubble tube must be parallel to the line of collimation.
2. The axis of the bubble tube must be perpendicular to the vertical axis.
3. The horizontal cross hair should lie in a plane perpendicular to the vertical axis.

12.3 Temporary Adjustments

Setting up The level is set up at a convenient point for ease of observation of stations whose elevations are to be measured. The instrument station has no other significance in levelling. The instrument is set up on a tripod. There is no need for centring the instrument in levelling. The tripod legs should be set well apart to ensure stability of the instrument. These should be firmly fixed to the ground so that the tripod does not move when the telescope is turned.

Levelling the bubble tube   Levelling is a basic operation and should be carefully performed at every setting of the instrument. The whole operation of finding elevations depends upon the accuracy of levelling the spirit level. Levelling has been discussed in Chapter 4. The operation is the same as in the case of levels. Instruments usually have a three- or four-screw levelling head.

Focusing the eyepiece   It is the process of bringing the cross hairs to focus. The focusing position varies with the eyesight of the observer. If the same observer takes all the readings, this has to be done only once. To focus the eyepiece proceed as follows.

1. Keep a piece of white paper in front of the telescope, or direct the telescope to a clear portion of the sky.
2. Looking through the telescope, adjust the vision by rotating the eyepiece till the cross hairs come in a sharp and clear view.
3. If the eyepiece has graduations, note the graduation at which you get a clear view of the cross hairs. This can help in later adjustment, if required.

Focusing the objective   The objective has to be focused whenever an object is sighted, because this depends upon the distance between the instrument and the object. There is a focusing screw on the side of the telescope, which is operated to focus the objective. This operation brings the image of object in the plane of the cross hairs. This helps to exactly bisect the object, say, a levelling staff. To focus the objective, swing the instrument to bring the object in view by looking over the telescope. Rotate the focusing knob till the object is in a sharp view along with the cross hairs.

12.4   Permanent Adjustments

Permanent adjustments are done to check that the desired relationships between the fundamental lines exist. If there is any error, it needs to be corrected. All adjustments involves the following three steps.

1. Testing whether the desired relationship exists
2. Adjusting the instrument by necessary corrective steps if it is not in adjustment
3. Repeating the above steps till the instrument is in adjustment

12.4.1   Dumpy Level

Dumpy levels were commonly used in earlier days due to their sturdy nature. The following adjustments may be required to adjust a dumpy level.

Test 1   To make the axis of the bubble tube perpendicular to the vertical axis

Principle of adjustment   Figure 12.3 shows the basic principle governing this adjustment. When the instrument is in adjustment, i.e., the axis of the bubble tube is horizontal (when the bubble traverses), the bubble tube axis is perpendicular to the vertical axis [Fig. 12.3(a)] for all positions of the telescope. If the instrument is not in adjustment, the vertical axis is inclined and the angle between the bubble tube axis and the vertical axis is not a right angle [Fig. 12.3(b)]. Next, the bubble tube is brought to the centre of its run as in temporary adjustment. The telescope is now rotated through 180°. The error in the angle, if any, between the bubble axis and the vertical axis is now doubled as per the principle of reversion.

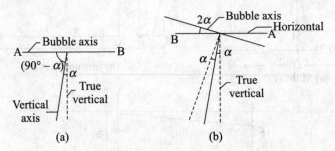

Fig. 12.3   First adjustment of level

The following steps are to be followed for testing.
1. To test whether the condition is satisfied, set up the instrument on firm, level ground.
2. Bring the bubble to the centre of its run in two positions at right angles.
3. When the bubble is over the third foot screw, rotate the telescope through 180° to bring the other end of the telescope over the third foot screw.
4. If the bubble traverses and is in the centre, the required condition is satisfied. Hence no adjustment is needed.
5. If the bubble does not traverse, adjustment is required.

The following steps are adopted for adjustment.
1. Adjust by noting down the number of divisions by which the bubble is off centre.
2. Correct half the error using the foot screw. Correct the remaining half by adjusting the capstan-headed nut at one end of the bubble tube. The bubble is in the centre of its run.
3. Repeat the test and adjustment till the instrument is in adjustment.

Test 2   To make the line of collimation parallel to the bubble tube axis

This is the most important test in the case of a level. When the bubble is centred, the bubble tube axis is horizontal. If the line of collimation is parallel to the bubble tube axis, then we have a horizontal line of sight

Principle of adjustment   When the level is kept exactly midway between two staff stations, the difference in elevations calculated will be correct even if the line of sight is inclined.

Test and adjustment are done by the two-peg test (see Fig. 12.4).

Two-peg test

First method   The procedure for the first method is as follows.
1. Drive two pegs A and B into clear ground about 50–100 m apart. Locate a point M exactly midway between A and B [see Fig. 12.4(a)].

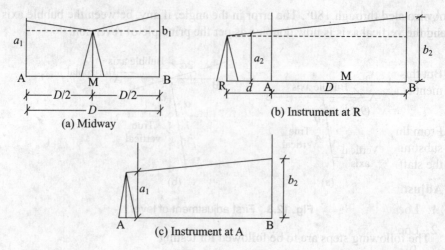

(a) Midway

(b) Instrument at R

(c) Instrument at A

Fig. 12.4 Two-peg test

2. Set up and level the instrument at M. Take readings on the staff held at A. Shift the staff to B and take a reading. The bubble should be at the centre of its run when the readings are taken. Let the readings obtained be a_1 and b_1, respectively.

3. Shift the instrument to position R. Point R can be within the stations A and B or outside the line AB. Let the distance between A and R be d [see Fig. 12.4(b)].

4. Again take readings on the staves held at P and Q. Let the readings be a_2 and b_2. The bubble should be brought to the centre of its run while taking readings.

5. The true difference in elevation between A and B is $(a_1 - b_1)$. Point B is higher if the true difference is positive $(a_1 > b_1)$; point B is lower if the difference is negative $(a_1 < b_1)$. The difference in elevation with the instrument at R is $(a_2 - b_2)$. If the two differences in elevation are the same, then the instrument is in adjustment. Otherwise, the correct readings with the instrument at R can be calculated and the instrument adjusted.

6. The corrections to the readings can be obtained as follows:
   True difference in elevation = $a_1 - b_1$
   Apparent difference in elevation = $a_2 - b_2$
   True reading in B, $b = a_1 \pm$ true difference

7. Compare b and b_2. If $b_2 > b$, then the line of collimation is inclined upwards. If $b > b_2$, then the line of collimation is inclined downwards. Now $b_2 - b$ is the collimation error in a distance D. The corrections to the readings on both pegs at A and B are
   Correction to the reading at A = $d(b_2 - b)/D$
   Correction to the reading at B = $(D + d)(b_2 - b)/D$

8. Another option is to assume that the line of collimation is inclined at an angle. If the line of collimation is inclined at an angle α, as in Fig. 12.4(c), then the adjustment to the readings taken from R can be calculated as follows:
   Correction to the reading at the near peg A = $a_2 \pm d \tan \alpha$
   Correction to the reading at the far peg B = $b_2 \pm (D + d) \tan \alpha$

A plus sign is used if the line of collimation is inclined downwards and a negative sign if the line of collimation is inclined upwards.

$$\text{Difference in elevation} = b_2 \pm (D + d) \tan \alpha - (a_2 \pm d \tan \alpha)$$

But this must be equal to the true difference in elevation obtained from the instrument at O, midway between A and B. Therefore,

$$b_2 \pm (D + d) \tan \alpha - (a_2 \pm d \tan \alpha) = a_1 - b_1$$

From this equation, the value of α can be calculated. Once α is known, it can be substituted in the equations for correct readings to obtain the reading to be set on the staff.

Adjustment Proceed as follows for the adjustment.

1. Locate the diaphragm screws at the top and bottom of the diaphragm.

2. Loosen one screw before tightening the other.

3. Turn the screws to set the correct reading on the staves held at the pegs at A and B.

4. Carry out the correction first at the farther peg B and then at the peg at A. It may be noted that the cross hair moves in the direction of the screw that is being tightened.

5. Repeat the test till the instrument is in adjustment.

Second method The procedure for the second method of adjustment is as follows [Refer Fig. 12.4(c)].

1. Set up the level midway between two points A and B on fairly level ground and level the instrument. Take readings on staves held at A and B. The true difference in elevation between points A and B is given by these two readings, even if the level is not in adjustment. If a_1 and b_1 are the readings, then $(a_1 - b_1)$ is the true difference in elevation between points A and B.

2. Shift the instrument very close to A (or B). Level the instrument and take readings on the staves placed at A and B. The reading at A may have to be taken through the objective, as the staff is very close to the instrument. Note down the readings. Let a_2 and b_2 be the readings on the staves held at A and B, respectively.

3. Compute the difference $(a_2 - b_2)$. If this is the same as $(a_1 - b_1)$, then the instrument is in adjustment. Otherwise, adjustment can be done as given below.

Adjustment

1. Take the reading a_2 on peg A to be correct, as the inclination of the line of collimation has little effect on the staff held so close. The correct reading on peg B is $a_1 \pm$ the true difference $= a_1 \pm (a_1 - b_1)$. Use a plus sign if there is a fall from A to B; use a negative sign, if there is a rise from A to B.

2. Using the diaphragm screws, bring the cross hair to the correct reading on B.

3. Repeat test and adjustment till the instrument is in adjustment.

Third method A third method is to find the true difference in elevation by reciprocal levelling (see Chapter 7 for reciprocal levelling). We then proceed as in the first or second method for adjustment.

12.4.2 Tilting Level

A tilting level is so made that the telescope axis can be independently tilted (independent of foot screws and the vertical axis) using a tilting screw. The only adjustment required is to make the line of collimation parallel to the bubble tube axis. The two-peg test is used in this case as well.

The correct reading on a peg is obtained by operating the tilting screw. The bubble is displaced from the centre when the tilting screw is turned. After getting the correct reading, bring the bubble to the centre using the capstan-headed screws of the bubble tube. The test and adjustment are repeated till the instrument is in adjustment.

12.4.3 Automatic Levels

Automatic levels also require adjustment from time to time. The test and adjustment are the same as for a dumpy level. The two adjustments required are to make the line of collimation parallel to the axis of the bubble tube and to make the vertical axis perpendicular to the axis of the bubble tube.

Example 12.1 A dumpy level was kept midway between two points A and B 100 m apart. The readings on staves held at A and B were 2.340 m and 1.795 m. The instrument was then kept at C, 20 m from A along BA produced. The readings on the staves held at A and B were 1.985 and 1.435, respectively. Calculate the correct readings at A and B with the instrument at C. Is the line of collimation inclined upward or downward?

Solution When the instrument is kept midway between the two points, we get the correct difference in elevation irrespective of the inclination of the line of sight. True difference in elevation between A and B = 2.340 − 1.795 = 0.545 m. Apparent difference in elevation when the instrument is at C = 1.985 − 1.435 = 0.55 m. Reading at A = 1.435 m. Add the true difference: 1.435 + 0.545 = 1.980 m. As the observed reading is more than the correct reading, the line of collimation is inclined upwards.

 Correction for reading at the closer peg = 0.005 × 20/100 = 0.001 m

 Correction to reading at the farther peg = 0.005 × 120/100 = 0.006 m

 Correct reading at the closer peg = 1.435 − 0.001 = 1.434 m

 Correct reading at the farther peg = 1.985 − 0.006 = 1.979 m

 Difference in elevation = 1.979 − 1.434 = 0.545 m

Example 12.2 Two points A and B were selected 90 m apart for testing a dumpy level. The following observations were recorded.

Instrument location	Reading at A	Reading at B
Midway between A and B	1.865	1.925
At A	1.405	1.460

Find the correct reading at the peg B. Also state whether the line of collimation is inclined upwards or downwards.

Solution The true difference in elevation is obtained when the instrument is midway between the stations.

 True difference in elevation = 1.925 − 1.865 = 0.06 m

Apparent difference in elevation = 1.460 – 1.405 = 0.055 m

With the level at A

Reading at A = 1.405

Adding true difference,

Reading at B = 1.405 + 0.06 = 1.465 m

Actual reading at peg B = 1.460 < 1.465 m

The line of collimation is thus inclined downwards.

Correct reading at B = 1.465 m

Example 12.3 A dumpy level was tested using reciprocal levelling. The following observations were recorded.

Instrument location	Reading at P	Reading at Q
P	1.860	2.660
Q	1.215	1.815

Determine whether the line of collimation is inclined upwards or downwards. Find the angle of inclination if the distance between P and Q is 80 m. Also find the correct readings on the staves held at P and Q.

Solution The true difference in elevation is the mean of the apparent difference in elevation in the two cases.

Instrument at P: Difference in elevation = 2.660 – 1.860 = 0.80 m

Instrument at Q: Difference in elevation = 1.815 – 1.215 = 0.60 m

True difference in elevation = (0.8 + 0.6)/2 = 0.70 m

Instrument at P: Reading at P = 1.860

True difference = 0.70 m

Correct reading at Q = 1.860 + 0.70 = 2.560 m

Since the actual reading at Q is more, the line of collimation is inclined upwards.

Instrument at Q: reading at Q = 1.815

True difference = 0.70 m

Correct reading at P = 1.815 – 0.70 = 1.115 m

Difference in the reading = 1.215 – 1.115 = 0.1 m

in a distance of 80 m. Inclination of the line of collimation is given by

$$\tan \alpha = 0.1/80 = 0.0125$$

Therefore, $\alpha = 6' 52''$ is the inclination of the line of collimation.

Example 12.4 For testing a dumpy level, two points P and Q 100 m apart were selected. The instrument was kept midway between the two points and readings were taken on staves held at P and Q. The instrument was then shifted to a point 20 m from P on QP produced and again readings were taken. The following readings were recorded.

Instrument location	Reading at P	Reading at Q
Midway between P and Q	1.885	2.435
At R (20 m from P)	1.635	2.085

Find the correct readings at P and Q when the instrument was at R. Also find the inclination of the line of sight.

Solution True difference in elevation = 2.435 – 1.885 = 0.55 m

Apparent difference (instrument at R) = 2.085 – 1.635 = 0.45 m

Reading at P = 1.635

True difference = 0.55

Reading at Q = 1.635 + 0.55 = 2.185 m

The reading at Q is 2.085 and is less than the reading to be obtained. The line of collimation is inclined downwards.

Correction in 100 m = 0.1 m

Correction to reading at P = 0.1 × 20/100 = 0.02 m

Corrections to reading at Q = 0.1 × 120/100 = 0.12 m

Correct reading at P = 1.635 + 0.02 = 1.655 m

Correct reading at Q = 2.085 + 0.12 = 2.225 m

Difference in readings = 2.225 – 1.655 = 0.55 m

Inclination of the line of collimation is given by

$$\tan \alpha = 0.1/100 = 0.001$$
$$\alpha = 3' \, 26''$$

Example 12.5 In testing a dumpy level, the following observations were recorded.

Instrument location	Reading at A	Reading at B
Midway between A and B	2.615	3.175
At 25 m from A	1.905	2.340
(between A and B)		

Find the correct readings at A and B when the instrument is at C. Also find the inclination of the line of collimation.

Solution True difference in elevation = 3.175 – 2.615 = 0.56 m

Apparent difference in elevation = 2.340 – 1.905 = 0.435 m

Reading at A = 1.905

True difference = 0.56

Reading at B = 1.905 + 0.56 = 2.465 m

As the observed reading at B is less, the line of collimation is inclined downwards. If α is the inclination of the line of collimation, we have

Correct reading at A = 1.905 + 25 tan α

Correct reading at B = 2.340 + 75 tan α

Difference in elevation = 2.340 + 75 tan α – (1.905 + 25 tan α)

$$= 0.435 + 50 \tan \alpha$$

But the true difference is 0.56. Therefore,

$$0.435 + 50 \tan \alpha = 0.56 \quad \Rightarrow \quad \alpha = 8' \, 36''$$

Correct reading at A = 1.905 + 25 tan (8′ 36″) = 1.967 m

Correct reading at B = 2.340 + 75 tan (8′ 36″) = 2.527 m

Check: Difference = 2.527 – 1.967 = 0.56 m

12.5 Adjustments of Theodolite

The fundamental lines of a theodolite are the following.

1. Vertical axis
2. Horizontal axis or trunnion axis
3. Line of collimation or line of sight
4. Axis of plate level
5. Axis of altitude level
6. Axis of the striding level, if provided

The desired relationships among these lines are (see Fig. 12.5) the following:

(a) The axis of the plate level must lie in a plane perpendicular to the vertical axis.

(b) The line of collimation must be perpendicular to the horizontal axis. The line of collimation, the vertical axis, and the horizontal axis must intersect at a point.

(c) The horizontal axis must be perpendicular to the vertical axis.

(d) The axis of the altitude bubble must be parallel to the line of collimation.

(e) The vertical circle should read zero when the line of collimation is horizontal.

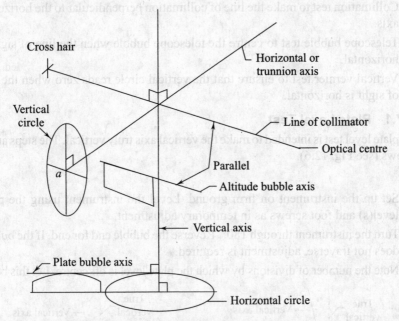

Fig. 12.5   Desired relationships between the fundamental lines of a theodolite

Each of these conditions (relationships) will result in a desirable performance of the instrument as follows. Condition 1 ensures that the vertical axis is truly vertical when the plate bubble is in the centre. Condition 2 ensures that when the telescope is moved about the horizontal axis or transited, it moves in a vertical plane. Condition 3 ensures that the vertical angle measured will be free from the error caused by axes not being parallel. Condition 4 ensures that the line of sight generates a vertical plane when the telescope is transited. Condition 5 ensures that vertical angles will be free from any index error due to the vernier.

12.6   Temporary Adjustments of Theodolite

The temporary adjustments of a theodolite are the same as that of a level. These have already been discussed in Chapter 4.

12.7   Permanent Adjustments of Theodolite

The permanent adjustments of a theodolite are done in a prescribed order so that one adjustment does not affect any other adjustment. The order in which the adjustments are to be done is the following.

1. Plate level test to make the plate level at the centre when the vertical axis is truly vertical.
2. Cross hair ring test to make the line of collimation coincide with the optical axis and also to ensure that the line of collimation generates a vertical plane when the telescope is transited.
3. Spire test to make the horizontal axis perpendicular to the vertical axis.
4. Collimation test to make the line of collimation perpendicular to the horizontal axis.
5. Telescope bubble test to centre the telescope bubble when the line of sight is horizontal.
6. Vertical vernier test to ensure that the vertical circle reads zero when the line of sight is horizontal.

12.7.1   Plate Level Test

The plate level test is intended to make the vertical axis truly vertical. The steps are as follows (see Fig. 12.6).

Test

1. Set up the instrument on firm ground. Level the instrument using the plate level(s) and foot screws as in temporary adjustment.
2. Turn the instrument through 180°. Reverse the bubble end for end. If the bubble does not traverse, adjustment is required.
3. Note the number of divisions by which the plate level is off centre. Let this be n.

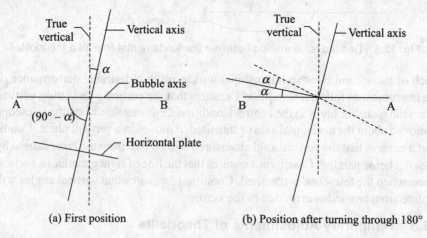

(a) First position   (b) Position after turning through 180°

Fig. 12.6   Plate level test

Adjustment

1. Shift the bubble towards the centre by $n/2$ divisions using the capstan-headed screws at the end of the bubble tube. Bring the bubble to the centre of its run using the foot screws.
2. Repeat the test and adjustment till the instrument is in adjustment.

Alternative method The altitude level on the telescope or on the T-frame is more sensitive than the plate level. This is used for adjustment. The procedure is as follows.

Test

1. Turn the telescope parallel to any two foot screws. Bring it to the centre of its run using the two screws. Turn the telescope through 90° and bring the level over the third foot screw. Bring it to the centre of its run. Repeat until the bubble remains at the centre in these two positions.
2. Turn the telescope through 180°. Reverse the bubble end for end. Note down the number of divisions by which the bubble is off centre. Let the bubble be off centre by n divisions.

Adjustment

1. Bring the bubble halfway ($n/2$ divisions) by means of the clip screw or the vertical circle tangent screw. Bring the bubble to the centre of its run (correct the remaining $n/2$ divisions) using the foot screws.
2. Bring the bubble over the third foot screw. Check whether the bubble traverses. If not, adjust with the third footscrew.
3. Repeat the test and adjustment till the altitude bubble traverses in all positions.
4. The vertical axis has now been made truly vertical. If the plate level is off centre, bring it to the centre by means of the capstan-headed screws. The bubble should now traverse in all positions.

12.7.2 Cross Hair Ring Adjustment

The cross hair ring adjustment test is done to make the line of collimation coincide with the optical axis of the telescope. The vertical and horizontal hairs are adjusted separately.

Adjustment of the horizontal hair

For the adjustment of the horizontal hair, proceed as follows [see Fig. 12.7(a)].

Test

1. Select two points P and Q about 100 m apart. Select a point R about 10 m from P in line with P and Q on the line PQ.
2. Set up the theodolite at P and level the instrument carefully.
3. Keep a levelling staff each at R and Q with face left and take readings. Let the readings be r_1 and q_1.
4. Transit the telescope and swing it through 180° to get the face right.
5. Set the horizontal hair to the earlier reading r_1 obtained on the staff at R.

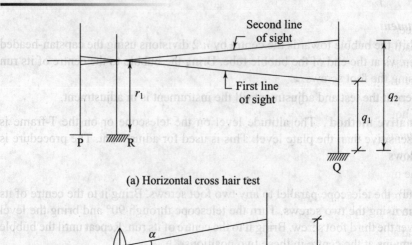

(a) Horizontal cross hair test

(b) Alternative horizontal cross hair text

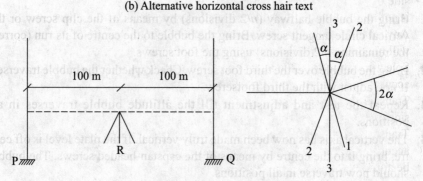

(c) Adjustment of vertical hair

Fig. 12.7 Cross hair ring test

6. Again take the reading on the staff at Q. If the reading is the same as that obtained earlier, q_1, the instrument is in adjustment as far as the horizontal hair is concerned. If not, let the reading be r_2. Adjustment is required for the hair.

Adjustment

1. With the top and bottom screws on the diaphragm, bring the reading to the mean reading $(r_1 + r_2)/2$.
2. Repeat the test till the instrument is in adjustment.

Alternative method   As an alternative, the horizontal angle can be noted while taking the reading. Proceed as follows [see Fig. 12.7(b)].

Test

1. Select two points P and Q about 100 m apart from each other. Set up and level the instrument at P.

2. Keep a staff vertically at Q and take the reading on the staff. Note down the reading of the vertical angle.
3. Transit the telescope and turn it through 180° to sight the staff again at Q with the alternative face. Set the reading on the vertical circle to the same angle. Read the staff. If the reading is the same as q_1, the instrument is in adjustment. If not, let the reading be q_2.

Adjustment

1. The adjustment is done using the diaphragm screws by setting the reading to the mean of the two readings, $(q_1 + q_2)/2$.
2. Repeat the test and adjustment until the instrument is in adjustment.

Adjustment of the vertical hair

Vertical adjustment is done to ensure that when the telescope is transited, it generates a vertical plane. If the adjustment is not done, it will generate a cone with its axis as the horizontal axis. We proceed as follows for this adjustment [see Fig. 12.7(c)].

Test

1. Select a point R such that about 100 m of level ground is available on either side of R. Set up and level the instrument at R. This is to ensure that the test is not affected by the difference in elevation between the points on either side. Any further adjustment of the horizontal axis, even if it is not strictly horizontal, will not affect the adjustment.
2. Select a point P about 100 m from R. Keep a ranging rod at R and bisect it accurately with the telescope. Keep the upper and lower motions clamped.
3. Transit the telescope. Keep a ranging rod at about 100 m from R on the other side of P in the line of sight. Place the ranging rod accurately in line with PR.
4. Transit the telescope again. Check whether P is in the line of sight. If P is sighted accurately, the instrument is in adjustment. Otherwise, the vertical hair has to be adjusted.

Adjustment

1. Place a ranging rod in the line of sight beside P. Let this position be P′. Measure PP′.
2. Using the diaphragm screws on the horizontal diameter, shift the vertical hair to a point P″ at one-fourth distance of PP′ from P.
3. Repeat the test till the instrument is in perfect adjustment.

It must be noted that the instrument is transited twice and the apparent error is four times the actual error.

12.7.3 Spire Test

In the spire test the horizontal axis is made perpendicular to the vertical axis when the instrument is levelled. This, together with the adjustment of the vertical hair, ensures that the line of collimation generates a vertical plane. Proceed with the test and adjustment as follows (see Fig. 12.8).

Test

1. Set up the instrument at 10–20 m
    from a building or any structure
    that has a clearly visible top such
    as a flag pole, lightning conductor,
    or a church or temple spire.

2. Set up and level the instrument
    at this point. With face left, sight
    the high point. Clamp the motions
    and depress the telescope. Mark a
    point on the ground or on the wall.
    It may be better to use a levelling

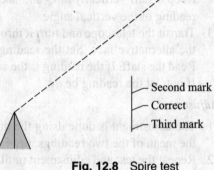

— Second mark
— Correct
— Third mark

Fig. 12.8 Spire test

staff placed horizontally below the high point. In such a case, note the reading
on the staff.

3. Transit the telescope, turn it through 180°, and again sight the point on the top.
    Clamp both motions and depress the telescope. Note the reading on the staff. If
    it is the same as the previous one, the instrument is in adjustment; otherwise,
    adjustment is required.

Adjustment

1. If points are marked, measure the distance between the points on the wall or
    ground. Find the point midway between the two marks. If a staff is used, take
    the mean of the two readings to get the point midway between the two readings.
    Sight this point with the telescope. Clamp both motions.

2. Raise the telescope and check whether the high point is sighted. It will not
    bisect the point.

3. With the horizontal axis adjustment screw on the standards, adjust the horizontal
    axis so that the point is sighted.

4. Repeat the test and adjustment till the instrument is in adjustment.

12.7.4 Collimation Test

The collimation test ensures that the line of sight coincides with the optical axis
of the telescope. The test procedure is the same as the two-peg test described for
a level [see Fig. 12.4(a)].

Alternative method The procedure is the same as in the two-peg test except
for the following.

1. With the vertical circle clamped, bring the bubble to the centre of its run using
    the clip screw.

2. In adjusting the reading on Q to the correct reading, use the clip screw. This
    eliminates any index error in the vertical circle and vernier.

12.7.5 Altitude Bubble Test

The objective of the altitude bubble test is to ensure that the vertical circle reads
zero when the telescope bubble is centred. If the telescope bubble is in the centre,
but the vertical circle does not read zero, then there is an index error. Corrections
can be applied to the vertical angles measured. However, the test and adjustment

are done to eliminate the index error so that confusion of signs in applying the corrections is avoided. Proceed as follows.

1. Set up the instrument on firm ground. Level it accurately with reference to the plate bubble.
2. Bring the altitude bubble to the centre of its run with the help of the vertical circle tangent screw.
3. Read the vertical circle. If it reads zero, the instrument is in adjustment.
4. If the vertical circle does not read zero, note down the reading. This is the index error, which should be added to or subtracted from the vertical angles measured.
5. To make the vertical circle read zero, the screws attaching the vernier arc to the standards are loosened and adjusted to make the vertical circle read zero.

12.8 Importance of Adjustments

In this section, we will discuss the important adjusments required for levels and theodolites.

12.8.1 Adjustments of the Level

In the case of levels, the most important adjustment is the two-peg test to make the line of collimation parallel to the bubble tube axis. The principle of spirit levelling is based on this condition. This adjustment should be carried out at intervals for accurate work.

12.8.2 Adjustments of the Theodolite

The error caused by many of the required conditions of the fundamental lines not being met can be eliminated by face-left and face-right observations.

The plate level adjustment is very important for all the operations of the theodolite. This adjustment should be done frequently. If the vertical axis is not truly vertical, the error cannot be eliminated by face-left and face-right observations. The plate level adjustment ensures that the horizontal circle and trunnion axis are truly horizontal. The measurement of horizontal angles is not affected except when using the method of repetition.

The second adjustment is that of cross hairs. The adjustment of the horizontal hair is important for measuring vertical angles and also when the theodolite is used as a level. If the horizontal hair does not lie on the optical axis of the telescope, its position will change during focusing. This, however, does not affect the measurement of horizontal angles. The adjustment of the vertical hair is important in many operations such as prolonging a line by transiting, measuring horizontal angles, and measuring with the telescope in inclined positions. The errors caused can be eliminated by face-left and face-right observations.

The third adjustment to make the horizontal axis perpendicular to the vertical axis becomes important in all operations involving the motion of the telescope in a vertical plane. The telescope should generate a vertical plane during such motion, which is ensured by this adjustment.

The fourth adjustment, of the altitude level, is important for accurate measurement of vertical angles and for using the theodolite as a level.

The index error adjustment of the vertical circle and vernier is done for convenience only. It is possible to apply corrections to the measured angles after noting down the index error.

Summary

Levels and theodolites are important instruments used extensively in surveying. Over a period of time, surveying instruments may not give accurate results. This is due to the desired relationships between the fundamental lines of the instruments getting altered because of mechanical wear and tear of the instrument with time. Regular adjustments should be done to ensure the existence of such desired relationships, the frequency depending upon the extent of use.

The principle of reversion states that the apparent error is double the actual error due to reversion of the axis or line. The vertical axis, the bubble tube axis, and the line of collimation are the three fundamental lines of a level. Two important conditions are necessary for levelling to be accurate, namely the line of collimation must be parallel to the bubble tube axis and the bubble tube axis must be perpendicular to the vertical axis. These two adjustments are very important for a level.

There are five fundamental lines in a theodolite: the plate bubble tube axis, altitude bubble tube axis, vertical axis, horizontal or trunnion axis, and the line of collimation. Sometimes a striding level is provided, in which case the axis of this level tube is also included. The vertical axis, horizontal axis, and the line of collimation must be perpendicular to each other at a point. The two bubble tube axes must be parallel to each other and perpendicular to the vertical axis.

The five adjustments to a theodolite are done in an order so that one adjustment does not affect any other adjustment made. The order of adjustment is as follows: plate level test, cross hair ring test, collimation test, spire test, and the altitude bubble test.

Exercises

Multiple-Choice Questions

1. The principle of reversion used in adjusting survey instruments states that–if there is error in a line with respect to another line, the error
   (a) remains the same when this line is reversed in position
   (b) gets doubled when this line is reversed in position
   (c) becomes zero when the position of the line is reversed
   (d) gets magnified four times when the position of this line is reversed
2. The three fundamental lines of a level are
   (a) the line of collimation and the lines joining the footscrews
   (b) the line of collimation, axis of the bubble tube and the horizontal axis of the instrument
   (c) the line of collimation, axis of the bubble tube and the vertical axis through the centre of the instrument
   (d) the line of sight, line of collimation and the axis of the bubble tube
3. The two-peg test is used to ensure that
   (a) the axis of the bubble tube is perpendicular to the vertical axis
   (b) the line of collimation is parallel to the bubble tube axis
   (c) the horizontal cross hairs is in a plane perpendicular to the vertical axis
   (d) the horizontal cross hair is parallel to the bubble tube axis
4. The principle of adjustment of the two-peg test is
   (a) the principle of reversion

(b) that the true difference in level is obtained with an inclined line of collimation when the level is kept exactly midway between the two points
(c) that when the level is kept closer to the staff it is read through the objective lens
(d) based on similarity of triangles

5. The horizontal axis of the theodolite is the line
   (a) tangential to the plate bubble
   (b) tangential to the altitude level
   (c) perpendicular to and passing through the centre of the vertical circle
   (d) the line joining the intersection of cross hairs and centre of objective lens

6. Plate level test of a theodolite is done to ensure that
   (a) horizontal axis is truly horizontal
   (b) the vertical axis is truly vertical
   (c) the plate bubble axis is horizontal
   (d) the horizontal circle is truly horizontal

7. The spire test for permanent adjustment of a theodolite is done to ensure that
   (a) the line of sight is perpendicular to the vertical axis
   (b) the two axes, of plate level and altitude level, are parallel
   (c) the vertical circle lies in a vertical plane
   (d) the horizontal axis is perpendicular to the vertical axis

8. The error that cannot be eliminated by face left and face right observations is when the
   (a) vertical axis is not truly vertical
   (b) line of collimation and trunnion axes are not perpendicular
   (c) axes of the bubble tubes are not parallel
   (d) line of collimation and axis of plate level are not parallel

Review Questions

1. State and explain the principle of reversion.
2. List the fundamental lines of a level and give the desired relationships among them.
3. Describe the permanent adjustment of a level to make the bubble tube axis perpendicular to the vertical axis. State what is achieved by this adjustment.
4. Describe the permanent adjustment of a level to make the line of collimation parallel to the bubble tube axis. What is achieved by this adjustment?
5. List the fundamental lines of a theodolite and give the desired relationship among them. State also what is achieved by each of the conditions of adjustment.
6. List the order in which the permanent adjustments of a theodolite is to be done. Why is this order necessary?
7. Describe the permanent adjustment of a theodolite to make the vertical axis truly vertical.
8. Describe the spire test for the permanent adjustment of a theodolite. Why is this test needed?
9. Describe the permanent adjustment of a theodolite to ensure that the vertical circle reads zero when the line of sight is horizontal.
10. Describe the permanent adjustment of a theodolite to make the line of collimation perpendicular to the horizontal axis.

Problems

1. In testing a level, the following observations are recorded:

Level at	Reading at P	Reading at Q
Midway	1.845	2.645
At P	1.375	2.060

Determine whether or not the level is in adjustment. Find the inclination of the line of collimation and the correct reading at Q when the level is at P. The distance between P and Q is 100 m.

2. In testing a dumpy level, the instrument is kept midway between two stations A and B and also at a point 20 m away from A in the line BA produced. The following is the record of observations made.

Level at	Reading at A	Reading at B
Midway	2.315	3.170
20 m from A	1.935	2.690

Determine whether the instrument is in adjustment. Find the inclination of the line of collimation and the correct reading at A and B when the instrument is at the second station. The distance between A and B is 80 m.

3. Reciprocal levelling is used in testing a dumpy level. The following observations are recorded.

Level at	Reading at P	Reading at Q
P	1.315	2.25
Q	1.280	2.20

Determine whether the instrument is in adjustment. Also, determine the inclination of the line of collimation and find the correct readings at P and Q when the instrument is kept at the other station. The distance between the stations is 50 m.

4. In testing a level, the following observations are recorded.

Instrument at	Reading at P	Reading at Q
Midway	2.315	3.345
10 m from P	1.355	2.355

The second station is 10 m from P on QP produced. Determine whether the instrument is in adjustment. Find the inclination of the line of collimation and the correct reading at P and Q.

5. In testing a dumpy level, the instrument is kept midway between two stations P and Q and at a point 20 m from P between P and Q, the distance between P and Q being 90 m. The following observations are recorded.

Instrument at	Reading at P	Reading at Q
Midway	1.985	2.830
20 m from P	1.345	2.215

Determine whether the level is in adjustment. Find the inclination of the line of collimation and the correct readings at P and Q.

Setting Out Works

Learning Objectives

After going through this chapter, the reader will be able to

- explain the purpose and importance of setting out works
- explain the methods used to achieve horizontal and vertical control
- explain the method, materials required, procedure, and checks for setting out a building
- explain the method used to set out a culvert
- explain the methods used to find the length of a bridge and locate its piers
- explain the methods used to transfer the alignment and levels from surface to underground in tunnelling
- explain the methods, tools required, and the procedure to set a sewer or pipeline to a given gradient

---

Introduction

In the preceding chapters, we studied several methods of surveying. Some other methods, such as hydrographical surveying and aerial surveying, will be discussed in the subsequent chapters. In survey work, distances, angles, and elevations of points are obtained. Using such data, plans and maps are prepared as per requirement.

Based on the plans and maps, various structures are designed. These structures can be small, such as simple buildings or culverts and bridges, or big projects such as dams. These structures have to be translated from the plan to the ground in their exact positions in all three dimensions. Setting out works is the important process of accomplishing this accurately.

The instruments and methods of surveying are used to accomplish this task. In the methods of surveying studied so far, we have discussed the survey of available ground features by conducting different types of surveys such as traversing, levelling and contouring, and the associated computations; the problem of setting out is slightly different. As an example, suppose we locate in the plan of an area a building to be constructed. The location drawing will indicate the location of the building, in terms of coordinates or other such parameters, with reference to some permanent ground features. The task now is to locate the building on the ground accurately using horizontal and vertical distances and angles.

In terms of instruments and methods, the procedure is the same. The only difference is that now we locate the specific points available in the plan on ground. Care should be taken in this regard, as accurate setting out is very important to avoid problems later on. In this chapter, we will discuss some relevant methods and techniques.

13.1 Prerequisites

The new structure is located with reference to some permanent features. This is an important aspect and needs some detailed consideration. The following points need to be kept in mind.

Horizontal control

As mentioned in Chapter 1, horizontal control means establishing some control points accurately for reference purposes. Depending upon the importance and extent of the project, different types of control points are established.

Control points

Triangulation (see Chapter 17) is a method of accurately locating control points (Fig. 13.1) that are far apart. Primary triangulation locates points large distances apart, to which secondary triangulation points are related. Secondary triangulation points can further be used to locate points by tertiary triangulation. The basic method remains the same, only the type of triangulation differs depending upon the methods used and the accuracy specified. Triangulation thus provides us with a large number of control points, which can be used for referencing our work. Such triangulation points are generally used in the case of large projects.

From such triangulation points, we estab-

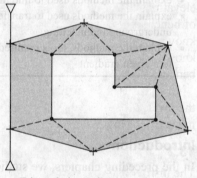

△ Triangulation points
+ Local control points
• Points of structure

Fig. 13.1   Control points

lish local control points by traversing or other means. Such points should be close to the structure and permanently marked. Care should be taken to ensure that they are located close to the structure, but sufficiently apart so as not to be disturbed during construction.

It is also possible to use a base reference line connecting the control points. If one or two such lines are laid out, it is possible to set out the structure using these lines. The points of the structure are located with the help of the base lines by using different methods (Fig. 13.2).

The points near the structure have to be permanently marked by some means—using pegs and nails, masonry, or a concrete structure with an embedded metal bolt. The location of the station should be clearly marked.

In the case of small works, permanent structures such as buildings, if located in the plan, can be used as control points. The new structure is laid out using distances and angles from the points on such structures.

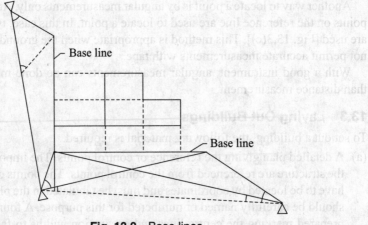

Fig. 13.2 Base lines

Vertical control

As in the case of horizontal control, it is necessary to have reference points of known elevation for setting out the structure in the vertical direction. A sufficient number of benchmarks is established near the structure for this purpose. A benchmark of known elevation is obtained near the structure by fly levelling (Chapter 7) from a known benchmark some distance away. In the case of small structures, arbitrary benchmarks may be used on some permanent point of an existing structure.

In all cases, the benchmarks are located very accurately and checked by closing the levels. These benchmarks should be established near the structure but located in such a way that they are not disturbed by the construction activity.

13.2 Instruments and Methods

The instruments and methods used for setting out works are same as for other surveys. Levels, theodolites, and tape are used to locate points. A point can be located with respect to a reference point or line by different means (Fig. 13.3).

A point can be located by two distances from two different points on a reference line as in Fig. 13.3(a). A reference line is a line joining two control points. A tape is usually enough to locate points by this method. This method can be used depending upon whether or not the ground situation facilitates accurate tape measurements.

A point can be located by its polar coordinates, i.e., one distance and the angle from the reference line. For this purpose, at least one theodolite and tape are required [Fig. 13.3(b)].

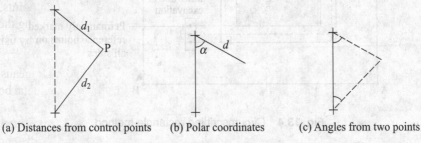

(a) Distances from control points   (b) Polar coordinates   (c) Angles from two points

Fig. 13.3 Locating points

Another way to locate a point is by angular measurements only. Angles from two points on the reference line are used to locate a point. In this case, two theodolites are used [Fig. 13.3(c)]. This method is appropriate when the ground conditions do not permit accurate measurements with tape.

With a good instrument, angular measurements can be done more accurately than distance measurement.

13.3 Laying Out Buildings

To set out a building, the following material is required.

(a) A detailed plan giving the reference or control points. The important points of the structure are referenced from the control points. The points of the structure have to be located by coordinates and must be tabulated in the plan. Each point should be carefully named or numbered for this purpose. A foundation plan is prepared marking the centre lines and excavation widths to facilitate setting out on ground.

(b) Depending upon the method used, the instruments are selected. For important buildings, a theodolite and tape are invariably used. One will also need a cord to stretch between the points to mark a straight line between them. Stakes, nails, hammer, and marking powder, such as lime, are the other items required.

Consider the layout of a rectangle shown in Fig. 13.4. One method is to reference the layout with a circumscribing rectangle. The procedure for the same is as follows:

1. Establish control points A and B near the structure but at sufficiently large distances such that they are not disturbed during the excavation. Erect perpendiculars at A and B to get points C and D of sufficient length so that they are away from the excavation limits. This can be done with a tape alone using the 3-4-5 principle, or more accurately using a prismatic compass or theodolite.

2. Check the rectangle set out by measuring the diagonals AC and BD, which should be equal to their calculated lengths. Correct any error before proceeding further.

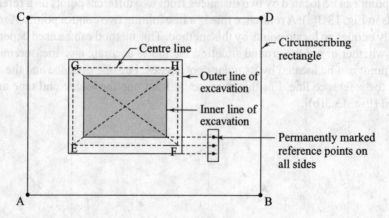

Fig. 13.4 Circumscribing rectangle method

3. The centre lines of the four walls of the building form the rectangle EFGH. Locate the four corners from the respective corners of ABCD. Check the lengths of the sides and the diagonals of EFGH. Set temporary stakes at points E, F, G, and H and mark these points with a nail.
4. Tie a cord between the nails and mark the centre line with dry lime powder.
5. Mark the inner and outer boundaries of the foundation widths in the same way. Again mark the four corners of these rectangles with temporary stakes. Note that these stakes will go away once the excavation starts.

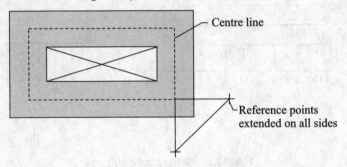

Fig. 13.5 Centre line method

Alternately, the centre lines of the four walls of the building themselves (Fig. 13.5) can be used as a reference rectangle. The corners of the centre line are accurately set out with respect to the control points near the site. The sides and diagonals are measured as a check. Temporary stakes are put as in the previous case. The inner and outer rectangles are then marked with respect to this rectangle. The stakes set up in this case also will be lost during excavation. It is therefore necessary to have these reference points transferred to outside locations for future work. This is accomplished by extending the centre lines to points outside the excavation limits. The minimum distance required is 2–4 m [Fig. 13.6(a)]. In many instances all the three lines are extended and permanent stakes are put for future reference.

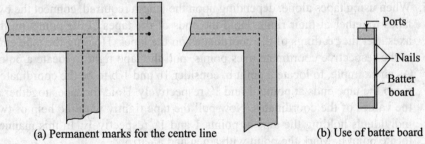

(a) Permanent marks for the centre line   (b) Use of batter board

Fig. 13.6 Extending the centre line

For larger buildings, batter boards are used along with stakes for setting out. Batter boards are wooden pieces of cross-sectional area 25 mm × 100 mm. These are nailed to the wooden stakes driven into the ground. The top of the batter boards carry nails, which can be used for extending the lines using string [Fig. 13.6(b)].

13.4 Setting Out Culverts

Culvert foundations, abutments, and wing walls can be set by using the coordinates of a number of points on the lines defining them. For this purpose, the origin is selected at the intersection of the centre lines of a waterway or a road or railway line passing over it [Fig. 13.7(a)]. The following procedure is adopted.

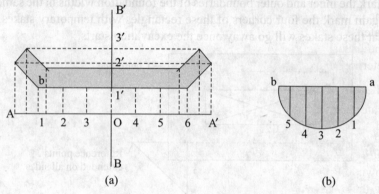

(a)   (b)

Fig. 13.7 Setting out a culvert

1. From the foundation plan of the culvert and the roadway, Fig. 13.7 locate the two centre lines AOA′ and BOB′, O being the origin. Locate these centre lines, from the control points available, near the site by coordinates. Check and verify that these two lines are at right angles.
2. Drive a peg at O and mark it carefully. Set up a theodolite at O, centre and level it.
3. Set up a number of points along both the lines. Assume that 1, 2, 3, 4, etc. are the points along AOA′ and 1′, 2′, 3′ 4′, etc. are the points along BOB′. Mark the points with pegs and arrows such that a cord tied along the arrows defines the lines and the points marked on them.
4. Set the points on the foundation lines using coordinates with the help of a tape or by using two theodolites placed at the ends of the lines.
5. When using tapes alone, depending upon the length required, connect the two tapes together at their rings. Hold the ends of the tapes at the points on the axes. Set the readings of the coordinates on the tapes. Holding the tape ends at the respective coordinate axes points, pull the tape tight to locate a point. As an example, to locate a point b, consider 1b and 1′b to be the coordinates. Keep the tape ends at points 1 and 1′, respectively. Hold the tapes together at the values of the coordinates. Now pull the tape tightly with the help of two individuals holding the tape at points 1 and 1′, respectively. In this manner, locate point b. Mark the point with a peg and an arrow.
6. As an alternative, calculate polar coordinates from O. Set the angle at the theodolite kept at O and measure the distance r with a tape to locate b.
7. Once all the points a, b, c, d, etc, are located, tie a string around the points located. Make a mark along the string with dry lime powder or make a line by nicking.
8. In case any of the walls have a curved outline, locate the end points first. Then locate the curved boundary by using coordinates from the chord of the curve [Fig. 13.7(b)].

13.5 Setting Out Bridges

In the case of bridges, the problem is different from that of culverts. Bridges generally have more than one span. Most bridges will be across perennial rivers or streams. The centre line method used in the case of culverts cannot be adopted for bridges. The difficulty in the case of bridges is twofold as given below.

(a) The centre line length of the bridge between two points on either side of the waterway has to be determined accurately.

(b) The locations of piers that lie between these points on the banks have to be determined.

13.5.1 Locating the Centre Line

To locate the centre line length of the bridge, two methods are commonly employed.

Triangulation

As shown in Fig. 13.8(a), A and B are two points on a banks of a river and the line joining them forms the centre line of the bridge. In order to find the length AB, proceed as follows.

1. Set out a line AC perpendicular to AB at A using a theodolite set up at A.
2. Measure a convenient length along AC and mark point C.
3. Shift the instrument to C, and centre and level it. Measure the angle ACB. This can be done by repetition. Repeat three times each with face left and face right. Find the angle ACB accurately from this data.
4. Length AB = AC tanα, where $\alpha = \angle$ACB.
5. As a check repeat the procedure at B. Set out a line BD perpendicular to AB at B. Measure a convenient length and mark point D.
6. Shift the instrument to D. Measure the angle BDA by repetition as before. Accurately determine the angle BDA.
7. Length AB = BD tanβ, where $\beta = \angle$BDA.
8. Compare the two values of AB obtained from the two measurements. If they match and are within the limits of the precision set, then find the length as the average of the two values obtained. Otherwise, repeat the procedure until the values obtained are within the limits of precision.

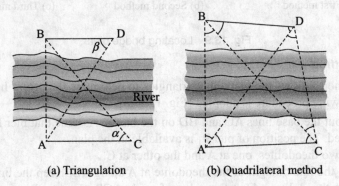

(a) Triangulation   (b) Quadrilateral method

Fig. 13.8 Centre line length of the bridge

Quadrilateral method

In this method, instead of one triangle, two triangles forming a quadrilateral are used. The procedure is as follows [Fig. 13.8(b)].

1. AB is the centre line of the bridge whose length AB has to be determined.
2. Set out a base line AC approximately at right angles to the line AB. Measure the length of the base line AC accurately.
3. Set out another line BD approximately at right angles to AB. Measure the length BD accurately.
4. Measure the angles in the quadrilateral ABCD accurately. There are two angles at each station, making a total of eight angles. Thus, measure angles BAC and DAC at A, angles ACB and ACD at C, angles ABC and CBD at B, and angles BDA and CDA at D.
5. Assuming one of the base line lengths as correct, calculate the length of the other base line. As an example, using the measured value of AC, calculate the length of BD. The sine rule can be applied to calculate this length.
6. If the calculated length matches the measured length of BD within the specified limits of precision, then calculate the length of AB using these values. The limit of precision in the case of large bridges is not less than 1 in 5000.
7. If not, repeat the procedure until the values match.

13.5.2 Locating Bridge Piers

Once the centre line length is known, the centre points of the bridge piers are located on this line. There are many methods to achieve this.

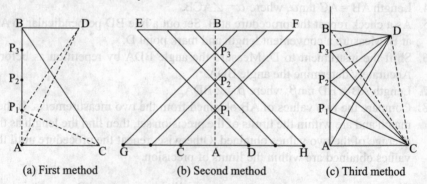

| (a) First method | (b) Second method | (c) Third method |

Fig. 13.9 Locating bridge piers

First method

This method follows the method of triangles to determine the centre line length. The following procedure is adopted [Fig. 13.9(a)].

1. Lay out the base lines AC and BD on the banks. Assume that pier P_1 is to be located. The position of pier P_1 is available in the plan.
2. Use two theodolites, one at A and the other at C.
3. Keep the line of sight of the theodolite at A along AB. Keep the line of sight of the theodolite at C at the value of angle ACP_1. Locate the pier P_1 at the intersection of the line of sights.

4. Calculate the angle ACP_1 from the triangle using the length of the base line AC and the length of AP_1.
5. Check the location of pier P_1 using the above procedure from stations B and D. Calculate the angle BDP_1 from the length of the base line BD and the length of BP_1.
6. Keep the line of sight of the theodolite at B along BA. Keep the line of sight of the theodolite at D such that angle BDP_1 is the calculated value of this angle.
7. The intersection of the lines of sight from B and D locates P_1. If the location of P_1 matches the earlier location from A and C within the precision limits, consider the location correct. Otherwise, repeat the procedure until it is found correct.
8. Locate and check the other piers in a similar way.

Second method

In this method, the position in the plan of the piers is obtained from the design marked in the plan. The distances AP_1, AP_2, etc. are thus known. The following procedure is adopted [Fig. 13.9(b)].

1. Lay out two lines, perpendicular to the base line AB, at A and B accurately.
2. To locate the centre of the pier P_1, knowing the length AP_1, measure it accurately on both the sides of A to get points C and D, respectively. $AC = AD = AP_1$. As line CAD is at right angles to A, both angles ACP_1 and ADP_1 are equal to $45°$.
3. Similarly at B, lines BE and BF are at right angles to AB and equal in length to BP_1. Both angles BEP_1 and BFP_1 are equal to $45°$.
4. It is clear from the figure that triangles EFP_1 and CDP_1 are similar. If we keep a theodolite at D and sight the signal at E, the centre of pier P_1 is along this line. If we keep a theodolite at C and sight the signal at F, the line of sight will again pass through P_1.
5. Thus, locate the position of pier P_1 at the intersection of the two lines DE and CF. Locate the other piers in a similar way.
6. Since the piers are located by the intersection of lines of sight by keeping the instruments at one bank and the signal at the other bank, this method has an advantage of giving accurate results if the right angles can be set precisely.

Third method

In the third method [Fig. 13.9(c)], the quadrilateral used for determining the centre line length is used. The following procedure is adopted.

1. Measure the base lines AC and BD accurately and check their lengths by calculation. Take these two lines as the base lines.
2. For pier P_1, calculate the angles ACP_1 and BDP_1. CAP_1 and DBP_1 are right angles.
3. Find the location of pier P_1 by intersection. Keep theodolites at A and C. Set the theodolite at A to have its line of sight along AB. Keep the theodolite at C to measure angle ACP_1 while keeping the line of sight along CP_1. Pier P_1 will be at the intersection of the two lines of sight.

4. To check this location further, shift the transits to B and D and locate the pier P_1 at the intersection of lines of sight from B towards AB and angle BDP_1 giving the line of sight DP_1. The intersection of the two lines of sight should be the same as the point P_1 obtained earlier for the work to be accurate. Otherwise repeat the procedure until the points are correctly located.
5. Locate the other piers in the same way.

13.6 Setting Out Tunnels

Tunnel surveying is discussed in Chapter 20. Tunnel surveying consists of two surveys—surface survey and underground survey. Surface survey is done in the usual way. As the terrain near the tunnel is likely to be mountainous and difficult, careful surveying is required to get the proper alignment of the tunnel. The alignment is marked on the surface. In setting out tunnels, a major problem is the transfer of surface alignments and levels to the underground tunnel base. The alignments and levels have to be transferred to points several metres below the surface.

13.6.1 Transferring Alignment

To transfer the surface alignment to a point inside the tunnel the following procedure is adopted [Fig. 13.10(a)].

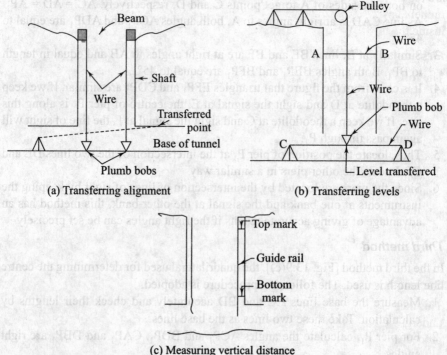

(a) Transferring alignment   (b) Transferring levels

(c) Measuring vertical distance

Fig. 13.10 Setting out tunnels

1. Make a vertical shaft from the surface to the tunnel. On the top of this shaft, lay two wooden beams at right angles to the alignment of the shaft.

2. Along the alignment available on the surface, mark points A and B on the wooden beams accurately. Check the alignment from both directions and mark the points. A low error margin is permissible in this work as any shift in the alignment can take the tunnel very far from the original direction.
3. Hang two heavy plumb bobs using steel wires from points A and B marked on the wooden beams. Use heavy plumb bobs weighing 5–10 kg so that they hang vertically.
4. You may also keep the plumb bobs in oil or water to keep them from swaying due to minor air flow.
5. Now shift a theodolite to the bottom of the shaft. Align the line of sight of the theodolite with the two wires after a number of trials.
6. Once the alignment is available, mark the points along this direction on the roof and the ground with permanent markers.

13.6.2 Transferring Benchmarks

Surface levelling is done in the usual way. Vertical control points are first established near the site. Local benchmarks are established near the shafts for transferring the levels underground. These levels can be transferred easily from the surface to the underground locations near the tunnel entry and exit. However, special procedures are required to transfer the levels near the shafts towards the inner portions of the tunnel. Two methods are used to transfer the levels or benchmarks to the points inside the tunnel from the surface.

First method

The following procedure is adopted [Fig. 13.10(b)] for the first method.
1. Mark two horizontal lines AB and CD, one at the top of the shaft and the other at the base of the tunnel, using fine wires tied between points marked on the shaft walls and heavy tripods set on the ground.
2. From the surface, hang a fine wire with a heavy plumb bob with the help of pulleys. Make the plumb bob touch the station mark on the ground. Adjust this wire so that it is in touch with the two horizontal wires.
3. Mark this hung wire at the level of the two horizontal wires accurately. Stretch the hung wire fully and keep the horizontal wires taut.
4. Now pull out the wire and measure the length between the marks.
5. Using this length and the other measured lengths, calculate the reduced level of the bottom of the plumb bob.

Second method

This method can be used when the shaft has guide rails and a cage can be transferred underground using it. The vertical distance is measured using a tape. The following procedure is adopted [Fig. 13.10(c)].
1. Make a mark at the top of one of the guides. Determine the reduced level of the position of this mark accurately by levelling from one of the benchmarks near the shaft.
2. With an assistant holding the zero end of the tape on the mark, descend with the tape until the end of the tape is reached. Hold the tape taut and mark the end of the tape on the guide rail accurately.

3. Repeat this procedure until the base of the tunnel is reached using temporary platforms, which are made in the shaft at suitable intervals.
4. Calculate the level of a point at the base of the tunnel or a mark on the guide rail using this procedure.

In tunnelling work, high precision needs to be maintained in transferring the alignment and levels from the surface to underground.

13.7 Setting out Sewer Lines

Underground pipelines and sewer pipes have to be laid to a particular gradient designed for the ease of flow of water or sewage. The inside bottom of the sewer pipe is known as invert. Some special tools (Fig. 13.11) are required to set the line in a gradient.

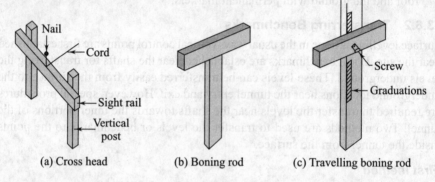

(a) Cross head   (b) Boning rod   (c) Travelling boning rod

Fig. 13.11 Tools

Cross head It consists of two posts and a horizontal bar. The posts are of suitable heavy cross section and the horizontal member, known as the sight rail, has an area of 50 mm × 150 mm. Cross heads are kept at the top and bottom of the gradient section and at 20–30 m intervals. A cross head is placed at every change of gradient. The top of the sight rail is kept at a convenient height, 2–5 m.

Boning rod It resembles a cross. A vertical wooden piece measuring 100 mm × 30 mm is used. The horizontal member is of the same area, 400 mm long, and is placed at a convenient height depending upon the setting of the sight rail. The top of the boning rod is a reference for determining the depth of excavation in a section. One boning rod is required for a particular gradient. Different boning rods are required for different gradients.

Travelling rod It is an adjustable boning rod. The horizontal piece can be moved over the vertical piece thus varying the height of the rod.

Setting out

To set out the sewer on the gradient, the following procedure is adopted.

1. Set the alignment of the line on the surface with reference to the control points brought near the site. Also establish the vertical control points for setting out the grade.
2. Lay a line parallel to the alignment for reference, as the first line will be lost once the excavation starts.

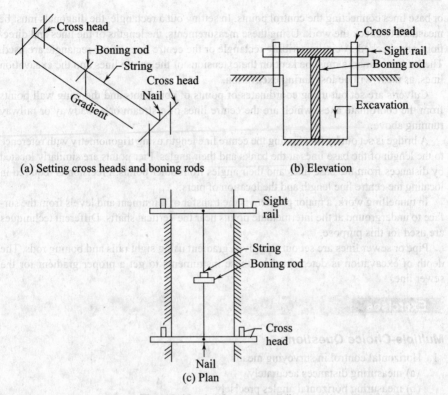

(a) Setting cross heads and boning rods   (b) Elevation

(c) Plan

Fig. 13.12 Laying a sewer line

3. Start work from the lower point. Place the sight rails at the bottom and top of the gradient line (Fig. 13.12). Drive nails at the centre of the cross piece of the cross heads. Stretch a string between the nails to act as a reference for vertical distances.
4. Drive the cross heads in such a way that the top of the sight rail is 2–5 m above the invert of the sewer. You may do this with a level and take the readings. Use a boning rod of the same length as the setting of the sight rail in the gradient section.
5. The string stretched between the nails on the sight rails should have the same gradient as that of the sewer line.
6. Check the proper depth of the excavation at any intermediate point using the boning rod.

Summary

Setting out works is an important task in surveying. The equipment and methods used are the same as that of surveying for ground features. The difference between setting out and surveying is that here we locate or transfer points from a plan to the ground.

In setting out works, control points for both horizontal alignment and vertical direction have to be established. These are determined near the site by surveying from the triangulation points. For a small structure, permanent features near the site can be used as control points.

In setting out buildings, rectangles formed by the centre lines of the four walls are set out. The corners of the rectangles are located on the ground by coordinates from control points or base lines connecting the control points. In setting out a rectangle, the diagonals must be measured to check the work. Using these measurements, the lengths of the lines and directions are checked. A circumscribing rectangle or the centre lines of the rectangle are used. The reference marks must be kept on the extensions of the centre lines and the excavation lines, as these will be lost during excavation.

Culverts are set out using coordinates of points of abutments and the wing wall points from the coordinate axes, which are the centre lines of a stream or a roadway or railway running above.

A bridge is set out by first finding the centre line length using trigonometry with reference to the length of the base lines at the banks and their angles. Pier points are similarly located by distances from the base lines and their angles from other points. Precision is required in locating the centre line length and the location of piers.

In tunnelling work, a major problem is the transfer of alignment and levels from the surface to underground at the intermediate points near the vertical shafts. Different techniques are used for this purpose.

Pipe or sewer lines are set out to a given gradient using sight rails and boning rods. The depth of excavation is determined by these instruments to get a proper gradient for the sewer line.

Exercises

Multiple-Choice Questions

1. Horizontal control in surveying means
    (a) measuring distances accurately
    (b) measuring horizontal angles precisely
    (c) having triangles as basic figures for surveying
    (d) locating some points accurately as reference points
2. Vertical control in surveying means
    (a) using precise levelling instruments
    (b) measuring vertical angles accurately
    (c) establishing accurately benchmarks
    (d) establishing contour lines
3. When setting out a rectangle using a tape, a check is done by measuring the
    (a) sides a second time using a different tape
    (b) distance from a control point
    (c) diagonals of the rectangle
    (d) angles using the 3-4-5 principle

Review Questions

1. Explain with an example, the importance of setting out works.
2. Explain the method of achieving horizontal and vertical control in setting out works.
3. Explain the basic procedure, instruments, and materials required to set out the foundation of a building on the ground as per plan.
4. Explain the procedure, instruments, and materials required to set out a culvert.
5. Explain the different methods used to determine the centre line length of a bridge.
6. Explain the different methods of locating piers while setting out a bridge.
7. Explain the procedure to transfer the alignment of a tunnel to an underground point.
8. Explain the methods used to transfer levels from surface to underground.
9. Using neat sketches, explain the purpose of a boning rod and sight rails.
10. Explain the procedure used to set out gradient stakes and setting out a sewer line on a given gradient.

14

Tacheometry

Learning Objectives

After going through this chapter, the reader will be able to

- define and explain the term tacheometry
- explain the basic principle of stadia tacheometry
- describe the different methods of stadia tacheometry
- explain the methods to find horizontal distances and elevations from stadia observations
- explain the subtense bar method
- explain the tangential method of tacheometry
- explain the advantages and disadvantages of tacheometry

Introduction

Tacheometry, also known as tachymetry or telemetry, is a branch of surveying in which both horizontal and vertical distances are measured without the use of a chain or tape. It is mostly used for contouring, in which horizontal distances and elevations are to be determined to give a complete relief map of the ground. It can also be used for checking measurements taken by a tape or other means. The method is suitable in rough terrains where chaining is difficult or impossible. In terms of precision, it is not very suitable. An accuracy of 1 in 1000 can be achieved with careful handling and reading of instruments, the normal range being 1 in 600 to 1 in 850. With the availability of electromagnetic distance-measuring instruments, conventional tacheometry may go out of use.

In this chapter we will discuss the instruments and methods of tacheometric surveying.

14.1 Basic Principle of Stadia Tacheometry

Tacheometry is based on the principle known as *stadia surveying*. The method is illustrated in Fig. 14.1. The term 'stadia' comes from the stadia diaphragm (of a theodolite) which has three horizontal cross hairs as shown in Fig. 14.1(a). The cross hairs may be arranged in several forms. Some of these are shown in Fig. 14.2(a). Most levels and theodolites have a stadia diaphragm. The readings on a staff are taken against all the three cross hairs in tacheometry. While the reading against the middle hair is used for finding differences in elevation, the readings against the top and bottom hairs are used to find horizontal distances. One method of tacheometry (the tangential method) does not use these hairs at all and reading from the middle hair is used to calculate the distance.

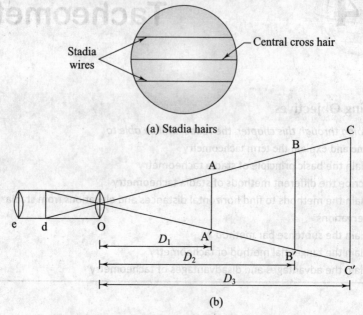

Fig. 14.1 Basic principle of stadia tacheometry

If, as shown in Fig. 14.1(b), e is the eyepiece, d is the diaphragm, and O is the objective lens of a theodolite, then it is evident that the staff readings at different distances from the instrument will be proportional to these distances. This is clear from the fact that the triangles shown in the figure are isosceles and similar. The distances D_1, D_2, D_3, etc. to the object from the instrument are proportional to the intercepts AA′, BB′, CC′, etc. In general, the distance from the instrument to the object can be proved to be equal to $Ks + C$, where K and C are constants. K is known as the multiplying constant and C is known as the additive constant. In the following sections, we will discuss the different methods of tacheometry used to find the horizontal distance of a point from the instrument station and the elevation of that point.

14.2 Tacheometric Methods

There are two basic methods of tacheometry: (i) Stadia method and (ii) tangential method. The stadia method is based on the principle outlined above. There are two forms of the stadia method.

1. Fixed hair stadia method In this method the distance between the stadia hairs is kept constant. The intercept varies as the distance varies.

2. Movable hair (subtense bar) method This method uses a special diaphragm that enables us to change the distance between the cross hairs. The intercept is kept constant even though the distance varies.

In the tangential method, vertical angles are measured from the central cross hair and distances are calculated using trigonometric formulae. The stadia hairs are not used in this method.

14.3 Instruments

The instrument used in tacheometry is a theodolite fitted with a stadia diaphragm or a tacheometer. A tacheometer is similar to a theodolite but has some special features. The following are some of the features necessary in a tacheometer.

(a) The telescope should be truly anallactic (making the additive constant zero).

(b) The telescope must have a magnification of about 30 diameters to read the staff accurately at large distances.

(c) The aperture of the objective should be 35 to 45 mm diameter to give a bright image.

(d) The multiplying constant should be 100 and the additive constant zero.

A theodolite also satisfies most of these requirements. A theodolite having an internal focusing telescope is very nearly anallactic with a very small additive constant. Old external focusing telescopes are fitted with an additional lens to make them anallactic.

14.3.1 Staff and Stadia Rod

For rough and small-scale work, a levelling staff can be used for measuring the intercept. For more accurate work, a stadia rod is used. A stadia rod is similar to a levelling staff but may be longer and more accurately and finely divided. Stadia rods should have bright, bold, and clear markings for ease of reading. Some of the marking forms are shown in Fig. 14.2(b).

Holding the staff

In the case of a horizontal line of sight, the staff is held vertical. In the case of an inclined line of sight, the staff may be held vertical or normal to the line of sight.

Holding the staff vertical   In ordinary work, the verticality of the staff can be judged visually. The staff must be held truly vertical in accurate work. For this purpose, the verticality can be checked by a suspended plumb bob. Often, for accurate work, the stadia rod may be provided with a circular level to check the verticality of the staff. Any deviation in verticality can result in serious error in the calculation of distances and elevations. Holding it vertical is easier than holding it perpendicular to an inclined line of sight. Though this involves a little more calculation, reduction of stadia notes to find distances and elevations can be done easily using stadia tables.

Holding the staff normal   Holding the staff normal to the line of sight is more difficult. Generally, either open sights with a cross hair or a telescope attached to the stadia rod is used to ensure that the rod is normal to the line of sight. While looking through the open sights or the telescope, the staff is tilted in both directions to obtain the position of bisection of the telescope. The reading is then taken at this position of the staff.

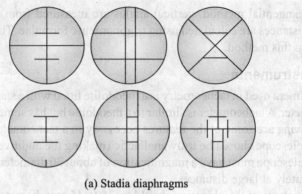

(a) Stadia diaphragms

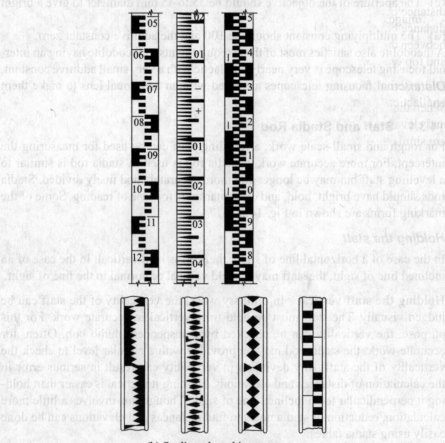

(b) Stadia rod marking patterns

Fig. 14.2   Stadia diaphragms and rods

14.4   Stadia Method

The stadia method, as mentioned earlier, is based on readings taken against the stadia cross hairs. The difference in the readings can be related to the horizontal distance. The two methods of stadia surveying are discussed in this section.

14.4.1 Fixed hair stadia method

In the fixed hair method, the stadia interval, i.e., the distance between stadia hairs, remains constant. There are two variations of this method depending upon the line of sight—horizontal or inclined.

The following symbols have been used here. O denotes the objective lens of the telescope and a-b the plane of cross hairs, 'a' being the top cross hair and 'b' the bottom one. The central cross hair is 'c'. i represents the stadia interval—the distance between the top and bottom cross hairs, f the focal length of the objective lens, f_1, f_2 the conjugate focal lengths—conjugate focal lengths are distances to the object and its image from the optical centre, v-v′ the vertical axis of the instrument, d the distance between the objective lens and the vertical axis, D the horizontal distance between the vertical axis and the staff, s the staff intercept, V the vertical distance between the optical centre and the target, α the parallactic angle—angle between the rays coming from the readings on the staff corresponding to the bottom and top cross hairs, and θ the inclination of the line of sight.

Distance and elevation formula for horizontal line of sight

When the line of sight is horizontal, as shown in Fig. 14.3, the horizontal distance and elevation can be determined as follows. In Fig. 14.3(a), triangles Oab and Oa′b′ are similar. This gives

$$s/i = f_1/f_2$$

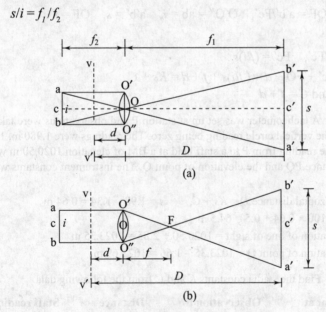

Fig. 14.3 Horizontal line of sight

From the lens formula,

$$\frac{1}{f} = \frac{1}{f_1} + \frac{1}{f_2}$$

Multiplying by f_1,

$$\frac{f_1}{f} = 1 + \frac{f_1}{f_2}$$

As $f_1/f_2 = s/i$, we get

$$\frac{f_1}{f} = 1 + \frac{s}{i}, \qquad f_1 = \frac{fi + sf}{i}$$

$$D = f_1 + d = (f/i)s + (f + d) = Ks + C$$

$K = f/i$ is the multiplying constant and $C = (f + d)$ is the additive constant. Horizontal distance $= Ks + C$.

If a reading is taken to a benchmark (BM) of known elevation, then the elevation of the line of sight can be determined and the reduced level (RL) of the staff station can be found.

Elevation of staff station P = RL of benchmark + $c_1 - c_2$

where c_1 and c_2 are the readings against the middle hair on the benchmark and the station P.

Alternative proof   An alternative proof can be given by considering Fig. 14.3(b). Here, rays of light parallel to the optical axis are considered. These rays pass through point F, the focal point of the lens. Triangles O′O″F and a′b′F are similar. Therefore,

$$O'O''/OF = a'b'/Fc', \quad O'O'' = ab = i, \quad a'b' = s, \quad OF = f$$

Substituting,

$$i/f = s/Fc', \quad Fc' = (f/i)s$$
$$D = Fc' + FO + d = (f/i)s + f + d = Ks + C$$

where $K = f/i$ and $C = f + d$.

Example 14.1   A tacheometer was set up at station P and observations were taken on a staff held at Q, the vertical circle reading being zero. The readings were 1.980 m, 1.660 m, and 1.340 m. The reading from P to a staff held at a BM of elevation 1020.50 m was 2.85 m. Find the distance PQ and the elevation of point Q. The instrument constants were 100 and 0.5.

Solution   Horizontal distance $D = Ks + C$, $s = 1.98 - 1.34 = 0.64$ m

$D = 100 \times 0.64 + 0.5 = 64.5$ m

Elevation of line of sight $= 1020.50 + 2.85 = 1023.35$ m

Elevation of point Q $= 1023.35 - 1.66 = 1021.69$ m

Example 14.2   Find the stadia constants K and C from the following data.

Instrument at	Observation to	Distance	Staff readings
P	Q	50 m	1.354, 1.603, 1.852
P	R	100 m	1.152, 1.65, 2.149

The line of sight was horizontal in both cases.

Solution   Two equations can be formulated with the known distances, and the constants can be determined. For the observation from P to Q,

$$50 = K(1.852 - 1.354) + C$$

For the observation from P to R,

$$100 = K(2.149 - 1.152) + C$$

Solving these equations,

$$K = 100 \text{ and } C = 0.3$$

Example 14.3 The readings on a staff held vertically 60 m from a tacheometer were 1.460 and 2.055. The line of sight was horizontal. The focal length of the objective lens was 24 cm and the distance from the objective lens to the vertical axis was 15 cm. Calculate the stadia interval.

Solution   Constant $C = f + d = 0.25 + 0.15 = 0.40$ m

Staff intercept $s = 2.055 - 1.46 = 0.0595$ m

Distance $D = 60$ m

$60 = 0.48K + 0.4$, $K = (60 - 0.4)/0.595 = 100.16$

$K = f/i$, $i = f/K = 250/100.16$ mm $= 2.5$ mm

Distance and elevation formula for inclined line of sight

In most cases, except in flat terrains, it is difficult to keep the line of sight horizontal. In the case of an inclined line of sight, the vertical angle to the staff is read, with respect to the central cross hair. The staff intercept is read against the top and bottom cross hairs.

There can be two cases of inclined line of sight—(i) the staff is held normal to the line of sight (by tilting the staff as the line of sight is inclined) and (ii) the staff is held vertical irrespective of the inclination of line of sight. In either case the vertical angle can be an angle of elevation or depression.

Staff held vertical   In this case, the vertical angle can be an angle of elevation or depression.

Angle of elevation: The situation is shown in Fig. 14.4(a). P is the point at which the staff is held. A, C, and B are the points on the staff against the cross hairs. We draw the line B′CA′ perpendicular to the line OC. $\angle BOA = B'OA' = \alpha$ and $\angle COF = \theta$. In triangle COA′, $\angle COA' = \alpha/2$; $\angle CA'O = 90° - \alpha/2$, and $\angle CA'A = 180° - (90° - \alpha/2) = 90° + \alpha/2$. Similarly, from triangle COB′, $\angle COB' = \alpha/2$, $\angle CB'O = 90° - \alpha/2$, and $CB'B = 90° + \alpha/2$. As angle α is very small, angles CB′B and CA′A are nearly equal to 90°. The triangles can be considered right-angled.

$$A'B' = A'C + CB' = AC \cos\theta + CB \cos\theta = s \cos\theta$$
$$\text{(since } AC + CB = s)$$

Inclined length $OC = L = KA'B' + C = K(s \cos\theta) + C$

Horizontal distance $D = L \cos\theta = K(s \cos^2\theta) + C \cos\theta$

$V = L \sin\theta = (Ks \cos\theta + C) \sin\theta = Ks \sin\theta \cos\theta + C \sin\theta$

$\qquad = 1/2\, Ks \sin 2\theta + C \sin\theta$

The elevation of P can be determined from this expression.

Angle of depression: In case the vertical angle is an angle of depression, referring to Fig. 14.4(b), it is clear that the formulae for D and V remain the same:

$$D = Ks \cos^2\theta + C \cos\theta, \qquad V = (1/2)\, Ks \sin 2\theta + C \sin\theta$$

In calculating the reduced level of the point, V and the distance corresponding to the central hair reading have to be subtracted.

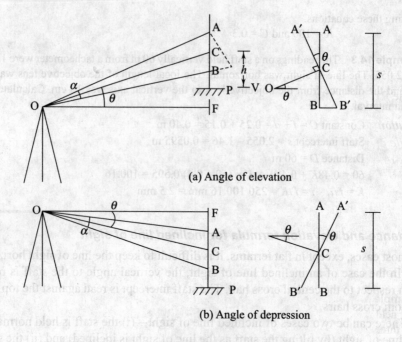

(a) Angle of elevation

(b) Angle of depression

Fig. 14.4 Inclined line of sight, staff held vertical

Staff held normal In this case also, the vertical angle can be an angle of elevation or depression. (Fig. 14.5).

Angle of elevation: Referring to Fig. 14.5(a), using the same notations,

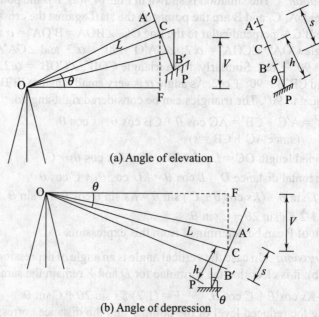

(a) Angle of elevation

(b) Angle of depression

Fig. 14.5 Inclined line of sight, staff held normal

$$OC = L = Ks + C$$
$$OF = L \cos \theta = (Ks + C) \cos \theta$$
$$D = OF + FF' = (Ks + C) \cos \theta + PC \sin\theta$$
Elevation of staff station $V = OC \sin \theta = (Ks + C) \sin \theta$

If the RL of the height of the instrument is known, then

$$\text{RL of } P = \text{RL of line of sight} + V - h \cos \theta$$

Angle of depression: If the vertical angle is an angle of depression, referring to Fig. 14.5(b) and using the same notations, we have

Inclined length $L = Ks + C$
$$OF = L \cos \theta = (Ks + C) \cos \theta$$
$$D = OF - FF' = (Ks + C) \cos \theta - h \sin \theta$$
$$V = OC \sin \theta = (Ks + C) \sin \theta$$
$$\text{RL of } P = \text{RL of instrument} - V - h \cos \theta$$

The following examples illustrate the use of these formulae.

Example 14.4 A tacheometer was set up at station P and observations were made to a staff held normal to the line of sight over point Q. The vertical angle measured was 6° 36'. The three hair readings were 1.905, 2.480, and 3.055. The reading from P, with the line of sight horizontal to a BM of RL 852.55 was 1.855. If the instrument constants are 100 and 0.5, find the RL of Q.

Solution Intercept $s = 3.055 - 1.905 = 1.15$ m

As the staff is held normal to the line of sight,

Horizontal distance $D = L \cos \theta$

where L is the inclined length

$$L = Ks + C, \quad D = (Ks + C) \cos \theta, \quad \theta = 6° 36'$$
$$D = (100 \times 1.15 + 0.5) \cos (6° 36') = 114.73 \text{ m}$$

V, the difference in height from the instrument height at P, is given by $L \sin \theta$:

$$V = (Ks + C) \sin \theta = (100 \times 1.15 + 0.5) \sin (6° 36') = 13.275 \text{ m}$$
$$\text{RL of BM} = 852.55 \text{ m}$$
$$\text{RL of line of sight} = 852.55 + 1.855 = 854.405 \text{ m}$$
$$\text{RL of Q} = 854.05 + 13.275 - 2.48 = 865.2 \text{ m}$$

Example 14.5 A tacheometer was set up at station P and observations were made to two stations Q and R. The vertical angles to Q and R were 5° 30' and 1° 08', respectively. The cross hair readings at Q were 2.105, 2.47, and 2.835 and those at R were 2.215, 2.56, and 2.905. The staff was held vertical in both cases. The instrument constants were 100 and 0.3. The reading from P to a BM of RL 285.35 m was 2.255. The horizontal angle QPR measured was 58° 30'. Find the distance QR, the gradient from Q to R, and the RLs of Q and R.

Solution $K = 100$ and $C = 0.3$. When the staff is held vertical,

Horizontal distance $D = Ks \cos^2\theta + C \cos \theta$
$$V = (1/2) Ks \sin 2\theta + C \sin \theta$$

Observation to Q: vertical angle $\theta = 5° 30'$, $s = 2.835 - 2.105 = 0.73$ m:

$$D = 100 \times 0.73 \times \cos^2(5° 30') + 0.3 \cos(5° 30') = 72.638 \text{ m}$$

$$V = (1/2) \times 100 \times 0.73 \times \sin (2 \times 5° 30') + 0.3 \times \sin (5° 30') = 6.992 \text{ m}$$

Observation to R: vertical angle $= 1° 08'$, $s = 2.905 - 2.215 = 0.69$ m:

$$D = 100 \times 0.69 \times \cos^2(1° \, 08') + 0.3 \times \cos(1° \, 08') = 69.273 \text{ m}$$

$$V = (1/2) \times 100 \times 0.69 \sin(2° \, 16') + 0.3 \times \sin(1° \, 08') = 1.367 \text{ m}$$

In triangle PQR, PQ = 72.638 m, PR = 69.273 m. Angle QPR = 58° 30′. The distance QR can be calculated using the cosine rule:

$$QR^2 = PQ^2 + PR^2 - 2PQPR \cos(58° \, 30')$$

$$= 72.638^2 + 69.273^2 - 2 \times 72.638 \times 69.273 \times \cos(58° \, 30')$$

$$= 4816.75$$

QR = 69.4 m

RL of BM = 285.35 m

RL of line of sight = 285.35 + 2.255 = 287.605 m

RL of Q = 287.605 + 6.992 − 2.47 = 292.127 m

RL of R = 287.605 + 1.367 − 2.56 = 286.412 m

Difference in elevation of Q and R = 292.127 − 286.412 = 5.715 m

Gradient from Q to R = 5.715/69.4 = 0.0823   (about 1 in 12)

Example 14.6   To determine the elevation of a point P, a tacheometer was set up at station A and observations were made to a staff held vertically at P. As a check, the instrument was set up at another point B and observations were taken to a staff held at P. The RL of the BM was 235.455. The instrument constants were 100 and 0.3. Determine the RL of P from the following data recorded:

Instrument at	Staff at	Vertical angle	Hair readings	Readings at BM
A	P	3° 45′	2.235, 2.795, 3.355	1.75
B	P	2° 30′	0.945, 1.490, 2.035	2.25

Solution   The horizontal distance between stations and their difference in elevation can be determined as follows:
A to P: staff intercept s = 3.355 − 2.235 = 1.12 m:

$$D = (K + s) \cos^2\theta + C \cos\theta = (100 + 1.12) \cos^2(3° \, 45') + 0.3 \cos(3° \, 45')$$

$$= 111.82 \text{ m}$$

$$V = (1/2)(K + s) \sin 2\theta + C \sin\theta$$

$$= (1/2)(100 + 1.12) \sin 7° \, 30' + 0.3 \sin(3° \, 45')$$

$$= 7.328 \text{ m}$$

RL of BM = 235.455 m

RL of line of sight = 237.205 + 1.75 = 237.205 m

RL of P = 237.205 + 7.328 − 2.795 = 241.718 m

Observation from B to P: staff intercept = 2.035 − 0.945 = 1.09 m:

$$D = (100 + 1.09) \cos^2(2° \, 30') + 0.3 \cos(2° \, 30') = 101.197 \text{ m}$$

$$V = (1/2)(100 + 1.09) \sin 5° + 0.3 \sin 2.5° = 5.5 \text{ m}$$

RL of line of sight at B = 235.455 + 2.25 = 237.705 m

RL of P = 237.705 + 5.5 − 1.490 = 241.715 m

RL of P = (1/2)(241.718 + 241.715) = 241.716 m

Example 14.7 Find the gradient from P to Q using the data given in Table 14.1.

Table 14.1 Data for Example 14.7

Instrument at	Staff at	Line	Bearing	Vertical angle	Cross hair readings
A	P	AP	84° 36′	3° 30′	1.35, 2.10, 2.85
A	Q	AQ	142° 24′	2° 45′	1.955, 2.875, 3.765

The staff was held normal to the line of sight in both cases.

Solution The horizontal and vertical distances from the instrument axis to the central cross hair can be found from the cross hair readings:

$$D = (Ks + C) \cos \theta, \quad V = (Ks + C) \sin \theta$$

Observation from A to P: staff intercept = 2.85 – 1.35 = 1.5 m:

$$D = AP = (100 \times 1.5 + 0.3) \cos(3° 30′) = 150 \text{ m}$$
$$V = (100 \times 1.5 + 0.3) \times \sin(3° 30′) = 9.175 \text{ m}$$

Observation from A to Q: staff intercept = 3.765 – 1.955 = 1.81 m:

$$D = AQ = (100 \times 1.81 + 0.3) \cos(2° 45′) = 101.992 \text{ m}$$
$$V = (100 \times 1.81 + 0.3) \sin(2° 45′) = 4.899 \text{ m}$$

In the triangle formed by the points A, P, and Q, the sides AP and AQ are known. The angle PAQ can be found from the bearings of the lines.

Angle PAQ = 142° 24′ – 84° 36′ = 57° 48′

Using the cosine rule,

$$PQ^2 = AP^2 + AQ^2 - 2 \times AP \times AQ \times \cos(57° 48′)$$
$$= 150^2 + 101.992^2 - 2 \times 150 \times 101.992 \times \cos(57° 48′)$$
$$PQ = 128.83 \text{ m}$$

Assuming the horizontal line of sight at A as datum,

Elevation of P = 9.175 – 2.10 = 7.075 m

Elevation of Q = 4.899 – 2.875 = 2.024 m

Difference in elevation = 7.075 – 2.024 = 5.051 m

P is higher than Q.

Gradient from P to Q = 5.051/128.83 = 0.0392   (1 in 25.5)

Example 14.8 As chaining was not possible, a traverse was conducted using tacheometry. The line of sight was horizontal in all cases. The data (Table 14.2) was obtained. Find the lengths of the sides and the length and bearing of line AC. Find also the gradient from A to C if the reading on a staff held at a benchmark is 2.415 from A and 0.645 from C. The instrument constants were 100 and 0.3.

Table 14.2 Data for Example 14.8

Line	Bearing	Instrument at	Staff at	Cross hair readings
AB	70° 30′	A	B	1.535, 2.214, 2.893
BC	120° 45′		D	2.018, 2.70, 3.708
CD	223° 30′		C	
DA	320° 47′	C	B	1.033, 1.733, 2.432
			D	1.363, 2.243, 3.123
			A	

Solution The included angles can be calculated from the bearings. Referring to Fig. 14.6,

Included angle A = 140° 47′ – 70° 30′

Angle at A = 70° 17′

Included angle B = 250° 30′ – 120° 45′ = 129° 45′

Included angle C = 300° 45′ – 223° 30′ = 77° 15′

Included angle at D = 320° 47′ – 43° 30′ = 277° 17′ (exterior angle)

= 82° 43′

Sum of angles = 70° 17′ + 129° 45′ + 77° 15′ + 82° 43′ = 360° 00′

The lengths of the sides can be found from the stadia readings. The line of sight is horizontal.

$$D = Ks + C$$

Length of AB : staff intercept s = 2.893 – 1.535 = 1.358:

AB = 100 × 1.358 + 0.3 = 136.1 m

Length of BC: staff intercept = 2.432 – 1.033 = 1.399

BC = 100 × 1.399 + 0.3 = 140.2 m

Length of CD: intercept = 3.123 – 1.363 = 1.76

CD = 100 × 1.76 + 0.3 = 176.3 m

Length of DA: staff intercept = 3.708 – 2.018 = 1.69

DA = 100 × 1.69 + 0.3 = 169.3 m

In triangle ABC, AB and BC are known and the angle at B has been found from the bearings. Applying the cosine rule, length AC can be calculated.

$$AC^2 = AB^2 + BC^2 - 2 \times AB \times BC \times \cos (\angle B)$$
$$= 136.1^2 + 140.2^2 - 2 \times 136.1 \times 140.2 \times \cos(129° 45′)$$
$$= 62{,}581.8$$
$$C = \sqrt{62{,}581.8} = 250.16 \text{ m}$$

The bearing of AC can be found from triangle ABC. From Fig. 14.6, 129° 45′ + α + β = 180°. We apply the sine rule to evaluate angles α and β:

$$\frac{250.16}{\sin(129°45′)} = \frac{136.1}{\sin \beta}$$

From this, sin β = 0.4183, β = 24° 44′
Similarly,

$$\frac{250.16}{\sin(129°45′)} = \frac{140.2}{\sin \alpha}$$

sin α = 140.2 sin(129° 45′)/250.16
= 0.4308, α = 25° 31′

Check: 129° 45′ + 24° 44′ + 25° 31′ = 180°

Bearing of AC = 70° 30 + 25° 31′
= 96° 01′

The difference in elevation of A and C
= 2.415 – 0.645 = 1.77 m. Gradient from A
to C = 1.77/250.16, approximately 1 in 141.

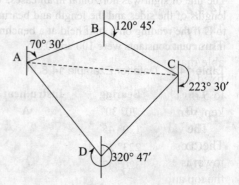

Fig. 14.6 Traverse (Example 14.8)

Example 14.9 To determine the tacheometric constants K and C, the instrument was set up at O. Distances of 30 m, 60 m, and 90 m were carefully measured and stations P, Q, and R were carefully marked. A stadia rod was kept at the three stations and the following readings (Table 14.3) were obtained.

Table 14.3 Data for Example 14.9

Instrument at	Staff at	Distance from O	Cross hair readings
O	P	30	1.135, 1.284, 1.433
O	Q	60	1.025, 1.325, 1.624

Determine the instrument constants.

Solution Two distance equations can be formed and solved for the constants:

$$30 = Ks + C = K(1.433 - 1.135) + C$$
$$60 = K(1.584 - 1.025) + C$$
$$30 = 0.298K + C, \quad 60 = 0.599K + C$$

Subtracting the first from the second,

$$30 = 0.261K, \quad K = 99.67, \quad C = 0.298$$

Example 14.10 Find the tacheometric constants from the observations given in Table 14.4.

Table 14.4 Data for Example 14.10

Instrument at	Staff at	Distance	Vertical angle	Cross hair readings
P	A	80	2° 30′	1.325, 2.122
	B	140	1° 36′	0.985, 2.382

Solution The horizontal distance is given by

$$D = Ks \cos^2 \theta + C \cos \theta$$
$$80 = K \times (2.122 - 1.325) \cos^2(2° 30') + C \cos(2° 30')$$
$$80 = 0.7955K + 0.999C$$
$$140 = K \times (2.382 - 0.985) \cos^2(1° 36') + C \cos(1° 36')$$
$$140 = 1.3959K + 0.9996C$$

Solving these equations, $K = 100$ and $C = 0.4$.

14.4.2 Movable Hair Stadia Method

The movable hair stadia method basically enables the stadia interval i to be varied. This requires a special, variable diaphragm, shown in Fig. 14.7(a). With the variable diaphragm, a fixed distance on the target can be measured all the time. The fixed staff intercept is called the *base*. The staff intercept is generally between 3 m to 5 m. The intercept is obtained from a subtense bar [Fig. 14.7(b)] which can be kept vertical or horizontal.

The variation of the stadia interval is obtained by a special subtense diaphragm. The cross hairs are attached to two different vertical legs which can be moved towards each other or away from each other by milled-headed screws provided at the top and bottom. A drum provided with a vernier enables readings to be obtained up to a 1000th of the pitch of the screw moving the legs. A comb scale seen in the field of view enables the complete turns to be read directly.

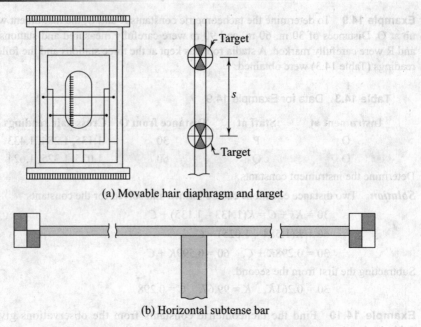

(a) Movable hair diaphragm and target

(b) Horizontal subtense bar

Fig. 14.7 Movable hair method and the subtense bar

The movable hair method rarely finds use nowadays. The many moving parts make it not a very reliable method compared to the fixed hair method.

Vertical subtense bar method

The distance formula for the movable hair method obviously is the same as that for the fixed hair method as the stadia interval i and the intercept s (which is kept constant) are involved in both the cases. Thus, for a horizontal line of sight,

Horizontal distance $D = Ks + C$

and for an inclined line of sight with the staff vertical,

Horizontal distance $D = Ks \cos^2 \theta + C \cos \theta$

and

$V = (1/2) Ks \sin 2\theta + C \sin \theta$

The stadia interval in such cases is varied by the use of the micrometer screw to get the desired staff intercept. If p is the pitch of the screw and m the number of revolutions of the screw, then the distance moved $= mp$. The stadia interval $i = mp$. As $K = f/i$,

$D = (f/mp)s + C$

for a horizontal line of sight. The constant K is taken as f/p. Thus,

$D = Ks/m + C$

where $C = f + d$

$D = Ks \cos^2\theta/m + C \cos \theta$ and $V = (1/2) Ks \sin2\theta/m + C \sin \theta$

for an inclined line of sight. If there is an index error in the micrometer screw, the denominator becomes $(m - e)$ instead of m.

Constants *K* and *C*   The value of constant *K* is kept at a convenient value of 1000 but may vary between 600 and 1000. The constant *K* can be determined as follows.

1. Keep the distance between the vanes on the staff at a convenient length between 3 and 6 m.
2. Measure two convenient distances *D* and *D'* on a fairly level ground. Keep the vertical angle zero.
3. Keep the movable hair instrument at one end of these lines. Measure the micrometer screw movement in terms of the number of complete turns of the screw required to get the intercept. Let these be *m* and *m'*.
4. From these observations, $D = Ks/m + C$ and $D' = Ks/m' + C$.
5. These two equations can be solved to get the values of *K* and *C*.
6. $K = \dfrac{(D - D')m\,m'}{s(m - m')}$, $C = \dfrac{Dm - D'm'}{m - m'}$

Horizontal subtense bar method

In the horizontal subtense bar method, the subtense bar is kept horizontal and the angle subtended by the distance between the vanes at the instrument is measured. Referring to Fig. 14.8, if the line PQ is perpendicular to the line IR, then the horizontal distance is given by

$$D = (1/2)\, s \cot(\alpha/2) = s/(2 \tan \alpha/2)$$

If α is very small, $\tan \alpha/2 = \alpha/2$. Then,

$$D = s/\alpha$$

where α is in radians.

Effect of angular error on horizontal distances   If there is an error in the measurement of angle α, there will be a corresponding error in the measurement of distance *D*. To determine this effect,

$$D = s/(2 \tan \alpha/2)$$

and if α is small, $\tan \alpha/2 = \alpha/2$ radians, and

$$D = s\alpha$$

Let the angle measured be $(\alpha - \delta\alpha)$ and the distance $D = D + \delta D$.

$$s = D\alpha = (D + \delta D)(\alpha - \delta\alpha)$$
$$D\alpha = (D + \delta D)(\alpha - \delta\alpha)$$

This gives

$$\frac{D + \delta}{D} = \frac{\alpha}{\alpha - \delta\alpha}$$
$$\delta D = \frac{D\,\delta\alpha}{\alpha - \delta\alpha}$$

Similarly, if α is greater by an amount $\delta\alpha$,

$$\delta D = \frac{D\,\delta\alpha}{\alpha + \delta\alpha}$$

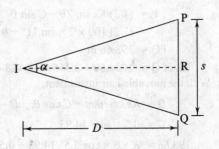

Fig. 14.8   Horizontal subtense bar method

Taking $\delta\alpha$ to be small in comparison to α, in both cases,

$$\delta D = D\ \delta\alpha/\alpha$$

The following examples illustrate these principles.

Example 14.11 Observations were made from a station P on a subtense bar held at station Q. The vertical angle was 8° 15′. The number of revolutions of the micrometer screw was 21.35. The instrument constants were 1000 and 0.4. The intercept was kept at 3 m. Find the horizontal distance between P and Q.

Solution The horizontal distance in the case of a movable hair instrument is given by

$$D = Ks\ \cos^2\theta/m + C\cos\theta$$

$$= 1000 \times 3 \times \cos^2(8°\ 15′)/21.35 + 0.4 \times \cos(8°\ 15′) = 138.01\ \text{m}$$

Example 14.12 The distance between two stations A and B was 258 m. A movable hair instrument was used to measure this distance again. The vertical angle was 6° 30′. The distance between the vanes on the subtense bar was 5 m. The constants of the instrument were 1000 and 0.5. Find the number of turns of the micrometer screw registered during this measurement.

Solution The horizontal distance is given by

$$D = Ks\ \cos^2\theta/m + C\cos\theta$$

$$= 258\ \text{m}$$

$$s = 5\ \text{m}, \quad \theta = 6°\ 30′, \quad K = 1000, \quad C = 0.5$$

$$258 = 1000 \times 5 \times \cos^2(6°\ 30′) + 0.5 \times \cos(6°\ 30′)$$

Solving this equation,

$$m = 19.13$$

Example 14.13 A distance PQ was measured with a tacheometer (constants 100 and 0.5) at P. The vertical angle was 5° 30′. The cross hair readings were 1.335, 2.335, and 3.335. Find the distance PQ and the RL of Q if the reading at the staff at a BM of RL 1030.50 was 2.355. A movable hair instrument was then set up over P and observations were made over the same distance. The vertical angle was the same. The intercept was 3 m and the number of turns of the micrometer screw was noted as 14.93. If $C = 0.5$, find the constant K of the instrument.

Solution From the tacheometer observations,

$$D = Ks\ \cos^2\theta + C\cos\theta, \quad s = 3.335 - 1.335 = 2\ \text{m}, \quad \theta = 5.5$$

$$D = 100 \times 2 \times \cos^2 5.5° + 0.5\cos 5.5° = 198.66\ \text{m}$$

$$V = (1/2)\ Ks\ \sin 2\theta + C\sin\theta$$

$$= (1/2)\ 100 \times 2 \times \sin 11° + 0.5\sin 5.5° = 19.128\ \text{m}$$

$$PQ = 198.66\ \text{m}$$

$$\text{RL of Q} = 1030.50 + 2.355 + 19.128 - 2.335 = 1049.648\ \text{m}$$

With the movable hair instrument,

$$D = Ks\ \cos^2\theta/m + C\cos\theta, \quad D = 198.66\ \text{m} \quad s = 3\ \text{m}, \quad \theta = 5.5°$$

$$C = 0.5, \quad m = 14.93$$

$$198.66 = K \times 3 \times \cos^2 5.5°/14.93 + 0.5 \times \cos 5.5°$$

Solving this equation,

$$K = 997.84$$

Example 14.14 To determine the constants K and C of a movable hair instrument, from a station O two distances of 60 and 120 m were laid out. The intercept was 1.5 m. The number of turns of the micrometer screw recorded was 22.5 for a distance of 60 m and 11.28 for a distance of 120 m. Find the constants K and C of the instrument.

Solution Two equations can be set up for the two measurements:

$$D = Ks/m + C$$
$$60 = K \times 1.5/22.5 + C, \quad 120 = K \times 1.5/11.28 + C$$

Solving these two equations,

$$K = 904.8 \quad \text{and} \quad C = 0.32$$

The constants can also be determined from the formula

$$K = (D - D')mm'/[s(m - m')]$$
$$= (120 - 60) \times 22.5 \times 11.28/[1.5(22.5 - 11.28)] = 904.8$$
$$C = (Dm - D'm')/(m - m')$$
$$= (120 \times 11.28 - 60 \times 22.5)/(22.5 - 11.28) = 0.32$$

14.5 Anallactic Lens

As mentioned earlier, in an external focusing telescope, the focusing is done by varying the distance between the object lens and the diaphragm. The additive constant C thus has a considerable value. In an internal focusing telescope, a third lens is provided between the object lens and the eyepiece and focusing is done by moving this lens. The additive constant of an internal focusing telescope thus has a small value. Internal focusing telescopes can thus be considered anallactic. A similar lens provided in the external focusing telescope to make the additive constant zero is known as an anallactic lens. An anallactic lens is a double convex lens provided in the telescope between the object lens and the eyepiece to make the additive constant zero. This makes the calculation simple. However, the additional lens also absorbs some light and reduces the brightness of the object.

The principle of the anallactic lens can be studied from Fig. 14.9. The essential principle is to make the vertex of the tacheometric triangle coincide with the vertical axis of the instrument. The following symbols are used: O denotes the objective lens, A the anallactic lens, P, Q the cross hair readings on the staff, p, q the image of P and Q, f_1, f_2 the conjugate focal lengths of the objective lens, f' the focal length of the anallactic lens, and $f_2 - n$, $m - n$ the conjugate lengths of the anallactic lens.

The following relationships can be obtained from the lens formula and the properties of similar triangles.

(1) From the lens formula,

$$\frac{1}{f} = \frac{1}{f_1} + \frac{1}{f_2}$$

(2) From triangle PQF and rsF (rs= pq)

$$\frac{s}{i'} = \frac{f_1}{f_2}$$

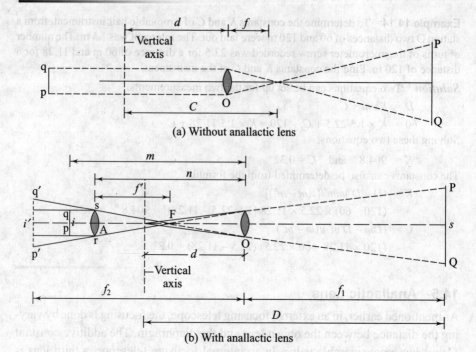

Fig. 14.9 Anallactic lens

(3) For the anallactic lens, f' is the focal length and $(f_2 - n)$ and $(m - n)$ are conjugate distances. Hence,

$$\frac{1}{f'} = \frac{-1}{f_2 - n} + \frac{1}{m - n}$$

(The negative sign is due to i' and i being on the same side of A.)

(4) Also

$$\frac{i'}{i} = \frac{f_2 - n}{m - n}$$

From (2) and (4), we get

$$\frac{s}{i} = \frac{f_1(f_2 - n)}{f_2(m - n)}$$

From (1),

$$\frac{f_1}{f_2} = \frac{f_1 - f}{f} \quad \text{or} \quad f_2 = \frac{f f_1}{f_1 - f}$$

From (3)

$$\frac{f_2 - n}{m - n} = \frac{f_2 - f + f'}{f'}$$

Substituting these values, we get

$$\frac{s}{i} = \frac{(f_1 - f)(f_2 - n + f')}{f f'}$$

Substituting for $f_2 = ff_1/(f_1 - f)$ and reducing, we get

$$\frac{s}{i} = \frac{f_1(f + f' - n)}{ff'} + \frac{f(n - f')}{ff'}$$

The distance f_1 can be derived from this equation as

$$f_1 = \frac{s\,ff'}{i(f + f' - n)} - \frac{f(n - f')}{f + f' - n}$$

The horizontal distance D is given by $D = f_1 + d$. Substituting for f_1,

$$D = \frac{ff's}{(f + f' - n)i} - \frac{f(n - f')}{f + f' - n} + d$$

This form is equivalent to $Ks + C$, where

$$K = \frac{ff'}{(f + f' - n)i}$$

$$C = d - \frac{f(n - f')}{f + f' - n}$$

The purpose of the anallactic lens is to make the additive constant zero. For this,

$$d - \frac{f(n - f')}{f + f' - n} = 0$$

$$\frac{f(n - f')}{f + f' - n} = d$$

To make this happen, the distance n between the anallactic lens and the objective lens is made equal to

$$n = f' = \frac{fd}{f + d}$$

Once this equation is satisfied, the additive constant $C = 0$ and the apex of the tacheometric triangle will be at the centre of the horizontal axis. Also, the multiplier constant should be some convenient value, say, 100. For this,

$$\frac{ff'}{(f + f' - n)i} = 100$$

The horizontal distance D then will be equal to

$$D = 100s$$

The following examples illustrate this principle.

Example 14.15 A theodolite is to be fitted with an anallactic lens. From the following data, find the distance from the objective lens at which the lens is to be fixed and the stadia interval required to give a multiplying constant of 100. Focal length of the objective lens = 25 cm, focal length of the anallactic lens = 11.6, distance from the objective lens to the vertical axis = 13.5.

Solution Let the distance between the anallactic lens and the objective be n. Then,

$$N = f' + \frac{fd}{f + d} = 1.6 + \frac{25 \times 13.5}{25 + 13.5} = 20.36 \text{ cm}$$

The stadia interval can be determined from the equation for:

$$K = \frac{ff'}{(f + f' - n)i}$$

from which

$$i = \frac{ff'}{K(f + f' - n)}$$

$$i = (25 \times 11.6)/[100 \times (25 + 11.6 - 20.36)] = 0.18 \text{ cm}$$

Example 14.16   Find the focal length of the anallactic lens and the distance at which it should be placed to get a multiplying constant of 100 from the following data: focal length of the objective lens = 24 cm, stadia interval = 0.17 cm, distance between the objective lens and the vertical axis = 10.5 cm.

Solution   The focal length of the anallactic lens can be found from the equation

$$N = f' + fd/(f + d), \quad f' = n - fd/(f + d) = n - 24 \times 10.5/(24 + 10.5) = n - 7.3$$

$$K = ff'/i(f + f' - n) = 24f'/0.17(24 + f' - n)$$

$$= 100$$

Solving these equations,

$$n = f' + 7.3$$

$$100 = 24 f'/0.17(24 + f' - f' - 7.3), \quad f' = 11.83 \text{ cm}$$

$$n = 11.83 + 7.3 = 18.6 \text{ cm}$$

Focal length of the anallactic lens = 11.83 cm

Distance from the objective lens = 18.6 cm

14.6   Tangential Method

The difference between the tangential method of tacheometry and the stadia method is that the stadia hairs are not used in the former. The distances and elevations are computed using the readings corresponding to the central hair only. Two vanes are fixed on the stadia rod or on another target a fixed distance apart. These vanes are bisected by the central cross hair and the vertical angles corresponding to each vane are measured. The tangential method is suitable if the theodolite does not have a stadia diaphragm. It involves the measurement of two angles at each station and this increases the chances of error.

14.6.1   Elevation and Distance Formulae

There can be three different cases depending upon the nature of the angles:

(i) Both angles of elevation

(ii) Both angles of depression

(iii) One angle of elevation and one angle of depression

The following symbols will be used in deriving the distance and elevation formulae: D denotes the horizontal distance between the stations, V the difference in elevation between the instrument axis and the lower vane, θ_1, θ_2 the vertical angles to the upper and lower vanes, respectively, s the distance between the vanes, i.e., the intercept, I the position of the instrument axis, r the staff reading at the lower vane or the height of the lower vane above the foot of the staff, h the height of the instrument axis, and P, Q the points corresponding to the upper and lower vanes.

Both angles of elevation

When both angles are those of elevation, the situation is as shown in Fig. 14.10(a). From triangle IPS,

$$PS = D \tan \theta_1, \quad V + s = D \tan \theta_1$$

From triangle IQS,

$$QS = D \tan \theta_2, \quad V = D \tan \theta_2$$

From these two relations,

$$s = D(\tan \theta_1 - \tan \theta_2), \quad D = s/(\tan \theta_1 - \tan \theta_2)$$
$$D = s \cos \theta_1 \cos \theta_2/[\sin(\theta_1 - \theta_2)] \quad \text{(by putting } \tan \theta = \sin \theta/\cos \theta)$$
$$V = D \tan \theta_2 = s \cos \theta_1 \sin \theta_2/[\sin(\theta_1 - \theta_2)]$$

Elevation of station point = elevation of instrument axis + $V - r$

Both angles of depression

When both angles are those of depression, the situation is as shown in Fig. 14.10(b). From triangle IPS,

$$PS = D \tan \theta_1, \quad V - s = D \tan \theta_1$$

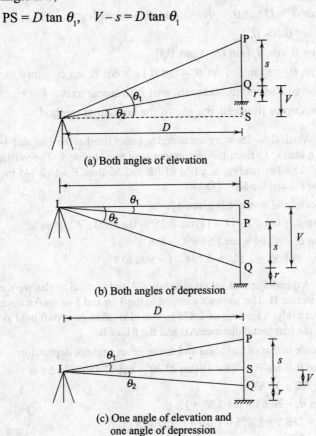

(a) Both angles of elevation

(b) Both angles of depression

(c) One angle of elevation and
one angle of depression

Fig. 14.10 Tangential method

From triangle IQS,

$$QS = V = D\tan\theta_2$$

From these two equations,

$$s = D(\tan\theta_2 - \tan\theta_1)$$
$$D = s/(\tan\theta_2 - \tan\theta_1)$$
$$D = s\cos\theta_1\cos\theta_2/[\sin(\theta_2 - \theta_1)] \quad \text{(putting } \tan\theta = \sin\theta/\cos\theta)$$
$$V = D\tan\theta_2 = s\tan\theta_2/(\tan\theta_2 - \tan\theta_1) = s\cos\theta_1\sin\theta_2/[\sin(\theta_2 - \theta_1)]$$

Elevation of staff point = elevation of instrument axis $- V - r$

One angle of elevation and one angle of depression

This situation is shown in Fig. 14.10(c). From triangle IPS,

$$PS = D\tan\theta_1, \quad s - V = D\tan\theta_1$$

From triangle IQS,

$$QS = V = D\tan\theta_2$$

From these two equations,

$$s = D\tan\theta_1 + D\tan\theta_2, \quad D = s/(\tan\theta_1 + \tan\theta_2)$$

Putting $\tan\theta = \sin\theta/\cos\theta$,

$$D = s\cos\theta_1\cos\theta_2/[\sin(\theta_1 + \sin\theta_2)]$$
$$V = D\tan\theta_2 = s\tan\theta_2/(\tan\theta_2 - \tan\theta_1) = s\cos\theta_1\sin\theta_2/[\sin(\theta_1 + \theta_2)]$$

Elevation of staff point = elevation of instrument axis $- V - r$

The following examples illustrate the use of the tangential method.

Example 14.17 Vertical angles were measured to vanes fixed at the 1-m and 4-m marks of a staff held at a station Q from the instrument kept at a station P. The vertical angles were 3° 30′ and 6° 15′. The reading at a BM of RL 985.55 from P was 2.345 m. Find the horizontal distance PQ and the RL of Q.

Solution　The horizontal distance is given by

$$D = s/(\tan\theta_1 - \tan\theta_2) = (4 - 1)/(\tan 6.25° - \tan 3.5°) = 62.04 \text{ m}$$
$$V = D\tan\theta_2 = 62.04 \times \tan 3.5° = 3.794 \text{ m}$$
$$\text{RL of Q} = 985.55 + 2.345 + 3.794 - 1 = 990.689$$

Example 14.18 A theodolite was set up at a station A and vertical angles were measured to vanes kept at a station B. The angles measured to the 1-m and 5-m marks were $- 2° 30′$ and $+ 3° 45′$, respectively. A reading of 1.875 m was also taken on a staff held at a BM of RL 258.5 m. Find the horizontal distance AB and the RL of B.

Solution　One angle is that of elevation and the other an angle of depression.

$$D = s/(\tan\theta_1 + \tan\theta_2) = (5 - 1)/(\tan 2° 30′ + \tan 3° 45′) = 36.63 \text{ m}$$

Distance AB = 36.63 m

$$V = D\tan\theta_2 = 36.63 \times \tan 2.5° = 1.6 \text{ m}$$
$$\text{RL of B} = 258.5 + 1.875 - 1.6 - 1 = 257.775 \text{ m}$$

14.7 Fieldwork

Tacheometry by its very nature is suitable for determining the heights and distances of points. It can be used for traversing and contouring as well as for filling in the details. Generally, tacheometry is used for the following:

(a) Surveys involving the location of a point on a plan only
(b) Surveys requiring the location of points on a plan as well as the elevation of these points
(c) Sectioning involving the location of points along a line with their elevations, and.
(d) Contouring to locate points on a plan and determine their elevations

Fieldwork involves the following activities.

Reconnaissance

Reconnaissance of the area should be conducted first to select the tachometric stations. Depending on the nature of the terrain, appropriate stations are fixed. Steep vertical angles should be avoided by selecting a large number of stations.

Fixing stations

As tacheometry involves finding horizontal distances and elevations, both horizontal and vertical control are involved. For horizontal control, already established triangulation stations may be used. In the absence of such stations, it is necessary to lay out a base line as accurately as possible. For vertical control, well-established benchmarks may be used. In the absence of such benchmarks, fly levelling is done to establish local benchmarks very accurately for the purpose of the survey.

Tacheometric stations should be so selected as to give a clear view of a wide area and cover a large tract of land from one station. The intervisibility of stations must be ensured.

Field observations

At every station, the orientation of the instrument with respect to the horizontal angles is ensured by setting the instrument to the bearing of a line. Once this setting is done, the instrument is kept undisturbed till all the observations from the stations are completed. This is important, as tacheometric observations should relate the points to a reference direction.

For elevation observations, the height of the line of sight must be known. Available benchmarks may be used or a benchmark established accurately by fly levelling from a point of known elevation.

Tacheometric fieldwork involves (a) setting up the instrument at a station, (b) taking observations of the horizontal angles, (c) taking stadia and central hair readings on the staff, and (d) observing vertical angles. Observations [(c) and (d)] are repeated for every point that can be seen from the instrument station.

Recording observations

The observations should be recorded in a field book, sample of which is shown in Table 14.5.

Table 14.5   Sample field book for tacheometric work

Instru-ment at	Staff at	Bearing	Cross hair readings			Distance	Elevation	RL
			Top	Middle	Bottom			
1	BM	35° 64' 10"	1.834	2.494	3.154			
	P	110 ° 34'55"	0.985	1.410	1.835			
	Q	165° 35' 36"	1.155	1.662	2.165			
2	Q	286° 34' 55"	2.355	1.76	3.165			
	R	47° 35' 36"	1.655	1.805	1.945			

14.8   Reduction of Stadia Observations

Reduction is the process of calculating horizontal distances and elevations from the stadia observations. The central hair reading is used to compute the reduced level of the station point. In case the number of observations is small, it is possible to calculate these quantities using the tacheometric formulae along with log tables or a calculator. In the case of vast areas involving a large number of observations, tacheometric tables or charts are resorted to. Using tacheometric tables is easier and more accurate than using charts from the point of view of reading values.

14.8.1   Tacheometric Tables

Many forms of tacheometric tables are available. A page from one such form is shown in Table 14.6. This table lists horizontal corrections and differences in elevation for various values of vertical angles. The first column lists minutes at a $2'$ interval. The next four columns list degrees at $1°$ intervals. It is easy to see from the values in the table that horizontal distance $= Ks - Ks \cos^2\theta = Ks \sin^2\theta$. The values listed are $K \sin^2\theta$. For difference in elevation, the values listed are $(1/2) K \sin 2\theta$. The value of K is assumed to be 100. As an example, when $\theta = 2° 30'$, $K \sin^2\theta = 100 \sin^2(2.5°) = 0.19$ and $(1/2) 100 \sin 5° = 4.36$. Similarly, in the three rows at the bottom, for different values of additive constant C ($C = 0.2, 0.3$, and 0.4), the values to be added to the value obtained in the main table are given. Thus, when $\theta = 2° 30'$ and $C = 0.2$,

Horizontal distance $= 0.19s + (0.2 - 0.0)$

Difference in elevation $= 4.36s + 0.01$

Table 14.7 shows another form of the tacheometric table. The basic principle remaining the same, the values listed are horizontal distances and differences in elevation. $K \cos^2\theta$ is listed under horizontal correction and $(1/2)K \sin \theta$ is listed under difference in elevation. The value of K is assumed to be 100. For non-zero values of C—0.2, 0.3, and 0.4—the additive values are listed in the bottom rows. The values against C are $C \cos \theta$ and $C \sin \theta$.

Table 14.6 Tacheometric table

Minutes	Vertical angle							
	0°		1°		2°		3°	
	Hor.	Diff.	Hor.	Diff.	Hor.	Diff.	Hor.	Diff.
	corr.	*elev.*	*corr.*	*elev.*	*corr.*	*elev.*	*corr.*	*elev.*
0	0.00	0.00	0.03	1.74	0.12	3.49	0.27	5.23
2	0.00	0.06	0.03	1.80	0.13	3.55	0.28	5.28
4	0.00	0.12	0.03	1.86	0.13	3.60	0.29	5.34
6	0.00	0.17	0.04	1.92	0.13	3.66	0.29	5.40
8	0.00	0.23	0.04	1.98	0.14	3.72	0.30	5.46
10	0.00	0.29	0.04	2.04	0.14	3.78	0.31	5.52
12	0.00	0.35	0.04	2.09	0.15	3.84	0.31	5.57
14	0.00	0.41	0.05	2.15	0.15	3.89	0.32	5.63
16	0.00	0.47	0.05	2.21	0.16	3.95	0.32	5.69
18	0.00	0.52	0.05	2.27	0.16	4.01	0.33	5.75
20	0.00	0.58	0.05	2.33	0.17	4.07	0.34	5.80
22	0.00	0.64	0.06	2.38	0.17	4.13	0.34	5.86
24	0.00	0.70	0.06	2.44	0.18	4.18	0.35	5.92
26	0.01	0.76	0.06	2.50	0.18	4.24	0.36	5.98
28	0.01	0.81	0.07	2.56	0.19	4.30	0.37	6.04
30	0.01	0.87	0.07	2.62	0.19	4.36	0.37	6.09
32	0.01	0.93	0.07	2.67	0.20	4.42	0.38	6.15
34	0.01	0.99	0.07	2.73	0.20	4.47	0.38	6.21
36	0.01	1.05	0.08	2.79	0.21	4.53	0.39	6.27
38	0.01	1.11	0.08	2.85	0.21	4.59	0.40	6.32
40	0.01	1.16	0.08	2.91	0.22	4.65	0.41	6.38
42	0.01	1.22	0.09	2.97	0.22	4.71	0.41	6.44
44	0.02	1.28	0.09	3.02	0.23	4.76	0.42	6.50
46	0.02	1.34	0.10	3.08	0.23	4.82	0.43	6.56
48	0.02	1.40	0.10	3.14	0.24	4.88	0.44	6.61
50	0.02	1.45	0.10	3.20	0.24	4.94	0.44	6.67
52	0.02	1.51	0.11	3.26	0.25	4.99	0.45	6.73
54	0.02	1.57	0.11	3.31	0.25	5.05	0.46	6.79
56	0.03	1.63	0.11	3.37	0.26	5.11	0.47	6.84
58	0.03	1.69	0.12	3.43	0.27	5.17	0.48	6.90
60	0.03	1.74	0.12	3.49	0.27	5.23	0.49	6.96
C = 0.2 m	0.00	0.00	0.00	0.01	0.00	0.01	0.00	0.01
C = 0.3 m	0.00	0.00	0.00	0.01	0.00	0.01	0.00	0.02
C = 0.4 m	0.00	0.00	0.00	0.01	0.00	0.02	0.00	0.02

Table 14.7 Tacheometric table

| Minutes | Vertical angle | | | | | | | |
| | 0° | | 1° | | 2° | | 3° | |
	Hor. dist.	Diff. elev.	Hor. dist.	Diff. elev.	Hor. dist.	Diff. elev.	Hor. dist.	Diff. elev.
0	100.00	0.00	99.97	1.74	99.88	3.49	99.73	5.23
2	100.00	0.06	99.97	1.80	99.87	3.55	99.72	5.28
4	100.00	0.12	99.97	1.86	99.87	3.60	99.71	5.34
6	100.00	0.17	99.96	1.92	99.87	3.66	99.71	5.40
8	100.00	0.23	99.96	1.98	99.86	3.72	99.70	5.46
10	100.00	0.29	99.96	2.04	99.86	3.78	99.69	5.42
12	100.00	0.35	99.96	2.09	99.85	3.84	90.69	5.57
14	100.00	0.41	99.95	2.15	99.85	3.89	99.68	5.63
16	100.00	0.47	99.95	2.21	99.84	3.95	99.68	5.69
18	100.00	0.52	99.95	2.27	99.84	4.01	99.67	5.75
20	100.00	0.58	99.95	2.33	99.83	4.07	99.66	5.80
22	100.00	0.64	99.94	2.38	99.83	4.13	99.66	5.86
24	100.00	0.70	99.94	2.44	99.82	4.18	99.65	5.92
26	99.99	0.76	99.94	2.50	99.82	4.24	99.64	5.98
28	99.99	0.81	99.93	2.56	99.81	4.30	99.63	6.04
30	99.99	0.87	99.93	2.62	99.81	4.36	99.63	6.09
32	99.99	0.93	99.93	2.67	99.80	4.42	99.62	6.15
34	99.99	0.99	99.93	2.73	99.80	4.47	99.61	6.21
36	99.99	1.05	99.92	2.79	99.79	4.53	99.61	6.27
38	99.99	1.11	99.92	2.85	99.79	4.59	99.60	6.32
40	99.99	1.16	99.92	2.91	99.78	4.65	99.59	6.38
42	99.99	1.22	99.91	2.97	99.78	4.71	99.58	6.44
44	99.98	1.28	99.91	3.02	99.77	4.76	99.58	6.50
46	99.98	1.34	99.90	3.08	99.77	4.82	99.57	6.56
48	99.98	1.40	99.90	3.14	99.76	4.88	99.56	6.61
50	99.98	1.45	99.90	3.20	99.76	4.94	99.55	6.67
52	99.98	1.51	99.89	3.26	99.75	4.99	99.55	6.73
54	99.88	1.57	99.89	3.31	99.74	5.05	99.54	6.79
56	99.97	1.63	99.89	3.37	99.74	5.11	99.53	6.84
58	99.97	1.69	99.88	3.48	99.73	5.17	99.52	6.90
60	99.97	1.74	99.88	3.49	99.73	5.23	99.51	6.96
$C = 0.2$ m	0.20	0.00	0.20	0.01	0.20	0.01	0.20	0.01
$C = 0.3$ m	0.30	0.00	0.30	0.01	0.30	0.01	0.30	0.01
$C = 0.2$ m	0.40	0.00	0.40	0.01	0.40	0.02	0.40	0.02

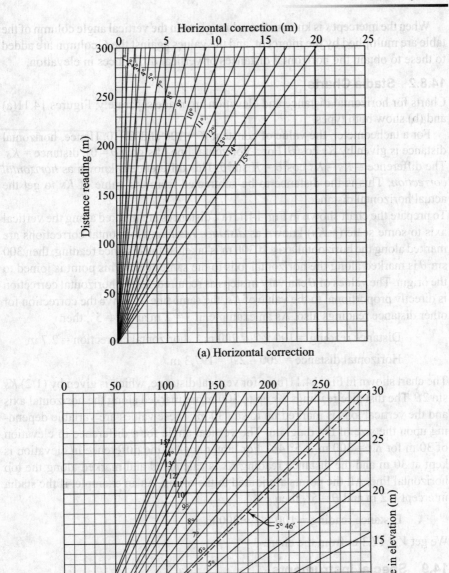

(a) Horizontal correction

(b) Elevation

Fig. 14.11 Stadia charts

When the intercept *s* is known, the values given in the vertical angle column of the table are multiplied by the intercept, and the values against the *C* column are added to these to obtain the horizontal distances or vertical differences in elevation.

14.8.2 Stadia Charts

Charts for horizontal distance and elevation are made separately. Figures 14.11(a) and (b) show both types.

For a tacheometer, the value of constant $K = 100$ and $C = 0$. Hence, horizontal distance is given by $Ks \cos^2\theta$. For a horizontal line of sight ($\theta = 0$), distance = Ks. The difference = $Ks - Ks \cos^2\theta = K \sin^2\theta$. This difference is known as *horizontal correction*. This is the distance to be subtracted from the value of Ks to get the actual horizontal distance.

To prepare the chart shown in Fig. 14.11(a), distances are marked along the vertical axis to some scale (= Ks) known as *distance reading*. Horizontal corrections are marked along the horizontal axis. If 300 m is taken as the distance reading, then 300 $\sin^2\theta$ is marked along the horizontal axis to the same scale. This point is joined to the origin. The values of θ can vary in steps as required. As the horizontal correction is directly proportional to the value of Ks, the same line will give the correction for other distance readings also. As an example, if $s = 2$ m and $\theta = 5°$, then

Distance reading = $100 \times 2 = 200$ m,   horizontal correction = 2.7 m

Horizontal distance = $200 - 2.7 = 197.3$ m

The chart shown in Fig. 14.11(b) is for vertical distance, which is given by $(1/2) Ks \sin 2\theta$. The distance readings Ks up to 300 m are marked along the horizontal axis and the vertical axis is marked up to say 30 m. These values are variable depending upon the size of the diagram. The vertical angle for a difference in elevation of 30 m for $Ks = 300$ m is 5° 46′. For larger angles, the difference in elevation is kept at 30 m and the distance reading Ks is calculated and marked along the top horizontal line. All the points are joined to the origin. As an example, if the stadia intercept = 2 m and $\theta = 5°$, then

Distance reading = $100 \times 2 = 200$ m

We get $V = 17.3$ m from the chart.

14.9 Special Instruments

When the number of observations is large, the calculations of distances and elevations with an inclined line of sight become laborious. To reduce this toil, many devices or techniques have been developed. Some of these are discussed in this section.

14.9.1 Beaman Stadia Arc

The Beaman stadia arc is a device that can be attached externally to a theodolite or tacheometer. A sketch of the Beaman stadia arc is shown in Fig. 14.12. The objective is to reduce the number of observations required for the distances and elevations directly. These observations depend upon $\cos^2\theta$ and $(1/2) \sin 2\theta$. $\cos^2\theta$ varies very slowly with the small angles generally used in tacheometry. The markings are thus based on $(1/2) \sin 2\theta$ and the elevations and distances can be worked out easily. Beaman stadia arc is based on the principle that for certain angles, the value of $(1/2) \sin 2\theta$ is a convenient figure. Such values are shown in Table 14.8.

Table 14.8 Beaman stadia arc angles

(1/2) sin 2θ	Angle	(1/2) sin 2θ	Angle
0.01	00° 34' 23"	0.06	03° 26' 46"
0.02	01° 08' 46"	0.07	04° 01' 26"
0.03	01° 43' 12"	0.08	04° 36' 12"
0.04	02° 17' 39"	0.09	05° 11' 06"
0.05	02° 52' 11"	0.10	0° 46' 07"

The arc has two scales—one for horizontal distance and the other for vertical distance or difference in elevation. The vertical scale is marked with 50 at the centre and increases on one side and decreases on the other side. Each graduation corresponds to the angle corresponding to the value shown in Table 14.8. Thus, if the reading is 54, it means that the vertical angle is equal to 2° 17' 39". Values less than 50, such as 48 or 44, will correspond to angles of depression.

For $K = 100$ and $C = 0$, $V = (1/2) Ks \sin 2\theta$. If $(1/2) \sin 2\theta = 0.01$, then $V = s$; if $(1/2) \sin 2\theta = 0.02$, then $V = 2s$; and so on.

For horizontal distances, the H-scale is marked for horizontal correction. The H-scale reading multiplied by the intercept gives the horizontal correction.

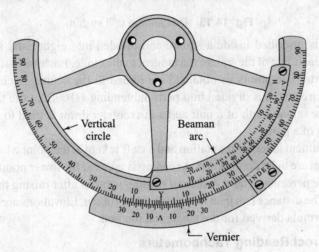

Fig. 14.12 Beaman stadia arc

Horizontal distance $D = 100\, s -$ horizontal arc reading $\times\, s$

In practice, the staff or stadia rod kept at the station is sighted and the staff intercept s is noted. If the index does not rest against a graduation of the Beaman stadia arc, the tangent screw is used to place the index against the nearest graduation of the stadia arc. There is no significant change in the intercept due to this. The arc reading is taken and the horizontal distance and elevation are calculated.

14.9.2   Fergusson's Percentage Unit System

The percentage system devised by J.C. Fergusson helps in the case of tangential tache-ometry. It essentially consists of markings of tangents of angles as percentages.

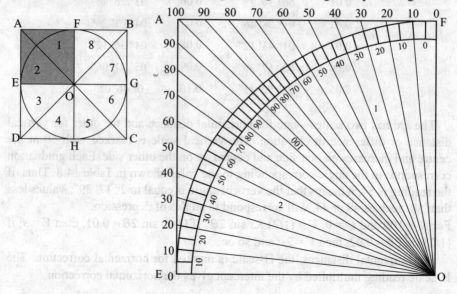

Fig. 14.13   Percentage unit system

A circle is inscribed inside a square and divided into eight parts as shown in Fig. 14.13. Each side of the square is a tangent to the circle. Each half-side is divided into 100 parts. The points so obtained are joined to the centre. Each one-eighth part of the circle is thus divided into parts subtending 1/100th of the tangent. The markings are in one-tenth of a unit and a micrometer drum is used to measure up to one-tenth of a part.

The instrument is set up at a station and a staff is kept at the point whose distance and elevation are to be determined. A reading is taken at the lower point of the staff with a whole percentage unit. A second reading is taken after raising the telescope by n units. The distance can then be calculated as $100s/n$. Elevations are determined using the formula derived for the tangential method.

14.9.3   Direct Reading Tacheometers

Direct reading tacheometers are specially designed to facilitate the calculation of distances and elevations. Once the intercept is read, one has to multiply it by a simple factor like 10 or 100 to get the required value.

Jeffcot direct reading tacheometer

The view from the eyepiece of this tacheometer is shown in Fig. 14.14(a). There is a fixed pointer and two movable pointers. The movable pointer on the left is for horizontal distance and that on the right is for vertical difference in elevation. The two movable pointers are activated by cams that synchronize with the vertical angle measured. The intercepts are read between the fixed pointer and the movable pointer. Generally, distances and elevations are obtained as

Distance = 100 × intercept between fixed pointer and left movable pointer

$V = 10$ × intercept between the fixed pointer and right movable pointer

Szepessy direct reading tacheometer

The view from the eyepiece of this tacheometer is shown in Fig. 14.14(b). This instrument uses the concept of percentage angles and the tangential method. It has a scale marked with percentage angles, meaning that a reading of 15 on the scale indicates an angle whose tangent is 0.15. This scale is aligned with the help of prisms, along with the staff seen through the eyepiece.

The following procedure is adopted to find the distance and difference in elevation using this tacheometer.

1. Set up the instrument at a station and perform the temporary adjustments.
2. Keep the staff over the point whose distance and elevation are to be determined.
3. Sight the staff and set the line of sight at a convenient position. With the vertical circle tangent screw, bring one of the main divisions of the tangent scale along the horizontal cross hair.
4. Read the staff intercept. Let it be s. Let the tangent scale reading be r. Then,

   Horizontal distance $D = 100s$

   Difference in elevation = rs

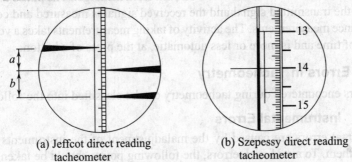

(a) Jeffcot direct reading tacheometer   (b) Szepessy direct reading tacheometer

Fig. 14.14   Direct reading tacheometers

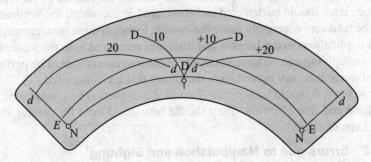

Fig. 14.15   Auto-reduction tacheometer (Hammer and Fennel)

14.9.4   Auto-reduction Tacheometers

Of the many auto-reduction tachometers conceived, the one designed by Hammer and Fennel is the most popular one. This instrument provides the distance and difference in elevation through a single reading of a rod held at a station. A special auto-reduction device is fitted onto the instrument. The view from the telescope is divided into two parts. One half gives the view of the staff whereas the other half shows the auto-reduction diagram (Fig. 14.15). Four curves can be seen, namely N, E, d, and D. Separate curves are seen for angles of elevation and angles of depression. The N-curve is the zero curve, the E-curve is the distance curve, the D-curve is used for vertical angles up to 14, and the d-curve for vertical angles greates 14 and up to 47. The D- and d-curves are height curves. The multiplying constant for distances is 100. For differences in elevation, this constant varies depending upon the angle.

To find distance and elevation, the N-curve (zero-curve) is made to bisect a specially marked zero point on the staff. At this point, the E-curve reading is taken and multiplied by 100 to obtain the distance. Similarly, the D- or d-curve is read (depending upon the vertical angle) and multiplied by 10 or 20 to obtain the difference of elevation.

14.9.5   Electronic Tacheometers

Electronic tacheometers work on the principle of electromagnetic distance measurement. Electromagnetic radiation waves are used with special receiving and transmitting targets at the station point. The horizontal and vertical angles are measured using special glass graduated circles as in an electronic theodolite. The ranging unit transmits the radiation which is received back from the target. The phase difference between the transmitted signal and the received signal is measured and converted into distance measurements. The activity of taking measurements takes a very small amount of time and is more or less automatic, at the press of a button.

14.10   Errors in Tacheometry

The errors encountered during tacheometry can be classified into the following.

14.10.1   Instrumental Errors

Instrumental errors are caused by the maladjustment of the instruments used or faults in them. To avoid such errors, the following points should be taken care of. The tachometer should be in perfect adjustment for taking observations. The altitude bubble should be at the centre of its run while reading the vertical circle for angles. Any index error should be detected and eliminated or accounted for. Vertical angles should be read very accurately using the vernier. Errors in these measurements have serious implications in the heights and distances measured with the instrument.

The multiplying and additive constants of the instrument should be periodically checked to see that they indeed have the values that are being used.

The stadia rod should be accurately divided into parts. The graduations should be uniform and free of errors. They should be marked bold for greater visibility from a large distance.

14.10.2   Errors due to Manipulation and Sighting

Errors due to manipulation and sighting include the following.
(a) Inaccurate levelling of the instrument
(b) Inaccurate reading of horizontal and vertical angles
(c) Focusing errors
(d) Inaccurate bisection of the target

(e) Inaccurate reading of the staff intercept
(f) Errors in holding the staff

14.10.3 Errors due to Natural Causes

Errors due to natural causes include errors due to, say, high winds or the refraction pattern changing during observation. During high winds it is difficult to keep the staff vertical and read it accurately. Unequal refraction is caused by the different densities of different layers of air. A line of sight passing too close to the ground should be avoided. Work should not be undertaken during the hot mid-day period. In very hot conditions, the instrument should be protected with an umbrella to avoid errors due to the unequal expansion of different parts. Work may also be hampered by bad visibility due to strong sunlight and glare.

14.10.4 Precision in Tacheometry

The accuracy attainable of tacheometry depends upon many factors such as the optical accuracy of the instrument, the precision of the stadia rod markings, length of sight, vertical angle, and the observer's eyesight. The error in the calculated horizontal distances should not exceed 1 in 500 for a single observation under average conditions. For vertical distances, the error should not exceed 0.1 m.

Accuracy up to 1 in 1000 can be obtained by taking great care during the various operations, and even greater accuracies can be obtained with better instruments. The average error in the measurement of horizontal distances varies from 1 in 600 to 1 in 850. In traverse survey, the closing error should not be greater than 0.055 m, where P is the perimeter of the traverse.

The error of closure in elevation varies from $0.08\sqrt{D}$ to $0.25\sqrt{D}$, where D is the distance in kilometres.

14.11 Effect of Errors on Distance and Elevation

In this section we will study the effect of common errors on calculated values. The following errors will be considered.
(a) Error due to staff not being normal to the line of sight
(b) Error due to non-verticality of staff
(c) Error due to misreading of vertical angle

14.11.1 Staff Not Being Normal

Consider the case shown in Fig. 14.16. PQ is normal to the line of sight at C. The staff is held in the position P'Q' at an inclination of α from this normal.

Let OC be the line of sight inclined at an angle θ to the horizontal. PQ is the position of the stadia rod normal to the line of sight. The rod is, however, held along P'Q' at an angle α to the correct position. P, C, and Q cor-

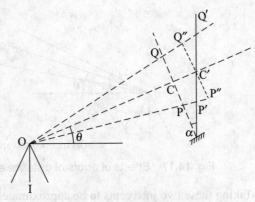

Fig. 14.16 Effect of errors of distance and elevations, staff not normal

respond to the three cross hairs in the normal position and P′, C′, and Q′ correspond to the incorrect position. Draw P″C′Q″ normal to the line of sight at C′. The staff intercept $s = PQ = P″Q″$. The staff intercept observed is $s′ = P′Q′$. Taking the angles at P″ and Q″ as 90° approximately, it is clear that

$$P″Q″ = PQ = P′Q′ \cos \alpha$$

Assuming that the multiplying constant is K and the additive constant $C = 0$,

Distance calculated = $Ks′$,   correct distance = $Ks = Ks′ \cos \alpha$

Error in distance = $Ks′ - Ks′ \cos \alpha = Ks′(1 - \cos \alpha)$

14.11.2 Non-verticality of Staff

If the staff is supposed to be held vertical but is kept out of plumb, an error is introduced in the measurement. To determine this error, consider the cases illustrated in Fig. 14.17. The correct position of the staff is PQ. The staff is, however, held along the direction P′Q′ at an angle α to the vertical. The staff can be inclined either away from the observer or towards the observer.

Let us consider the case in which the staff is held away from the observer [Fig. 14.17(a)].

$$P_1Q_1 = PQ \cos \theta = s \cos \theta \quad \text{and} \quad P_2Q_2 = P′Q′ \cos(\theta + \alpha) = s′ \cos(\theta + \alpha)$$

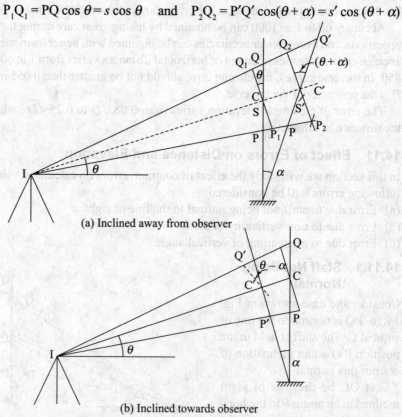

(a) Inclined away from observer

(b) Inclined towards observer

Fig. 14.17 Effects of errors of distance and elevation, staff not vertical

Taking these two intercepts to be approximately equal, we have

$$s \cos \theta = s′ \cos(\theta + \alpha)$$

from which

$$s = s'\cos(\theta + \alpha)/\cos\theta$$

If the staff was inclined at an angle α towards the observer, then

$$s = s'\cos(\theta - \alpha)/\cos\theta$$

Therefore, in either case,

$$s = s'\cos(\theta \pm \alpha)/\cos\theta$$

(Use the plus sign when the staff is tilted away from the observer.) Here s is the true intercept and s' is the observed intercept.

True distance $= Ks\cos^2\theta = K[s'\cos(\theta \pm \alpha)/\cos\theta]\cos^2\theta$

Incorrect distance $= Ks'\cos^2\theta$

Error $= Ks'\cos^2\theta\{[\cos(\theta \pm \alpha) - \cos\theta]/\cos\theta\}$

14.11.3 Misreading the Vertical Angle

If the staff is held vertical and the line of sight is inclined at an angle θ, the horizontal distance D (assuming $C = 0$) is given by

$$D = Ks\cos^2\theta$$

Differentiating,

$$\delta D = -2Ks\cos\theta\sin\theta\,\delta\theta$$

where $\delta\theta$ is the error in the measurement of the angle θ.

$$\delta D/D = [2Ks\cos\theta\sin\theta\,\delta\theta]/Ks\cos^2\theta = 2\tan\theta\,\delta\theta$$

Depending upon the desired value of $\delta D/D$, the accuracy of angle measurement can be determined.

$$\delta\theta = [(\delta D/D)\cot\theta]/2$$

If $\delta D/D = 1$ in 850 and $\theta = 30$, then

$$\delta\theta = \cot 30° \times 206,265/1700'' = 3'\,30''$$

Thus, if the overall precision in distance required is 1 in 850, then the angle close to 30° should be measured to a precision of $3'\,30''$.

Example 14.19 A tacheometer was kept at a station P and observations were made to a staff kept at Q. The staff was supposed to be kept normal but was kept 3° 30′ out of normal, away from observer. The vertical angle was 4° 30′ and the cross hair readings were 1.235, 2.2, and 3.165. Find the percentage error in the horizontal distance.

Solution   Staff intercept $= 3.165 - 1.235 = 1.93$ m

Horizontal distance calculated $= 100 \times 1.93 = 193$ m

Intercept normal to the line of sight $= 1.93 \times \cos 3° 30' = 1.926$ m

Correct distance $= 100 \times 1.926 = 192.6$ m

Error $= 193 - 192.6 = 0.4$ m

Error % $= 0.4 \times 100/193 = 0.21\%$

Example 14.20 The vertical angle of the line of sight to B from a tacheometer kept at A was 8° 30′. The staff was supposed to be held vertical at B. The readings taken at the outer cross hairs were 2.385 m and 1.063 m. It was found that the staff, 4 m long, was kept out of plumb by 120 mm. Find the error in the horizontal distance calculated due to the non-verticality of the staff.

Solution The inclination of the staff is given by

$$\tan \alpha = 0.12/4 = 0.03, \quad \alpha = 1°\ 43'\ 06''$$

Staff intercept $= 2.385 - 1.063 = 1.322$ m

Correct value of intercept $= 1.322 \cos (8°\ 30' + 1°\ 43'\ 06'')/\cos (8°\ 30')$

$$= 1.315 \text{ m}$$

Correct horizontal distance $= 100 \times 1.315 \cos^2 (8°\ 30') = 128.627$ m

Error in distance $= 100 \cos^2 (8°\ 30') (1.322 - 1.315) = 0.685$ m

14.12 Precision in Tacheometry

The accuracy that can be obtained with a tacheometer depends on the instrument features, such as optical power and precision, graduations on the stadia rod, length of sight, and the vertical angle in addition to natural conditions of temperature, wind, etc.

The average error in horizontal distance is from 1 in 600 to 1 in 850, but an accuracy of 1 in 1000 is normally attainable. Higher precision can be obtained with a shorter length of the line of sight. In a stadia traverse, the closing error should not be greater than $0.055 \sqrt{P}$, where P is the perimeter of the traverse.

The error in closure of elevation varies from $0.08 \sqrt{K}$ to $0.25 \sqrt{K}$ for plane and hilly areas, where K is the distance in kilometres.

Summary

Tacheometry or telemetry is a method of surveying in which distances are measured indirectly. Tacheometry is done with a theodolite, a stadia diaphragm, or a tachometer. The appropriate instrument is used in conjunction with a levelling staff in the case of ordinary work and a stadia rod in the case of precise work. A stadia diaphragm has three horizontal cross hairs. The difference in readings obtained with the top and bottom hairs is known as *stadia intercept*. The distance between the top and bottom cross hairs is known as *stadia interval*.

The basic principle of tacheometry is the similarity of the triangles formed by the object and its image as bases.

With a horizontal line of sight, the vertical circle reading zero, the horizontal distance is given by $D = Ks + C$, where K and C are instrument constants. The constants are obtained in terms of f (the focal length of the objective lens) and i (the stadia interval). $K = f/i$ is known as the *multiplying constant* and $C = f + d$ is known as the *additive constant*. K is generally kept at a value of 100. Here d is the distance between the objective lens and the vertical axis. When the line of sight is inclined, the distance and elevation formulae, respectively, become

$$D = Ks \cos^2\theta + C \cos \theta \quad \text{and} \quad V = (1/2) Ks \sin 2\theta + C \sin \theta$$

In a tacheometer, the constant C is zero. Internal focusing telescopes have a very small value of C. In an external focusing telescope C is made zero by using an anallactic lens. An anallactic lens is a double convex lens introduced between the objective lens and the diaphragm such that the vertex of the tacheometric triangle coincides with the vertical axis, thus making the additive constant C zero.

In *movable hair tacheometry*, a special diaphragm is used in which the stadia interval i can be varied using a micrometer screw. Using a special subtense bar which gives a constant stadia intercept, the stadia interval is varied and the distance and elevations can be calculated using the same formulae.

A third method of tacheometry, known as the *tangential method*, does not use the stadia wires but calculates the distances and elevations from the vertical angles measured to two points on the staff.

Stadia readings are reduced in the case of large-scale work using stadia tables or charts.

The Beaman stadia arc is a device that facilitates the reduction of stadia readings for horizontal and vertical distances. Direct reading tacheometers enable us to take direct readings and multiply the intercept by 100 or 10 for distances and elevations. The auto-reduction tacheometer is an instrument that facilitates the direct reading of distances and elevations. Electronic tacheometers are instruments with digital display and the facility of storage of distances and elevations. They measure and display horizontal and vertical angles as in electronic theodolites. For distance measurement, a ranging unit using electromagnetic radiation is used.

Exercises

Multiple-Choice Questions

1. Stadia tacheometry is based on the principle that
   (a) trigonometrical formulae can be used to calculate distances from vertical angles
   (b) intercepts on measuring rods are proportional to the distance
   (c) horizontal distances vary linearly as vertical angles
   (d) knowing the side and two angles of a triangle, another side can be calculated

2. A stadia diaphragm has
   (a) a single cross wire at the centre
   (b) two cross wires at equal distances from centre
   (c) two cross wires, one at the centre and another above or below it
   (d) three cross wires, one at the centre and another two at equal distances on either side

3. Movable hair tacheometry is based on the principle that
   (a) intercepts on a stadia rod are proportional to the distance
   (b) trigonometrical formulae can be used to find distance from vertical angles
   (c) a constant intercept on the staff obtained by moving stadia wires is a measure of distance
   (d) by measuring two vertical angles distance can be calculated

4. A theodolite can be used as a stadia tacheometer if it has
   (a) an external focusing telescope
   (b) a Ramsden's eyepiece
   (c) an internal focusing telescope and stadia diaphragm
   (d) an external focusing telescope with stadia diaphragm

5. The distance formula for finding distances using a theodolite, for a horizontal line of sight, is (K is multiplying constant, s is intercept and C is additive constant)
   (a) $Ks + C$ (b) $K + Cs$ (c) $K/s + C$ (d) $K + Cs$

6. The multiplying constant in the distance formula by tacheometry is given by
   (a) focal length of objective lens divided by the distance between stadia wires
   (b) focal length of objective lens multiplied by the distance between stadia wires
   (c) stadia intercept divided by the focal length of objective lens
   (d) stadia intercept multiplied by the focal length of the objective

7. The additive constant in the distance formula by tacheometry is given by
   (a) focal length of objective lens divided by the distance from the objective lens to the vertical axis

    (b) focal length of the objective lens multiplied by the distance from the objective lens to the vertical axis

    (c) sum of focal length of objective lens and the distance from objective lens to vertical axis

    (d) the distance between the diaphragm and the objective lens

8. The additive constant of a theodolite, used as a tacheometer, is nearly zero if it
    (a) has a shorter telescope   (c) is of the internal focusing type
    (b) has an external focusing telescope   (d) the objective lens diameter is small

9. By using a tacheometer with a horizontal line of sight, the three hair readings were recorded as 1.980, 1.660, and 1.340 m. If the multiplying constant is 100 and, additive constant is zero, then the distance between the instrument and stadia rod station is
    (a) 198 m   (b) 166 m   (c) 64 m   (d) 32 m

10. When the line of sight is inclined, having an angle of elevation, the tacheometer formula for distance with staff held vertical is
    (a) $Ks + C$ (c) $Ks \cos \theta + C \cos \theta$
    (b) $Ks \cos \theta + C$ (d) $Ks \cos^2 \theta + C \cos \theta$

11. The distance formula with an inclined line of sight, for an angle of depression, of a tacheometer when the staff is held perpendicular to the line of sight is
    (a) $Ks + C - h$ (c) $Ks \cos \theta + C \cos \theta - h$
    (b) $Ks \cos \theta + C - h$ (d) $(Ks + (C) \cos \theta - h \sin \theta$

12. In vertical subtense bar (movable hair) method,
    (a) the staff intercept is made a constant value
    (b) the distance between cross hairs is kept constant
    (c) the multiplying constant is made 100
    (d) the additive constant is made zero

13. By using an anallactic lens, in a theodolite
    (a) the multiplying constant is made 100
    (b) the additive constant is made zero
    (c) we get a clearer image of the staff
    (d) the cross wires are more clearly seen

14. The tangential method of tacheometry uses
    (a) the readings against all the three cross hairs
    (b) the readings against the top and bottom cross hairs only
    (c) the reading against the middle cross hair only
    (d) a constant intercept on the staff

Review Questions

1. Explain the basic principle of tacheometry.
2. State the tacheometric formula for horizontal and inclined lines of sight and explain what each term depicts.
3. Derive the distance and elevation formulae for an inclined line of sight with an angle of elevation and an angle of depression.
4. What is an anallactic lens? In which telescope is it used? What is the condition under which the additive constant is zero with an anallactic lens?
5. Explain the subtense bar method of tacheometry. Explain the difference between the fixed hair method and the movable hair method of tacheometry.
6. Draw a neat sketch of a movable hair stadia diaphragm and label its parts.
7. Explain the tangential method of tacheometry.
8. Derive the formulae for distance and elevation in the tangential method.
9. What is a tacheometric table? Explain the use of such a table.
10. Draw a neat sketch and explain the method of using the Beaman stadia arc.

11. Explain the working of a direct reading tacheometer.
12. What is an auto-reduction tacheometer? Explain its functioning.
13. Explain the features of an electronic tacheometer and its functioning.
14. What are the common errors and mistakes encountered in tacheometry. Explain the precautions to be taken to eliminate them.
15. Derive a formula for the correction to the distance value calculated due to the staff not being normal to the line of sight.

Problems

1. A tacheometer was kept at a station P and observations were made to a stadia rod kept at station Q. The stadia readings were 1.135, 2.05, 2.965. The reading on a staff held at a BM of RL 110.95 was 2.135 m. Find the distance PQ and the RL of point P.
2. To determine the tacheometric constants, the following observations were made:

Distance	50 m	100 m
Stadia readings	1.235, 1.483, 1.731	1.345, 1.843, 2.341

Find the constants of the instrument.
3. The stadia readings obtained with a horizontal line of sight from an instrument were 1.365, 0.996, and 2.361 at a distance of 100 m. If the focal length of the objective lens was 25 cm and the distance between the objective lens and the vertical axis was 15 cm, find the stadia interval. The tacheometric constants are $K = 100$ and $C = 0$.
4. A tacheometer was kept at station A and observations were taken on a stadia rod kept over station B. The vertical angle was 6° 30′ and the stadia readings were 1.375, 2.003, 2.631, the staff being held normal to the line of sight. Also the reading on a staff held at a benchmark 878.55 m was 1.875 m. Find the distance AB and the RL of point B. $K = 100$, $C = 0$.
5. A tacheometer was kept at a station P and observations were made to a staff held vertically at Q. The cross hair readings were 1.835, 1.92, and 3.755. The vertical angle of depression was 8° 06′. From the same set-up, the reading on a staff held at a BM of RL 962.55 was 2.035 m. Find the horizontal distance PQ and the RL of point Q. $K = 100$, $C = 0$.
6. Find the difference in elevation between stations P and Q from the data given below The stadia constants are $K = 100$, $C = 0.3$.

Instrument at	Staff at	Vertical angle	Stadia readings
A	P	+ 3° 15′	1.355, 2.58, 3.935
	Q	− 1° 45′	0.985, 1.66, 2.335

7. Find the horizontal distance PQ and the gradient from P to Q from the data given below ($K = 100$, $C = 0.3$).

Instrument at	Staff at	Vertical angle	Cross hair readings
A	P	6° 50′	
	Q	3° 30′	

Horizontal angle PAQ = 68° 30′
8. In a plot of land in the shape of a hexagon, a tacheometer was kept at one of the vertices and readings were taken on a staff held at three consecutive points from that station. The data collected is as follows.

Instrument at	Staff at	Vertical angle	Cross hair readings
A	B	2° 30′	1.256, 1.404, 1.552
A	C	1° 45′	1.865, 2.125, 2.385
	D	0° 56′	1.634, 1.932, 2.230

Check whether the plot is a regular hexagon.

9. Find the stadia constants from the data given below.

Instrument at	Staff at	Cross hair readings	Distance
O	P	1.135, 1.285, 1.435	OP = 30 m
O	Q	1.025, 1.324, 1.625	OQ = 60 m

10. Find the stadia constants of a theodolite from the data given below.

Instrument at	Staff at	Vertical angle	Intercept	Distance
P	A	2° 30′	0.999	100
	B	1° 36′	1.998	200

11. Determine the horizontal distance to station Q from station P if the vertical angle observed from P was 6° 30′. The distance between the target vanes was 3 m and the number of revolutions of the micrometer screw noted was 18.55. The instruments constants are 1000 and 0.4.

12. A tacheometer was kept at station A and the target at station B. The intercept was kept at 5 m. The vertical angle was 4° 45′. The instrument constant was 1000. If the distance between A and B is known to be 365 m, find the number of turns of the micrometer screw on the movable hair diaphragm.

13. A distance PQ was measured with a tacheometer (constants 100 and 0.3). The vertical angle was 5° 30′ and the intercept was 2.15 m. If a movable hair instrument was used to measure the same distance, find the constant K of the instrument if the number of turns of the micrometer screw was observed to be 15.64 for an intercept of 4 m and the same angle of elevation. C = 0.4.

14. A tacheometer was kept at A and the stadia rod was held at B. The vertical angles measured to the 1-m and 4-m marks on the rod were 3° 45′ and 5° 30′, respectively. If the RL of the instrument axis is 1050 m, find the distance AB and the RL of station B.

15. From the instrument kept at station P, observations were made to a staff held at Q. The vertical angles measured to the 1-m and 6-m marks were − 1° 45′ and 3° 30′. Find the horizontal distance PQ.

16. From the instrument kept at A, the following vertical angles were observed: staff at P: 1° 30′ to the 1-m mark and 6° 45′ to the 4-m mark; staff at Q: 0° 50′ to the 0.5-m mark and 4° 30′ to the 4-m mark. The horizontal angle PAQ was measured as 62° 30′. The reading at a BM (RL 905 m) was 2.325 m. Determine the distance PQ and the gradient from P to Q.

17. A staff supposed to be held normal was held inclined at an angle to the line away from the observer. The line of sight was inclined at 20° (elevation) to the horizontal. Find the maximum value of the angle such that the error in the distance measured is not more than 0.1%. K = 100 and C = 0.

18. The vertical angle to a station B was measured as 16° 45′. The staff intercept was 3.15 m. The staff was supposed to be held vertical but was out of plumb by 50 mm in 4 m (away from the observer). Find the error in the horizontal distance. K = 100, C = 0.

CHAPTER

15

Curve Surveying

Learning Objectives

After going through this chapter, the reader will be able to

- state the purpose of and explain the classification of curves
- describe the methods of setting out curves in the field
- calculate the necessary quantities for setting out curves by linear methods
- calculate the necessary quantities for setting out curves by angular methods
- state the purpose and explain the profile of transition curves
- calculate the quantities required to set out a transition curve
- state the purpose and explain the profile of vertical curves
- calculate the quantities required to set out vertical curves

Introduction

Curves are used in roads and railway tracks to change the direction of motion of vehicles. When a moving vehicle has to change its direction, it can do so only by gradual directional changes. It is not possible to effect a direct change at a single point as shown in Fig. 15.1(a). To provide a smooth change of direction, a curve is introduced between two straight lines [Fig. 15.1(b)]. Curves need to be carefully designed and prepared for the safety of vehicles and comfort of drivers and passengers. Very sharp and blind curves are sometimes necessary in hill roads.

A vehicle must move at a lower speed along curves, as the kinetics of its motion gives rise to a force perpendicular to the direction of motion. Vehicles moving at higher speeds are in danger of being thrown out of the road due to this force. The road is also banked on the outer side, called superelevation, for safety.

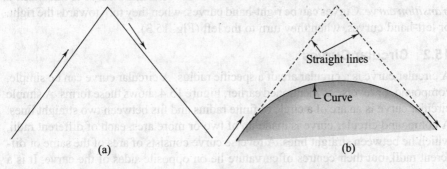

Straight lines

Curve

(a)   (b)

Fig. 15.1 Curve for smooth change of direction

In this chapter, we will study about curves, their geometry, and the methods of laying them out.

15.1 Types of Curves

Curves can be basically classified into two types—horizontal curves, (curves in plan), and vertical curves (curves in a vertical section). Of these two types, horizontal curves are more common and they often lie along a gradient. Vertical curves are used in hilly terrain to move from one elevation to another smoothly. Figure 15.2 shows both these types of curves.

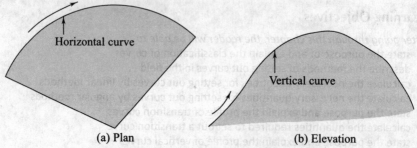

(a) Plan   (b) Elevation

Fig. 15.2 Horizontal and vertical curves

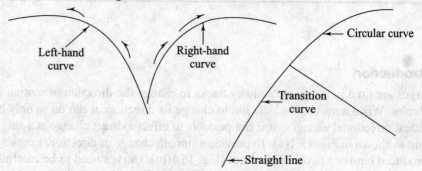

Fig. 15.3 Classification of curves

Curves can be classified based on their profile into circular curves, parabolic curves, spirals, etc. A curve, of finite radius, has to be attached to a straight line, which is a curve of infinite radius. There is a sharp change in the radius at the meeting point. To smoothen out this sharp change of radius, a curve of varying radius is introduced between the straight part and the curve. Such a curve is known as a *transition curve*. Curves can be right-hand curves, when they turn towards the right, or left-hand curves, when they turn to the left (Fig. 15.3).

15.2 Circular Curves

A circular curve is a circular arc of a specific radius. A circular curve can be simple, compound, or reverse as mentioned earlier. Figure 15.4 shows these forms. A simple circular curve is an arc of a circle of finite radius and fits between two straight lines. A compound circular curve is made up of two or more arcs, each of different radii, which lie between straight lines. A reverse curve consists of arcs of the same or different radii, but their centres of curvature lie on opposite sides of the curve. It is a combination of a right-hand and a left-hand curve or a reverse combination.

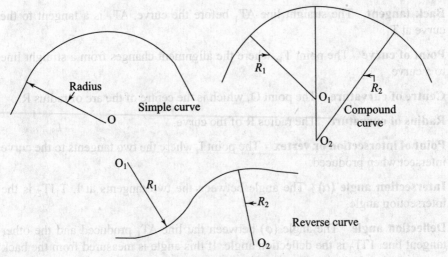

Fig. 15.4 Classification of curves

15.2.1 Terminology

We lay out the terminology of circular curves with the example of a simple curve. The notations of a simple circular curve are shown in Fig. 15.5. A simple circular curve lies between two straight lines. The curve starts from the end T_1 of a straight line AT_1 and ends at the beginning, T_2, of another straight line, T_2B. Both the straight lines are tangents to the curve. The following terminology with reference to circular curves needs to be understood.

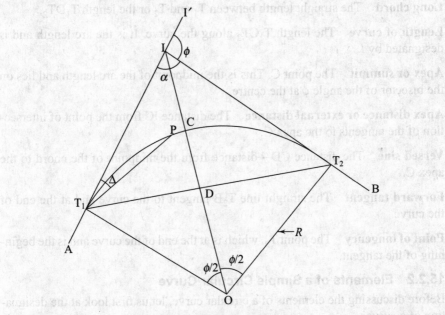

Fig. 15.5 Terminology of curves

Back tangent   The straight line AT_1 before the curve. AT_1 is a tangent to the curve at T_1.

Point of curve   The point T_1, where the alignment changes from a straight line to a curve.

Centre of curvature   The point O, which is the centre of the arc of radius R.

Radius of curvature   The radius R of the curve.

Point of intersection or vertex   The point I, where the two tangents to the curve intersect when produced.

Intersection angle (α)   The angle between the two tangents at I. T_1IT_2 is the intersection angle.

Deflection angle   The angle (ϕ) between the line AT_1 produced and the other tangent line. $I'IT_2$ is the deflection angle. If this angle is measured from the back tangent in the clockwise direction, the curve is a right-hand curve. If it is measured in the anticlockwise direction from the back tangent, the curve is a left-hand curve. The deflection angle is also the angle between the radii OT_1 and OT_2, from the properties of circles.

Deflection angle to a point   The deflection angle to any point like P on the curve is the angle by which the chord to that point, T_1P, deflects from the back tangent. The angle IT_1P is the deflection angle at P.

Tangent lengths   The lengths IT_1 or IT_2 are known as tangent lengths or tangent distances.

Long chord   The straight length between T_1 and T_2 or the length T_1DT_2.

Length of curve   The length T_1CT_2 along the curve. It is the arc length and is designated by l.

Apex or summit   The point C. This is the midpoint of the arc length and lies on the bisector of the angle ϕ at the centre.

Apex distance or external distance   The distance IC from the point of intersection of the tangents to the apex C.

Versed sine   The distance CD—distance from the midpoint of the chord to the apex C.

Forward tangent   The straight line T_2B tangent to the curve and at the end of the curve.

Point of tangency   The point T_2, which is at the end of the curve and is the beginning of the tangent.

15.2.2   Elements of a Simple Circular Curve

Before discussing the elements of a circular curve, let us first look at the designation of a curve.

Designation of a curve There are two methods of designating a curve, either by the degree of the curve (e.g., a 2° curve) or by the radius of the curve (e.g., a 300-m curve). In India, designation by the degree of the curve is commonly adopted. The degree of a curve is the angle subtended at the centre by a specified length of chord or arc. In metric units, the length can be 20 or 30 m.

Chord definition If D is the degree of the curve, then from Fig. 15.6(a),

$$\sin(D/2) = T_1 D/R, \quad R = T_1 D/\sin(D/2)$$

$T_1 D = 15$ or 10 m. The values of radius for some degree values are given in Table 15.1.

When D is small, $\sin(D/2)$ can be taken equal to $D/2$ radians. Then,

$$\text{Radius} = 15/(D \times \pi/2 \times 180) \text{ for a 30-m chord}$$
$$= 10/(D \times \pi/2 \times 180) \text{ for a 20-m chord}$$

These values are approximate and are listed in the last two columns of Table 15.1.

Table 15.1 Chord definition of curves—radius values for various degrees of curve

Degree of curve	Radius (l = 30 m)	Radius (l = 20 m)	Radius (l = 30 m) (approx.)	Radius (l = 20 m) (approx.)
1	1718.89	1145.93	1718.87	1145.91
2	859.48	572.98	859.44	572.96
3	573.02	382.01	572.95	381.97
4	429.8	286.54	429.72	286.48
5	343.88	229.25	343.77	229.18
6	286.61	191.07	286.48	190.99

Arc definition If the curve is designated by an arc in place of a chord, then the

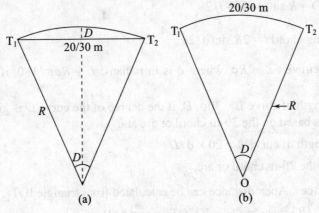

Fig. 15.6 Designation of curves

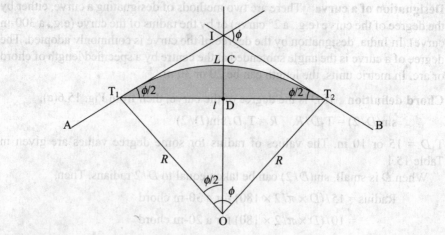

Fig. 15.7 Elements of a simple curve

arc length is 30 or 20 m. The degree of the curve is the angle subtended by an arc of length 30 or 20 m at the centre. From the properties of circles and proportional lengths of arcs, we can write

$$D:30::360:2\pi R, \quad R = 30 \times 360/(2\pi D) = 1718.87/D$$

For a 20-m arc,

$$D:20::360:2\pi R, \quad R = 20 \times 360/(2\pi D) = 1145.91/D$$

Figure 15.7 shows the elements of a simple circular curve.

Tangent length Length T_1I or T_2I. Considering the triangle OT_1I with the right-angle at T_1, if ϕ is the central angle, then

Tangent length = $R \tan(\phi/2)$

Length of the long chord $T_1T_2 = l$. From the triangle OT_1D with the right-angle at D,

$$T_1D = R \sin(\phi/2) = l/2$$

Thus,

Long chord $l = 2R \sin(\phi/2)$

Length of curve $L = R\phi$, where ϕ is in radians; $L = R\phi\pi/180$, if ϕ is in degrees.

Length of curve $L = 30\phi/D$, if the degree of the curve D is given instead of the radius based on the 30-m chord or arc and

Length of curve $L = 20 \times \phi/D$

if based on the 20-m chord or arc.

Apex distance Apex distance can be calculated from triangle IOT_1.

$$OT_1/IO = \cos(\phi/2), \quad IO/OT_1 = \sec(\phi/2)$$

Apex distance = $IO - OC = IO - R, \quad IC + R = IO$

$(IC + R)/OT_1 = \sec \phi/2$; $IC = R \sec \phi/2 - R$

$$= R (\sec \phi/2 - 1)$$

$$= R \operatorname{exsec}(\phi/2)$$

Mid-ordinate Mid-ordinate $DC = OC - OD$. From triangle ODT_1,

$$OD = R \cos(\phi/2) \quad \text{and} \quad OC = R$$

Mid-ordinate $= R - R \cos(\phi/2) = R(1 - \cos \phi/2) = R \operatorname{versine}(\phi/2)$

The following examples illustrate the calculation of the elements of a circular curve.

Example 15.1 Determine the radius of a curve if it is designated as a 3° curve on a 30-m arc.

Solution Radius × degree of curve in radians = 30 m. Therefore,

Degree of curve $= 3° = 3\pi/180$ rad

Radius of curve $= 30 \times 180/3\pi = 573$ m

Example 15.2 If the radius of a curve is 380 m, what is its degree designation on a 20-m arc?

Solution $RD = 20$, D (in radians) $= 20/380 = 0.0526$

D (in degrees) $= 0.0526 \times 180/\pi = 3.01$

Degree designation of the curve $= 3.01$

Example 15.3 If a curve is designated as a 4° curve on a 30-m arc, find the tangent distance, length of long chord, length of arc, apex distance, and mid-ordinate if the deflection angle is 36°.

Solution The radius R of the curve is given by

$R = 30 \times 180/(4\pi) = 429.72$ m

Tangent distance $= R \tan(\phi/2) = 429.72 \times \tan 18° = 139.62$ m

Length of long chord $l = 2R \sin(\phi/2) = 2 \times 429.72 \times \sin 18° = 265.85$ m

Length of arc $L = R\phi = 429.72 \times 36\pi/180 = 270$ m

Apex distance $= R[\sec(\phi/2) - 1] = 429.72(\sec 18° - 1) = 22.13$ m

Mid-ordinate (versine distance) $= R[1 - \cos(\phi/2)] = 429.72(1 - \cos 18°)$

$$= 21.03 \text{ m}$$

Example 15.4 If the tangent distance of a curve is 235.6 m and the deflection angle is 42°, calculate the other parameters of the curve, such as length of long chord, length of curve, apex distance, and mid-ordinate.

Solution Let the length of long chord be l.

Tangent distance $= R \tan(\phi/2)$ and $l = 2R \sin(\phi/2)$

$R \tan(\phi/2) = 235.6$, $R \tan 21° = 235.6$

$2R \sin(21°) = l$

Dividing one by the other,

$\tan(21°)/2 \sin(21°) = 236.6/l$, $l = 220$ m

$R = 235.6/\tan 21° = 613.8$ m

Apex distance = $R[\sec(\phi/2) - 1] = 613.8 \times (\sec 21° - 1) = 43.67$ m

Mid-ordinate = $R[1 - \cos(\phi/2)] = 613.8 \times (1 - \cos 21°) = 40.77$ m

15.2.3 Methods of Ranging Simple Circular Curves

Curves are set on ground by two methods.

Linear methods These methods use only a tape or chain. A curve is set by knowing the distances (offsets) to the curve from main lines such as tangents or the long chord. This is done when the curve is short and when an angle-measuring instrument is not available. The accuracy of setting is not very high.

Angular methods These methods use a theodolite with or without a tape and are more accurate. One or more theodolites are used for setting out a curve. It must be recognized that when linear measurements are used, the measurements have to be along a straight line and cannot be along a curve. The curve set out may be a series of chords and not arcs. In the case of flat curves, with large radii, the difference between arc and chord distances may be insignificant. In the case of curves of short radius, equivalent chord distances can be calculated for setting out the curves. The difference between arc and chord distances can be seen in Table 15.2. The table shows the arc lengths for 100 m chord length. It can be seen that the difference between the arc length and the chord length becomes smaller as the radius becomes larger (as the curve becomes flatter).

Table 15.2 Arc length for 100 m chord length for different radii (all in metres)

Radius	Arc length	Chord length	Radius	Arc length	Chord length
100	104.7197	100	500	100.1674	100
200	101.0721	100	600	100.1161	100
300	100.489	100	700	100.0852	100
400	100.2622	100	800	100.0652	100

The stake-out interval or peg interval is the distance between the pegs set out on the ground. For ease of calculation and setting out, the peg interval is kept constant. Obviously, neither the chord length nor the arc length can be a certain number of chain lengths, either 30 m or 20 m. The first length and the last length, thus, may be shorter than a chain length. A chain length is known as a *full chord* and the shorter ones are known as *subchords*.

Locating main points Before the curve can be set out on the ground, the main points of the curve have to be located. These are the point of intersection and the tangent points. A control survey in the form of a traverse will be available, to which the curve elements can be related. The tangent lines are located by offsets from the control traverse lines. The point of intersection can be located by continuing the tangent lines. The tangent lengths can now be measured on either side to locate the tangent points. If a theodolite is available, the deflection angle can be set out to locate the forward tangent point. These linear measurements must be done very accurately, as the curve setting depends upon the accuracy of these points. Once the point of intersection and the tangent points are located, curves can be set by linear or angular methods.

Linear methods

Linear methods make use of only a chain or tape. Essentially, the points on the curve are located by offset distances from any of the major lines—tangents or the long chord. The following methods are commonly employed.

(a) Offsets from the long chord
(b) Radial offsets from tangents
(c) Perpendicular offsets from tangents
(d) Offsets from the chord produced
(e) Successive bisection of arcs

   Since circular arcs are symmetrical about the centre line—the line joining the point of intersection to the centre of curvature—many methods require only calculating the values for half the arc. The other half will have equal values of offsets. We will now discuss the calculations involved and the procedure for setting out the curve using the values calculated.

   In this discussion, the following notations will be used: R is the radius of the curve, ϕ is the deflection angle or the central angle subtended by the arc, x is the distance along the chord or tangent, Y or O is the offset or ordinate, l is the length of the long chord, and L is the length of the arc.

Offsets or ordinates from the long chord   The geometrical construction is shown in Fig. 15.8. The ordinate will be symmetrical about the mid-ordinate. The length of the long chord is first calculated. If a 20-or 30-m peg interval is chosen, then the length of the long chord may not give exactly an integral number of ordinates at the interval chosen. It will be necessary to use a subchord in the beginning and another subchord at the end. The mid-ordinate, as we have seen, is the versine distance and is given by $R[1 - \cos(\phi/2)] = \text{versine}(\phi/2)$, where ϕ is the deflection angle. The ordinates are calculated from the midpoint d of the long chord. Let the ordinate be Y at point P at a distance x from D. Let P′ be the point on the curve corresponding to the ordinate at P. Then PP′ = Y. Then from triangle OP′E,

$$(OP')^2 = OE^2 + P'E^2, \quad R^2 = OE^2 + x^2$$

Now, OE = OD + DE. DE = Y. OD can be calculated from triangle OT$_1$D:

$$OD^2 = OT_1^2 - T_1D^2, \quad OD^2 = R^2 - (l/2)^2, \quad OD = \sqrt{R^2 - (1/2^2)}$$

$$OE^2 = R^2 - x^2, \quad OE = \sqrt{(R^2 - x^2)}, \quad OD + Y = \sqrt{R^2 - x^2}$$

$$\sqrt{R^2 - (1/2^2)} + Y = \sqrt{R^2 - x^2}$$

From this, the ordinate at x from D is

$$Y = \sqrt{R^2 - x^2} - \sqrt{R^2 - (1/2^2)}$$

When $x = 0$, this gives the mid-ordinate as

$$\text{Mid-ordinate} = R - \sqrt{R^2 - (1/2^2)}$$

The Y values need to be calculated only for one half of the curve. The other half will have exactly the same values for the same distances x.

In the case of very flat curves, where the radius is very large compared to the length of the long chord, an approximate formula can be derived by measuring distances from T_1, the tangent point. This assumes that the vertical ordinate is approximately equal to the radial ordinate. Referring to Fig. 15.8(b), the offset $Y = PP'$, is taken equal to $P'P_1$, the radial offset to the same point P'. From the properties of circles,

$$P'P_1 \times 2R = T_1P' \; P'T_2$$

We take $T_1P' = x$ and $P'T_2 = (l-x)$. Therefore,

$$P'P_1 = x(l-x)/2R$$

While in the previous case, x is measured from D, in this case, x is measured from T_1. It will be necessary to calculate the subchord length for the first x value, ensuring that a full chain length falls at D.

Procedure to set out the curve

1. Set the curve by first locating the intersection point of the tangents.

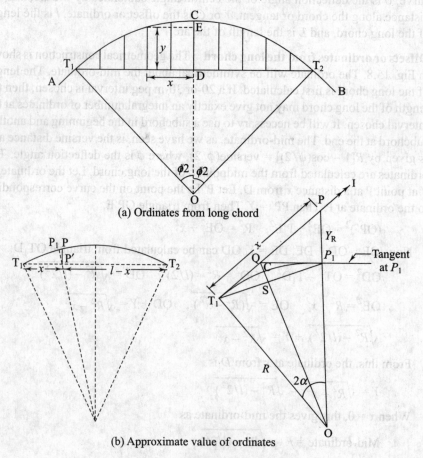

(a) Ordinates from long chord

(b) Approximate value of ordinates

Fig. 15.8 Offsets from long chord

2. Measure the tangent length backward and forward from I to locate the tangent points T_1 and T_2, respectively.

3. Once the tangent points are located, measure the distance of the long chord T_1T_2. Bisect T_1T_2 to get the point D, the midpoint of the chord.

4. If the distance x is measured from midpoint D, then measure a chain length 20 or 30 m from D. Set out a right-angle offset of the calculated length from this point.

5. Locate further points by measuring 20 m or 30 m lengths and setting off the calculated lengths of offsets.

6. Repeat the procedure on the other side of D.

7. If the approximate method is used, then meaure the lengths from T_1.

8. It will be necessary to calculate the first subchain length, shorter than a chain length, so that a full chain length falls on D. To do this, halve the long chord length and calculate the number of full chain lengths possible in the length T_1D.

9. Set off the ordinates as in the previous case.

10. Set the other half of the curve from the second tangent point T_2 in the same way.

Radial offsets from tangents   The geometrical construction for this method is shown in Fig. 15.9. A radial offset is an offset from the tangent to the curve but in the direction of the radius of the curve from that point. While the offsets can be set off at chain lengths along the tangent, the points obtained on the curve will not be at equal intervals.

Referring to Fig. 15.9(a), T_1I is the tangent and Y_R is the radial offset. Point P is chosen to be 20 m or 30 m from T_1. OT_1P is a right-angled triangle. $T_1P = x$, OT_1 is the radius R of the curve, and PO is the hypotenuse. $PO = Y_R + R$. We can then write

$$OP^2 = OT_1^2 + T_1P_2, \quad (Y_R + R)^2 = R^2 + x^2$$

from which

$$Y_R = \sqrt{R^2 - x^2} - R$$

This is the exact value of the radial offset at any distance x from T_1.

For an approximate relationship, we expand $\sqrt{R^2 - x^2}$.

$$\sqrt{R^2 - x^2} = R\sqrt{1 + x^2/R^2} = R(1 + x^2/2R^2 - x^4/8R^4 + \cdots)$$

Neglecting the higher powers of x, we have

$$\sqrt{R^2 - x^2} = R + x^2/2R - R = x^2/2R$$

This relation can also be obtained from the properties of a circle. From Fig. 15.9(a),

$$T_1P^2 = PP_1(2R + PP_1), \quad x^2 = Y_R(2R + Y_R)$$

Considering Y_R to be small compared to $2R$, we get

$$Y_R = x^2/2R$$

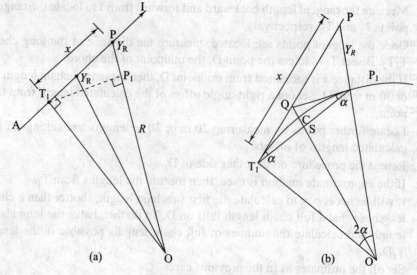

(a)   O   (b)   O

Fig. 15.9 Radial offsets

The approximate formula for offsets will not give an exactly circular arc but the profile of a parabola. When the radius is large, this will follow very nearly a circular profile.

Procedure for setting out the curve

1. Locate the point of intersection and tangent points as before.
2. Select a length equal to that of a full chain of 20 m or 30 m or half-chain lengths for calculating the offsets.
3. Set out the selected length from T_1 and obtain point P. Measure the offset distance Y_R along the radial direction from P to get the first point on the curve.
4. Set out the second length from P. Measure the offset length from this point in the radial direction to get the second point on the curve.
5. Continue this process till the half-point of the curve is reached.
6. Set out the other half on similar lines from the second tangent point.

The offset length can sometimes become too long and, in such a case, it is possible to get a second tangent at a suitable point. The procedure is illustrated in the next section on perpendicular offsets, as setting a perpendicular and measuring long offsets are prone to errors.

In the case of radial offsets, we can find the tangent direction from a point on the curve as follows: Referring to Fig. 15.9(b), T_1I is the backward tangent. At a distance x from T_1, the radial offset is Y_R, giving the point P_1 on the curve. $PP_1 = Y_R$. From triangle PT_1O, $\angle T_1OP = 2\alpha$; α is given by $(1/2)\tan^{-1}(x/R)$. If we join Q to O to intersect the chord at S, then SQ is the sum of the versine of the arc and the apex distance of arc T_1P_1. Therefore,

$$SQ = SC + CQ = R[1 - \cos(\alpha/2)] + R[\sec(\alpha/2) - 1]$$

$$T_1S = R\sin(\alpha/2)$$

The distance $T_1Q = \sqrt{T_1S^2 + SQ^2}$. This distance can be measured along the tangent to locate Q. The line QP_1 can be ranged and extended to get the new tangent direction, from which further offsets can be measured to set out the curve.

Perpendicular offsets from tangents Offsets can also be set at right-angles to the tangent to locate points on the curve. In Fig. 15.10, T_1I is the tangent to the curve, $PP_1 = Y_P$ is perpendicular to the tangent, and OT_1 is the radius of the circle with centre at O. T_1P is the length along the tangent and can be conveniently taken as a chain length of 20 m or 30 m. From P_1 draw a line parallel to the tangent to intersect OT_1 at P_2. $P_1P_2 = x$ and $T_1P_2 = Y_P$. From the right-angled triangle OP_1P_2, we can write

$$(OP_2)^2 = (OP_1)^2 - (P_1P_2)^2$$

$$(R - Y_P)^2 = R^2 - x^2$$

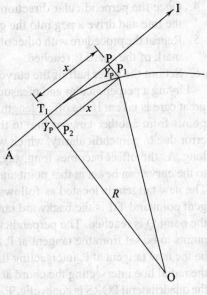

Fig. 15.10 Perpendicular offset from tangents

From this, the perpendicular offset is

$$Y_P = R - \sqrt{R^2 - x^2}$$

This is the exact relationship for the perpendicular offset from the tangent. To obtain an approximate relationship, we can expand the term $(R^2 - x^2)^{1/2}$ as

$$\sqrt{R^2 - x^2} = R(1 - x^2/2R^2 - x^4/8R^4 - \cdots)$$

Neglecting higher power of x,

$$\sqrt{R^2 - x^2} = R - x^2/2R$$

and the perpendicular offset then becomes

$$Y_P = x^2/2R$$

The comments given in the earlier section apply here as well. The approximate values of offsets , when used, will not lie on a circle but on a parabola. In flat curves the difference will be insignificant.

Procedure for setting out the curve The procedure for setting out the curve is as follows.

1. Locate the point of intersection I and the tangent points T_1 and T_2 by measurement.
2. From the tangent point T_1, measure the length of a chain or half-chain as per calculation.
3. Lay the offset perpendicular to the tangent. This can be done using an optical square or other instruments.

4. Once the perpendicular direction is laid out, measure the offset length along the line and drive a peg into the ground to locate the point on the curve.
5. Repeat the procedure with other offset lengths to get more points till the halfway mark of the curve is reached.
6. Set out the other half of the curve from the second tangent point T_2.

Laying a perpendicular and measuring the offset are both prone to errors unless great care is taken. If the offset length becomes large, it is preferable to set further points from another tangent line to the curve. This should be done to avoid any error due to perpendicularity, which will result in appreciable error if the offset is long. As the offset becomes long, say, a chain length of 20 m, then a new tangent to the curve can be set at that point and further offsets measured from this tangent. The new tangent is located as follows. Consider Fig. 15.11(b). T_1 is the first tangent point and T_1I is the backward tangent. A number of offsets have been set till the point Q is reached. The perpendicular offset Y_Q is large enough for the further points to be set from the tangent at P, the point set with the offset from Q. Let PR be the new tangent at P, intersecting the tangent T_1I at R. T_1P is the chord and RO the radial line intersecting the chord at S. The angles at Q and S being right-angles, the quadrilateral PQRS is concyclic. T_1RQ and T_1SP are chords from external points and we can therefore write the relation

$$T_1R \times T_1Q = T_1S \times T_1P$$

But $T_1S = T_1P/2$. Also from the right-angled triangle T_1QP, $T_1P^2 = T_1Q^2 + QP^2$. From these, we get

$$T_1R = T_1S \times T_1P/T_1Q = T_1P^2/2 \times T_1Q, \quad T_1R = (T_1Q^2 + QP^2)/2 \times T_1Q$$

Setting $T_1Q = x$, the distance along the tangent, and $QP = Y_Q$,

$$T_1R = (x^2 + Y_Q^2)/2x$$

The direction of the tangent at P can thus be obtained as RP and can be extended to set further offsets from this tangent to the curve.

Offsets from the chord produced   Another method to set out circular curves, particularly long ones, is the method of offsets from the chord produced. A chord is taken of the arc and produced for a certain length. The offset from the chord to the curve is then calculated. The chord must be continuous and can be of a chain length of 20 or 30 m. Subchords will have to be taken at the beginning and at the end.

Consider the case shown in Fig. 15.11(c). T_1 is the first tangent point and T_1I is the direction of the tangent. P_1, P_2, P_3, etc. are points to be established on the curve. The first chord T_1P_1 (as also the last chord) will be shorter than a chain length and the other chords P_1P_2, P_2P_3, P_3P_4, etc. will be equal to a full chord of length 20 m or 30 m. The points on the curve are established by swinging an arc equal to the length of the chord with the previous point as the centre. For the first subchord, T_1 is the centre; for the second subchord, P_1 is the centre; and so on. Therefore,

$$T_1P_1' = T_1P_1 = c_1, \quad P_1P_2' = P_1P_2 = c_2, \quad P_2P_3' = P_2P_3 = c_3, \quad \text{and so on}$$

Generally, $c_2 = c_3 = c_4 = \cdots = c_{n-1} =$ a chain length, 20 or 30 m. c_1 and c_n will be subchords of smaller length.

To establish the points on the curve, the length of swing or the ordinate at each point is to be calculated.

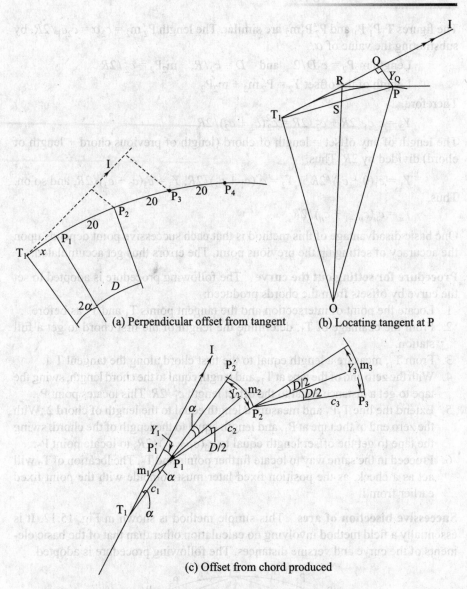

(a) Perpendicular offset from tangent

(b) Locating tangent at P

(c) Offset from chord produced

Fig. 15.11 Setting out simple circular curves

To calculate the offset distances, we proceed as follows. Considering the first chord c_1, which may be shorter than a full chain length, let the angle subtended by it at the centre be 2α. Then, $R \times 2\alpha = c_1$. From this, $\alpha = c_1/2R$. Now consider the figure $T_1 P_1' P_1$. $T_1 P_1' = c_1$, $\angle P_1' T_1 P_1 = \alpha$. The offset Y_1 is then given by $Y_1 = c_1 \alpha$. Substituting the value of α,

$$Y_1 = c_1 c_1 / 2R = c_1^2 / 2R$$

For the second and subsequent offsets except the last offset, the angle subtended at the centre is the degree of the curve on a 20- or 30-m arc or chord. Let this angle be D. Let us take 20 m chord lengths. Let the tangent drawn at P_1 meet the second offset Y_2 at m_2. Then $\angle P_2' P_1 m_2 = \alpha$, being the angle opposite to $\angle T_1 P_1 m_1$.

The figures $T_1P_1'P_1$ and $P_2'P_1m_2$ are similar. The length $P_2'm_2 = c_2\alpha = c_2c_1/2R$, by substituting the value of α.

$$\text{Length } m_2P_2 = c_2D/2 \quad \text{and} \quad D = c_2/R, \quad m_2P_2 = c_2^2/2R$$

$$\text{Length of the offset } Y_2 = P_2'm_2 + m_2P_2$$

Therefore,

$$Y_2 = c_2c_1/2R + c_2^2/2R = c_2^2(c_1 + c_2)/2R$$

The length of any offset = length of chord (length of previous chord + length of chord) divided by $2R$. Thus,

$$Y_1 = c_1(0 + c_1)/2R, \quad Y_2 = c_2(c_1 + c_2)/2R, \quad Y_3 = c_3(c_2 + c_3)/2R, \text{ and so on.}$$

Thus,

$$Y_n = c_n(c_{n-1} + c_n)/2R$$

One basic disadvantage of this method is that each successive point depends upon the accuracy of setting up the previous point. The errors thus get accumulated.

Procedure for setting out the curve The following procedure is adopted to set the curve by offsets from the chords produced.

1. Locate the point of intersection and the tangent points T_1 and T_2 as before.
2. From the chainage of T_1, determine the length of the first chord to get a full station.
3. From T_1, measure a length equal to the first chord along the tangent T_1I.
4. With the zero end of the tape at T_1, and length equal to the chord length, swing the tape to get a distance equal to the offset length $c_1^2/2R$. This locates point P_1.
5. Extend the line T_1P_1 and measure a length equal to the length of chord 2. With the zero end of the tape at P_1, and length equal to the length of the chord, swing the tape to get the offset length equal to $c_2(c_1 + c_2)/2R$, to locate point P_2.
6. Proceed in the same way to locate further points and T_2. The location of T_2 will act as a check, as the position fixed later must coincide with the point fixed earlier from I.

Successive bisection of arcs This simple method is shown in Fig. 15.12. It is essentially a field method involving no calculation other than that of the basic elements of the curve and versine distances. The following procedure is adopted.

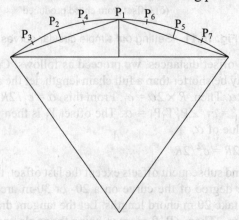

Fig. 15.12 Successive bisection of arcs

1. Locate the point of intersection and the tangent points as before.
2. Measure the long chord carefully with a tape and bisect it at point D.
3. Range the line ID and locate the point C by measuring DC equal to the mid-ordinate.
4. Measure the line T_1C with a tape and bisect it at D_1. Measure D_1C_1 = mid-ordinate of the arc T_1C. This of course is equal to $R[1 - \cos(\phi/4)]$. Locate point C_1 by laying out an offset equal to this distance.
5. Continue the same way for locating more points by bisection of arcs till a sufficient number of points are located.
6. Each mid-ordinate for a successive arc will be based on half the previous angle; e.g., $\phi/2$, $\phi/4$, $\phi/8$, and so on.

Example 15.5   A curve is designated as a 4° curve (20-m arc). The deflection angle is 40°. Calculate the offsets from the long chord at 20-m intervals.

Solution   We calculate the radius of the curve from the degree of the curve:

$$\text{Radius} = 20 \times 180/4\pi = 286.5 \text{ m}$$

The length of the long chord is given by $2R \sin(\phi/2)$. In this case,

$$\text{Length of long chord } l = 2 \times 286.5 \times \sin 20° = 196 \text{ m}$$

The length of the offset y at a distance x from the centre is given by

$$Y = \sqrt{R^2 - x^2} - \sqrt{R^2 - (l/2)^2} = \sqrt{R^2 - x^2} - \sqrt{286.5^2 - 98^2}$$

$$= \sqrt{286.5^2 - x^2} - 269.2$$

The values of x are 20, 40, 60, 80, and 98 m. If we use the approximate formula, measuring x from the tangent point,

$$Y = x(l - x)/2R = x(196 - x)/(2 \times 286.5)$$

To have the ordinates at the same point, the values for x are 18, 38, 58, 78, and 98 m. These values are calculated and given in Table 15.3.

Table 15.3   Calculation of ordinates (Example 15.5)

Distance from midpoint (m)	Offset (m)	Distance from tangent point, x	l − x	Offset
20	16.6	98	98	16.76
40	14.49	78	118	16.06
60	10.95	58	138	
80	5.90	38	158	
98	0	18	178	

Example 15.6   A simple circular curve has a radius of 800 m and a deflection angle of 36°. Tabulate the ordinates from the chord to set out the curve.

Solution   Length of long chord = $2R \sin(\phi/2) = 2 \times 800 \times \sin 18° = 494.42$ m

$$\text{Mid-ordinate} = R[1 - \cos(\phi/2)] = 800(1 - \cos 18°) = 39.15 \text{ m}$$

As the offsets are long, we can work with a divided chord subtending an angle of 18°. As shown in Fig. 15.13, DC is the mid-ordinate. We take T_1C as the chord and calculate ordinates from this chord. The point C is located as follows. The long chord is bisected at D.

The intersection point I of the tangents is located. The line ID is ranged and a distance of the mid-ordinate equal to 39.15 m is set out from D. This locates point C.

Length $T_1C = 2 \times 800 \times \sin 9° = 250.3$ m

The mid-ordinate to the curve from this chord $= 800(1 - \cos 9°) = 9.85$ m. The ordinates are tabulated as shown in Table 15.4.

Table 15.4 Ordinates to set out the curve (Example 15.6)

From D (x)	Ordinate (exact)	From T_1 (x)	l – x	Ordinate (approx.)
0	9.85	125.15	125.15	9.79
20	9.60	105.15	145.15	9.54
40	8.85	85.15	165.15	8.79
60	7.60	65.15	185.15	7.54
80	5.84	45.15	205.15	5.79
100	3.58	25.15	225.15	3.54
120	0.80	5.15	245.15	0.8

In the second column of Table 15.4, the exact ordinates are calculated as follows. The ordinate y at any distance x from D_1 is given by

$$Y = \sqrt{R^2 - x^2} - \sqrt{R^2 - (l/2)^2} = \sqrt{800^2 - x^2} - \sqrt{800^2 - (125.15)^2}$$

$$= \sqrt{R^2 - x^2} - 790.15$$

By setting the values of $x = 20, 40, 60$, etc. we get the values of y.

By the approximate method, the values of x are measured from T_1. To get coordinates at equal intervals, a smaller peg distance is introduced in the beginning. This distance is obtained from the length of the half-chord, which is 125.15. At the 20-m peg interval, we take 120 m for six offset distances. The balance is 5.15 m. The sublength offset is thus at $x = 5.15$ m. The offset is given by $y = x(l - x)/2R$. The ordinate values of y from this equation are given in the fifth column of Table 15.4. For comparison purposes, these are listed in the reverse order. The difference in the ordinates can be seen clearly by comparing the values in columns 2 and 5.

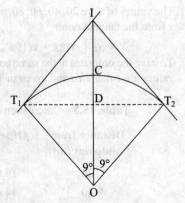

Fig. 15.13

The values for the other half of this chord will be identical and the values for the other half of the curve will also be the same.

Example 15.7 A curve of radius 300 m has a deflection angle of 32°. Calculate the radial and perpendicular offsets from the tangent to locate points on the curve. Calculate and tabulate the offsets. The number of offsets must be such that the offset length is less than 20 m.

Solution Radial offsets: The radial offset length is given by

$$Y_R = \sqrt{R^2 + x^2} - R^2 \text{ (for exact value)} \quad \text{and} \quad Y_R = x^2/2R \text{ (approximately)}$$

We calculate the offsets at 20-m intervals so that $x = 20, 40, 60, 80$, etc.

The perpendicular offset at x from the tangent point is given by $Y_P = R - (R^2 + x^2)^{1/2}$, exactly, and $Y_P = x^2/2R$, approximately. The approximate values are the same in both the cases. The values are listed in Table 15.5.

Table 15.5 Radial and perpendicular offsets (Example 15.7)

Distance along tangent, x	Radial offset (exact)	Perpendicular offset (exact)	Offset (approx.)
20	0.666	0.667	0.666
40	2.655	2.678	2.667
60	5.941	6.061	6.000
80	10.483	10.863	10.667
100	16.227	17.157	16.667
120	23.11	25.045	24.000

For example, if $x = 20$,

$$\text{Radial offset} = \sqrt{300^2 + 20^2} - 300 = 0.667 \text{ m}$$

$$\text{Perpendicular offset} = 300 - \sqrt{300^2 + 20^2} = 0.667 \text{ m}$$

$$\text{Approximate value} = 20^2/(2 \times 300) = 0.666 \text{ m}$$

The other values in the table are calculated similarly. As the value for 120 m exceeds 20 m, further values are not required.

Example 15.8 A curve of radius 800 m has a deflection angle of 40° between tangents. Calculate the radial and perpendicular offsets at 20-m intervals up to 100 m.

Solution The radial offset is $Y = \sqrt{R^2 + x^2} - R$. The perpendicular offset is given by $Y_P = R - (R^2 - x^2)^{1/2}$. The approximate value of the offset in both cases is $x^2/2R$. These values are calculated and tabulated in Table 15.6.

Table 15.6 Radial and perpendicular offsets (Example 15.8)

Distance along tangent x	Radial offset, Y_R (exact)	Perpendicular offset, Y_P (exact)	Approximate value of offset
20	0.25	0.25	0.25
40	1.00	1.00	1.0
60	2.246	2.25	2.25
80	4.00	4.01	4.0
100	6.225	6.27	6.25

For example, with $x = 20$ m,

$$\text{Radial offset } Y_R = \sqrt{800^2 + 20^2} - 800 = 0.25 \text{ m}$$

$$\text{Perpendicular offset } Y_P = 800 - \sqrt{800^2 - 20^2} = 0.25 \text{ m}$$

$$\text{Approximate value of } Y = 20^2/(2 \times 800) = 0.25 \text{ m}$$

The remaining values are calculated in the same way. One can see that with a large radius, the radial, perpendicular, and approximate values of the offsets are nearly the same.

Example 15.9   A curve of 400 m radius and 42° deflection angle is to be set up by offsets from the chords produced. Calculate the offset distances for the first five chords. The offsets are to be at 20-m intervals.

Solution   We assume that the offsets are so calculated that one of the offsets will give the maximum ordinate of the curve.

Degree of curve with 20-m chord $= 2\sin^{-1}[20/(2 \times 400] = 2.864°$

Half the central angle $= 21°$

This will give seven full chords of 20 m.

Angle for seven chords $= 7 \times 2.864 = 20.048°$

Balance angle $= 21 - 20.048 = 0.952°$

This will give a first chord length of

$c_1 = 20 \times 0.952/2.864 = 6.65$ m

Offset $= c_n(c_{n-1} + c_n)/2R$

The values are calculated and tabulated in Table 15.7.

Table 15.7   Offsets from chords produced (Example 15.9)

Chord length (m)	c_{n-1} (m)	c_n (m)	Offset Y (m)
6.65	0	6.65	0.055
20	6.65	20	0.66
20	20	20	1.00
20	20	20	1.00
20	20	20	1.00

First offset $= 6.65(0 + 6.65)/(2 \times 400) = 0.055$ m

Second offset $= 20(6.65 + 20)/(2 \times 400) = 0.66$ m

Third offset $= 20(20 + 20)/(2 \times 400) = 1.00$ m

The fourth and fifth will be the same. The other half of the curve can be set with similar values.

Example 15.10   A curve of radius 300 m was being set with radial offsets from tangents. The curve was set with a distance of 100 m along the tangent when it was decided to have a new tangent for further offsets at the last point on the curve. Find the point at which the new tangent will intersect the first tangent.

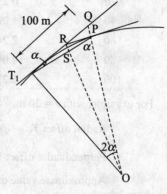

Solution   The situation is shown in Fig. 15.14. $T_1Q = 100$ m and QP is the radial offset giving point P on the curve. T_1P is the chord. The tangent to the curve at P intersects the first tangent at R. RSO is the radial line from R intersecting the chord T_1P at S. The distance T_1R has to be calculated so that the new tangent RP can be ranged and extended for further offsets.

Radial offset $Y_R = \sqrt{R^2 + x^2} - R$

$= \sqrt{300^2 + 100^2} - 300$

$= 16.23$ m

Fig. 15.14

From the right-angled triangle OT_1Q, if 2α is the central angle for chord T_1P, $\cos 2\alpha = 300/316.22$, from which $2\alpha = 18.43°$.

Chord length $T_1P = 300 \times 18.43\pi/180 = 96.5$ m

$T_1R = 96.5/2 \cos(9.215°) = 48.9$ m (from triangle RT_1S)

The direction of tangent RP can now be ranged after locating R.

Angular methods

Angular methods of laying curves use an angle-measuring instrument such as the theodolite. The points on the curve are set with angles measured with or without measuring distances. There are three angular methods:
1. Rankine's method of deflection angles
2. Two-theodolite method
3. Tacheometric method

The first method uses measurements of angles and distances with a tape while the latter methods do not use any distance measurement. The points are set with angle measurements only. This is particularly suitable in rough ground, where distance measurement is difficult or impossible.

Rankine's method of deflection angles The deflection angle to any point on the curve is the angle between the tangent and the chord from the tangent point to the point on the curve. Also, from the properties of a circle, the deflection angle is half the angle subtended by the arc at the centre. Thus, in Fig. 15.15(a), T_1P is the chord and T_1I is the tangent at T_1. The angle IT_1P is δ and the angle subtended at the centre by the arc T_1P is 2δ. Assuming the arc length and chord length to be nearly equal, the value of the angle δ can be derived as follows. Let c_1 be the length of the chord. If R is the radius of the circle, then

$$c_1 = R \times 2\delta\pi/180 \quad \text{and} \quad \delta = 90c_1/\pi R \text{ (degrees)}$$

Since the angles will be small, we convert them into minutes. Thus,

$$\delta = (90/\pi) \times 60c_1/R \text{ minutes} = 1718.9c_1/R \text{ minutes}$$

Table 15.8 Values of tangential angles for different chord lengths

Radius (m)	10-m chord	15-m chord	20-m chord	30-m chord
200	1° 26″	2° 8′ 55″	2° 52′	4° 17′ 50″
300	57′ 18″	1° 26′	1° 54′ 35″	2° 52′
400	42′ 58″	1° 04′ 27″	1° 26′	2° 08′ 55″
500	34′ 23″	51′ 34″	1° 08′ 45″	1° 43′ 08″
600	28′ 39″	43′	57′ 18″	1° 25′ 57″
700	24′ 33″	36′ 50″	49′ 07″	1° 13′ 40″
800	21′ 29″	32′ 14″	43′ 00″	1° 04′ 27″

Now, in practice, the first chord from T_1 will be smaller than a full chain or half chain length to make P a full station of a certain number of chains. The remaining chords will be full or half chain lengths except the last one, which will also be shorter. If chord values are 10, 15, 20, or 30 m, the value of the angles will be as given in Table 15.8.

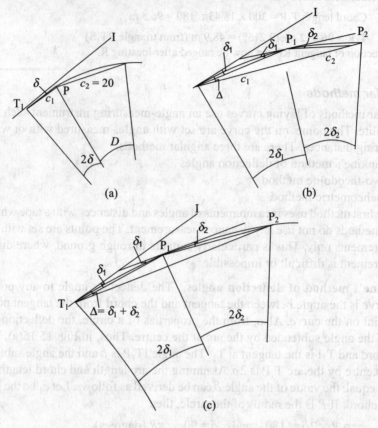

Fig. 15.15 Method of deflection angles

We designate the tangential angle, i.e., the angle between the tangent at the end of the chord and the tangent, δ. The angle between the back tangent and the line joining T_1 and P, the deflection angle, is designated Δ.

In Fig. 15.15(b), if the first chord has a tangential angle of δ_1, this is $\angle IT_1P_1$. The deflection angle for this chord is the same as the tangential angle. For the second chord c_2, the angle from the tangent P_1 is δ_2. If we have to locate point P_2 from T_1, we require the $\angle IT_1P_2$ (deflection angle of point P_2). This angle can be seen to be $\delta_1 + \delta_2$. This is so because the angle subtended by the chord P_1P_2 at the centre is $2\delta_2$. The same chord will subtend an angle of δ_2 at T_1. Thus, $\angle P_1T_1P_2$ is δ_2 and $\angle IT_1P_2$ is $\delta_1 + \delta_2$. A similar procedure is applied to all the angles.

$\delta_1, \delta_2, \delta_3$, etc. are known as tangential angles, i.e., the angles between the tangents at the points on the curve and the chord. The deflection angle is the angle between the tangent IT_1 and the chord T_1P to any point. This is the angle by which the chord deviates or deflects from the tangent. Points on the curve are located by measuring the deflection angles from the line IT_1. For chord c_1, δ_1 is the tangential angle. It is also the deflection angle for this chord. For chord c_2, δ_2 is the tangential angle. The deflection angle is $\delta_1 + \delta_2$, the sum of the deflection angles up to this chord. It can also be stated as the sum of the deflection angle for the previous chord and the tangential angle of this chord. So, the deflection angle for c_2 is $\Delta_1 + \delta_2$. For chord c_3, the tangential angle is δ_3 and the deflection angle is $\delta_1 + \delta_2 + \delta_3$. $(\delta_1 + \delta_2)$ is the deflection angle for the previous chord c_2. So the deflection angle for chord $c_3 = \Delta_2 + \delta_3$. For any chord c_n, deflection angle $= \Delta_{n-1} + \delta_n$.

In general, we can say that the deflection angle of a chord = deflection angle of the previous chord + tangential angle of the chord itself. In practice, tangential angles and deflection angles are calculated for half the curve from T_1. From this the angle to be set in the theodolite to obtain the direction of T_1P (P being any point on the curve) is obtained. The length T_1P has to be measured with a tape.

Procedure for setting the curve   The procedure to be followed for setting points on the curve with deflection angles is as follows.

1. Locate the point of intersection I and the tangent points.
2. From the chainage of I and the tangent length IT_1, calculate the chainage of T_1.
3. From the chainage of T_1, calculate the length of the first subchord.
4. Calculate the tangential for this subchord, and designate it δ_1. This is also Δ_1 for this chord.
5. Depending upon the peg interval 10 m, 15 m, 20 m, or 30 m, calculate the deflection angles for the remaining chords. The tangential angle δ remains the same for any length of constant peg interval.
6. Calculate the deflection angles for all the pegs to be set out from T_1.
7. From these values, work out and tabulate, in the format given in Table 15.9, the angle to be set in the theodolite for obtaining the direction of the pegs (radius = 300 m).

Table 15.9   Format for recording angles in Rankine's method

Point	Chainage	Chord	Tangential angle	Deflection angle	Theodolite reading	Direction
T_1	1812.65	–	–	–	–	–
P_1	1820	7.35	42′ 7″	0° 42′ 7″		
P_2	1840	20	1° 54′ 36″	2° 36′ 43″		
P_3	1860	20	1° 54′ 36″	4° 31′ 19″		
P_4	1880	20	1° 54′ 36″	7° 25′ 55″		

8. Set up the theodolite on the ground at T_1. Both the verniers reading zero, sight the signal at I and bisect it accurately. Measure all angles from this line but chordwise.
9. Release the upper clamp and set the vernier to read the deflection angle of the first chord. Set the theodolite to read the angle accurately. Measure a distance equal to the chord length accurately along this line, in the line of sight of the instrument, and fix the first peg.
10. Set the theodolite to read the second reading calculated and set the angle accurately. With the zero end of the tape on P_1 and a length equal to the second chord, move the tape till the arrow at the end of the chord length is bisected by the line of sight. Fix the point P_2.

11. Proceed similarly to fix more points and T_2. The position of T_2 will act as a check, as the position fixed earlier should coincide with the point obtained later.

One disadvantage of this method is that points are fixed with reference to the previous point. Thus, any error in fixing any one point is carried forward.

Two-theodolite method   In the two-theodolite method, measuring distances with a chain or tape is dispensed with. This method is particularly useful in rough ground, where chainage is impossible or difficult. The method requires two theodolites, and two people to handle them, thus making it more expensive. However, it is very accurate. The geometrical construction is shown in Fig. 15.16(a) and the steps are given below.

The principle of the method can be understood from Fig. 15.16(b). T_1 and T_2 are the two tangent points. Consider any point P_1 on the curve. T_1P_1 is a chord and subtends an angle 2δ at the centre. From the properties of a circle, the angle between the tangent and the chord is δ, half the angle subtended at the centre. A chord will also subtend half the angle subtended at the centre at another point on

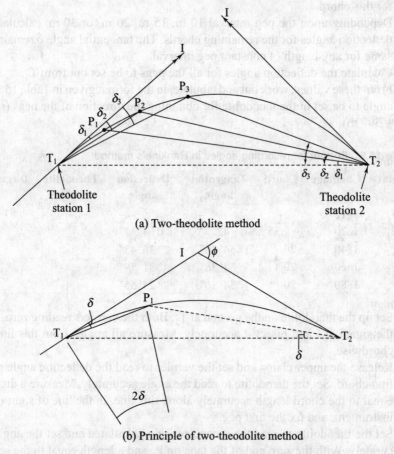

(a) Two-theodolite method

(b) Principle of two-theodolite method

Fig. 15.16   Setting out curve by deflection angles

the curve. Thus, angle $T_1T_2P_1$ is also δ. Thus, if we set angle δ from T_1I and the same angle δ from T_2T_1, we get point P_1 as their intersection. This is the principle on which the two-theodolite method is based.

1. Fix the intersection point I and tangent points T_1 and T_2 accurately.
2. Set up one theodolite at T_1 and a second one at T_2.
3. Set the theodolite at T_1 to zero and sight the signal at the point of intersection I. Direct the theodolite at T_2 to sight T_1 with the vernier set to zero.
4. Turn the theodolites clockwise to set the deflection angle Δ_1 accurately. Have a person with a ranging rod move such that he comes in the line of sight of both the theodolites. The point at which this person stands is a point on the curve for the deflection angle set.
5. Repeat the procedure with the other deflection angles with both the instruments reading the same angles. Locate a number of such points on the curve. The smaller the distance between such points, the closer one gets to the actual profile of the curve.

This method is very accurate as no linear measurements are involved and the angles can be set accurately with the theodolite. If T_1 is not visible from T_2, then the angles at T_2 can be set with respect to the point of intersection. The theodolite at T_2 is set to zero and the signal at I is sighted. The angle to be set in the theodolite at T_1 is Δ as before. The angle to be set in the instrument at T_2 is $360° - \Delta$, as the instrument is rotated anticlockwise to locate the point from I. The data for setting out the curve can be recorded as shown in Table 15.10.

Table 15.10 Format for recording data for setting a curve using the two-theodolite method

Point	Chord length (m)	Tangential angle	Deflection angle	Theodolite reading at T_1	Theodolite reading at T_2
P_1	7.35	0° 31′ 35″	0° 31′ 35″	0° 31′ 35″	359° 28′ 25″
P_2	20	1° 26′	1° 57′ 35″	1° 57′ 35″	358° 02′ 25″
P_3	20	1° 26′	3° 23′ 35″	3° 23′ 35″	356° 36′ 25″
P_4	20	1° 26′	4° 49′ 35″	4° 49′ 35″	355° 10′ 25″

Tacheometric method　Again in the tacheometric method, chaining is dispensed with. Distance is measured using the principle of tacheometry. A theodolite with a stadia diaphragm or a tacheometer is required for conducting the survey.

In tacheometry the distance is obtained as

$$\text{Distance} = K_1 S + K_2$$

where K_1 is the multiplying constant, S is the staff intercept, and K_2 is the additive constant. Most of the instruments will have $K_1 = 100$; K_2 is generally zero or very small. The procedure here is the reverse of finding the distance. The distance is known and the staff intercept is to be determined. The staff location is then determined to obtain the calculated staff intercept.

The basic principle is the same as that of Rankine's method. The procedure is as follows.

1. Locate the point of intersection I and tangent points T_1 and T_2 as before.
2. Set up the tacheometer or a theodolite with a stadia diaphragm at T_1.
3. Direct the instrument to I with the instrument set to zero.
4. Set the instrument to the first deflection angle Δ_1. The line of sight will be in the direction of T_1P_1.
5. Obtain the calculated staff intercept for the chord length c_1 on the staff by moving it along the line of sight. When the required staff intercept is obtained, the point is fixed.
6. For the second point P_2, the distance is not chord c_2 but the length T_1P_2. This length is easily calculated as $2R \sin(\Delta_2/2)$.
7. Set the instrument to the angle Δ_2 to get the direction of T_1P_2. Move the staff along this line of sight till the required staff intercept for this distance is obtained. The point P_2 is thus fixed.
8. Locate further points similarly.

It will be necessary to tabulate the calculated values for facilitating the fieldwork. This can be done in the format given in Table 15.11, which has been prepared assuming $K = 100$ and $C = 0$.

Table 15.11 Format for tabulating calculated values in the tacheometric method

Point	Tangential angle	Deflection angle	Chord length (from T_1; m)	Staff intercept
P_1	0° 31′ 35″	0° 31′ 35″	7.35	0.073
P_2	1° 26′	1° 57′ 35″	13.681	0.137
P_3	1° 26′	3° 23′ 35″	23.684	0.237
P_4	1° 26′	4° 49′ 35″	33.684	0.337

Example 15.11 A curve has a radius of 400 m and a deflection angle of 40°. The chainage of T_1 is 1804.25 m. Compute and tabulate the angles and theodolite readings to set out the curve using Rankine's method.

Solution In Rankine's method, tangential and deflection angles are calculated. The points on the curve are set with deflection angles from tangent point T_1 and distances are measured with a chain or tape. The calculations are given in Table 15.12.

Radius of the curve = 400 m

Length of long chord = $2 \times 400 \times \sin 20° = 273.31$ m

Length of arc = $400 \times 40\pi/180 = 279.25$ m

Length of first chord = $1820 - 1804.25 = 15.75$ m

There will be 12 chords of 20 m each. The last chord will be taken as 23.5 m. For the first chord,

Tangential angle = $1718.88 \times 15.75/400$ min = 1° 07′ 40″

For the next 12 chords,

Tangential angle = $1718.88 \times 20/400$ min = 1° 25′ 56″

Table 15.12 Calculations for setting out the curve of Example 15.11

Point	Chord length	Tangential angle	Deflection angle	Theodolite reading	Distance from T₁
1	15.75	1° 07′ 40″	1° 07′ 40″	1° 07′ 40″	15.75
2	20	1° 25′ 56″	2° 33′ 37″	2° 33′ 37″	35.736
3	20	1° 25′ 56″	3° 59′ 34″	3° 59′ 34″	55.704
4	20	1° 25′ 56″	5° 25′ 31″	5° 25′ 31″	75.638
5	20	1° 25′ 56″	6° 51′ 28″	6° 51′ 28″	95.524
6	20	1° 25′ 56″	8° 17′ 25″	8° 17′ 25″	115.35
7	20	1° 25′ 56″	9° 43′ 22″	9° 43′ 22″	135.105
8	20	1° 25′ 56″	11° 09′ 19″	11° 09′ 19″	154.775
9	20	1° 25′ 56″	12° 35′ 16″	12° 35′ 16″	174.348
10	20	1° 25′ 56″	14° 01′ 13″	14° 01′ 13″	193.812
11	20	1° 25′ 56″	15° 27′ 10″	15° 27′ 10″	213.155
12	20	1° 25′ 56″	16° 53′ 07″	16° 53′ 07″	232.365
13	20	1° 25′ 56″	18° 19′ 04″	18° 19′ 04″	251.429
14	23.5	1° 40′ 57″	20° 00′ 01″	20° 00′ 00″	273.61

For the last chord,

$$\text{Tangential angle} = 1718.88 \times 23.5/400 \text{ min} = 1° 40′ 57″$$

Deflection angle = deflection angle for the previous chord + tangential angle for the present chord. For the first chord,

$$\text{Deflection angle} = \text{tangential angle} = 1° 07′ 40″$$

For the second chord,

$$\text{Deflection angle} = 1° 07′ 40″ + 1° 25′ 56″ = 2° 33′ 37″$$

and so on.

The reading to be set on the theodolite is the deflection angle. The distance to be measured along any direction is the chord length from the tangent point T_1. This is obtained as

$$\text{Distance} = 2 \times 400 \times \sin(\text{deflection angle})$$

For the first chord,

$$\text{Distance} = 800 \times \sin(1° 07′ 40″) = 15.75 \text{ m}$$

For the second chord,

$$\text{Distance} = 800 \times \sin(2° 33′ 37″) = 35.736 \text{ m}$$

and so on.

Example 15.12 A curve of radius 300 m is to be set out with two theodolites. The deflection angle of the curve is 36°. Tabulate the angles and theodolite readings, assuming that both the theodolites are turned clockwise to locate points. If the angles at the instrument at the second tangent point T_2 are to be measured from the point of intersection, as T_1 cannot be clearly seen from T_2, list the angles to be set in the instrument at T_2. The chainage of tangent point T_1 is 1083.65 m.

Solution With two theodolites, the points on the curve are fixed by the intersection of lines of sight from the instruments, and distances need not be measured. From the first tangent point, angles are measured from the tangent line (line joining the tangent point and the point of intersection), and from the second tangent point, angles are measured from the long chord (line joining the two tangent points). If for some reason the latter measurement is not possible, then the angle at the second tangent point is also measured from the tangent. In the first case both angles are equal, whereas in the second case the angle at the forward tangent is $360° - \Delta$. The calculations can be done in tabular form as shown in Table 15.13.

Table 15.13   Deflection angles (Example 15.12)

Point	Length	Tangential angle	Deflection angle	Deflection angle at T_2 (read from tangent)
1	16.35	1° 33′ 40″	1° 33′ 40″	358° 26′ 20″
2	20	1° 54′ 36″	3° 28′ 16″	356° 31′ 44″
3	20	1° 54′ 36″	5° 22′ 52″	354° 37′ 08″
4	20	1° 54′ 36″	7° 17′ 28″	352° 42′ 32″
5	20	1° 54′ 36″	9° 12′ 04″	350° 47′ 56″
6	20	1° 54′ 36″	11° 06′ 40″	348° 53′ 20″
7	20	1° 54′ 36″	13° 01′ 16″	346° 58′ 44″
8	20	1° 54′ 36″	14° 55′ 52″	345° 04′ 08″
9	20	1° 54′ 36″	16° 50′ 28″	343° 09′ 32″
10	12.15	1° 09′ 36″	18° 00′ 04″	342° 00′ 00″

Example 15.13   A curve of radius 400 m and deflection angle 30° is to be set using a tacheometer. If the tacheometer constants are 100 and 0, list the staff intercepts to locate points on the curve at 20-m intervals. The chainage of tangent point T_1 is 1326.78 m.

Solution   Radius = 400 m, deflection angle = 30°, length of arc = 209.44 m, length of chord = 207.05 m.

Table 15.14   Solution to Example 15.14

Point	Chord length	Tangential angle	Deflection angle	Distance	Staff intercept
1	13.22	0° 56′ 48″	0° 56′ 48″	13.22	0.132
2	20	1° 25′ 57″	2° 22′ 45″	33.21	0.332
3	20	1° 25′ 57″	3° 48′ 42″	53.18	0.532
4	20	1° 25′ 57″	5° 14′ 39″	73.12	0.731
5	20	1° 25′ 57″	6° 40′ 36″	93.01	0.93
6	20	1° 25′ 57″	8° 06′ 33″	112.85	1.128
7	20	1° 25′ 57″	9° 32′ 30″	132.61	1.326
8	20	1° 25′ 57″	10° 58′ 27″	152.30	1.523
9	20	1° 25′ 57″	12° 24′ 24″	171.88	1.719
10	20	1° 25′ 57″	13° 50′ 29″	191.39	1.914
11	16.22	1° 09′ 40″	15° 00′ 01″	207.05	2.070

Chainage of T_1 = 1326.78 m,   length of first chord = 13.22 m

There will be eight chords of 20 m and a shorter chord at the end. The solution is shown in Table 15.14.

The tangential angles and deflection angles are calculated as before. For the 13.22-m chord,

Tangential angle = $\sin^{-1}(13.22/800)$ = 0° 56′ 48″

For the 20-m chord,

Tangential angle = $\sin^{-1}(20/800)$ = 1° 25′ 56″

For the 16.22-m chord,

Tangential angle = $\sin^{-1}(16.22/800)$ = 1° 09′ 40″

Deflection angle = deflection angle for the previous chord + tangential angle of the present chord

For the 13.22-m chord,

Deflection angle = 0° 56′ 48″

For the 20-m chord,

Deflection angle = 0° 56′ 48″ + 1° 25′ 57″ = 2° 22′ 45″

and so on. The distances to be measured along these directions are chord lengths from T_1, the first tangent point. For the first chord

Distance = 13.22 m

For the next chord,

Distance = $2 \times 400 \times \sin(2° 22′ 45″)$ = 33.21 m

and so on. Since the multiplying constant is 100 and the additive constant is zero, the staff intercept is obtained as

Distance = 100 × staff intercept + 0,   staff intercept = distance / 100

These are listed in the last column of Table 15.14.

Obstacles to curve ranging

While laying out curves in the field many obstacles may be encountered. Ways to set curves despite these obstacles have to be figured out. The actual curve can be laid out later when the obstacle is removed. The following is a discussion on some of the common obstacles to curve ranging and viable solutions to such problems.

Point of intersection not accessible   When the point of intersection is not accessible, either because of some obstacle or because it falls in a lake or river, the problem will be to locate the tangent points T_1 and T_2. The procedure (refer to Fig. 15.17) is as follows.

1. In Fig. 15.17, the point of intersection I falls behind the obstacle. The direction of the straight lines is known.
2. Take any two points on the tangent lines clear of the obstacle such that the line joining them can be conveniently measured. The points can be P and Q on the outside towards the obstacle or P′ and Q′ before the curve. If a line cannot be found, a traverse can be run with the line as an omitted quantity, and the line can be determined from traverse calculations.

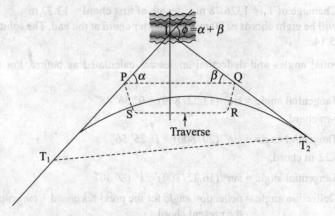

Fig. 15.17 Point of intersection inaccessible

3. Measure the angles made by this line with the tangents with a theodolite.
4. If these angles are α and β, then the deflection angle $\phi = \beta + \alpha$.
5. Measure the line PQ accurately.
6. Solve the triangle IPQ to find the lengths of IP and IQ from the sine formula for triangles:

$$IP/\sin \beta = IQ/\sin \alpha = PQ/\sin(180° - \phi)$$

From this,

$$IP = PQ \times \sin \beta/\sin(180° - \phi), \quad IQ = PQ \sin \alpha/\sin(180° - \phi)$$

$$\text{Tangent length } T_1I = R \sin(\phi/2) = T_2I$$

Therefore,

$$T_1P = T_1I - IP \quad \text{and} \quad T_2Q = T_2I - IQ$$

7. Knowing these distances, measure them along the tangents to locate both the tangent points.
8. Set the curve from either of the tanget points.

One or both tangent points not accessible   These are temporary obstacles and normally get cleared later. However, to proceed with the survey work, the following solutions may be useful. There are three cases under this head.

(a) Backward tangent point not accessible: When the backward tangent point is not accessible (see Fig. 15.18), the chainage of the tangent point has to be obtained to proceed further. The following procedure is adopted.

1. Since the tangent directions are known, select two points on either side of the obstacle on the tangent line. P and Q are the two points.
2. Measure the distance IP accurately.
3. Select a point R as shown. The distances PR and QR can be measured along with the angle PRQ. Solve triangle PQR to get the distance PQ.

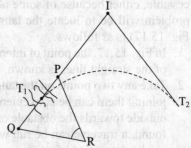

Fig. 15.18 Backward tangent point not accessible

4. Calculate the tangent length from the radius and the deflection angle, which are known.
5. The chainage of point Q is known. Obtain the chainage of I as chainage of Q + PQ + IP.
6. Chainage of T_1 = chainage of I – tangent length. Then, chainage of T_2 = chainage of T_1 + length of the curve.
7. Set the curve from the forward tangent point T_2.

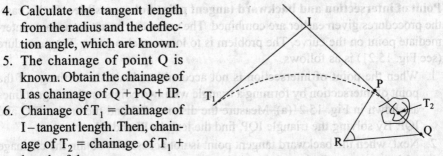

Fig. 15.19 Forward tangent point inaccessible

(b) Forward tangent point not accessible: When the forward tangent point is not accessible, the problem is to find the chainage of point Q so that the work can proceed further. The solution is similar to the previous case (see Fig. 15.19).

1. Select a point P on the forward tangent and measure the length IP.
2. Select another point Q on the tangent line and any arbitrary point R.
3. Solve the triangle PQR to get the length PQ. For this, measure the distances PR and QR and the angle PRQ.
4. Chainage of Q = chainage of I + length PQ. Chainage of T_2 = chainage of I + tangent length.

The curve can be set from T_1 even without these calculations. To proceed further along the tangent line T_2, the chainage of Q is required.

(c) Both tangent points not accessible: When both the tangent points are not accessible, the solution is to get an intermediate point on the curve from which the curve can be set. The best location of this point is the apex or summit of the curve (see Fig. 15.20). The procedure is as follows.

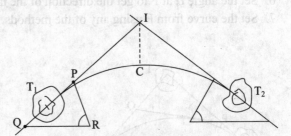

Fig. 15.20 Both tangent points not accessible

1. Find the chainage of T_1 as before by selecting three points P, Q, and R and solving the triangle PQR.
2. Chainage of T_2 = chainage of T_1 + length of curve.
3. Chainage of apex point C = chainage of T_1 + half the curve length.
4. Obtain the direction of IC by bisecting the angle between the tangents at I.
5. The distance IC is equal to = $R[\sec(\phi/2) - 1]$. Measure this distance along the direction IC to get point C. Set the curve from C in either direction using any of the methods discussed earlier, offsets from tangents or offsets from the chord produced.
6. To proceed beyond the curve at T_2, adopt the procedure suggested for the case when the forward tangent point is not accessible.

Point of intersection and backward tangent point not accessible   In this case, the procedures given earlier are combined. The curve can be set from some intermediate point on the curve. The problem is to locate such a point. The procedure (see Fig. 15.21) is as follows.

1. When the point of intersection is not accessible, obtain the chainage of the point of intersection by forming a triangle with two points on the tangent lines as shown in Fig. 15.21(a). Measure the distance PQ and the angles IPQ and IQP. By solving the triangle IQP, find the lengths PI and QI.
2. Next, when the backward tangent point is not accessible, obtain the chainage of the tangent point by solving a triangle as shown in Fig. 15.21(b). PMN is a triangle formed with arbitrary points M and N. Solve this triangle by measuring the angle at M and the distances PM and MN. Calculate the length PN from this triangle. Then calculate the chainage of T_1.
3. As shown in Fig. 15.21(c), draw a line parallel to the tangent from point E, AE being at right-angles to the tangent. The length PF = $R[1 - \cos(\alpha/2)]$. Obtain angle α from triangle PT_1O using $\sin \alpha = PT_1/OT_1$. $OT_1 = R$. The distance EF = twice the distance PT_1. F is a point on the curve and EF is a chord. This is evident if we consider that the curve is extended to E.
4. If we measure the angle at P, the angle T_1EF, this is half the angle subtended at the centre. The same angle is formed at F, between the tangent at F and the chord EF.
5. Set up the theodolite at F and take a backsight on E. Transit the telescope and the direction of the chord EF extended will be obtained.
6. Set the angle α at F to get the direction of the tangent at F.
7. Set the curve from F using any of the methods.

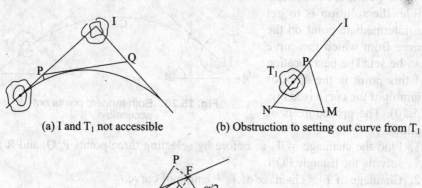

(a) I and T_1 not accessible   (b) Obstruction to setting out curve from T_1

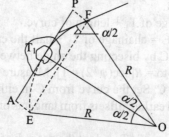

(c) Procedure for setting out

Fig. 15.21   Obstacles to setting out curves

Obstacles to setting out the curve from tangent point   Sometimes, an obstacle may prevent the setting of the curve from the backward tangent point. In such a case, the procedure involves obtaining the direction of the next chord at the last point set from T_1. The instrument is shifted to the last point set and the procedure continued as before. Two cases may arise:
(a) Intermediate point is visible from the tangent point.
(b) Intermediate point is not visible from the tangent point.

(a) Intermediate point visible from the tangent point: Referring to Fig. 15.22(a), let the curve be set up to point P from T_1. Further points cannot be set from T_1 due to the obstacle. The following procedure is adopted.
1. Shift the instrument to the last point P set from T_1.
2. To locate the direction of the next chord from P, do one of the following.
    *   Shift the instrument to P. After making the temporary adjustments, set the instrument to zero and take a backsight to T_1. Transit the telescope. Now set the deflection angle of the arc T_1P from this line of sight and the direction of the next chord from P will be obtained.
    *   If the instrument is not in adjustment and transiting may not give good results, locate a point P_1 in line with T_1P with the instrument at T_1. Shift the instrument to P. Set up and level the instrument. With vernier A set to zero, take a sight to P_1 and bisect it. Now set the deflection angle of the chord T_1P to get the direction of the next chord from P.
    *   A third method is to set the vernier to read 180° with the instrument set up at P. Take a backsight to T_1. Now rotate the instrument in azimuth (in the horizontal plane) through 180°. The instrument is in the direction T_1P. Set the deflection angle to get the direction of the next chord from P.
3. Proceed as if the instrument is at T_1. Use the deflection angles calculated earlier for setting further points. This can be proved as follows.

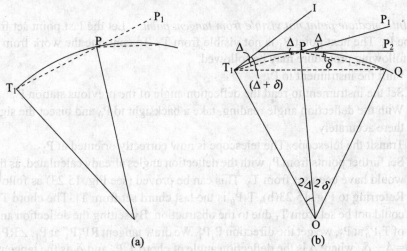

Fig. 15.22   Obstacles to setting out curve

Referring to Fig. 15.22(b), point P on the curve has been set from the tangent at T_1. Let Δ be the deflection angle for chord T_1P:

$\angle T_1OP = 2\Delta$, $\angle IT_1P = \Delta$

PP_1 is the extension of the chord T_1P. The deflection angle for the chord $T_1Q = \Delta + \delta$, where δ is the tangential angle for the chord PQ. Draw the tangent PR at P. RP_2 is the tangent at P, intersecting T_1I at R. $\angle T_1PR = \Delta$. $\angle P_1PP_2 = \Delta$, being the opposite angle of intersecting lines. $\angle P_2PQ = \delta$, the tangential angle for chord PQ. Thus $\angle P_1PQ = \Delta + \delta$, same as the deflection angle from tangent T_1I. Thus, when the chord P is extended, with the instrument at P, the deflection angles can be measured as if the instrument were at T_1.

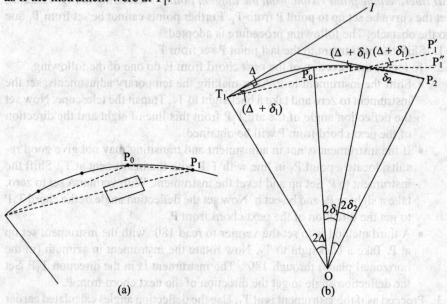

Fig. 15.23 Obstacles to setting out curves

(b) Intermediate point not visible from tangent point: Let the last point set from T_1 be P_0. The next point P_1 is not visible from T_1. To continue the work from P_1, the following procedure may be followed.

1. Shift the instrument to P_1.
2. Set the instrument to read the deflection angle of the previous station P_0.
3. With the deflection angle reading, take a backsight to P_0 and bisect the signal there accurately.
4. Transit the telescope. The telescope is now correctly oriented at P_1
5. Set further points from P_1 with the deflection angles already calculated, as they would have been set from T_1. This can be proved (see Fig. 15.23) as follows. Referring to Fig. 15.23(b), T_1P_0 is the last chord set from T_1. The chord T_1P_1 could not be set from T_1 due to the obstruction. By setting the deflection angle of T_1P_0 at P_1, we get the direction P_1P_1'. We draw tangent RP_1P_1'' at P_1. $\angle RP_1T_1 = \Delta + \delta_1$, where Δ is the deflection angle of chord T_1P_0 and δ_1 is the tangential angle of chord P_0P_1. Also $\angle P_1'P_1P_1'' = \Delta + \delta_1$, being the opposite angle. It is clear that $\angle P_1''P_1P_2 = \delta_2$, the tangential angle of chord P_1P_2. Therefore,

$$\angle P_1'P_1P_2 = \Delta + \delta_1 + \delta_2$$

which is the deflection angle of chord P_1P_2 from tangent IT_1. For setting purposes, the same deflection angles, as calculated from T_1, may be used to set points from P_1P_1'.

Obstacles to linear measurements Let P_0 be the last point set from T_1 and P_n the next point visible from it. One or more points on the curve cannot be set due to obstacles to chaining the distance from T_1 (Fig. 15.24).

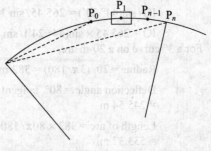

One simple solution is to set the remaining points on the curve from the forward tangent point T_2. A second method is to set out the point P_n independently of the previously set points. For this, do the following.

Fig. 15.24 Obstacles to chaining

1. Set off the deflection angle for P_n from T_1.
2. Along this line, measure the length of the chord T_1P_n. The length of this chord is equal to $(1/2)R \sin(\Delta_n/2)$, where Δ_n is the deflection angle to P_n.
3. Points beyond P_n can be set out without difficulty.
4. Points in between P_0 and P_n can be set out once the obstacle is removed.

The following examples illustrate the principles outlined above.

Example 15.14 A $3°$ curve was to be set out on the ground. The point of intersection was not accessible. The following measurements were made. A line PQ was measured, P and Q lying on the tangent lines. Length PQ = 265.45 m. $\angle T_1PQ = 135°\ 36'$ and $\angle T_2QP = 144°\ 24'$. Find the tangent lengths and chainage of the forward tangent point if the chainage of the backward tangent point is 1342.65 m.

Solution The situation is shown in Fig. 15.25. The angles IPQ and IQP can be found from the given data.

$$\angle IPQ = 180° - 135°\ 36' = 44°\ 24',$$

$$\angle IQP = 180° - 144°\ 24' = 35°\ 36'$$

Triangle IPQ can be solved to get the lengths of sides IP and IQ.

$$\angle PIQ = 180° - (44°\ 24' + 35°\ 36') = 100°$$

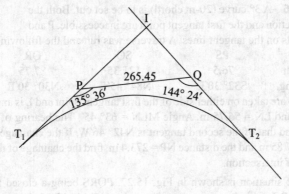

Fig. 15.25

From the sine rule,

$$IP/\sin(35° 36') = PQ/\sin 100°, \quad PQ = 265.45 \text{ m}$$

$$IP = 265.45 \times \sin(35° 36')/\sin 100° = 156.91 \text{ m}$$

$$IQ/\sin(44° 24') = 265.45/\sin 100°$$

$$IQ = 265.45 \times \sin(44° 24')/\sin 100° = 188.59 \text{ m}$$

For a 3° curve on a 20-m arc,

Radius = $20/(3\pi/180) = 382$ m

Deflection angle = 80°, tangent length = 382 sin 40°
= 245.54 m

Length of arc = $382 \times 80\pi/180$
= 533.37 m

Chainage of tangent point T_2
= 1342.65 + 533.37 = 1876.02 m

Example 15.15 A curve of radius 350 m was to be set out. The first tangent point is not accessible. The following measurements were made. P and Q are points on the backward tangent line on either side of the obstacle. R is any arbitrary point on the side of the curve clearing the obstacle. PR = 50 m, QR = 68 m, and ∠PRQ = 110° 00'. IQ = 133.48 m. If the chainage of P before the obstacle is 1138.535 m, find the chainage of both the tangent points and the point of intersection. The deflection angle of the curve is 48°.

Solution The situation is shown in Fig. 15.26. From triangle PQR,

$$PQ^2 = PR^2 + RQ^2 - 2 \times PR \times RQ \times \cos 110° = 85.77 \text{ m}$$

Chainage of P = 1138.535 m

Tangent length = 350 tan 24° = 155.83 m

QT_1 = 155.83 − 133.48 = 22.35 m

Chainage of T_1 = 1138.535 + 85.77 − 22.35

= 1201.955 m

Length of arc = $350 \times 48\pi/180 = 251.327$ m

Chainage of T_2 = 1201.955 + 251.327 = 1453.282 m

Chainage of I = 1201.955 + 155.83 = 1357.785 m

Example 15.16 A 3° curve (20-m chord) is to be set out. Both the **Fig. 15.26**
point of intersection and the first tangent point are inaccessible. P and
Q are two points on the tangent lines. A traverse was run and the following data obtained.

Line	PS	SQ	QR
Length	76.5	121.35	87.75
Bearing	S32° 30'E	N84° 45'E	N30° 30'E

Points M and N are taken on either side of the first tangent point and L is any arbitrary point. LM = 34.6 m and LN = 56.75 m. Angle MLN = 63° 45'. The bearing of the first tangent is N36° 32'E and that of the second tangent is N42° 46'W. If the chainage of M before the obstacle is 1267.85 m and the distance NP = 273.4 m, find the chainage of the tangent points and the point of intersection.

Solution The situation is shown in Fig. 15.27. PQRS being a closed traverse, we can find the length PQ from the properties of a traverse. Table 15.15 shows the latitudes and departures of lines whose lengths and bearings are known.

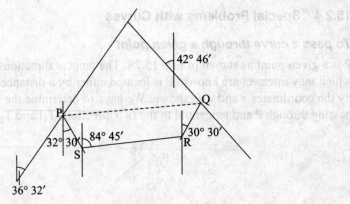

Fig. 15.27

Table 15.15 Calculating length PQ (Example 15.16)

Line	Length (m)	Bearing	Latitude (m)	Departure (m)
PS	76.5	S32° 30′E	– 64.52	41.10
SR	121.35	N84° 45′E	11.10	120.84
RQ	87.75	N30° 30′E	5.61	44.54

Sum of latitudes = 22.19 m, sum of departures = 206.48 m

Latitude of QP = – 22.19, departure of QP = – 206.48

Lenth of QP = $\sqrt{(22.19^2 + 206.482^2)}$ = 207.67 m

Bearing of QP = $\tan^{-1}(-2219/-206.48)$ = 6.134° = 6° 08′ = S83° 52′W

∠IPQ = 83° 52′ – 36° 32′ = 47° 20′

∠IQP = 96° 08′ – 42° 46′ = 53° 22′

The lengths IP and IQ can be obtained by applying the sine rule to triangle IPQ:

IP = 207.67 × sin(53° 22′)/sin(79° 18′) = 169.6 m

IQ = 207.67 × sin(47° 20′)/sin(79° 18′) = 155.4 m

Triangle LMN can be solved to get the length MN. Applying the cosine rule,

$$MN^2 = LM^2 + LN^2 - 2 \times LM \times LN \times \cos 63° 45'$$

$$= 34.62 + 56.75^2 - 2 \times 34.6 \times 56.75^2 - 2 \times 34.6 \times 53.75 \times \cos 63° 45'$$

$$= 51.78 \text{ m}$$

For a 3° curve on a 20-m arc,

Radius = 20/(3π/180) = 381.97 m

Tangent length = 381.7 tan(50° 21′) = 460.90 m

Chainage of point of intersection = 1267.85 + 51.78 + 273.4 + 169.6

$$= 1762.63 \text{ m}$$

Chainage of T_1 = 1762.63 – 460.9 = 1301.73 m

Length of arc = 460.9 × 100° 42′π/180 = 810.05 m

Chainage of T_2 = 1301.73 + 810.05 = 2111.78 m

15.2.4 Special Problems with Curves

To pass a curve through a given point

P is a given point as shown in Fig. 15.28. The tangent directions and the angle at which they intersect are known. P is located either by a distance r and angle α or by the coordinates x and y as shown. We have to determine the radius of a curve passing through P and tangential to the two directions T_1I and T_2I.

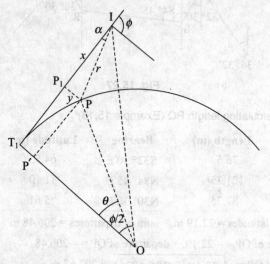

Fig. 15.28 Passing the curve through a given point

(i) r and α are given. From triangle T_1OI, $\angle T_1IO = 90° - \phi/2$. $\angle T_1IP = \alpha$. Therefore, $\angle PIO = 90° - \phi/2 - \alpha$. Let $\angle T_1OP = \theta$. In triangle IPO, $\angle POI = \phi/2 - \theta$. From these two angles,

$$\angle IPO = 180° - (90° - \phi/2 - \alpha) - (\phi/2 - \theta) = 90° + \alpha + \theta$$

We apply the sine rule to triangle IPO:

$$OI = R + R(\sec \phi/2 - 1) = R \sec \phi/2, \quad OP = R$$

Therefore,

$$R \sec(\phi/2)/R = \sin(90° + \alpha + \theta)/\sin[90° - (\phi/2 + \alpha)]$$

From which

$$\sec \phi/2 = \cos(\alpha + \theta)/\cos(\phi/2 + \alpha)$$

from which θ can be calculated. If PP_1 is the perpendicular on the tangent and PP' is a perpendicular on OT_1, then $P'T_1 = PP_1 = R(1 - \cos \theta) = r \sin \alpha$, where r is the distance of P from I. From this,

$$R = r \sin\alpha/(1 - \cos \theta)$$

The radius R can thus be determined.

(ii) If P is located by coordinates x and y from I, we calculate α from $\tan \alpha = y/x$. Knowing α, the procedure is the same as before.

To pass a curve tangential to three given lines

As shown in Fig. 15.29, the directions of lines T_1P and T_2Q are known. The angles at which a third line PQ intersects these lines are also known along with the length of line PQ. We have to find the radius of the curve that will be tangential to all the three points. In Fig. 15.29, angle $\phi = \alpha + \beta$, PC = R tan $\alpha/2$, and QC = R tan $\beta/2$.

$$PQ = PC + QC = R \tan \alpha/2 + R \tan \beta/2$$

Therefore,

$$R = PQ/(\tan \alpha/2 + \tan \beta/2)$$

A curve of this radius will be tangential to all the three lines.

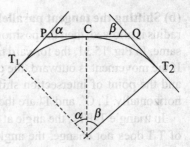

Fig. 15.29 Passing the curve tangential to three lines

Shifting the tangent parallel to itself

If during fieldwork, it is found necessary to shift any of the tangent points, keeping the direction of the tangent same, it can be done either without changing the radius of the curve already decided or with a new radius. Both these cases can be tackled as follows.

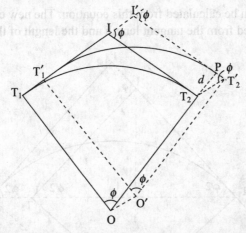

Fig. 15.30 Shifting the tangent parallel to itself

(a) Shifting the tangent parallel to itself without change in radius Figure 15.30 shows the case of shifting the forward tangent parallel to itself without changing the radius of the curve. Let the curve of radius R have two tangent points T_1 and T_2. The tangents intersect at I. The deflection angle is ϕ. It is now required to shift, by a distance d, the forward tangent parallel to itself outward to a new position. The radius has to remain the same. The point of intersection I, tangent point T_2, and the centre of curvature move to new positions. Let I′, $T_1′$, $T_2′$, and O′ be the new positions of these points. All these points move parallel to their new positions.

From triangle $T_2PT_2′$, this movement is equal to $d/\sin\phi$. Note that angle $T_2T_2′P = \phi$. If the old chainage of T_1 is C, then new chainage of T_1 will be $T_1′ = C + d/\sin\phi$. The point of intersection and the centre O will move by the same amount. The chainage I′ and $T_2′$ can be calculated now, as the length of the curve does not change.

(b) Shifting the tangent parallel to itself with change in radius If a change of radius is allowed, then the position of T_1 (or the other tangent point) can be kept the same. In Fig. 15.31, the forward tangent is shifted parallel to itself by an amount d. If the movement is outward, the curve becomes longer, the radius becomes larger, and the point of intersection shifts. We assume that the main chord is extended horizontally. I', O', and T_2' are the new positions of the corresponding points.

In triangle T_2PT_2', the angle at T_2' is $\phi/2$. (From triangle IT_1T_2', as the direction of T_1I does not change, the angle between T_1I and the new long chord is $\phi/2$.) From triangle PT_2T_2',

$$\sin(\phi/2) = d/T_2T_2', \quad T_2T_2' = d/\sin(\phi/2)$$

From triangle $I'IM$,

$$\sin \phi = d/II', \quad II' = d/\sin\phi$$

New tangent length $T_1I' = T_2I' = T_1I + II' = R\tan(\phi/2) + d/\sin\phi$

Since the deflection angle does not change, the new radius R' can be found as

$$R'\tan(\phi/2) = T_1I' = R\tan(\phi/2) + d/\sin\phi$$

$$R' = R + d/[\sin\phi + \tan(\phi/2)]$$

The new radius can be calculated from this equation. The new chainages of I' and T_2' can be calculated from the tangent length and the length of the arc.

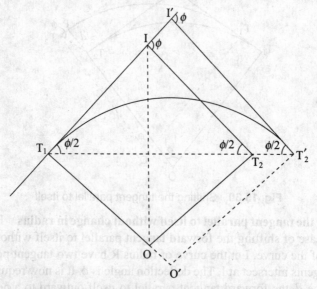

Fig. 15.31 Shifting the tangent parallel to itself

Changing the direction of tangent

If one of the tangents is rotated about the tangent point, the point of intersection changes and the other tangent point shifts. The radius of the curve also changes. There may also be situations in which the tangent direction changes along with the shifting of the tangent point.

Changing the direction of tangent; tangent point remaining the same In this case, the tangent line is rotated about the tangent point. Both the point of intersection and the other tangent point get shifted. Referring to Fig. 15.32, let the forward tangent be rotated by an angle θ. The point of intersection I shifts to I′, the other tangent point T_1 shifts to T_1' and the radius R changes to R'. The new deflection angle $= \phi' = \phi + \theta$. From triangle II′T_2, applying the sine rule,

$$I'T_2 = IT_2 \times \sin \phi / \sin \phi'$$

$$IT_2 = R \tan \phi/2, \quad I'T_2 = R'\tan \phi'/2$$

Therefore,

$$R' \tan \phi'/2 = (R \tan \phi/2) \times \sin \phi/ \sin \phi'$$

$$R' = R \times [\tan(\phi/2) \sin \phi]/(\tan \phi'/2 \times \sin \phi')$$

$$\tan \phi/2 = (1 - \cos \phi)/ \sin \phi$$

Therefore,

$$R' = R(1 - \cos \phi)/(1 - \cos \phi')$$

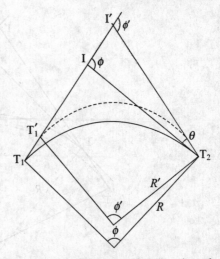

Fig. 15.32 Changing the direction of tangent

The distance by which the point of intersection shifts, II′, can be found from triangle II′T_2 by applying the sine rule:

$$II' = IT_2 \sin \theta/ \sin \phi'$$

To find the distance by which the tangent point T_1 has shifted,

$$T_1I = R \sin \phi, \quad T_1'I' = R' \sin \phi'$$

$$T_1T_1' = T_1I - T_1'I = T_1I - (T_1'I' - II')$$

$$= R \sin \phi - (R'\sin \phi' - T_1I \sin \theta/ \sin \phi')$$

$$= R \sin \phi(1 + \sin \theta/ \sin \phi') - R' \sin \phi'$$

But $R' \sin \phi' = R \sin \phi/ \sin \phi'$. Therefore,

$$T_1T_1' = R \sin \phi[(1 + \sin \theta/ \sin \phi') - (\sin \phi/ \sin \phi')]$$

$$= R \sin \phi[1 - (\sin \phi - \sin \theta)/ \sin \phi']$$

From these equations all the required values can be calculated.

Changing the direction of tangent; shifting tangent point   The case is illustrated in Fig. 15.33.

T_1I and T_2I are the tangents intersecting at I, deflection angle ϕ, of a curve of radius R. The tangent T_1I is rotated by an angle θ, keeping T_2 the same. The new point of intersection is I′. The angle of deflection is ϕ'. The new radius is R' and the centre of curvature is O′ on the line OT$_2$. $\phi' = \phi + \theta$. By applying the sine rule to triangle II′ T_1, II′ $= T_1I \times \sin \theta/ \sin \phi'$. T_2I' is the new tangent length and T_2I $(= T_1I)$ is the tangent length of the original curve.

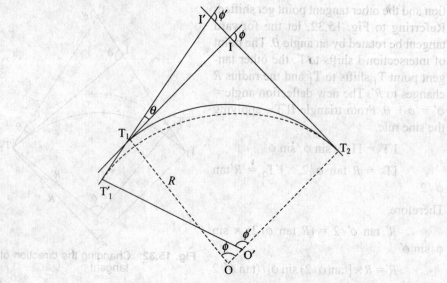

Fig. 15.33 Shifting the direction of tangent

$$T_2I' = T_2I + II' = T_2I + T_1I \sin \theta / \sin \phi' = T_2I[1 + (\sin \theta / \sin \phi')]$$
$$T_2I' = R' \tan \phi'/2 \quad \text{and} \quad T_2I = R \tan(\phi/2)$$

Therefore,

$$R' \tan(\phi'/2) = R \tan(\phi/2)[1 + (\sin \theta / \sin \phi')]$$
$$R' = R[\tan(\phi/2)/\tan(\phi'/2)][1 + (\sin \theta / \sin \phi')]$$

R can be calculated from this equation. The tangent point T_1 shifts to T_1'. To find this shift, we apply the sine rule to triangle $II' T_1$:

$$I' T_1 = IT_1 \sin \phi / \sin \phi'$$
$$T_1' T_1 = T_1' I' - I' T_1$$
$$= R \tan(\phi/2)[1 + (\sin \theta / \sin \phi') - R \tan(\phi/2)\sin \theta / \sin \phi' \quad (\text{since}$$
$$T_1' I' = T_2 I')$$
$$= R \tan(\phi/2)[1 - (\sin \phi - \sin \theta)/\sin \phi']$$

The new tangent point T_1' can now be located.

The following examples illustrate the principles outlined above.

Example 15.17 AT_1 and BT_2 are two straight lines meeting at point I at an angle of 110°. Find the radius and tangent lengths of a curve touching the two lines and passing through a point P. Angle $IPT_1 = 30°$ and the distance $IP = 70$ m.

Solution The situation is shown in Fig. 15.34. The deflection angle $\phi = 180° - 110° = 70°$. If $\angle T_1OP = \theta$, then we have, in triangle IPO,

$$\angle PIO = 110°/2 - 30° = 25°, \quad \angle POI = \phi/2 - \theta = 35° - \theta$$
$$\angle IPO = 180° - \angle PIO - \angle POI = 180° - 25° - 35° + \theta = 120° + \theta$$

Applying the sine rule to triangle IPO,

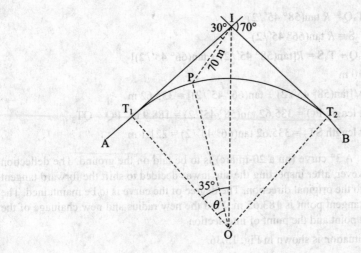

Fig. 15.34

$$R/\sin 25° = OI/\sin(120° + \theta)$$

Now,

$$OI = R + R(\sec 35° - 1) = R \sec 35°$$

$$\sin(120° + \theta) = \cos(30° + \theta)$$

Therefore,

$$R/\sin 25° = R \sec 35°/\cos(30° + \theta), \quad \cos(30° + \theta) = \sin 25°/\cos 35°$$

$$30° + \theta = 58.94, \quad \theta = 28.94° = 28° 56' 27''$$

$$R(1 - \cos \theta) = 70 \sin 30° = 35, \quad R = 35/(1 - \cos 28.94°) = 280.28 \text{ m}$$

Tangent length = $280.28 \times \tan 35° = 196.25$ m

Example 15.18 The following are the bearings of three lines: PQ = N21° 45′E, QS = N80° 30′E, SQ = S32° 45′E. Find the radius of a curve tangential to the three lines. Length QS = 410 m. Also find the tangent lengths.

Solution The situation is shown in Fig. 15.35. T_1, T_2, and T_3 are the tangent points.

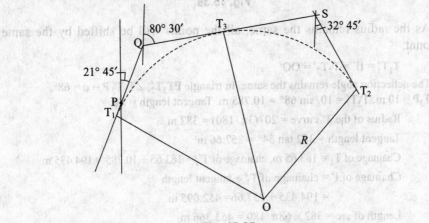

Fig. 15.35

$T_1Q = T_3Q = R \tan(58° \ 45'/2)$

$T_3S = T_2S = R \tan(66° \ 45'/2)$

$QS = T_3Q + T_3S = R[\tan(58° \ 45'/2) + \tan(66° \ 45'/2)]$

$QS = 410$ m

$R = 410/[\tan(58° \ 45'/2) + \tan(66° \ 45'/2)] = 335.62$ m

Tangent length $PQ = 335.62 \tan(58° \ 45'/2) = 188.9$ m, $PQ = QT_3$

Tangent length $ST_2 = 335.62 \tan(66° \ 45'/2) = 221.1$ m

Example 15.19 A 3° curve (on a 20-m arc) is to be laid on the ground. The deflection angle is 68°. However, after inspecting the site it was decided to shift the forward tangent by 10 m parallel to the original direction. The degree of the curve is to be maintained. The chainage of first tangent point is 183.65 m. Find the new radius and new chainage of the backward tangent point and the point of intersection.

Solution　The situation is shown in Fig. 15.36.

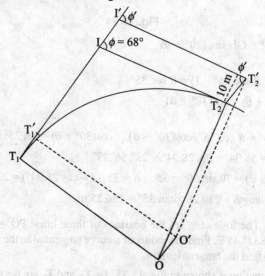

Fig. 15.36

As the radius remains the same, all the points will be shifted by the same amount:

$$T_1T_1' = II' = T_2T_2' = OO'$$

The deflection angle remains the same. In triangle PT_2T_2', $\angle T_2T_2'P = \phi = 68°$, $T_2P = 10$ m, $T_2T_2' = 10/\sin 68° = 10.785$ m. Tangent length $= R\phi$.

Radius of the 3° curve $= 20/(3\pi/180) = 382$ m

Tangent length $= 382 \tan 34° = 257.66$ m

Chainage of $T_1 = 183.65$ m, chainage of $T_1' = 183.65 + 10.785 = 194.435$ m

Chainage of $I' =$ chainage of $T_1' +$ tangent length

$\qquad = 194.435 + 257.66 = 452.095$ m

Length of arc $= 382 \times 68\pi/180 = 453.366$ m

Chainage of $T_2' = 194.435 + 453.366 = 647.801$ m

Example 15.20 A curve of radius 400 m and deflection angle 75° was to be set out. It became necessary to rotate the forward tangent by 10° about the tangent point. Find the new radius and chainages of the tangent points and point of intersection if the chainage of the backward tangent point was 986.45 m.

Solution The situation is shown in Fig. 15.37.

Tangent length $IT_2 = R \tan(75°/2) = 400 \times \tan 37.5° = 306.93$ m

From triangle IT_2I',

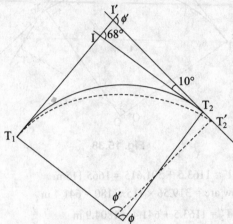

Fig. 15.37

$II' = IT_2 \sin 10°/\sin 85° = 306.93 \times \sin 10°/\sin 5° = 53.5$ m

$I'T_2 = 306.93 \times \sin 75°/\sin 85° = 297.6$ m

New radius $R' = 297.6/\tan 42.5° = 324.77$ m

$T_1T_1' = T_1I + II' - T_1'I' = 306.93 + 53.5 - 297.6 = 62.83$ m

Chainage of $T_1 = 986.45$ m

Chainage of $T_1' = 986.45 + 62.83 = 1049.28$ m

Chainage of $I' = 1049.28 + 297.6 = 1346.88$ m

Length of new arc $= 324.77 \times 85\pi/180 = 481.80$ m

Chainage of $T_2 = 1049.28 + 481.80 = 1531.08$ m

Example 15.21 A 3.5° curve (on a 20-m arc) has a deflection angle of 100°. The forward tangent is rotated by 15°, keeping the backward tangent point fixed. Find the new radius, and the chainages of the new point of intersection and the tangent point.

Solution The situation is shown in Fig. 15.38.

Radius for a 3.5° curve $= 20 \times 180/3.5\pi = 327.4$ m

$\phi = 100, \quad \theta = 15, \quad \phi' = 115$

$IT_2 = 327.4 \times \tan 50° = 390.18$ m

From triangle $II'T_2$,

$II' = 390.18 \times \sin 15°/\sin 115° = 111.435$ m

$T_1I' = 390.18 + 111.435 = 501.615$ m

$R' \tan(115°/2) = 501.615, \quad R' = 319.56$ m

Chainage of $T_1 = 1163.5$ m

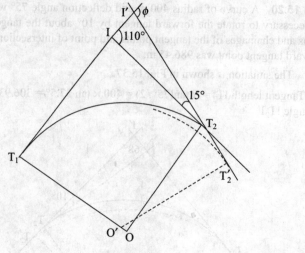

Fig. 15.38

Chainage of I' = 1163.5 + 501.615 = 1665.115 m
Length of new arc = 319.56 × 115π/180 = 641.4 m
Chainage of T'$_2$ = 1163.5 + 641.4 = 1804.9 m

15.3 Compound Curves

A compound curve is made up of two or more simple circular arcs of different radii. Here, we restrict the discussion to compound curves made up of arcs of two different radii only. Figure 15.39 shows such a compound curve. The following symbols are used in the discussion and derivation. R_1 is the radius of the first arc, R_2 is the radius of the second arc, T_1 is the first (or backward) tangent point, T_2 is the second (or forward) tangent point, I is the point of intersection of tangents at the end of the curve, EF is the tangent at the common point of the two arcs, ϕ is the deflection angle of the compound curve, ϕ_1 is the deflection angle of the first arc, ϕ_2 is the deflection angle of the second arc, Lt_1 is the length of tangent T_1I to the combined curve at the backward tangent point, LT_2 is the length of tangent T_2I to the combined curve at the forward tangent point, LT_1 is the length of tangent T_1E to the first arc, and Lt_2 is the length of tangent T_2F to the second arc.

A compound curve has seven elements—R_1, R_2, ϕ, ϕ_1, ϕ_2, LT_1, and LT_2. Four of the seven quantities must be known to evaluate the other three. The other three elements can be evaluated from the three equations to be derived.

15.3.1 Elements of a Compound Curve

Elements of a compound curve are related by the equation $\phi = \phi_1 + \phi_2$. This equation is clear from triangle IEF, where ϕ is the external angle equal to the sum of the interior opposite angles. Also, from triangles ET_1O and FT_2O,

$$T_1E = Lt_1 = R_1\tan(\phi_1/2), \quad T_2F = Lt_2 = R_2\tan(\phi_2/2)$$

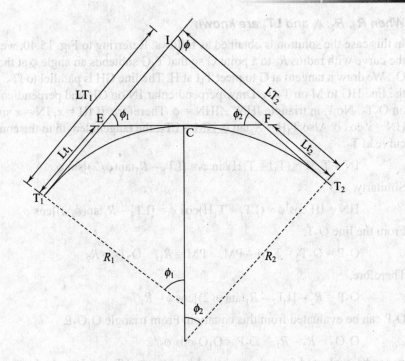

Fig. 15.39 Compound curve

From triangle IEF, we find the lengths IE and IF using the sine rule. Thus,

$$IE = \sin \phi_2[EF/\sin(180° - \phi)] = EF \sin \phi_2/\sin \phi$$

$$IF = \sin \phi_1[EF/\sin(180° - \phi)] = EF \sin \phi_1/\sin \phi$$

Now $T_1I = T_1E + EI$ and $T_2I = T_2F + FI$. Substituting for these quantities,

$$T_1I = LT_1 = R_1\tan(\phi_1/2) + EF \sin \phi_2/\sin \phi$$
$$= R_1\tan(\phi_1/2) + (R_1\tan \phi_1/2 + R_2\tan \phi_2/2)\sin \phi_2/\sin \phi$$
$$T_2I = LT_2 = R_2\tan(\phi_2/2) + EF \sin \phi_1/\sin \phi$$
$$= R_2\tan(\phi_2/2) + [R_1 \tan(\phi_1/2) + R_2 \tan(\phi_2/2)] \sin \phi_1/\sin \phi$$

The three equations for ϕ, LT_1, and LT_2 are sufficient to solve for three unknowns. Generally R_1, R_2, and ϕ are known. If one of the other quantities are known, the remaining three quantities can be evaluated from the above equations. However, the solution may become complex as angles and their trigonometric ratios are involved. The following procedure is adopted in different cases.

When R_1, R_2, ϕ, and ϕ_1 or ϕ_2 are known

This is the simplest case and can be solved easily. $\phi = \phi_1 + \phi_2$ and ϕ_1 or ϕ_2 can be obtained from this equation. Once the two angles are known, triangle IEF can be solved to get the lengths IE and IF. The tangent lengths are

$$LT_1 = IE + ET_1 = IE + R_1\sin(\phi_1/2)$$
$$LT_2 = IF + FT_2 = IF + R_2\sin(\phi_2/2)$$

When R_1, R_2, ϕ, and LT_1 are known

In this case the solution is obtained as follows. Referring to Fig. 15.40, we extend the curve with radius R_1 to a point G so that T_1G subtends an angle ϕ at the centre O_1. We draw a tangent at G to meet T_1I at H. The line GH is parallel to IT_2. Extend the line HG to M on T_2O_2. Draw perpendicular IN on GH and perpendicular O_1P on O_2T_2. Now, in triangle IHN, $\angle IHN = \phi$. Therefore, if IH = x, IN = $x \sin \phi$ and HN = $x \cos \phi$. Also $T_1H = R_1 \tan \phi$. NM = LT_2, the tangent length to the compound curve at T_2.

$$IN = T_2M = (T_1I - T_1H)\sin \phi = [LT_1 - R_1\tan(\phi/2)]\sin \phi$$

Similarly,

$$HN = IH \cos \phi = (LT_1 - T_1H)\cos \phi = [LT_1 - R_1\tan(\phi/2)]\cos \phi$$

From the line O_2T,

$$O_2P = O_2T_2 - T_2M - PM, \quad PM = R_1, \quad O_2T_2 = R_2$$

Therefore,

$$O_2P = R_2 - [LT_1 - R_1\tan(\phi/2)]\cos \phi - R_1$$

O_2P can be evaluated from this equation. From triangle O_1O_2P,

$$O_1O_2 = R_2 - R_1, \quad O_2P = O_1O_2\cos \phi_2/2$$

ϕ_2 can be evaluated from the equation $\phi_1 = \phi - \phi_2$. LT_2 can be calculated as $O_2P + R_1 + T_2M$.

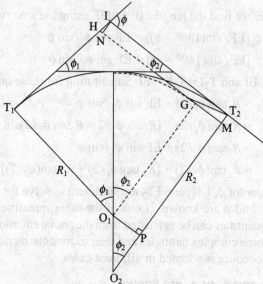

Fig. 15.40 Elements of compound curve

When R_1, LT_1, LT_2, and ϕ are known

The solution in this case is found in the same way as in the previous case. We have, from Fig. 15.40,

$$T_1H = R_1\tan \phi \times T_2M = [LT_1 - R_1\tan(\phi/2)]\sin \phi$$

Also,

$$HN = [LT_1 - R_1\tan(\phi/2)]\cos\phi$$

We also have

$$O_1P = GM = MN + NH - GH = LT_2 + HI\cos\phi - R_1\tan\phi/2$$

Extend the line CG to pass through T_2. In triangle T_2GM,

$$\angle T_2GM = \phi_2/2,$$
$$\tan(\phi_2/2) = T_2M/GM$$

ϕ_2 can be evaluated from the equation $\phi_1 = \phi - \phi_2$. The remaining values can be computed easily.

When R_1, R_2, ϕ, and LT_2 are known

In this case we extend the arc of larger radius to subtend an angle ϕ at the centre as shown in Fig. 15.41. The tangent to this arc at G will be parallel to T_1I. $T_2H = GH = R_2\tan(\phi/2)$. Draw a perpendicular IN from I to GH or GH extended. Draw also a perpendicular O_1P from O_1 to O_2G. From triangle IHN,

$$IH = T_2H - T_2I = R_2\tan(\phi/2) - LT_2$$
$$GM = IN = [R_2\tan(\phi/2) - LT_2]\sin\phi$$
$$HN = IH\cos\phi = [R_2\tan(\phi/2) - LT_2]\cos\phi$$

In triangle O_1O_2P,

$$O_2P = (R_2 - R_1)\cos\phi_1$$
$$= R_2 - GM - MP = R_2 - [R_2\tan(\phi/2) - LT_2]\sin\phi - R_1$$
$$\cos\phi_1 = O_2P/(R_2 - R_1) = R_2 - [R_2\tan(\phi/2) - LT_2]\sin\phi - R_1$$

$\cos\phi_1$ can be calculated from the equation $\phi_2 = \phi - \phi_1$. All the remaining values can be calculated after ϕ_1 and ϕ_2 are known.

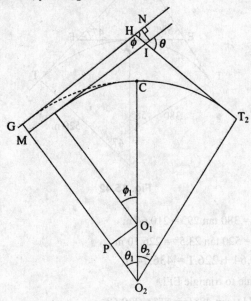

Fig. 15.41 Elements of compound curve

When R_2, LT_1, LT_2, and ϕ are known

The procedure is the same as in the previous case. Extend the larger arc to G so that it subtends an angle of ϕ at the centre. We have, as calculated earlier in the previous case,

$$IH = R_2\tan(\phi/2) - LT_2$$
$$GM = IN = [R_2\tan(\phi/2) - LT_2]\sin\phi$$
$$HN = [R_2\tan(\phi/2) - LT_2]\cos\phi$$

Extend the chord CT_1 to pass through G. In triangle GT_1M, $\angle GT_1M = \phi_1/2$.

$$\tan(\phi_1/2) = GM/MT_1 = IN/O_1P$$
$$O_1P = MT_1 = MI - T_1I = GN - LT_1 = GH + HN - LT_1$$

ϕ_1 can be calculated from the equation $\phi_2 = \phi - \phi_1$. From triangle O_1O_2P, $R_1 = R_2 - O_1O_2 = R_2 - O_2P/\sin\phi_1$.

The following examples illustrate the procedures explained here.

Example 15.22 A compound curve is made up of two arcs of radii 380 m and 520 m. The deflection angle of the combined curve is 105° and that of the first arc of radius 380 m is 58°. The chainage of the first tangent point is 848.55 m. Find the chainages of the point of intersection, common tangent point, and forward tangent point.

Solution Figure 15.42 represents the compound curve. Deflection angle of the compound curve = 105°, deflection angle of the first arc = 58°, deflection angle of the second arc = 105° – 58° = 47°. In Fig. 15.42,

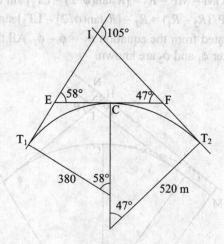

Fig. 15.42

$$EC = ET_1 = 380 \tan 29° = 210.64 \text{ m}$$
$$CF = FT_2 = 520 \tan 23.5° = 226.10 \text{ m}$$
$$EF = 210.64 + 226.1 = 436.74 \text{ m}$$

Applying the sine rule to triangle EFI,

$$EI = 436.74 \times \sin 47°/\sin 75° = 330.68 \text{ m}$$

FI = 436.74 × sin 58°/ sin 75° = 383.44 m

$T_1I = T_1E + EI = 210.64 + 330.68 = 541.32$ m

$T_2I = T_2F + FI = 226.1 + 383.44 = 609.54$ m

Chainage of T_1 = 848.55 m

Chainage of I = 848.55 + 541.32 = 1389.87 m

Length of first arc = 380 × 58°π/180 = 384.67 m

Length of second arc = 520 × 47°π/180 = 426.58 m

Chainage of C = 848.55 + 384.67 = 1233.32 m

Chainage of T_2 = 1233.32 + 426.58 = 1659.8 m

Example 15.23 A compound curve is made up of two arcs of radii 280 m and 400 m. The deflection angle of the combined curve is 76°. The length of tangent LT_1 = 260 m. Find the angles subtended by the two arcs at their centres and the chainages of the point of intersection, common tangent point, and forward tangent point if the chainage of the backward tangent point is 986.45 m.

Solution We extend the curve of 280 m radius to G so that the central angle is equal to 76°, as shown in Fig. 15.43. GH is the tangent to the curve at G. $\angle T_1O_1G = 76°$.

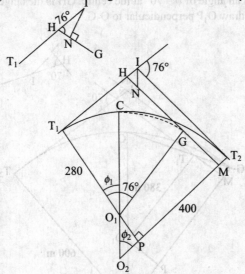

Fig. 15.43

GH = T_1H = 280 × tan 38° = 218.76 m

Draw IN perpendicular to GH. Angle IHN = 76°:

IH = 260 – 218.76 = 41.24 m

IN = IH sin ϕ = 41.24 sin 76° = 40.01 m

HN = 41.24 cos 76° = 9.98 m

Extend HG to M to intersect O_2T_2. T_2M = IN = 40.01 m. Draw O_1P perpendicular to O_2T_2.

$O_2P = O_2T_2 - PM - T_2M = 400 - 280 - 40.01 = 80$ m

$O_1O_2 = 400 - 280 = 120$ m

$\cos \phi_2 = O_2P/O_1O_2 = 80/120 = 0.6665$, $\phi_2 = 48.196° = 48° 12'$

$\phi_1 = 76° - 48° 12' = 27° 48'$, $O_1P = GM = 120 \sin 48° 12'$

$LT_2 = T_2I = GM + GH - HN = 89.46 + 218.76 - 9.98 = 298.24$ m

Length of arc $l_1 = 280 \times (27° 48') \times \pi/180 = 135.86$ m

Length of arc $l_2 = 400 \times (48° 12') \times \pi/180 = 336.47$ m

Chainage of $T_1 = 986.45$ m

Chainage of $I = 986.45 + 260 = 1246.45$ m

Chainage of $C = 986.45 + 135.86 = 1122.31$ m

Chainage of $T_2 = 1122.31 + 336.47 = 1458.78$ m

Example 15.24 A compound curve consists of two arcs of radius 380 m and 600 m with a deflection angle of 70°. If the length of the forward tangent is 370 m, find the angles subtended by the two arcs at their centre and the length of the backward tangent. If the chainage of the point of intersection is 1248.55 m, find the chainage of the tangent points and the common point of the two curves.

Solution The situation is shown in Fig. 15.44. Extend the arc of larger radius to G such that the arc subtends an angle of $\phi = 70°$ at the centre. GH is the tangent. Draw IN perpendicular to GH. Also draw O_1P perpendicular to O_2G.

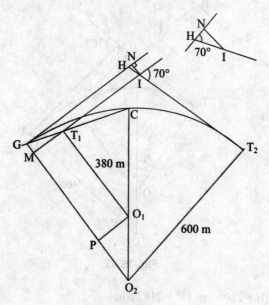

Fig. 15.44

$GH = T_2H = 600 \times \tan 35° = 420.12$ m

$T_2I = 370$ m, $IH = 420.12 - 370 = 50.12$ m

From triangle IHN,

$IN = 50.12 \sin 70° = 47.09$ m, $HN = 50.72 \cos 70° = 17.14$ m

In triangle O_1PO_2,

$$O_2P = O_2G - PM - MG, \quad O_2G = 600$$

$$PM = 380 \quad \text{and} \quad MG = IN = 47.09$$

Thus,

$$O_2P = 600 - 380 - 47.09 = 172.91 \text{ m}$$

In triangle O_1PO_2, $\angle PO_2O_1 = \phi_1$

$$\cos \phi_1 = O_2P/O_1O_2 = 172.91/220 = 0.8645, \quad \phi_1 = 30° \ 10'$$

$$\phi_2 = 70° - 30° \ 10' = 39° \ 50'$$

$$O_1P = T_1M = 220 \times \sin 39° \ 10' = 110.56$$

$$LT_1 = GH + HN - T_1M = 420.12 + 17.14 - 110.56 = 326.7 \text{ m}$$

Chainage of $T_1 = 1248.55$ m

Chainage of $I = 1248.55 + 326.7 = 1575.25$ m

Length of arc $l_1 = 380 \times (30° \ 10') \times \pi/180 = 200.07$ m

Length of arc $l_2 = 600 \times (39° \ 50') \times \pi/180 = 417.13$ m

Chainage of $C = 1248.55 + 200.07 = 1448.62$ m

Chainage of $T_2 = 1448.62 + 417.13 = 1865.75$ m

Example 15.25 A compound curve consists of two circular arcs, the first arc having a radius of 400 m. The deflection angle is 72°. The lengths of the two tangents are 260 m and 480 m. The chainage of the first tangent point is 1020.5 m. Find the radius of the second arc, the angles subtended by the arcs at the centres, and the chainages of the point of intersection, the common point of the arcs, and the second tangent point.

Solution As shown in Fig. 15.45, we extend the curve T_1C to G so that the arc T_1G subtends an angle of ϕ at the centre O_1. Draw the tangent at G to this arc to intersect the tangent T_1I at H. Draw a perpendicular IN from I onto GH. Draw a perpendicular O_1P from O_1 to O_2T_2. Extend the arc CG to pass through T_2.

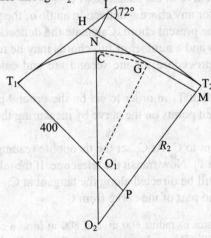

Fig. 15.45

Length of $T_1H = GH = 400 \times \tan 36° = 290.62$ m

In triangle IHN,

$$\angle IHN = \phi = 72°, \quad HI = 320 - 290.62 = 29.38$$

From triangle IHN,

$$HN = IH \cos \phi = 29.38 \times \cos 72° = 9.08 \text{ m}$$
$$IN = IH \sin \phi = 29.38 \times \sin 72° = 27.94 \text{ m}$$
$$T_2I = MN = 400 \text{ m}$$
$$MH = HN + NM = HN + IT_2 = 9.08 + 360 = 369.08 \text{ m}$$
$$GM = MH - HG = 369.08 - 290.62 = 78.46 \text{ m}$$

In triangle T_2GM, the angle at G is $\phi_2/2$. From triangle T_2GM,

$$\tan(\phi_2/2) = T_2M/GM = IN/GM = 27.94/78.46 = 0.3561, \quad \phi_2 = 39° \, 12'$$
$$\phi_1 = 72° - 39° \, 12' = 32° \, 48'$$

From triangle O_1O_2P,

$$O_1O_2 = O_1P/\sin \phi_2 = GM/\sin \phi_2 = 78.46/\sin(39° \, 12') = 124.14 \text{ m}$$
$$R_2 = R_1 + O_1O_2 = 400 + 124.14 = 524.14 \text{ m}$$

15.3.2 Setting Out a Compound Curve

A compound curve can be set on the ground in two stages. The method of deflection angles can be easily adopted for this purpose. The following procedure is adopted for this purpose.

1. Knowing the chainage of I, calculate the two tangent lengths of the compound curve from the equations given in the preceding section. Obtain the chainage of tangent point T_1 by subtracting the tangent length from the chainage of I.
2. Obtain the chainage of C, the apex point, or the common point of the two arcs by adding the length of the arc of the first curve to the chainage of T_1.
3. Obtain the chainage of the second tangent point T_2 by adding the length of the arc of the second curve to the chainage of C.
4. Tangential angle for any chord = deflection angle of the previous chord + tangential angle of the present chord. Calculate the deflection angles up to point C. Two subchords and a number of full chords may be required.
5. Adopt a similar procedure for the second arc and calculate the deflection angles.
6. Set up a theodolite at T_1 in order to set on the ground points on the first arc of the curve. Obtain points on the curve by measuring the deflection angles at T_1.
7. Shift the instrument to C. At C, set the theodolite reading to $360° - \phi_1/2$ and take a backsight to T_1. Now transit the telescope. If the telescope is now turned through $\phi_1/2$, it will be directed along the tangent at C.
8. Now set the second part of the curve from C.

Example 15.26 Two arcs of radius 600 m and 800 m form a compound curve with a deflection angle of 74° 30'. The angle subtended by the arc of radius 600 m at its centre is 40°. Calculate the tangential and deflection angles required to set out the first arc of the compound curve; the first arc is to be set out with the theodolite at the backward tangent point. The peg interval is 20 m and the chainage of the first tangent point is 863.5 m.

Solution The situation is shown in Fig. 15.46. The tangential angles are calculated as $1718.9 \times$ chord length/radius. First, the chord length is chosen to make a full station:

First chord length = 880 − 863.5

= 16.5 m

Length of first arc = 600 × 40π/180

= 418.37 m

Second arc subtends an angle = 74°
30′ − 40 = 34° 30′

Length of second arc = 800 × 34°
30′ π/180 = 481.87 m

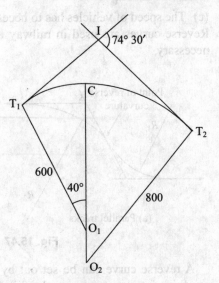

Fig. 15.46

The first arc of 600 m radius will be divided into
the first arc of length 16.5 m, 19 arcs of length
20 m, and the last arc of length 22.37 m.

First arc: tangential angle for 16.5-m
chord = 16.5 × 1718.9/600

= 47′ 16″

Tangential angle for 20-m chords
= 1718.9 × 20/600 = 57′ 18″

Tangential angle for last chord of
22.37 m = 1718.9 × 22.7/600

= 1° 04′ 05″

The deflection angles are shown in Table 15.16.

Table 15.16 Deflection angles required to set out the first arc (Example 15.26)

Chord length (m)	Deflection angle	Chord length (m)	Deflection angle
16.5	47′ 16″	20	11° 17′ 34″
20	1° 44′ 34″	20	12° 14′ 52″
20	2° 41′ 52″	20	13° 12′ 10″
20	3° 39′ 10″	20	14° 09′ 28″
20	4° 36′ 28″	20	15° 06′ 46″
20	5° 33′ 46″	20	16° 04′ 04″
20	6° 31′ 04″	20	17° 01′ 22″
20	7° 28′ 22″	20	17° 58′ 40″
20	8° 25′ 40″	20	18° 55′ 58″
20	9° 22′ 58″	22.37	20° 00′ 03″
20	10° 20′ 16″		

15.4 Reverse Curves

A reverse curve, as explained earlier, consists of two arcs of the same or different
curvature (Fig. 15.47) but with their centres of curvature on opposite sides of the
curve. They have a common tangent point known as the *point of reverse curvature*.
A reverse curve is used when the straight lines which the two curves connect, are
nearly parallel or intersect at a small angle. A reverse curve should be avoided for
the following reasons.

(a) The sudden change of curvature causes difficulty in movement.

(b) Superelevation cannot be provided conveniently at the intersection point of the
two arcs.

(c) The speed of vehicles has to necessarily be low due to different curvatures. Reverse curves are used in railway lines and roads, where they are absolutely necessary.

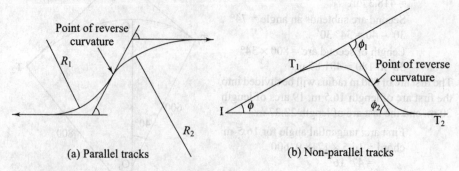

(a) Parallel tracks   (b) Non-parallel tracks

Fig. 15.47 Reverse curve

A reverse curve can be set out by calculating its elements only when some condition is given—e.g., equal radii of the two arcs or equal deflection angles for both the arcs.

15.4.1 Elements of a Reverse Curve

The elements of a reverse curve can be determined as follows. In Fig. 15.48(a), AT_1 and BT_2 are the two straight lines meeting at an angle ϕ. T_1 and T_2 are the two tangent points. The tangents intersect at I, the point of intersection of the tangents. The common tangent of the two curves is HJ. ϕ_1 and ϕ_2 are the angles made by the common tangent with the straight lines. The long chord connecting the two ends of the curve is T_1T_2. α_1 is the angle made by this chord with the straight line AT_1 and α_2 is the angle made with the other straight. O_1 and O_2 are the centres of the two curves of radii R_1 and R_2, respectively.

$$\angle T_1O_1D = \alpha_1 \quad \text{and} \quad \angle T_2O_2E = \alpha_2$$
$$\angle DO_1E = \phi_1 - \alpha_1 \quad \text{and} \quad \angle EO_2G = \phi_2 - \alpha_2$$

From triangle HIJ,

$$\phi_1 = \phi + \phi_2 \quad \text{or} \quad \phi = \phi_1 - \phi_2$$

From triangle T_1IT_2,

$$\alpha_1 = \phi + \alpha_2 \quad \text{or} \quad \phi = \alpha_1 - \alpha_2$$

This gives

$$\phi_1 - \phi_2 = \alpha_1 - \alpha_2$$

The same result is obtained by considering the two parallel lines O_1D and O_2E. Angles DO_1G and EO_2G are equal, being alternate angles. Therefore, $\phi_1 - \alpha_1 = \phi_2 - \alpha_2$. Now,

$$T_1T_2 = T_1D + DE + ET_2$$

$$T_1D = R_1\sin\alpha_1$$

$$DE = O_1K = (R_1 + R_2)\sin(\phi_2 - \alpha_2) \quad \text{and} \quad ET_2 = R_2\sin\alpha_2$$

$$T_1T_2 = R_1\sin\alpha_1 + R_2\sin\alpha_2 + (R_1 + R_2)\sin(\phi_2 - \alpha_2)$$

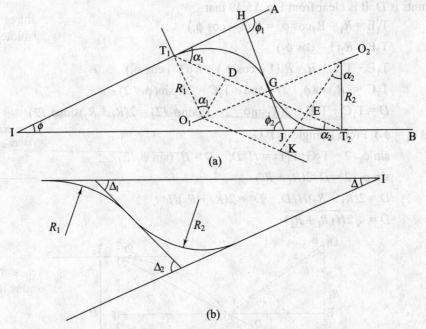

Fig. 15.48 Elements of a reverse curve

Similarly,

$$O_1D = KE = R_1\cos \alpha_1, \quad O_2E = R_2\cos \alpha_2$$

$$O_2K = KE + EO_2 = R_1\cos \alpha_1 + R_2\cos \alpha_2$$

From triangle O_2KO_1,

$$O_2K = (R_1 + R_2)\cos(\phi_2 - \alpha_2)$$

Equating the two expressions for O_2K, we have

$$R_1\cos \alpha_1 + R_2\cos \alpha_2 = (R_1 + R_2)\cos(\phi_2 - \alpha_2)$$

This gives

$$\cos(\phi_2 - \alpha_2) = (R_1\cos \alpha_1 + R_2\cos \alpha_2) / (R_1 + R_2) = \cos(\phi_1 - \alpha_1)$$

These equations are sufficient to determine the various quantities if some condition is available. Generally, $R_1 = R_2$ or $\phi_1 = \phi_2$ is the condition available. In the above equations, it is assumed that $\phi_1 > \phi_2$. This will be the case when the point of intersection I occurs before the reverse curve. If, as in Fig. 15.48(b), the point of intersection occurs after the reverse curve, then $\phi_2 > \phi_1$. In general, $\phi = \pm (\phi_1 - \phi_2)$.

The most common use of the reverse curve is when the two straight lines are parallel. The following is a discussion on the most common cases with the conditions required for evaluating the elements.

Straight lines are parallel

In the case of parallel straight lines, a reverse curve is a common method to go from one line to the other. Figure 15.49 represents the situation.

Let the distance along the direction of the tangents between T_1 and T_2 be d. The distance between the parallel straight lines is H and the distance between the tangent points is D. It is clear from Fig. 15.49 that

$$T_1E = R_1 - R_1\cos \phi_1 = R_1(1 - \cos \phi_1)$$

$$T_2F = R_2(1 - \cos \phi_2)$$

$$T_1E + T_2F + H = R_1(1 - \cos \phi_1) + R_2(1 - \cos \phi_2)$$

$$T_1C = 2R_1\sin(\phi_1 / 2) \quad \text{and} \quad T_2C = 2R_2\sin(\phi_2 / 2)$$

$$D = T_1C + T_2C = 2(R_1\sin\phi_1/2 + R_2\sin\phi_2/2) = 2(R_1 + R_2)\sin(\phi_1/2)$$

(as $\phi_1 = \phi_2$). From triangle T_1T_2G,

$$\sin \phi_1/2 = T_2G/T_1T_2 = H/D, \quad D = H/(\sin \phi_1/2)$$

$$\sin \phi_1/2 = D/2(R_1 + R_2)$$

$$D = 2(R_1 + R_2)H/D, \quad D^2 = 2(R_1 + R_2)H$$

$$D = \sqrt{2H(R_1 + R_2)}$$

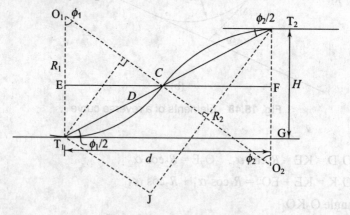

Fig. 15.49 Parallel tracks

These equations can be used to solve for different values if $R_1 = R_2 = R$. Then,

$$EF = 2R(1 - \cos \phi_1), \quad D = 2\sqrt{(RH)}, \quad d = 2R \sin \phi$$

Straight lines are non-parallel

In the case, there can be four different situations depending upon the given quantities, as follows.

ϕ_1, ϕ_2, and length of common tangent known and $R_1 = R_2$ In Fig. 15.50,

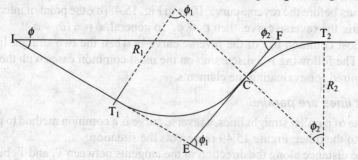

Fig. 15.50 Non-parallel track

$$EF = EC + CF, \quad EC = T_1E = R_1\tan \phi_1/2, \quad CF = FT_2 = R_2\tan \phi_2/2$$
$$EF = R_1\tan \phi_1/2 + R_2 \tan \phi_2/2$$

If $R_1 = R_2$, then

$$R = EF/(\tan \phi_1/2 + \tan \phi_2/2)$$

In triangle IFE, $\phi_2 = \phi + \phi_1$. Applying the sine rule to triangle IFE, $EI = EF \sin \phi_2/\sin \phi$. The distance between T_1 and I can be calculated as

$$T_1I = T_1E + EI = R \tan \phi_1/2 + EF \sin \phi_2/\sin \phi$$

The required quantities can be calculated from these expressions.

α_1, α_2, and T_1T_2 **(distance between tangent points) known and** $R_1 = R_2$ In Fig. 15.51, α_1 and α_2 are the angles made by the tangents with the line T_1T_2. T_1I and T_2I are the straight lines meeting at an angle ϕ. R_1 and R_2 are the radii of the two arcs and generally $R_1 = R_2$.

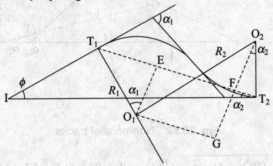

Fig. 15.51 Non-parallel tracks

$$\angle T_1O_1E = \alpha_1, \quad \angle T_2O_2F = \alpha_2$$

In triangle T_1O_1E,

$$O_1E = R_1\cos \alpha_1 = FG$$

In triangle T_2O_2F,

$$O_2F = R_2 \cos \alpha_2, \quad O_1O_2 = R_1 + R_2$$

From triangle O_1GO_2,

$$O_1G = (R_1 + R_2)\cos \theta = EF$$
$$O_2G = (R_1 + R_2)\sin \theta, \quad O_2G = O_2F + FG = R_1\cos \alpha_1 + R_2\cos \alpha_2$$
$$\sin \theta = (R_1\cos \alpha_1 + R_2\cos \alpha_2)/(R_1 + R_2)$$

From triangle T_1O_1E,

$$T_1E = R_1\sin \alpha_1$$

From triangle O_2FT_2,

$$FT_2 = R_2\sin \alpha_2$$
$$T_1T_2 = T_1E + EF + FT_2 = R_1\sin \alpha_1 + (R_1 + R_2)\cos \theta + R_2\sin \alpha_2$$

T_1T_2 is known. If $R_1 = R_2 = R$, R and θ can be calculated. The central angles of the two arcs can be calculated as follows.

For arc of radius R_1: Central angle = $\alpha_1 + 90° - \theta$

For arc of radius R_2: Central angle = $\alpha_2 + 90° - \theta$

ϕ_1, ϕ_2, and distance between tangent points, and one of the radii known This case is the same as the previous case except that R_1 or R_2 is given. Referring to Fig. 15.52,

$$T_1E = R_1\sin \alpha_1, \quad T_2F = R_2\sin \alpha_2$$

From triangle O_2O_1G,

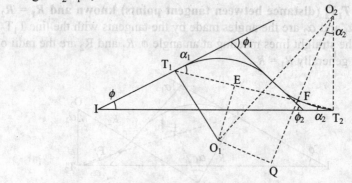

Fig. 15.52 Non-parallel tracks

$$O_1G = \sqrt{O_1O_2^2 - O_2G^2}, \quad O_1O_2 = R_1 + R_2, \quad O_2G = GF + FO_2 = O_1E + FO_2$$

$$O_1E = R_1\cos \alpha_1, \quad FO_2 = R_2\cos \alpha_2$$

$$O_2G = R_1\cos \alpha_1 + R_2\cos \alpha_2$$

$$O_1G = EF \ \sqrt{(R_1 + R_2)^2 - (R_1\cos\alpha_1 + R_2\cos\alpha_2)^2}$$

The distance between tangent points, T_1T_2, is given. Let this be D. Then

$$\begin{aligned} D &= T_1E + EF + FT_2 \\ &= R_1\sin \alpha_1 + \sqrt{(R_1 + R_2)^2 - (R_1\cos\alpha_1 + R_2\cos\alpha_2)^2} + R_2\sin \alpha_2 \end{aligned}$$

$$[D - (R_1\sin\alpha_1 + R_2\sin \alpha_2)]^2 = (R_1 + R_2)^2 - (R_1\cos \alpha_1 + R_2\cos \alpha_2)^2$$

One of the two radii is known. The other radius can be calculated from this equation, as all the other quantities are known. Once R_1 and R_2 are known, θ can be calculated from triangle O_2O_1G.

R_1, R_2, angle of intersection f, and one tangent length known The case is illustrated in Fig. 15.53. AI and BI are the two straight lines intersecting at I. T_1 and T_2 are the tangent points of arcs of radii R_1 and R_2. The central angles of these arcs are ϕ_1 and ϕ_2. Draw O_2C perpendicular to AI and O_2E parallel to AI to intersect O_1T_1 at E. Angle $T_2DI = 90 - \phi = CDO_2$ (opposite angles).

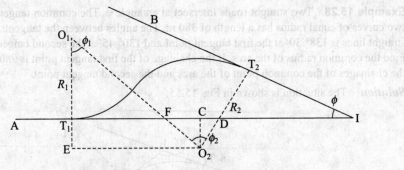

Fig. 15.53 Elements of reverse curves

$$T_2O_2C = \phi, \quad CO_2F = \phi_2 - \phi = \phi_1 \quad \text{(as } O_2C \text{ and } O_1T_1 \text{ are parallel)}$$

$\phi = \phi_2 - \phi_1$. Let $IT_1 = \angle T_1$ and $IT_2 = \angle T_2$. From triangle DIT_2,

$$DT_2 = LT_2 \tan \phi, \quad DI = LT_2 / \cos \phi$$

If LT_2 is given, these lengths can be calculated. From triangle O_2CD,

$$O_2C = O_2D \cos \phi \quad \text{and} \quad CD = O_2D \sin \phi, \quad O_2D = R_2 - T_2D$$

These values can be calculated. In triangle O_1EO_2,

$$O_1E = R_1 + T_1E = R_1 + O_2C$$

From this equation O_1E can be calculated. The central angle ϕ_1 is given by $\cos \phi_1 = O_1E/(R_1 + R_2)$

$$O_2E = O_1E \tan \phi_1$$

$$\text{Tangent length } LT_1 = ID + DC + CT_1 = ID + CD + O_2E$$

Example 15.27 A reverse curve is to be set between parallel tracks 20 m apart. The arcs have the same radius. The distance between the tangent points is 200 m. Determine the common radius of the arcs.

Solution The situation is shown in Fig. 15.54. In this figure, $T_1T_2 = 200$ m and $T_2G = 20$ m. From ΔT_1T_2G,

$$\sin \phi/2 = T_1G/T_2G = 20/200 = 0.1, \quad \phi = 11° 28' 42''$$

$$\text{Radius} = (20/2)(1 - \cos \phi) = 500 \text{ m}$$

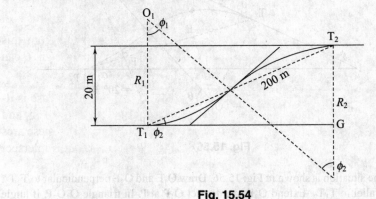

Fig. 15.54

Example 15.28 Two straight roads intersect at an angle ϕ. The common tangent of the two curves of equal radius has a length of 380 m. The angles between the tangents and the straight lines is 138° 30′ at the first tangent point and 130° 45′ at the second tangent point. Find the common radius of the arcs. If the chainage of the first tangent point is 980 m, find the chainages of the common point of the arcs and the second tangent point.

Solution The situation is shown in Fig. 15.55.

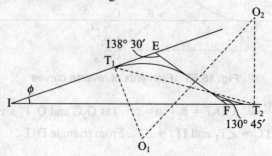

Fig. 15.55

$\phi_1 = 180° - 138° 30′ = 41° 30′$, $\phi_2 = 180° - 130° 45′ = 49° 15′$

As the two arcs have the same radius,

Length of EF = 380 = R(tan 41.5°/2 + tan 49.25°/2), $R = 453.9$ m

Length of arc $l_1 = 453.9 \times 41.5° \times \pi/180 = 328.76$ m

Length of arc $l_2 = 453.9 \times 49.25° \times \pi/180 = 390.16$ m

Chainage of $T_1 = 980$ m

Chainage of C = 980 + 328.76 = 1308.76 m

Chainage of $T_2 = 1308.76 + 390.16 = 1698.92$ m

Example 15.29 A reverse curve has to be set between two arcs intersecting at an angle ϕ. The distance between the tangent points is 680 m. The angle between the line joining the two tangent points is 30° 15′ at the first tangent point and 20° 30′ at the second tangent point. Find the common radius of the arcs and the angle of intersection.

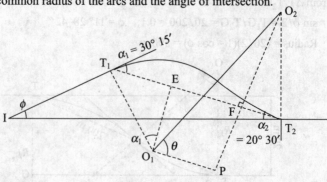

Fig. 15.56

Solution The situation is shown in Fig. 15.56. Draw O_1E and O_2F perpendicular to T_1T_2. Draw O_1P parallel to T_1T_2. Extend O_2F to intersect O_1P at P. In triangle O_1O_2P, if angle

$O_2O_1P = \theta$, then $\sin \theta = O_2P/O_1O_2 = O_2P/2R$.

$O_2P = O_1E + O_2F = R \cos \phi_1 + R \cos \phi_2 = R \cos 30° 15' + R \cos 20° 30'$

$\sin \theta = R(\cos 30° 15' + \cos 20° 30')/2R = 0.9002, \quad \theta = 64° 11'$

$T_1T_2 = T_1E + EF + FT_2 = R \sin \phi_1 + 2R \cos \theta + R \sin \phi_2$

$$R = T_1T_2/(\sin 30° 15' + 2 \sin 64° 11' + \sin 20° 30')$$

$$= 680/(0.5038 + 0.8711 + 0.9367)$$

$$= 294.17 \text{ m}$$

Angle of intersection $= 30° 15' - 20° 30' = 9° 45'$

Example 15.30 A reverse curve consists of two arcs of radii 280 m and 400 m. The reverse curve lies between tangents intersecting at an angle of 20°. If the tangent length is 180 m at the second tangent point, find the other tangent length and central angles of the arcs. If the chainage of the first tangent point is 1020 m, find the chainages of the point of intersection, the second tangent point, and the common point of the arcs.

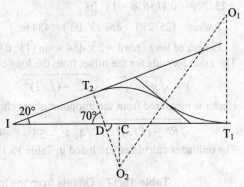

Fig. 15.57

Solution The situation is shown in Fig. 15.57. In this figure, in triangle IT_2D, $\angle T_2ID = 20°$, $\angle IDT_2 = 70° = \angle ODC$. In triangle CO_1D, $\angle CO_1D = 20°$. If ϕ_2 is the central angle of the arc of radius R_2, then $\angle CO_2O_1 = \phi_2 - 20° = \angle T_1O_1O_2$ (O_2C and O_1T_1 are parallel). In triangle IT_2D, $IT_2 = 180$ m. $T_2D = 180 \sin 20° = 61.56$ m, $ID = 180 \cos 20° = 169.14$ m. $O_2T_2 = R_2 = 400$ m. $O_2D = 400 - 61.56 = 338.44$ m. In triangle O_2CD,

$O_2C = 338.44 \cos 20° = 318.03$ m

In triangle O_1O_2E,

$O_1E = R_1 + T_1E = R_1 + O_2C = 318.03 + 280 = 598.03$ m

$\cos \phi_1 = 598.03/(400 + 280) = 0.8794, \quad \phi_1 = 28.42$

$\phi_2 = 28.42 + 20 = 48.42$

$IT_1 = ID + DC + DT_1 = ID + DC + O_2E$. $ID = 169.14$ m, $DC = 338.44 \sin 20° = 115.75$ m, and $O_2E = 680 \sin 28.42° = 363.63$ m. $IT_1 = 608.52$ m.

Length of arc $l_1 = 280 \times 28.42 \pi/180 = 138.88$ m

Length of arc $l_2 = 400 \times 48.42 \pi/180 = 338.03$ m

Chainage of $T_1 = 1020$ m

Chainage of $C = 1020 + 138.88 = 1158.88$ m

Chainage of $T_2 = 1158.88 + 338.03 = 1196.91$ m

15.4.2 Setting Out Reverse Curves

Reverse curves can be set out with offsets from tangents. One arc can be set from one tangent and the other arc from the other tangent. Deflection angles can also be used for setting out the curve. A theodolite can be set up at a tangent point for one arc and the second arc can be set from the common point of the arcs or the other tangent points. The following examples illustrate the necessary calculations.

Example 15.31 A reverse curve lying between parallel straight lines 25 m apart consists of two arcs of equal radius. If the distance between the tangent points is 220 m, find the radius. Find the chainages and offsets at 10-m intervals from the long chord to set out the curve if the chainage of the first tangent point is 901.8 m.

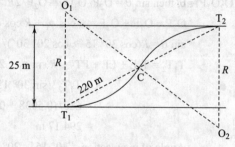

Fig. 15.58

Solution The curve is shown in Fig. 15.58. Let ϕ be the central angle. $\sin \phi/2 = 25/220 = 0.1136$, $\phi = 13° 03'$.

Radius $= (25/2)(1 - \cos 13° 03') = 484$ m

Length of long chord $= 2 \times 484 \times \sin (13° 03'/2) = 110$ m

The exact formula for the offset from the long chord is

$$Y = \sqrt{R^2 - x^2} - \sqrt{R^2 - (L/2)^2}$$

Here x is measured from the midpoint of the chord.

$$\sqrt{R^2 - (L/2)^2} = \sqrt{484^2 - 55^2} = 480.86$$

The ordinates calculated are listed in Table 15.17.

Table 15.17 Offsets from the long chord (Example 15.31)

Distance from D, x (m)	Offset from long chord, Y (m)
0	3.14
10	3.04
20	2.73
30	2.21
40	1.48
50	0.55
55	0.0

The offsets will be the same for the other half of the chord and for the second curve.

Example 15.32 If the curve of Example 15.31 is to be set by tangential or deflection angles, list the deflection angles with the theodolite at the first tangent point for the first curve.

Solution Length of arc $= 484 \times 13.05 \times \pi / 180 = 110.24$ m. The tangential angle is given by $1718.9c/R$ minutes. At 10-m intervals, the length of first chord $= 8.2$ m. There will be nine chords of 10 m each and the length of the last chord $= 12.04$ m.

Tangential angle of the first chord $= 1718.9 \times 8.2/484 = 29' 07''$

Tangential angle of the 10-m chords $= 1718.9 \times 10/484 = 35' 31''$

Tangential angle of the 12.04-m chord $= 1718.9 \times 12.04/484 = 42' 46''$

Deflection angle = tangential angle of the chord + deflection angle of the previous chord. These are listed in Table 15.18.

The angles given in Table 15.18 are for the first arc of the reverse curve. From the tangent at the common point, the angles will be the same. The theodolite is then shifted to C, the common point of the arcs. Half the central angle is set and the telescope transited. The direction of the tangent at C is obtained.

Table 15.18 Deflection angle for setting the curve (Example 15.32)

Point	Chord length (m)	Tangential angle	Deflection angle
1	8.2	29' 07"	29' 07"
2	10	35' 31"	1° 04' 38"
3	10	35' 31"	1° 40' 09"
4	10	35' 31"	2° 15' 40"
5	10	35' 31"	2° 51' 11"
6	10	35' 31"	3° 26' 42"
7	10	35' 31"	4° 02' 13"
8	10	35' 31"	4° 37' 44"
9	10	35' 31"	5° 13' 15"
10	10	35' 31"	5° 48' 46"
11	12.04	42' 46"	6° 31' 32"

15.5 Transition curves

Transition curves are used in road and railway curves to avoid the suddenness of change from a straight line to a curve of finite radius. On a curve, the vehicle is subjected to forces that tend to throw the vehicle outward due to centrifugal action. Such forces are resisted either by the frictional forces developed between the tyres of vehicles and the road surface or by the cant or superelevation provided on the outer edge of the road. In the case of railway curves the forces are resisted by the outer flanges pressing against the outer rails. A transition curve, also known as the *easement curve*, serves the purpose of reducing the effect of a sudden change in curvature from zero to a finite value.

Transition curves serve the following purposes.

(a) They effect the change in radius gradually by a curve of varying radius.
(b) They provide superelevation gradually from zero to a stipulated value at the beginning of the curve.

15.5.1 Requirements of a Transition Curve

A transition curve should satisfy the following conditions.

(a) It should meet the straight line part of the road tangentially. This means that its curvature should be zero at the beginning.
(b) It should meet the circular curve tangentially. This means that the radius of curvature at the meeting point with the circular curve must be the same as that of the circular curve.
(c) The length of the transition curve must be such that the cant or superelevation can be provided conveniently to its maximum value at the beginning of the circular curve.
(d) The rate of increase of the curvature should be such that it matches with the rate of increase of cant.

It is obvious that transition curves need to be provided both at the beginning and the end of a circular curve. Superelevation is an important aspect for the safety of vehicles moving along a curve. The basic principles, which come from mechanics, are explained below,

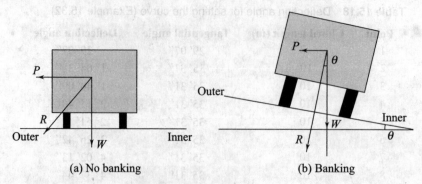

(a) No banking (b) Banking

Fig. 15.59 Superelevation

Superelevation

Superelevation (or cant or banking) is the increase in height of the outer edge of a curved path over the inner edge. It provides a means of balancing the forces due to the movement of a vehicle over a curve.

Let us consider the case of a vehicle moving along a curved path as shown in Figs 15.59(a) and (b). In Fig. 15.59(a) the surface is level and in (b) it is banked on the outer edge. The vehicle is acted upon by the forces shown. W is the weight of the vehicle and P is the centrifugal force. In case (a), P is parallel to the surface on which the vehicle moves. The centrifugal force is equal to mass × acceleration. The acceleration is equal to v^2/R, where v is the velocity and R is the radius. Since W is vertical and $P = ma = (W/g)v^2/R$ is horizontal, the horizontal force has to be balanced by the frictional force developed between the tyres of the vehicle and the surface of the road. The frictional force that can be developed has a maximum value and the speed of the vehicle is thus limited. Beyond a certain value of the velocity, the frictional force required will be more than what can be developed and the vehicle will slip and be thrown outwards.

In case (b), the forces acting on the vehicle are the same. However, the surface is banked by an angle θ. If R is the net force acting normal to the surface, $P = R \sin \theta$ and $W = R \cos \theta$. W will be inclined at an angle θ to R.

$$P = Wv^2/gR \quad \text{and} \quad P/W = v^2/gR$$

It is easy to see that $\tan \theta = P/W = v^2/gR$. If h is the cant or superelevation and B is the width of the road,

$$h = B \tan \theta = Bv^2/gR$$

In the case of railway lines, if G is the gauge,

$$h = Gv^2/gR$$

The effect of superelevation is to reduce the dependence on friction for the stability of the vehicle. The force P is no longer parallel to the surface, as it is always horizontal. It is perpendicular to the axis of rotation, which is vertical.

For the maximum speed of the vehicle allowed on the track, if sufficient cant is provided, the vehicle will move with stability and the passengers will not feel the centrifugal force. If the banking height is less, the passengers will feel the force and sway outwards. Such a case has cant deficiency.

Cant $h = Bv^2/gR$

The cant thus depends upon the velocity and the width of the track.

Centrifugal ratio

Centrifugal ratio is the ratio of the centrifugal force to the weight of the vehicle.

Centrifugal ratio $= P/W = v^2/gR$

Generally, the maximum value of the centrifugal ratio is taken as $1/4$ for roads and $1/8$ for railways. Thus, for roads,

$$1/4 = v^2/g \quad \text{or} \quad v = \sqrt{gR/4}$$

and for railways,

$$1/8 = v^2/gR \quad \text{or} \quad v = \sqrt{gr/8}$$

These equations decide the minimum radius of the curve for moving with a given velocity.

Friction

Friction between the vehicle and the surface opposes the effect of centrifugal force. Consider the case shown in Fig. 15.60. NR is the reaction normal to the surface, W is the weight of the vehicle, and P is the horizontal centrifugal force. The frictional force will be μ'NR, where μ' is the friction factor and is less than μ, the frictional coefficient between the vehicle and the road surface. We resolve the forces W and P parallel to the pavement surface. Note that NR, the normal reaction is inclined at θ to the vertical, where θ is the angle of banking.

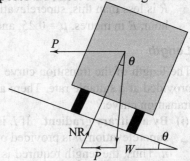

Fig. 15.60 Friction

Forces normal to the pavement $= W \cos\theta + P \sin\theta$

Forces parallel to the pavement $= P \cos\theta - W \sin\theta$

Forces parallel to the pavement are resisted by the frictional force. Therefore,

$$W \sin\theta + P \cos\theta = \mu'\text{NR}$$

But NR = forces normal to the pavement $= W \cos\theta + P \sin\theta$

$$P \cos\theta - W \sin\theta = \mu'(P \sin\theta + W \cos\theta)$$

Rearranging the terms,

$$P(\cos\theta - \mu' \sin\theta) = W(\sin\theta + \mu'\cos\theta)$$

$$P/W = (\sin\theta + \mu'\cos\theta)/(\cos\theta - \mu'\sin\theta)$$

Dividing the numerator and denominator by $\cos\theta$,

$$P/W = (\tan\theta + \mu')/(1 - \mu' \tan\theta)$$

$$= v^2/gR \quad (\text{as } P/W = v^2/gR)$$

The condition for the safety of the vehicles is

$$v^2/gR < (\tan\theta + \mu')/(1 - \mu'\tan\theta)$$

The frictional coefficient μ has a maximum value, which is generally taken as 0.25. The value of μ' in the above equation is limited to 0.25. For safe design of highways this equation must be satisfied. It is clear from the above that the effect of centrifugal forces is balanced by superelevation and friction. If superelevation is increased (increase in θ), there will be less dependence on friction. On the other hand, if there is cant deficiency, there will be more reliance on friction. How much of the centrifugal force should be resisted by superelevation or friction is a matter of debate. In general, there are two methods—(i) relying entirely on superelevation and (ii) relying entirely on friction.

(i) **Method of maximum superelevation** If we rely entirely on superelevation for resisting the centrifugal forces, then $P/W = \tan\theta = v^2/gR$; $R = v^2/g\tan\theta$. This is the minimum value of R. If the radius is less than this value, friction has to come into play. This can also be obtained by putting $\mu' = 0$ in the equation for $P/W = (\tan\theta + \mu')/(1 - \mu'\tan\theta)$.

(ii) **Method of maximum friction** In this case, we rely entirely on frictional forces to resist centrifugal forces. Thus, $Wv^2/gR < \mu W$, from which $R > v^2/\mu g$. For a given velocity and friction coefficient, this is the minimum value of R. If R is less than this, superelevation has to be provided. If v is in kilometres per hour, R in metres, $\mu = 0.25$, and $g = 9.8$ m s^{-2}, then $R = 0.03143v^2$.

Length

The length of the transition curve should be such that the superelevation can be provided at a suitable rate. There are three methods of deciding the length of the transition curve.

(i) **By arbitrary gradient** If L is the length of the transition curve and h is the superelevation, h is provided over the length L at an arbitrary gradient of 1 in n. Thus, the length required is $L = hn$. L is the length of the transition curve in metres, h in metres, and 1 in n is the rate of banking. The value of n lies between 300 to 1200.

(ii) **By time rate** In this method, the cant is provided at an arbitrary rate of r metres per second. If v is the speed of the vehicle in metres per second, then the time taken by the vehicle to pass the length of the transition curve is given by time $= hv/r$, where h is the maximum cant and r is the time rate of banking. If L is the length of the transition curve in metres and v is the speed in metres per second, then time taken $T = L/v$ seconds. As the time rate of canting is r per second, the superelevation in this time, $h = Lr/v$ metres, or $L = hv/r$ metres.

(iii) **By rate of change of radial acceleration** In this method, the rate of change of radial acceleration acceptable for the comfort of passengers is considered. This rate of change of radial acceleration is generally taken as 0.3 m/s^2. Let L be the length of transition curve, V the velocity in m/s, α the rate of change of radial acceleration, and h the maximum superelevation provided. The time taken to pass over the transition curve $= L/v$ seconds. Radial acceleration gained in this time $= L\alpha/v$. The radial acceleration is also equal to v^2/R, where R is the radius of the curve. Therefore,

$$L\alpha/v = v^2/R \quad \text{or} \quad L = v^3/\alpha R = v^3/(0.3R) \quad (\text{if } \alpha = 0.3 \text{ m/s}^2)$$

If V is the speed in kilometres per hour, then

$$L = (V \times 1000/3600)^3/(0.3R) = V^3/14R$$

This method is most commonly used for finding the length of the transition curve.

15.5.2 Fundamental Requirement of a Transition Curve

We have the basic equation

$$P = Wv^2/gr$$

where P is the centrifugal force, W is the weight of the vehicle, g is the acceleration due to gravity, and r is the radius of curvature at a point. Both P and L, the length of the transition curve, must vary with time. If l is the length to any point where the radius of curvature is r, then

$$P \propto l \propto Wv^2/gr$$

W, v, and g are constants and hence

$$P \propto l \propto 1/r$$

This gives

$$lr = \text{constant}$$

Also,

$$LR = \text{constant}$$

where L is the length of the transition curve and R is the minimum radius at its junction with the circular curve.

Thus, the fundamental requirement of a transition curve is that its radius of curvature r must vary as the length l from the beginning of the transition curve. The beginning of the transition curve is the point of intersection with the straight part of the road or railway. The product of length measured along the curve from the tangent point and the radius of curvature at that point has a constant value.

15.5.3 Types of Transition Curves

A type of curve satisfying the fundamental requirement of a transition curve is the *clothoid* or *true spiral*. This curve is somewhat modified in practice by reducing it to a cubic spiral or cubic parabola.

Ideal transition curve–Clothoid

A clothoid is shown in Fig. 15.61. T is the tangent point at the beginning of the curve, where it meets the straight line AT tangentially. T_1 is the tangent point of the circular curve at the end of the transition curve, and the tangent to the circular curve is also tangential to the clothoid. M is any point on the curve at l from T. The tangent to the curve at M intersects the tangent TB at T at point M′. The following symbols are used in this section: r is the radius of the curve at M, ϕ the angle made by the tangent at M with the initial tangent TB, ϕ_d the deflection angle or spiral angle, i.e., angle between the tangents to the clothoid at its ends, R the radius of the circular curve, and L is the total length of the transition curve. From the fundamental requirement of a transition curve, we have

$$lr = LR$$

which gives

$$1/r = l/RL$$

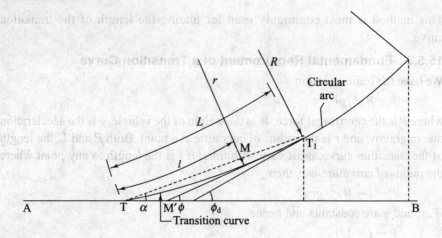

Fig. 15.61 Ideal transition curve

But $1/r$ = curvature = $d\phi/dl$. Therefore,

$$d\phi/dl = l/RL \quad \text{or} \quad d\phi = (l/RL)\, dl$$

Integrating,

$$\phi = l^2/2RL + C$$

where C is the constant of integration. At $l = 0$, $\phi = 0$. Therefore $C = 0$. Thus,

$$\phi = l^2/2RL$$

This is the intrinsic equation of an ideal transition curve. This equation can also be expressed in the form

$$l = \sqrt{2RL\phi} = K\sqrt{\phi}$$

where $k = \sqrt{2RL}$. Also note that when $l = L$, $\phi d = L^2/2RL = L/2R$.

Cartesian coordinates for points on the curve   It will be easier to set out the curve from tangent point T. To do this, rectangular coordinates of a point on the curve are required. Referring to Fig. 15.62, T is the origin of the coordinates, (x, y) are the coordinates of point M, N is any point having coordinates $(x + \delta x)$ and $(y + \delta y)$, δl is the distance between M and N along the curve, ϕ is the angle made by the tangent at M with TB, and $\phi + \delta\phi$ is the angle made by the tangent at N with TB.

Coordinate x of point M

$$dx/dl = \cos\phi, \quad dx = dl\cos\phi$$

Expanding $\cos\phi$,

$$\cos\phi = (1 - \phi^2/2! + \phi^4/4! - \cdots)$$

$$dx = dl(1 - \phi^2/2! + \phi^4/4! - \cdots)$$

$$l = K\sqrt{\phi} \qquad dl = K\, d\phi/2\sqrt{\phi}$$

Substituting this value of dl,

$$dx = (K/2)(\phi^{-1/2} - \phi^{3/2}/2! + \phi^{7/2}/4! - \cdots)d\phi$$

Integrating,

$$x = K(\phi^{1/2} - \phi^{5/2}/10 + \phi^{9/2}/216 - \cdots)$$

$$= K\phi^{1/2}(1 - \phi^2/10 + \phi^4/216 - \cdots)$$
$$= l(1 - \phi^2/10 + \phi^4/216 - \cdots)$$

This equation is in terms of ϕ, the deviation angle of the tangent. It can be expressed in terms of K and l using $\phi = l^2/K^2$ as follows:

$$x = l(1 - l^4/10K^4 + l^8/216K^8 - \cdots)$$

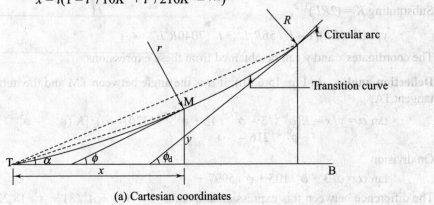

(a) Cartesian coordinates

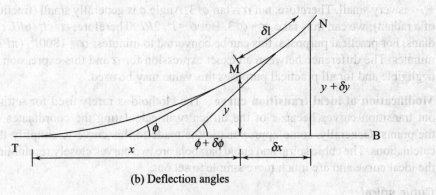

(b) Deflection angles

Fig. 15.62 Points on transition curves

This can be expressed in terms of R and L as, from $K = \sqrt{2RL}$,

$$x = l(1 - l^4/40\,R^2L^2 + l8/3456R^4L^4 - \cdots)$$

Coordinate y of point M

$$dy/dl = \sin\phi, \quad dy = dl\sin\phi$$

Expanding $\sin\phi$,

$$dy = dl(\phi - \phi^3/3 + \phi^5 5 - \cdots)$$

$l = K\sqrt{\phi}$, $dl = K\,d\phi/2\phi^{1/2}$. Substituting the value of dl,

$$dy = (K/2)[\phi^{1/2} - \phi^{5/2}/6 + \phi^{9/2}/120 - \cdots)$$

Integrating,

$$y = K[\phi^{3/2}/3 - \phi^{7/2}/42 + \phi^{9/2}/120 - \cdots)$$

Taking $\phi^{3/2}/3$ as the common term,

$$y = (K\phi^{3/2}/3)[1 - \phi^2/14 + \phi^4/440 - \cdots)$$

Putting $K\phi^{1/2} = l$ and $\phi = l^2/K^2$,

$$y = (l^3/3K^2)[1 - \phi^2/14 + \phi^4/440 - \cdots)$$
$$= (l^3/3K^2)[1 - l^4/14K^4 + l^8/14K^8 - \cdots)$$

Substituting $K = (2RL)^{1/2}$,

$$y = (l^3/6RL)[1 - l^4/56R^2L^2 + l^8/7040R^4L^4 - \cdots]$$

The coordinates x and y can be obtained from these expressions.

Deflection angles In Fig. 15.62(b), if α is the angle between TM and the initial tangent TA,

$$\tan \alpha = y/x = K(\phi^{1/2}/3 - \phi^{7/2}/42 + \phi^{11/2}/1320 - \cdots)/[K(\phi^{1/2} - \phi^{5/2}/10 + \phi^{9/2}/216 - \cdots)$$

On division

$$\tan \alpha = \phi/3 + \phi^3/105 + \phi^5/5997 + \cdots$$

The difference between this expression and $\tan A/3 = A/3 + A^3/81 + A^5/18{,}225 + \cdots$ is very small. Therefore, $\tan \alpha \cong \tan \phi/3$. Angle ϕ is generally small (fraction of a radian); we can thus take $\alpha = \phi/3$. But $\phi = l^2/2RL$. Therefore, $\alpha = l^2/6RL$ radians. For practical purposes, this can be converted to minutes: $\alpha = 1800l^2/(\pi PR)$ minutes. The difference between an exact expression for α and this expression is negligible and for all practical purposes this value may be used.

Modification of ideal transition curve The clothoid is rarely used for setting out transition curves because of the difficulty in calculating the coordinates of the points. Generally, some approximation is made to this curve to simplify the calculations. The cubic spiral and cubic parabola are two curves closely resembling the ideal curve and are much more simple to set out.

Cubic spiral

We had the following expressions while deriving the equation for the clothoid:

$$dy = dl \sin \phi, \quad l = K\phi^{1/2}$$

where $K = \sqrt{2RL}$.

$$dy = dl \sin \phi = dl(\phi - \phi^3/3! + \phi^5/5! - \cdots)$$

Neglecting all the terms except the first, we make the assumption that $\sin \phi = \phi$. Then,

$$dy = dl\, \phi$$

From $l = K\phi^{1/2}$, we have $dl = (K/2)\phi^{1/2}$. Substituting this for dl,

$$dy = K\phi/2\phi^{1/2} = K\phi^{1/2}/2$$

Integrating,

$$y = K\phi^{3/2}/3 = K\phi^{1/2}(\phi/3) = l\phi/3$$

Putting $\phi = l^2/K^2$, we have

$$y = l^3/3K^2$$

$K = \sqrt{2RL}$. We thus have

$$y = l^3/6RL$$

which is the equation for the cubic spiral. One assumption, $\sin \phi = \phi$, is made in this derivation.

The cubic spiral can be set from offsets from tangents or chord lengths. The curve can also be set by deflection angles. For deflection angles, we have $\alpha = \phi/3 = l^2/6RL$ radians.

Cubic parabola

The most widely used transition curve is the cubic parabola because it can easily be set with coordinates.

We have $dx = dl \cos \phi = 1$, by taking $\cos \phi = 1$, as the angle ϕ is very small. We therefore get $x = l$. For the cubic spiral we have derived $y = l^3/6RL$. Thus we get for the cubic parabola $x = l$ and $y = x^3/6RL$. Both these coordinates are used in setting the curve.

It may be noted that the cubic parabola involves two assumptions ($\cos \phi = 1$ and $\sin \phi = \phi$) as against one made in the case of the cubic spiral ($\sin \phi = \phi$). Greater error is involved in the approximations made in the case of the cubic parabola— the assumption that $\cos \phi = 1$ introduces a greater error, as the cosine series is less rapidly converging. However, the cubic parabola is preferred over the cubic spiral due to the ease with which it can be set with coordinates.

15.5.4   Characteristics of a Transition Curve

Transition curves are introduced at the ends of a circular curve. In order to accommodate the transition curve, it becomes necessary to shift the circular curve inwards to get the required space for the transition curve. The inward movement of the circular curve is known as *shift*.

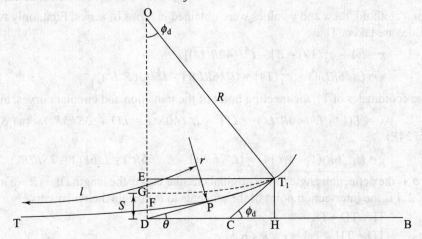

Fig. 15.63   Characteristics of transition curves

In Fig. 15.63, TT_1 is the transition curve. AT is the straight line and TA is tangential to the curve at T. T_1 is the meeting point of the transition and circular curves and the common tangent is T_1C. The deflection angle, also known as the *spiral angle*, is ϕ_d. We consider any point P on the curve at a distance l from T. The length of the transition curve is L. The angle made by the tangent at P with the initial tangent at T is θ. The radius of the transition curve at T is infinity and the radius at T_1 is R. Let the radius at P be r.

OD is perpendicular to the initial tangent. T_1E is perpendicular to OD and T_1H is perpendicular to the initial tangent. The transition curve intersects OD at F. If the circular curve were to be extended, it will intersect the line OD at G. DG is known as the *shift of the transition curve*.

$$\angle GOT_1 = \phi_d, \quad GT_1 = R\phi_d = RL/2R = L/2 \quad (\text{as } \phi_d = L/2R)$$

FC is very nearly equal to GC and $GC = L/2R$. The line OD bisects the transition curve at G.

$$\text{Shift } S = GD = ED - EG = y - (OG - OE)$$
$$OG - OE = R(1 - \cos\phi_d) = 2R\sin^2\phi_d/2 = 2R\phi_d^2/4 = R\phi_d^2/2$$
$$\text{Shift } S = y - R\phi_d^2/2$$
$$y = L^3/6RL = L^2/6R \quad \text{and} \quad \phi_d = L/2R$$
$$\text{Shift } S = L^2/6R - RL^2/8R^2 = L^2/24R$$

F is a point on the transition curve at $L/2$. Its y-coordinate is

$$y = (L/2)^3/6RL = L^2/48RL$$

which is half the value of y at C. Hence the shift S is bisected by the transition curve at F.

15.5.5 Elements of a Transition Curve

For the three types of transition curves, the various values for setting up the curve are obtained as follows.

True spiral or clothoid

For a clothoid, the x and y values were obtained in terms of series. First, only two values are taken. Thus,

$$x = l(1 - \phi_d^2/10) = l[1 - l^4/(40R^2L^2)]$$
$$y = (l^3/6RL)[1 - \phi_d^2/14] = (l^3/6RL)[1 - l^4/(56R^2L^2)]$$

The coordinates of T_1, the meeting point of the transition and circular curves, are

$$x = L(1 - L^4/40R^2L^2) = L(1 - L^2/40R^2) = L(1 - 3S/5R^2) \quad (\text{as } S = L^2/24R)$$
$$y = (L^2/6R)(1 - \phi_d^2/14) = (L^2/6R)(1 - L^2/56R^2) = L^26R(1 - 3S/7R^2)$$

If ϕ' is the deflection angle of the original circular curve, the length $DI = (R + S)\tan \phi'/2$. I is the intersection point of the tangents to the original circular curve.

$$TI = TD + DI$$
$$TD = TH - T_1J = x - R\sin\phi_d$$
$$\text{Total tangent length} = (R + S)\tan\phi'/2 + X - R\sin\phi_d$$
$$TI = (R + S)\tan\phi'/2 + L(1 - \phi_d^2/10) - R(\phi_d/3 - \phi_d^3/6)$$

(substituting the value of X and expanding sin ϕ_d). Substituting $\phi_d = L/2R$, we get

$$TI = (R + S)\tan \phi'/2 + L(L^2/40R^2) - R(L/2R - L^3/48R^3)$$
$$= (R + S)\tan \phi'/2 + (L/2)(1 - L^2/120R^2)$$
$$= (R + S)\tan \phi'/2 + (L/2)(1 - S/5R) \quad (S = L^2/24R)$$
$$= R \tan \phi'/2 + S \tan \phi'/2 + L(1 - S/5R)$$

$R \tan \phi'/2$ is the length of the tangent of the original circular curve. When the transition curve is inserted, this length increases. The increase $= S \tan \phi'/2 + L(1 - S/5R)$. $S \tan \phi'/2$ is called the *shift increment* and the second factor $(X - R \sin \phi_d)$ is called the *spiral extension*.

Cubic spiral

In the case of a cubic spiral,

$$y = l^3/6RL$$

l is measured along the curve. The total length is the same as in the case of the true spiral.

$$TI = (R + S)\tan \phi'/2 + L(1 - S/5R)$$

If the deflection angle ϕ_d of the transition curve is small, as is the usual case, the formula for the tangent length of the cubic parabola (derived below) may be used.

$$TI = (R + S)\tan \phi'/2 + L/2$$

Cubic parabola

In the case of a cubic parabola, the length is measured along the tangent (Fig. 15.64).

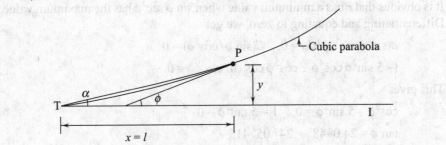

Fig. 15.64 Cubic parabola

$$L = x, \quad y = l3/6RL = x^3/6RL$$
$$\sin \phi_d = \phi_d = L/2R \text{ radians}$$
$$\text{Tangent length} = (R + S)\tan \phi'/2 + L - R\phi_d$$
$$= (R + S)\tan \phi'/2 + L - RL/2R$$
$$= (R + S)\tan \phi'/2 + L/2$$
$$\tan \alpha = y/x = x^2/6RL = l^2/6RL$$
$$\phi_d = l^2/6RL \text{ radians}$$

Therefore,

$$\alpha = (1/3)\phi_d \text{ radians}$$

The angle α has the same value in all cases.

$$\alpha = l^2/6RL \text{ radians} = (l^2/6RL) \times 180/\pi \text{ degrees} = 9.549l^2/6RL \text{ degrees}$$
$$= 9.549 \times 60 \ (l^2/6RL) \text{ minutes} = 573(l^2/6RL) \text{ minutes}$$

Note that α is the angle between TI and TP, where P is any point on the curve, and is called the *polar angle*.

In the case of a cubic parabola, which is the most widely used curve, the minimum radius of curvature is obtained at a particular value of the angle α. If the curve is made longer than this, the radius of curvature increases, and it is not useful as a transition curve. The equation for the cubic parabola is $y = x^3/6RL$, where x is measured along the tangent. Now,

$$dy/dx = x^2/2RL \quad \text{and} \quad d2y/dx^2 = x/RL$$

Also,

$$dy/dx = \tan\phi \quad \text{and } x = \sqrt{2RL\tan\phi}$$

Then,

$$d^2y/d^2 = \sqrt{2RL\tan\phi/R^2L^2} = \sqrt{2\tan\phi/RL}$$

The radius of curvature r in terms of Cartesian coordinates is given by

$$r = [(1 + (dy/dx)^2]^{3/2}/d^2y \, dx^2$$
$$= (1 + \tan^2\phi)^{3/2}/\sqrt{2\tan\phi/RL}$$
$$= \sec^3\phi/\sqrt{2\tan\phi/RL}$$
$$= 1/(2\sin\phi\sec^5\phi/RL)$$

It is obvious that r has a minimum value when $\sin\phi\sec^5\phi$ has the maximum value. Differentiating and equating to zero, we get

$$d(\sin\phi\sec^5\phi)/d\phi = 0, \quad d(\sin\phi/\cos^5\phi) = 0$$
$$(-5\sin^2\phi\cos^4\phi + \cos^5\phi\cos\phi)/\cos^{10}\phi = 0$$

This gives

$$\cos^2\phi - 5\sin^2\phi = 0, \quad 1 - 5\tan^2\phi = 0$$
$$\tan\phi = 24.0948° = 24° \ 05' \ 41''$$

Substituting this value of ϕ in the expression for r, we get

$$R = 1/\sqrt{2\sin(24.0948°)\cos^5(24.0948°)/RL} = 1.39\sqrt{RL}$$

This is the minimum value of the radius. Angle ϕ should not be more than this value to take full advantage of the transition curve.

Length of the combined curve Transition curves are introduced between the tangents and the two ends of the circular curve. To accommodate the transition curve, the circular curve has to be shifted. Referring to Fig. 15.65, ϕ_d is the angle subtended at the centre by the two transition curves and ϕ_c is the angle subtended by the circular curve. ϕ' is the deflection angle of the total curve. The total (combined) length of the curves can be calculated as

Combined length = length of circular arc + length of two transition curves

Length of one transition curve = L

Angle subtended by circular arc at centre = $\phi_c = \phi' - 2\phi_d$

Length of the circular arc = $(\phi' - 2\phi_d)R\pi/180$

Total length = $(\phi' - 2\phi_d)R\pi/180 + 2L$

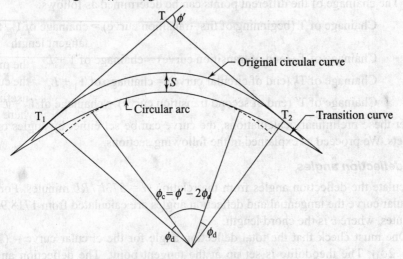

Fig. 15.65 Length of combined curve

Table 15.19 summarizes the various elements of transition curves.

Table 15.19 Summary of transition curve elements

Elements	True spiral (clothoid)	Cubic spiral	Cubic parabola
Intrinsic equation	$\phi = l^2/2RL$	$y = l^3/6RL$	$y = x^3/6RL$
Coordinate x	$l(1 - \phi^2/10)$	–	$x = l$
Coordinate y	$(l^3/6RL)(1 - \phi^2/14)$	$l^3/6RL$	$x^3/6RL$
Tangent length	$(R + S)\tan \phi'/2 +$ $(L/2)(1 - S/5R)$	$(R + S)\tan \phi'/2 +$ $(L/2)(1 - S/5R)$	$(R + S)\tan \phi'/2 +$ $L/2$
Deflection angle α	$\phi/3 = l^2/6RL$ radians $[(573l^2/RL)$ minutes]	$l^2/6RL$ radians $[(573l^2/RL)$ minutes]	$l^2/6RL$ radians $(573L/R$ minutes)
Nature	Ideal curve	Very close to ideal	Not very ideal
Assumption	None	$\sin \phi = \phi$	$\sin \phi = \phi$, $\cos \phi = 1$

15.5.6 Setting Out Transition and Combined Curves

Transition and combined curves can be set out by offsets or angles. The procedure is as follows.

1. The following data will be available: deflection angle ϕ' of the original circular curve, radius R of the circular curve, length L of the transition curve, and chainage of point of intersection I of the tangents.
2. Calculate the shift S of the circular curve from the given data as $L^2/24R$.

3. Calculate the deflection angle or spiral angle from $L/2R$.
4. Calculate the total tangent length as $(R + S)\tan \phi'/2 + (L/2)(1 - S/5R)$ for spirals and $(R + S)\tan \phi'/2 + L/2$ for cubic parabolas.
5. Calculate the length of the curves as for the circular curve, $L_c = \pi R(\phi_c - 2\phi_d)/180$, and the total length as L (combined) $= 2L + \pi R(\phi_c - 2\phi_d)/180$.
6. The chainage of the different points can be determined as follows:

Chainage of T (beginning of first transition curve) = chainage of I – total tangent length

Chainage of T_1 (end of transition curve) = chainage of T + L

Chainage of T_2 (end of circular curve) = chainage of $T_1 + L_c$

Chainage of T' (end of second transition curve) = chainage of $T_2 + L$

After these preliminary calculations, the curve can be set either by angles or by offsets. We proceed as explained in the following sections.

By deflection angles

Calculate the deflection angles from the relation $\alpha = 573 l^2 / RL$ minutes. For the circular curve the tangential and deflection angles are calculated from $1718.9c/R$ minutes, where c is the chord length.

One must check that the total deflection angle for the circular curve = $(1/2)$ $(\phi - 2\phi_d)$. The theodolite is set up at the tangent point. The deflection angles are measured from the tangent to locate the points on the transition curve. After setting out pegs for the transition curve, the theodolite can be shifted to the common point of the transition and circular arcs. The circular curve is set out from this point after locating the tangent direction. This can be done by setting (an angle equal to) $360°$ – the spiral angle in the instrument and sighting the tangent point T. The instrument is then made to read zero by turning it in azimuth, giving the direction of the tangent. The circular curve can be set by deflection angles after transiting the telescope.

By linear measurements

The linear measurements made can be those of coordinates or offsets. For Cartesian coordinates we use the following relations.

True spiral: $y = (l^3/6RL)(1 - \phi^2/10) = (l^3/6RL)(1 - l^4/56R^2L^2)$

(Here l is measured along the curve and y perpendicular to the tangent).

Cubic spiral: $y = l^3/6RL$

(l measured along the curve and y perpendicular to the tangent).

Cubic parabola: $y = x^3/6RL$

(x measured along the tangent and y perpendicular to it).

Points on the curve are obtained by keeping the zero end of the tape at the tangent point and taking a length l (along the curve) for the clothoid and cubic parabola. The tape is swung with this length till the required offset is obtained. In the case of the cubic parabola, the length is measured along the tangent. Convenient lengths are taken for the value of x and the corresponding y values are calculated. Tangent offsets are then laid at right angles to the tangent to obtain points on the curve.

Example 15.33 The deflection angle between the tangents of a circular curve is 60° and it is proposed to have a transition curve at its ends. The maximum speed of vehicles is assumed to be 100 kmph and the centrifugal ratio is $1/4$. The rate of change of radial acceleration is 0.3 m/s^3. Find the radius of the circular curve, length of the transition curve, and chainages of the points at the beginning and end of the curves if the chainage of the point of intersection of the tangents is 1850 m.

Solution Maximum speed of vehicles $= 100$ kmph $= 100 \times 1000/3600 = 27.78$ m/s. We calculate the radius from the centrifugal ratio.

Centrifugal ratio $= v^2/gR = 1/4$, $R = 4v^2/g = 4 \times 27.782/9.81 = 314.67$ m

Length of transition curve $= v^3/\alpha R$

$\alpha =$ rate of change of radial acceleration $= 0.3$ m/s^2

Length $L = 27.78^3/(0.3 \times 314.67) = 227$ m

Spiral angle $\phi_d = L/2R$ radians $= (227/2 \times 314.67) \times 180/\pi = 20°\ 40'$

Central angle of circular arc $= 60° - 2 \times 20°\ 40' = 18°\ 40'$

Length of circular arc $= 314.67 \times (\pi \times 18°\ 40'/180) = 102.52$ m

Shift $S = L^2/24R = 227^2/24 \times 314.67 = 6.82$ m

Tangent length $= (R + S)\tan \phi'/2 + L/2$

$= (314.67 + 6.82) \tan 30° + 227/2 = 299.11$ m

Length of combined curve $= 102.52 + 2 \times 227 = 556.52$ m

Chainage of I $= 1850$ m

Chainage of beginning of first transition curve $= 1850 - 299.11 = 1550.89$ m

Chainage of beginning of circular arc $= 1550.89 + 227 = 1777.83$ m

Chainage of end of circular arc $= 1777.89 + 102.52 = 1880.41$ m

Chainage of end of combined curve $= 1880.41 + 227 = 2107.41$ m

Example 15.34 A circular curve of radius 300 m on a railway line of gauge 1.5 m is to be provided with transition curves at both ends. The superelevation is restricted to 20 cm. The design should be such that the rate of change of radial acceleration is 0.3 m/s^3. Calculate the length of the transition curve and design speed of vehicles if no lateral pressure is to be exerted on the rails. Also calculate the shift and spiral angle of the transition curve.

Solution If θ is the angle of banking, $\tan \theta = 15/150 = v^2/gR$. The velocity v can be calculated as follows:

$$v = \sqrt{15gR/150} = \sqrt{15 \times 9.81 \times 300/150} = 17.155 \text{ m/s}$$

$= 17.155 \times 3600/1000$ km/h $= 61.76$ kmph

Rate of change of radial acceleration $= 0.3$ m/s^3

Length of transition curve $L = v^3/\alpha R = 17.155^3/(0.3 \times 300) = 56.1$ m

Spiral angle ϕ_d $= L/2R$ radians $= (L/2R) \times 180/\pi$

$= (56.1/2 \times 300) \times 180/\pi = 5°\ 21'$

Shift $= L^2/24R = 56.1^2/(24 \times 300) = 44$ cm

Example 15.35 A combined curve has a circular arc of radius 300 m with 75-m-long transition curves at both ends. The deflection angle between the tangents is 60°. The circular curve is to be set with 20-m peg intervals and the transition curve with 10-m peg intervals. Find the necessary angles for obtaining the points on the first transition curve.

Solution With the given data,

Spiral angle $= L/2R = 75/2 \times 300$ rad $= (1/8) \times 180/\pi = 7° \ 10'$

Shift $= L^2/24R = 75^2/24 \times 300 = 0.78$ m

Tangent length $= (R + S) \tan \phi/2 + L/2 = 300.78 \times \tan 30° + 75/2 = 187.89$ m

Central angle for circular arc $= 60° - 7° \ 10' - 7° \ 10' = 45° \ 40'$

Length of circular arc $= 300 \times (45° \ 40') \times \pi/180 = 239.11$ m

Chainage of I $= 2050$ m

Chainage of T $= 2050 - 187.89 = 1862.11$ m

Chainage of C $= 1862.11 + 75 = 1937.11$ m

Chainage of D $= 1937.11 + 239.11 = 2176.22$ m

Chainage of end point $= 2176.22 + 75 = 2251.22$ m

To set points on the transition curve, angles are calculated from

$$\alpha = 1800 l^2 / \pi RL \text{ minutes} = 0.0254 l^2 \text{ minutes}$$

The angles are listed in Table 15.20.

Table 15.20 Angles for setting curve (Example 15.35)

Peg interval	Chord length l (m)	Angle
7.89	7.89	1′ 35″
10.	17.89	8′ 08″
210	27.89	19′ 45″
10	37.89	36′ 28″
10	47.89	58′ 15″
10	57.89	1° 25′ 07″
10	67.89	1° 57′ 04″
7.11	75	2° 22′ 52″

Example 15.36 Two straight roads intersect at chainage 2050 m. The angle of deflection is 40° 30′. The combined curve consists of a circular arc of radius 600 m and two spirals 120 m long. Find the chainages at the beginning and end of the three curves.

Solution Deflection angle between tangents $= 40° \ 30'$

Shift $S = L^2/24R = 1202/24 \times 600 = 1$ m

Total tangent length $= (R + S)\tan \phi'/2 + (L/2)(1 - S/5R)$

$$= (600 + 1) \tan 20° \ 15' + (120/2)(1 - 1/5 \times 600)$$

$$= 281.72 \text{ m}$$

Deflection angle $\phi_d = L/2R$ radians $= 120/(2 \times 600) \times 180/\pi = 5° \ 44'$

Central angle for the circular arc $= 40° \ 30' - 2 \times (5° \ 44') = 29° \ 2' \ 30''$

Length of circular arc $= 600\pi \times 29.042/180 = 304$ m

Chainage of I $= 2050$ m

Chainage of beginning of first transition curve $= 2050 - 281.72 = 1768.28$ m

Chainage of beginning of circular curve $= 1768.28 + 120 = 1888.28$ m

Chainage of end of circular curve = 1888.28 + 304 = 2192.28 m

Chainage of end of combined curve = 2192.28 + 120 = 2312.28 m

Example 15.37 Two straight roads intersect at a deflection angle of 60° 30' at chainage 3030 m. The maximum speed of vehicles is 120 kmph. The centrifugal ratio is 1/4 and the rate of change of radial acceleration is to be 0.3 m/s³. Design the transition and circular curves and find the chainages of points at the beginning and end of the curves. If the transition curve is a cubic parabola, find offsets from the tangent to set up points on the curve.

Solution   Maximum velocity = 120 kmph = 33.33 m/s

Centrifugal ratio = $1/4 = v^2/gR$

from which

$R = 4v^2/g = 4 \times 33.33^2/9.81 = 453$ m

Length of transition curve $L = v^3/\alpha R$ (from rate of change of radial acceleration)

$$= 33.33^3/(0.3 \times 453) = 272.41 \text{ m}$$

Spiral angle $\phi_d = L/2R = 272.41/2 \times 453 = 17° \ 14'$

Central angle of circular curve = 60° 30' − (2 × 17° 14') = 26° 02'

Length of circular curve = 453 × 26° 02'π/180 = 205.84 m

Shift $S = L^2/24R = 272.41^2/(24 \times 453) = 6.825$ m

Total tangent length = $(R + S) \tan \phi' + L/2$

$$= 459.87 \times \tan 60.5° + 272.41/2$$

$$= 949.03 \text{ m}$$

Chainage of I = 3030 m

Chainage of first tangent point = 3030 − 949.03 = 2080.97 m

Chainage of beginning of circular arc = 2080.97 + 272.41 = 2353.38 m

Chainage of end of circular arc = 2353.38 + 205.84 = 2559.22 m

Chainage of end of curve = 2559.22 + 272.41 = 2831.63 m

In the case of a cubic parabola, the length is measured along the tangent. The ordinate from the tangent, $y = x^3/6RL$. Table 15.21 gives the offsets.

Table 15.21   Offsets (Example 15.37)

Length along tangent, x	Offset y	Length along tangent, x	Offset y
20	0.01	180	7.88
40	0.086	200	10.81
60	0.29	220	14.38
80	0.69	240	18.67
100	1.35	260	23.74
120	2.33	272.41	27.30
140	3.71		
160	5.532		

15.6 Spiralling Compound Curves

A compound curve having two circular arcs of different radii has three points of abrupt change of radius. At the beginning of the first arc, the radius changes from infinity to R_1; at the common point, the radius changes from R_1 to R_2; and at the end, the radius changes from R_2 to infinity. To smoothen out these changes, transition curves are introduced in the beginning, at the common point, and at the end. The procedure to calculate the values required is as follows (Fig. 15.66).

1. Shift both segments of the circular curves by different amounts. The shifts can be calculated if the length of the transition curve is known.
2. With the radii R_1 and R_2 and the maximum speed, calculate the superelevation for the two arcs ($h = v^2G/gR$). Let h_1 and h_2 be the superelevations.
3. Find the lengths of the transition curves at the ends either from the rate of application of cant or from comfort conditions ($L = nh$). Let L_1 and L_2 be the lengths of the transition curves at the beginning and the end.
4. Obtain the shifts (S_1 and S_2) from $L^2/24R$. It is clear that the distance between the shifted tangents (distance E_1E_2) $= S_1 - S_2$.

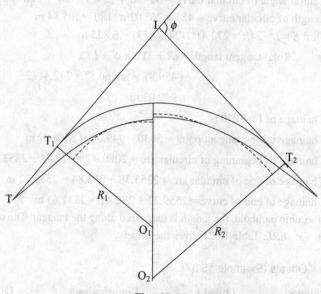

Fig. 15.66

5. The transition curve at the common tangent point will have a length $= n(h_1 - h_2)$. The length can also be fixed arbitrarily or for the comfort condition.
6. Calculate the lengths of the tangents and circular arcs. The chainages of the points can now be obtained.
7. Carry out the necessary calculations for setting out the curve, computing the offsets or angles from the equations already derived. An approximate expression for offsets is $y = 4(S_1 - S_2)x^3/L^3$.

15.7 Spiralling Reverse Curves

In the case of a reverse curve, even if the radii of the two arcs are the same, there is an abrupt change in the direction of the curve from right hand to left hand or vice versa. A transition curve has to be introduced between these two arcs. The transition curve itself will be a reverse transition curve to match the curvature of the two arcs.

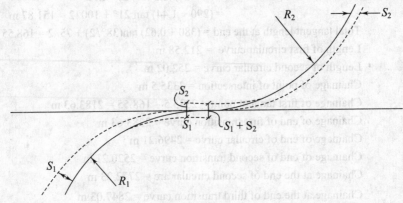

Fig. 15.67

A reverse curve is shown in Fig. 15.67. The procedure for setting out such a curve is the following.

1. Calculate the superelevation to be provided to the two arcs from the expression $h = (v^2/gR) \times$ width of track. If R_1 and R_2 of the reverse curve are different, calculate the superelevation from two values of radii.
2. Greatest change of cant $= h_1 + h_2$. If n is the rate of cant, the length of transition curve $L = n(h_1 + h_2)$. Provide half this length on each side of the midpoint of the reverse curve.
3. The distance between the shifted circular arcs will be equal to the sum of shifts S_1 and S_2, each given by $(L/2)/24R$.
4. The offsets to the transition curve are calculated from the generally used approximate expression $y = 4(S_1 + S_2)x^3/L^3$.

The following example illustrates the procedure.

Example 15.38 A compound curve consists of two circular arcs of radii 290 m and 380 m. The deflection angle is 80°. The central angle of the 290-m-radius arc is 42°. The maximum speed of vehicles is 90 kmph. The distance between the rails is 1.5 m. The chainage of the point of intersection is 2335.5 m. Find the chainages of the points at which transition curves are to be introduced at the beginning, end, and the common junction of the curves. Find the lengths of the transition curves at the beginning, common point, and end. The rate of application of cant is 1 in 300.

Solution   Refer to Fig. 15.66. Speed $= 90$ kmph $= 90 \times 1000/3600 = 25$ m/sec

Cant $h_1 = (v^2/gR)G = 25 \times 25 \times 1.5/(9.81 \times 290) = 0.33$ m

Cant $h_2 = 25 \times 25 \times 1.5/(380 \times 9.81) = 0.25$ m

Length of transition curve $L_1 = 300 \times 0.33 = 100$ m

Spiral angle $= L/2R = 100/2 \times 290 = \times 0.1724$ rad $= 9°\ 53'$

Length of transition curve at the end $= 300 \times 0.25 = 75$ m

Spiral angle $= 0.0987$ rad $= 5°\ 39'$

Length of transition curve at the common point $= 300(0.33 - 0.25) = 24$ m

Shift $S_1 = 100 \times 100/(24 \times 290) = 1.44$ m

Shift $S_2 = 75 \times 75/(24 \times 380) = 0.62$ m

Total tangent length $= (R + S_1) \tan(42°/2) + L/2$

$= (290 + 1.44) \tan 21° + 100/2 = 151.87$ m

Total tangent length at the end $= (380 + 0.62) \tan(38°/2) + 75/2 = 168.55$ m

Length of first circular curve $= 212.58$ m

Length of second circular curve $= 252.02$ m

Chainage of point of intersection $= 2335.5$ m

Chainage of first tangent point $= 2335.5 - 168.55 = 2183.63$ m

Chainage of end of first transition curve $= 2283.63$ m

Chainage of end of circular curve $= 2496.21$ m

Chainage of end of second transition curve $= 2520.21$ m

Chainage at the end of second circular arc $= 2772.23$ m

Chainage at the end of third transition curve $= 2847.03$ m

15.8   Lemniscate

Bernoulli's lemniscate is a curve used in modern roads. The lemniscate is a symmetrical curve and is preferred over other transition curves due to the following reasons.

(a) The radius of curvature decreases more gradually.

(b) The rate of increase of curvature decreases towards the circular curve, which is a desirable property.

(c) It corresponds more closely to an autogenous curve, the path traced by an automobile moving freely.

Figure 15.68 shows a full lemniscate. OX and OY are tangents to the curve. The major axis OA makes an angle of 45° with the axis OX. The minor axis is MM' and the tangent at M makes an angle of 45° with OX and is parallel to the major axis. The polar angle, the angle made by a ray from the origin to any point, P is α. The polar angle of point M is 15°. The tangent at P makes an angle ϕ with OX and the angle between the tangent and ray OP is β. It is clear that $\phi = \alpha + \beta$.

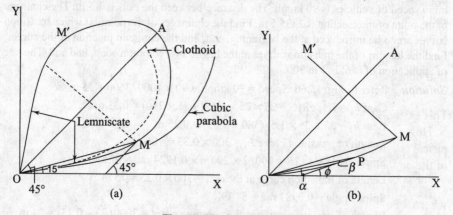

Fig. 15.68   Lemniscate

15.8.1 Elements of a Lemniscate

It is easier to work with polar coordinates to derive the various properties of Bernoulli's lemniscate. The equation describing the lemniscate is

$$l = K \sqrt{\sin 2\alpha}$$

where l = OP. From the property of polar coordinates, $\tan \beta = l(d\alpha/d_1)$.
From $l = K \sqrt{\sin 2\alpha}$, $dl/d\alpha = K \cos 2\alpha/\sin 2\alpha$. Substituting this value in $\tan \beta$,

$$\tan \beta = K\sqrt{\sin 2\alpha} \ \sqrt{\sin 2\alpha}/\cos 2\alpha$$

$$\tan \beta = \tan 2\alpha, \quad \beta = 2\alpha$$

From $\phi = \alpha + \beta$

$$\phi = 3\alpha$$

This shows that for a lemniscate, the deflection angle is exactly three times the polar angle. This relation is approximately true for the spiral and cubic parabolas discussed earlier.

The radius of curvature r at a point of a lemniscate is given in polar coordnates as

$$r = \frac{[l^2 + (dl/d\alpha)^2]^{3/2}}{[l^2 + 2(dl/d\alpha)^2 - l\, d^2 l/d\alpha^2]}$$

Now, from

$$dl/d\alpha = K \cos 2\alpha / \sqrt{\sin 2\alpha}$$

and

$$d^2 l/d\alpha^2 = -K(1 + \sin^2 2\alpha)/(\sin 2\alpha)^{3/2}$$

we get

$$r = \frac{[K^2 \sin 2\alpha + K^2 \cos^2 2\alpha/\sin 2\alpha]^{3/2}}{[K^2 \sin 2\alpha + 2K^2 \cos^2 2\alpha/\sin 2\alpha + K\sqrt{\sin 2\alpha}\,K(1+\sin^2 2\alpha)/(\sin 2\alpha)]^{3/2}}$$

This reduces to $r = (K/3)\sqrt{\sin 2\alpha}$. Substitutng $K = l/\sqrt{\sin 2\alpha}$, we get

$$r = l/3 \sin 2\alpha$$

For small polar angles, $\sin 2\alpha = 2\alpha$ and $l = 6r\alpha$ (approximately) if α is in radians and $l = r\alpha/9.55$ if α is in degrees. From the expression for $r = (K/3\sqrt{\sin 2\alpha}$, $K = 3r\sqrt{\sin 2\alpha}$. Substituting $\sqrt{\sin 2\alpha} = l/K$,

$$K = 3rl/K \quad \text{or} \quad K = \sqrt{3lr}$$

If at point P, the deviation angle is ϕ and the radius of curvature is r, we get

$$dl/d\phi = r = (K/3)\sqrt{\sin 2\alpha}$$

This expression can be integrated to get the value of l.

However, this does not result in a convenient expression to calculate l. An empirical formula suggested by R. Dawson is commonly used to calculate the length of the curve:

$$L = 2K\alpha \cos \alpha/\sqrt{\sin 2\alpha} = 6r\alpha \cos k\alpha \sqrt{\cos \alpha}$$

where k is a constant and its values for different values of α are as given in Table 15.22.

Table 15.22 Values of k and α

Polar angle α	Constant k	Polar angle α	Constant k
5	0.191	30	0.173
10	0.187	35	0.168
15	0.184	40	0.163
20	0.181	45	0.159
25	0.177		

As mentioned earlier, the minor axis is obtained by drawing a polar ray at 15°. The tangent at this point M will make $3 \times 15° = 45°$ with OX and hence is parallel to the major axis. Join O to M and M'. OMM' is an equilateral triangle. $l = K\sqrt{\sin 2\alpha}$; at M, $\alpha = 15°$, $2\alpha = 30°$, and OM = $K\sqrt{\sin 30°}$; at N, $\alpha = 45°$ and $l = K\sqrt{\sin 90°} = K$.

Minor axis MM' $= K / \sqrt{2} = ON / \sqrt{2}$

Therefore, for a lemniscate,

Major axis = minor axis × $\sqrt{2}$

Radius of curvature = $K / (3\sqrt{\sin 2\alpha})$

This decreases with increasing value of α. Maximum value of $\alpha = 45°$. Hence,

Minimum radius = $(K/3)\sqrt{\sin(2 \times 45°)} = K/3$

As K is the length of the major axis, this means that the minimum radius is equal to major axis/3. This is the radius of curvature at N.

Lengh $l = 6r\alpha \sqrt{\cos\alpha} \cos k\alpha$, $r = K/3 = ON/3$, $k = 0.159$

$= 6 (ON/3)(\pi/4) \sqrt{\cos 45°} \times \cos(0.159 \times 45°) = 1.31115$ ON

This is the length of the curve OMN. There are two cases to be considered here—(i) the lemniscate used as a transitional curve throughout without a circular arc and (ii) the lemniscate used at the ends of a circular curve.

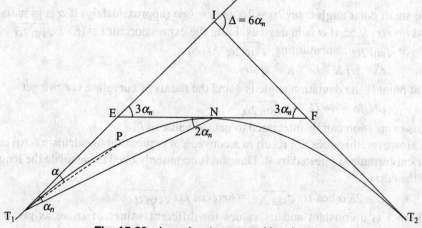

Fig. 15.69 Lemniscate as transitional curve

Lemniscate used as transitional curve throughout

In Fig. 15.69, T_1 and T_2 are two tangent points intersecting at I. T_1NT_2 is a lemniscate tangential to the lines T_1I and T_2I. EF is the tangent at N. To any point P on the curve, the polar angle is α. If $\angle IT_1N$ is equal to α_n, then $\angle IEN = 3\alpha_n$. If the deflection angle of the tangents is Δ, as the curve is symmetrical about N, then $\angle EIN = 90 - \Delta/2$. From triangle IEN, $\angle IEN = 180° - 90° - (90° - \Delta/2)$ $= \Delta/2$. But $\angle IEN = 3\alpha_n$. Therefore $3\alpha_n = \Delta/2$ or $\alpha_n = \Delta/6$. This is the condition required if the lemniscate is transitional throughout. The maximum polar deflection angle must be equal to one-sixth of the deflection angle between the tangents. Also from triangle ENT_1, $\angle ENT_1 = 3\alpha_n - \alpha_n = 2\alpha_n$. For different values of α, l can be calculated for setting out the curve.

Lemniscate as transition curve at the end of circular curves

If the polar deflection angle is less than one-sixth of the deflection angle between tangents, then the lemniscate cannot be transitional throughout. A circular curve is introduced between the two arms of the lemniscate.

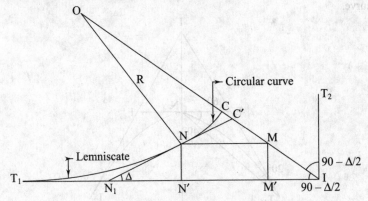

Fig. 15.70 Lemniscate at the end of circular curves

In Fig. 15.70, I is the point of intersection of tangents. A lemniscate is attached symmetrically to a circular curve at its ends. T_1N is the lemniscate and NC is half the circular curve. Δ is the deflection angle between the tangents. α_n is the polar deflection angle of point N and ϕ is the deflection angle of the tangent at N with the tangent T_1I. ON is the radius of the circular curve. Because of symmetry, the ray OC will pass through I. The tangent N_1N is extended to meet OI at C'. Draw NM parallel to the tangent and MM' and NN'' perpendicular to the tangent T_1I.

$$\text{Tangent length } T_1I = T_1N' + N'M' + M'I$$

Now,

$$\alpha_n = \phi/3, \quad T_1N' = T_1N \cos \alpha_n = T_1N \cos \phi/3$$

$$\angle T_1IC = \angle T_2IC = 90° - \Delta/2$$

Let the half central angle for the circular arc be β. From triangle $N_1C'I$, $\angle N_1C'O$ (external angle) $= \phi + 90° - \Delta/2$. From triangle NC'O, $\angle ONC' = 90°$. Therefore, $\beta = 90° - \angle NC'O = 90° - (\phi + 90° - \Delta/2) = \Delta/2 - \phi$. In triangle ONM, by applying the sine rule,

$$NM/\sin \beta = ON/\sin(90° - \Delta/2) \quad \text{or} \quad NM = R \sin \beta/\sin (90° - \Delta/2)$$
$$NM = R \sin(\Delta/2 - \phi)/\sin(90° - \Delta/2) = R \sin(\Delta/2 - \phi)/(\cos \Delta/2)$$
$$= R(\sin\Delta/2 \cos \phi - \cos \Delta/2 \sin \phi)/(\cos \Delta/2)$$
$$= R(\cos \phi \tan \Delta/2 - \sin \phi)$$

Also
$$M'I = MM' \cot(90° - \Delta/2) = MM' \tan \Delta/2 \quad \text{(from triangle MM'I)}$$
$$MM' = NN' = T_1N \sin \alpha_n = l_n \sin \alpha_n$$

Total tangent length $= l_n \cos \phi_n/3 + R(\cos \phi \tan \Delta/2 - \sin \phi) + l_n \sin \phi \tan \Delta/2$

The chainages and other data required can now be computed.
The following examples illustrate the above principles.

Example 15.39 A lemniscate has to be fitted between two tangents intersecting at an angle of 120°. It is used as transition curve throughout. If the apex distance is 40 m, determine the length of the transition curve and the lengths for different values of the polar angle to set the curve.

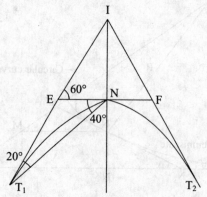

Fg. 15.71

Solution Refer to Fig. 15.71. Deflection angle $\Delta = 120°$. IN = 40 m. $\angle EIN = (180° - 120°)/2 = 30°$. $\angle INE = 90°$ and hence $\angle IEN = 60°$. Maximum polar angle $\alpha_n = 1/6$th of deflection angle $= 120°/6 = 20°$. From triangle IT_1N, applying the sine rule,

$$T_1N/\sin 30° = IN/\sin 20°, \quad T_1N = 40 \times \sin 30°/\sin 20° = 58.46 \text{ m}$$
$$L = K \sqrt{\sin 2\alpha}, \quad K = 1/\sqrt{\sin 2\alpha} = l_n/\sqrt{\sin 2\alpha_n} = 40/\sqrt{\sin 40°} = 72.92$$

Therefore, $L = 72.92 \sqrt{\sin 2\alpha}$ is the polar equation. For different values of α, the values of L can be calculated. These are listed in Table 15.23.

Table 15.23 α and L values (Example 15.39)

α	L	α	L	α	L
30'	9.63	4° 00'	27.2	10° 00'	42.64
1° 00'	13.62	5° 00'	30.39	12° 00'	46.5
1° 30'	16.68	6° 00'	33.25	14° 00'	49.96
2° 00'	19.25	7° 00'	35.87	16° 00'	53.08
2° 30'	21.53	8° 00'	38.28	18° 00'	55.90
3° 00'	23.57	9° 00'	40.53	20° 0'	58.46

Example 15.40 The deflection angle between two straight lines is 120. The maximum polar deflection angle is 15° for the lemniscate. A circular curve of 100 m is used between the lemniscates. Find the tangent length and length of the lemniscate and the circular curve.

Solution Referring to Fig. 15.72, T_1N is the lemniscate and NC is half the circular arc. N_1NC' is the common tangent at N. T_1I is the tangent to the transition curve at the beginning. As 15° < 120°/6, a circular curve is required between the lemniscates. The ray OC will pass through I due to symmetry.

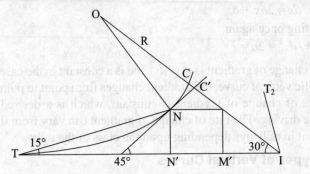

Fig. 15.72

As $\phi = 3\alpha_n$, deflection angle $\phi = 3 \times 15 = 45°$. In triangle N_1IC, $\angle C'N_1I = 45°$. $\angle C'IN_1 = (180° - 120°)/2 = 30°$. External $\angle N_1C'O = 45 + 30 = 75$. In triangle ONC', $\angle ONC' = 90°$, $\angle OC'N = 75°$. Therefore, half the central angle = $180° - 75° - 90° = 15°$. Central angle = 30°, $R = 100$ m, length of circular arc = $100 \times 30\,\pi/180 = 52.35$ m. In triangle ONM, $\angle NOM = 15°$, $\angle NMO = 180° - 90° - 60° = 30°$. From the sine rule,

$$NM = R \sin 15°/\sin 30° = 40 \times \sin 15°/\sin 30° = 51.76 \text{ m}$$

Length of lemniscate = $6R\alpha \sqrt{\cos \alpha} \cos k\alpha$ (for $\alpha = 15$, $k = 0.184$)

Length $l_n = 6 \times 100 \times (15\pi/180) \sqrt{\cos 15°} \cos(0.184 \times 15°) = 154.23$ m

Tangent length $T_1I = T_1N' + N'M' + M'I$

$T_1N' = l_n \cos 15° = 154.23 \times \sin 15° = 148.97$ m

$N'M' = NM = 51.76$ m

$M'I = MM'\cot 30° = NN' \cot 30° = 154.23 \sin 15° \times \cot 30° = 69.14$ m

Tangent length = $148.97 + 51.76 + 69.14 = 269.87$ m

15.9 Vertical Curves

Vertical curves are used to connect two gradient lines to smoothen out the change from one gradient to another. An abrupt change of gradient is likely to subject the vehicles to impact that could be dangerous. Vertical curves are thus essential for safety and comfort. Circular or parabolic curves can be used in vertical curves but a parabola is mostly preferred. A parabolic curve produces a constant rate of change of gradient.

A gradient can be expressed as a percentage or a ratio. A 2 per cent gradient means that the ground rises by 2 m over a length of 100 m. A gradient of 1 in n means that the ground rises by 1 m in n metres; e.g., 1 in 200, 1 in 400, etc. A gradient is considered positive (upgradient when the elevations increase as we go along the gradient and negative (downgradient) when the elevations decrease as we go along the gradient.

Rate of change of gradient   The general equation for a parabola with a vertical axis is

$$y = ax^2 + b$$

Differentiating,

$$dy/dx = 2ax + b$$

Differentiating once again,

$$d^2y/dx^2 = 2a = r$$

The rate of change of gradient is d^2y/dx^2 and is a constant in the case of a parabola. In a parabolic vertical curve, the gradient changes from point to point on the curve. But the rate of change of gradient is constant, which is a desired condition for comfortable driving. The rate of change of gradient can vary from 0.03 per cent to 0.12 per cent in railways depending upon the class of the railway.

15.9.1   Types of Vertical Curves

There can be six different cases of vertical curves depending upon their gradients. These are shown in Fig. 15.73. It is necessary to designate appropriate signs to the gradient, taking the upward gradient as positive as mentioned earlier.

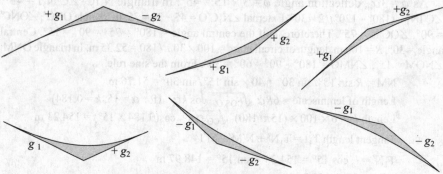

Fig. 15.73   Types of vertical curves

15.9.2   Length of a Vertical Curve

The length of a vertical curve can be obtained from gradients and the rate of change of gradient specified for the particular curve. Thus,

> Length L of vertical curve = change in gradient/rate of change of gradient
>
> $= (g_1 - g_2)/r$

$g_1 - g_2$ is the algebraic change in gradient over the whole length. Appropriate signs should be given to compute the total change of gradient. If $g_1 = 0.05$ per cent and $g_2 = -0.03$ per cent, then change of gradient $= 0.05 - (-0.03) = 0.08$ per cent. In the case of road curves the minimum length of the curve is determined from the point of sight distance as will be explained later.

15.9.3   Setting Out Vertical Curves

Vertical curves can be set by offsets from tangents or by deflection angles. The distances are measured horizontally and the offsets measured vertically. This is acceptable, as the difference in length between the distance along the horizontal and that along the curve is very small because vertical curves are generally very flat.

15.9.4 Tangent Correction

In Fig. 15.74, let O be the origin of the X- and Y-axis. The tangent T_1I has a slope of g_1 and the tangent IT_2 has a slope of g_2 (negative). Let P be any point on the curve with coordinates (x, y). Draw a vertical line through P intersecting the tangent at P' and the X-axis at R. The equation for the parabola is

$$y = ax^2 + bx, \quad dy/dx = 2ax + \beta$$

From the condition at $x = 0$, $dy/dx = + g_1$, we get $g_1 = b$. The equation then becomes

$$y = ax^2 + g_1x$$

Let $P'P = h = P'R - PR$. Now, $P'R = g_1x$ and $PR = y$. Therefore,

$$PP' = g_1x - y = -ax^2 \quad \text{(from } y = ax^2 + g_1x\text{)}$$
$$= Cx^2$$

where C is a constant.

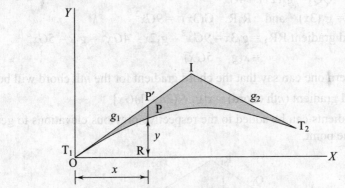

Fig. 15.74 Tangent correction

The difference in elevation between a point on the tangent and the corresponding point on the curve varies as the square of its distance (x) from the tangent point. This difference in elevation is also known as the *tangent correction*.

If g_1 and g_2 are numerically equal, the tangent correction can be measured vertically. As these are generally different, the axis is slightly inclined to the vertical. Tangent correction should thus strictly be measured parallel to the tilted axis. It is generally measured vertically, as the difference is negligible for practical purposes.

The value of C will have to be determined to calculate the tangent corrections. This can be done as follows. In Fig. 15.74(b), y-coordinate of $T_2 = (g_1 - g_2)l/100$, where l is the length on either side of the apex. Also, $x = 2l$. Therefore,

$$y = (g_1 - g_2)l/100 = C \times 4l^2$$

Therefore,

$$C = (g_1 - g_2)/400l$$

15.9.5 Chord Gradients

In Fig. 15.75, a vertical curve is shown with the two tangents meeting at I. The chord gradient is the difference in elevation between two consecutive points on the curve.

Equal chord lengths are taken for convenience. If P_1, P_2, etc. are points at chord lengths c, then (elevation of P_2) – (elevation of P_1) is called the chord gradient. It can be determined as follows: In Fig. 15.75, P, Q, and R are points on the curve. It is clear that the chord gradients are the distances PP_2, QQ_3, RR_3, etc. A general expression for chord gradient can be derived as follows.

First chord: $P_1P_2 = g_1x$, $\quad P_1P = $ tangent correction $= Gx^2$

where $G = (g_1 - g_2)/400l$.

Chord gradient $= P_1P_2 - P_1P = g_1x - Gx^2 = x(g_1 - Gx)$

Second chord: Chord gradient $= QQ_3 = Q_1Q_2 - Q_1Q - Q_3Q_2$

$$= Q_1Q_2 - Q_1Q - PP_2$$

$Q_1Q_2 = g_1(2x)$, $\quad Q_1Q = $ tangent correction for $Q = G(2x)^2$

$$= 4Gx^2, \quad PP_2 = g_1x - Gx^2$$

Chord gradient $= g_12x - 4Gx^2 - g_1x + Gx^2 = g_1x - 3Gx^2 = x(g_1 - 3Gx)$

Third chord: Chord gradient $= RR_3 = R_1R_2 - R_1R - R_3R_2$

$R_3R_2 = QQ_2 = g_1(2x) - 4Gx^2$

$R_1R_2 = g_1(3x)$ $\quad$ and $\quad R_1R = G(3x)^2 = 9Gx^2$

Chord gradient $RR_3 = g_13x - 9Gx^2 - g_12x - 4Gx^2 = g_1x - 5Gx^2$

$$= x(g_1 - 5Gx)$$

From this pattern one can say that the chord gradient for the nth chord will be

Chord gradient (nth chord) $= x[g_1 - (2n - 1)Gx]$

The chord gradients can be added to the respective previous elevations to get the elevation of the point.

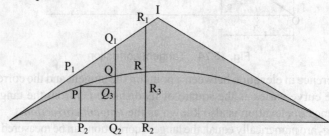

Fig. 15.75 Chord gradients

15.9.6 Method of Length of Vertical Curves

The length of the vertical curve is decided based upon the centrifugal effect or the sight distance. The sight distance criterion will give more length than the effect of centrifugal force. Taking the centrifugal effect into account, the equation for a vertical curve is

$$y = ax^2 + bx, \quad dy/dx = 2ax + b, \quad \text{and} \quad d^2y/dx^2 = 2a = 1/R$$

$$1/R = 2a = 2(g_1 - g_2)/400l = (g_1 - g_2)/200l$$

Length of curve $= 2l = R(g_1 - g_2)/100$

In general, the length calculated this way is less than that required for visibility or sight distance.

15.9.7 Sight Distance

Sight distance is the distance required for a vehicle to perceive an object, react, and stop the vehicle if required. This distance should be sufficient in order to avoid a collision. The required sight distance depends upon the speed of the vehicle or the speed allowed on a vertical curve. There can be two cases—one where the sight distance is less than the length of the curve and the other where the sight distance is more than the length of the curve.

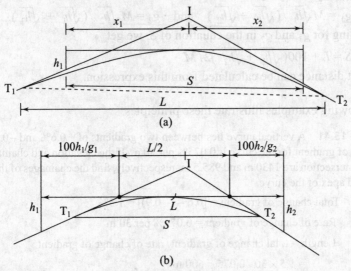

Fig. 15.76 Sight distance

Sight distance less than length of curve

Consider the case shown in Fig. 15.76(a). If h_1 is the height of the driver's eye level and h_2 is the height of the object at the instant of visibility, the line joining h_1 and h_2 will be tangential to the vertical curve. If x_1 and x_2 are the distances to the vehicle and the object, then

$$h_1 = Cx_1^2 \quad \text{and} \quad h_2 = Cx_2^2 \quad [\text{where } C = (g_1 - g_2)/400l]$$

If $2l = L$, the length of the curve, then the length of the curve for a given sight distance can be calculated from the following expression:

$$\text{Sight distance } S = x_1 + x_2 = (1/\sqrt{C})(\sqrt{h_1} + \sqrt{h_2})$$
$$= [\sqrt{h_1} + \sqrt{h_2}]\sqrt{200L} / \sqrt{g_1 - g_2}$$

Squaring both sides, we get

$$S = 200L/(g_1 - g_2)[\sqrt{h_1} + \sqrt{h_2}]^2$$

or

$$L = S^2(g_1 - g_2)/[200(\sqrt{h_1} + \sqrt{h_2})^2]$$

Sight distance greater than length of curve

When the sight distance is greater than L (the length of the curve), referring to Fig. 15.76(b),

$$S = L + 100 \ (h_1/g_1 + h_2/g_2) \quad \text{(using scalar values of } g_1 \text{ and } g_2\text{)}$$

For the minimum value of S, differentiating and equating to zero, we get

$$h_1/g_1^2 - h_2/g_2^2 = 0$$

This gives

$$g_1 = g_2 \sqrt{h_1/h_2} \quad \text{and} \quad g_2 = g_1 \sqrt{h_2/h_1}$$

Setting $g_1 + g_2 = M$, we get

$$g_1 = M\sqrt{h_1} \ / (\sqrt{h_1} + \sqrt{h_2}) \quad \text{and} \quad g_2 = M\sqrt{h_2} \ / (\sqrt{h_1} + \sqrt{h_2})$$

Substituting for g_1 and g_2 in the equation of S, we get

$$S = L + 100(\sqrt{h_1} + \sqrt{h_2})2/M$$

The sight distance can be calculated from this expression.

The following examples illustrate these principles.

Example 15.41 A vertical curve lies between two gradients of $+0.6\%$ and -0.9%. Rate of change of gradient for the curve is 0.075% per 30 m. If the elevation and chainage of the point of intersection are 1430 m and 985.5 m, respectively, find the chainages of the tangent points and apex of the curve.

Solution Total change of gradient $= [0.6 - (-0.9)] = 1.5\%$

Rate of change of gradient $= 0.075\%$ per 30 m

Length $=$ total change of gradient/rate of change of gradient

$= 1.5 \times 30 \ /0.075 = 600$ m

The curve will be in equal lengths on either side of the apex. The difference in elevation between the point of intersection and the tangent point is given by

Difference $= 0.6 \times 300/100 = 1.8$ m

The difference in elevation between the point of intersection and the second tangent point is

Difference in elevation $= 0.8 \times 300/100 = 2.4$ m

Reduced level (RL) of point of intersection $= 1430$ m

RL of first tangent point $= 1430 - 1.8 = 1428.2$ m

RL of second tangent point $= 1430 - 2.4 = 1427.6$ m

Chainage of point of intersection $= 985.5$ m

Chainage of apex of curve $= 1285.5$ m

Chainage of second tangent point $= 1585.5$ m

Example 15.42 A vertical curve connects two gradients of 0.6% and -0.8%. Rate of change of gradient is 0.05% per 20 m. The chainage and elevation of the point of intersection are 950.5 m and 858.75 m, respectively. Find the length of the vertical curve, chainages of points on the curve, and the elevations of the points on the curve at 20-m intervals for the first half.

Solution Total change of gradient $= 0.6 + 0.8 = 1.4\%$

Length of curve $= 1.4 \times 20/0.05 = 560$ m

Chainage of point of intersection $= 950.5$ m

Chainage of $T_1 = 950.5 - 280 = 670.5$ m

Chainage of $T_2 = 670.5 + 560 = 1230.5$ m

RL of I = 858.75 m

Height of I above $T_1 = 0.6 \times 280/100 = 1.68$ m

RL of $T_1 = 858.75 - 1.68 = 857.07$ m

Height of I above $T_2 = 0.8 \times 280/100 = 2.24$ m

RL of $T_2 = 858.75 - 2.24 = 856.51$ m

Tangent corrections   The tangent correction is given by (see Table 15.24)

$$\text{Tangent correction} = (g_1 - g_2)x^2/400l$$

Table 15.24   Tangent corrections (Example 15.42)

x	Tangent correction	Elevation of point	x	Tangent correction	Elevation of point
0	0.000	857.07	160	0.320	857.71
20	0.005	857.185	180	0.405	857.745
40	0.020	857.290	200	0.500	857.77
60	0.045	857.385	220	0.605	857.785
80	0.080	857.470	240	0.720	857.79
100	0.125	857.545	260	0.845	857.785
120	0.180	857.61	280	0.980	857.77
140	0.245	857.665			

Example 15.43   A vertical curve lies between two gradients of 0.4% and – 0.6%. The RL of the point of intersection is 385.5 m. A vertical parabolic curve has to be introduced in the gradients. The rate of change of gradient is 0.05% per 20 m. Find the length of the curve and chord gradients for half of the curve (rising curve) and the elevation of points on the curve at 20-m intervals.

Solution   Total change of gradient = $0.4 - (- 0.6) = 1.0$

Length = total change of gradient/rate of change = $(1.0/0.05)20 = 400$ m

Elevation of point of intersection = 385.5 m

Height of I above tangent point 1 = $0.4 \times 200/100 = 0.8$ m

Elevation of first tangent point = $385.5 - 0.8 = 384.7$ m

Elevation of I above second tangent point = $0.6 \times 200/100 = 1.2$ m

Elevation of second tangent point = $385.5 - 1.2 = 384.3$ m

The chord gradient is given by $x[g_1 - (2n - 1)Gx]$. The chord gradients and elevations of points are tabulated at 20-m intervals in Table 15.25.

Table 15.25   Chord gradients and elevations (Example 15.43)

Point x	Chord gradient	Elevation	Point x	Chord gradient	Elevation
T_1	–	384.7	6	0.025	385
1	0.075	384.775	7	0.015	385.015
2	0.065	384.84	8	0.005	385.020
3	0.055	384.895	9	– 0.005	385.015
4	0.045	384.94	10	– 0.015	385.000
5	0.035	384.975	11		

Example 15.44 Calculate the length of a vertical curve if the sight distance is 1.5 times the length of the curve and the gradients are 0.8% followed by –1.2%. The height of the eye level of the driver is 1.13 m.

Solution Sight distance, when it is more than the length of the curve, is given by

$$S = L + 100(\sqrt{h_1} + \sqrt{h_2})^2 / M$$

where $M = g_1 + g_2$.

$$M = 0.8 - (-1.2) = 2, \quad S = 1.5\,L, \quad h_1 = h_2 = 1.13 \text{ m}$$

Therefore,

$$1.5L = L + 100 \times 4 \times 1.13/2$$

from which $L = 452$ m.

Summary

Curves are used in roads and railway lines to change the direction of motion smoothly. Without curves, the change in the direction of motion will be abrupt, will not be comfortable, and can be dangerous. Circular curves are commonly used but the curves can be parabolic as well. Circular curves can be simple (having one arc of a single radius), compound (having two or more arcs of different radii) or reverse or serpentine (having two or more arcs of the same or different radii). Curves can also be right-hand or left-hand depending upon whether they bend towards the right or left of the tangent direction.

Curves lie between tangents or straight lines which meet at a point known as the *point of intersection*. The angle by which one tangent deflects from the other is known as the *deflection angle*.

The central angle subtended by a circular arc is the same as the deflection angle. The tangent length is given by $R \tan \phi/2$ and the length of the curve by $R\phi$, where ϕ is in radians. The length of the long chord is $2R \sin \phi/2$.

A curve can be set by linear measurements, which can be offsets from the long chord, radial or perpendicular offsets from the tangent, offsets from the chords produced or the successive bisection of arcs. Angular methods of setting out curves include Rankine's method of deflection angles, the two-theodolite method, and the method of tacheometry, the last two methods not requiring linear measurements at all.

Compound curves are made up of two arcs of different radii and are used in situations where a single-arc curve may not be suitable. There is a common tangent to the two curves. A compound curve is a curve tangential to three lines. It has seven elements, namely, the two radii, deflection angle, two tangent lengths, and the two central angles of the two arcs. If any four of them are known, the remaining three can be determined from equations for the compound curve. A compound curve can be set by linear and angular methods as in the case of simple circular curves.

A reverse or serpentine curve consists of two or more arcs of the same or different radii, but curves in opposite directions. The centres of curvature lie on opposite sides of the curve. The calculations for a reverse curve require the conditions of equal radii, equal deflection angles, etc. A reverse curve can also be set by linear or angular methods.

With a circular curve, there is an abrupt change in radius, from the straight line having zero curvature to a curve of finite curvature. To ease this change of curvature and make the change smooth, a transition or easement curve is introduced between the straight line and the circular arc. The transition curve may be a true spiral (clothoid), a cubic spiral, or a cubic parabola. Though a cubic parabola is not the closest to a true spiral, it is preferred because of the ease with which it can be set on the ground using Cartesian coordinates. A vehicle moving along a curved path is subjected to centrifugal force, which tend to throw the vehicle outwards. This is resisted by friction between the tyres and the surface of the road. As the frictional forces developed have an upper limit, the general practice is not to depend only upon friction for safety. The outer ends of a road or the outer track of a railway are banked or given a cant or superelevation to provide additional forces to resist lateral motion. The cant is provided from a zero value at the tangent point to the value required at the circular curve. It is provided on the basis of either an arbitrary gradient or the time rate or rate of change of radial acceleration.

The requirements of a transition curve are (i) to have sufficient length to provide a smooth cant, (ii) to provide a smooth change of radius from infinity to the radius of the circular curve, and (iii) to be tangential to the straight line as well as to the circular curve. The basic requirement of a transition curve is that its radius should change inversely proportional to its length from the tangent point. A clothoid or true spiral is the ideal transition curve. By making some assumption(s), the cubic spiral or cubic parabola is obtained from a true spiral. A transition curve is set by an offset from the tangent or by polar angles. The length is measured along the curve in the case of a clothoid and a cubic spiral and along the tangent in the case of a cubic parabola.

A transition curve can be provided to a compound curve, making it a spiralling compound curve. It can also be provided to reverse curves, making them spiralling reverse curves, which are used to smoothen out the change in curvature or direction.

Bernoulli's lemniscate is another form of transition curve which can be used either as a transition curve without a circular curve or along with a circular curve. Bernoulli's lemniscate is preferred in roads due to the autogenous nature of the curve and the smoother change in radius along the length.

Vertical curves are used when the road has to pass from one gradient to another to smoothen out the change in gradient. These are set out using tangent corrections or chord gradients. The length of vertical curves on roads is governed by centrifugal force considerations or the sight distance. Sight distance generally requires a longer length than centrifugal force. Sight distance, both for passing or stopping, depends upon the speed and the height of the observer and the object. It can be more or less than the curve length.

Exercises

Multiple-Choice Questions

1. A 2° curve of chord length 20 m has a radius of
   - (a) 573 m
   - (b) 286.5 m
   - (c) 143 m
   - (d) 72.5 m
2. The radius of a 20-m arc length, 3° curve, is
   - (a) 1146 m
   - (b) 764 m
   - (c) 573 m
   - (d) 382 m
3. The ratio of radii, of a 20-m chord, 2° curve to 20-m arc, 3° curve is
   - (a) 0.9988
   - (b) 0.9999
   - (c) 1.0000
   - (d) 1.5000
4. For a 3° curve (20-m chord) having a deflection angle of 36°, the length of long chord is
   - (a) 118 m
   - (b) 236 m
   - (c) 449 m
   - (d) 572 m
5. For a 2° curve (30-m chord), of deflection angle 30°, the apex distance is
   - (a) 15.15 m
   - (b) 30.31 m
   - (c) 60.62 m
   - (d) 121.24 m
6. If the degree of curve is 3° on 20-m chord and deflection angle is 36°, then the length of the curve is
   - (a) 60 m
   - (b) 120 m
   - (c) 240 m
   - (d) 480 m
7. The ratio of lengths of curves of a 2° curve on 20-m chord to that of a 3° curve on 30-m chord is
   - (a) 0.25
   - (b) 0.50
   - (c) 0.75
   - (d) 1.00

8. For a 2° curve on 20-m chord having a deflection angle of 36°, the tangent length is
    (a) 118 m   (b) 177 m   (c) 248.2 m   (d) 363.27 m
9. If the tangent distance of a curve is 242 m and deflection angle is 36°, then the length
    of the curve is
    (a) 246 m   (b) 492 m   (c) 632 m   (d) 783 m
10. If the length of the long chord of a curve is 341.6 m and deflection angle is 42°, then
    the length of the curve is
    (a) 175 m   (b) 349.4 m   (c) 438 m   (d) 524 m
11. If the tangent distance of a simple circular curve of radius 578 m is 190 m, then the
    length of the curve is
    (a) 193.6 m   (b) 387.2 m   (c) 538 m   (d) 714 m
12. The length of the long chord of a simple circular curve is
    (a) twice the apex distance   (c) twice the tangent length
    (b) twice the mid ordinate   (d) twice the radius of the curve
13. If the chainage of the first tangent point is 1100 m and tangent length is 245 m of a 2°
    (20-m chord) curve, then the chainage of second tangent point is
    (a) 1672.95 m   (b) 1606.31 m   (c) 1483.15 m   (d) 1353.15 m
14. In the case of curves of very large radius, the ordinates at x from the centre of long
    chord for setting points on the curve can be calculated from the approximate formula
    (a) $x/^2R$ (b) $x^2/2R$ (c) $x(1-x)/2R$ (d) $x^2(1-x)/2R^2$
15. In the case of curves of very large radius, the radial offsets from tangents can be ap-
    proximately be found as
    (a) $x/2R$ (b) $x^2/2R$ (c) $x(1-x)/2R$ (d) $x^2(1-x)/2R^2$
16. In setting a curve by angular method using a theodolite, the general rule for deflection
    angle of a chord is
    (a) deflection angle = 1718.9 × chord length × radius
    (b) deflection angle = tangential angle
    (c) deflection angle = 2 × tangential angle
    (d) deflection angle = deflection angle of the previous chord + tangential angle of the
        chord
17. In using the two-theodolite method for setting curves, the principle used is
    (a) deflection angle is equal to tangential angle for any chord to the point
    (b) angle of intersection is the same as the angle subtended at the centre
    (c) deflection to any point P from the first tangent is the same as the angle between
        the long chord and the direction to P from the second tangent point
    (d) equal chords subtend equal angles at the centre
18. A compound curve has
    (a) a simple circular curve and a transition curve at one end
    (b) a simple circular curve and transition curves at both ends
    (c) the equation of a clothoid
    (d) two or more simple circular curves of different radii
19. A reverse curve is one
    (a) with a simple circular curve and a transition curve
    (b) where the simple circular curve is set from the second tangent point in the reverse
        direction
    (c) having two simple circular curves with centres in opposite directions
    (d) having have circular and half cubic parabola as a compound curve
20. A transition curve is essentially used to
    (a) generate more frictional forces for stability
    (b) allow vehicles to have increased speed while driving

(c) negate the effect of centrifugal forces

(d) avoid abrupt change in radius from a straight line to a finite radius curve

21. Superelevation is
    (a) when the circular curve lies on a gradient
    (b) provided on the inner side of the circular curve
    (c) provided at the centre of the circular curve
    (d) provided on the outer side of the curved track

22. Centrifugal ratio is given by, using usual notations
    (a) $v/(g\,r)$ (b) v^2/g (c) v/r (d) $v^2/(gr)$

23. In designing curves based on maximum superelevation, the minimum radius of curve is given by, using usual notations,
    (a) v^2/g (b) $v^2/\tan\theta$ (c) $v/\mu\tan\theta$ (d) $v^2/g\tan\theta$

24. In designing curves based on maximum friction, the minimum radius is given by
    (a) v^2/g (b) $v^2/\tan\theta$ (c) $v/\mu\tan\theta$ (d) $v^2/\mu g$

25. In designing transition curves, one fundamental requirement is that
    (a) there must be enough frictional force generated for stability
    (b) superelevation must be provided on the outer side of the track
    (c) the circular curve following the transition curve must have large radius
    (d) the radius of curvature must vary as the length from the beginning

26. The ideal transition curve is called a
    (a) clothoid   (b) cubic spiral   (c) cubic parabola (d) hyperbola

27. The equation for an ideal transition curve is, using usual notations,
    (a) $1=2RL\,\phi$ (b) $l^2=2RL\,\phi$ (c) $1=2RL\sqrt\phi$ (d) $1=\phi\sqrt{2RL}$

28. The equation for cubic spiral is obtained from that of an ideal transition curve by making an assumption that
    (a) $\sin\phi=\phi$ (b) $\cos\phi=1$ (c) $\tan\phi=\phi$ (d) $\cos\phi=0$

29. The equation for a transition curve in the form of a cubic spiral is
    (a) $y=l^2/6R$ (b) $y=l^3/6RL$ (c) $y^3=6Rl$ (d) $y^2=6R\sqrt{l}$

30. The equation of a cubic parabola as a transition curve is obtained by making another assumption on the equation of cubic spiral that
    (a) $\sin\phi=\phi$ (b) $\cos\phi=1$ (c) $\tan\phi=\phi$ (d) $\cos\phi=0$

31. The equation of a transition curve in the form of a cubic parabola is given by
    (a) $y=x^2/6R$ (b) $y=x^3/6RL$ (c) $y^3=6RL$ (d) $y^2=6R\sqrt{l}$

32. The term shift used in transition curves is
    (a) the movement of the centre of the circular due to introduction of transition curve
    (b) the movement of the tangent of the circular curve because of the transition curve
    (c) the shift in the point of intersection to accommodate the transition curve
    (d) the movement of the circular curve inwards to accommodate the transition curve

33. The equation for Bernoulli's lemniscate is given by
    (a) $1=k\sqrt{\cos}\,\alpha$ (b) $1=k\sqrt{\sin}\,2\alpha$ (c) $1=k\sqrt{\tan}\,\alpha$ (d) $12=k\sqrt{\sin}\,\alpha$

34. Bernoulli's lemniscate is preferred over other forms of transition curves because it is
    (a) easier to set the curve   (c) easier to calculate the coordinates
    (b) more like an autogenous curve   (d) a symmetrical curve

35. In the case of Bernoulli's lemniscate, the deflection angle is equal to
    (a) the polar angle   (c) three times the polar angle
    (b) twice the polar angle   (d) four times the polar angle

36. In the case of vertical curves, a 3 per cent gradient means that the ground rises
    (a) 1m for every 3 m   (c) 03 m for every 100 m
    (b) 0.03 m for every m   (d) 3m for every 100 m

37. In the case of vertical curves, the curves are generally in the form of
    (a) simple circular curves   (c) compound curves
    (b) simple reverse curves   (d) parabolic curve
38. In the case of vertical parabolic curves, the rate of change of gradient is
    (a) always negative   (c) always positive
    (b) zero   (d) constant
39. In a vertical curve, with an upward gradient of 0.4% followed by a downward gradient of 0.3%, the length of the curve for a rate of change of gradient 0.075% for 30 m is
    (a) 20 m   (b) 160 m   (c) 180 m   (d) 280 m
40. Tangent correction in vertical curves varies
    (a) as the distance along the horizontal
    (b) as the square of the distance along the horizontal
    (c) linearly with the vertical distance to the point
    (d) linearly with the distance along the tangent
41. Sight distance requirement in the case of vertical curves depends upon
    (a) the length of the vertical curve
    (b) the gradients on the upward and downward sides
    (c) the rate of change of gradient
    (d) the speed of the vehicle moving on the curve

Review Questions

1. Explain the term degree of a curve.
2. Give the classification of curves along with neat sketches of each.
3. Explain the terms apex distance and versine distance of a curve.
4. Explain briefly the linear methods of setting out a circular curve.
5. Explain briefly the method of setting out a curve with (a) one theodolite and (b) two theodolites.
6. Explain briefly the tacheometric method of setting out circular curves.
7. Explain the procedure to set out a curve if the point of intersection is not accessible.
8. Explain the procedure to set out a curve if the tangent point is not accessible.
9. Explain the term compound curve and its use.
10. What is a reverse or serpentine curve? Explain the situations in which it is used.
11. Derive the expressions for the elements of a compound curve.
12. Derive the expressions for the elements of a reverse curve.
13. Explain the requirements of a transition curve. Derive the elements of a clothoid.
14. Explain the difference between a clothoid and a cubic parabola.
15. Why does a vehicle tend to move out while being driven along a curve? Explain the term superelevation.
16. Explain the basis on which the length of a transition curve is decided.
17. Why is a lemniscate preferred over other curves in modern roads? Show in a sketch all the transition curves for comparison.
18. What is the purpose of a vertical curve? Should they be circular or parabolic?
19. Explain the terms tangent correction and chord gradient.
20. Explain the term sight distance and explain how it is taken care of in designing vertical curves.

Problems

1. Determine the radius of a 3° circular curve if it is on a (a) 20-m arc and (b) 20-m chord.
2. A 4° curve is tangential to two lines intersecting at an angle of 108°. Calculate the tangent length, length of the long chord, apex distance, and versine distance.

3. The apex distance of a 3° degree circular curve is 82.45 m. Determine the deflection angle, tangent length, and length of the long chord.

4. A 2.5° circular curve is tangential to two lines AI and BI. The bearing of AI is N36° 45′E and the bearing of BI is N48° 30′W. Find the tangent length, length of the long chord, apex distance, and versine distance.

5. The maximum distance of a 3° curve (on a 30-m arc) is 86.75 m from the long chord. Determine the tangent length, length of the long chord, and apex distance.

6. The chainages of the point of intersection and tangent point of a curve are 1083.585 m and 829.665 m, respectively. The deflection angle of the curve is 48°. Compute the radius, the length of the long chord, the apex distance, and the versine distance.

7. A curve of radius 420 m is to be set out by offsets from the long chord. The deflection angle is 60°. Tabulate the offsets from the tangent point at 20-m intervals for half the curve.

8. A curve of radius 280 m was to be set by offsets from the long chord. The deflection angle is 60°. Tabulate the offset values at 20-m intervals from the midpoint of the long chord. Tabulate also the chainages of the points obtained on the curve if the chainage of the backward tangent point is 1200 m.

9. A 4° curve (on a 20-m arc) has a deflection angle of 75°. Tabulate the radial offsets from the tangents to set the left half of the curve. Also tabulate the chainages of the points obtained if the chainage of the point of intersection is 1300.5 m.

10. A curve of radius 800-m (deflection angle = 100°) is to be set with perpendicular offsets from a tangent. Tabulate the values of the offsets for points at 20-m intervals along the tangent, till the value of offsets $\geq$ 20 m.

11. Line AB is along the north direction and line BC has a bearing of 102°. A curve has to be set tangential to a point 250 m from B along BA and also tangential to BC. Tabulate the perpendicular offsets from the tangents to set out the curve.

12. A curve of radius 600 m with deflection angle 98° is to be set out by perpendicular offsets from the tangent. Tabulate the offset values till the offset is not more than 20 m. The remaining curve has to be set from a new tangent from the last point set. Determine the direction of the new tangent or the point at which this tangent intersects the backward tangent point.

13. A curve of radius 400 m and deflection angle 85° was to be set from offsets from the chords produced. The chainage of the first tangent point is 1002.35 m. Calculate the first five offsets from the chords produced to set out the curve.

14. A curve of radius 300 m has a deflection angle of 76°. It has to be set with deflection angles from the backward tangent point. Tabulate the angles and theodolite readings for locating points on the curve at 20-m peg intervals. The chainage of the tangent point is 1008.65 m.

15. A curve of radius 400 m and deflection angle 82° is to be set in the field with two theodolites at the two tangent points. The chainage of the first tangent point is 228.25 m and the peg interval is 20 m. Tabulate the angles and theodolite settings for setting out the curve.

16. A curve has to lie between two tangents having bearings of 30° 30′ and 300° 30′. The point of intersection was not accessible due to an obstacle. A line PQ, P and Q lying on the tangents, was measured and found to be 182.5 m long. The bearing of PQ was 70° 45′. If $T_1P = 200$ m and the chainage of T_1 is 1002.50 m, find the chainages of the point of intersection and the forward tangent point.

17. To set up a curve of radius 300 m on the ground, it was found that the point of intersection of the curve was not accessible. Two points P and Q were chosen on the tangent lines and a traverse was run as follows:

Line	PS	SR	RQ
Length (m)	101.8	88.5	145.3
Bearing	135° 30′	95° 15′	48° 30′

The bearings of the tangents were 40° 50′ and 300° 30′. If the chainage of the first tangent point is 850 m, find the chainages of the point of intersection and the second tangent point.

18. During fieldwork to set up a 3° curve, deflection angle 85°, it was found that the first tangent point was not accessible. A point Q, chainage 380 m, was selected before the tangent point. Another point P was selected on the tangent line after the obstacle. An arbitrary point R was selected and the lengths PR and QR were 78.5 m and 101.5 m, respectively. Angle PRQ was measured and found to be 38° 45′. The distance of P from the point of intersection was 315 m. Find the chainages of the point of intersection and the tangent points.

19. In setting up a curve of radius 300 m in the field, it was found that both the point of intersection and the first tangent point were not accessible. The bearing of the backward tangent was 38° 30′, forward tangent 318° 45′, and of a line PQ (P and Q lying on tangents) was 85° 30′. Length PQ = 372.5 m. A triangle PNM was measured with N lying on the tangent and before the obstacle. The lengths PM and MN were 52.5 m and 80.6 m, respectively. Angle PMN was found to be 80° 50′. If the chainage of N was 785 m, find the chainage of the point of intersection and the tangent points.

20. Three lines AB, BC, and CD have the following bearings: AB = 18° 30′, BC = 88° 30′, and CD = 140° 45′. If BC is 235 m long, find the radius of a circular curve tangential to all the three lines.

21. A circular curve is tangential to lines AI and BI intersecting at I at an angle of 110°. Find the radius of a curve tangential to the lines and passing through a point P. P is located at an angle of 26° from the line IA and at a distance of 120 m from I along this direction.

22. Two straight lines with bearings 41° 30′ and 158° 15′ intersect at I. A third line with a bearing of 98° 30′ intersects these lines at E and F. Two arcs of radii 300 m and 400 m are tangential to EF and form a compound curve, each tangential to one of the other lines. If the chainage of I is 1230.5 m, find the chainage of the three tangent points.

23. A compound curve consists of two arcs of radii 600 m and 800 m. The deflection angle is 95°. If the distance between the first tangent point and the point of intersection is 770 m and the chainage of the point of intersection is 1620 m, find the chainage of the second tangent point and the point of common curvature.

24. A compound curve consists of two arcs, a 3° curve followed by a 4° curve, with the tangent intersecting at a deflection angle of 85°. The chainage of point of intersection is 1020.65 m. If the central angle of the first arc is 38° 30′, find the chainage of the tangent points. List the deflection angles to set out the first five points of the arc from the first tangent point at 20-m peg interval.

25. A compound curve consists of two arcs, with the first arc having a radius of 400 m. The deflection angle is 75°. The two tangent lengths are $T_1I = 350$ m and $T_2I = 400$ m. Calculate the lengths of the arcs and the chainages of the tangent points if the chainage of the point of intersection is 985 m.

26. Two straight lines T_1I (bearing 52° 40′) and T_2I (bearing 312° 40′) meet at I. A compound curve is a tangent to both the lines. The common tangent to the arcs of the compound curve lies along the bearing 98° 40′. If the length of the common tangent is 232.5 m and the radius of the first arc is 274 m, find the radius of the second arc and the length of the curve.

27. Two parallel railway tracks, 15 m apart, are to be connected by a reverse curve of two arcs of equal radius. If the distance between the tangent points is 120 m, find the common radius. Calculate the offsets required to set out the first arc.

28. AB and CD are two straight lines with bearings 0° 00′ and 130° 30′. BC is 400 m long, lying along the bearing 80° 00′, and is the common tangent of a reverse curve. If the two arcs are of the same radius, find the common radius.

29. A reverse curve consists of two arcs, one of radius 210 m followed by another arc of a different radius. The line joining the tangents is 485 m long, inclined at 54° 30′ to the first tangent and 24° 15′ to the second tangent. Find the radius of the second arc of the reverse curve.

30. The maximum permissible speed on a roadway is 120 kmph. The rate of change of radial acceleration is 0.3 m/s^3. The centrifugal ratio is 1/4. If the deflection angle is 75°, find the minimum radius of a transition curve, its length, shift, and tangent length.

31. A circular curve of radius 300 m is to be provided with transition curves at the ends. The deflection angle between the tangents is 80°. The maximum permissible speed is 100 kmph. The rate of cant is 1 in 300. Find the length of the transition curve.

32. A railway curve lies between tangents intersecting at 85°. The maximum speed is limited to 120 kmph. The radial acceleration is 0.4 m/s^3 and the centrifugal ratio is 1/4. The circular arc is to be provided with transition curves in the form a cubic parabola at the ends. Find the minimum radii and lengths of the transition and circular curves, the shift, and the total tangent length.

33. A circular curve is provided with clothoid spirals at the ends. If the maximum speed is 108 kmph, radial acceleration is 0.3 m/s^3, and centrifugal ratio is 1/4, find the radius and length of the transition curve. Deflection angle = 40°.

34. The deflection angle between two tangents to a lemniscate is 75°. If the minimum radius is to be 120 m, find the necessary data to set up the curve if it is to be transitional throughout.

35. A circular arc of radius 120 m is to be provided with lemniscates at the ends. If the deflection angle is 80°, find the equation of the curve and the length of the lemniscates.

36. Find the length of a vertical curve if it is tangential to two gradients 0.8% and −1.0%. The rate of change of gradient is 0.05%/chain (20 m).

37. Calculate the necessary data to set up a vertical curve joining two gradients 0.6% and −0.8%. The rate of change of gradient is 0.1% per 30 m. The elevation and chainage of point of the intersection are 385 m and 1025 m, respectively.

38. A downgradient of 1% is followed by an upgradient of 0.9%. Find the elevations of five points on the curve at 20-m intervals. The rate of change of gradient is 0.05% per 20 m. The elevation of the point of intersection is 1230 m and its chainage is 895 m.

39. A 2% upgradient meets a 1% downgradient at a chainage of 1150 m. The reduced level of the point of intersection is 100. The sight distance is 400 m. The eye level of the driver can be assumed to be 1.12 m above the road surface and the height of the obstacle as 0.1 m. Find the length of the vertical curve and the chainages and RLs of the tangent points.

40. An upgradient of 1 in 25 is followed by a downgradient of 1 in 200. The sight distance is 300 m. If the level of the eye of the driver is 1.2 m above the road surface and the height of the obstacle is 0.1 m, find the length of the vertical curve.

Trigonometric Levelling

16

Learning Objectives

After going through this chapter, the reader will be able to

- find the reduced level of a point from the vertical angles and the horizontal distance
- determine the horizontal distance and height of an object from the vertical angles measured from two stations
- apply (in angular measure) curvature, refraction, and axis-signal corrections to vertical angles
- apply (in linear measure) curvature, refraction, and axis-signal corrections to computed elevations
- find the difference in elevation, after applying necessary corrections, between points using vertical angles
- determine the difference in elevation between points from reciprocal observations

Introduction

Trigonometric levelling is the process of determining the elevations of stations from observed vertical angles and horizontal distances. Vertical angles are measured with a theodolite. The distances may be measured accurately with a tape or may be computed.

Heights and distances can be determined using the principles of plane surveying when it can be safely assumed that the distances are not very large. Corrections to the observed measurements for curvature and refraction (see Chapter 7) may be omitted as being small or applied linearly to the calculated elevations. Such plane surveying methods are used to find the heights of objects or the differences in elevation between two points. When the distances involved are large, geodetic distances or distances reduced to the mean sea level are involved. The corrections for curvature and refraction are applied in angular measure before calculating the heights and distances.

To have vertical control in precise surveying, precise levelling or trigonometric levelling is used. Precise levelling is the same as levelling discussed in Chapter 7 but uses more precise instruments and techniques. Trigonometric levelling is used by accurately determining the vertical angles after the necessary corrections for various factors. These angles, together with the distances calculated or measured, are used to determine accurately the heights of points on the surface of the earth.

We start with the first of the methods of trigonometric levelling based on the principles of plane surveying.

16.1 Basic Principles of Trigonometric Levelling

Consider the case shown in Fig. 16.1. The instrument is set up at station I. The elevation of point P is to be determined. The distance between P and I can be measured and is equal to D. The reduced level (RL) of the line of sight of the instrument can be determined by observations to a benchmark (BM). The vertical angle measured to point P is α. In practice, a staff is held at P and the reading is taken to a fixed mark (say, the 3-m mark) on the staff.

Height $H = D \tan \alpha$

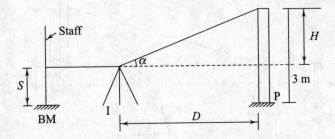

Fig. 16.1 Basic principle

If the RL of the benchmark is M and the reading of the staff kept at the benchmark with the instrument at I is S, then

RL of line of sight $= (M + S)$ m

RL of point P $= (M + S + D \tan\alpha - 3)$ m

Thus, by measuring the vertical angle to a mark on the staff held at the point, the elevation of point P can be obtained.

In the above discussion, it is assumed that the distance D can be measured. If it is not possible to measure the distance D, then both D and the elevation of point P can be calculated from the observation of vertical angles from two instrument positions. There may arise two distinct cases depending upon whether the instrument stations are coplanar with the object or not. These cases are explained in the following section.

16.2 Calculation of Heights and Distances

By observing the vertical angles to a point at an elevation, the height of the point above the instrument axis and the horizontal distance to the point can be calculated. The following cases may arise.

(i) Both the instrument stations are in the same vertical plane as the object.

(ii) The instrument stations are not in the same vertical plane as the object.

16.2.1 Instrument Stations and Object in Same Vertical Plane

There are two cases in this category as illustrated in Fig. 16.2.

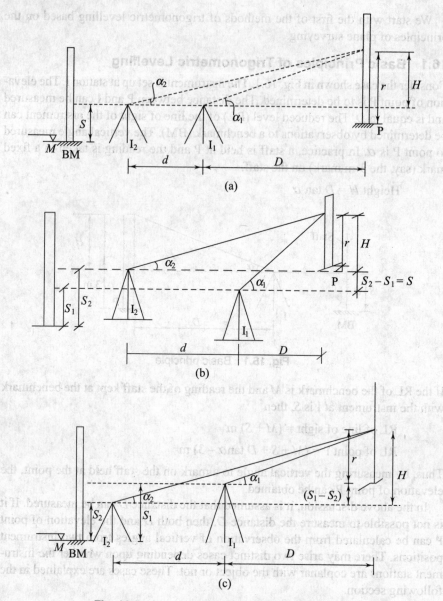

Fig. 16.2 Instrument stations in the same plane

Instrument heights equal

This situation is not usually encountered in practice. In Fig. 16.2(a), I_1 and I_2 are the two instrument stations. The instrument stations and the object lie in the same vertical plane. Let D be the unknown horizontal distance between the object and station I_1 and d be the distance between the instrument stations. Let the reading taken to a benchmark of RL M be S.

RL of line of sight $= M + S$

The staff is held at station P and vertical angles are read to the same mark, say, the 3-m mark, from both the instrument positions. r is the reading or mark on the staff. The vertical angle to the mark is α_1 from I_1 and α_2 from I_2. Now,

$$D \tan \alpha_1 = H, \quad (D + d)\tan \alpha_2 = H$$

Equating the two,

$$D \tan \alpha_1 = (D + d)\tan \alpha_2$$

From this, $D = d \tan \alpha_2/(\tan \alpha_1 - \tan \alpha_2)$

$$H = D \tan \alpha_1$$

Substituting for D,

$$H = d \tan \alpha_1 \tan \alpha_2/(\tan \alpha_1 - \tan \alpha_2)$$

This can be reduced to (by $\tan \alpha = \sin \alpha/\cos \alpha$)

$$H = d \sin \alpha_1 \sin \alpha_2/\sin (\alpha_1 - \alpha_2)$$

RL of station P $= M + S + H - r$

Instrument heights different

There can be two cases under this category. The height of the instrument at I_2 is greater than the height of that at I_1 and vice versa.

Height of instrument at I_2 is greater than that at I_1 In Fig. 16.2(b), I_1 and I_2 are the two instrument stations and α_1 and α_2 are the observed vertical angles to a fixed mark on a staff held at point P. The difference in the instrument heights can be determined from observations to a staff held at a benchmark from the two stations. If S_1 and S_2 are the two readings, then $S_1 - S_2 = S$ is the difference in the heights of the instrument axes at I_1 and I_2. If H is the height of the mark on the staff at station P above the instrument axis at station I_2, then $H + S$ will be the height above the instrument axis at I_1 to the same point. D is the unknown horizontal distance of P from I_1 and $D + d$ is the distance from station I_2, d being the distance between the instrument stations. We then have

$$H = (D + d) \tan \alpha_2 \quad \text{and} \quad H + S = D \tan \alpha_1$$

From these equations,

$$(D + d) \tan \alpha_2 + S = D \tan \alpha_1$$

$$D = (S + d \tan \alpha_2)/(\tan \alpha_1 - \tan \alpha_2)$$

$$= (S \cot \alpha_2 + d) \tan \alpha_2/(\tan \alpha_1 - \tan \alpha_2)$$

$$H + S = D \tan \alpha_1 = (S \cot \alpha_2 + d) \tan \alpha_1 \tan \alpha_2/(\tan \alpha_1 - \tan \alpha_2)$$

$$= (S \cot \alpha_2 + d) \sin \alpha_1 \sin \alpha_2/\sin (\alpha_1 - \alpha_2)$$

H can be calculated as S is known. The RL of P can then be calculated.

Height of instrument at I_1 is greater than that at I_2 In Fig. 16.2(c), S_1 and S_2 are the readings on the staff held at the benchmark. The difference in height of the instrument axis is $S_2 - S_1 = S$.

$$H = (D + d)\tan \alpha_2 \quad \text{and} \quad H - S = D \tan \alpha_1 \, H = S + D \tan \alpha_1$$

Equating the two,

$$D \tan \alpha_2 + d \tan \alpha_2 = S + D \tan \alpha_1$$

$$D = (d \tan \alpha_2 - S)/(\tan \alpha_1 - \tan \alpha_2)$$

$$= (d - S \cot \alpha_2) \tan \alpha_2 / (\tan \alpha_1 - \tan \alpha_2)$$

$$H - S = D \tan \alpha_1 = (d - S \cot \alpha_2) \tan \alpha_1 \tan \alpha_2 / (\tan \alpha_1 - \tan \alpha_2)$$

$$= (d - S \cot \alpha_2) \sin \alpha_1 \sin \alpha_2 / \sin (\alpha_1 - \alpha_2)$$

H can be calculated from this equation, as S is known. The RL of P can then be calculated. If S is taken as $S_1 - S_2$, then in both the cases,

$$H = S + (d \pm S \cot \alpha_2) \sin \alpha_1 \sin \alpha_2 / \sin (\alpha_1 - \alpha_2)$$

16.2.2 Instrument Stations Not in Same Vertical Plane

This situation is shown in Fig. 16.3. I_1 and I_2 are the instrument stations and P is the point whose elevation is to be determined. The measurements taken include the vertical angles, α_1 and α_2, to the target at P and the horizontal angles, θ_1 and θ_2, between the line joining the instrument stations and lines from the stations to P. Let the distance from I_1 to P be D and the distance between the instrument stations be d; d can be measured. $H = D \tan \alpha_1$. Triangle $I_1 I_2 P$ can be solved, as all the three angles are known. By applying the sine rule,

$$\frac{D}{\sin \theta_2} = \frac{d}{\sin [180° - (\theta_1 + \theta_2)]}$$

$$D = d \sin \theta_2 / \sin (\theta_1 + \theta_2)$$

$$H = D \tan \alpha_1 = d \sin \theta_2 \tan \alpha_1 / \sin (\theta_1 + Q_2)$$

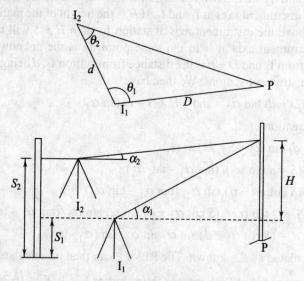

Fig. 16.3 Instruments in different planes

Knowing the height H and the reading on a staff held at the benchmark from I_1, the RL of P can be calculated.

The following examples illustrate the application of these principles.

Example 16.1 A theodolite was set up at station A and the staff reading on a benchmark of elevation 380.355 was 2.785 m. The staff was then placed at station P and the vertical angle reading at the 3.0 mark of the staff was 8° 28′ 40″. Find the reduced level of station P if the distance between the instrument and P was 185 m.

Solution The RL of station P is determined as follows:

RL of benchmark = 380.355 m

Staff reading = 2.785 m

RL of line of sight (horizontal) = 380.355 + 2.785 = 383.14 m

Height of staff mark above line of sight = $D \tan \alpha$ = 185 tan(8° 28′ 40″)

$$= 27.575 \text{ m}$$

RL of station P = 383.14 + 27.575 − 3 = 407.715 m

Example 16.2 A theodolite was set up at point I and the reading on a benchmark of RL 1583.55 was 1.875 m. The staff was then held at two stations P and Q and the vertical angle (of depression) readings on the staff at the 4.0 m mark were 4° 30′ and 7° 34′ 30″. The distances between the instrument and the stations P and Q were 300 m and 2850 m, respectively. Find the RLs of the station points P and Q.

Solution The distance to P being small (300 m), no corrections for curvature and refraction need to be made. Station Q is at a large distance and hence corrections for curvature and refraction are required.

RL of benchmark = 1583.55 m

Staff reading = 1.875 m

RL of line of sight (horizontal) = 1583.55 + 1.875 = 1585.425 m

For P, height of 4-m mark above line of sight = 300 tan(4° 30′) = 23.61 m

RL of station P = 1585.425 + 23.61 − 4.0 = 1605.035 m

For Q, height of 4-m mark above line of sight = 2850 tan(7° 34′ 30″) = 379 m

The combined correction for curvature and refraction = $0.06375D^2$, where D is in kilometres.

Correction = $0.06375 (2.85)^2$ = 0.518 m

This should be added to the reading obtained. Therefore,

RL of Q = 1585.425 + 23.61 + 0.518 − 4 = 1605.553 m

Example 16.3 A theodolite was kept at two stations I_1 and I_2 150 m apart and in a vertical plane containing a station P. The height of the instrument in both cases was the same. The vertical angle observed was 30° 30′ from I_1 and 18° 30″ from I_2. The readings were taken to the 4-m mark on the staff in both the cases. Find the horizontal distance of P from the stations and the elevation of P if the RL of the instrument axis is 1355.765 m.

Solution Let the distance of the instrument station I_1 from P be D. The distance of I_2 from P = D + 150 m. The height of the instrument being the same, the height h of the 4-m mark on the staff from the line of sight is given by

$D \tan(30° 30′)$ (from I_1)   and $(D + 150)\tan(18° 30′)$ (from I_2)

These two heights are equal. Therefore,

$D \tan 30.5° = (D + 150) \tan 18.5°$

$D = 150 \tan 18.5° / (\tan 30.5° − \tan 18.5°)$ = 197.288 m

Height h of the 4-m mark = 197.288 tan 30.5° = 116.2 m

The corrections for curvature and refraction need not be applied, as the distance is small.

$$\text{RL of station } P = 1355.765 + 116.2 - 4 = 1467.965 \text{ m}$$

Example 16.4 To determine the height of a chimney, a theodolite was kept at two stations I_1 and I_2 200 m apart, I_1 being nearer to the chimney. The readings at the benchmark (of RL 1020.375) were 1.35 m from station I_1 and 2.15 m from I_2. The vertical angles to the top of the chimney were 19° 30′ and 8° 15′ from stations I_1 and I_2, respectively. Find the horizontal distance and RL of the top of the chimney.

Solution The difference in the heights of the instrument can be obtained from the staff readings to the benchmark.

$$\text{Difference in height of instrument axis} = 2.15 - 1.35 = 0.8 \text{ m}$$

If D is the distance between instrument station 1 and the object, then

$$D \tan(19° \, 30′) = H + S$$

where H is the height from station 2 and S is the difference between the staff intercepts. Instrument station 2 is at $D + 200$ m from the object. Therefore,

$$(D + 200) \tan(8° \, 15′) = H$$

Comparing the two equations and equating the value of H,

$$(D + 200) \tan(8° \, 15′) = D \tan(19° \, 30′) - S$$

$$D = [200 \tan(8° \, 15′) + S] / (\tan 19° \, 30′ - \tan 8° \, 15′), \quad D = 142.5 \text{ m}$$

$$H = 342.5 \tan(8° \, 15′) = 49.66 \text{ m}$$

Check: $H = 142.5 \tan(19° \, 30′) + 0.8 = 49.66$ m

$$\text{RL of the top of chimney} = 1020.375 + 2.15 + 49.66 = 1072.185 \text{ m}$$

Example 16.5 Two stations I_1 and I_2, 180 m apart, were selected for making observations to find the elevation of a point P on a hillock. The following horizontal angles were measured: $\angle PI_1I_2 = 58° \, 30′$ and $\angle PI_2I_1 = 50° \, 50′$. The vertical angles at the staff held at point P to the 3-m mark were 10° 50′ and 9° 27′ from stations I_1 and I_2, respectively. To find the RL of the instrument axis, readings were taken to a benchmark of RL 1085.65 m. The readings from stations I_1 and I_2 were 1.65 m and 2.85 m, respectively. Find the RL and distance from I_1 to point P.

Solution In triangle I_1I_2P, the distance between the stations = 180 m and all the angles are known. The sine rule can be applied to determine the other sides.

$$\angle I_1 P I_2 = 180° - 50° \, 50′ - 58° \, 30′ = 70° \, 40′$$

By the sine rule,

$$180 / \sin(70° \, 40′) = I_1P / \sin(58° \, 30′) = I_2P / \sin(50° \, 50′)$$

$$I_1P = 180 \times \sin(58° \, 30′) / \sin(70° \, 40′) = 147.9 \text{ m}$$

$$I_2P = 180 \times \sin(58° \, 30′) / \sin(70° \, 40′) = 162.647 \text{ m}$$

$$\text{Height} = 147.9 \times \tan(10° \, 50′) = 28.3 \text{ m}$$

$$\text{RL of P} = 1085.65 + 1.65 + 28.3 = 1115.6 \text{ m}$$

Check: As a check, find the RL from the observation from I_2:

$$\text{Height} = 162.647 \times \tan(9° \, 27′) = 27.1 \text{ m}$$

$$\text{RL} = 1085.65 + 2.85 + 27.1 = 1115.6 \text{ m}$$

16.3 Refraction

Terrestrial refraction is the bending of rays of light as they pass through layers of different densities of air. It affects levelling, as the direction in which the line of sight is kept is not the true direction from the station to the object. As we have seen in Chapter 7, the effect of refraction is to reduce the reading on the staff. In the linear correction adopted in this situation, the effect of refraction is thus added to the staff reading. This is opposite to the effect of curvature, which increases the staff reading and is hence subtracted from the reading.

The effect of refraction does not remain constant during the day, as it depends upon the atmospheric conditions. If observations from two stations are taken at different times, it is quite possible that the effect of refraction is different for the two readings. Hence for very precise levelling, observations must be taken at the same time with two instruments.

Here, we consider the angular error in the vertical angles measured by a theodolite. In Fig. 16.4, P and Q are two stations, Q at a higher plane than P. Observations were made from P to Q and the angle measured is α (angle of elevation). The angle measured from Q to P is β (angle of depression).

The true line of sight from P to Q or vice versa is along PQ. Refraction makes the line of sight tangential to a curve at both the stations. From P, the line of sight is along PP', as Q is apparently seen in that direction due to refraction. Similarly, at Q, the line of sight is along QQ'.

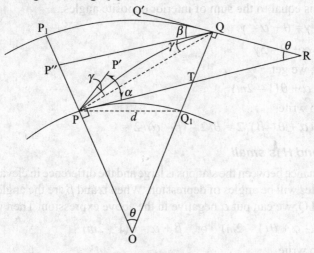

Fig. 16.4

The angle between PP' and PQ is γ, the angle of refraction. Assuming it to be the same at Q also, $\angle Q'QP = \gamma$. The vertical angle measured at P, $\angle P'PT = \alpha$. The angle measured at Q, $\angle Q'QP = \beta$. Both are measured with respect to the horizontal lines through the respective stations.

The correct angle of elevation (from the horizontal) at $P = \angle QPT = \alpha - \gamma$. Similarly, the correct angle of depression at $Q = \angle Q_1QP = \beta + \gamma$.

The correction for refraction is negative (subtractive) for angles of elevation and positive (additive) for angles of depression.

16.3.1   Coefficient of Refraction

The coefficient of refraction is generally related to the angle subtended at the centre by the line joining the two stations. The ratio of the angle of refraction to the angle subtended at the centre is known as the coefficient of refraction. If θ is the angle subtended at the centre and γ is the angle of refraction, then

$$\text{Coefficient of refraction } m = \gamma / \theta$$

The coefficient of refraction is variable and has a value between 0.06 and 0.08. An average value of 0.07 is taken when the correct value is not known.

The magnitude of correction will depend upon the nature of the vertical angle, whether both are angles of depression or one is an angle of elevation and the other an angle of depression. This will depend upon the horizontal distance between stations P and Q and the difference in height H between them. The following two cases can arise.

d is small and H is large

In such a case, one will be an angle of elevation and the other will be an angle of depression. Referring to Fig. 16.4, in triangle QPT, $\angle QRP = \theta$ (this is clear from the fact that OP and OQ are radial lines and PT and QQ' are perpendicular to them).

$$\angle QPT = \alpha - \gamma$$

The external angle of this triangle, $\angle PQQ' = \beta + \gamma$. From the property that an external angle is equal to the sum of interior opposite angles,

$$\beta + \gamma = \theta + \alpha - \gamma$$
$$\beta = \alpha + \theta - 2\gamma$$

As $m = \gamma / \theta$, we get

$$\beta = \alpha + \theta(1 - 2m)$$

One can also write

$$\gamma = (\alpha + \theta - \beta)/2 = \theta/2 - (\beta - \alpha)/2$$

d is large and H is small

When the distance between the stations is large and the difference in elevation is small, both the angles will be angles of depression. When α and β are the angles of depression at P and Q, we can put α negative in the above expression. Then we get

$$\beta = -\alpha + \theta(1 - 2m) \quad \text{or} \quad \beta + \alpha = \theta(1 - 2m)$$

One can also write

$$\gamma = (\theta - \alpha - \beta)/2 = \theta/2 - (\beta + \alpha)/2$$

16.3.2   Determining the Coefficient of Refraction

Refraction is a variable quantity and depends on atmospheric conditions. It is maximum in the morning hours, reduces and remains constant during the day, and increases again in the evening hours. The coefficient of refraction is a measure of the curvature of the line of sight which is curved concave to the surface. It is a ratio of the radius of curvature of the earth to that of the line of sight. $m = R/R'$, where R is the radius of the earth and R' is the radius of the refraction curve.

To determine m, several reciprocal observations are made and angles β and α are observed. If θ is the central angle, then

Curvature correction $= \theta/2$

Refraction correction $\gamma = m\theta$

The net correction is $\theta/2 - m\theta$ and is additive to the angle of elevation α and subtractive to the angle of elevation β. Obviously,

$$\alpha + \theta/2 - m\theta = \beta - \theta/2 + m\theta$$

$$m\theta = (1/2)(\alpha - \beta + 2\theta/2)$$

$$= (1/2)(\alpha - \beta + \theta)$$

$$m = (\alpha - \beta + \theta)/2\theta$$

The value of m obtained usually lies between 0.06 and 0.08. An average value of 0.07 is usually assumed.

16.4 Curvature

The curvature correction, as we have seen earlier in Chapter 7, tends to increase the reading on the staff, thus making the point appear lower than it actually is. From Fig. 16.5, it is apparent that the actual reading, as per the level line through the instrument, increases because of curvature and reduces due to refraction. The curvature correction when linearly applied is

Curvature correction $= 0.0785D^2$ m   (in kilometres)

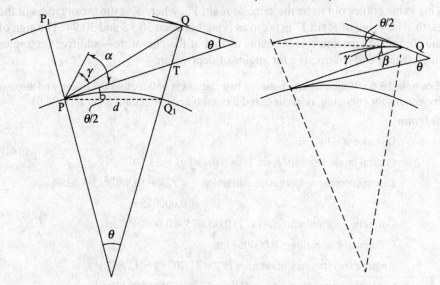

Fig. 16.5

The curvature correction in the angular measure can be calculated as follows. In measuring the vertical angle with the instrument at P, the angle is measured with respect to the horizontal line PT. This angle should actually be measured with respect to chord PQ_1. Angle TPQ_1 is the angle between the tangent and chord PQ_1. From the properties of a circle this is equal to $\theta/2$ (half the central angle subtended by the chord). Therefore, curvature correction in angular measure $= \theta/2$ and should be added to the angle of elevation measured at P. The correct angle of elevation should be $\alpha + \theta/2$. The correction is thus positive.

At Q, similarly, angle β is measured with respect to the horizontal through Q, while it should be measured with respect to the chord through Q, QP_1. The correction is again $\theta/2$, and should be subtracted from angle β. As β is an angle of depression, it is clear from the figure that the correct vertical angle from Q to P is $\beta - \theta/2$.

The curvature correction is $\theta/2$ (half the central angle) and is positive (added) in the case of angle of elevation and negative (subtracted) in the case of angle of depression.

The combined correction for curvature and refraction in angular measure is

Curvature correction $= \theta/2 = d/2R$ radians $= (d/2R) \times 180 \times 3600/\pi$ seconds
This is equal to $d/(2R \sin 1'')$ seconds.

Correction for refraction $= m\theta = md/R$ radians $= md/(R \sin 1'')$ seconds
The curvature correction is greater than the refraction correction. Hence,

Combined correction $= d/(2R \sin 1'') - md/(R \sin 1'')$ seconds
$$= (1 - 2m)d/(R \sin 1'') \text{ seconds}$$

In converting from radians to degrees, we multiply the radian measure by $(180/\pi)$ to get the value in degrees. This value can be converted into seconds by multiplying it further by 3600. The value in seconds $= (180 \times 3600/\pi)$ seconds. This value comes out to be the same as $R \sin 1''$, where R is the mean radius of the earth. The value of $R \sin 1''$ is taken as lying between 30.88 and 30.94. The sign of the combined correction is the same as that of the curvature—additive for angles of elevation and subtractive for angles of depression.

Example 16.6 If the distance between two stations is 980 m, find the linear and angular corrections for curvature, refraction, and the combined correction. $R \sin 1'' = 30.91$.

Solution

Distance $d = 980$ m

Central angle $\theta = 980/R \sin 1'' = 980/30.91 = 31.70''$

Linear correction: Curvature correction $= d^2/2R = (0.980)^2/2 \times 6380$

$$= 0.000075 \text{ m}$$

Correction for refraction $= (1/7) \, 0.000075 = 0.0000107$ m

Combined correction $= 0.0000643$ m

Angular correction: Curvature $= \theta/2 = 31.70''/2 = 15.85''$

Correction for refraction $= md/R \sin 1'' = 0.075 \times 980/30.91 = 2.38''$

Combined correction $= 15.85 - 2.38 = 13.47''$

Example 16.7 The distance d between two stations P and Q is 2860 m. The vertical angle measured at station P, while sighting to Q, is 48'' (angle of depression). Determine the angular corrections for curvature and refraction. What will be the vertical angle measured from Q while sighting to P? $R \sin 1'' = 30.90$ m and $m = 0.07$.

Solution

Central angle $\theta = d/R \sin 1'' = 2860/30.90 = 92.55'' = 1' \, 32.55''$

Curvature correction $= \theta/2 = 46.28''$

Refraction correction $= 0.07 \times 92.55'' = 6.48''$

Combined correction $= 46.28 - 6.48 = 39.8''$

This is subtractive, as it is an angle of depression.

Corrected angle at $P = 68'' - 39.8'' = 28.2''$

From $r = \theta/2 - (\alpha + \beta)/2$,

$(\alpha + \beta)/2 = q/2 - r = 46.28 - 6.48 = 39.8''$

$\alpha + \beta = 79.6''$ and $\beta = 79.6 - 28.2 = 51.4''$

16.5   Axis-Signal Correction

Axis-signal correction, or *eye-object correction*, is the correction to be made to the vertical angle measured because of the difference between the height of the instrument at the observer station and the height of the signal at the target station. The sign of correction depends upon the nature of the vertical angle, whether an angle of elevation or of depression.

16.5.1   Observed Angle is Angle of Elevation

Refer to Fig. 16.6(a). The instrument is set up at P and the target is at Q. PT is the tangential (horizontal) line at P. The vertical angle measured is α, which is $\angle$TPQ. The following symbols are used: h_1, h_2 are the heights of instruments at stations P and Q; s_1, s_2 are the heights of signals at P and Q; θ is the central angle subtended by the horizontal distance PQ; α is the vertical angle observed at P to the target at Q; QQ′ is the difference between the height of the instrument and the height of the signal erected at Q, i.e., $s_2 - h_1$; and $\angle$QPQ′$= \delta$ is the axis-signal correction to the angle observed at Q. Draw QR perpendicular to PQ. Angle TPQ$_1 = \theta/2$, half the central angle subtended by the chord d. From triangle QPO,

$\angle POQ = \theta$, $\angle OPQ = 90° + \alpha$

$\angle PQO = 180° - \theta - 90° - \alpha = 90° - (\alpha + \theta)$

Angle PQR $= 90°$ by construction. Therefore, $\angle Q_1 QR = (\alpha + \theta)$. As angle δ is very small, $\angle QRP$ can be approximately taken equal to a right angle. From triangle QRQ′,

$QR = QQ' \cos(\alpha + \theta) = (s_2 - h_1) \cos(\alpha + \theta)$

Again from triangle QPQ$_1$,

$\angle QPQ_1 = (\alpha + \theta/2)$, $\angle PQQ_1 = 90° - (\alpha + \theta)$

$\angle PQ_1Q = 180° - (\alpha + \theta/2) - 90° - (\alpha + \theta) = 90° + \theta/2$

Applying the sine rule,

$PQ_1 / \sin [90° - (\alpha + \theta)] = PQ / \sin(90° + \theta/2)$

$PQ = PQ_1 \cos(\theta/2) / \cos(\alpha + \theta)$

$PQ_1 = d$, the horizontal distance between the stations. From triangle PQR, right-angled at R,

$$\tan \delta = QR/PQ$$

Substituting the values derived for PQ and QR,

$$\tan \delta = (s_2 - h_1)\cos(\alpha + \theta)/[d \cos(\theta/2)/\cos(\alpha + \theta)]$$
$$= (s_2 - h_1)\cos^2(\alpha + \theta)/d \cos \theta/2$$

δ can be calculated from this expression and is the angular axis-signal correction.

Depending upon the distance d, θ varies but is generally small compared to α. An approximate expression for the axis-signal correction is

$$\tan \delta_1 = (s_2 - h_1)\cos^2(\alpha)/d$$

As δ is generally very small, $\tan \delta = \delta$. Therefore,

$$\delta = (s_2 - h_1)\cos^2(\alpha)/d$$

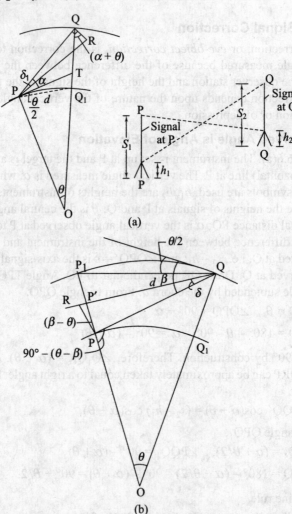

(a)

(b)

Fig. 16.6

The angle obtained will be in radians. This can be further simplified to

$$\delta = (s_2 - h_1)/(d \sin 1'') \text{ seconds} \quad \text{(if } \alpha \text{ is small)}$$

It is evident that this correction is subtractive for an angle of elevation.

16.5.2 Observed Angle is Angle of Depression

This case is illustrated in Fig. 16.6(b). In this case, the correction is additive. If β is the observed angle of depression and δ is the axis-signal correction, then

$$\tan \delta = (s_1 - h_2)\cos^2(\beta + \theta)/(d \cos \theta/2)$$

This can be taken equal to δ radians, as δ is generally very small. Also, θ is generally small compared to β and can be ignored, and $\cos \theta/2 = 1$. Then

$$\delta_2 = (s_1 - h_2)\cos^2\beta/d \text{ radians}$$
$$= (s_1 - h_2)\cos^2\beta/(d \sin 1'') \text{ seconds}$$

The axis-signal correction is negative (subtractive) for angles of elevation and positive (additive) for angles of depression.

A summary of these corrections is given in Table 16.1.

Table 16.1 Summary of corrections

Type of correction	Exact value	Approximate correction	Linear correction	Sign of correction
Axis-signal	$\dfrac{(s-h)\cos^2(\alpha+\theta)}{d\cos(\theta/2)}$	1. $(s-h)\cos^2\alpha/d$ 2. $(s-h)/d \sin 1''$ (seconds)	$(s-h)$	– for elevation
Curvature	$d/(2R \sin 1'')$ seconds	$d/(2R \sin 1'')$ seconds	$d^2/2R$	+ for elevation
Refraction	$md/(2R \sin 1'')$	$md/(2R \sin 1'')$	$(1/7)d^2/2R$	– for elevation

Example 16.8 Find the axis-signal correction to the observed angle from the following data: Instrument station P, height of instrument $= 1.35$ m, target station Q, target height $= 5$ m, vertical angle of elevation from P to Q $= 2°\ 34'\ 35''$, horizontal distance between P and Q $= 2574$ m.

Solution As there is a difference between the height of the instrument and the height of the signal, axis-signal correction is made to the vertical angle. The angle being an angle of elevation, the correction is subtractive.

Axis-signal correction $= (s - h)\cos^2(\alpha + \theta)/d \cos(\theta/2)$

$\alpha = 2°\ 34'\ 35''$, $\quad \theta = d/R \sin 1'' = 2574/(6,380,000 \times \sin 1'') = 83.22''$

$\alpha + \theta = 2°\ 34'35'' + 83.22'' = 2°\ 35'\ 58.22''$

$s = 5$ m, $\quad h = 1.35$ m, $\quad s - h = 3.65$ m

Axis-signal correction $= 3.65 \cos^2(2°\ 35'\ 58.22'')/2574 \cos(41.61'')$

$\qquad\qquad\qquad = 0.0014$ rad $= 4'\ 52.18''$

Corrected angle of elevation $= 2°\ 34'\ 35'' - 4'\ 52.18'' = 2°\ 29'\ 42.82''$

If we assume θ to be a small angle compared to α,

Approximate value of correction $= (s - h)\cos^2\alpha/d$

$$= 3.65\cos^2(2° 34' 35'')/2574$$

$$= 0.0014 \text{ rad (same as before)}$$

As α is also a small angle,

Correction $= (s - h)/(d \sin 1'')$ seconds $= 3.65/(2574 \sin 1'') = 292.489''$

$$= 4' 52.5'' \quad \text{(same as before)}$$

16.6 Determining Differences in Elevation

The difference in elevation between two stations is obtained from observations of vertical angles and horizontal distances. The vertical angles measured have to be corrected for curvature, refraction, and axis-signal height difference. There are two methods of finding the difference in elevation between two points—by single observation and by reciprocal observations.

16.6.1 Elevation from Single Observation

The sign of corrections depends upon the sign of the vertical angle—positive for angle of elevation and negative for angle of depression.

Angle of elevation

The situation is shown in Fig. 16.7. P is the instrument station and Q is the target station, α is the observed angle of elevation at P, α_1 is the elevation angle corrected for axis-signal difference $= (\alpha - \delta_1)$, where δ_1 is the axis-signal correction, which is negative for an angle of elevation, γ is the refraction angle $= m\theta$, and θ is the central angle for the chord PQ_1 of length d.

We have to find the height H. $H = QQ_1$. For this we solve the triangle PQQ_1. $PQ_1 = d$. The angles of the triangle are the following:

$$\angle QPQ_1 = \alpha_1 - \gamma + \theta/2 = \alpha_1 - m\theta + \theta/2$$

$$\angle PQ_1Q = 90° + \theta/2$$

$$\angle Q_1QP = 180° - \alpha_1 + m\theta - \theta/2 - 90° - \theta/2 = 90° - \alpha_1 + m\theta - \theta$$

Applying sine rule to triangle PQQ_1,

$$QQ_1/\sin QPQ_1 = PQ_1/\sin PQQ_1$$

$$QQ_1 = H = d \sin QPQ_1/\sin PQQ_1$$

$$H = d \sin(\alpha_1 - m\theta + \theta/2)/\sin[90° - (\alpha_1 - m\theta + \theta)]$$

$$= d \sin(\alpha_1 - m\theta + \theta/2)/\cos(\alpha_1 - m\theta + \theta)$$

$$R\theta = d, \quad \theta = d/R \text{ radians} = d/R \sin 1'' \text{ seconds}$$

Substituting this value of θ,

$$H = \frac{d \sin[\alpha_1 - md/R\sin 1'' + md/2R\sin 1'']}{\cos[\alpha_1 - md/R\sin 1'' + d/R\sin 1'']}$$

$$= \frac{d \sin[\alpha_1 - (1 - 2m)d/2R\sin 1'']}{\cos[\alpha_1 + (1 - m)d/R\sin 1'']}$$

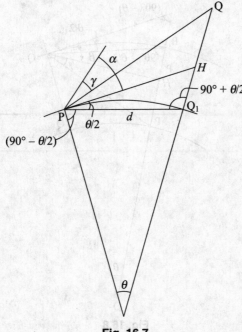

Fig. 16.7

This is the exact expression for H. When θ is small, which is usually the case, $(90° + \theta/2)$ can be approximately taken to be equal to $90°$. Angle $PQ_1Q = 90°$. Then

$$H = d \tan(PQ_1Q) = d \tan\{[\alpha_1 + (1 - 2m)d]/2R \sin 1''\}$$

Angle of depression

The situation is shown in Fig. 16.8. β is the vertical angle of depression as observed from Q to P. β_1 is the vertical angle corrected for axis-signal difference, $\beta_1 = \beta - \delta$.

$$\beta_1 = \beta + \tan^{-1}(s_2 - h_1) \cos^2 \beta / d$$

γ is the refraction angle $= m\theta$. θ is the central angle subtended by the chord $PQ_1 = d$. Consider triangle P'QO. $\angle$ P'QO = $90°$, $\angle$ QOP' = θ. Therefore,

$$\angle OP'Q = 180° - 90° - \theta = 90° - \theta$$

$$\angle P'QP_1 = \theta/2$$

being the angle between the tangent and the chord. Consider triangle P_1QO. $\angle P_1OQ = \theta$, $\angle P_1QO = 90° - \theta/2$. Therefore,

$$\angle QP_1O = 180° - \theta - 90° + \theta/2 = 90° - \theta/2$$

(Even otherwise, triangle P_1OQ is isosceles, as $P_1O = QO$, and the base angles are equal.) Now consider triangle P_1PQ:

$$\angle PP_1Q = 90° - \theta/2, \quad \angle P_1QP = \beta_1 - \gamma - \theta/2 = \beta_1 - m\theta - \theta/2$$

$$\angle QPP_1 = 180° - (90° - \theta/2) - (\beta_1 + m\theta - \theta/2) = 90° - (\beta_1 + m\theta - \theta)$$

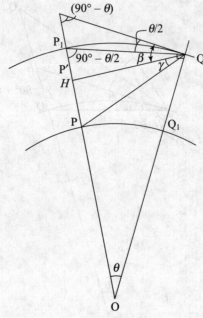

Fig. 16.8

$P_1P = H$, the difference in elevation between P and Q. $P_1Q = d$, the horizontal distance between stations P and Q. Applying the sine rule to triangle P_1PQ,

$$\frac{P_1P}{\sin(\beta_1 + m\theta - \theta/2)} = \frac{d}{\sin[90° - (\beta_1 + m\theta - \theta)]}$$

$$H = \frac{d\sin(\beta_1 + m\theta - \theta/2)}{\cos[\beta_1 + m\theta - \theta]}$$

Putting $\theta = d/(R\sin 1'')$ seconds and simplifying,

$$H = \frac{d\sin[\beta_1 - (1 - 2m)d/(2R\sin 1'')]}{\cos[\beta_1 - (1 - m)d/(R\sin 1'')]}$$

This is the exact expression for H. When θ is small, angle PP_1Q (= 90° − θ/2) can be considered equal to 90°. Then,

$$H = d\tan[\beta_1 - (1 - 2m)d/(2R\sin 1'')]$$

Corrections in linear measure

In the case of small distances, the corrections can be applied in linear measure. The corrections to be applied in linear measure are the following.

Axis-signal correction   This is applied in linear terms as $s - h$, where s is the height of the signal and h is the height of the instrument axis.

Curvature correction   This is applied as $d^2/2R$, where d is the distance between the stations and R is the radius of the earth.

Refraction correction   This is applied as refraction angle × distance = rd. This is equal to $m\theta d$. $\theta = d/R$.

Refraction correction $= m(d/R)d = m\, d^2/R$

Combined correction for curvature and refraction $= (1 - 2m)d^2/2R$

The corrections are applied linearly to the height calculated as $d\tan\alpha$. If the vertical angle is an angle of elevation, then

$$H = d\tan\alpha - (s - h) + d^2/2R - md^2/R$$

If the vertical angle is an angle of depression β, then

$$H = d\tan\beta + (s - h) + d^2/2R - md^2/R$$
$$= d\tan\beta + (s_1 - h_2) - (1 - 2m)d^2/2R$$

The following examples illustrate the above principles.

Example 16.9 The following data refers to observations from station A to station B to find the height of the stations: angle (elevation) $= 1° 54' 30''$, height of instrument at $A = 1.25$ m, height of signal at $B = 3.50$ m, distance between A and B $= 4350$ m. Coefficient of refraction $= 0.07$. Find the corrected vertical angle by applying axis-signal, curvature, and refraction corrections. Find the elevation of B if the elevation of A is 1150.55 m. If all the corrections are applied linearly, what is the elevation of B?

Solution Central angle $\theta = d/R \sin 1'' = 4350/6{,}380{,}000 \sin 1'' = 140.6'' = 2' 20.6''$

Height difference between signal and instrument $= 3.5 - 1.25 = 2.25$ m

Axis-signal correction $= 2.25 \cos^2(\alpha + \theta)/(d\cos\theta/2)$

$\qquad = 2.25 \cos^2(1° 54' 30'' + 0° 2' 20.6'')/\cos(1' 10.3'')$

$\qquad = 1' 46''\ (-)$

$\alpha_1 = \alpha - 1' 46'' = 1° 54' 30'' - 0° 1' 46'' = 1° 52' 44''$

Curvature correction $= \theta/2 = 1' 10.3''\ (+)$

Refraction correction $= \gamma = m\theta = 0.07 \times 2' 20.3'' = 9.84''\ (-)$

Corrected vertical angle $= 1° 54' 30'' + 1' 103'' - 1' 46'' - 9.84'' = 1° 53' 44.6''$

The difference in elevation H between A and B is given by

$$H = d\sin(1° 53' 44.6'')/\cos(1° 52' 44'' + 2' 20.6'' - 0.07 \times 2' 20.66'')$$

$\qquad = 143.97$ m

RL of B $= 1150.55 + 143.97 = 1294.52$ m

Linear corrections:

Axis-signal correction $= 3.5 - 1.25 = 2.25$ m

Curvature $= d^2/2R = 4350^2/(2 \times 6{,}380{,}000) = 1.483$ m

Refraction $= (1/7)$ (curvature correction) $= 0.212$ m

$d\tan\alpha = 4350 \tan(1° 54' 30'') = 144.937$ m

Difference in elevation $= 144.937 - 2.25 + 1.483 - 212 = 143.958$ m

RL of B $= 1150.55 + 143.958 = 1294.508$ m

Example 16.10 The following data refers to observations taken from station Q to station P: vertical angle (depression) $= 2° 24' 50''$, height of instrument $= 1.3$ m, height of signal $= 2.75$ m, distance between stations $= 7416$ m. If the elevation of Q is 2345.65 m, find the elevation of P.

Solution If δ is the axis-signal correction, then

$$\tan \delta = (s - h)\cos^2(\alpha + \theta)/(d \cos\theta/2)$$

Central angle $\theta = d/R \sin 1'' = 7416/30.9 = 240 \text{ sec} = 4 \text{ min}$

$\tan \delta = (2.75 - 1.3)\cos^2(2° 24' 50'' + 4')/[7416 \cos (2')] = 40.25''$

[By an approximate formula, $\delta = (s - h)/d \sin 1'' = 1.45/(7416 \sin 1'') = 40.32''$.]

Vertical angle corrected for axis-signal difference = $2° 24' 50'' + 0° 0' 40.25''$

$\beta_1 = 2° 25' 30.25''$

Curvature correction = $\theta/2 = 2 \text{ min}$

Refraction correction = $0.07 × 4 = 0.28 \text{ min} = 17''$

Corrected vertical angle = $2° 25' 30.25'' - 2' + 17'' = 2° 23' 47.25''$

The difference in level is given by

$$H = \frac{d \sin[\beta_1 - (1 - 2m)d/(2R\sin 1'')]}{\cos[\beta_1 - (1 - m)d/(R\sin 1'')]}$$

$= 7416 \sin[2° 25' 30.25'' - 2' + 17'']/\cos[2° 25' 30.25'' + 17'' - 4']$

$= 309.133 \text{ m}$

Elevation of P = $2345.65 - 309.133 = 2036.517 \text{ m}$

Example 16.11 The distance between two stations P and Q is 5436 m. The angle of elevation from P to Q was measured as $2° 34' 46''$. The height of instrument at P was 1.25 m and the height of the signal at Q was 4 m. If the RL of P is 1325.75 m, find the RL of Q, if we (a) neglect all corrections, (b) neglect the axis-signal correction, (c) neglect curvature and refraction corrections, and (d) take all corrections into account. Take $R \sin 1'' = 30.9$ m and $m = 0.07$.

Solution Central angle $\theta = d/R \sin 1'' = 5436/30.9 = 176'' = 2' 56''$

Curvature correction = $\theta/2 = 2' 56''/2 = 1' 28''$

Refraction correction = $m\theta = 0.07 × 2' 56'' = 12.3''$

Axis-signal correction = $(s - h)/d \sin 1'' = (4 - 1.25)/(5436 \sin 1'')$

$= 104.35'' = 1' 44.35''$

Or more accurately,

Axis-signal correction = $(s - h)\cos^2(\alpha + \theta)/[d \cos (\theta/2)]$

$= 2.75 \cos^2(2° 37' 32'')/[5436 \cos (1' 28'')]$

$= 1' 44''$

(a) *All corrections neglected* $H = d \tan \alpha = 5436 \tan(2° 34' 46'') = 244.63 \text{ m}$

Elevation of Q = $1325.75 + 244.63 = 1570.38 \text{ m}$

(b) *Axis-signal correction neglected* The angle is corrected for curvature and refraction.

$\alpha = 2° 34' 36'' + 1' 28'' - 12.3'' = 2° 35' 51.7''$

$H = d \tan \alpha = 5436 × \tan (2° 35' 51.7'') = 246.63 \text{ m}$

Elevation of Q = $1325.75 + 246.63 \text{ m} = 1572.38 \text{ m}$

(c) *Curvature and refraction neglected* The angle is corrected for axis-signal difference only.

$$\alpha_1 = 2° \ 34' \ 36'' - 0° \ 1' \ 44.35'' = 2° \ 32' \ 51.65''$$

$$H = d \tan (2° \ 32' \ 51.65'') = 241.87 \text{ m}$$

Elevation of Q = 1325.75 − 241.87 = 1567.62 m

(d) *All corrections applied*

$$\alpha_1 = 2° \ 32' \ 51.65'', \quad \theta = 2' \ 56'', \quad m\theta = 12.3''$$

$$H = d \sin[\alpha_1 + \theta/2 - m\theta]/[\cos\alpha_1 - m\theta + \theta]$$

$$= 5436 \sin[2° \ 32' \ 51.65'' + 1' \ 28'' - 12.3'']/\cos [2° \ 32' \ 51.65''$$

$$- 12.3'' + 2' \ 56'']$$

$$= 243.78 \text{ m}$$

Elevation of Q = 1325.75 + 243.78 = 1569.53 m

Example 16.12 If the distance between two stations is 4758 m and the vertical angle observed from one station is 1° 23′ 44″ (depression) and height of the signal above the instrument axis is 3.2 m, find the RL of the station if the RL of the instrument axis (when horizontal) is 2314.5mR sin 1″ = 30.9 m and m = 0.07.

Solution   s − h = 3.2 m

Central angle θ = 4758/30.9 sec = 154″ = 2′ 34″

Curvature correction = $\theta/2$ = 1′ 17″,   refraction = 0.07θ = 10.78″

Axis-signal correction = 3.2 $\cos^2(\alpha + \theta)/d \cos(\theta/2)$

$$= 3.2 \cos^2(1° \ 26' \ 18'')/4758 \cos(1' \ 17'')$$

$$= 0.000672 \text{ rad} = 2' \ 18.6''$$

Observed vertical angle β = 1° 23′ 44″

Angle corrected for axis-signal difference β_1 = 1° 23′ 44″ + 00° 2′ 18.6″

$$= 1° \ 26' \ 2.6''$$

$$H = \frac{d \sin[\beta_1 + m\theta - \theta/2]}{\cos[\beta_1 + m\theta - \theta]}$$

$$= \frac{4758 \sin[1°26'2.6'' + 10.78'' - 1'17'']}{\cos[1°26'2.6'' + 10.78'' - 2'34'']}$$

$$= 117.58 \text{ m}$$

Elevation of P = RL of Q − H = 2314.5 − 117.58 = 2196.92 m

16.6.2   Elevation from Reciprocal Observations

Another method to find the difference in level between two stations is to make observations from both the stations. Ideally, the observations should be made simultaneously to avoid any change in refraction between observations. If two instruments cannot be deployed simultaneously, observations are taken at the same time of the day, on two days. Let P and Q be two stations, as shown in Fig. 16.9. At station P,

$$\angle P'PT = \alpha_1 \text{ (vertical angle measured at P corrected for axis-signal}$$
$$\text{difference)}$$

$$= \alpha - (s_2 - h_1)/(d \sin 1'')$$

$\angle P'PQ$ = refraction angle, $\quad \gamma = m\theta$

$\angle TPQ_1 = \theta/2$ [angle between tangent and chord = $(1/2) \times$ central angle]

$$= \text{curvature correction}$$

Similarly at Q,

$SQQ' = \beta_1$ (vertical angle of depression corrected for axis-signal difference)

$$= \beta + (s_1 - h_2)/(d \sin 1'')$$

$\angle Q'QP$ = angle of refraction $\gamma = m\theta$

$\angle SQP_1 = \theta/2$ (curvature correction

$\angle QPQ_1 = \alpha_1 - \gamma + \theta/2 = \alpha_1 - m\theta + \theta/2$

$\angle PQP_1 = \beta_1 + \gamma - \theta/2 = \beta_1 + m\theta - \theta/2$

As PQ_1 is parallel to QP_1, these two angles are equal and each is equal to half their sum.

Each corrected angle = $(\alpha_1 + \beta_1)/2$

In triangle QPQ_1, $\angle QPP_1 = (\alpha_1 - m\theta + \theta/2) = (\alpha_1 + \beta_1)/2$

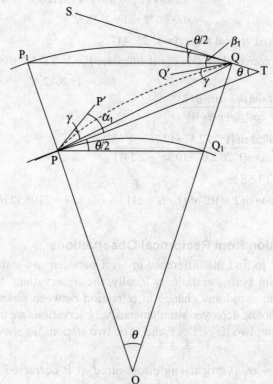

Fig. 16.9

$\angle PQQ_1 = 90° - (\beta_1 + m\theta)$

$\beta_1 + m\theta - \theta/2 = (\alpha_1 + \beta_1)/2, \quad \beta_1 + m\theta = (\alpha_1 + \beta_1)/2 + \theta/2$

Applying the sine rule,

$$\frac{QQ_1}{\sin QPQ_1} = \frac{PQ_1}{\sin PQQ_1}, \quad QQ_1 = H, \quad \text{and} \quad PQ_1 = d$$

$QQ_1 = H = d\,[\sin(\alpha_1 + \beta_1)/2]/\sin[90° - (\beta_1 + m\theta)]$

$H = d[\sin(\alpha_1 + \beta_1)/2]/[\cos(\alpha_1 + \beta_1)/2 + \theta/2]$

As θ is generally small, it can be neglected. Then,

$H = d[\sin(\alpha_1 + \beta_1)/2]/[\cos(\alpha_1 + \beta_1)/2] = d\tan(\alpha_1 + \beta_1)/2$

In the case discussed, it was assumed that one is an angle of elevation and the other an angle of depression. Another possibility is that both may be angles of depression. In such a case, the sign of α has to be changed (made negative).

$H = d[\sin(\beta_1 - \alpha_1)/2]/[\cos(\beta_1 - \alpha_1)/2] = d\tan(\beta_1 - \alpha_1)/2$

To cover both cases, we can write

$H = d[\sin(\beta_1 \pm \alpha_1)/2]/[\cos(\beta_1 \pm \alpha_1)/2] = d\tan(\beta_1 \pm \alpha_1)/2$

In the above expression, use the positive sign when α is an angle of elevation and the negative sign when α is an angle of depression.

Note that when reciprocal observations are made, the refraction correction is eliminated. If angle θ is small, angles are corrected for axis-signal difference only and averaged. The following examples illustrate the application of this formula.

Example 16.13 Reciprocal observations made from stations A and B are as follows. Distance between stations = 7224 m, angle of elevation at A to B = 1° 34′ 56″ , angle of depression at B to A = 0° 56′ 42″, height of instrument at A = 1.2 m, height of signal at B = 3 m, height of instrument at B = 1.3 m, height of signal at A = 4.2 m. Find the difference in level between the stations.

Solution Reciprocal observations are available and hence refraction is eliminated. Angles are corrected for axis-signal difference and each angle is equal to the average corrected angle.

Instrument at A: Height of instrument = 1.2 m, height of signal at B = 3 m

$(s_2 - h_1) = 3 - 1.2 = 1.8$ m

Axis-signal correction at $\delta_1 = 1.8/(d\sin 1″) = 1.8/(7224\sin 1″) = 51.39″$

Corrected angle $\alpha_1 = 1° 34′ 56″ - 51.39″ = 1° 34′ 4.61″$

Instrument at B: Height of instrument = 1.3 m

Height of signal at A = 4.2 m

$s_1 - h_2 = 4.2 - 1.3 = 2.9$ m

Axis-signal correction $\delta_2 = 2.9/(7224\sin 1″) = 1′ 22.8″$

Corrected angle $\beta_1 = 0° 56′ 42″ + 1′ 22.8″ = 0° 58′ 4.8″$

Central angle $\theta = 7224/30.9 = 3′ 53.78″$

$(\alpha_1 + \beta_1)/2 = (1° 34′ 4.61″ + 0° 58′ 4.8″)/2 = 1° 16′ 4.7″$

$$(\alpha_1 + \beta_1)/2 + \theta/2 = 1° \ 16' \ 4.7'' + (3' \ 53.78'')/2 = 1° \ 18' \ 15.9''$$

$$H = \frac{d \sin(\alpha_1 + \beta_1)/2}{\cos[(\alpha_1 + \beta_1)/2 + \theta/2]}$$

$$= 7224 \sin(1° \ 16' \ 4.7'')/\cos(1° \ 18' \ 15.9'')$$

$$= 159.89 \ \text{m}$$

Alternately, if θ is considered small, then

$$H = d \tan(\alpha_1 + \beta_1)/2 = 7224 \tan(1° \ 16' \ 4.7'') = 159.89 \ \text{m} \quad \text{(as before)}$$

Example 16.14 The horizontal distance between two stations P and Q is 8254 m. The following data was recorded: vertical angle from P to Q $= -54' \ 34''$, vertical angle from Q to P $= -1° \ 23' \ 34''$, height of instrument at P $= 1.25$ m, height of instrument at Q $= 1.32$ m, height of signal at P $= 3.54$ m, height of signal at Q $= 4.56$ m. Find the difference in level between the two stations.

Solution We find the axis-signal corrections for observations made at both stations. When the instrument is at P,

$$(s_2 - h_1) = 3.54 - 1.25 = 2.29 \ \text{m}$$

Axis-signal correction $= 2.29/(d \sin 1'') = 2.29/(8254 \sin 1'') = 57.22''$

$$\alpha_1 = \alpha + 57' \ 22'' = 54' \ 34' + 57.22'' = 55' \ 31.22''$$

When the instrument is at Q,

$$(s_1 - h_2) = 4.56 - 1.32 = 3.24 \ \text{m}$$

Axis-signal correction $= 3.24/(8254 \sin 1'') = 1' \ 20.96''$

$$\beta_1 = 1° \ 23' \ 34'' + 1' \ 20.96'' = 1° \ 24' \ 54.96''$$

Central angle $\theta = 8254/30.90 = 267.12'' = 4' \ 27''$

$$(\beta_1 - \alpha_1)/2 = (1/2)(1° \ 24' \ 54.96'' - 55' \ 31.22'') = (1/2)(29' \ 23.74'') = 0° \ 14' \ 42''$$

$$H = d \tan(\beta_1 - \alpha_1)/2 = 8254 \tan(0° \ 14' \ 42'') = 35.29 \ \text{m}$$

$$\text{(neglecting } \theta \text{ as small)}$$

Example 16.15 The following data refers to reciprocal observations made from three stations:

Data	Instrument at P	Instrument at Q	Instrument at R
Horizontal distance	PQ = 13,200 m	QR = 8542 m	RP = 6548 m
Vertical angle	$- 19' \ 34''$ to Q	$- 13' \ 26''$ to P	$+ 1° \ 23' \ 36''$ to P
	$- 0° \ 42' \ 50''$ to R	$- 0° \ 23' \ 56''$	$+ 0° \ 34' \ 56''$ to Q
Height of instrument	1.35 m	1.2 m	1.27 m
Height of signal	4.2 m at Q	3.85 m at P	3.5 m at P
	3.8 m at R	4.25 m at R	3.95 m at Q

$$H = d \ [\sin(\beta_1 - \alpha_1)/2]/[\cos(\beta_1 - \alpha_1)/2 + \theta/2]$$

$$H = 8254 \sin(0° \ 14' \ 42'')/\cos(0° \ 14' \ 42'' + 2' \ 13.5'') = 35.29 \ \text{m}$$

$$\text{(taking } \theta \text{ into account)}$$

If the RL of P is 956.75 m, find the RL of Q and R.

Solution *Instrument at P, observation to Q*

Axis-signal correction = $(4.2 - 1.35)/(13,200 \sin 1'') = 44.53''$

$\alpha_1 = 19' \, 34'' + 44.53'' = 20' \, 8.53''$

Instrument at Q, observation to P

$s - h = 3.85 - 1.2 = 2.65$ m

Axis-signal correction = $2.65/(13,200 \sin 1'') = 41.41''$

$\beta_1 = 13' \, 26'' + 41.41'' = 14' \, 7.41''$

Difference in elevation = $d \tan(\beta_1 - \alpha_1)/2$

$= 13,200 \tan[(14' \, 7.41'' - 20' \, 8.53'')/2]$

$= -23.11$ m

P is at a higher level due to the negative sign of *H*.

Instrument at P, observation to R As this is a single observation, all corrections are to be made.

$D = 6548$ m, $(s - h) = 3.5 - 1.35 = 2.15$ m

$\alpha = -42' \, 50''$ (angle of depression)

$\theta = 6548/30.9 = 211.9'' = 3' \, 31.9''$

Axis-signal correction = $2.15/(6548 \sin 1'') = 67.35'' = 1' \, 7.35''$

Curvature correction = $211.9''/2 = 1' \, 50''$

Refraction correction = $0.07 \times 211.9'' = 14.8''$

$\alpha_1 = 42' \, 50'' + 1' \, 7.35'' = 43' \, 57.35''$

Height difference $H = d \sin[\alpha_1 + m\theta - \theta/2]/\cos[\alpha_1 + m\theta - \theta]$

$H = 6548 \sin(42'22.3'')/\cos(40' \, 32.3'') = 80.7$ m

R is 80.7 m below P.

Instrument at Q, observation to R This is again a single observation and all corrections have to be made to the vertical angle.

$\alpha = 23' \, 56''$, $s - h = 3.5 - 1.2 = 2.3$ m

$\theta = 8542/30.9 = 276.44'' = 4' \, 36.44''$

Curvature correction = $\theta/2 = 276.44''/2 = 2' \, 18.22''$

Refraction correction = $0.07 \times 276.44'' = 19.35''$

Axis-signal correction = $2.3/(8542 \sin 1'') = 55.53''$

$\alpha_1 = 23' \, 56'' + 55.53'' = 24' \, 51.53''$

$H = 8542 \sin[24' \, 51.53'' + 19.35'' - 2' \, 18.22'']/\cos[24' \, 51.53'' + 19.35'' - 4' \, 36.44'']$

$= 56.8$ m

R is below Q by this distance, as the angle is a depression angle. P is above Q and R and Q is above R. Thus,

H (P to R) $- H$ (Q to R) $= H$ (P to Q)

$80.7 - 56.8 = 23.9$ m

while H (from P to Q) $= 23.11$ m

Difference = $23.9 - 23.11 = 0.79$ m

This difference can be distributed among the three heights equally, as no weightage has been given to the observations.

Summary

Trigonometric levelling is a precise form of levelling used to determine vertical and/or horizontal distances. It is most useful in tough terrains, where measuring distances is difficult or impossible.

In determining heights and distances, the instrument is set up at a station and a target is set at the station whose elevation is to be determined. The vertical angle to the target is measured accurately. If possible, the horizontal distance to the object is also measured. The height of the target above the instrument axis can be calculated. If observations to a benchmark are also made with the same instrument set-up, then the reduced level of the station can also be found.

If the horizontal distance cannot be measured, then vertical angles to the target from two instrument stations are measured. Knowing the distance between the instrument stations and the two vertical angles, the horizontal distance and the height of the target above the instrument axis can be calculated. If the two instrument stations lie in the same plane as the object, only vertical angles are required. If the two stations do not lie in the same plane as the object, the horizontal angles to the object are also measured to find the horizontal distance.

If the distances are large, curvature and refraction corrections are applied linearly to the calculated height or difference in elevation. In geodetic observations, where the distances involved are large, elevations are determined from horizontal distances and vertical angles. The corrections are applied in the angular measure to the observed vertical angles.

Terrestrial refraction tends to bend light rays concave to the ground surface. The angle between the horizontal and the tangent to the refraction curve is known as the refraction angle (r). The coefficient of refraction, m, is the ratio of the refraction angle r to the angle subtended at the centre: $r = m\theta$. The effect of refraction does not remain constant and varies during the day, remaining constant from 10 a.m. to 4 p.m.

If the horizontal distance is d and the central angle is θ, then the curvature correction is $\theta/2$ and has a sign opposite to that of refraction.

Another correction made is due to the difference between the height of the instrument at one station and the height of the signal at the other. This correction is negative for angles of elevation and subtractive for angles of depression.

Elevations are determined either from one vertical angle observed from one station or by reciprocal observations. Reciprocal observations, made from each station to the other, eliminate the effect of refraction. If the distances involved are not large (θ is small), the height can be determined as distance multiplied by the tangent of the average reciprocal angles.

Exercises

Multiple-Choice Questions

1. Trigonometric levelling is
   (a) the same as ordinary levelling
   (b) levelling where precise instruments are used
   (c) levelling where precise methods are used
   (d) where elevations are obtained from vertical angles and staff readings
2. The correction for refraction is
   (a) always subtractive
   (b) always additive
   (c) subtractive for angles of depression
   (d) additive for angles of depression
3. Angle of refraction is
   (a) a constant fraction of the vertical angle observed from the station
   (b) angle between the line joining the two stations to the line of sight

    (c) the angle between tangent to the line of sight with the horizontal

    (d) the angle between the line of sight and plumb line through the point

4. Coefficient of refraction is defined as the ratio of

    (a) angle of refraction to the angle subtended at the centre by the line joining the two stations

    (b) difference in height between the two stations to the distance between them

    (c) distance between the two stations to the radius of the earth

    (d) angle of elevation or depression to the angle subtended at the centre by the line joining the two stations

5. Curvature correction in angular measure is, if d is the distance between the stations and R is the radius of the earth,

    (a) $d/2$ radians   (b) $d/2R$ radians   (c) $d/3R$ radians   (d) $2d/R$ radians

6. Curvature correction in angular measure for two stations d apart subtending an angle θ at the centre is equal to

    (a) θ (b) $\theta/2$ (c) $\theta/3$ (d) 2θ

7. In trigonometric levelling, curvature correction in angular measure is

    (a) always subtractive

    (b) always additive

    (c) additive in the case of angle of elevation

    (d) subtracted in the case of angle of elevation

8. Axis-signal correction is applied to correct for

    (a) the inclination of the Earth's axis

    (b) the tilt of the instrument axis

    (c) the tilt in the horizontal axis of the instrument

    (d) difference in height between the instrument and the signal height.

9. Axis-signal correction is

    (a) always subtractive

    (b) always additive

    (c) subtractive for angles of elevation

    (d) subtractive for angles of depression

10. Axis-signal correction in linear measure is,

    (a) $(s-h)$ (b) $(s-h)/2$ (c) $(s+h)$ (d) $(s+h)/2$

11. Axis-signal correction in angular measure is

    (a) $(s-h)/(\sin 1'')$ sec   (b) $(s-h)/d$ min

    (c) $(s-h)/R$ min   (d) $(s-h)/(d \sin 1'')$ sec

12. While finding elevations from reciprocal observations

    (a) curvature correction is eliminated

    (b) refraction correction is eliminated

    (c) curvature and refraction corrections are eliminated

    (d) both curvature and refraction corrections are to be taken into account

13. Curvature correction in linear measure can be applied as

    (a) $d/2R$ (b) d/R (c) d^2/R (d) $d^2/2R$

14. Refraction correction if applied in linear measure is given by

    (a) $d^2/7R$ (b) $d^2/2R$ (c) $d^2/14R$ (d) d^2/R

Review Questions

1. Explain the basic principle of trigonometric levelling and the difference between plane and geodetic methods.

2. Describe the procedure to determine the elevation when the distance to the object cannot be measured.

3. Explain the term geodetic distance.
4. Explain the term coefficient of refraction.
5. Does the effect of refraction remain constant? Explain your answer.
6. State the formulae for curvature and refraction when they are applied in linear measure.
7. Explain the term axis-signal correction. Derive the formula for axis-signal correction when the observed vertical angle is (a) an angle of elevation and (b) an angle of depression.
8. Show that the curvature correction in angular measure is half the central angle.
9. Explain the term reciprocal observation in trigonometric levelling and state its advantages.
10. State the exact and approximate formulae for difference in elevation when reciprocal observations are made. What is the assumption in deriving the approximate formula?

Problems

1. The horizontal distance between two stations P and Q is 920 m. The vertical angle observed from P to the vane at Q at a height of 4 m is 8° 24′ 30″. An observation made to a staff held at a benchmark of RL 385.55 m from P was 2.85 m. Find the RL of Q.
2. A theodolite was kept at two stations A and B. The instrument height at the two stations was the same. The distance between the two instrument stations was 200 m and in the same vertical plane containing the object. The vertical angles observed were 12° 30′ and 8° 50′ from A and B, respectively. The staff reading on a benchmark of RL 985.55 m was 2.35 m from A. If the sighting was done to a staff held at the object station to a mark 4 m from the ground, find the RL of the object station.
3. Two instrument stations were selected 150 m apart in the same vertical plane as the object. The vertical angles observed to the two stations were 12° 13′ 30″ and 9° 10′ 40″. The readings on a staff held at a benchmark of RL 1080.00 m were 2.35 and 1.05 m from the stations, the instrument reading being higher for the station further from the object. Find the RL of the object station if the vertical angles were read on a mark 5 m from the ground.
4. To find the height of a chimney, two stations P and Q were selected, not lying in the same plane as the object. The vertical angles observed were 12° 40′ 20″ and 8° 30′ 40″. The distance between P and Q was 380 m. The horizontal angle between the line joining the chimney to P and the line PQ was 70° 40′ and the horizontal angle at Q, between QP and the line joining Q to the chimney, was 50° 30′. Find the height of the chimney if the height of the instrument at P was 1.3 m.
5. Two stations A and B were used to measure the height of an object P; the stations do not lie in the same vertical plane as the object. The distance between the stations was 250 m and the vertical angles observed to the object were 10° 30′ from A and 6° 45′ from B. The horizontal ∠PAB was 82° 30′ and ∠PBA = 38° 40′. The staff readings on a benchmark of RL 2345.5 m were 2.35 m and 1.85 m from A and B, respectively. Find the RL of station P if the readings were to a mark 4 m above the point P.
6. The distance between two stations P and Q was 5145 m. The vertical angle observed from P to Q was 1° 24′ 30″ (elevation) and from Q to P 2° 12′ 40″ (depression). Find the axis-signal correction to the angles if the instrument and signal heights at P and Q were 1.2 m, 3.5 m and 1.3 m, 4.2 m, respectively.
7. The distance between stations A and B was 6450 m. The instrument was set up at A and the vertical angle to Q measured was 1° 30′ 40″. The instrument height was 1.25 m and the signal height at B was 4 m. Correct the angle and find the difference in elevation between the points.

8. The distance between two stations P and Q was 8145 m. Reciprocal observations were made from P and Q. At P, the vertical angle to Q = 2° 30′ 40″, height of instrument = 1.2 m, and height of signal at Q = 4 m. At Q, vertical angle = 1° 10′ (depression), height of instrument = 1.3 m, and height of signal at P = 3.5 m. If the RL of P was 1035.5 m, find the RL of Q.

9. The distance between two stations A and B was 11,150 m. The following data reciprocal observations is made from the two stations: Instrument at A: vertical angle = (−) 0° 30′ 40″, height of instrument = 1.2 m, height of signal at B = 3 m. Instrument at B: vertical angle = (−) 0° 46′ 20″, height of instrument = 1.3 m, height of signal at A = 4 m. Find the difference in elevation between A and B.

10. The following observations refer to trigonometric levelling data between three stations. Determine the difference in the heights of the stations and find their reduced levels if the reduced level of A = 1050 m.

Instrument at	Height of instrument (m)	Signal at	Height of signal (m)	Vertical angle	Distance (m)
A	1.35	B	4.0	+1° 10′	AB = 3820
B	1.5	A	4.5	− 56′ 50″	BC = 3255
		C	5	− 1° 31′	
C	1.25	A	4.5	− 50′ 15″	CA = 2550
		B	4	− 58′ 30″	

CHAPTER

17

Geodetic Surveying

Learning Objectives

After going through this chapter, the reader will be able to

- distinguish between geodetic surveying and plane surveying
- explain the basic principle of triangulation
- explain triangulation systems with sketches
- explain the precision and specifications for triangulation surveys
- explain the concept of strength of figure and calculate the same for a given network
- explain signals, towers, intervisibility, and phase correction with sketches
- describe briefly the equipment and methods used to measure the base line and angles
- describe the equipment and methods for precise levelling

Introduction

The distinction between geodetic surveying and plane surveying has been explained in Chapter 1. The objective of geodetic surveying is the precise location of a set of control points. Essentially, this survey provides a set of points for horizontal control. Geodetic surveying does not purport to obtain details of the topography of an area. The objective is only to get a set of control points whose locations are absolutely and relatively precise. Relative positions are obtained in terms of bearings or angles and lengths of the lines joining the points. Absolute location includes latitudes and longitudes of the points. These points can be further used for more detailed surveys such as locating secondary and tertiary sets of control points. Such points are known as triangulation stations.

Geodetic surveys cover a very large area such as a continent or a country. Thus, geodetic surveying is important and is generally carried out by state survey agencies. In India, the Survey of India conducts such surveys and is responsible for preparing maps.

The methods of geodetic surveying include triangulation, trilateration, and precise traversing. Of the three, triangulation is the most popular method.

In triangulation, a base line connecting two points or stations is measured. The line is measured very accurately taking all corrections into account. A third point is now located to form a triangle. The angles between the lines are measured accurately by precise methods and an average value is found. Using the length of one line and the angles of the triangle, the lengths of the other lines can be found. More triangles are added to this triangle by locating further points and measuring only the angles.

In trilateration, the procedure involves selecting three stations and measuring the distances between them. The three sides of the triangle are measured. The method is not very popular because of the distances involved and difficulties in precise measurement of the sides.

Precise traversing is another method that can be used for conducting surveys of large areas. As the number of checks available is small in a polygon, it is inferior to triangulation. Precise traversing is conducted only in special situations where triangulation is not possible, for example, in a heavily wooded area. However, triangulation is preferred over the other two methods.

Even in triangulation, a quadrilateral with its diagonals is preferred because of more stringent checks available to determine the accuracy of work.

The methods of geodetic surveying are changing fast. With the availability of total stations, EDMs (electromagnetic distance measuring instruments), distomats, and smart stations with GPS (global positioning system), such surveys can be done much more precisely and in a shorter period of time.

Along with horizontal control, one also needs to locate geodetic points with known elevations above a datum. This is done by using the methods of precise levelling. The principle remaining the same, precise levelling uses precise instruments and methods to accurately determine the elevations of points with reference to a datum such as mean sea level. The methods of precise levelling are discussed at the end of this chapter.

17.1 Basic Concepts of Triangulation

In triangulation, one side and three angles of a triangle are measured. The side measured is known as the *base line*. The basic principle of triangulation is that if one side and three angles are known, the remaining sides can be calculated using the sine rule (Fig. 17.1). A, B, and C are used to denote angles and a, b, and c are used to denote the sides opposite these angles. Thus, if one side a is measured and the angles A, B, and C are measured, then

$$\frac{a}{\sin A} = \frac{b}{\sin B} = \frac{c}{\sin C}$$

This gives $b = a \sin B/\sin A$ and $c = a \sin C/\sin A$. This triangle can be extended by adding two sides to any of the sides of the base triangle as shown in the figure. The three angles of the added triangle are measured and the two unknown sides calculated from the sine formula with the help of the known side.

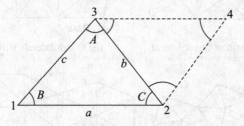

Fig. 17.1 Basic concept of triangulation

The base line is precisely measured. Theoretically, it is not necessary to measure any other line to have the complete data required for plotting. However, if the computations are based upon the measurement of a single line, error accumulation could be high. Therefore, as in any other work requiring precision, some intermediate lines are measured as check bases. The number of such lines depends upon the extent of the survey. Measuring a line precisely is a tedious task and the computations required to find the horizontal length are lengthy. The method of measuring the base line is explained later.

The sum of the angles in a triangle is 180°. The sum of the angles can be checked but there is no check on the accuracy of the individual angles measured. The angles are measured accurately using the repetition or reiteration method.

17.1.1 Triangulation Systems

Triangulation systems describe the manner in which a base triangle is extended. A number of triangulation systems are shown in Fig. 17.2. It must be noted that in all the systems, the basic figure is a triangle.

In Fig. 17.2(a), the base triangle is extended to cover a larger area by adding triangles in one or both directions resulting in a chain of triangles. This system is economical and is amenable to rapid survey work. However, this can lead to accumulation of errors; also the number of checks available is small. This form is suitable for narrow strips of land to be covered by the survey. It will be necessary to have a sufficient number of check bases as a check on the work. In Fig. 17.2(b), the triangle is extended on both sides such that a set of diamond-shaped figures or a double chain of triangles is formed. This system can cover a much larger area but suffers from the same infirmities as mentioned before. Check bases are required to prevent the accumulation of errors.

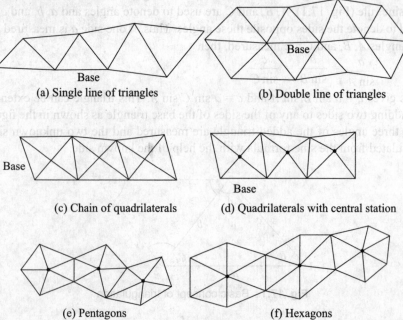

(a) Single line of triangles   (b) Double line of triangles

(c) Chain of quadrilaterals   (d) Quadrilaterals with central station

(e) Pentagons   (f) Hexagons

Fig. 17.2 Triangulation systems

In Fig. 17.2(c), we have a set of quadrilaterals with their diagonals being calculated by measuring all the eight angles of the quadrilateral. This system is the most accurate, as the number of checks available are more. Note that there is no station at the intersection of the diagonals.

The other systems, although basically consisting of triangles, use polygonal figures, such as quadrilaterals, pentagons, hexagons, etc., with a station inside. In Fig. 17.2(d), there is a set of quadrilaterals with a central station, which may not necessarily be at the intersection of the diagonals. Figures 17.2(e) and 17.2(f) show a set of pentagons and hexagons, respectively, with central stations.

Grid iron system   In surveying very large areas or countries such as India, a chain of triangles or quadrilaterals is set out in two approximately perpendicular directions. The area left out is surveyed from the available stations of these triangles or quadrilaterals [Fig. 17.3(a)].

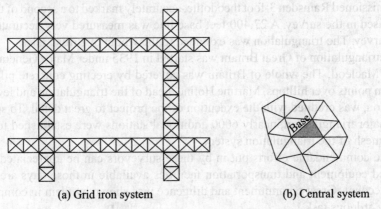

| (a) Grid iron system | (b) Central system |

Fig. 17.3   Triangulation systems

Central system   In surveying small areas or countries, the survey is extended from a base triangle or quadrilateral at the centre of the area by adding figures in radial directions. This is known as the central system [Fig. 17.3(b)].

Selection of triangulation systems

In selecting a system for triangulation, the points to be kept in mind are as follows.
(a) It must be possible to compute the sides of the triangles through two independent routes.
(b) The independent routes available should be such that both the routes go through well-conditioned figures.
(c) All the lines forming the figure should be of comparable length. Very long lines and very short lines should be avoided.
(d) As triangulation is expensive, the system should enable, for the amount of work required or put in, maximum coverage of the surface.

Great Trigonometrical Survey of India   The Great Trigonometrical Survey (GTS) of India was carried out in the 19th century. The survey was a mammoth effort by William Lambton and George Everest. The survey was mainly conducted for demarcation of territories belonging to the British regime. In addition to conducting the tedious and long winding operations with the limited equipment available at that time, the survey is also hailed for its achievement in measuring the heights of mountain ranges such as Mount Everest, K2, and Kanchenjunga.

The GTS started in 1802 and was not an operation of a small period. It took 60 years to complete the survey.

The base line for the survey was measured at Madras and was nearly 7 miles long (12.1 km). The base line was very accurately measured and various corrections applied.

Survey of Great Britain Two major surveys were conducted in Great Britain. The principal triangulation of Great Britain was conducted between 1783 and 1853. The survey was conducted under the leadership of General William Roy. A specially commissioned Ramsden 3-foot theodolite, accurately marked to a second of the arc, was used in the survey. A 27,400 feet base line was measured very accurately for the survey. The triangulation was extended to the whole of British Isles.

Retriangulation of Great Britain was started in 1935 under Major General Malcolm Macleod. The whole of Britain was covered by erecting concrete pillars as station points over hilltops. Martine Hotine, head of the triangulation and levelling division, was credited with the execution of the project to great detail. To support the major triangulation, nearly 6000 additional stations were established to get a finer mesh of the triangulation system.

The commendable efforts put in by these surveyors can be appreciated if the limited equipment and transportation facilities available in those days are taken into account. Great commitment and diligence was shown by them in completing a very arduous task.

With the geopositioning systems available today, what was done in years by these great surveyors can be accomplished in days to get exact location of points within a metre of accuracy. Technological developments have helped surveyors to undertake accurate surveys with minimum effort.

17.2 Order of Triangulation

Order of triangulation refers to the classification of triangulation into different types. Essential principles remaining the same, triangulation is carried out with different types of instruments and methods. The different types of triangulations differ only in the accuracy stipulated for the work. Thus we have first-order or primary triangulation, second-order or secondary triangulation, and third-order or tertiary triangulation.

First-order triangulation

First-order triangulation is done on extensive areas, such as a country for which the highest level of accuracy is stipulated. The lengths of the sides of the triangle are large. Triangulation stations are precisely located using first-order triangulation. A stringent control is exercised in all the measurements. The instruments used must be precise and should be tested and adjusted daily.

In primary triangulation, care is taken to minimize errors by using very high precision equipment, methods used are commensurate with the equipment, and the criteria for permissible errors are equally stringent. This is necessary because all the other surveys are based on such a triangulation.

Second-order triangulation

Second-order triangulation is carried out within the primary triangulation stations. The extent of area covered is small and the sides of the triangle are also small. The instruments and methods used are not as precise as in first-order triangulation. The accuracy limits specified are not as stringent as in the first case.

Third-order triangulation

Third-order or tertiary triangulation is performed within the area covered by second-order triangulation stations. This triangulation gives a set of control points which are normally used by agencies conducting engineering surveys. The equipment and methods used are of lower precision. The accuracy limits are also not as stringent as in the case of the other two methods of triangulation.

The requirements and the extent of area for the three types of triangulation are given in Table 17.1. These will change with the availability of new and more precise equipment.

Table 17.1 Specifications for the different types of triangulation

Specification	Triangulation type		
	Primary	Secondary	Tertiary
Length of the base line	5–20 km	2–5 km	0.5–3 km
Length of the sides of triangles	30–160 km	8–70 km	1.5–10 km
Error in base	1 in 300,000	1 in 150,000	1 in 75,000
Probable error of base	1 in 1000,000	1 in 500,000	1 in 250,000
Probable error in the computed side	1 in 60,000 to 1 in 250,000	1 in 20,000 to 1 in 50,000	1 in 5,000 to 1 in 20,000
Discrepancy in two computations* (mm)	$10\sqrt{K}$	$20\sqrt{K}$	$25\sqrt{K}$
Average triangle of closure	$< 1''$	$3''$	$6''$
Maximum error in triangle of closure	$< 3''$	$8''$	$12''$
Probable azimuth error	$0.5''$	$2''$	$5''$
Instrument used	Least count $0.1''$	Least count $1''$	Least count $5''$
Limiting strength of figure	25	40	50
Sets of angle observations	16	8	3

*K is the distance in kilometres of the section surveyed.

17.3 Size and Shape of Triangles

The size of the side of a triangle depends upon many factors. While large triangles reduce the cost of a survey, many ground factors limit the maximum size that can be used. For geodetic surveying, the size of the triangles is large, as maximum area has to be covered with minimum amount of labour involved in measuring the angles. However, the sizes are limited by the intervisibility of stations. There are many instances where, to have intervisibility, towers are erected at a station. The visibility and height of the towers will limit the size of the triangle. In topographic survey, small triangles are used since less precise equipment can be used for running the traverses.

Similarly, the shape of triangles also depends on a few factors. It may be assumed that an equilateral triangle with equal sides and angles would be the best choice. However, often it may be difficult to set an equilateral triangle. In principle, the shape of the triangle should be such that the sides of the triangle computed with the measured angles are least affected by any error in the angles. The sides are computed using the sine rule. Sine values change by larger amounts if the angle is small. The side opposite to a small angle will get affected more due to an error in the measurement of angles as compared to the side lying opposite a larger angle.

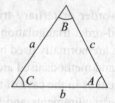

Fig. 17.4 Optimum angles of a triangle

To find the optimum angles of a triangle for this purpose, consider the triangle shown in Fig. 17.4. A, B, and C are angles lying opposite to sides a, b, and c, respectively. Let δA, δB, and δC be the errors in the measurement of these angles. Let δa_A, δa_B, and δa_C be the errors in the measurement of side a due to the error in the measurement of the angles.

Side a is calculated by the sine formula from side c and angle C as

$$a = c \sin A/\sin C$$

Differentiating partially with respect to A

$$\delta a_A = c \cos A \, \delta A/\sin C$$

Dividing this by the first equation

$$\delta a_A/a = \cos A \, \delta A/\sin A = \cot A \, \delta A$$

Similarly, partially differentiating with respect to C

$$\delta a_C = - c \sin A \cos C \, \delta C/\sin^2 C$$

and by division, we get

$$\delta a_C/a = - \cos C \, \delta C/\sin C = - \cot C \, \delta C$$

Assuming that both angles A and C are measured with the same precision, let the error in their measurement be $\pm \beta$.

$$\delta a_A = \delta a_C = \pm \beta$$

The probable error in the computed side a is given by

$$\delta a/a = \sqrt{(\delta a_A/a)^2 + (\delta a_C/a)^2} = \beta \sqrt{\cot^2 A + \cot^2 C}$$

This is minimum when $\cot^2 A + \cot^2 C$ is minimum.

$$C = 180° - A - B = 180° - 2A$$

taking $A = B$ as a condition to evaluate A. Here $\cot^2 A + \cot^2 2A$ is minimum. Differentiating with respect to A and equating to zero, we get

$$4 \cos^4 A + 2 \cos^2 A - 1 = 0$$

The solution of this quadratic equation yields

$$A = 56° \ 13' \ 40''$$

Therefore, an isosceles triangle with base angles equal to 56° 13′ 40″ is the correct shape of the triangle to give the least error in computing the sides from the angles. In practice, however, it is more difficult to set this angle in the field and, hence, an equilateral triangle is preferred. The triangulation stations should be located such that no angle is less than 30° and more than 120°.

17.4 Strength of Figure

In a triangulation system, the base line and angles of the triangles forming the system are measured. These are subjected to errors, and the computed length of a side may be in error because of errors in the measurement. The triangulation system may be such that the length of a side can be computed by going through many paths of the triangulation system. It is important to choose the path that gives the least error.

The strength of figure of a triangulation system may be defined as the number that gives the least error in the computed length of the last line of the system due to the shape and composition of figures.

When we compute the length of an unknown side using the sine rule, we use the sines of two angles. These are known as the distance angles. Thus, if a is the known side and b is the computed side, then

$$b = a \ \sin B / \sin A$$

B and A are known as the distance angles. When computed this way, the angle C has no effect on computations. It is advisable to have the distance angles large and non-effective angle small for greater accuracy in the computation of the side.

17.4.1 Strength of Figure Criterion

In order to develop a criterion for the strength of figure, US Coast and Geodetic Survey developed a method to calculate the same. This is based upon the assumption that all angles are measured with equal precision. The strength of figure is, then, assumed to depend upon many other factors such as the size of distance angles, number of observed angles, number of geometric conditions, and so on. The US Coast and Geodetic Survey method is based upon the expression for the square of the most probable error, L^2, which would occur in the sixth decimal place of the logarithm of any side. The expression is

$$L^2 = \frac{4d^2}{3} \frac{D-C}{D} \sum (\delta A^2 + \delta A \delta B + \delta C^2)$$

$$= \frac{4d^2 R}{3}$$

where $R = \dfrac{D-C}{D} \sum (\delta A^2 + \delta A \delta B + \delta C^2)$

$$= \frac{(D-C)a}{D}$$

In these equations, R is the strength of figure, d is the probable error in an observed direction in seconds, D is the number of angles observed, excluding those for the known side, δA and δB are the tabulated differences in the sixth place in the logarithmic sine of the respective distance angles, $a = \Sigma(\delta A^2 + \delta A\ \delta B + \delta C^2)$ is the value of the probable error of an observed direction in seconds for a particular chain of angles (Table 17.2 gives these values for different values of angles), C is the number of geometric conditions—sum of side and angle conditions—and is calculated from $C = (n' - s' + 1) + (n - 2s + 3)$, where n is the total number of lines, n' is the number of lines observed in both directions, s is the total number of stations, and s' is the number of occupied stations. Here, $(n' - s' + 1)$ is the number of angle conditions and $(n - 2s + 3)$ is the number of side conditions.

The relative strength of figure can be computed in terms of R. The lower the value of R, the stronger the figure. R can be calculated for different routes for computations and the figure having the lowest R can be chosen for computation to get the least error in the computed value. The following examples illustrate these concepts.

Example 17.1   In a triangle ABC, AB is the base line and the three angles A, B, and C are 70°, 45°, and 65°, respectively. Find the strength of figure.

Solution   In a single triangle, $n = 3$, $s = 3$, $n' = 3$, $s' = 3$.

$$C = (n - s + 1) + (n' - 2s' + 3) = (3 - 3 + 1) + (3 - 2 \times 3 + 3) = 1$$
$$D = 6 - 2 = 4 \quad \text{(there are six angles and sides; two known base lines}$$
$$\text{are not to be counted)}$$

For $A = 70°$ and $B = 45°$, $a = 7$ (from Table 7.2)

$$R = (D - C)a/D = (4 - 1) \times 7/4 = 5.25$$

Example 17.2   The probable error in the measurement of an angle is 1″. Find the maximum value of R if the maximum probable error in computing the sides is to be limited to 1 in 20,000.

Solution   Probable error stipulated = 1 in 20,000 = 0.00005
L is the difference in the sixth place of the logarithm.

log 1 = 0.000000

log(1 + 0.00005) = 0.0000217

Difference in the sixth place = 21, $L = 21$, $L^2 = 441$

Now

$$L^2 = (4/3)d^2R, \quad R = (3/4)L^2/d^2, \quad d = 1″$$

Therefore

$$R = (3/4)441 = 331$$

Example 17.3   A triangulation system consists of two quadrilaterals as shown in Fig. 17.5, where AC is the base line. Find the value of $(D - C)/D$ for the system.

Solution   Total number of lines, $n = 11$. Total number of stations, $s = 6$. Number of occupied stations = 6. Number of stations observed in both directions is 11.

$$D = (11 \times 2 - 2) = 20$$
$$C = (11 - 6 + 1) + (11 - 2 \times 6 + 3) = 8$$
$$(D - C)/D = (20 - 8)/20 = 0.6$$

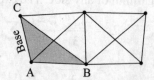

Fig. 17.5   Quadrilaterals for Example 17.3

Example 17.4   In the triangulation network shown in Fig. 17.6, AB is the base line. The length of DC is to be computed. Find the route that gives the most accurate value in computing the side DC.

Table 17.2 Table for determining relative strength of triangulation figures (values of $\delta A^2 + \delta A\,\delta B + \delta B^2$)

	10°	12°	14°	16°	18°	20°	22°	24°	26°	28°	30°	35°	40°	45°	50°	55°	60°	65°	70°	75°	80°	85°	90°
10°	428																						
12°	359	359																					
14°	315	295	253																				
16°	284	253	214	187																			
18°	262	225	187	162	143																		
20°	245	204	168	143	126	113																	
22°	232	189	153	130	113	100	91																
24°	221	177	142	119	103	91	81	74															
26°	213	167	134	111	95	83	74	67	61														
28°	206	160	126	104	89	77	68	61	56	51													
30°	199	153	120	99	83	72	63	57	51	47	43												
35°	188	148	115	94	79	68	59	53	48	43	40	33											
40°	179	137	106	85	71	60	52	46	41	37	33	27	23										
45°	172	129	99	79	65	54	47	41	36	32	29	23	19	16									
50°	167	124	93	74	60	50	43	37	32	28	25	20	16	13	11								
55°	162	119	89	70	57	47	39	34	29	26	23	18	14	11	9	8							
60°	159	115	86	67	54	44	37	32	27	24	21	16	12	10	8	7	5						
65°	155	112	83	64	51	42	35	30	25	22	19	14	11	9	7	5	4	4					
70°	152	109	80	62	49	40	33	28	24	21	18	13	10	7	6	5	4	3	2				
75°	150	106	78	60	48	38	32	27	23	19	17	12	9	7	5	4	3	2	2	1			
80°	147	104	76	58	46	37	30	25	21	18	16	11	8	6	4	3	2	2	1	1	1		
85°	145	102	74	57	45	36	29	24	20	17	15	10	7	5	3	2	2	1	1	1	0	0	
90°	143	100	73	55	43	34	27	23	19	16	14	9	7	4	3	2	1	1	1	0	0	0	0
95°	140	98	71	54	42	33	26	22	19	16	13	9	6	4	3	2	1	1	0	0	0	0	0

Table 17.2 (Contd)

	10°	12°	14°	16°	18°	20°	22°	24°	26°	28°	30°	35°	40°	45°	50°	55°	60°	65°	70°	75°	80°	85°	90°
100°	138	95	68	51	40	31	25	21	17	14	12	8	6	4	3	2	1	1	0	0	0		
105°	136	93	67	50	39	30	25	20	17	14	12	8	5	4	2	2	1	1	0	0			
110°	134	91	65	49	38	30	24	19	16	13	11	7	5	3	2	2	1	1	1				
115°	132	89	64	48	37	29	23	19	15	13	11	7	5	3	2	2	1	1					
120°	129	88	62	46	36	28	22	18	14	12	10	7	5	3	2	2	1						
125°	127	86	61	45	35	27	22	18	14	12	10	7	5	4	3	2							
130°	125	84	59	41	34	26	21	17	14	12	10	7	5	4	3								
135°	122	82	58	43	33	26	21	17	14	12	10	7	6	4									
140°	119	80	56	42	32	25	20	17	15	12	10	8	6										
145°	116	77	55	41	32	25	21	17	16	13	11	9											
150°	112	75	54	40	32	26	21	18	17	15	13												
152°	111	75	53	40	32	26	22	19	19	16													
154°	110	74	53	41	33	27	23	21	19														
156°	108	74	54	42	34	28	25	22															
158°	107	74	54	43	35	30	27																
160°	107	74	56	45	38	33																	
162°	107	76	59	48	42																		
164°	109	79	63	54																			
166°	113	86	71																				
168°	122	98																					
170°	143																						

Solution There are many methods to calculate the length of line CD. AB is the known base line and all calculations have to make use of this line. These are as follows.

1. From triangle ABD, calculate BD and then calculate CD from triangle BCD.
2. From triangle ABD, calculate AD and then calculate CD from triangle ACD.
3. From triangle ABC, calculate AC and then calculate CD from triangle ACD.
4. From triangle ABC, calculate BC and then calculate CD from triangle BCD.

1. In triangle ABD, the distance angles are 55° and 58°. $\Sigma(\delta A^2 + \delta A\ \delta B + \delta B^2) = a$ can be read from Table 17.2, $a = 5.8$. In triangle BCD, the distance angles are 28° and 125°.

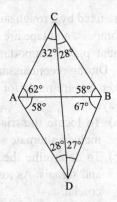

Fig. 17.6 Network for Example 17.4

$$a = 12 \quad \text{(from Table 17.2)}$$

$$\Sigma a = 12 + 5.8 = 17.8$$

There are two triangles with one base line.

$$n = 6, \quad s = 4, \quad n' = 6, \quad s' = 4 \quad (s = 4 \text{ because two parameters of the base line are not counted)}$$

$$C = (n - s + 1) + (n' - 2s' + 3) = (6 - 4 + 1) + (6 - 2 \times 4 + 3) = 4$$

$$D = (6 \times 2) - 2 = 10$$

Strength of figure, $R = (D - C)/D \times a = (10 - 4)/10 \times 17.8 = 10.68$

2. In triangle ABD, the distance angles are 55° and 68°, $a = 4.4$. In triangle ACD, the distance angles are 32° and 120°, $a = 8.8$.

$$\Sigma a = 13.2$$

We note that the value of $(D - C)/D$ does not change for any of the combinations.

$$(D - C)/D = (10 - 4)/10 = 0.6$$

Strength of figure, $R = 0.6 \times 13.2 = 7.92$

3. In triangle ABC, the distance angles are 60° and 58°, $a = 4.4$. In triangle ACD, the distance angles are 28 and 120, $a = 12$.

$$\Sigma a = 16.4$$

Strength of figure, $R = 0.6 \times 16.4 = 9.84$

4. In triangle ABC, the distance angles are 60° and 62°, $a = 4$. In triangle BCD, the distance angles are 27° and 125°, $a = 14$.

$$\Sigma a = 18$$

Strength of figure, $R = 0.6 \times 18 = 10.8$

The best method to calculate CD is from route 2, as the value of R is the least in that route.

17.5 Triangulation Fieldwork

Reconnaissance

Reconnaissance forms an important part of triangulation work. Thorough reconnaissance will save time, effort, and money. In reconnaissance, triangulation stations are fixed by a rough survey of the terrain. Instruments such as a vernier theodolite, compass, and tape are carried by the reconnaissance party. Experience and judgement play an important role in taking decisions during reconnaissance.

During reconnaissance, many important things are recorded and field notes and sketches are made for further analysis. The data collected helps in the following ways.

(a) To locate the triangulation stations, which is convenient for surveys that include alternate stations.

(b) To determine the intervisibility of stations and the requirement of towers and signals. As towers are expensive to construct, station selection becomes crucial.

(c) To determine the location for the base line. Measuring a base line is also expensive; hence, a good location for the base line is important.

(d) To find permanent marks or features that can reference the triangulation stations.

Hilly areas are easy to survey as stations can be located over peaks having intervisibility. In the plains, the selection of stations and base line has to be carefully considered. After reconnaissance, the data collected is analysed and the final location for triangulation stations and base line decided.

Fixing and marking stations

Depending upon the nature and purpose of triangulation, stations are fixed. Triangulation stations are permanently marked. The marking is done on a copper or brass plate with the station identification clearly marked. The plates are set both underground and on the surface. Each station is also identified by local reference marks which are the permanent features of that area.

In the case of rocky terrains, a hole is drilled into the rock and the underground mark is set inside. The hole is filled and a surface mark is also established. If the soil is soft, it is excavated and a bolt or copper plate is set in a concrete base inside. A similar mark is also made on the surface. In either case, the station should be clearly identifiable and a way to reach the station should be sketched clearly and maintained.

Intervisibility of stations

The intervisibility of stations located on peaks or ridges does not pose a problem. Intervisibility is to be checked in the case of stations having small differences in elevation in plains or when the distances are large. As triangulation stations are generally separated by large distances, the intervening ground, if it is high, also affects intervisibility. Curvature and refraction are also to be taken into account. In addition, the line of sight should not run too close to the ground anywhere. A minimum distance of 3 m above ground must be maintained.

The line of sight remains horizontal while the ground surface follows a curved path. The effect of curvature is opposite to that of refraction. The effect of refraction is about one-seventh of that due to curvature. The net effect is

$$h = (1 - 2r)D^2/2R$$

where r is the refraction coefficient, D is the distance between the station and the point of tangency, and R is the radius of the earth [Fig. 17.7(a)]. The value of r is between 0.07 and 0.08. Therefore, $h = 0.06735D^2$, where h is in metres and D is in kilometres.

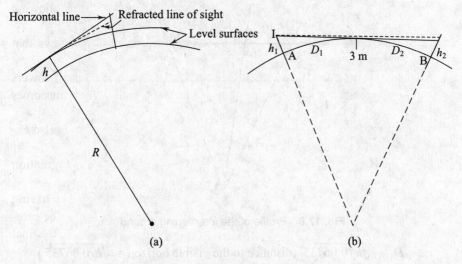

Fig. 17.7 Intervisibility of stations

Distance between stations

If the instrument is at a height h_1 and the signal at the other station is at a height h_2 [Fig. 17.7(b)], then

$$D_1 = \sqrt{h_1/0.06735}$$

and $\quad D_2 = \sqrt{h_2/0.06735}$

where D_1 and D_2 are distances to the instrument station and the signal station, respectively.

$$\text{Height of signal} = 0.06735D_2^2$$

Knowing the value of h and the RL of the signal station, it is possible to determine if the signal at B needs to be further elevated for visibility. The signal height is kept at least 3 m above the calculated value to avoid the line of sight passing too close to the ground.

Profile of the intervening ground

If the intervening ground between the instrument station and the signal station has peaks (Fig. 17.8), then the distances from the instrument and the elevations of these peaks need to be determined. At each of these points, the position of the line of sight has to be checked using the above formula.

The procedure for checking intervisibility can be explained with the help of Fig. 17.8. Let A be the instrument station and B be the signal station. Let C be an intervening peak. The horizontal line of sight from A will strike the mean level surface through a at e. If h_1 is the height of point A above e, then the distance D_1 will be calculated as

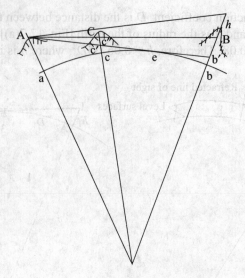

Fig. 17.8 Profile of the intervening ground

$D_1 = \sqrt{h_1/0.06735}$ (distance to the visible horizon $= \sqrt{h/0.06735}$)

Knowing D_1, D_2 can be calculated as

$$D_2 = (D - D_1)$$

a, c, and b are points on the level surface through e on the plumb lines through A, C, and B.

$$ec = D_1 - AC$$

If c′ is a point on the line ae, then

$$Cc' = 0.06735 (D_1 - AC)^2$$

We consider the line of sight AB. Let this intersect the plumb line through c at c″. Then, from triangle Ab′B,

$$cc'' = Bb' \times AC/D$$

Cc′ + c′c″ is the elevation of the line of sight over ab. If the elevation of C is more than this, then the line of sight AB fails to clear the peak at C. At C, the line of sight has to go up by Cc″ + any minimum clearance specified. The normal clearance is 2 to 3 m. Let this be h'. The line of sight AB should go up by Cc″ + h' at C. At B, this will give (Cc″ + h')D/AC. This is the height of signal at B above the point B.

McCaw's method Captain McCaw developed a simple solution to the problem of intervisibility in the case of an intervening high ground. Referring to Fig. 17.9, the height h of the line of sight at the peak point is given by

$$h = 1/2(h_2 + h_1) + 1/2(h_2 - h_1)x/s - (s^2 - x^2) \operatorname{cosec}^2 \xi (1 - 2r/R)$$

where h_1 is the height of station A above datum, h_2 is the height of station B above datum, $2s$ is the distance between A and B, $(s + x)$ is the distance of obstruction from station A, $(s - x)$ is the distance of obstruction from station B, ξ is the zenith distance from A to B, and $(1 - 2r)/R = 0.06735$ is the coefficient required to adjust the effects of curvature and refraction. X, s, and R are in kilometres and the elevations h, h_1, and h_2 are in metres. In most cases, $\text{cosec}^2\xi = 1$. Hence, MacCaw's solution can be written as

$$h = 1/2(h_2 + h_1) + 1/2 (h_2 - h_1)x/s - (s^2 - x^2)(1 - 2r/R)$$

The following examples illustrate the computation of intervisibility.

Example 17.5 Two stations A and B are 72 km apart. The elevations of the stations A and B are 372 m and 458 m, respectively. The intervening ground has a uniform elevation of 328 m. Find the height of the signal required at B if the line of sight has to pass at least 3 m above the ground at all points.

Solution Minimum elevation of line of sight = 328 + 3 = 331 m

If this elevation of line of sight is taken as datum, then

   Elevation of A = 372 – 331 = 41 m

   Elevation of B = 418 – 331 = 87 m

The line of sight from A will strike the ground at a distance of D_1.

$$D_1 = \sqrt{h/0.06735} = 3.833\sqrt{h_1} = 3.8533\sqrt{41} = 24.673 \text{ km}$$

The distance D_2 from C to B is given by

$$D_2 = 72 - 24.673 = 47.327 \text{ km}$$

For this distance, the height of the point in the line of sight at B is given by
$H = 0.06735 D_2^2 = 0.06735 \times (47.327)^2 = 150.85 \text{ m}$

   Elevation of signal at B = 331 + 150.85 = 481.85 m

   Height of signal at B = 481.85 – 418 = 23.85 m

Example 17.6 Two stations A and B are 100 km apart. The elevation of A is 185 m and that of B is 885 m. In the line of sight between A and B, there are two intervening high points C and D. C is 42 km from A and D is 81 km from A. The elevations of peaks C and D are 318 m and 750 m. Check whether the line of sight from A to B clears the peaks with a minimum clearance of 3 m above ground level. Determine the height of the signal at B for intervisibility.

Solution The situation is shown in Fig. 17.9.
Let the distance to the visible horizon from A be equal to D_1. Then,
$D_1 = 3.8533\sqrt{185} = 52.41 \text{ km}$

   Ae = 52.41 km, Ce = 52.41 – 42 = 10.41 km, De = 81 – 52.41 = 28.59 km

   Be = 100 – 52.41 = 47.59 km

If the horizontal line of sight Ae intersects the verticals through C, D, and B at c', d', and b', then

   Cc' = 0.06735 × 10.41² = 7.2986 m

   Dd' = 0.06735 × 28.59² = 55.05 m

   Bb' = 0.06735 × 47.59² = 152.022 m

   Bb' = 885 – 152.022 = 732.978 m

Taking the line of sight as AB, from triangle Ab'B,

   c'c" = 732.978 × 42/100 = 307.85 m

   D'd" = 732.978 × 81/100 = 593.71 m

   Cc" = 307.85 + 7.298 = 315.143 m

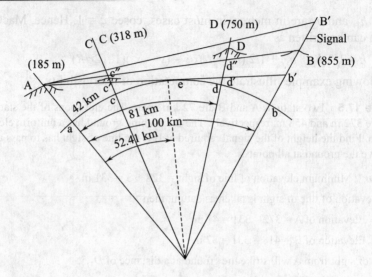

Fig. 17.9   Figure for Example 17.6

Dd" = 593.71 + 55.05 = 648.76 m

The line of sight clears peak C by

315.143 – 310 = 5.143 m

but fails to clear peak D by

655 – 648.76 = 6.24 m

Considering that the line of sight must pass 3 m clear of the ground surface, the height of the line of sight at D must be

6.24 + 3 = 9.24 m

Height of signal at B = 9.24 × 100/81 = 11.41 m

Example 17.7   Two triangulation stations A and B are 90 km apart. The elevation of A is 418.85 m and that of B is 702.63 m. An intervening peak is 66 km from A and has an elevation of 524.6 m. Check the intervisibility from A to B and also the required height of the signal at B for clear visibility. The line of sight must pass at least 2.5 m above ground at all points.

Solution   The profile is shown in Fig. 17.10.

Distance of visible horizon from A = 3.8533 $\sqrt{418}$ = 78.78 km

Distance ce = 78.78 – 66 = 12.78 km

Height from c to line of sight cc' = 0.06735 × (12.78)^2 = 11 m

Distance eb = 90 – 78.78 = 11.22 km

Distance bb' = 0.06735 × $\sqrt{11.22}$ = 8.48 m

Bb' = 702.63 – 8.48 = 694.15 m

From similar triangles Ab'B and Ac'c",

c'c" = 694.15 × 66/90 = 509.04 m

Height of line of sight at C = cc" = 509.04 + 11 = 520.04 m

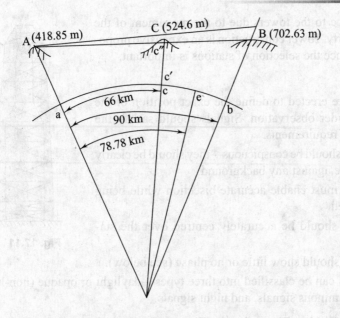

Fig. 17.10 Profile of ground (Example 17.7)

The line of sight fails to clear the peak by 524.6 – 520.04 = 4.56 m. Taking a minimum of 2.5 m clearance from the ground, the height of the line of sight at C is 4.56 + 3 = 7.56 m above the line of sight AB.

Height of the signal at B = 7.56 × 90/66 = 10.31 m

Captain McCaw's solution

According to Captain McCaw, the height of the line of sight at C, *h*, is given by

$$h = 1/2(h_1 + h_2) + 1/2(h_2 - h_1)x/s - (s^2 - x^2) \times 0.06735 \text{ (taking cosec}^2\xi = 1)$$

Here

$$h_1 = 418.85 \text{ m}, \quad h_2 = 702.63, \quad s = 45 \text{ km}, \quad x = 66 - 45 = 21 \text{ km}$$

Substituting these values,

$$h = 1/2(418.85 + 702.63) + 1/2(702.63 - 418.85) \, 21/45 - (45^2 - 21^2) \times 0.06735$$

$$= 520.273 \text{ m}$$

The signal fails to clear the peak by

$$524.6 - 520.273 = 4.327 \text{ m}$$

Considering the 2.5 m clearance from the ground,

Height of line of sight at C = 6.827 m

Height of the signal at B = 6.827 × 90/66 = 9.31 m

Towers

Towers (Fig. 17.11) are temporary structures erected for the purpose of surveying to elevate instrument and signals. Short and medium-height towers may be constructed of timber. Tall towers are made of steel. In the case of large towers, two independent towers are constructed. The inner tower with a platform is made for the instruments only and the outer tower is meant for the survey party. This is done to avoid any disturbance to the towers due to the movement of the survey party. Tower construction is an expensive proposition; hence the selection of stations is important.

Signals

Signals are erected to define the exact position of the station under observation. Signals should satisfy the following requirements.

(a) They should be conspicuous—they should be clearly visible against any background.

(b) They must enable accurate bisection while being sighted.

(c) They should be accurately centred over the station.

(d) They should show little or no phase (see below).

Fig. 17.11 Tower

Signals can be classified into three types—daylight or opaque (non-luminous) signals, luminous signals, and night signals.

Opaque signals These signals [Fig. 17.12(a)] can be used for distances up to 30 km. They can be various forms of masts, targets, or tin-cone types. For small distances, they can be round poles mounted on tripods or quadrupods with the pole painted bright with alternate dark and white bands. For larger distances, target signals with two rectangular panels placed at right angles and mounted on poles over tripods are used. The panels are made of cloth stretched between the sides of a wooden frame.

To make the signal conspicuous, a flag is tied to the top of the signal. The signal should be dark in colour to be visible against the sky. The diameter of the signal is about $1.3D$ to $1.9D$ (cm), where D is the distance in kilometres. The height of the signal is about $13.3D$ (cm), the vertical angle to the signal being at least $30°$.

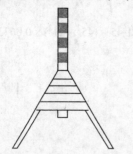

Fig. 17.12(a) Opaque signal

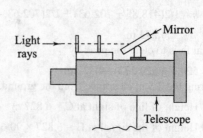

Fig. 17.12(b) Heliotrope

Luminous signals These are also known as sun signals as they are based upon the reflection of the sun's rays from a mirror. They can be used for long distances. Heliotrope and heliograph are special cases of luminous signals. A heliotrope [Fig. 17.12(b)] essentially consists of a mirror for reflecting the sun's rays. It may have a telescope attached to it. Luminous signals are generally used in higher order triangulations. Their limitation is their dependence on the availability of bright sunlight. Luminous signals work better towards the end of the day. As the sun moves constantly over the sky, the signal has to be adjusted frequently to get a better reflection, which requires additional manpower. The signal has to be directed towards the station using the instrument. This is achieved by sending light signals from the station towards the signal.

Night signals   These are used for functioning at night. They consist of oil lamps with reflectors or optical collimators. They can be used for large distances up to 80 km. An acetylene lamp, designed by Captain McCaw, is commonly used. Various types of electrical lamps can also be used as night signals.

Phase correction

The phase of a signal is caused due to the signal being partly illuminated due to lateral illumination. The observer sees the bright part of the signal and bisects it. This leads to a shift of the exact bisection point from the intended centre. There can be two cases in practice: (a) when the bright portion seen by the observer is bisected; (b) when the visible bright line is bisected.

When the visible bright portion is bisected   The instrument (observer) is at A and the signal is at B. The observer sees only the bright portion DE and bisects it at

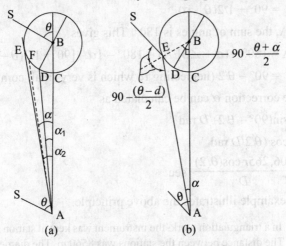

Fig. 17.13   Phase correction

F [Fig. 17.13(a)]. C is the centre of the signal in line with AB. θ is the inclination of the sun's rays to the line AB. α is the angle to the centre of the observed portion. α_1 and α_2 are the angles made by the lines AD and AE with AB. AE is the line of sight and r is the radius of the signal. The correction to the angle α is measured as

$$\alpha = \alpha_1 + (\alpha_2 - \alpha_1)/2 = (\alpha_2 + \alpha_1)/2$$

If D is the distance between the stations

$$\alpha_2 = r/D \text{ (in rad)}$$
$$\alpha_1 = r \sin(90° - \alpha)/D \text{ (in rad)} = r \cos\alpha D$$

$$\alpha = \frac{1}{2}\left(\frac{r}{D} + \frac{r + \cos\alpha}{D}\right) = \frac{r(1 + \cos\alpha)}{2D}$$

$$= \frac{r\cos^2(\alpha/2)}{D} \text{ rad} = \frac{r\cos^2(\alpha/2)}{D\sin 1''} \text{ sec}$$

$$= \frac{206,265r\cos^2(\alpha/2)}{D} \text{ sec}$$

When the observation is made on the bright line In this case the line of sight is AE [Fig. 17.13(b)]. The bright line formed by the reflected rays is SE. With the same notations used earlier,

$$\alpha = \text{phase correction} = \angle EAB$$

The lines SE and S'E are parallel and

$$\angle SEA + \angle S'EA = 180°$$

This gives

$$\angle SEA = 180° - (\theta - \alpha)$$

As the line BE bisects $\angle$SEA,

$$\angle BEA = 180° - (1/2) \angle SEA$$
$$\angle BEA = 180° - 1/2[180° - (\theta - \alpha)]$$
$$= 90° + 1/2(\theta - \alpha)$$

In triangle BEA, the sum of angles is 180°. This gives

$$\angle EBA = 180° - (\alpha + \angle BEA) = 180° - \{\alpha - [90°+ 1/2(\theta - \alpha)]\}$$
$$= 90° - \theta/2 \text{ (neglecting } \alpha, \text{ which is very small compared to } \theta)$$

Now, the phase correction α can be calculated as

$$\alpha = r \sin(90° - \theta/2)/D \text{ rad}$$
$$= r \cos (\theta/2/D \text{ rad}$$
$$= \frac{206,265r\cos(\theta/2)}{D} \text{ sec}$$

The following example illustrates the above principle.

Example 17.8 In a triangulation work, the instrument was kept at station A and the signal was at station B. The distance between the stations was 8560 m. The diameter of the signal at B was 14 cm. The observations were taken when the sun made an angle of 50° with the line AB. Find the phase correction to the angle if (a) the observation was made on the bright portion of the signal and (b) the observation was made on the bright line.

Solution (a) When the bright portion of the signal is observed, the phase correction is given by

$$\alpha = 206,265r \cos^2(\theta/2)/D$$

where r, the radius of the signal = 7 cm, D = 8560 m = 856,000 cm, and $\theta = 50°$.

$$\alpha = 206,265 \times 7 \times \cos^2(50°)/856,000 = 1.4''$$

(b) When the bright line is observed,

$$\alpha = 206,265r \cos(\theta/2)/D$$
$$= 206,265 \times 7 \times \cos 50°/856,000 = 1.53''$$

17.6 Base Line Measurement

Measuring the base line is the most important part of a triangulation survey. The base line must be measured very accurately so that the other sides calculated from the base line and the angles are accurate. The technique of measuring base lines has changed over the years and continues to change with the availability of better and high precision equipment. Base lines are measured using the following.

(a) Rigid bars

(b) Steel and invar tapes and wires

(c) Electronic tacheometry or electromagnetic distance measurement

Rigid bars

Rigid bars were used earlier in the GTS (Great Trigonometrical Survey) of India. This is an obsolete technique now. Colby's apparatus (Fig. 17.14) was used in this technique with the following procedure.

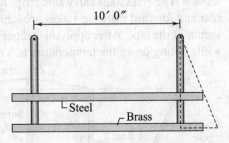

Fig. 17.14 Colby's apparatus

The apparatus consisted of two 10-ft-long bars, one of steel and the other of brass. The two bars were riveted together at the centre and held by two short bars at each end. The coefficient of thermal expansion being known, the distance between the bars was so adjusted that even with a change in temperature, the distance between two marks on the short bars did not change. The distance between the two marks was kept exactly at 10′. A number of such bars were employed to measure the base line. When two bars were placed together, the gap between the two marks was kept constant by a framework similar to that of the main bar. Microscopes were used to set the marks on the rods with their cross hairs. A number of such bars were used together to measure the base line by repeating the measurement with the rods.

Flexible apparatus

The flexible apparatus consists of steel or invar tapes or wires. Invar tapes are more popular due to their low coefficient of expansion. Tapes are longer compared to bars and can measure a larger length at one time. They speed up the work and are hence less expensive.

Steel tapes are 30–100 m long. Their coefficient of thermal expansion is $11.6 \times 10^{-6}/°C$. As the temperature of the tape cannot be accurately measured, it is best used in overcast conditions. The air and tape temperatures remain the same in such conditions.

Invar tapes and wires are more suitable, as the coefficient of thermal expansion is very small. An invar tape is a steel alloy containing 36 per cent nickel. Its coefficient of thermal expansion is about $0.64 \times 10^{-6}/°C$. The coefficient of thermal expansion for a tape is accurately determined before and during use at frequent intervals. They also come in varying lengths of 30–100 m. The ends of the tapes have millimetre divisions at both the ends. Invar has the disadvantage that it is very soft and hence, can get easily kinked. It is wound upon large reels for safe carriage.

To measure base lines, at least two tapes are required. One tape is used for field-work while the other is used for checking the field tape. Tripods for holding the tape, stakes, and a number of thermometers are required. A spring for measuring the tension in the tapes is also required. Many methods are used for measuring base lines with tapes. Of these, Wheeler's apparatus and Jaderin's method are popular.

Wheeler's apparatus   It consists of two end stakes—of these two, one is a strain-ing pole. Stakes are also kept in between, at about 5–15 m intervals, depending upon the length of the tape. The poles are either kept at the same level or on a uniform slope. The stakes carry hooks on which the tape is supported. The straining pole carries a spring balance to measure the tension as well as a straining weight to apply tension. The end stakes carry zinc strips to accurately mark the ends of the tape. The rear and forward marking stakes are kept at a distance approximately equal to the length of the tape. After applying proper tension, the ends of the tapes are marked, while noting down the temperature at a number of points. (see Fig. 17.16)

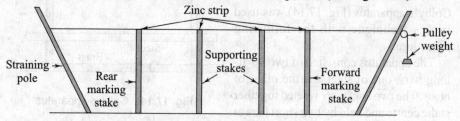

Fig. 17.15   Wheeler's apparatus

Jaderin's apparatus   It consists of two straining tripods and two measuring tri-pods. The ends of the tape are kept on the measuring tripods and tension is applied using weights at both the ends and measured with a spring balance. The temperature is measured at the time of tape measurement. The levels of the measuring tripods are measured using a level. The two ends of the tape are marked on the measuring tripods. (see Fig. 17.16)

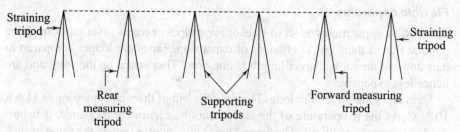

Fig. 17.16   Jaderin's method

EDM instruments

EDM instruments with much greater accuracy are now available for geodetic distance measurements (Chapter 23). EDM instruments come either independently or along with electronic theodolites. They use different types of waves in the electromagnetic spectrum including visible waves, microwaves, and radio waves.

They measure distances using the phase difference method. Geodimeters with a range of about 25 km, tellurometers, and higher-range distomats can be used for distance measurements. The accuracy of distance measurement varies from 5 mm ($\pm$ 1 mm/km) to 10 m ($\pm$ 3 mm/km).

17.6.1 Tape Corrections

Once the base line is measured, the measured length has to be corrected for various factors, which are different from the parameters on which the tape is standardized. Some of these have been discussed in Chapter 2. The essential corrections are as follows.

Correction for absolute length   This correction is applied when the length of the tape is different from its designated length. If the actual length is more, then the measured length will be recorded as too short and the correction will be positive. The correction will be subtractive if the actual length is less than the designated length. The correction is given as

$$C_a = Lc/l$$

where L is the measured length, c is the correction per tape length, and l is the designated length of the tape.

Correction for slope   If the two ends of the tape are not in the same horizontal plane, then the correction for slope is applied. This correction is always subtractive, as the measured length is always more than the equivalent horizontal length. If the difference between the ends of the tape is h (Fig. 17.17), then

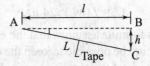

Fig. 17.17   Correction for vertical alignment

$$\text{Correction} = AC - AB = L - \sqrt{L^2 - h^2} = -L - L\left(1 - \frac{h^2}{2L^2} - \frac{h^4}{8L^4} - \cdots\right)$$

$$= \frac{h^2}{2L} + \frac{h^4}{8L^3}$$

Correction for vertical alignment

$$C_v = \frac{h^2}{2L} + \frac{h^4}{8L^3}$$

For flatter slopes, the second term can be neglected, giving the value of C_v as

$$C_v = h^2/2L$$

If the slope is given in terms of angle α, then

$$C_v = L(1 - \cos\alpha) = 2L\sin^2(\alpha/2)$$

Correction for temperature   The correction for temperature is given by

$$C_t = L\alpha T$$

where L is the measured length, α is the coefficient of thermal expansion for the material of the tape or wire, and $T = T_m - T_s$. T_m is the average temperature during field measurement and T_s is the temperature of tape standardization. The correction can be positive or negative depending upon the signs of T_m and T_s.

Correction for sag Flexible materials, such as tape and wires, tend to sag when supported between stakes during measurement. The measured length is always more than the true length due to the sag (Fig. 17.18). Sag correcton is given as

$$C_s = L - l = \frac{8h^2}{3l}$$

Assuming the curve to be a flat parabola,

$$L = l + 8h^2/3l^2$$

From Chapter 2, we have seen that

$$h = wLl/8P$$

where w is the weight of the tape per unit length and P is the tension applied to the ends of the tape. From this

$$C_s = L(wL)^2/24P^2$$

If the length L of the tape is suspended over n spans, then

$$C_s = L(wL)^2/24\, n^2P^2$$

This formula assumes that the ends of the tape are at the same level. Sometimes, it may not be possible to keep the ends of the tape at the same level. If the line joining the ends of the tape is at an inclination θ to the horizontal, then

$$C_s' = C_s\cos^2\theta\,(1 \pm wL\,\sin\theta/P)$$

The positive sign is used if the pull is applied at the higher end and the negative sign is used when the pull is applied at the lower end. If the tape is standardized in the form of a catenary, then the sag correction is given by

$$C_s = (w^2L^3/24P^2) - (w^2L^3/24P_0^2)$$

where P_0 is the standardizing tension and P is the tension applied in the field.

Correction for pull or tension Tapes are standardized at a certain pull P_s. If the pull P used during the measurement is greater than P_s, then the length of the tape is more and the correction is positive. If the pull P is less than P_s, then the length of the tape is less and the correction is negative. Correction for the pull is given by

$$C_p = (P - P_s)L/AE$$

where P is the pull applied during the field measurement, P_s is the pull of standardization, A is the area of cross section of the tape, and E is the modulus of elasticity of the material of the tape.

Reduction to sea level The horizontal distances calculated are reduced to mean sea level distances. Such distances are known as geodetic distances. This is done by (Fig. 17.19) reducing the distance from AB to A′B′ using the property of a circle.

$$L = (R + h)\theta \quad \text{and} \quad L' = R\theta$$
$$L' = LR/(R + h) = L/(1 + h/R)$$

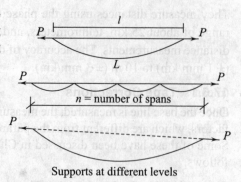

Supports at different levels

Fig. 17.18 Correction for sag

$$L' = L(1 + h/R)^{-1} = L(1 - h/R)$$

where L is the measured horizontal distance, θ is the angle subtended at the centre, h is the mean equivalent height above the MSL (mean sea level) of the line AB, L' is the geodetic distance A′B′, and R is the radius of the earth.

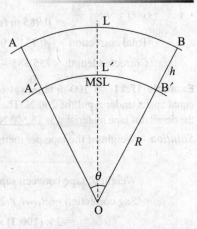

$$\text{Correction} = L - L' = L - L$$
$$(1 - h/R) = Lh/R$$

This is always subtractive.

The following examples illustrate the tape corrections discussed above.

Fig. 17.19 Reduction to sea level

Example 17.9   A 30-m-long tape has an actual length of 29.995 m. The tape was standardized at a temperature of 38°C. During measurement in the field, the temperature was 22°C. Find the true length of a line recorded as 8560 m, if the coefficient of thermal expansion of the material of the tape is $11.6 \times 10^{-6}/°C$.

Solution   Correction for absolute length = 29.995 − 30 = − 0.005 m

Correction for temperature = $30 \times 11.6 \times 10^{-6} \times (22 - 38) = -0.0055$

Total correction = 0.0105 m for a tape length

Total correction for the line = 0.0105 × 8560/30 = 2.396 m

True length of the line = 8560 − 2.396 = 8557.604 m

Example 17.10   A tape was used to measure a line and the length was recorded as 755.385 m. The tape was standardized at 27°C. The temperature during the measurement was 12°C. The line was measured along the slopes as follows:

  1° 30′ 120 m
  2° 10′ 248 m
  3° 30′ 136 m
  2° 45′ 135 m

and the remaining length was measured along a slope of 4° 30′. Find the true length of the line.

Solution   The correction for temperature is $L\alpha(T_m - T_s)$, where L is the length of the wire, α is the coefficient of its thermal expansion, T_m is the mean temperature during measurement, and T_s is the standardization temperature.

  Temperature correction = $755.385 \times 11.6 \times 10^{-6} \times (12 - 27)$

         = − 0.131 m (subtractive)

  Correction for slope = $L(1 - \cos \alpha)$

  Length for 4° 30′ slope = 755.385 − (120 + 248 + 136 + 135) = 116.385 m

  Slope correction = 120 (1 − cos 1° 30′) + 248 (1 − cos 2° 10′)

      + 136 (1 − cos 3° 30′) + 135 × (1 − cos 2° 45′)

      + 116.385 × (1 − cos 4° 30′)

$$= 0.985 \text{ m (subtractive)}$$

Total correction $= -0.131 - 0.985 = -1.116$ m

Corrected length $= 755.385 - 1.116 = 754.269$ m

Example 17.11   A 100-m tape was used to measure a line. The tape was hung in three equal spans under a pull of 200 N. The area of cross section of the tape was 8 sq. mm and the density of tape material was 78,500 N/cu. m. Find the sag correction.

Solution   Weight of the tape per metre length $=$ area $\times 1 \times 78,500$

$$= 8 \times 10^{-6} \times 78,500 = 0.628 \text{ N/m}$$

Weight of tape between supports $= 0.628 \times 100/3 = 20.933$ N

Sag correction $= nl_1(wl_1)^2/24P^2$

$$= 3 \times (100/3) \times (20.933)^2/(24 \times 2002)$$

$$= 0.046 \text{ m}$$

Example 17.12   A 50-m-long tape has been standardized at 25°C under a pull of 100 N. During field measurement, the tape was supported at two points A and B. The elevations of A and B were 110.385 and 110.12 m with respect to a local benchmark. Elevation of A above the mean sea level is 1163.853 m. The temperature and pull during the measurements were 42°C and 150 N, respectively. Find the corrected length of a tape length reduced to mean sea level.

Solution   Corrections have to be done for slope, sag, pull, temperature, and then reduced to the geodetic distance:

Correction for slope $= h^2/2L = (110.385 - 110.12)^2/2 \times 50 = 0.0026$ m (negative)

Correction for temperature $= L\alpha(T_m - T_s) = 50 \times 11 \times 10^{-6} \times (42 - 25)$

$$= 0.0038 \text{ m (positive)}$$

Correction for pull $= (P - P_0)^2/AE = (150 - 100)^2/(8 \times 10^{-6} \times 2 \times 10^{11})$

$$= 0.0015 \text{ m (positive)}$$

Correction for sag $= LW^2/24P^2 = 50 \times 31.4^2/(24 \times 150^2)$

$$= 0.091 \text{ m (negative)}$$

Total correction $= -0.0883$ m

Corrected length $= 50 - 0.0883 = 49.912$ m

Reduction at mean sea level $= Lh/R = 50 \times 1163.853/6370 \times 10^3$

$$= 0.0098 \text{ m}$$

Geodetic length $= 49.912 - 0.0098 = 49.903$ m

Example 17.13   A 50-m-long tape is suspended between level supports. The area of cross section of the tape is 0.08 sq. cm, the density is 78,500 N/cu. m, and Young's modulus is 200 GN/m². The standardization pull for the tape is 100 N. Find the normal tension of the tape. (Normal tension is the tension to be applied to balance the effect of sag.)

Solution   Correction for pull $= (P - P_0)L/AE$

Correction for sag $= l(wl)^2/24P^2$

The two corrections are opposite in nature. Equating the two, we get the pull to be applied to balance the effect of sag:

$$(P - P_0)l/AE = lW^2/24P^2$$

where $W = wl$

$$P = 0.2041W \sqrt{AE} / \sqrt{P - P_0}$$

As P appears on both sides of the equation, we solve this equation by trial and error.

$$A = 0.08 \text{ sq. cm} = 8 \times 10^{-6} \text{ m}^2$$

$$W = 8 \times 10^{-6} \times 50 \times 78,500 \text{ N} = 3.14 \text{ N}, E = 2 \times 10^{11} \text{ N/m}^2$$

Substituting these values,

$$P = 810.647/ \sqrt{P - 100}$$

Solving by trial and error,

$$P = 135.7 \text{ N}$$

17.7 Problems in Base Line Measurement

Some of the problems in measuring base lines are given below.

Extending a short base line   The base line is generally the shortest line in the triangulation system; the reason is the difficulty and expenses involved in measuring a long base line. Occasionally, some obstacles may also prevent the measurement of a long base line. The base line can be extended using the principles of triangulation itself. Figure 17.20 shows some of the methods used.

In Fig. 17.20(a), the base line AB is extended to CD by the triangles ABD and ABC. This can be extended to EF by the triangles CED and CFD and further to GH by triangles EGF and EHF. Any number of such extensions is possible.

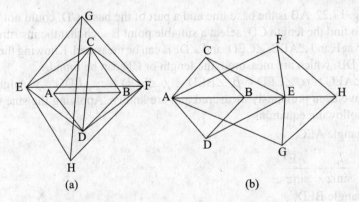

Fig. 17.20   Extending the base line

In Fig. 17.20(b), another method is shown for extending the base line in the same direction. The short base line AB is measured accurately. Two points C and D are selected such that the triangles ABC and ABD are well-conditioned. Select a point E along the line AB, by prolonging the line AB such that triangles ACE and ADE are also well-conditioned. AE can be calculated and it becomes the new, longer base line. Further extension to AF can be done, as shown in the figure, using a similar procedure.

Some more methods of extending the base line are shown in Fig. 17.21.

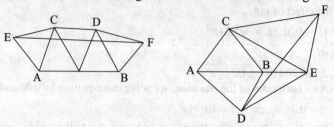

Fig. 17.21 Extending the base line

Determining the immeasurable part of the base Sometimes, a part of the base cannot be measured due to some other reason such as an obstruction. This can happen if no other base length is available nearby and one is compelled to select a base line with an obstruction. The length of this part of the base can be found using trigonometry.

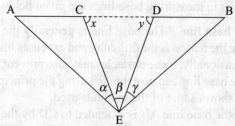

Fig. 17.22 Determining the length of a part of the base

In Fig. 17.22, AB is the base line and a part of the base, CD, could not be measured. To find the length CD, select a suitable point E such that the instrument can be set up at E and ∠AEC, ∠CED, and ∠DEB can be measured. Knowing the lengths AC and DB, which are measured, the length of CD can be found.

Let ∠AEC = α, ∠CED = β, ∠BED = γ, ∠ECD = z, ∠EDC = y. Angles α, β, and γ have been previously measured and are known. Applying the sine rule, we get the following equations.

From triangle AEC,

$$\frac{l_1}{\sin\alpha} = \frac{AE}{\sin z} \tag{17.1}$$

From triangle BED,

$$\frac{l_2}{\sin\gamma} = \frac{BE}{\sin y} \tag{17.2}$$

From triangle AED,

$$\frac{l_1 + x}{\sin(\alpha + \beta)} = \frac{AE}{\sin y} \tag{17.3}$$

From triangle BEC

$$\frac{l_2 + x}{\sin(\beta + \gamma)} = \frac{BE}{\sin z} \tag{17.4}$$

Dividing Eqn (17.1) by Eqn (17.3), we get

$$\frac{l_1 \sin(\alpha + \beta)}{(l_1 + x)\sin\alpha} = \frac{\sin y}{\sin z} \tag{17.5}$$

From Eqns (17.2) and (17.4),

$$\frac{l_2 \sin(\beta + \gamma)}{(l_2 + x)\sin\gamma} = \frac{\sin z}{\sin y} \tag{17.6}$$

Multiplying Eqns (17.5) and (17.6), we get

$$\frac{l_1 \sin(\alpha + \beta) l_2 \sin(\beta + \gamma)}{(l_1 + x)\sin\alpha(l_2 + x)\sin\gamma} = 1$$

Let

$$\frac{\sin(\alpha + \beta)\sin(\beta + \gamma)}{\sin\alpha\sin\gamma} = M$$

We then get

$$l_1 l_2 M = (l_1 + x)(l_2 + x)$$

This reduces to

$$x^2 + (l_1 + l_2)x + l_1 l_2(1 - M) = 0$$

Solving this equation,

$$x = -\frac{l_1 + l_2}{2} \pm \frac{\sqrt{(l_1 + l_2)^2 - 4(l_1 l_2)(1 - M)}}{2}$$

$$= -\frac{l_1 + l_2}{2} + \frac{\sqrt{(l_1 - l_2)^2 + l_1 l_2 M}}{2}$$

Thus,

$$x = -\frac{l_1 + l_2}{2} + \frac{\sqrt{[(l_1 - l_2)/2]^2 + l_1 l_2 \sin(\alpha + \beta)\sin(\beta + \gamma)}}{\sin\alpha\sin\gamma}$$

Length from broken base If due to any reason, the direct length AC of the base line (Fig. 17.23) cannot be measured, then lengths AB and BC are measured, where B is a point near the base line. Thus, length AC can be determined from triangle ABC, knowing the lengths AB and BC and ∠ ABC. By the cosine rule,

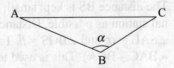

Fig. 17.23 Broken base

$$b^2 = a^2 + c^2 - 2ac \cos\alpha$$

$$a^2 + c^2 - b^2 + 2ac = 2ac - 2ac \cos\alpha \quad \text{(adding } 2ac \text{ to both sides and rearranging)}$$

$$(a + c - b)(a + c + b) = 2ac(1 - \cos\alpha)$$

$$(a + c - b) = \frac{2ac(1 - \cos\alpha)}{(a + c + b)}$$

$$= \frac{4ac(\sin^2\alpha/2)}{(a + c + b)}$$

$$b = (a + c) - \frac{4ac(\sin^2\alpha/2)}{(a + c + b)}$$

$(a + c + b)$ can be taken approximately equal to $2(a + c)$ and hence b can be calculated from the above equation.

Satellite station and reduction to centre Sometimes, a triangulation is fixed, for convenience, at a point where the instrument cannot be kept for taking measurements, such as a church spire. In such cases, a station near the triangulation station is selected. Measurements are made from this station and the angle at the original triangulation station is calculated. Such a station is known as a *satellite or false station*. The process of finding the angle at the original station is known as reduction to centre.

In Fig. 17.24, B is the station where the instrument cannot be kept. A and C are two other stations of the triangulation system. The length AC is calculated. S is the satellite station. ∠ABC is to be found so that the lengths AB and BC can be calculated.

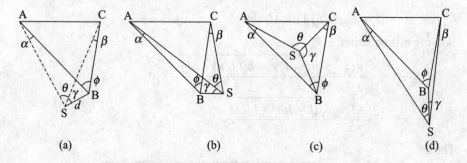

(a)   (b)   (c)   (d)

Fig. 17.24 Satellite station and reduction to centre

To find ∠ABC, the instrument is kept at S and ∠ASC and ∠BSC are measured. The distance BS is kept small; the satellite station is selected as close to the original station as possible. Distance BS = d. Let ∠ASC = θ, ∠BSC = γ, ∠ABC = ϕ, ∠SAB = α, and ∠BCS = β. To start with ∠ABC is approximately taken as (180° − BAC − BCA). This is used to calculate the lengths of the unknown sides. From triangle ASB,

$$d/\sin\alpha = c/\sin(\theta + \gamma), \quad \sin\alpha = d\sin(\theta + \gamma)/c$$

From triangle BCS,

$$d/\sin\beta = a/\sin\gamma, \quad \sin\beta = d\sin\gamma/a$$

If the distance d is small, then the values of α and β will be small. The value of their sines will be equal to the value of their respective angles in radians. Thus,

$$\sin\alpha = \alpha \text{ rad} = \alpha \times 180° \times 3600/\pi \text{ sec} = 206{,}265\alpha \text{ sec}$$

Similarly,

$$\sin\beta = 206,265\beta \sec$$

Therefore,

$$\alpha = 206,265d \sin(\theta + \gamma)/c \sec \quad \beta = 206,265d \sin\gamma/a \sec$$

If AB and SC intersect at P, then from triangle ASP,

$$\angle APC = \alpha + \theta \quad \text{(sum of the interior angles)}$$

On similar lines, considering triangle BPC,

$$\angle APC = \angle ABC + \beta$$

From these two equations,

$$\angle ABC = \alpha + \theta - \beta$$

This derivation has been carried out for the position of S shown in Fig. 17.24(a). The position of S may vary as shown in Fig. 17.24(b), (c), or (d). On the same principles it can be shown that

$$\angle ABC = \theta + \beta - \alpha \qquad \text{[from Fig. 17.24(b)]}$$
$$\angle ABC = \theta - \beta - \alpha \qquad \text{[from Fig. 17.24(c)]}$$
$$\angle ABC = \theta + \beta + \alpha \qquad \text{[from Fig. 17.24(d)]}$$

Satellite stations should be avoided, as they tend to introduce errors and involve more calculations. They should be used only in unavoidable circumstances.

Eccentricity of signal If the signal is not placed exactly over the station, then the instrument sights an eccentric signal. This is the same as having a satellite station as described and the corrections are the same. In Fig. 17.25, let A, B, and C be the triangulation stations. The signal at B has been kept eccentrically at S. In triangle ABC, ∠A and ∠B are measured. The distances AC and BA are known.

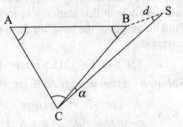

Fig. 17.25 Eccentricity of signal

$$BC = AC \sin A/\sin B$$

Knowing distance BS = d and ∠BSC, the correction to ∠C = α is

$$\alpha \text{ (in sec)} = d \sin(\angle BSC)/BC \sin 1''$$
$$= 206,265 \sin(\angle BSC)/BC$$

The following examples illustrate these principles.

Example 17.14 The side AB of a triangle (Fig. 17.26) could not be measured fully due to obstructions. Another station E was chosen and the angles at E were measured as ∠AEC = 20°, ∠CED = 30°, and ∠DEB = 15°. Length AC = 80 m and length DB = 60 m. Find the length CD.

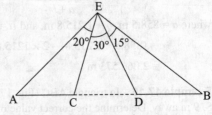

Fig. 17.26 Figure for Example 17.14

Solution Let the length CD be x m. From triangle AEC,

$$80/\sin 20° = AE/\sin z$$

From triangle BED,

$$60/\sin 15° = BE/\sin y$$

From triangle AED,

$$(80 + x)/\sin 50° = AE/\sin y$$

From triangle BEC,

$$(60 + x)/\sin 45° = BE/\sin z$$

$$(80 \sin 50°)/[(80 + x)\sin 20°] = \sin y/\sin z$$

$$(60 \sin 45°)/[(80 + x)\sin 15°] = \sin z/\sin y$$

$$(80 \sin 50° × 60 \sin 45°)/(80 + x)(\sin 20°)(60 + x)(\sin 15°) = 1$$

$$4{,}800 × \sin 50° × \sin 45° = (80 + x)(\sin 20°)(60 + x)(\sin 15°)$$

$$2{,}600.04 = (4800 + 140\, x + x^2)0.0885$$

$$x^2 + 140x + 4800 = 29{,}379$$

This quadratic equation on solving gives

$$x = 101.67 \text{ m}$$

Example 17.15 A base line AB could not be measured due to some obstructions. A station S by the side (Fig. 17.27) was chosen and the angle at that station was measured. Find the length of the base line from the following data:

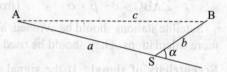

Fig. 17.27 Figure for Example 17.15

$$a = 858.5 \text{ m}, b = 1215.8 \text{ m}, \alpha = 10° 06' 30''.$$

Solution From triangle ASB, length of AB (side c) can be calculated as follows:

$$c^2 = a^2 + b^2 + 2ab \cos\alpha, \quad a^2 + b^2 - c^2 = -2ab \cos\alpha$$

$$a^2 + b^2 + 2ab - c^2 = 2ab - 2ab \cos\alpha$$

$$(a + b)^2 - c^2 = 2ab(1 - \cos\alpha)$$

$$a + b - c = 2ab(1 - \cos\alpha)/(a + b + c)$$

$$c = (a + b) - \frac{2ab(1 - \cos\alpha)}{a + b + c}$$

$(a + b + c)$ is approximately equal to $2(a + b)$.

$$(1 - \cos\alpha) = 2 \sin^2(\alpha/2)$$

$$c = (a + b) - \frac{2ab\sin^2(\alpha/2)}{a + b}$$

where $a = 858.5$ m, $b = 1{,}215.8$ m, and $\alpha = 10° 06' 30''$. Substituting,

$$c = (858.5 + 1{,}215.8) - 2 × 1215.8 × 858.5 \sin^2(5° 03' 15'')/(1{,}215.8 + 858.5)$$

$$= 2{,}066.573 \text{ m}$$

Example 17.16 In triangle ABC (Fig. 17.28), the signal at B was placed eccentrically at S, 9 m away. Determine the correct value of $\angle ABC$. Here BA = 3,896.8 m, BC = 4,470.72 m, $\angle BSC = 33° 15' 40''$, and $\angle ASC = 67° 56' 20''$.

Solution Eccentricity of signal = 9 m.

$$BA = 3896.8 \text{ m}, \quad BC = 4470.72 \text{ m}$$

$$\angle BSC = 33° \ 15' \ 40'',$$

$$\angle ASC = 67° \ 56' \ 20''$$

Let $\angle SAB = \alpha$ and $\angle SCB = \beta$. α and β can be calculated using the sine rule. From triangle ASB,

$$d/\sin\alpha = AB/\sin(\angle BSA)$$

Here

$$d = 9 \text{ m}$$

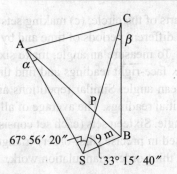

Fig. 17.28 Figure for Example 17.16

and

$$\angle BSA = 33° \ 15' \ 40'' + 67° \ 56' \ 20'' = 101° \ 12' \ 00''$$

$$\sin\alpha = 9 \times \sin(101° \ 12')/3,896.8$$

As these angles are small, $\sin\alpha$ (or $\sin\beta$) = α (or β) in radians.

$$\alpha = 9 \times \sin(101° \ 12') \times 206,265/3,896.8 = 467.3 \text{ sec}$$

From triangle BSC,

$$d/\sin\beta = BC/\sin(33° \ 15' \ 40'')$$

$$\sin\beta = 9 \times \sin (33° \ 15' \ 40'') \times 206,265/4470.72 = 227.7''$$

From triangle SPA,

$$\angle CPA = 67° \ 56' \ 20'' + \alpha = 69° \ 04' \ 07.3''$$

From triangle BPC,

$$\angle PBC + \beta = \angle CPA$$

$$\angle ABC = 69° \ 04' \ 07.3'' - 227.3'' = 69° \ 00' \ 19.6''$$

17.8 Measurement of Angles

After the base line is measured, the next major task in triangulation is to measure angles. Angles are measured using a theodolite. The common vernier theodolite is not useful, as it is not accurate enough. Special or precise theodolites are used for triangulation.

In earlier days, greater precision was obtained by making the horizontal plate larger, as much as 30 or 36 in. in diameter. These instruments were bulky and difficult to carry. With the advent of optical as well as electronic theodolites, smaller instruments with the same or greater precision are available to surveyors. Microoptic theodolites, electronic theodolites, and total stations are now available for angle measurements. These instruments are very accurate, measuring 1 in. or less, and have a much higher range, which is required for geodetic surveying. Greater accuracy in centring is obtained using an optical plummet.

Whichever instrument is used, angles are measured a number of times using the method of repetition or reiteration. These have been discussed in Chapter 4.

Method of repetition

In the method of repetition [Fig. 17.29(a)], angles are measured repeatedly by (a) taking face-left and face-right observations, (b) reading the angle from different parts of the circle, (c) making sets of observations, and (d) making measurements at different periods of time and by a different survey party.

To measure an angle, make six face-left readings and find the average. Make six face-right readings and find the average. The average of these two sets is the mean angle. Similar repetitions are made by measuring the angle with different initial readings. The average of all the set values is taken as the mean value of the angle. Sixteen sets (each set consisting of six face-left and face-right readings) are used in precise (first-order) triangulation while only three such sets may be used in third-order triangulation work.

Method of reiteration

The method of reiteration or the direction method is employed when a number of angles are to be measured at a station or figures with a central station are used in the triangulation system. With a central station [Fig. 17.29(b)], the reiteration method uses the principle of closing the horizon while measuring angles. After the closure of horizon, the reading should be the same as the initial reading.

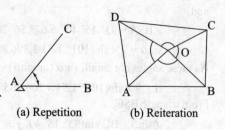

(a) Repetition (b) Reiteration

Fig. 17.29 Measuring angles

Any error observed is distributed among the angles. Sets are made, as in the case of repetition, by taking face-left and face-right observations, measuring in the reverse direction, and reading from different parts of the circle plate. Mean values of the angles are obtained as the average values obtained from the number of sets measured.

In case a number of angles are to be measured, as in Fig. 17.29(c), one of the directions, say OA, is taken as the reference direction. Successive angles are measured as $\angle AOB$, $\angle BOC$, and $\angle COD$. The same angles are now measured in the reverse direction. The face is changed and the angles are measured in the same way first in the forward direction and then in the reverse direction. Four values of the angles are now available and this is taken as one set.

Further sets are measured in the same way but with a different reading in the initial direction OA. A number of such sets are obtained. The format for recording the observations is already discussed in Chapter 4.

Field checks on angles

While measuring angles and before removing the instrument, the surveyor should check the following.

(a) The sum of the angles in a triangle is equal to 180°.
(b) In the reiteration method, if the horizon is closed, the initial reading is obtained when coming back to the starting direction.
(c) Face-left and face-right observations differ by 180°.
(d) In the case of a quadrilateral as in Fig. 17.30(a), eight angles are measured. The product of the sine of the alternate angles is equal to the product of the sine of the other set of angles. The alternate angles are $\angle CAB$, $\angle DBC$, $\angle ACD$,

and ∠ADB. The other set of angles are also alternate such as ∠ABD, ∠BCA, ∠CDB, and ∠DAC. Thus,

$$\sin(\angle CAB)\sin(\angle DBC)\sin(\angle ACD)\sin(\angle ADC)$$
$$= \sin(\angle ABD)\sin(\angle BCA)\sin(\angle CDB)\sin(\angle DAB)$$

This can be easily verified by applying the sine rule.

$$BC = AB\sin(\angle BAC)/\sin(\angle BCA),\ CD = BC\sin(\angle DBC)/\sin(\angle BDC)$$

Therefore, substituting for BC,

$$CD = \frac{AB\sin(\angle BAC)\sin(\angle DBC)}{\sin(\angle BCA)\sin(\angle BDC)}$$

CD can also be calculated as, in the same way,

$$AD = AB\sin(\angle ABD)/\sin(\angle ADB),\ CD = AD\sin(\angle DAC)/\sin(\angle DCA)$$

Putting the value of AD,

$$CD = \frac{AB\sin(\angle ABD)\sin(\angle DAC)}{\sin(\angle ADB)\sin(\angle DCA)}$$

Equating the two values of CD,

$$\frac{AB\sin(\angle BAC)\sin(\angle DBC)}{\sin(\angle BCA)\sin(\angle BDC)} = \frac{AB\sin(\angle ABD)\sin(\angle DAC)}{\sin(\angle ADB)\sin(\angle DCA)}$$

Thus,

$$\sin(\angle BAC)\sin(\angle DBC)\sin(\angle ADB)\sin(\angle DCA)$$
$$= \sin(\angle ABD)\sin(\angle DAC)\sin(\angle ADB)\sin(\angle DCA)$$

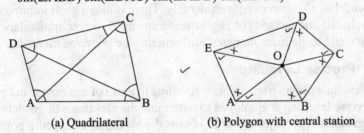

(a) Quadrilateral   (b) Polygon with central station

Fig. 17.30 Field checks on angles

If a polygon with a central station is surveyed, the same condition will apply. Thus, in Fig. 17.30(b), the product of sines of angles marked √ if equal to the product of sines of angles marked with ×. Thus,

$$\sin(\angle OAB)\sin(\angle OBC)\sin(\angle OCD)\sin(\angle ODE)\sin(\angle OEA)$$
$$= \sin(\angle OBA)\sin(\angle OCB)\sin(\angle ODC)\sin(\angle OED)\sin(\angle OAE)$$

This can be easily verified by calculating OB from triangle OAB, OC from triangle OBC, etc. Thus,

$$OB = OA\sin(\angle OAB)/\sin(\angle OBA)\quad\text{(from triangle OBA)}$$
$$OC = OB\sin(\angle OBC)/\sin(\angle OCB)\quad\text{(from triangle OBC)}$$

$$= OA \sin(\angle OAB)\sin(\angle OBC)/\sin(\angle OBA)\sin(\angle OCB)$$

From triangle ODC,

$$OD = OC \sin(\angle OCD)/\sin(\angle ODC)$$

$$= \frac{OA \sin(\angle OCD)\sin(\angle OAB)\sin(\angle OBC)}{\sin(\angle ODC)\sin(\angle OBA)\sin(\angle OCB)}$$

From triangle OED,

$$OE = OD \sin(\angle ODE)/\sin(\angle OED)$$

$$= \frac{OA \sin(\angle ODE)\sin(\angle OCD)\sin(\angle OAB)\sin(\angle OBC)}{\sin(\angle OED)\sin(\angle ODC)\sin(\angle OBA)\sin(\angle OCB)}$$

Finally, from triangle OEA,

$$OA = OE \sin(\angle OEA)/\sin(\angle OAE)$$

$$= \frac{OA \sin(\angle OEA)\sin(\angle ODE)\sin(\angle OCD)\sin(\angle OAB)\sin(\angle OBC)}{\sin(\angle OAE)\sin(\angle OED)\sin(\angle ODC)\sin(\angle OBA)\sin(\angle OCB)}$$

Therefore,

$$\sin(\angle OAE) \sin(\angle OED) \sin(\angle ODC) \sin(\angle OCB) \sin(\angle OBA)$$

$$= \sin(\angle OEA) \sin(\angle ODE) \sin(\angle OCD) \sin(\angle OAB) \sin(\angle OAB)$$

17.9 Trilateration

Trilateration is a method of geodetic surveying wherein the three sides of a triangle are measured. The disadvantage of this method is the difficulty in precise measurement of long distances. With the advent of EDM instruments, the measurement of distances has become easier. The accuracy attainable with EDM instruments has also improved considerably. There is also a great improvement in the instruments and methods of angle measurement. However, triangulation, with a base line and angle measurements, still remains the preferred method.

17.10 Precise Levelling

Precise levelling differs from ordinary levelling in terms of equipment and methods used. Precise levelling is employed to determine the elevations of geodetic points in triangulation. The objective is to obtain the elevation of specific points very accurately. Some of the features of precise levelling that make it different from ordinary levelling are as follows.

(a) Precise levels are used which are expensive and specially made for precise levelling work.

(b) The lines of sight are generally longer than in ordinary levelling but are limited to about 100 m for the purpose of precise reading of the staff.

(c) Backsight and foresight distances are made nearly equal to eliminate errors due to collimation and curvature.

(d) For more precise work, an invar levelling staff is used.

(e) Efforts are made to take readings in quick succession to eliminate errors due to temperature and refraction.

(f) The instruments are tested and adjusted daily.

Precise levels

Many manufacturers make highly precise levels. Modern automatic and digital levels are very precise. Some of the features of such levels are as follows.

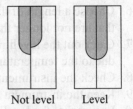

(a) The telescope aperture is not less than 40 mm and the focal length of objective is kept at a distance of 37.5–50 mm or more.

Not level   Level

(b) The telescope magnification is about 50 diameters, as long sights are involved.

Fig. 17.31 Coincidence reading of bubble tube

(c) The bubble tube is made of a larger radius of curvature to have a higher sensitivity. The value of the 2-mm division is 3–5 sec.

(d) The position of the bubble is seen either through an auxiliary telescope or in the main telescope itself.

(e) Only automatic, tilting, or digital levels are used. A parallel plate micrometer is generally fitted in the tilting level to read the staff accurately.

(f) The telescope is set on a wider tribrach for stability.

(g) Most precise levels have a coincidence bubble tube (Fig. 17.31). The two ends of the bubble tube are seen simultaneously through an auxiliary telescope at the eyepiece end of the main telescope. This is achieved through an optical arrangement to bring both the ends together. This coincidence can be achieved accurately.

Precise levelling staff   A levelling staff of superior quality wood is used and the graduations are made accurately. The staff invariably has a bubble tube to check its verticality at the time of reading. For more precise work, an invar staff is used. An invar staff is a flexible plate of invar, with graduations up to 3 m, which is fixed onto a wooden piece. It is fixed at the bottom but is hung from a spring at the top. This eliminates any error due to the expansion of the wooden piece that holds the invar plate.

Precise methods   The procedure for precise levelling has many built-in features to eliminate errors.

1. Make the foresights and backsights equal by stadia observations to eliminate errors due to curvature and collimation.
2. Take the foresight and backsight readings in quick succession to eliminate any errors due to the settlement of the tripod or staff.
3. Change the order of readings in alternate set-ups. From one instrument set-up, read the foresight followed by the backsight. From the next set-up, read the backsight first and then the foreseght.
4. Take all the three cross hair readings and record them to average the central cross hair reading.
5. Use two levelling rods and two staff men. Move the staff read first at any setting, leaving the other staff in position for the next reading from the next set-up. This will ensure that the rod read first is always read first from all set-ups irrespective of the changing order of readings. This will eliminate errors due to refraction changes, level, or the staff being disturbed.

6. Choose a length of the line of sight not greater than 100 m. Level the sections that are not longer than 1200 m.
7. Carry out the levelling operation during that part of the day when the changes due to the temperature and refraction are the least.
8. Check the instrument used for levelling every day.
9. Keep the instrument and the staff protected from the sun.
10. Ensure that the levelling done by a party in a section is checked on a different day, by a different party, using different turning points.
11. Close the level and keep the error of closure within the prescribed limits.

Field book The format of the field book used for recording levels is the same as for ordinary levelling. However, all the three hairs are read from every sight and recorded. A page from a field record is shown in Fig. 17.32.

Instrument station	Backsight	Difference	Foresight	Difference	Height of instrument	Reduced level
1	1.846	0.61				
(to BM)	2.456				189.842	187.385
	3.068	0.612				
	2.457					
2	2.466	1.196	2.628	1.196		
(to TP 1)	3.662		3.824		187.984	186.019
	4.856	1.194	5.018	1.194		
	3.661		3.823			
3	1.965	0.98	2.115	0.999		
(to TP 2)	2.936		3.114		187.802	184.871
	3.896	0.96	4.112	0.998		
	2.931		3.113			

Notes

1. All the three hair readings are taken. The difference between the middle hair reading and the two outer hair readings is determined. This difference is limited to 0.005 m. If the difference is more, the readings are taken again.

2. The first reading is taken to a benchmark of known elevation. The height of the instrument is then calculated as 187.835 + 2.457 = 189.842. With this instrument position, the next reading taken is the foresight of 3.823. The RL of the turning point is 189.842 − 3.823 = 186.019.

3. With this RL, the second height of the instrument is determined from the backsight reading as 186.019 + 1.965 = 187.984.

Fig. 17.32 Sample of field book for precise levelling

Limits of precision The limits of precision used in precise levelling are stringent. The permissible error of closure is limited to the range $(4K^{1/2} - 12K^{1/2})$mm, where K is the distance in kilometres, of the section surveyed. The lower limit is to be used for obtaining the elevations of triangulation points and the higher limits are used for transferring benchmarks.

Summary

The objective of geodetic surveying is to accurately establish points for horizontal control. Geodetic surveying differs from plane surveying in different ways: the distances involved are large, the curvature of the earth has to be taken into account, and precise instruments and methods have to be used.

The general method used for geodetic surveying is triangulation. In this method, one base line is measured, which is the side of a triangle, followed by the measurement of all the three angles of the triangle. The remaining sides are calculated using the sine rule. Theoretically, only one base line has to be measured; however, in practice, one or more lines are measured as check bases.

The triangulation system can be built up of triangles, quadrilaterals, or polygons with central stations. In all these cases, the system consists of triangles only. The order of triangulation depends upon the purpose of the survey. First-order triangulation is done in the most precise way with a stringent control over instruments and methods. Triangulation points that are very far apart are connected by secondary triangulation points. These in turn are connected for more localized points by third-order or tertiary triangulation. Such points can be used as control points for more detailed surveys.

Triangles have to be well conditioned with no angle less than 30° or more than 120°. Practically, an equilateral triangle is the best figure.

Fieldwork in triangulation consists of reconnaissance and then fixing the stations and the base line. Intervisibility between the stations has to be checked. To achieve intervisibility, stations and signals are raised above the ground level using towers and other means.

Base lines can be measured using rods, tapes, and wires or with the help of optical and electronic means of measuring distance. The Colby apparatus used earlier in Indian surveys is obsolete; invar tapes and wires are used now. In addition, total stations and EDM instruments are often used in geodetic surveying.

Angles are measured using precise theodolites. Optic and electronic theodolites and total stations are used to measure angles. Angles can be measured using the repetition or reiteration method. A number of sets are made using face-right and face-left observations and by taking the readings from different parts of the circle plate. The average of all the sets is taken as the mean value of the angle.

Precise levelling is undertaken to provide points for vertical control. Precise levelling uses refined instruments, methods, and a stringent control to establish benchmarks of known elevation.

Exercises

Multiple-Choice Questions

1. Geodetic surveying is used for
   (a) detailing the topography of a large area
   (b) getting control points for horizontal control
   (c) finding elevations of points precisely
   (d) finding the latitude and longitude of points
2. Triangulation is based on the principle that knowing
   (a) three sides, the angles can be calculated precisely
   (b) three angles, the sides can be calculated precisely
   (c) two sides and one angle, the remaining side and angles can be determined precisely
   (d) one side and three angles, the remaining sides can be calculated precisely
3. Trilateration is a method of triangulation where
   (a) three angles of a triangle are accurately measured
   (b) three sides of a triangle are accurately measured

(c) one side aand two angles of a triangle are accurately measured

(d) one angle and two sides of a triangle are accurately measured

4. Order of triangulation is based on the

(a) instruments used for triangulation (c) triangulation system adopted

(b) method used for triangulation   (d) accuracy prescribed for measurements

5. The optimum angle of a triangle for triangulation is

(a) $75°\,30'\,20''$ (b) $60°\,15'\,30''$ (c) $66°\,45'\,40''$ (d) $56°\,13'\,40''$

6. The strength of figure in triangulation system is more

(a) when the angles of a triangle are very nearly equal to 60

(b) when any angle of a triangle is not less than 30 or more than 120

(c) in the method of trilateration

(d) when the error is the least when computing the length of last line.

7. The distance angles in a triangulation system are

(a) the angles calculated using distances measured

(b) the angles used in calculating distances

(c) the angles measured during triangulation

(d) the lines measured during triangulation

8. If the measured length of a line is L and the mean height of the stations is h, then the distance reduced to mean sea level is

(a) R/Lh   (b) Lh/R   (c) L/h   (d) R/L

9. Phase correction in triangulation is due to the inaccuracy in

(a) the measurement of base line

(b) the instruments used

(c) bisecting an object due to lateral illumination

(d) the verticality of signal

10. In the case of a quadrilateral used as a basic figure, divided into triangles by diagonals, then,

(a) sum of sines of alternate angles = sum of sines of remaining angles

(b) sum of log sines of alternate angles = sum of log sines of remaining angles

(c) sum of cosines of alternate angles = sum of cosines of remaining angles

(d) product of sines if alternate angles = product of remaining angles

11. In the case of any polygon with a central station, the rule governing the angles measured at the vertices of the polygon can be stated as

(a) sum of sines of alternate angles = sum of sines of remaining angles

(b) sum of log sines of alternate angles = sum of log sines of remaining angles

(c) sum of cosines of alternate angles = sum of cosines of remaining angles

(d) product of sines if alternate angles = product of remaining angles

Review Questions

1. Briefly explain the difference between plane surveying and geodetic surveying.
2. Explain the objective and basic principle of triangulation.
3. Explain, with sketches, the different triangulation systems.
4. Explain the concept of order of triangulation and the specifications for different orders of triangulation.
5. Explain the concept of strength of figure and the method used to calculate it.
6. Explain briefly the different aspects of fieldwork in triangulation.
7. Explain the importance of reconnaissance before triangulation.
8. Explain the method used to decide the intervisibility of stations.
9. Using sketches, explain how towers are built for stations.
10. Explain the different types of signals used in triangulation.

11. Explain the different methods for base line measurement.
12. Describe briefly the corrections to be made to field measurements with a tape.
13. What is geodetic length? Explain how it is derived.
14. Explain the method of making sets while measuring angles with the repetition and reiteration methods.
15. Explain the equipment and methods used for precise levelling.

Problems

1. In a triangle PQR, PR is the base line. $\angle P$, $\angle Q$, and $\angle R$ were measured and found to be 63°, 72°, and 45°, respectively. Find the strength of figure.
2. If the probable error in the measurement of angles is 1.5″ and the probable error in the distance computed is to be limited to 1 in 10,000, find the maximum value of the strength of figure.
3. Find the value of $(D - C)/D$ from Fig. 17.33.
4. In the network shown in Fig. 17.34, find the most suitable route to calculate the length of DC.
5. Two triangulation stations A and B are 50 km apart. The elevation of A is 205.5 m and that of B is 232.2 m. The intervening ground may be assumed to have a uniform elevation of 175 m. Determine the height of the signal at B if the line of sight is required to pass at least 3 m above ground.
6. Two stations A and B are 130 km apart. The elevation of A is 500 m and that of B is 790 m. An intervening peak is at 75 km from A and has an elevation of 392 m. Check the intervisibility of stations A and B. Find the height of signal required at B if the line of sight is required to pass at least 2.5 m above ground. Compare this with Captain McCaw's solution.
7. Two stations P and Q are 100 km apart. The elevations of P and Q are 375 m and 1025 m, respectively. Two intervening peaks R and S are at 42 km and 84 km, respectively, from A. The elevation of R is 490 m and that of S is 835 m. Check the intervisibility between the stations P and Q. Find the height of the signal required at Q if the line of sight must pass 3 m above ground.
8. A 30-m-long tape is standardized at 22°C. It is used to measure the distance along a slope of 1°. At what temperature will the effect of the slope and the temperature neutralize?

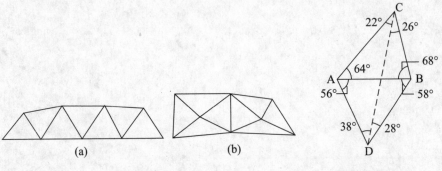

| (a) | (b) | |
| Fig. 17.33 | Fig. 17.34 | |

9. A tape of nominal length 30 m was standardized at a temperature of 27°C. The mean temperature during measurement was 12°C. The total length measured was 185.5 m and along the slopes as follows:

1° 30′   30 m
2° 45′   60 m

3° 20'   60 m

4° 10'   Balance length of the wire

Find the corrected length of the line.

10. A 100-m-long tape was suspended over two equal spans. The weight of the tape was 9.8 N and the pull applied during measurement was 120 N. Find the sag correction for the tape.

11. A 30-m tape (standardized at 22°C with a pull of 100 N in catenary) was used to measure a base line. The mean temperature during measurement was 38°C and the pull applied was 150 N. For the tape, the coefficient of thermal expansion $= 11.6 \times 10^{-6}/°C$, $E = 2 \times 10^{11}$ N/m², area of cross section $= 8$ mm², and density $= 78,500$ N/m³. Find the corrections per tape length due to the temperature, pull, and sag and reduce this distance to mean sea level if the elevation of the points of support is 4,500 m.

12. Show that, for a tape suspended between supports at the same level, the horizontal distance between the supports is given by $T_c/w \log[(T + W)/(T - W)]$, where T_c is the tension at the centre of the tape, w is the weight per metre, T is the tension at the supports, and $2W$ is the total weight of the tape.

13. Show that for a tape suspended between supports at different levels, the sag and slope correction $= w^2 l^3/24P^2 + h^2/2l$, where w is the weight per metre length, l is the distance between supports, P is the tension applied, and h is the difference in level between the supports.

14. While measuring a line PQ, the whole length could not be measured due to some obstructions. A station T was chosen and the angles to stations R, S, and Q were measured. $\angle PTR = 18°$, $\angle RTS = 34° 30'$, and $\angle STQ = 16°$. The length PR = 32.8 m and SQ = 28.4 m. Find the length SR.

15. A, B, and C are the stations of a triangle. As station B could not be occupied, a satellite station S, 21.4 m from B, was occupied. S and C were on the opposite sides of the line AB. The angles measured were $\angle BSC = 61° 18' 24''$ and $\angle ASC = 69° 38' 12''$. Length BC = 7,180.55 m and length BA = 10,357.67 m. Find the angle ABC.

Theory of Errors and Survey Adjustments

Learning Objectives

After going through this chapter, the reader will be able to

- distinguish between systematic errors and accidental errors
- state the laws of weights
- state and prove the principle of least squares
- define and explain the terms probable value and probable error, and explain the methods used to determine their values from given data
- describe the methods used to find the probable value by the method of normal equations and correlates
- explain the terms station adjustment and figure adjustment
- adjust triangles, quadrilaterals, and polygons with central stations using different methods
- adjust levels and level nets

Introduction

A brief discussion on errors was taken up in Chapter 1. Subsequently, various types of errors have been mentioned. Errors, in general, can be classified into instrumental errors, personal errors, and errors due to natural causes. Despite using the most precise instruments and methods, errors still remain in field measurements. Errors that remain in spite of taking all possible precautions are known as *accidental errors*. Such errors come within the realm of probability. One looks for the most probable values of errors and observed quantities. As the true value of an observed quantity is not known, the objective of adjustments is to compute a value, from a number of observations, nearest to the true value.

18.1 Errors

Errors can be classified into instrumental errors, personal errors, and errors due to natural causes. This classification has been discussed earlier. Errors can also be classified into systematic errors and accidental errors.

Systematic errors These are errors caused by factors that can be accounted for and corrections applied. For example, variation in length due to temperature can be determined and corrected. Systematic errors can be positive or negative. These errors follow definite physical laws and are generally cumulative.

Accidental errors These are errors that remain after the systematic errors and mistakes have been accounted for. These errors are beyond the control of the observer. They do not follow any specific laws and may be positive or negative. There is an equal chance of the quantity having a higher or lower value than the true value. Accidental errors follow the law of probability. The discussion in this chapter is limited to accidental errors.

Random errors that continue to remain after the removal of systematic errors in observations are governed by the law of probability. The precision with which the measurement is done can be known by plotting the probability curve from a large number of observations. The spread of the curve is an indication of the care with which the measurements are made. If systematic errors and mistakes are not eliminated, we may have precise but inaccurate measurements (the difference between the true value and the most probable value remains high).

Law of accidental errors

Accidental errors are dealt with based on the probability error curve shown in Fig. 18.1(a). This curve has been plotted between the size of errors and their frequency of occurrence. The following observations are pertinent from this curve.

(a) Very small errors and very large errors (in magnitude) have a small chance of occurring.

(b) Positive and negative errors have equal chances of occurring. The curve is thus symmetrical about the mean error value.

(c) The values close to the mean error have a greater chance of occurrence. Their frequency is higher.

(d) When the mean and true values are very close [Fig. 18.1(b)], an absence of systematic errors is implied. When systematic errors are present, the mean and true values will be very different, as shown in Fig. 18.1(c).

The theory of errors is derived from the probable error curve. The curve is plotted from a large number of observations. Probable values are expressed at a desired level of confidence as mean value ± error. This gives the limit, at a certain level of confidence, within which the true value lies.

From the probability curve, the probable error in a single observation, E_s, and the probable error in the mean of a number of observations, E_m, are given by

$$E_s = \pm\, 0.674 \sqrt{\Sigma v^2/(n-1)}$$

and

$$E_m = [\sqrt{n} \times 0.6745\sqrt{\Sigma v^2/(n-1)}]/n \quad E_s/\sqrt{n}$$

Thus, the probable error in the mean is $1/\sqrt{n}$ times the probable error in a single observation. These are explained in the later sections for their application.

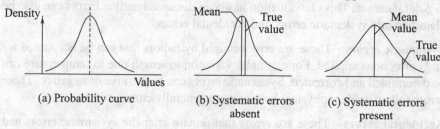

| (a) Probability curve | (b) Systematic errors absent | (c) Systematic errors present |

Fig. 18.1 Probability curves

18.2 Definitions

The following definitions are pertinent to the discussions in this chapter.

Observed value The observed value of a quantity is the value derived from an observation after correcting the same for all the known errors.

True value The true value of a quantity is the value of the quantity absolutely free from all errors. Since the true error is not known, the true value is never known.

Most probable value The most probable value of a quantity is the value that has a greater chance of being the true value than any other value. This value is derived from several measurements of the quantity.

Independent quantity An independent quantity is a quantity whose value does not depend upon the value of other quantities. Reduced levels of benchmarks are independent quantities. They are not affected by changes in the values of other quantities.

Dependent or conditioned quantity A dependent or conditioned quantity is a quantity whose value depends upon the value of one or more quantities. 'The sum of the angles in a triangle is 180°' or 'the sum of the angles around a point is 360°' are conditions imposed on angles. In the case of a triangle, there are three angles. Any two angles can be independent quantities while the third is dependent on the value of these quantities because of the imposed condition.

Direct observation A direct observation is the measured numerical value of a quantity (length, angle), made directly on the quantity whose value is being determined. A base line, a single measurement of an angle, etc. are examples.

Indirect observation An indirect observation of a quantity is a value derived from measuring some related quantity. When an angle is measured by repetition, the observed value of the quantity is indirectly derived from a number of observations.

Weight The weight of an observation is indicative of the worth, trustworthiness, precision, or confidence placed upon the observed value. It is also a measure of the reliability of the observed value. A value with a weight of 5 is (5/3) times as reliable as a quantity with a weight of 3. When different weights are not assigned to quantities, they are said to be of the same weight. Observations are assigned different weights depending upon the accuracy with which the observations are made and the conditions under which they are made. In general, weight is proportional to the number of times a quantity is measured.

True error It is the difference between the true value and the observed value of a quantity. It is not known.

Most probable error It is the quantity which, when added to and subtracted from the most probable value, defines the limits within which there is a greater probability that the true value of the quantity may lie.

Residual error It is the difference between the most probable value and the observed value of a quantity.

Observation equation   It is an equation relating a quantity to its numerical value.

Conditional equation   It is an equation expressing the relation between several dependent quantities.

Normal equation   It is obtained by multiplying each equation by the coefficient of the unknown quantity whose normal equation is desired. The number of normal equations is the same as the number of unknowns.

18.3   Laws of Weights

As mentioned earlier, weights are assigned to observations based upon the confidence one has in the observed values. All observations are not taken at the same time with equal care. Many conditions change in the course of survey work. Observations taken with greater care and under very favourable conditions are likely to be more accurate than observations taken under unfavourable conditions.

The reliability of an observation is indicated by its weight. The weights assigned are relative. They are established by either measurement conditions or judgement. The following points should be considered while assigning weights.
(a) The weight of a quantity is proportional to the number of measurements made.
(b) If a measurement is repeated a large number of times, the weight is proportional to the square of the probable error.
(c) In the case of levels, the weight varies inversely as the length of the path.

The following laws of weights are valid in the cases dealt in this chapter.
(a) The weight of the arithmetic mean of a number of observations of equal weight is equal to the number of observations.
(b) The weight of the weighted arithmetic mean of observations is equal to the sum of the weights of the observations.

$$\text{Weight } W \text{ of mean} = w_1 + w_2 + w_3 + \cdots + w_n$$

(c) The weight of the algebraic sum of two or more quantities is equal to the reciprocal of the sum of the reciprocals of the individual weights. The same applies to the difference in quantities too.
(d) If a weighted observation is multiplied by a factor, the weight of the product is obtained by dividing the weight by the square of the factor.
(e) If a weighted observation is divided by a factor, the weight of the result is equal to the weight multiplied by the square of the factor.
(f) If an equation is multiplied by its own weight, the weight of the resulting quantity is the reciprocal of the weight of the equation.
(g) The weight of an equation remains the same if the signs of all the terms of the equation are changed or the equation is added to or subtracted from a constant.

The following examples illustrate the concept of weights.

Example 18.1   The following six observations are of equal weight: 42° 22′ 32″, 42° 22′ 30″, 42° 22′ 33″, 42° 22′ 34″, 42° 22′ 29″, 42° 22′ 30″. Find the arithmetic mean and the weight of arithmetic mean.

Solution   Mean value of angle = 42° 22′ + (32″ + 30″ + 33″ + 34″ + 29″ + 30″)/6

$$= 42° \ 22′ \ 31.33″$$

Weight of the mean = number of observations = 6

Example 18.2   From the following six weighted observations of A, find the weight of the mean of A.

42° 22′ 32″	– 2	42° 22′ 34″	– 2
42° 22′ 30″	– 4	42° 22′ 29″	– 3
42° 22′ 33″	– 3	42° 22′ 30″	– 4

Solution   Sum of weights = 2 + 4 + 3 + 2 + 3 + 4 = 18

Mean value of A = 42° 22′ + (32 × 2 + 30 × 4 + 33 × 3 + 34 × 2 + 29 × 3
+ 30 × 4)/18

$$= 42° \ 22′ \ 31″$$

Weight of the mean = sum of the weights = 18

Example 18.3   Two angles A and B were measured. A = 32° 16′ 18″ of weight 3 and B = 26° 14′ 12″ of weight 2. Find the values and weights of $(A + B)$ and $(A - B)$.

Solution   Sum of angles $(A + B)$ = 32° 16′ 18″ + 26° 14′ 12″ = 58° 30′ 30″

Difference $(A - B)$ = 6° 02′ 06″

For sums and differences, the weight is the reciprocal of the sum of the reciprocals.

$$\text{Weight} = \frac{1}{\dfrac{1}{3} + \dfrac{1}{2}} = 6/5 = 1.2$$

Example 18.4 A = 42° 22′ 31.33″ and has a weight of 6. Find the weight of $3A$ and $A/4$.

Solution   When a quantity with a weight w is multiplied by a constant C, the weight of the product is w/C^2.

Weight of $3A = 6/3^2 = 2/3$

When a quantity of weight w is divided by a constant C, the weight of the divided quantity is wC^2.

Weight of $A/4 = 6 \times 4^2 = 96$

Example 18.5   Two angles of a triangle are A = 42° 32′ 40″ (weight 3) and B = 51° 29′ 20″ (weight 2). Find the value and weight of C.

Solution $C = 180° - (A + B) = 180° - 42° \ 32′ \ 40″ - 51° \ 29′ \ 20″ = 85° \ 58′ \ 00″$

Weight of $A + B$ = reciprocal of $(1/3 + 1/2) = 6/5$

Weight of C = weight of $[180° - (A + B)] = 6/5$

18.4   Principle of Least Squares

The principle of least squares is a very powerful tool in the theory of errors. The general principle of least squares can be stated as 'The most probable value of a quantity evaluated from a number of observations is the one for which the squares of the residual errors is minimum.' Let X_1, X_2, X_3, etc., be the observed values. If

M is the arithmetic mean, then

$$M = (X_1 + X_2 + X_3 + \cdots + X_n)/n = \Sigma X/n$$

If Z is the most probable value, then the residual errors, e, can be calculated as

$$Z - X_1 = e_1$$
$$Z - X_2 = e_2$$
$$Z - X_3 = e_3$$
$$\cdots \cdots \cdots \cdots$$
$$Z - X_n = e_n$$

Adding,

$$nZ - \Sigma X = \Sigma e$$

or

$$Z = \Sigma X/n + \Sigma e/n$$

As $\Sigma X/n = M$,

$$Z = M + \Sigma e/n$$

If the number of observations is large, n is very large, and the second term is nearly equal to zero. Thus

$$Z = M$$

If the number of observations is large, the arithmetic mean is the true value or the most probable value

$$Z \approx M$$

Now we calculate the residual errors from the mean value. If r_1, r_2, r_3, etc. are the residual errors, then

$$M - X_1 = r_1$$
$$M - X_2 = r_2$$
$$M - X_3 = r_3$$
$$\cdots \cdots \cdots \cdots$$
$$M - X_n = r_n$$

Adding,

$$nM = \Sigma X + \Sigma r$$
$$M = \Sigma X/n + \Sigma r/n$$

As $M = \Sigma X/n$, $\Sigma r/n = 0$ and therefore

Sum of residuals, $\Sigma r = 0$

Squaring the equations of residuals and adding,

$$\Sigma r^2 = nM^2 + \Sigma X^2 - 2M\Sigma X$$
$$nM = \Sigma X$$

Substituting,

$$\Sigma r^2 = M\Sigma X - 2M\Sigma X + \Sigma X^2$$
$$= \Sigma X^2 - \Sigma X/n$$

Since $M = \Sigma X / n$,

$$\Sigma X^2 = \Sigma r^2 + \Sigma X^2 / n$$

We can find a similar equation by finding the residual d from any other value V

$$V - X_1 = d_1$$
$$V - X_2 = d_2$$
$$V - X_3 = d_3$$
$$\cdots \cdots \cdots \cdots$$
$$V - X_n = d_n$$

Squaring and adding, we get

$$\Sigma d^2 = nV^2 + \Sigma X^2 - 2V\Sigma X$$

If we substitute

$$\Sigma X^2 = \Sigma r^2 + \Sigma X^2 / n$$

we have

$$\Sigma d^2 = nV^2 + \Sigma r^2 + \Sigma X^2 / n - 2V\Sigma X$$
$$= \Sigma r^2 + n(V^2 - 2V\Sigma X / n + \Sigma V^2 / n^2)$$
$$= \Sigma r^2 + n(V - \Sigma X / n)^2$$

The second term, being a square of a quantity, is always positive. This shows that Σd^2 is always more than Σr^2. Hence, the sum of the squares of the errors from the mean is the minimum, which is the principle of least squares.

18.5 Most Probable Value

The most probable value of a quantity can be determined from the principle of least squares. We use the following notations.

(a) $X_1, X_2, X_3, \ldots, X_n$ are the observed values.

(b) P is the most probable value.

(c) $e_1, e_2, e_3, \ldots, e_n$ are the residual errors.

The quantity may have been measured directly or indirectly. For direct observation, the observed values may have equal or unequal weights. For indirect observation, the observed values may or may not be weighted. These cases are solved by forming normal equations. There may also be conditioned quantities. The observation and conditional equations are used to solve such cases.

We take up these cases separately.

Direct observations of equal weight   The fundamental principle may be stated as 'In observations of equal weight, the most probable value of the quantity is the one that yields the minimum result as the sum of the squares of residual errors.' Therefore, from the law of least squaes, $e_1^2 + e_2^2 + e_3^2 + \cdots + e_n^2$ is minimum when the residual errors are calculated from the mean. Also $(P - X_1)^2 + (P - X_2)^2 + (P - X_3)^2 + \cdots + (P - X_n)^2$ is minimum. Differentiating the above equation,

$$(P - X_1) + (P - X_2) + (P - X_3) + \cdots + (P - X_n) = 0$$

or

$$P = (X_1 + X_2 + X_3 + \cdots + X_n)/n = M$$

where M is the arithmetic mean. Therefore, the most probable value in the case of observations of equal weight is the arithmetic mean. The weight of the arithmetic mean is n, the number of observations.

Direct observations of unequal weight This principle can be stated as 'If the observations are assigned weights, the most probable value is the one that gives the minimum result for the sum of the weighted squares of residual errors.'

Let $X_1, X_2, X_3, \ldots, X_n$ be observations having weights $w_1, w_2, w_3, \cdots, w_n$. Then, by the principle stated above, $w_1e_1^2 + w_2e_2^2 + w_3e_3^2 + \cdots + w_ne_n^2$ is minimum. If P is the most probable value, then the above equation can be written as

$$w_1(P - X_1)^2 + w_2(P - X_2)^2 + w_3(P - X_3)^2 + \cdots + w_n(P - X_n)^2 = 0$$

Differentiating with respect to P and equating to zero, we get

$$w_1(P - X_1) + w_2(P - X_2) + w_3(P - X_3) + \cdots + w_n(P - X_n) = 0$$

$$P = (w_1X_1 + w_2X_2 + w_3X_3 + \cdots + w_nX_n)/n$$

$$= \text{weighted arithmetic mean}$$

Therefore, the most probable value of a set of weighted observations is the weighted arithmetic mean of the observations. The weight of the most probable value is the sum of the weights of the observations $= w_1 + w_2 + w_3 + \cdots + w_n = \Sigma w$.

Example 18.6 The following values are the observed values of $\angle A$: 36° 22′ 40″, 36° 22′ 44″, 36° 22′ 42″, 36° 22′ 50″, 36° 22′ 38″, 36° 22′ 46″. Find the most probable value of the angle.

Solution The most probable value is the arithmetic mean. The number of observations is 6. The most probable value P is given by

$$P = 36° 22′ + (40″ + 44″ + 42″ + 50″ + 38″ + 46″)/6 = 36° 22′ 43.33″$$

Example 18.7 The following values are the observed values of $\angle A$: 36° 22′ 40″ (1), 36° 22′ 44″ (2), 36° 22′ 42″ (4), 36° 22′ 50″ (3), 36° 22′ 38″ (2), 36° 22′ 46″ (3). The weights assigned to the measurements are given in brackets. Find the most probable value of $\angle A$.

Solution The most probable value in the case of weighted observations is given by the weighted arithmetic mean of the observations.

Weighted arithmetic mean $= \Sigma wX/\Sigma w$

$$\Sigma wX/\Sigma w = 36° 22′ + (40 \times 1 + 44 \times 2 + 42 \times 4 + 50 \times 3 + 38 \times 2 + 46 \times 3)/15$$

$$= 36° 22′ 44″$$

Indirect observation of independent quantities The most probable value of an indirectly observed quantity is found from a normal equation. A normal equation is a conditional equation from which we find the most probable value of any one quantity by assigning a particular set of values to the remaining quantities. Normal equations are formed for each of the unknowns from their observation equations.

Normal equations The observation equation is a simple equation expressing the relationship between a quantity and its value. If we measure the angles around a point a point by the reiteration method, closing the horizon, then the sum of the angles is 360°.

$$\angle A + \angle B + \angle C = 360° = -k$$

This equation results in

$$\angle A + \angle B + \angle C + k = 0$$

The error e in the sum can be distributed equally amongst the three angles to get their correct values if the measurements are done with equal precision. On the other hand, if one or more angles are measured directly and the others indirectly, then the equation takes the form

$$aX + bY + cZ + k = e$$

where e is the error.

If a number of measurements of the unknowns are made, then the equations may take the form

$$a_1X_1 + b_1Y_1 + c_1Z_1 = e_1$$
$$a_2X_2 + b_2Y_2 + c_2Z_2 = e_2$$
$$a_3X_3 + b_3Y_3 + c_3Z_3 = e_3$$

or

$$\Sigma(aX + bY + cZ) = \Sigma e$$

According to the principle of least squares, if e_1, e_2, e_3, etc. are errors in the sum, then $\Sigma e^2 = \Sigma(aX + bY + cZ)^2$ is minimum. Differentiating and equating to zero with respect to each of the unknowns,

$$\Sigma a(aX + bY + cZ) = 0 \quad \text{(normal equation for } X)$$
$$\Sigma b(aX + bY + cZ) = 0 \quad \text{(normal equation for } Y)$$
$$\Sigma c(aX + bY + cZ) = 0 \quad \text{(normal equation for } Z)$$

These equations are obtained by multiplying the equations with the coefficient of the unknown which is to be evaluated. These equations are known as normal equations. As the number of equations is equal to the number of unknowns, the solutions of these equations yield the probable value of the quantities.

If the observations are weighted, a similar procedure is adopted to form equations. Let w_1, w_2, w_3, etc. be the weights of observations $(X_1, Y_1, Z_1), (X_2, Y_2, Z_2)$, etc. Then, the equations take the form

$$\Sigma w_1(aX_1 + bY_1 + cZ_1) = e_1$$
$$\Sigma w_2(aX_2 + bY_2 + cZ_2) = e_2$$
$$\Sigma w_3(aX_3 + bY_3 + cZ_3) = e_3$$

From the principle of least squares,

$$\Sigma w(aX + bY + cZ)^2 = \Sigma we^2$$

Differentiating with respect to each of the unknowns as before,

$$\Sigma wa(aX + bY + cZ) = 0 \quad \text{(normal equation for } X)$$
$$\Sigma wb(aX + bY + cZ) = 0 \quad \text{(normal equation for } Y)$$
$$\Sigma wc(aX + bY + cZ) = 0 \quad \text{(normal equation for } Z)$$

In both the cases discussed above, the normal equations for an unknown are added to get one equation. Similarly, the normal equations for the other unknowns can be added to get equations. These equations are then solved to determine the probable values of the unknowns. The following examples illustrate the formulation of normal equations.

Example 18.8 From the equations of equal weight given below, form the normal equations for each of the unknowns.

$$2X + 3Y + 5Z - 41 = 0 \tag{i}$$
$$3X + 5Y + 4Z - 46 = 0 \tag{ii}$$
$$4X + 2Y + 7Z - 51 = 0 \tag{iii}$$

Solution To form the normal equation for any unknown, each equation is multiplied by the coefficient of that unknown in that equation. Thus, the normal equations in X are obtained as follows:

$$4X + 6Y + 10Z - 82 = 0 \quad \text{[by multiplying Eqn (i) by 2]}$$
$$9X + 15Y + 12Z - 138 = 0 \quad \text{[by multiplying Eqn (ii) by 3]}$$
$$16X + 8Y + 28Z - 204 = 0 \quad \text{[by multiplying Eqn (iii) by 4]}$$

Adding the three equations,

$$29X + 29Y + 50Z - 424 = 0 \tag{iv}$$

The normal equations in Y are obtained as follows:

$$6X + 9Y + 15Z - 123 = 0 \quad \text{[by multiplying Eqn (i) by 3]}$$
$$15X + 25Y + 20Z - 230 = 0 \quad \text{[by multiplying Eqn (ii) by 5]}$$
$$8X + 4Y + 14Z - 102 = 0 \quad \text{[by multiplying Eqn (iii) by 2]}$$

Adding,

$$29X + 38Y + 49Z - 455 = 0 \tag{v}$$

The normal equations in Z are obtained as follows:

$$10X + 15Y + 25Z - 205 = 0 \quad \text{[by multiplying Eqn (i) by 5]}$$
$$12X + 20Y + 16Z - 184 = 0 \quad \text{[by multiplying Eqn (ii) by 4]}$$
$$28X + 14Y + 49Z - 574 = 0 \quad \text{[by multiplying Eqn (iii) by 7]}$$

Adding,

$$50X + 49Y + 90Z - 1418 = 0 \tag{vi}$$

Equations (iv), (v), and (vi) can be solved to get the probable values of X, Y, and Z.

Example 18.9 The equations given below have the weights specified. Form the normal equations for each of the unknowns.

$$2X + 3Y + 5Z - 41 = 0 \quad \text{(weight 2)} \tag{i}$$
$$3X + 5Y + 4Z - 46 = 0 \quad \text{(weight 3)} \tag{ii}$$
$$4X + 2Y + 7Z - 51 = 0 \quad \text{(weight 3)} \tag{iii}$$

Solution In the case of weighted quantities, the normal equations are formed by multiplying the equations by the coefficient of the unknown and the weight of that equation. Thus, the normal equations in X are obtained as follows:

$$8X + 12Y + 20Z - 164 = 0 \quad \text{[by multiplying Eqn (i) twice by 2]}$$
$$27X + 45Y + 36Z - 414 = 0 \quad \text{[by multiplying Eqn (ii) twice by 3]}$$

$48X + 24Y + 24Z - 612 = 0$ [Eqn (iii) multiplied by 4 and 3]

Adding,

$83X + 81Y + 80Z - 1190 = 0$ (iv)

The normal equations in Y are obtained as follows:

$12X + 18Y + 30Z - 246 = 0$ [Eqn (i) multiplied by 3 and 2]

$45X + 75Y + 60Z - 640 = 0$ [Eqn (ii) multiplied by 5 and 3]

$24X + 12Y + 42Z - 306 = 0$ [Eqn (iii) multiplied by 2 and 3]

Adding,

$81X + 105Y + 132Z - 1192 = 0$ (v)

The normal equations in Z are obtained as follows:

$20X + 30Y + 50Z - 410 = 0$ [Eqn (i) multiplied by 5 and 2]

$36X + 60Y + 48Z - 552 = 0$ [Eqn (ii) multiplied by 4 and 3]

$84X + 42Y + 147Z - 1071 = 0$ [Eqn (iii) multiplied by 7 and 3]

Adding,

$140X + 132Y + 245Z - 2033 = 0$ (vi)

Equations (iv), (v), and (vi) are solved to get the probable values of X, Y, and Z. The two cases that arise (of equal weight observations and weighted observations) are solved to get the probable value as explained below.

Indirect observations of equal weight   The normal equation for an unknown is formed by multiplying each equation by the coefficient of the unknown in that equation. These equations are then solved to get the most probable value of the unknown. Thus, if the observation equations are given as

$$A = \alpha \tag{18.1}$$

$$2A = \beta \tag{18.2}$$

$$6A = \gamma \tag{18.3}$$

then the normal equations are formed as follows:

$1 \times A = 1 \times \alpha$ [Eqn (18.1) is multiplied by 1, the coefficient of
                      unknown A]

$2 \times 2A = 2\beta$ [Eqn (18.2) is multiplied by 2, the coefficient of A]

$6 \times 6A = 6\gamma$ [Eqn (18.3) is multiplied by 6, the coefficient of A]

Adding all the equations,

$41A = \alpha + 2\beta + 6\gamma$

This equation is solved to get the probable value of A.

Indirect observations of unequal weight   In case the observations are weighted, the normal equation is formed using the following rule: 'The normal equation is formed for any unknown by multiplying the equation by the product of the coefficient of the unknown and the weight of that observation.'

$A = \alpha$ (weight 2)   (18.4)

$2A = \beta$ (weight 4)   (18.5)

$6A = \gamma$ (weight 6)   (18.6)

The equations are multiplied by the coefficient and the weight to form the normal equations:

$$2A = 2\alpha \quad \text{[Eqn (18.4) multiplied by 1 and 2]}$$

$$16A = 8\beta \quad \text{[Eqn (18.5) multiplied by 2 and 4]}$$

$$36 \times 6A = 36\gamma \quad \text{[by multiplying Eqn (18.6) twice by 6]}$$

Adding the above equations, we get

$$234\,A = 2\alpha + 8\beta + 36\gamma$$

This equation will give the probable value of A.

Conditioned quantities In the case of conditioned quantities, one or more conditional equations are available in addition to the observation equations. For example, $A = \alpha$, $B = \beta$, $C = \gamma$, $A + B = C$ (Fig. 18.2) has three observation equations and one conditional equation.

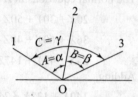

Fig. 18.2 Conditioned quantities

In one method of solution, we avoid the conditional equation by writing $A = \alpha$, $B = \beta$, $A + B = \gamma$. These equations are solved for A and B by forming normal equations using the rules stated above. The normal equations in A are obtained as follows:

$$A = \alpha$$
$$A + B = \gamma$$

Therefore, the normal equation for A is

$$2A + B = \alpha + \gamma$$

The normal equations in B are obtained as follows:

$$B = \beta$$
$$A + B = \gamma$$

Therefore, the normal equation for B is

$$A + 2B = \beta + \gamma$$

These two equations can be solved to get probable values of A and B. The probable value of C is given as

$$C = A + B$$

The second method of solution is by the method of correlates. This is explained later.

Method of differences In forming normal equations, the values may become very large. The equations are, therefore, formed in terms of the corrections to the angles, which makes it easier to solve the equations. For example, in Fig. 18.2,

$$A = \alpha, \ B = \beta, \ C = A + B = \gamma$$

Let the corrections to the angles A and B be c_1 and c_2. Then

$$A = \alpha + c_1$$

$$B = \beta + c_2$$
$$A + B = C = \gamma$$
$$\alpha + c_1 + \beta + c_2 = \gamma$$

Subtracting the given equations from the respective derived equations,

$$c_1 = 0$$
$$c_2 = 0$$
$$c_1 + c_2 = \gamma - (\alpha + \beta)$$

If the equations have equal weight, the equations are multiplied by the coefficients of the terms. In the above case the coefficients of c_1 and c_2 are unity. If the observations are weighted, then the equations are multiplied by the coefficients and the respective weights. The normal equations in c_1 and c_2 are formed as follows:

$$2c_1 + c_2 = \gamma - (\alpha + \beta) \quad \text{(normal equation in } c_1)$$
$$c_1 + 2c_2 = \gamma - (\alpha + \beta) \quad \text{(normal equation in } c_2)$$

These two equations are solved to get the values of c_1 and c_2. Knowing c_1 and c_2, the probable values of A, B, and C ($= A + B$) can be calculated.

The following examples illustrate the application of these principles.

Example 18.10 Find the probable values of A, B, and C given that $A = 46° 12' 34''$, $B = 22° 18' 36''$, and $C = A + B = 68° 31' 13''$.

Solution We eliminate the conditional equation by writing the equations as

$$A = 46° 12' 34'' \tag{i}$$
$$B = 22° 18' 36'' \tag{ii}$$
$$A + B = 68° 31' 13'' \tag{iii}$$

To form a normal equation in A, multiply (i) by 1 and (iii) by 1 and add. Thus, the normal equation in A is

$$2A + B = 114° 43' 47''$$

Similarly, the normal equation in B is obtained by multiplying (ii) by 1 and (iii) by 1 and adding them. Therefore, the normal equation in B is

$$A + 2B = 90° 49' 49''$$

Solving these two equations, we get

$$A = 46° 12' 35'', \quad B = 22° 18' 37''$$

Probable value of $A = 46° 12' 35''$

Probable value of $B = 22° 18' 37''$

Probable value of $C = 68° 31' 12''$

Solution by the method of differences

Let c_1 and c_2 be the corrections to A and B.

$$A = 46° 12' 34'' + c_1$$

By subtracting Eqn (i) from the above equation, we get

$$c_1 = 0$$
$$B = 22° 18' 37'' + c_2$$

By subtracting Eqn (ii) from the above equation, we get

$$c_2 = 0$$

$$46° 12' 34'' + c_1 + 22° 18' 36'' + c_2 = 68° 31' 13''$$

This gives

$$c_1 + c_2 = 3''$$

The three difference equations are

$$c_1 = 0, \ c_2 = 0, \ c_1 + c_2 = 3''$$

On forming normal equations for c_1 and c_2, we get

$$2c_1 + c_2 = 3'' \ \text{(normal equation in } c_1)$$

$$c_1 + 2c_2 = 3'' \ \text{(normal equation in } c_2)$$

Solving these two equations, we get

$$c_1 = 1'', \ c_2 = 1''$$

The probable values of A, B, and C are calculated as follows:

$$A = 46° 12' 34'' + 1'' = 46° 12' 35''$$

$$B = 22° 18' 36'' + 1'' = 22° 18' 37''$$

$$C = A + B = 68° 31' 12''$$

Example 18.11   The angles A, B, and C of a triangle are $A = 59° 32' 46''$, $B = 56° 12' 18''$, and $C = 64° 02' 15''$. Find the probable values of A, B, and C if (a) the values are of equal weight and (b) the values of A, B, and C have weights 2, 4, and 3, respectively.

Solution   The angles of a triangle are governed by the condition $A + B + C = 180°$. This condition can be written as $A + B = 180° - C = 180° - 64° 02' 15'' = 115° 44' 58''$. Thus, we have three equations:

$$A = 59° 32' 46''$$

$$B = 56° 12' 18''$$

$$A + B = 115° 44' 56''$$

(a) When the observations are of equal weight, the coefficients being 1,

$$A = 59° 32' 46'' \tag{i}$$

$$B = 56° 12' 18'' \tag{ii}$$

$$A + B = 115° 44' 56'' \tag{iii}$$

The normal equation in A is the sum of equations (i) and (iii) as given below.

$$2A + B = 175° 17' 44''$$

The normal equation in B is the sum of equations (ii) and (iii) as given below.

$$A + 2B = 171° 57' 16''$$

Solving these two equations, we get

$$A = 59° 32' 44''$$

$$B = 56° 12' 16''$$

$$C = 180° - (A + B) = 64° 15' 00''$$

These are the probable values of A, B, and C.

(b) When the observations are weighted: In this case, when forming the normal equation, the weights of the equations are taken into account. As in the previous case,

$$A = 59° 32' 46'' \ \text{(weight} = 2) \tag{iv}$$

$$B = 56° 12' 18'' \ \text{(weight} = 4) \tag{v}$$

$$A + B = 115° \ 44' \ 56'' \quad \text{(weight = 3)} \tag{vi}$$

$$2A = 119° \ 05' \ 32'' \quad \text{[by multiplying Eqn (iv) by 2]}$$

$$4B = 224° \ 49' \ 12'' \quad \text{[by multiplying Eqn (v) by 2]}$$

$$3A + 3B = 347° \ 14' \ 54'' \quad \text{[by multiplying Eqn (vi) by 3]}$$

The normal equation in A is given as

$$5A + 3B = 466° \ 20' \ 26''$$

The normal equation in B is given as

$$3A + 7B = 572° \ 04' \ 06''$$

Solving these equations, we get

$$A = 59° \ 32' \ 43.23'', \quad B = 56° \ 12' \ 16.62''$$

$$C = 180° - (59° \ 32' \ 43.23'' + 56° \ 12' \ 16.62'') = 64° \ 15' \ 0.15''$$

These are the probable values of A, B, and C.

Method of differences

In the above solution it is seen that the right-hand side values are very large. The method of differences makes the calculations simple. We take A and B as independent quantities. If c_1 and c_2 are the corrections in A and B, then

$$A = 59° \ 32' \ 46'' + c_1 \quad \text{(weight = 2)}$$

$$B = 56° \ 12' \ 18'' + c_2 \quad \text{(weight = 4)}$$

$$A + B = 180° - 64° \ 15' \ 02'' = 115° \ 44' \ 58'' \quad \text{(weight = 3)}$$

$$59° \ 32' \ 46'' + c_1 + 56° \ 12' \ 18'' + c_2 = 115° \ 44' \ 58''$$

$$c_1 + c_2 = -6''$$

Also from the first two equations, we get $c_1 = 0$ and $c_2 = 0$.

$$c_1 = 0 \quad \text{(weight 2)}$$

$$c_2 = 0 \quad \text{(weight 4)}$$

$$c_1 + c_2 = -6'' \quad \text{(weight 3)}$$

We form the normal equations in c_1 and c_2:

$$5c_1 + 3c_2 = -18''$$

$$3c_1 + 7c_2 = -18''$$

Solving these two equations,

$$c_1 = 2.77'', \quad c_2 = 1.38''$$

The probable values of A, B, and C are

$$A = 59° \ 32' \ 46'' - 2.77'' = 59° \ 32' \ 43.23''$$

$$B = 56° \ 12' \ 18'' - 1.38'' = 56° \ 12' \ 16.62''$$

$$C = 180° - (59° \ 32' \ 43.23'' + 56° \ 12' \ 16.62'') = 64° \ 15' \ 0.15''$$

Example 18.12   Four angles are measured at a station closing the horizon. The values of the angles are $A = 102° \ 48' \ 51''$ (weight = 3), $B = 85° \ 42' \ 37''$ (weight = 2), $C = 108° \ 36' \ 47''$ (weight = 4), and $D = 62° \ 51' \ 50''$ (weight = 1). Find the probable values of these angles.

Solution   The following condition holds for the angles (Fig. 18.3):

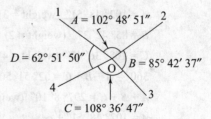

Fig. 18.3   Angles of Example 18.12

$$A + B + C + D = 360°$$

Three angles are independent quantities and the fourth angle is subject to the condition stated above. The observation equations can be written as

$A = 102° \, 48' \, 51''$ (weight = 3) (i)

$B = 85° \, 42' \, 37''$ (weight = 2) (ii)

$C = 108° \, 36' \, 47''$ (weight = 4) (iii)

$360° - A + B + C = 62° \, 51' \, 50''$

$A + B + C = 297° \, 8' \, 10''$ (weight = 1) (iv)

To obtain the normal equation in A, we multiply the equation containing A with its coefficient and the weight.

$3A = 308° \, 26' \, 23''$

$A + B + C = 297° \, 8' \, 10''$

Adding these two equations, we get the normal equation in A as

$4A + B + C = 605° \, 34' \, 43''$

Similarly, for the normal equation in B, we multiply the equation containing B with its coefficient and the weight.

$2B = 171° \, 25' \, 14''$

$A + B + C = 297° \, 08' \, 10''$

Adding these two equations, we get the normal equation in B as

$A + 3B + C = 468° \, 33' \, 24''$

To obtain the normal equation in C, we multiply the equation containing C with its coefficient and the weight.

$4C = 434° \, 27' \, 8''$

$A + B + C = 297° \, 8' \, 10''$

Adding these two equations, we get the normal equation in C as

$A + B + 5C = 731° \, 35' \, 18''$

Solving the three normal equations in C, we get the most probable values of the angles as

$A = 102° \, 48' \, 50.2''$

$B = 85° \, 42' \, 35.8''$

$C = 108° \, 36' \, 46.4''$

$D = 62° \, 51' \, 47.6''$

Method of differences

As there are three independent quantities, let c_1, c_2, and c_3 be the errors in A, B, and C. The observation equations are as follows:

$A = 102° \, 48' \, 51''$ (weight = 3) (v)

$B = 85° \, 42' \, 37''$ (weight = 2) (vi)

$C = 108° \, 36' \, 47''$ (weight = 4) (vii)

$360° - A + B + C = 62° \, 51' \, 50''$

$A + B + C = 297° \, 8' \, 10''$ (weight = 1)

$A = 102° \, 48' \, 51'' + c_1$ (viii)

$$B = 85° 42' 37'' + c_2 \tag{ix}$$

$$C = 108° 36' 47'' + c_3 \tag{x}$$

$$102° 48' 51'' + c_1 + 85° 42' 37'' + c_2 + 108° 36' 47'' + c_3 = 297° 8' 10''$$

$c_1 = 0$ (weight = 3) [subtracting Eqn (v) from Eqn (viii)]

$c_2 = 0$ (weight = 2) [subtracting Eqn (vi) from Eqn (ix)]

$c_3 = 0$ (weight = 4) [subtracting Eqn (vii) from Eqn (x)]

$c_1 + c_2 + c_3 = -5''$ (weight = 1)

We form the normal equations in c_1, c_2, and c_3 as follows:

$$4c_1 + c_2 + c_3 = -5$$

$$c_1 + 3c_2 + c_3 = -5$$

$$c_1 + c_2 + 5c_3 = -5$$

Solving these three equations, we get

$$c_1 = -0.8'', \quad c_2 = -1.2'', \quad c_3 = -0.6''$$

From these corrections, the probable values of the angles are

$$A = 102° 48' 51'' - 0.8'' = 102° 48' 50.2''$$

$$B = 85° 42' 37'' - 1.2'' = 85° 42' 35.8''$$

$$C = 108° 36' 47'' - 0.6'' = 108° 36' 46.4''$$

$$D = 360° - 102° 48' 50.2'' - 85° 42' 35.8'' - 108° 36' 46.4'' = 62° 51' 47.6''$$

Example 18.13 Find the most probable values of the angles A, B, and C from the following measurements: $A = 68° 12' 40.8''$, $A + B = 140° 40' 48.2''$, $B + C = 109° 16' 29.2''$, $A + B + C = 177° 29' 8.2''$. All measurements have equal weights.

Solution The four observation equations are

$$A = 68° 12' 40.8'' \tag{i}$$

$$A + B = 140° 40' 48.2'' \tag{ii}$$

$$B + C = 109° 16' 29.2'' \tag{iii}$$

$$A + B + C = 177° 19' 8.2'' \tag{iv}$$

Since the weights are the same for all equations and the coefficients are unity, we form normal equations by adding the equations. The normal equations in A are obtained as follows:

$$A = 68° 12' 40.8''$$

$$A + B = 140° 40' 48.2''$$

$$A + B + C = 177° 19' 8.2''$$

Adding,

$$3A + 2B + C = 386° 12' 37.2'' \tag{v}$$

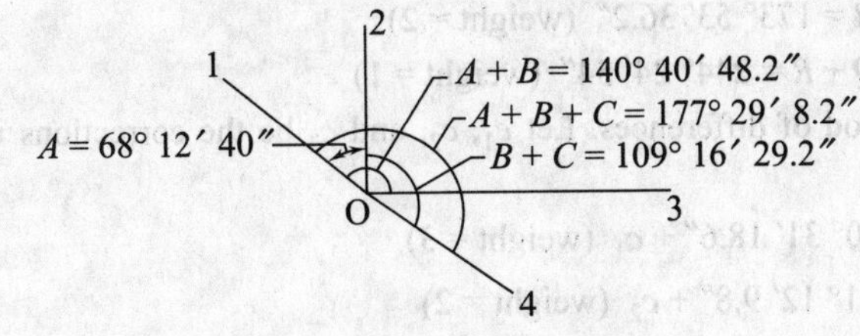

Fig. 18.4 Angles of Example 18.13

The normal equations in B are obtained as follows:

$$A + B = 140° \, 40' \, 48.2''$$

$$B + C = 109° \, 16' \, 29.2''$$

$$A + B + C = 177° \, 19' \, 8.2''$$

Adding,

$$2A + 3B + 2C = 427° \, 16' \, 25.6''$$ (vi)

The normal equations in C are obtained as follows:

$$B + C = 109° \, 16' \, 29.2''$$

$$A + B + C = 177° \, 19' \, 8.2''$$

Adding,

$$A + 2B + 2C = 286° \, 45' \, 37.4''$$ (vii)

Solving equations (v), (vi), and (vii) simultaneously, we get the probable values of A, B, and C as

$$A = 68° \, 12' \, 40.2''$$

$$B = 72° \, 28' \, 8''$$

$$C = 36° \, 48' \, 20.6''$$

Example 18.14 Find the probable values of angles P, Q, and R using the given data (Fig. 18.5). The weights of the observations are given in brackets. $P = 70° \, 31' \, 18.6''$ (3), $Q = 61° \, 12' \, 9.8''$ (2), $R = 112° \, 41' \, 31.6''$ (4), $P + Q = 131° \, 43' \, 20.6''$ (2), $Q + R = 173° \, 53' \, 36.2''$ (2), $P + Q + R = 244° \, 24' \, 54''$ (1).

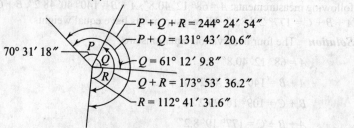

70° 31′ 18″ —

$P + Q + R = 244° \, 24' \, 54''$
$P + Q = 131° \, 43' \, 20.6''$
$Q = 61° \, 12' \, 9.8''$
$Q + R = 173° \, 53' \, 36.2''$
$R = 112° \, 41' \, 31.6''$

Fig. 18.5 Angles of Example 18.14

Solution The six observation equations are

$$P = 70° \, 31' \, 18.6'' \quad (\text{weight} = 3)$$

$$Q = 61° \, 12' \, 9.8'' \quad (\text{weight} = 2)$$

$$R = 112° \, 41' \, 31.6'' \quad (\text{weight} = 4)$$

$$P + Q = 131° \, 43' \, 20.6'' \quad (\text{weight} = 2)$$

$$Q + R = 173° \, 53' \, 36.2'' \quad (\text{weight} = 2)$$

$$P + Q + R = 244° \, 24' \, 54'' \quad (\text{weight} = 1)$$

We use the method of differences. Let c_1, c_2, and c_3 be the corrections in P, Q, and R, respectively.

$$P = 70° \, 31' \, 18.6'' + c_1 \quad (\text{weight} = 3)$$ (i)

$$Q = 61° \, 12' \, 9.8'' + c_2 \quad (\text{weight} = 2)$$ (ii)

$$R = 112° \, 41' \, 31.6'' + c_3 \quad (\text{weight} = 4)$$ (iii)

$$P + Q = 131° 43' 28.4'' + c_1 + c_2 \text{ (weight = 2)} \qquad \text{[adding (i) and (ii)]}$$
$$Q + R = 173° 53' 41.4'' + c_2 + c_3 \text{ (weight = 2)} \qquad \text{[adding (ii) and (iii)]}$$
$$P + Q + R = 244° 25' 00'' + c_1 + c_2 + c_3 \text{ (weight = 1)} \qquad \text{[adding (i), (ii)}$$
$$\text{and (iii)]}$$

Subtracting the above equations with the help of the observation equations, we get

$$c_1 = 0 \text{ (weight = 3)}$$
$$c_2 = 0 \text{ (weight = 2)}$$
$$c_3 = 0 \text{ (weight = 4)}$$
$$c_1 + c_2 = -7.8 \text{ (weight = 2)}$$
$$c_2 + c_3 = -5.2 \text{ (weight = 2)}$$
$$c_1 + c_2 + c_3 = -6 \text{ (weight = 1)}$$

We form normal equations in c_1, c_2, and c_3 and multiply them with their coefficients and weights. The normal equations in c_1 are obtained as follows:

$$3c_1 = 0$$
$$2c_1 + 2c_2 = -15.6$$
$$c_1 + c_2 + c_3 = -6$$

Adding these three equations, the normal equation in c_1 is

$$6c_1 + 3c_2 + c_3 = -21.6 \qquad \text{(iv)}$$

The normal equations in c_2 are obtained as follows:

$$2c_2 = 0$$
$$2c_1 + 2c_2 = -15.6$$
$$2c_2 + c_3 = -10.4$$
$$c_1 + c_2 + c_3 = -6$$

Adding these equations, the normal equation in c_2 is

$$3c_1 + 7c_2 + 2c_3 = -32 \qquad \text{(v)}$$

The normal equations in c_3 are obtained as follows:

$$4c_3 = 0$$
$$2c_2 + 2c_3 = -10.4$$
$$c_1 + c_2 + c_3 = -6$$

Adding these equations, the normal equation in c_3 is

$$c_1 + 3c_2 + 7c_3 = -16.4 \qquad \text{(vi)}$$

Equations (iv), (v), and (vi) are simultaneously solved to get the values of c_1, c_2, and c_3:

$$c_1 = -1.74, \quad c_2 = -3.58, \quad c_3 = -0.56$$

With these corrections, the probable values of the angle are

$$P = 70° 31' 16.86'', \quad Q = 112° 41' 31.05'', \quad R = 61° 12' 6.22''$$

Example 18.15 $A, B,$ and C were given as 70° 31' 18.6'' (weight = 3), 61° 12' 9.8'' (weight = 2), and 112° 41' 13.6'' (weight = 4), respectively. $(A + B)$ was also measured and recorded as 131° 43' 20.6'' (weight = 2). Find the probable values of A, B, and C.

Solution We use the method of differences. As there is one conditional equation, let the errors in the values be c_1, c_2, and c_3 in A, B, and C, respectively. The observation equations are

$$A = 70° 31' 18.6'' \text{ (weight = 3)}$$
$$B = 61° 12' 9.8'' \text{ (weight = 2)}$$
$$C = 112° 41' 13.6'' \text{ (weight = 4)}$$
$$A + B = 131° 43' 20.6'' \text{ (weight = 2)}$$

With the corrections, the equations become

$$A = 70° 31' 18.6'' + c_1$$
$$B = 61° 12' 9.8'' + c_2$$
$$C = 112° 41' 13.6'' + c_3$$
$$70° 31' 18.6'' + c_1 + 61° 12' 9.8'' + c_2 = 131° 43' 20.6''$$

Subtracting the observation equations, we get

$$c_1 = 0, \quad c_2 = 0, \quad c_3 = 0, \quad c_1 + c_2 = -7.8$$

To form normal equations, we multiply these equations by their weights and add them.

$$3c_1 = 0$$
$$2c_2 = 0$$
$$4c_3 = 0$$
$$2c_1 + 2c_2 = -15.6$$

The normal equations obtained are

$$5c_1 + 2c_2 = -15.6$$
$$2c_1 + 4c_2 = -15.6$$
$$4c_3 = 0$$

Solving the first two equations, we get

$$c_1 = -1.95, \quad c_2 = -2.92, \quad c_3 = 0$$

The corrected probable values of the angle are

$$A = 70° 31' 18.6'' - 1.95'' = 70° 31' 16.65''$$
$$B = 61° 12' 9.8'' - 2.92'' = 61° 12' 6.88''$$
$$C = 112° 41' 13.6''$$
$$A + B = 131° 43' 23.53''$$

18.6 Method of Correlates

With many conditional equations, the problem of determining the probable values becomes lengthy. In the case of complicated problems, the method of correlates is used to find the probable values.

Correlates are unknown constants which are derived from the principle of least squares and conditional equations. The number of correlates required is equal to the number of conditions.

Consider the case of angles measured around a point, closing the horizon. A, B, C, and D are the angles and the condition relating these angles is that $A + B + C + D = 360°$. Let w_1, w_2, w_3, and w_4 be the weights of the angles A, B, C, and D, respectively. Let e_1, e_2, e_3, and e_4 be the corrections in the angles A, B, C, and D. If E is the residual error in the sum of angles, we have

$$e_1 + e_2 + e_3 + e_4 = \Sigma e = E \tag{18.7}$$

Also from the principle of least squares

$$w_1 e_1^2 + w_2 e_2^2 + w_3 e_3^2 + w_4 e_4^2 = \Sigma w e^2 \text{ is minimum} \tag{18.8}$$

Differentiating these two equations,

$$\Sigma \delta e = \delta e_1 + \delta e_2 + \delta e_3 + \delta e_4 = 0 \tag{18.9}$$

$$w_1 e_1 \delta e_1 + w_2 e_2 \delta e_2 + w_3 e_3 \delta e_3 + w_4 e_4 \delta e_4 = \Sigma w e\ \delta e = 0 \tag{18.10}$$

We multiply Eqn (18.9) by a correlate $-\lambda$ and add the two equations:

$$-\lambda \delta e_1 - \lambda \delta e_2 - \lambda \delta e_3 - \lambda \delta e_4 = 0 \tag{18.11}$$

$$w_1 e_1 \delta e_1 + w_2 e_2 \delta e_2 + w_3 e_3 \delta e_3 + w_4 e_4 \delta e_4 = 0 \tag{18.12}$$

Adding Eqns (18.11) and (18.12), we get

$$\delta e_1 (w_1 e_1 - \lambda) + \delta e_2 (w_2 e_2 - \lambda) + \delta e_3 (w_3 e_3 - \lambda) + \delta e_4 (w_4 e_4 - \lambda) = 0$$

As δe_1, δe_2, δe_3, and δe_4 are definite independent quantities, the multiplying factors must independently vanish. Thus, we get

$$w_1 e_1 - \lambda = 0, \quad w_2 e_2 - \lambda = 0, \quad \text{etc.}$$

$$\lambda = w_1 e_1 = w_2 e_2 = w_3 e_3 = w_4 e_4$$

Also,

$$e_1 = \lambda / w_1, \quad e_2 = \lambda / w_2, \quad e_3 = \lambda / w_3, \quad e_4 = \lambda / w_4$$

The above values of corrections show that the corrections are inversely proportional to the weights of the quantities.

$$\frac{\lambda}{w_1} + \frac{\lambda}{w_2} + \frac{\lambda}{w_3} + \frac{\lambda}{w_4} = E$$

$$\lambda \left(\frac{1}{w_1} + \frac{1}{w_2} + \frac{1}{w_3} + \frac{1}{w_4} \right) = E$$

λ can be calculated from this equation, as the weights and E are known. The corrections can then be calculated from the four equations of corrections. Once the corrections are known, the probable values can be calculated.

The above illustration had only one conditional equation and hence one correlate λ was needed. In case there is more than one conditional equation, then a number of correlates such as λ_1, λ_2, λ_3 are used. The values of λ's can be calculated in the same way. The following examples illustrate the procedure to find the probable values.

Example 18.16 Four angles were measured at a station closing the horizon. The values of the angles are $A = 102° 48' 51''$ (weight = 3), $B = 85° 42' 37''$ (weight = 2), $C = 108° 36' 47''$ (weight = 4), $D = 62° 51' 50''$ (weight = 1). Find the probable values of the angles.

Solution There is one condition for the angles. The observation equations are

$$A = 102° 48' 51''$$

$$B = 85° 42' 37''$$

$$C = 108° 36' 47''$$

$$D = 62° 51' 50''$$

Sum of the angles = 360° 00' 05''

The conditional equation is

$$A + B + C + D = 360°$$

Let e_a, e_b, e_c, and e_d be the corrections to A, B, C, and D.

$$e_a + e_b + e_c + e_d = -5''$$

From the principle of least squares, $3e_a^2 + 4e_b^2 + 2e_c^2 + e_d^2$ is minimum. Differentiating the two equations,

$$\delta e_a + \delta e_b + \delta e_c + \delta e_d = 0 \tag{i}$$

$$3e_a\delta a + 2e_b\delta b + 4e_c\delta c + e_d\delta d = 0 \tag{ii}$$

There is one conditional equation. Therefore, one correlate $(-\lambda)$ is required. Multiplying Eqn (i) by $(-\lambda)$, we get

$$-\lambda\delta e_a - \lambda\delta e_b - \lambda\delta e_c - \lambda\delta e_d = 0$$

Adding the above equation and Eqn (ii), we get

$$\delta e_a(3e_a - \lambda) + \delta e_b(2e_b - \lambda) + \delta e_c(4e_c - \lambda) + \delta e_d(e_d - \lambda) = 0$$

The coefficients must separately vanish:

$$3e_a - \lambda = 0, \ e_a = \lambda/3$$

$$2e_b - \lambda = 0, \ e_b = \lambda/2$$

$$4e_c - \lambda = 0, \ e_c = \lambda/4$$

$$e_d - \lambda = 0, \ e_d = \lambda$$

Substituting these values in the equation $e_a + e_b + e_c + e_d = -5''$, we get

$$\frac{\lambda}{3} + \frac{\lambda}{2} + \frac{\lambda}{4} + \frac{\lambda}{1} = -5$$

From the above equation, $\lambda = -2.4$

$$e_a = \lambda/3 = -2.4/3 = 0.8'', \quad e_b = \lambda/2 = -2.4/2 = -1.2''$$

$$e_c = \lambda/4 = -2.4/4 = 0.6'', \quad e_d = \lambda = -2.4''$$

The corrected angles (probable values) are as follows:

$$A = 102° 48' 51'' - 0.8'' = 102° 48' 50.2''$$

$$B = 85° 42' 37'' - 1.2'' = 85° 42' 35.8''$$

$$C = 108° 36' 47'' - 0.6'' = 108° 36' 46.4''$$

$$D = 62° 51' 54'' - 2.4'' = 62° 51' 47.6''$$

Example 18.17 Find the most probable values of the angles A, B, and C of triangle ABC from the following measurements: $A = 76° 14' 25''$ (weight = 3), $B = 54° 34' 35''$ (weight = 4), $C = 49° 10' 48''$ (weight = 2).

Solution Sum of given angles = 76° 14' 25'' + 54° 34' 35'' + 49° 10' 48'' = 179° 59' 48''

Theoretical sum = 180° 00' 00''

Error = 179° 59' 48'' − 180° 00' 00'' = −12''

Total correction to angles = + 12''

This can be written as

$$e_a + e_b + e_c = 12''$$

From the principle of least squares, taking the weights into account, $3e_a^2 + 4e_b^2 + 2e_c^2$ is minimum. Differentiating both the equations,

$$\delta e_a + \delta e_b + \delta e_c = 0 \tag{i}$$

$$3e_a \delta e_a + 4e_b \delta e_b + 2e_c \delta e_c = 0 \tag{ii}$$

$$-\lambda \delta e_a - \lambda \delta e_b - \lambda \delta e_c = 0 \quad \text{[by multiplying (i) by correlate } -\lambda] \tag{iii}$$

Adding Eqns (ii) and (iii), we get

$$\delta e_a(3e_a - \lambda) + \delta e_b(4e_b - \lambda) + \delta e_c(2e_c - \lambda) = 0$$

The coefficients must separately vanish:

$$3e_a - \lambda = 0, \quad \lambda = 3e_a, \quad e_a = \lambda/3$$

$$4e_b - \lambda = 0, \quad \lambda = 4e_b, \quad e_b = \lambda/4$$

$$2e_c - \lambda = 0, \quad \lambda = 2e_c, \quad e_c = \lambda/2$$

We substitute these values in the equation

$$e_a + e_0 + e_c = + 12''$$

Thus, we obtain

$$\frac{\lambda}{3} + \frac{\lambda}{4} + \frac{\lambda}{2} = + 12''$$

$$\lambda = 11.08'', \quad e_a = 11.08/3 = 3.69'', \quad e_b = 11.08/4 = 2.77'', \quad e_c = 11.08/2 = 5.54''$$

The corrected and probable values of the angles are as follows:

$$A = 76° 14' 28.9''$$

$$B = 54° 34' 37.77''$$

$$C = 49° 10' 53.34''$$

Example 18.18 Find the most probable values of the angles P, Q, and R of a triangle PQR from the following measurements: $P = 70° 31' 18.6''$ (weight = 3), $Q = 61° 12' 11.8''$ (weight = 2), $R = 48° 16' 36.6''$ (weight = 4), $P + Q = 131° 43' 34.6''$ (weight = 2).

Solution There are two conditions in this case, $P + Q + R = 180°$ and $P + Q = \overline{P+Q}$. Let e_1, e_2, e_3, and e_4 be the corrections to angles P, Q, R, and $(P + Q)$.

$$P + Q + R = 70° 31' 18.6'' + 61° 12' 11.8'' + 48° 16' 36.6'' = 180° 00' 07''$$

$$P + Q = 131° 43' 30.4''$$

We can write the following equations:

$$e_1 + e_2 + e_3 = - 7''$$

$$e_1 + e_2 = - 4.2''$$

From the principle of least squares, $3e_1^2 + 2e_2^2 + 4e_3^2 + 2e_4^2$ is minimum. Differentiating these three equations,

$$\delta e_1 + \delta e_2 + \delta e_3 = 0 \tag{i}$$

$$\delta e_1 + \delta e_2 - \delta e_4 = 0 \tag{ii}$$

$$3e_1 \delta e_1 + 2e_2 \delta e_2 + 4e_3 \delta e_3 + 2e_4 \delta e_4 = 0 \tag{iii}$$

We use two correlates $-\lambda_1$ and $-\lambda_2$:

$$-\lambda_1 \delta e_1 - \lambda_1 \delta e_2 - \lambda_1 \delta e_3 = 0 \quad \text{[by multiplying (i) by } -\lambda_1]$$

$$-\lambda_2 \delta e_1 - \lambda_2 \delta e_2 + \lambda_2 \delta e_4 = 0 \quad \text{[by multiplying (ii) by } -\lambda_2]$$

Adding Eqn (iii) and the above two equations, we get

$$\delta e_1(3e_1 - \lambda_1 - \lambda_2) + \delta e_2(2e_2 - \lambda_1 - \lambda_2) + \delta e_3(4e_3 - \lambda_1) + \delta e_4(2e_4 + \lambda_2) = 0$$

As the coefficients of each of the δe terms should vanish independently,

$$3e_1 - \lambda_1 - \lambda_2 = 0 \text{ gives } e_1 = (\lambda_1 + \lambda_2)/3$$
$$2e_2 - \lambda_1 - \lambda_2 = 0 \text{ gives } e_2 = (\lambda_1 + \lambda_2)/2$$
$$4e_3 - \lambda_1 = 0 \text{ gives } e_4 = \lambda_1/4$$
$$2e_4 + \lambda_2 = 0 \text{ gives } e_4 = -\lambda_2/2$$

Substituting the values of e's into these equations, we get

$$(\lambda_1 + \lambda_2)/3 + (\lambda_1 + \lambda_2)/2 + \lambda_1/4 = -7$$
$$(\lambda_1 + \lambda_2) + (\lambda_1 + \lambda_2) + \lambda_2/2 = -4.2$$
$$13\lambda_1 + 10\lambda_2 = -84$$
$$5\lambda_1 + 8\lambda_2 = -25.2$$

Solving,

$$\lambda_1 = -7.78'', \quad \lambda_2 = 1.71''$$

Substituting these values back into the equations for e_1, e_2, e_3, and e_4, we get

$$e_1 = -2.02'', \quad e_2 = -3.03'', \quad e_3 = -1.95''$$

The correct (probable) values of the angles are

$$P = 70° \, 31' \, 16.58'', \quad Q = 61° \, 12' \, 8.77'', \quad R = 48° \, 16' \, 34.65''$$

18.7 Probable Error

Probable error in measurements is calculated from the probability curves of errors. In a series of sufficiently large number of observations made for a quantity, the probable error is of such a value that the number of errors larger than this value is equal to the number of errors smaller than this value. The probable error is a mathematical quantity derived from statistical principles and gives an idea of the precision in the measurements. The precision of observations can be gauged from their probable errors. The probable error is indicated along with the value of the quantity with a ± sign. For example, $A = 38° \, 12' \, 34'' \pm 3.15''$. The following cases are considered.

Direct observations of equal weight If a number of observations of equal weight are made for a quantity, then the probable error E_s of any single observation is given by

$$E_s = \pm 0.6745 \sqrt{\Sigma e^2 (n-1)}$$

where Σe^2 = sum of the squares of residual error = $e_1^2 + e_2^2 + e_3^2 + \cdots + e_n^2$, e is the residual error, and n is the number of observations.

Probable error in a single observation of weight w If an observation has weight w, the probable error E_{sw} in that observation is given by

$$E_{sw} = E_s / \sqrt{w}$$

Probable error in arithmetic mean The probable error in the arithmetic mean of observations, E_{sm}, is given by

$$E_{sm} = \pm 0.6745 \sqrt{\Sigma e^2 / n(n-1)}$$

Observations of unequal weights In the case of observations of unequal weights, the probable error is given by

$$E_s = 0.6745 \sqrt{\Sigma(we^2)/(n-1)}$$

for observations of unit weight,

$$E_{sw} = E_s / \sqrt{w}$$

for observations with weight w, and

$$E_m = E_s / \sqrt{\Sigma w}$$

for the weighted arithmetic mean.

Independent observations in independent quantities

Probable error in observations of unit weight,

$$E_s = \pm 0.6745 \sqrt{\Sigma we^2/(n-q)}$$

Probable error in an observation of weight $w = E_s / \sqrt{w}$

where n is the number of observation equations and q is the number of unknown quantities.

Indirect observations involving conditional equations Probable error in observations of unit weight, $E_s = \pm 0.6745 \sqrt{\Sigma we^2/(n-q+p)}$, where p is the number of conditions.

Probable error in an observation of weight $w = E_s / \sqrt{w}$

Probable error in computed quantities There can be a number of cases.

(i) The quantity is the sum or difference of a computed quantity and a constant. Let X be the computed quantity, e_X is the probable error in X, A is the observed quantity, e_A is the observed error in A, and K is a constant.

$$X = \pm A \pm K$$

$$e_X = e_A$$

Rule If a quantity is the sum or difference of an observed quantity and a constant, then the probable error is the same as that in the observed quantity.

(ii) The computed quantity is the product of an observed quantity and a constant:

$$X = KA$$

$$e_X = Ke_A$$

Rule If the computed quantity is the product of an observed quantity and a constant, then the probable error in the computed quantity is equal to the constant multiplied by the probable error in the observed quantity.

(iii) The computed quantity is the sum of two or more observed quantities. Let X be the computed quantity, A_1, A_2, A_3, etc. be the observed quantities, and e_1, e_2, e_3, etc. be the probable errors in A_1, A_2, A_3, etc.

$$X = A_1 + A_2 + A_3 + \cdots$$

$$e_X = \sqrt{e_1^2 + e_2^2 + e_3^2 + \cdots}$$

Rule If the computed quantity is obtained by adding two or more observed quantities, then the probable error is equal to the square root of the sum of the squares of the probable errors in the observed quantities.

(iv) The computed quantity is a function of the observed quantity. X is a computed quantity and A is the observed quantity,

$$X = f(A)$$

$$e_X = (dX/dA)e_A$$

Rule If the computed quantity is a function of the observed quantity, then the probable error is the product of the error in the observed quantity and the differential coefficient of the quantity with respect to the observed quantity.

(v) The computed quantity is a function of two or more quantities:

$$X = f(A_1, A_2, A_3, \ldots)$$

$$e_X = \sqrt{(e_{A_1}dX/dA_1)^2 + (e_{A_2}dX/dA_2)^2 + (e_{A_3}dX/dA_3)^2 + \cdots}$$

Rule If the computed quantity is a function of two or more observed quantities, its probable error is equal to the square root of the sum of the squares of the probable errors in the observed quantities and the differentiation with respect to the corresponding observed quantity.

The following examples illustrate the computation of probable errors.

Example 18.19 The following six observations of A were made: 62° 32′ 16.84″, 62° 32′ 18.22″, 62° 32′ 17.34″, 62° 32′ 15.95″, 62° 32′ 19.32″, 62° 32′ 14.84″. Find the mean, probable error in the mean, and the probable value.

Solution The solution is shown in Table 18.1.

Table 18.1 Solution of Example 18.19

Observed value	Error e	e^2
62° 32′ 16.84″	0.24	0.0576
62° 32′ 18.22″	− 1.14	1.300
62° 32′ 17.34″	− 0.26	0.0676
62° 32′ 15.95″	1.13	1.2769
62° 32′ 19.32″	− 2.24	5.0176
62° 32′ 14.84″	2.24	5.0176
Total = 375° 12′ 102.51″		Total = 12.7273

Mean angle = 375° 12′ 102.51″/6 = 62° 32′ 17.08″

Error in a single observation, $E_s = \pm 0.6745 \sqrt{12.7273/5} = 1.076″$

Error in the mean, $E_m = E_s / \sqrt{n}$ 1.076/ $\sqrt{6}$ = 0.44″

Probable value = 62° 32′ 17.08″ ± 0.44″

Example 18.20 A base line was measured and the following eight values were recorded: 5869.835 m, 5869.828 m, 5869.842 m, 5869.838 m, 5869.830 m, 5869.844 m, 5896.836 m, 5869.832 m. Find the probable value and the error in a single observation as well as the error in the mean.

Solution The solution is given in Table 18.2.

Mean = 46,958.685/8 = 5869.835 m

Error in a single observation = ± 0.6745 $\sqrt{0.000223/7}$ = ± 0.0038

Error in the mean = $E_s / \sqrt{8} = 0.001$

Probable value = 5869.835 ± 0.001 m

Table 18.2 Solution to Example 18.20

Observed value	Difference from the mean e	e^2
5869.835	0.000	0.0000
5869.828	0.007	0.000049
5869.842	− 0.007	0.000049
5869.838	− 0.003	0.000009
5869.830	0.005	0.000025
5869.844	− 0.009	0.000081
5869.836	− 0.001	0.000001
5896.832	0.003	0.000009
Total = 46,958.685		Total = 0.000223

Example 18.21 If the sides of a right-angled triangle are 90 ± 0.015 m and 120 ± 0.02 m, determine the probable error in the area of the triangle.

Solution Area $A = ab/2$

where a and b are the sides. When a quantity is computed as the product of two quantities, the probable error is given by

$$e_A = \sqrt{(e_a b/2)^2 + (e_b a/2)^2}$$

$e_a = 0.015, \ e_b = 0.02$

$$e_A = \sqrt{(0.015 \times 60/2)^2 + (0.02 \times 90/2)^2} = 1.272 \text{ m}^2$$

Area = $90 \times 120 = 10,800 \text{ m}^2$

Probable value = $10,800 \pm 1.272 \text{ m}^2$

18.8 Distribution of Error

Field data collected in terms of angles, lengths, or levels has to be checked for closing errors or other conditions. The error is distributed to the quantities. The following rules govern the distribution of error to the quantities.

(a) The correction to be applied to an observation is inversely proportional to the weight of the observation.

(b) The correction to be applied to an observation is directly proportional to the square of the probable error.

(c) In the case of levels, the correction to be applied is proportional to the length.

The following examples illustrate the distribution of error.

Example 18.22 The probable values of the angles of a triangle with the probable errors are as follows: $A = 70° 31' 18.6'' \pm 4''$, $B = 61° 12' 9.8'' \pm 6''$, $C = 48° 16' 22.2'' \pm 2''$. Find the error in the angle and distribute the error.

Solution The sum of the angles = $179° 59' 50.6''$

Discrepancy = $9.4''$

This is distributed among the angles in proportion to the square of the probable errors. If c_1, c_2, and c_3 are the corrections to the angles, then

$$c_1 : c_2 : c_3 :: (4)^2 : (6)^2 : (2)^2 = 16 : 36 : 4$$

Therefore,

$$c_1 = 9.4 \times 16/56 = 2.69''$$

$$c_2 = 9.4 \times 36/56 = 6.04''$$

$$c_3 = 9.4 \times 4/56 = 0.67''$$

The corrected angles are

$$A = 70° 31' 16.6'' + 2.69'' = 70° 31' 2.69''$$

$$B = 61° 12' 9.8'' + 6.04'' = 61° 12' 15.84''$$

$$C = 48° 16' 222.2'' + 0.67'' = 48° 12' 22.87''$$

Example 18.23 The four angles measured around a station have the following values and weights: $A = 115° 32' 18.8''$ (weight = 3), $B = 122° 46' 21.4''$ (weight = 4), $C = 62° 21' 16.8''$ (weight = 2), $D = 59° 20' 15.8''$ (weight = 1). Find the corrected values of the angles.

Solution Corrections to the angles are inversely proportional to their weights. If c_1, c_2, c_3, and c_4 are the corrections to the angles A, B, C, and D, respectively, then

$$c_1 : c_2 : c_3 : c_4 : \frac{1}{3} : \frac{1}{4} : \frac{1}{2} : \frac{1}{1} = 4 : 3 : 6 : 12$$

Sum of four angles = $360° 00' 12''$

Discrepancy = $-12''$

Then

$$c_1 = -12 \times 4/25 = -1.92'', \quad c_2 = -12 \times 3/25 = -1.44$$

$$c_3 = -12 \times 6/25 = -2.88, \quad c_4 = -12 \times 12/25 = -5.76$$

Check: $-1.92 - 1.44 - 2.88 - 5.76 = -12''$.

The corrected angles are

$$A = 115° 32' 18.8'' - 1.92'' = 115° 32' 16.88''$$

$$B = 122° 46' 21.4'' - 1.44'' = 122° 46' 19.96''$$

$$C = 62° 21' 14.8'' - 2.88'' = 62° 21' 13.92''$$

$$D = 59° 20' 15'' - 5.76'' = 59° 21' 9.24''$$

Check: $115° 32' 16.88'' + 122° 46' 19.96'' + 62° 21' 13.92'' + 59° 21' 9.24'' = 360° 00' 00''$.

18.9 Triangulation Adjustment

After the base line and angles of a triangulation system are measured, the data collected from the field has to be analysed and corrected. It is desirable to adjust the network in one operation, correcting all the angles simultaneously. However, the operation is very laborious. Generally, the operation of adjustment is divided into different parts. First, the angles are adjusted from their many measurements. The second part is the station adjustment, in which the angle measured at a station is corrected. The third part is figure adjustment, in which the figure is adjusted to satisfy certain conditions. The single angle adjustment begins with several observations of the angle. The observations for the same angle may be of the same or different weights. The correction to an angle is inversely proportional to the weight and directly proportional to the square root of the probable errors. In such cases, the most probable value of the angle is the arithmetic mean or the weighted arithmetic mean. These adjustments have already been discussed.

18.9.1 Station Adjustment

In station adjustment, the angles measured at the station are corrected. The angles may be measured by closing the horizon or otherwise. Three cases may arise.

(a) *Closing the horizon with angles of equal weight* In Fig. 18.6(a) four angles A, B, C, and D are measured, closing the horizon. If the angles are measured with equal care, the weights of all the angles will be equal or unity. The error of closure is the difference between the sum of the four angles calculated and their theoretical sum, i.e., 360°. The error is distributed equally among all the angles.

(b) *Closing the horizon with angles of unequal weight* The closing error is the difference between the theoretical and actual sum of the four angles. The corrections to the angles are inversely proportional to their weights. The angles are corrected by finding the error of closure and correction to each angle, inversely proportional to the weights of the angles.

(c) *When the angles and sum of angles are measured without closing the horizon* As shown in Fig. 18.6(b), the four angles as well as their sum are measured at the triangulation station. Their measurements may have different weights. For example, A, B, C, and D are measured individually. Also angles $(A + B)$, $(B + C)$, $(C + D)$, and $(A + B + C + D)$ are measured as combined angles. The values of the angles need to satisfy these four conditions. The most probable values of the angles are then determined by the method of normal equations. If, however, two or more angles and their sum are measured, the following rules may be observed while correcting the angles and their sum.

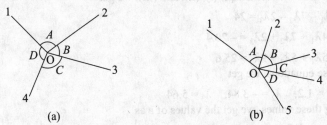

(a)   (b)

Fig. 18.6 Station adjustments

Rule 1 If the angles and their sum have equal weights, distribute the error equally among the angles. The sign of correction to the sum is opposite to that of the correction to the individual angles.

Rule 2 If the angles and their sum have different weights, the error distributed is inversely proportional to the weight of the angles. The sign of correction to the sum is opposite to that of the correction to individual angles.

The following example illustrates the method of station adjustment.

Example 18.24 P, Q, and R are angles measured at a station closing the horizon. Combined angles $(P + Q)$ and $(Q + R)$ were also measured. Find the probable values of the angles P,

Q, and R from the following data: $P = 42° 34' 26.2''$ (weight = 3), $Q = 64° 22' 14.8''$ (weight = 3), $R = 253° 03' 15''$ (weight = 2), $(P + Q) = 106° 56' 44.4''$ (weight = 2), $(Q + R) = 317° 25' 25.2''$ (weight = 1).

Solution We use the method of correlates to find the probable values. Let e_1, e_2, e_3, e_4, and e_5 be the corrections to angles P, Q, R, $(P + Q)$, and $(Q + R)$, respectively. Then,

$$42° 34' 26.2'' + e_1 + 64° 22' 14.8'' + e_2 + 253° 03' 15'' + e_3 = 360°$$

$$e_1 + e_2 + e_3 = 4'' \qquad \text{(i)}$$

$$42° 34' 26.4'' + e_1 + 64° 22' 14.8'' + e_2 = 106° 56' 44.4'' + e_4$$

$$e_4 - (e_1 + e_2) = -3.4'' \qquad \text{(ii)}$$

$$64° 22' 14.8'' + e_2 + 253° 03' 25.2'' + e_3 = 317° 25' 25.2'' + e_5$$

$$e_5 - (e_2 + e_3) = 4.6'' \qquad \text{(iii)}$$

From the principle of least squares,

$$w_1 e_1^2 + w_2 e_2^2 + w_3 e_3^2 + w_4 e_4^2 + w_5 e_5^2 \text{ is minimum} \qquad \text{(iv)}$$

Differentiating the four equations partially,

$$\delta e_1 + \delta e_2 + \delta e_3 = 0$$

$$\delta e_4 - \delta e_1 - \delta e_2 = 0$$

$$\delta e_5 - \delta e_2 - \delta e_3 = 0$$

$$3e_1 \delta e_1 + 3e_2 \delta e_2 + 2e_3 \delta e_3 + 2e_4 \delta e_4 + e_5 \delta e_5 = 0$$

The first three equations are multiplied by $-\lambda_1, -\lambda_2, -\lambda_3$ (correlates) and all the equations are added. We get

$$\delta e_1 (3e_1 - \lambda_1 + \lambda_2) + \delta e_2 (3e_2 - \lambda_1 + \lambda_2 + \lambda_3) + \delta e_3 (2e_3 - \lambda_1 + \lambda_3)$$
$$+ \delta e_4 (2e_4 - \lambda_2) + \delta e_5 (e_5 - \lambda_3) = 0$$

The coefficients in brackets should independently vanish. Thus,

$$e_1 = (\lambda_1 - \lambda_2)/3, \quad e_2 = (\lambda_1 - \lambda_2 - \lambda_3)/3, \quad e_3 = (\lambda_1 - \lambda_2)/2, \quad e_4 = \lambda_2/2, \quad e_5 = \lambda_3$$

Substituting these values in Eqns (i), (ii), and (iii), we get

$$7\lambda_1 - 7\lambda_2 - 2\lambda_3 = 24$$

$$-4\lambda_1 + 7\lambda_2 + 2\lambda_3 = -20.4$$

$$-5\lambda_1 + 5\lambda_2 + 9\lambda_3 = 25.6$$

Solving these equations, we get

$$\lambda_1 = 1.2, \quad \lambda_2 = -3.84, \quad \lambda_3 = 5.64$$

Substituting these values, we get the values of e as

$$e_1 = 1.68'', \quad e_2 = -0.2'', \quad e_3 = 2.52''$$

The corrected values of the angles are

$$P = 42° 34' 27.88'', \quad Q = 64° 22' 14.6'', \quad R = 253° 03' 17.52''$$

Check: $P + Q + R = 360°$.

18.9.2 Figure Adjustment

A triangulation network essentially consists of triangles. A quadrilateral with its two diagonals (without a station at their intersection) is another figure. Being a quadrilateral, it contains triangles and additionally other conditions that are imposed on it. A polygon with a central station is another configuration employed sometimes.

The determination of the most probable values of the angles in order to satisfy a geometrical condition is known as figure adjustment. We have already seen the methods of adjustment of angles with conditions. The method of normal equations, or preferably the method of correlates, is used for such adjustments.

Adjustment of a triangle The following notations are used in deriving the rules for adjusting angles:

(a) A, B, C are the angles of the triangle.

(b) c_A, c_B, and c_C are corrections to the observed angles.

(c) w_A, w_B, and w_C are the weights of the respective angles.

(d) d is the error of closure or discrepancy.

(e) n_A, n_B, and n_C are the number of observations of the respective angles.

(f) E_A, E_B, and E_C are the probable errors of the angles.

The following rules are observed in making corrections to the angles.

Rule 1 When the angles are of equal weight, distribute the error equally among the angles.

$$c_A = c_B = c_C = d/3$$

Rule 2 When the angles are unequally weighted, distribute the error among the angles inversely proportional to their weight. Thus,

$$c_A{:}c_B{:}c_C = \frac{1}{w_A} : \frac{1}{w_B} : \frac{1}{w_C}$$

Therefore,

$$c_A = \frac{d(1/w_A)}{1/w_A + 1/w_B + 1/w_C}$$

$$c_B = \frac{d(1/w_B)}{1/w_A + 1/w_B + 1/w_C}$$

$$c_C = \frac{d(1/w_C)}{1/w_A + 1/w_B + 1/w_C}$$

Rule 3 Corrections are made proportional to the number of observations. Thus,

$$c_A = \frac{d(1/n_A)}{1/n_A + 1/n_B + 1/n_C}$$

Similarly, corrections are calculated for other angles.

Rule 4 Corrections are made inversely proportional to the square of the number of observations. Thus,

$$c_A = \frac{d(1/n_A^2)}{1/n_A^2 + 1/n_B^2 + 1/n_C^2}$$

Similarly, corrections are calculated for other angles.

Rule 5 The correction to an angle is made proportional to the squares of the respective probable errors. This is derived from Rule 2, as the weight of an angle is inversely proportional to the square of its probable error.

$$c_A = \frac{dE_A^2}{1/E_A^2 + 1/E_B^2 + 1/E_C^2}$$

Similarly, the corrections for other angles are calculated.

Rule 6 This is known as Gauss's rule. This method is employed when the weights are not given but the number of observations is known. Gauss's rule provides a procedure to determine the weight of an angle from the number of observations and the residual errors.

$$W = (1/2)n^2/\Sigma e^2$$

Here Σe^2 is the sum of the squares of the difference between the mean value of the angle and the observed values. If M is the mean value and $A_1, A_2, A_3, \cdots, A_n$ are the different observations, then

$$M = (A_1 + A_2 + A_3 + \cdots + A_n)/n$$
$$\Sigma e^2 = (M - A_1)^2 + (M - A_2)^2 + (M - A_3)^2 + \cdots + (M - A_n)^2$$

Knowing the weights, the corrections are applied using Rule 2.

Of the many methods, Rules 1 and 2 are more commonly used. If the number of observations is known, Rule 6 is used along with Rule 2. The following example illustrates the procedure of angle adjustment.

Example 18.25 Several measurements of angles A, B, and C of a triangle were made. Find the probable value of the angles.

$A = 55°\ 36'$	15"	17"	18"	19"	16"	14"	15"
$B = 67°\ 14'$	30"	32"	35"	31"	28"	30"	
$C = 57°\ 09'$	15"	14"	17"	16"			

Solution The mean value of each angle can be calculated as

Mean value of $A = 55°\ 36' + (15" + 17" + 18" + 19" + 16" + 14" + 15")/7$
$$= 55°\ 36'\ 16.28"$$

Mean value of $B = 67°\ 14' + (30" + 32" + 35" + 31" + 28" + 30")/6$
$$= 67°\ 14'\ 31"$$

Mean value of $C = 57°\ 09' + (15" + 14" + 17" + 16")/4 = 57°\ 09'\ 15.5"$

Sum of the mean values of angles $= 180°\ 0'\ 2.78"$

Weight of a mesurement $= \dfrac{(1/2)n^2}{\Sigma e^2}$

Here n is the number of observations and the e's are the differences from the mean.

Weight of $A = 1/2(7)^2/(1.28^2 + 0.72^2 + 1.72^2 + 2.72^2 + 0.28^2 + 2.28^2 + 1.28^2)$
$$= 3.54$$

Weight of $B = 1/2(6)^2/(1^2 + 2^2 + 5^2 + 0 + 2^2 + 1^2)$
$$= 0.545$$

Weight of $C = 1/2(4)^2/(0.5^2 + 1.5^2 + 1.5^2 + 0.5^2)$
$$= 1.6$$

The corrections to the angles are inversely proportional to the weights. If c_1, c_2, and c_3 are the corrections to A, B, and C, respectively, then

$$c_1:c_2:c_3 = \dfrac{1}{3.54} : \dfrac{1}{0.545} : \dfrac{1}{1.6}$$
$$= 0.28:1.83:0.625$$

Discrepancy in the sum of angles $= -2.78"$

Therefore,

$$c_1 = (-2.78) \times 0.28/2.735 = -0.28$$
$$c_2 = (-2.78) \times 1.83/2.735 = -1.86$$
$$c_3 = (-2.78) \times 0.625/2.735 = -0.64$$

The probable values are as follows:

$$A = 55° \ 36' \ 16.28'' - 0.28'' = 55° \ 36' \ 16''$$
$$B = 67° \ 14' \ 31'' - 1.86'' = 67° \ 14' \ 29.14''$$
$$C = 57° \ 09' \ 15.5'' - 0.64'' = 57° \ 09' \ 14.86''$$

Check: Sum of $A + B + C = 180° \ 00' \ 00''$

18.10 Adjustment of a Triangle

In the case of triangles, if the sides of the triangles are small, they can be adjusted as plane triangles. Triangles with side lengths up to 3 km can be considered as plane triangles. If the sides are longer than 3 km, then the triangle has to be considered a geodetic triangle and will have to be adjusted like a spherical triangle.

Adjustment of a plane triangle

In the case of a plane triangle, the following steps are involved in adjustment.

1. Measure the three angles of the triangle. Determine the length of one of the sides, either by measuring it as a base line or by calculating it from another triangle.
2. Adjust individual angles first from the number of observations available. The arithmetic mean of the values will be the most probable value of the angle.
3. In a plane triangle, check the condition that the sum of angles is 180°.
4. If the sum is not 180°, adjust the three angles from the observation equations of individual angles and the conditional equation that their sum is 180°.
5. The method of normal equations or the method of correlates is more commonly used for adjusting the angles.
6. Knowing one side and three angles, calculate the remaining sides using the sine rule.

Adjustment of a geodetic triangle

In the case of a geodetic triangle (spherical triangle), the sides are considered as arcs of a sphere rather than straight lines. The condition with regard to the angles in a spherical triangle is that the sum of the angles must be equal to 180° in addition to the spherical excess.

Spherical excess

Spherical excess ϕ is the amount by which the sum of the angles of a spherical triangle exceeds 180°. It depends upon the area of the spherical triangle. If the sides of the triangle are less than 3.5 km in length, spherical excess is ignored. However, in most triangulation operations, the sides are longer than this value and the spherical excess should be taken into account.

Spherical excess is approximately 1'' for every 196.75 sq. km. The exact value of spherical excess may be calculated from the following formula.

$$\text{Spherical excess } \phi = \frac{\text{area} \times 180°}{\pi R^2}$$

$$= \frac{64,800 \times \text{area}}{\pi R^2} \sec$$

$$= \frac{\text{area}}{R^2 \sin 1''} \sec$$

where area is the area of the spherical triangle and R is the radius of the earth (6370 km). Area and R should be expressed in the same units. For works of higher precision,

$$\text{Spherical excess } \phi = \frac{\text{area} \times (1 - e^2 \sin^2 \theta)^2}{a^2 (1 - e^2) \sin 1''}$$

where θ is the mean of the latitude of the bounding stations, a is the equatorial semi-axis, e is the earth's eccentricity, and R is the radius of curvature of a meridian section at latitude θ.

To calculate the spherical excess, the area is required, which can be calculated when the angles are known accurately. To calculate the area, a plane triangle is considered and its area is calculated from a side and angles as

$$\text{Area } \Delta = \frac{a^2 \sin B \sin C}{2 \sin A}$$

or, if two sides and the included angle are known,

$$\text{Area } \Delta = 1/2 (ab) \sin C$$

where a, b, and c are the sides of the triangle and A, B, and C are the angles opposite these sides.

Sides of a spherical triangle

Knowing one side and the angles of a spherical triangle, the remaining sides can be calculated using any of the following methods.

Spherical trigonometry Let a be the known side and A_s, B_s, and C_s be the adjusted spherical angles [Fig. 18.7(a)]. Let α, β, and γ be the central angles subtended by the sides a, b, and c. α is the angle subtended at the centre by the side a. As a is known, α can be calculated as follows:

$$a = R \times \text{central angle in radians} = R\alpha\pi/180 \quad \text{(where } \alpha \text{ is in degrees)}$$

$$= \frac{180a}{\pi R}$$

where R is the radius of the earth (6370 km). The remaining central angles are found from the sine rule.

$$\sin \beta = \sin \alpha \times \sin B / \sin A$$

and

$$\sin \gamma = \sin \alpha \times \sin C / \sin A$$

Knowing $\sin \beta$ and $\sin \gamma$, b and c are calculated as follows:

$$b = \pi R\beta/180, \quad c = \pi R\gamma/180$$

Delambre's method The three stations A, B, and C are joined by chords. The plane angles are the angles between these chords. Let A_p, B_p, and C_p be the plane angles. The following steps are adopted [Fig. 18.7(b)] to determine the length.

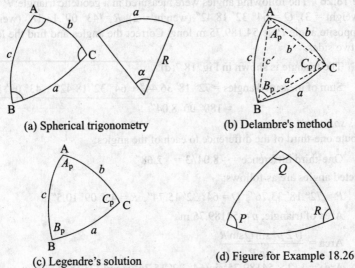

(a) Spherical trigonometry (b) Delambre's method

(c) Legendre's solution (d) Figure for Example 18.26

Fig. 18.7 Determination of sides of a spherical triangle

1. Knowing the arc length of any known side (say BC = a), calculate the central angle α:

   $$\alpha = 180a / \pi R$$

2. With the central angle α, calculate the chord length a' as

   $$a' = 2R \sin(\alpha/2)$$

3. From the chord length a' and the plane angles, compute the other two chord lengths as

   $$b' = a' \sin B_p / \sin A_p, \quad c' = a' \sin C_p / \sin A_p$$

4. From the chord lengths, calculate the corresponding central angles, β and γ, as

   $$\beta = 2\sin^{-1}(b'/2R), \quad \gamma = 2\sin^{-1}(c'/2R)$$

5. Compute the length of the arcs (or sides of the spherical triangle) as

   $$b = \pi R b' / 180, \quad c = \pi R c' / 180$$

Legendre's method This method is based upon Legendre's theorem, which states that 'In any spherical triangle, if each of the angles is diminished by one-third of the spherical excess, the sine of these angles is proportional to the length of the opposite sides and the triangle may be treated as a plane triangle, provided the lengths of the sides are small compared to the radius of the sphere.'

In Fig. 18.7(c), the spherical angles are corrected first. Each angle is then diminished by one-third of the spherical excess so that the sum is equal to 180°. These are then treated as plane angles and the sides are calculated using the sine rule for a plane triangle. If A_p, B_p, and C_p are the reduced spherical angles and a is a known side, then

$$b = a \sin B_p / \sin A_p, \quad c = a \sin C_p / \sin A_p$$

The following example illustrates these principles.

Example 18.26 The following angles were measured in a geodetic triangle: $P = 72° 18' 33.76''$ (weight = 3), $Q = 64° 32' 18.42''$ (weight = 2), $R = 43° 09' 13.18''$ (weight = 1). Side p, opposite angle P, is 54,189.75 m long. Correct the angles and find the lengths of the other two sides.

Solution The triangle is shown in Fig. 18.7(d).

$$\text{Sum of measured angles} = 72° 18' 36.44'' + 64° 32' 18.42'' + 43° 09' 13.18''$$
$$= 180° 00' 8.04''$$

Legendre's solution

We distribute one-third of the difference to each of the angles:

$$\text{One-third difference} = -8.04/3 = -2.68''$$

The corrected angles are as follows:

$$P = 72° 18' 33.76'', \quad Q = 64° 32' 15.74'', \quad R = 43° 09' 10.5''$$

Area of triangle, $p = 54189.75$ m

$$\text{Area} = \frac{1/2\, p^2 \sin Q \sin R}{\sin P}$$

$$\text{Area} = 1/2 \times 54189.75^2 \sin(64° 32' 15.74'')/\sin(72° 18' 33.76'')$$
$$= 951.55 \times 106 \text{ m}^2$$

Spherical excess $\phi = \text{area} \times 64,800/\pi R^2$
$$= 951.55 \times 10^6 \times 64,800/(\pi \times 6,370,290^2)$$
$$= 4.83''$$

$$\text{Error} = 8.04'' - 4.83'' = 3.21''$$

This is distributed among the angles in inverse proportion to their weights.

Ratio $= 1/3 : 1/2 : 1/1 = 2 : 3 : 6$

Correction to $P = 3.21 \times 2/11 = 0.58''$

Correction to $Q = 3.21 \times 3/11 = 0.88''$

Correction to $R = 3.21 \times 6/11 = 1.75''$

Corrected spherical angles are

$$P = 72° 18' 35.86'', \quad Q = 64° 32' 17.54'', \quad R = 43° 09' 11.43''$$

The corrected plane angles are obtained by distributing the spherical excess equally to the spherical angles. The corrected plane angles are

$$P = 72° 18' 34.25'', \quad Q = 64° 32' 15.93'', \quad R = 43° 09' 9.82''$$

Sides q and r are calculated using the sine rule as follows:

$$q = 54,189.75 \times \sin 64° 32' 15.93''/\sin 72° 18' 34.25''$$
$$= 51,354.698 \text{ m}$$
$$r = 54,189.75 \times \sin 43° 09' 9.82''/\sin 72° 18' 34.25''$$
$$= 38,902.453 \text{ m}$$

Delambre's method

On the earth's surface, 1 min of the arc subtends 1,853.79 m or 1'' of the arc subtends 30.896 m. If α_p, α_q, and α_r are the central angles subtended by the arcs p, q, and r, respectively, then

$\alpha_p = 54{,}189.75/1{,}853.79$ min $= 29'\ 13.91''$

Knowing the central angle of arc p, the chord p' can be calculated as

Chord $p' = 2R \sin(\alpha_p/2) = 2 \times 6{,}370{,}290 \sin(29'\ 13.91''/2) = 54{,}167.728$ m

Chords q' and r' can be calculated using the sine rule.

Chord $q' = 54{,}167.728 \times \sin 64°\ 32'\ 15.93''/\sin 72°\ 18'\ 34.25''$

$= 51{,}333.28$ m

Chord $r' = 54{,}167.728 \times \sin 43°\ 09'\ 9.82''/\sin 72°\ 18'\ 34.25''$

$= 38{,}886.643$ m

From the chord values, the central angles can be calculated. If α_q and α_r are the central angles of chords q' and r', then

$\alpha_q = 2 \sin^{-1}(\text{chord } q/2R) = 27.7022'$

$\alpha_r = 2 \sin^{-1}(\text{chord } r/2R) = 20.9853'$

The lengths of the arcs q and r can now be calculated as

Length $q = 1853.79 \times \alpha_q$ in minutes $= 1853.79 \times 27.7022$

$= 51{,}354.061$ m

Length $r = 1853.79 \times \alpha_r$ in minutes $= 1853.79 \times 20.9853$

$= 38{,}902.339$ m

By spherical trigonometry

If P_s, Q_s, and R_s are the corrected spherical angles, then by principles of spherical trigonometry

$\sin \alpha_q = \sin \alpha_p \times \sin Q_s/\sin P_s$

$= \sin(29°.2319') \times \sin 64°\ 32'\ 17.54''/\sin 72°\ 18'\ 35.86''$

$= 0.008058$

$\sin \alpha_r = \sin (29°.2319') \times \sin 43°\ 09'\ 11.43''/\sin 72°\ 18'\ 35.86''$

$= 0.006104$

$\alpha_q = 0.4617°, \ \alpha_r = 0.3497°$

Sides q and r can be calculated as $R \times \alpha$ in rad.

Side $q = \alpha_q R\pi/180° = 0.4617 \times 6{,}370{,}290 \times \pi/180° = 51{,}332.976$ m

Side r $= \alpha_r R\pi/180° = 0.3497 \times 6{,}370{,}290 \times \pi/180° = 38{,}880.532$ m

18.11 Adjustment of Chain of Triangles

In adjusting a chain of triangles (Fig. 18.8), station and figure adjustments are done in that order. In Fig. 18.8, ABC is the base triangle and AB is the base line measured, ABC, BCD, and CDE are the triangles. The angles measured are marked in the figure.

Station adjustment

The angles around a station have to satisfy the condition that their sum is 360°. The conditions are

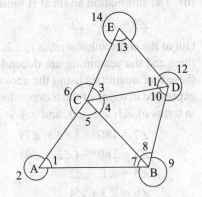

Fig. 18.8 Chain of triangles

$$\angle 1 + \angle 2 = 360°$$
$$\angle 3 + \angle 4 + \angle 5 + \angle 6 = 360°$$
$$\angle 7 + \angle 8 + \angle 9 = 360°$$
$$\angle 10 + \angle 11 + \angle 12 = 360°$$
$$\angle 13 + \angle 14 = 360°$$

The discrepancy, if any, is distributed equally among all the angles.

Figure adjustment

In figure adjustment, the condition that the sum of angles in each triangle should be 180° is to be satisfied. With corrected angles from station adjustment, the following conditions are to be fulfilled.

$$\angle 1 + \angle 5 + \angle 7 = 180°$$
$$\angle 8 + \angle 4 + \angle 10 = 180°$$
$$\angle 3 + \angle 11 + \angle 13 = 180°$$

If the angles have equal weight, the corrections are applied equally to each of the angles. If the angles are weighted, the corrections applied are inversely proportional to their weights. Adjustment of two triangles: As the process is very laborious, we illustrate the principle with two triangles as shown in Fig. 18.9.

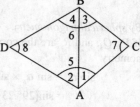

Fig. 18.9 Two connected

The eight angles measured in the triangles are $\angle 1$, $\angle 2$, $\angle 5$ ($\angle 1 + \angle 2$), $\angle 3$, $\angle 4$, $\angle 6$ ($\angle 3 + \angle 4$), $\angle 7$, and $\angle 8$. There are four conditions governing these angles.

(a) The sum of the angles in triangle ABC must equal 180°:
$$\angle 1 + \angle 3 + \angle 7 = 180°$$

(b) The sum of angles in triangle ADB = 180°:
$$\angle 2 + \angle 8 + \angle 4 = 180°$$

(c) The summation angle at A must be equal to its parts:
$$\angle 1 + \angle 2 = \angle 5$$

(d) The summation angle at B must be equal to its parts:
$$\angle 3 + \angle 4 = \angle 6$$

Out of the eight unknowns, $\angle 1$, $\angle 2$, $\angle 3$, and $\angle 4$ are considered independent quantities and the remaining are dependent quantities, as they can be calculated from these four quantities using the above conditions. The dependent unknowns can be expressed in terms of the independent unknowns. $\angle 5$ and $\angle 6$ are already expressed in terms of $\angle 1$, $\angle 2$, $\angle 3$, and $\angle 4$, $\angle 7$ and $\angle 8$ can be expressed as

$$\angle 7 = 180° - (\angle 3 + \angle 1)$$
$$\angle 8 = 180° - (\angle 2 + \angle 4)$$
$$\angle 5 = \angle 1 + \angle 2$$
$$\angle 6 = \angle 3 + \angle 4$$

Substituting these values, new observation equations in the four unknowns are obtained. By forming normal equations with these values, the four unknown independent quantities are determined. The solution is illustrated with an example.

Example 18.27 Two interconnected triangles ABC and ABD have a common base AB. The angles measured are $A_1 = 61° 10' 50''$, $A_2 = 66° 07' 34''$, $A = 127° 18' 26''$, $B_1 = 50° 01' 20''$, $B_2 = 62° 01' 50''$, $B = 112° 03' 14''$, $C = 58° 47' 47''$, $D = 51° 50' 34''$. Determine the most probable values of the angles.

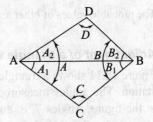

Fig. 18.10 Triangle of Example 18.27

Solution The triangles are shown in Fig. 18.10. We take A_1, A_2, B_1, B_2 as independent quantities. Other quantities can be expressed in terms of these four. Let c_1, c_2, c_3, and c_4 be the corrections to A_1, A_2, A_3, and A_4, respectively. We can write

$$A_1 = 61° 10' 50'' + c_1$$
$$A_2 = 66° 07' 34'' + c_2$$
$$B_1 = 50° 01' 20'' + c_3$$
$$B_2 = 62° 01' 50'' + c_4$$
$$A_1 + c_1 + A_2 + c_2 = A$$
$$c_1 + c_2 = 127° 18' 26'' - 61° 10' 50'' - 66° 07' 34'' = +2''$$
$$B_1 + c_3 + B_2 + c_4 = B$$
$$c_3 + c_4 = 112° 03' 14'' - 50° 01' 20'' - 62° 01' 50'' = +4''$$

The sum of the angles in triangles ABC and ABD is equal to 180°.

$$A_1 + c_1 + B_1 + c_3 + C = 180°$$
$$c_1 + c_3 = 180° - 61° 10' 50'' - 50° 01' 20'' - 58° 47' 47'' = +3''$$
$$A_2 + c_2 + B_2 + c_4 + D = 180°$$
$$c_2 + c_4 = 180° - 66° 07' 34'' - 62° 01' 50'' - 51° 50' 34'' = +2''$$

The new equations are

$$c_1 = 0, \quad c_2 = 0, \quad c_3 = 0, \quad c_4 = 0, \quad c_1 + c_2 = 2, \quad c_3 + c_4 = 4, \quad c_1 + c_3 = 3$$
$$c_2 + c_4 = 2$$

The following normal equations are formed from the above equations by multiplying them by their coefficients and the weights of equations:

$$3c_1 + c_2 + c_3 = 5$$
$$c_1 + 3c_2 + c_4 = 4$$
$$c_1 + 3c_3 + c_4 = 7$$
$$c_2 + c_3 + 3c_4 = 6$$

Solving these four equations, we get

$$c_1 = 0.93'', \quad c_2 = 0.6'', \quad c_3 = 1.6'', \quad c_4 = 1.27''$$

Applying these corrections to the angles, we get the probable values of the angles as

$$A_1 = 61° 10' 50.93''$$

$$A_2 = 66° 07' 34.6''$$
$$B_1 = 50° 01' 21.6''$$
$$B_2 = 62° 01' 51.27''$$

The probable values of other angles can be obtained from these angles.

Adjustment of a triangle with a central station

Figure 18.11 shows a triangle with a central station. The angles measured are indicated in the figure. Angles 7, 8, and 9 are angles measured at the central station O.

Angles 1, 3, and 5 are known as right angles and angles 2, 4, and 6 are known as left angles. We can easily prove the following condition with respect to these angles:

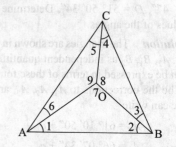

Fig. 18.11 Triangle with a central station

$$\sin 1 \times \sin 3 \times \sin 5 = \sin 2 \times \sin 4 \times \sin 6$$

In triangle ABO,

$$AO = BO \times \sin 3 / \sin 2$$

From triangle BCO,

$$BO = CO \times \sin 5 / \sin 4$$

Therefore,

$$AO = CO \times \sin 5 \times \sin 3 / \sin 2 \times \sin 4$$

From triangle ACO,

$$CO = AO \times \sin 1 / \sin 6$$

Therefore,

$$AO = CO \times \sin 6 / \sin 1$$

Equating the two expressions,

$$CO \times \frac{\sin 5 \times \sin 3}{\sin 2 \times \sin 4} = \frac{CO \times \sin 6}{\sin 1}$$

Thus, we get

$$\sin 1 \times \sin 3 \times \sin 5 = \sin 2 \times \sin 4 \times \sin 6$$

When we look from stations A, B, and C towards the central station, $\angle 1$, $\angle 3$, and $\angle 5$ are on our right and $\angle 2$, $\angle 4$, and $\angle 6$ are to our left. This condition is known as the side equation and is expressed generally in a summation form in terms of the logarithms of the sines of the angles. Thus,

$$\log \sin 1 + \log \sin 3 + \log \sin 5 = \log \sin 2 + \log \sin 4 + \log \sin 6$$

$$\Sigma \log \sin(\text{right angles}) = \Sigma \log \sin(\text{left angles})$$

The conditions to be satisfied by the measured angles are the following.

(a) The sum of angles around O is equal to 360°.

$$\angle 7 + \angle 8 + \angle 9 = 360°$$

(b) The sum of the angles in each of the three triangles must be equal to 180°.

(c) Side equation or log sine condition:

$$\log \sin 1 + \log \sin 3 + \log \sin 5 = \log \sin 2 + \log \sin 4 + \log \sin 6$$

Let $c_1, c_2, ..., c_9$ be the corrections to angles 1, 2, ..., 9. Let $d_1, d_2, ..., d_9$ be the tabular differences for one second for log sin 1, log sin 2, ..., log sin 9. The following equations can be written:

$$c_7 + c_8 + c_9 = \pm A_1 \tag{18.13}$$

$$c_1 + c_7 + c_6 = \pm A_2 \tag{18.14}$$

$$c_2 + c_3 + c_8 = \pm A_3 \tag{18.15}$$

$$c_4 + c_5 + c_9 = \pm A_4 \tag{18.16}$$

where A_1, A_2, A_3, A_4 are the differences between the theoretical sum and the sum of measured angles.

$$d_1 c_1 + d_3 c_3 + d_5 c_5 - d_2 c_2 - d_4 c_4 - d_6 c_6 = \pm M \tag{18.17}$$

Here M is in units of the seventh decimal place of the log sines. From the principle of least squares, $c_1^2 + c_2^2 + ... + c_9^2$ is minimum [let this be Eqn (18.18)]. We have three more conditions in terms of the angles or their differences. Angles 1–6 are taken as independent quantities and 7, 8, and 9 can be found from the given conditions. The six equations can [Eqns (18.13) – (18.18)] be solved to get the six corrections. The following example illustrates the procedure.

Example 18.28 In a triangle PQR with a central station O (Fig. 18.12), the angles have the following values: $\angle 1$ = 23° 18′ 16″, $\angle 2$ = 25° 10′ 20″, $\angle 3$ = 36° 22′ 19″, $\angle 4$ = 34° 38′ 28″, $\angle 5$ = 30° 45′ 16″, $\angle 6$ = 29° 45′ 31″. Adjust the angles of the triangle.

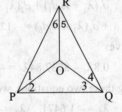

Solution The following conditions govern the angles.
 (a) $\angle 1 + \angle 2 + \angle 3 + \angle 4 + \angle 5 + \angle 6 = 180°$
 (b) Σlog sin(left angles) = Σlog sin(right angles)
 (c) $c_1^2 + c_2^2 + \cdots + c_6^2 = 0$ (least square principle), where c is the correction to an angle.

Fig. 18.12 Triangle of Example 18.28

Condition (b) can be written as

$$d_1 c_1 - d_2 c_2 + d_3 c_3 - d_4 c_4 + d_5 c_5 + d_6 c_6 = \Sigma\text{log sin(left angles)} - \Sigma\text{log sin(right angles)}$$
The calculation of the log sines is shown in Table 18.3.

Table 18.3 Calculation of log sines (Example 18.28)

Left angle	log sine + 10	Difference for 1″	Right angle	log sine +10	Difference for 1″
23° 18′ 16″	9.59727466	48.9	25° 10′ 20″	9.62873681	44.7
36° 22′ 19″	9.77307277	28.6	34° 38′ 28″	9.75468021	30.5
30° 45′ 16″	9.70872654	35.3	29° 45′ 31″	9.69578538	36.8
	29.07907397			29.07920240	

Sum of angles = 180° 00′ 10″

Σlog sin(right angles) – Σlog sin(left angles) = – 1,285

If $c_1, c_2, ..., c_6$ are the corrections to the six angles, then

$$c_1 + c_2 + c_3 + c_4 + c_5 + c_6 = -10'' \qquad \text{(i)}$$

$$d_1 c_1 - d_2 c_2 + d_3 c_3 - d_4 c_4 + d_5 c_5 - d_6 c_6 = -1,285 \qquad \text{(ii)}$$

$$c_1^2 + c_2^2 + c_3^2 + c_4^2 + c_5^2 + c_6^2 \text{ is minimum} \qquad \text{(iii)}$$

Differentiating the three equations partially

$$\delta c_1 + \delta c_2 + \delta c_3 + \delta c_4 + \delta c_5 + \delta c_6 = 0 \qquad \text{(iv)}$$

$$d_1 \delta c_1 + d_2 \delta c_2 + d_3 \delta c_3 + d_4 \delta c_4 + d_5 \delta c_5 + d_6 \delta c_6 = 0 \qquad \text{(v)}$$

$$\delta c_1 c_1 + \delta c_2 c_2 + \delta c_3 c_3 + \delta c_4 c_4 + \delta c_5 c_5 + \delta c_6 c_6 = 0 \qquad \text{(vi)}$$

Multiplying Eqn (iv) by $-\lambda_1$ and Eqn (v) by $-\lambda_2$ and adding the three equations, we get

$$\delta c_1(c_1 - \lambda_1 - \lambda_2 d_1) + \delta c_2(c_2 - \lambda_1 + \lambda_2 d_2) + \delta c_3(c_3 - \lambda_1 - \lambda_2 d_3)$$
$$+ \delta c_4(c_4 - \lambda_1 + \lambda_2 d_4) + \delta c_5(c_5 - \lambda_1 - \lambda_2 d_5) + \delta c_6(c_6 - \lambda_1 + \lambda_2 d_6) = 0$$

The coefficients of the terms should independently vanish.

$$c_1 = \lambda_1 + \lambda_2 d_1, \quad c_2 = \lambda_1 - \lambda_2 d_2, \quad c_3 = \lambda_1 + \lambda_2 d_3, \quad c_4 = \lambda_1 - \lambda_2 d_4$$
$$c_5 = \lambda_1 + \lambda_2 d_5, \quad c_6 = \lambda_1 - \lambda_2 d_6$$

Substituting these values back into the conditional equations, we get

$$6\lambda_1 - \lambda_2(d_1 - d_2 - d_3 - d_4 - d_5 - d_6) = -10''$$

or

$$6\lambda_1 + 0.8\lambda_2 = -10''$$

$$d_1(\lambda_1 + \lambda_2 d_1) + d_2(\lambda_1 - \lambda_2 d_2) + d_3(\lambda_1 + \lambda_3 d_3) + d_4(\lambda_1 - \lambda_2 d_4) + d_5(\lambda_1 + \lambda_2 d_5)$$
$$+ d_6(\lambda_1 - \lambda_2 d_6) = -1285$$

This gives

$$0.8\lambda_1 - 8737.84\lambda_2 = -1285$$

Solving these two equations, we get

$$\lambda_1 = -1.6471, \quad \lambda_2 = -0.1469$$

From these values, we get the corrections as

$$c_1 = -8.83'', \quad c_3 = 5.83'', \quad c_5 = -6.83'', \quad c_2 = 4.91, \quad c_4 = 2.83, \quad c_6 = 3.75$$

Check: Total $= -10''$. The corrected angles are

$$\angle 1 = 23° 18' 16'' - 8.83'' = 23° 18' 7.17''$$

$$\angle 2 = 25° 10' 20'' + 4.91'' = 25° 10' 24.91''$$

$$\angle 3 = 36° 22' 19'' - 5.83'' = 36° 22' 13.17''$$

$$\angle 4 = 34° 38' 28'' + 2.83'' = 34° 28' 30.83''$$

$$\angle 5 = 30° 45' 16'' - 6.83'' = 30° 45' 9.17''$$

$$\angle 6 = 29° 45' 31'' + 3.75'' = 29° 45' 34.75''$$

Check: Sum of angles $= 360°$.

18.12 Adjustment of a Quadrilateral

Figure 18.13 shows a quadrilateral with angular measurements taken along both diagonals. The diagonals are considered non-intersecting. Eight angles are measured, two each at each of the stations A, B, C, and D.

The conditions that the angles have to satisfy are as follows.

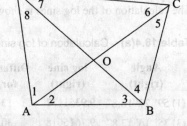

Fig. 18.13 Quadrilateral adjustment

(a) The sum of eight angles is equal to 360°.

$$\angle 1 + \angle 2 + \angle 3 + \angle 4 + \angle 5 + \angle 6 + \angle 7 + \angle 8 = 360°$$

(b) Considering triangles ABO and CDO,

$$\angle AOB = \angle COD$$
$$\angle 2 + \angle 3 = \angle 6 + \angle 7$$

(c) Considering triangles ADO and BCO and $\angle AOD = \angle BOC$,

$$\angle 1 + \angle 8 = \angle 4 + \angle 5$$

(d) The side equation or log sine equation is

$$\sin 1 \times \sin 3 \times \sin 5 \times \sin 7 = \sin 2 \times \sin 4 \times \sin 6 \times \sin 8$$

This is generally written as

$$\log \sin 1 + \log \sin 3 + \log \sin 5 + \log \sin 7 = \log \sin 2 + \log \sin 4$$
$$+ \log \sin 6 + \log \sin 8$$

$$\Sigma \log \sin(\text{left angles}) = \Sigma \log \sin(\text{right angles})$$

If $c_1, c_2, \ldots, c_8$ are the corrections to the observed angles, we can formulate the following equations:

$$c_1 + c_2 + \cdots + c_8 = \pm V_1$$
$$c_2 + c_3 = c_6 + c_7$$
$$c_1 + c_8 = c_4 + c_5$$
$$d_1 c_1 - d_2 c_2 + d_3 c_3 - d_4 c_4 + d_5 c_5 - d_6 c_6 + d_7 c_7 - d_8 c_8 = 0$$

By the theory of least squares, Σc^2 is minimum. With these equations, the values of corrections can be obtained by the method of correlates. The most probable values of angles can be found by applying these corrections to the angles. The following example illustrates the procedure.

Example 18.29 In a quadrilateral PQRS, the angles measured were as follows (Fig. 18.14): $\angle 1 = 59° 27' 16.8''$, $\angle 2 = 40° 17' 47.6''$, $\angle 3 = 35° 16' 12.8''$, $\angle 4 = 44° 58' 48.8''$, $\angle 5 = 45° 23' 28.6''$, $\angle 6 = 54° 21' 27.9''$, $\angle 7 = 41° 17' 15.5''$, $\angle 8 = 38° 57' 54.8''$. Adjust the angles.

Solution Sum of the angles = 360° 00′ 12.8″
The conditions are the following.

(a) Sum of interior angles = 360°

$$\angle 1 + \angle 2 + \angle 3 + \angle 4 + \angle 5 + \angle 6 + \angle 7 + \angle 8 = 360°$$

(b) $\Sigma \log \sin(\text{left angles}) = \Sigma \log \sin(\text{right angles})$

$$d_1 c_1 - d_2 c_2 + \cdots - d_8 c_8 = \text{difference of } \Sigma \log$$
sine of angles.

(c) $c_1 + c_2 - c_5 - c_6 = -7.9$

(d) $c_3 + c_4 - c_7 - c_8 = +8.7$

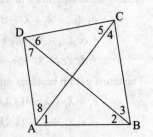

Fig. 18.14 Quadrilateral for Examples 18.29 and 18.30

(e) $c_1^2 + c_2^2 + \cdots + c_8^2 = $ minimum

The calculation of the log sines is shown in Table 18.4(a).

Table 18.4(a) Calculation of log sines (Example 18.29)

Angle (right)	log sine (right)	Difference for 1″	Angle (left)	log sine (left)	Difference for 1″
(1) 59° 27′ 16.8″	9.9351177	13.4	(2) 40° 17′ 47.6″	9.8107323	24.8
(3) 35° 16′ 12.8″	9.7615018	30.6	(4) 44° 58′ 48.8″	9.8493350	19.4
(5) 45° 23′ 28.6″	9.8524306	21.5	(6) 54° 21′ 27.9″	9.9099149	15.2
(7) 41° 17′ 15.5″	9.8194383	24.5	(8) 38° 57′ 54.8″	9.7985460	26.7
Total	39.3684884			39.3685282	

Σlog sin(left angles) $-$ Σlog sin(right angles) $= 398$

$$c_1 + c_2 + \cdots + c_8 = -12.8''$$
$$c_1 + c_2 - c_5 - c_6 = -7.9$$
$$c_3 + c_4 - c_7 - c_8 = +8.7$$
$$d_1 c_1 - d_2 c_2 + \cdots - d_8 c_8 = -398$$
$$c_1^2 + c_2^2 + \cdots + c_8^2 \text{ is minimum}$$

Differentiating the above equations partially,

$$\delta c_1 + \delta c_2 + \cdots + \delta c_8 = 0 \tag{i}$$
$$\delta c_1 + \delta c_2 - \delta c_5 - \delta c_6 = 0 \tag{ii}$$
$$\delta c_3 + \delta c_4 - \delta c_7 - \delta c_8 = 0 \tag{iii}$$
$$d_1 \delta c_1 - d_2 \delta c_2 + \cdots - d_8 \delta c_8 = 0 \tag{iv}$$
$$c_1 \delta c_1 + c_2 \delta c_2 + \cdots + c_8 \delta c_8 = 0$$

Using correlates $-\lambda_1$, $-\lambda_2$, $-\lambda_3$, and $-\lambda_4$, multiplying (i) by $-\lambda_1$, (ii) by $-\lambda_2$, (iii) by $-\lambda_3$, and (iv) by $-\lambda_4$, and adding all the equations,

$$-\lambda_1 \delta c_1 - \lambda_2 \delta c_2 - \cdots - \lambda_1 \delta c_8 = 0$$
$$-\lambda_2 \delta c_1 - \lambda_2 \delta c_2 + \lambda_2 \delta c_5 + \lambda_2 \delta c_6 = 0$$
$$-\lambda_3 \delta c_3 - \lambda_3 \delta c_4 + \lambda_3 \delta c_7 + \lambda_3 \delta c_8 = 0$$
$$-\lambda_4 d_1 \delta c_1 + \lambda_4 d_2 \delta c_2 - \lambda_4 d_3 \delta c_3 + \cdots + \lambda_4 d_8 \delta c_8 = 0$$

Adding these equations to the least square equation, we get

$$\delta c_1 (c_1 - \lambda_1 - \lambda_2 - d_1\lambda_4) + \delta c_2(c_2 - \lambda_1 - \lambda_2 + d_2\lambda_4) + \delta c_3(c_3 - \lambda_1 - \lambda_3 - d_3\lambda_4)$$
$$+ \delta c_4(c_4 - \lambda_1 - \lambda_3 + d_4\lambda_4) + \delta c_5(c_5 - \lambda_1 + \lambda_2 - d_5\lambda_4) + \delta c_6(c_6 - \lambda_1 + \lambda_2 + d_6\lambda_4)$$
$$+ \delta c_7(c_7 - \lambda_1 + \lambda_3 - d_7\lambda_4) + \delta c_8(c_8 - \lambda_1 + \lambda_3 - d_8\lambda_4) = 0$$

The terms within brackets must individually vanish. We get

$$c_1 = \lambda_1 + \lambda_2 + 13.4\lambda_4$$
$$c_2 = \lambda_1 + \lambda_2 - 24.8\lambda_4$$
$$c_3 = \lambda_1 + \lambda_3 + 30.6\lambda_4$$
$$c_4 = \lambda_1 + \lambda_3 - 19.6\lambda_4$$

$$c_5 = \lambda_1 - \lambda_2 + 21.5\lambda_4$$
$$c_6 = \lambda_1 - \lambda_2 - 15.2\lambda_4$$
$$c_7 = \lambda_1 - \lambda_3 + 24.5\lambda_4$$
$$c_8 = \lambda_1 - \lambda_3 - 26.7\lambda_4$$

Table 18.4(b) Corrections and final values

Angle	Value	Correlates	Corrections	Corrected angles
1	59° 27′ 16.8″	$\lambda_1 = -1.55$	$c_1 = -1.55 - 2.46 - 13.4$ $\times 0.11 = -5.48$	59° 27′ 11.32″
2	40° 17′ 47.6″	$\lambda_2 = -2.46$	$c_2 = -1.55 - 2.46 + 24.8$ $\times 0.11 = -1.28$	40° 17′ 46.32″
3	35° 16′ 12.8″	$\lambda_3 = +2.538$	$c_3 = -1.55 + 2.538 - 30.6$ $\times 0.11 = -2.38$	35° 16′ 10.42″
4	44° 58′ 48.8″	$\lambda_4 = -0.11$	$c_4 = -1.55 + 2.538 + 19.6$ $\times 0.11 = 3.14$	44° 58′ 51.94″
5	45° 23′ 28.6″		$c_5 = -1.55 + 2.46 - 21.5$ $\times 0.11 = -1.45$	45° 23′ 27.15″
6	54° 21′ 27.9″		$c_6 = -1.55 + 2.46 + 15.2$ $\times 0.11 = 2.58$	54° 21′ 30.48″
7	41° 17′ 15.5″		$c_7 = -1.55 - 2.538 - 24.5$ $\times 0.11 = -6.78$	41° 17′ 8.72″
8	38° 57′ 54.8″		$c_8 = -1.55 - 2.538 + 26.7$ $\times 0.11 = -1.15$	38° 57′ 53.65″

Substituting these back into the conditional equations, we get four equations in λ.

$$8\lambda_1 + 3.7\lambda_4 = -12.8$$
$$4\lambda_2 - 17.7\lambda_4 = -7.9$$
$$4\lambda_3 + 13.2\lambda_4 = 8.7$$
$$3.7\lambda_1 - 17.7\lambda_2 + 13.2\,\lambda_3 + 4113.75\lambda_4 = -398$$

We find λ_1, λ_2, and λ_3 from the first three equations in terms of λ_4. Substitute these values in the last equation to get the value of λ_4. λ_1, λ_2, and λ_3 can then be calculated from the first three equations. The solution of these four equations gives

$$\lambda_1 = 1.55,\ \lambda_2 = -2.46,\ \lambda_3 = +2.538,\ \lambda_4 = -0.11$$

The values of the corrections and corrected angles can be calculated as shown in Table 18.4(b).

Approximate solution

The following solution is sufficiently accurate for quadrilaterals of moderate size when very high precision is not required. The following steps are required.

1. Adjust the angles for spherical excess, if necessary.
2. The following conditions are to be satisfied.

$$\angle 1 + \angle 2 + \angle 3 + \angle 4 + \angle 5 + \angle 6 + \angle 7 + \angle 8 = 360°$$

$\angle 2 + \angle 3 = \angle 6 + \angle 7$ (considering triangles ABO and CDO, $\angle AOB = \angle COD$)

∠1 + ∠8 = ∠4 + ∠5 (considering triangles ADO and BCO, ∠AOD = ∠BOC)

The sine equation or log sine equation is as follows:

sin 1 × sin 3 × sin 5 × sin 7 = sin 2 × sin 4 × sin 6 × sin 8

This is generally written as

log sin 1 + log sin 3 + log sin 5 + log sin 7 = log sin 2 + log sin 4
+ log sin 6 + log sin 8

Σlog sin(left angles) = Σlog sin(right angles)

3. The adjustments can be made as follows.
   (i) Distribute the error equally among all the angles measured at a station for the station adjustment.
   (ii) Find the sum of these eight corrected angles of the quadrilateral. Correct the angles by distributing the error (by subtracting it from 360°) equally among all the angles.
   (iii) From the corrected angles obtained as above, find the difference between the sum of angles 1 + 3 and that of angles 6 + 7. Correct each angle by one-fourth of the difference. If (1 + 3) > (6 + 7), the corrections for 1 and 3 will be negative and those for 6 and 7 will be positive and vice versa.
   (iv) Similarly, correct the angles 1, 8, 4, and 5 by finding the difference between 1 + 8 and 4 + 5.
   (v) The corrected angles obtained from step (iv) are corrected for equality of the side or log sine equation. This is done as follows.
       (a) Determine the numerical value of the difference between the sum of the log sines of the left angles and the sum of the log sines of the right angles.
       (b) Write the difference d of the log sines of the various angles for 1″.
       (c) Apply the correction to the angles as

Correction to any angle $A = d_A \times m / \Sigma d^2$

where d_A is the tabular difference for 1″ for angle A, Σd^2 is the sum of the squares of tabular differences for 1″ for the log sines of all the angles, and m is the numerical difference between the log sines of the left and right angles.
       (d) Decide the sign of corrections by checking which of the sums is greater. If the left angle sum is greater, then the corrections will be negative for left angles, and vice versa.

The following example illustrates the procedure.

Example 18.30 In a quadrilateral PQRS, the angles measured were as follows (Fig. 18.14): ∠1 = 59° 27′ 16.8″, ∠2 = 40° 17′ 47.6″, ∠3 = 35° 16′ 12.8″, ∠4 = 44° 58′ 48.8″, ∠5 = 45° 23′ 28.6″, ∠6 = 54° 21′ 27.9″, ∠7 = 41° 17′ 15.5″, ∠8 = 38° 57′ 54.8″. Adjust the angles using the approximate method.

Solution The solution is shown in Table 18.5(a) and is explained below.
The stepwise procedure is as follows.
1. Satisfy the first condition of the angles that their sum should be 360°. Find the sum of the angles. The sum of angles in this case is 360° 00′ 12.8″.

Table 18.5(a) Adjustment of angles (Example 18.30)

Given angles	Adjustment 1	Corrected angles	Adjustment 2	Corrected angles
1. 59° 27′ 16.8″	− 1.6″	59° 27′ 15.2″	− 1.97″	59° 27′ 13.22″
2. 40° 17′ 47.6″	− 1.6″	40° 17′ 46″	− 1.97″	40° 17′ 44.02″
3. 35° 16′ 12.8″	− 1.6″	35° 16′ 11.2″	+ 2.17″	35° 16′ 13.37″
4. 44° 58′ 48.8″	− 1.6″	44° 58′ 47.2″	+ 2.17″	44° 58′ 49.37″
5. 45° 23′ 28.6″	− 1.6″	45° 23′ 27″	+ 1.97″	45° 23′ 28.97″
6. 54° 21′ 27.9″	− 1.6″	54° 21′ 26.3″	+ 1.97″	54° 21′ 28.27″
7. 41° 17′ 15.5″	− 1.6″	41° 17′ 13.9″	− 2.17″	41° 17′ 11.72″
8. 38° 57′ 54.8″	− 1.6″	38° 57′ 53.2″	− 2.17″	38° 57′ 51.02″

Table 18.5(b) Adjustment of angles (Example 18.30)

Left	log sine	d	d^2	Right	log sine	D	d^2
(1) 59° 27′ 13.22″	9.9351133	13.4	179.56	(2) 40° 17′ 40.2″	9.8107234	24.8	615.04
(3) 35° 16′ 13.37″	9.7615041	30.6	936.36	(4) 44° 58′ 49.37″	9.8493362	19.4	376.36
(5) 45° 23′ 28.97″	9.8524314	21.5	462.25	(6) 54° 21′ 28.27″	9.9099154	15.2	231.04
(7) 41° 17′ 11.72″	9.8194292	24.5	600.25	(8) 38° 57′ 51.02″	9.7985362	26.7	712.89
	39.3684780	90	2178.42		39.3685112	86.1	1935.33

Table 18.5(c)

Angle	Value	Correction	Corrected angle
1	59° 27′ 13.22″	1.08	59° 27′ 14.3″
3	35° 16′ 13.37″	2.47	35° 16′ 15.84″
5	45° 23′ 28.97″	1.73	45° 23′ 30.7″
7	41° 17′ 11.72″	1.98	41° 17′ 13.7″
2	40° 17′ 40.2″	− 2.00	40° 17′ 42.02″
4	44° 58′ 49.37″	− 1.56	44° 58′ 47.81″
6	54° 21′ 28.27″	− 1.23	54° 21′ 27.04″
8	38° 57′ 51.02″	− 2.15	38° 57′ 42.87

2. Distribute the error of 12.8″ equally among the angles. Correct each angle by subtracting − 12.8 / 8 = − 1.6″ from each angle.
3. There are two more conditions as follows:

$$\angle 1 + \angle 2 = \angle 5 + \angle 6, \quad \angle 3 + \angle 4 = \angle 7 + \angle 8$$

Therefore,

$$\angle 1 + \angle 2 = 59° 27′ 15.2″ + 40° 17′ 46″ = 99° 45′ 1.2″$$
$$\angle 5 + \angle 6 = 45° 23′ 27″ + 54° 21′ 26.3″ = 99° 44′ 53.5″$$

The difference is 7.9″. Take one-fourth of this as the correction to each angle. As the first sum is greater, subtract it from 1 and greater 2 and add it to 5 and 6. Similarly,

$$\angle 3 + \angle 4 = 35° \; 16' \; 11.2'' + 44° \; 58' \; 47.2'' = 80° \; 14' \; 58.4''$$
$$\angle 7 + \angle 8 = 41° \; 17' \; 13.9'' + 38° \; 57' \; 53.2'' = 80° \; 15' \; 7.1''$$

Distribute the difference of 8.7″ equally among the angles. Add 2.17″ to $\angle 3$ and $\angle 4$ and subtract it from $\angle 7$ and $\angle 8$.

4. The third condition is the side equation or log sine condition.

$$\Sigma \log \sin(\text{left angles}) = \Sigma \log \sin(\text{left angles})$$

Calculate the log sines of the angles, the difference d for 1″ for these angles, and d^2, and show them in the table.

Difference in the log sines of left and right angles = 332

Distribute the corrections as follows:

Correction for $\angle 1 = (d_1 / \Sigma d^2) \times 332 = (13.4 / 4113.75) \times 332 = 1.08''$
Correction for $\angle 2 = (30.6 / 4113.75) \times 332 = 2.47''$, and so on

List these corrections in the last table. Show the final corrected angles in the last column. As each of the corrections disturbs the earlier correction, the final values may not exactly satisfy all the conditions. But the difference, if any, will be small.

Sum of angles = 360° 00′ 00.28″
Difference between $\angle 1 + \angle 2$ and $\angle 5 + \angle 6 = 0$
Difference between $\angle 3 + \angle 4$ and $\angle 7 + \angle 8 = 0$

Adjustment of quadrilateral with central station

A quadrilateral with a central station is shown in Fig. 18.15. There are 12 angles measured—eight angles at the vertices of the quadrilateral and four angles at the central station. The angle conditions are as follows.

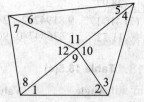

Fig. 18.15 Quadrilateral with a central station

(a) In each of the four triangles, the sum of angles must be 180°.

$$\angle 1 + \angle 2 + \angle 9 \; = 180°$$
$$\angle 3 + \angle 4 + \angle 10 = 180°$$
$$\angle 5 + \angle 6 + \angle 11 = 180°$$
$$\angle 7 + \angle 8 + \angle 12 = 180°$$

(b) The sum of angles around the central station is 360°.

$$\angle 9 + \angle 10 + \angle 11 + \angle 12 = 360°$$

(c) The side or log sine equation is given as

$$\Sigma \log \sin(\text{left angles}) = \Sigma \log \sin(\text{right angles})$$

18.13 Adjustment of Polygon with Central Station

A polygon with a central station is treated similarly to a quadrilateral with a central station. The central station may be within or outside the polygon and may or may not be occupied (Fig. 18.16). When the central station is not occupied, the angles at the central station are not measured. The following procedure is adopted.

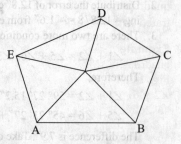

Fig. 18.16 Polygon with a central station

When the central station is not occupied

1. If n is the number of vertices of the polygon, the number of angles measured is $2n$. The angle equation is

    Sum of interior angles = $(2n - 4)$ right angles

2. The log sine equation is

    Σlog sin(left angles) = Σlog sin (right angles)

3. If $c_1, c_2, \ldots, c_n$ are the corrections to angles, we have, as in previous cases,

    $c_1, c_2, \ldots, c_n$ = discrepancy in the sum of angles

    and

    $d_1c_1 - d_2c_2 + \cdots + d_nc_n = M$

    By the theory of least squanes, $c_1^2 + c_2^2 + \cdots + c_n^2$ is minimum.

4. The method of correlates is used to find the corrections.

When the central station is occupied

1. Measure the angles at the central station and the angles at the vertices of the polygon. The number of angles measured is $2n + n = 3n$.

2. The following conditions have to be fulfilled.

(i) The sum of the angles measured at the central station is equal to 360°.

(ii) There are n triangles. The sum of the angles in each of the n triangles is equal to 180°.

(iii) The side or log sine equation is

   Σlog sin(left angles) = Σlog sin(right angles)

(iv) Determine the solution of these equations using the method of correlates.

18.14 Method of Equal Shifts

In adjusting any polygon with a central station, we have the following conditions.

(a) Sum of angles in each triangle = 180° (figure equation)

(b) Sum of angles measured at a station closing the horizon = 360° (station equation)

(c) Side equation or log sine equation

The method of equal shifts assumes that the correction or shift, necessary to the angles to satisfy the local equation, is the same for each triangle of the polygon. Similarly, any correction or shift necessary to satisfy the side equation is the same for each triangle. The method of adjusting quadrilaterals with a central station is illustrated with the following example.

Example 18.31 Adjust the angles of the quadrilateral with a central station shown in Fig. 18.17. The angles are $\angle 1 = 58° 10' 36''$, $\angle 2 = 61° 40' 50''$, $\angle 3 = 36° 2' 46''$, $\angle 4 = 28° 46' 26''$, $\angle 5 = 57° 6' 20''$, $\angle 6 = 62° 31' 48''$, $\angle 7 = 26° 12' 20''$, $\angle 8 = 29° 29' 04''$, $\angle 9 = 60° 8' 37''$, $\angle 10 = 115° 10' 54''$, $\angle 11 = 60° 21' 57''$, $\angle 12 = 124° 18' 40''$.

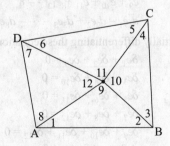

Fig. 18.17 Figure for Example 18.31

Solution   The solution is shown in Table 18.6(a).

Table 18.6(a)   Solution to Example 18.31

Right angle	log sine	d	d^2	Left angle	log sine	d	d^2
58° 10′ 36″	9.9292543	13.1	171.61	61° 40′ 50″	9.9446387	11.4	129.96
36° 2′ 46″	9.7696993	28.9	835.21	28° 46′ 26″	9.6824648	38.4	1474.56
57° 6′ 20″	9.9241099	13.1	171.61	62° 31′ 48″	9.9480472	10.9	118.89
26° 12′ 20″	9.6450220	43.1	1857.61	29° 29′ 04″	9.6921303	37.7	1421.29
Total	39.2680855	98.2	3036.04		39.2672810	98.4	3144.7

The conditions are as follows:
   (a) Sum of angles in each triangle = 180°
   (b) Sum of angles around central station = 360°
   (c) Σlog sin(left angles) = Σlog sin(right angles)
   (d) Least square equation
These conditions give

$$\angle 1 + \angle 2 + \angle 9 = 180°$$
$$\angle 3 + \angle 4 + \angle 10 = 180°$$
$$\angle 5 + \angle 6 + \angle 11 = 180°$$
$$\angle 7 + \angle 8 + \angle 12 = 180°$$
$$\angle 9 + \angle 10 + \angle 11 + \angle 12 = 360°$$

The side equation is as follows:

$$\log \sin 1 + \log \sin 3 + \log \sin 5 + \log \sin 7 = \log \sin 2 + \log \sin 4 + \log \sin 6$$
$$+ \log \sin 8$$

The least square equation is given as

$$\Sigma de^2 = \text{minimum}$$

where d is the difference for 1″ in the seventh figure of the seven figure log sine and e is the correction to the angles. In terms of corrections, $e_1, e_2, \ldots, e_{12}$, these equations can be written as

$$e_1 + e_2 + e_9 = -3″$$
$$e_3 + e_4 + e_{10} = -6″$$
$$e_5 + e_6 + e_{11} = -5″$$
$$e_7 + e_8 + e_{12} = -4″$$
$$e_9 + e_{10} + e_{11} + e_{12} = -8″$$
$$d_1 e_1 - d_2 e_2 + d_3 e_3 - \cdots - d_8 e_8 = 8045$$

Partially differentiating these equations, we get

$$\delta e_1 + \delta e_2 + \delta e_9 = 0 \tag{i}$$
$$\delta e_3 + \delta e_4 + \delta e_{10} = 0 \tag{ii}$$
$$\delta e_5 + \delta e_6 + \delta e_{11} = 0 \tag{iii}$$
$$\delta e_7 + \delta e_8 + \delta e_{12} = 0 \tag{iv}$$
$$\delta e_9 + \delta e_{10} + \delta e_{11} + \delta e_{12} = 0 \tag{v}$$
$$d_1 \delta e_1 - d_2 \delta e_2 + d_3 \delta e_3 - \cdots - d_8 \delta e_8 = 0 \tag{vi}$$
$$e_1 \delta e_1 + e_2 \delta e_2 + e_3 \delta e_3 + \cdots + e_{12} \delta e_{12} = 0$$

Multiplying Eqns (i) to (vi) by $-\lambda_1, -\lambda_2, -\lambda_3, -\lambda_4, -\lambda_5,$ and $-\lambda_6$ and adding these to the side equation, we get an equation in which the coefficients must independently vanish. This gives us

$$e_1 = \lambda_1 + 13.1\lambda_6$$
$$e_2 = \lambda_1 - 11.4\lambda_6$$
$$e_3 = \lambda_2 + 28.9\lambda_6$$
$$e_4 = \lambda_2 - 38.4\lambda_6$$
$$e_5 = \lambda_3 + 13.1\lambda_6$$
$$e_6 = \lambda_3 - 10.9\lambda_6$$
$$e_7 = \lambda_4 + 43.3\lambda_6$$
$$e_8 = \lambda_4 - 37.4\lambda_6$$
$$e_9 = \lambda_1 + \lambda_5$$
$$e_{10} = \lambda_2 + \lambda_5$$
$$e_{11} = \lambda_3 + \lambda_5$$
$$e_{12} = \lambda_4 + \lambda_5$$

Substituting these values back into the six conditional equations, we get six equations in λ_1 to λ_6. These can be solved to get the values of λ's and then the corrections. The six equations are as follows:

$$3\lambda_1 + \lambda_5 + 1.7\lambda_6 = -3$$
$$3\lambda_2 + \lambda_5 - 9.5\lambda_6 = -6$$
$$3\lambda_3 + \lambda_5 + 2.2\lambda_6 = -5$$
$$3\lambda_4 + \lambda_5 - 6.1\lambda_6 = -4$$
$$\lambda_1 + \lambda_2 + \lambda_3 + \lambda_4 + 4\lambda_5 = -8$$
$$13.1(\lambda_1 + 13.1\lambda_6) - 11.4(\lambda_1 - 11.4\lambda_6) + 28.9(\lambda_2 + 28.9\lambda_6) - 36.4(\lambda_2 - 36.4\lambda_6)$$
$$+ 13.1(\lambda_3 + 13.1\lambda_6) - 10.9(\lambda_3 - 10.9\lambda_6) + 43.3(\lambda_4 + 43.3\lambda_6) - (\lambda_4 - 37.2\lambda_6)$$
$$= -8,045$$
$$1.7\lambda_1 - 7.5\lambda_2 + 2.2\lambda_3 + 6.1\lambda_4 + 6180.7\lambda_6 = -8045$$

Solving these equations we get the λ values, from which the corrections are obtained. These are shown in Table 18.6(b).

Table 18.6(b) Corrections to angles (Example 18.31)

Values of λ	Corrections e_1 to e_{12}	Angles	Corrected angles
$\lambda_1 = +0.0196$	$e_1 = \lambda_1 + 13.1\lambda_6 = -17.1414$	58° 10′ 36″	58° 10′ 18.86″
$\lambda_2 = -5.871$	$e_2 = \lambda_1 - \lambda_1.4\lambda_6 = 14.9536$	61° 40′ 50″	61° 41′ 4.95″
$\lambda_3 = -0.4288$	$e_3 = \lambda_2 + 28.9\lambda_6 = -43.73$	36° 2′ 46″	36° 2′ 2.27″
$\lambda_4 = 1.6076$	$e_4 = \lambda_2 - 38.4\lambda_6 = 44.433$	28° 46′ 26″	28° 47′ 10.43″
$\lambda_5 = -0.8318$	$e_5 = \lambda_3 + 13.1\lambda_6 = -17.5898$	57° 6′ 20″	57° 6′ 2.41″
$\lambda_6 = -1.31$	$e_6 = \lambda_3 - 10.9\lambda_6 = 13.8502$	62° 31′ 48″	62° 32′ 1.85″
	$e_7 = \lambda_4 + 43.3\lambda_6 = -55.1154$	26° 12′ 20″	26° 11′ 24.88″
	$e_8 = \lambda_4 - 37.2\lambda_6 = 50.3396$	29° 29′ 04″	29° 29′ 54.34″
	$e_9 = \lambda_1 + \lambda_5 = -0.8122$	60° 8′ 37″	60° 8′ 36.19″
	$e_{10} = \lambda_2 + \lambda_5 = -6.7028$	115° 10′ 54″	115° 10′ 47.3″
	$e_{11} = \lambda_3 + \lambda_5 = -1.2606$	60° 21′ 57″	60° 21′ 55.74″
	$e_{12} = \lambda_4 + \lambda_5 = 0.7758$	124° 18′ 40″	124° 18′ 40.78″

Solution by the method of equal shifts

The same problem is solved by the method of equal shifts by adopting the following procedure.

1. Calculate the log sines, d, and d^2 of the angles as before. These are shown in Table 18.7(a).

Table 18.7(a) Calculation of log sines, d^2, and for angles

Right angle	log sine	d	d^2	Left angle	log sine	d	d^2
58° 10′ 36″	9.9292543	13.1	171.61	61° 40′ 50″	9.9446387	11.4	129.96
36° 2′ 46″	9.7696993	28.9	835.21	28° 46′ 26″	9.6824648	38.4	1474.56
57° 6′ 20″	9.9241099	13.1	171.61	62° 31′ 48″	9.9480472	10.9	118.89
26° 12′ 20″	9.6450220	43.1	1,857.61	29° 29′ 04″	9.6921303	37.7	1421.29
Total	39.2680855	98.2	3036.04		39.2672810	98.4	3144.7

The corrections to be applied are the following.

Corrections to angles of triangle ABO = – 3″

Corrections to angles of triangle BCO = – 6″

Corrections to angles of triangle CDO = – 5″

Corrections to angles of triangle DAO = – 4″

Corrections to angle at the central station = – 8″

2. Correct the central angles by distributing the error equally among the four angles. This is shown in column 1–3 of Table 18.7(b).

Table 18.7(b) Corrections to central and vertex angles (Example 18.31)

Central angle	Correc-tion	Corrected central angle	Right angles	Left angles	Sum	Correc-tions
60° 8′ 37″	– 2″	60° 8′ 35″	58° 10′ 36″	61° 40′ 50″	180° 00′ 01″	– 1″
115° 10′ 54″	– 2″	115° 10′ 52″	36° 2′ 46″	28° 46′ 26″	180° 00′ 04″	– 4″
60° 21′ 57″	– 2″	60° 21′ 55″	57° 6′ 20″	62° 31′ 48″	180° 00′ 03″	– 3″
124° 18′ 40″	– 2″	124° 18′ 38″	26° 12′ 20″	29° 29′ 04″	180° 00′ 02″	– 2″
360° 00′ 08″	– 8″	360° 00′ 00″				

3. To obtain the sums and the corresponding corrections, find the sum of the angles of the four triangles.
4. Distribute the corrections in the sum of the angles, one-third to each angle, to further correct the central angles.
5. This changes the sum of the central angles to deviate from 360°. Add + 3.33/4 to each of the first set of corrections to obtain a second set of corrections. Use this second set of corrections to correct the sum of the central angles.
6. Correct angles 1–8 in the same way by distributing one-third of the corrections to each angle. Further correct the central angles as they change. The sum of the angles will become 0.83″ more than 180°. Further apply this correction (0.83″/3) to each angle and obtain the corrected angles.

Table 18.7(c) Corrections to central angles

Central angles	Total corrections	First set	Second set	Final corrections	Corrected central angle
60° 8′ 35″	− 1″	− 0.33	+ 0.5	0.5″	60° 8′ 35.5″
115° 10′ 52″	− 4″	− 1.33	− 0.5	− 0.5″	115° 10′ 52.5″
60° 21′ 55″	− 3″	− 1.00	− 0.17	− 0.17″	60° 21′ 54.83″
124° 18′ 38″	− 2″	− 0.67	+ 0.17	+ 0.17″	124° 18′ 38.17″
360° 00′ 00″					360° 00′ 00″

Table 18.7(d) Corrections to the left angles

Right angles	Total corrections	First set	Second set	Final corrections	Corrected angles
58° 10′ 36″	− 1″	− 0.33	− 0.27″	− 0.6	58° 10′ 35.4″
36° 2′ 46″	− 4 ″	− 1.33	− 0.27″	− 1.6	36° 2′ 44.4″
57° 6′ 20″	− 3″	− 1.00	− 0.27″	− 1.27	57° 6′ 18.73″
26° 12′ 20″	− 2″	− 0.67	− 0.27″	− 0.94	26° 12′ 19.06″

Table 18.7(e) Corrections to right angles

Left angles	Total corrections	First set	Second set	Final corrections	Corrected angles
61° 40′ 50″	− 1″	− 0.33	− 0.27″	− 0.6″	61° 40′ 49.4″
28° 46′ 26″	− 4″	− 1.33	− 0.27″	− 1.6″	28° 46′ 24.4″
62° 31′ 48″	− 3″	− 1.00	− 0.27″	− 1.27″	62° 31′ 46.73″
29° 29′ 04″	− 2″	− 0.67	− 0.27″	− 0.94″	29° 29′ 3.06″

7. The difference in the log sines of the corrected left and right angles is 8050. The log sines of the left angles are greater than those of the right angles.

$$\Sigma \log \sin(\text{left angles}) - \Sigma \log \sin(\text{right angles}) = 39.2680769 - 39.2672719$$
$$= 8050$$

The left angles have to be decreased and the right angles have to be increased to make the sum of the log sines equal. The sum of d values for all the angles is 201.3.

  1″ shift = 201.3 for 8050

  Shift = 8050/201.3 = 40″

Decrease each of the left angles by 40″ and increase each of the right angles by 40″.

8. The process can be repeated once again if the log sine sums are still not equal. This is shown in Table 18.7(f). The log sine differs by 202 and hence the correction for the angles is 202/203.1 = 1″.

We can check the angles by summing them up as follows:

(a) 61° 41′ 30.38″ + 58° 9′ 54.39″ + 60° 8′ 35.23″ = 180° 00′ 00″

(b) 36° 2′ 3.39″ + 28° 47′ 5.38″ + 115° 10′ 51.23″ = 180° 00′ 00″

(c) $57° 5' 37.72'' + 62° 32' 27.72'' + 60° 21' 54.56'' = 180° 00' 00''$

(d) $26° 11' 38.05'' + 29° 29' 44.05'' + 124° 18' 37.9'' = 180° 00' 00''$

Table 18.7(f) Correction for side equation

Angle	First set	Second set	Final angle	Corrected
1. 58° 10′ 35.4″	− 40	− 1″	− 41″	58° 9′ 54.39
3. 36° 2′ 44.4″	− 40	− 1″	− 41″	36° 2′3.39″
5. 57° 6′ 18.73″	− 40	− 1″	− 41″	57° 5′ 37.72″
7. 26° 12′ 19.06″	− 40	− 1″	− 41″	26° 11′ 38.05″
2. 61° 40′ 49.4″	+40	+ 1″	+ 41″	61° 41′ 30.38″
4. 28° 46′ 24.4″	+40	+ 1″	+ 41″	28° 47′ 5.38″
6. 62° 31′ 46.73″	+40	+ 1″	+ 41″	62° 32′ 27.72″
8. 29° 29′ 3.06″	+40	+ 1″	+ 41″	29° 29′ 44.05″

18.15 Three-point Problem

The three-point problem is the same as the one we studied in plane table surveying (Chapter 5). While with the plane table we saw the graphical or mechanical solution, here we will look at the analytical solution.

A, B, and C (Fig. 18.18) are three triangulation stations and P is a point whose location is to be determined with respect to the three known stations. Let $\angle APC = \theta_1$, $\angle BPC = \theta_2$, $\angle PAC = \alpha$, $\angle PBC = \beta$.

$$\angle A + \angle P + \angle B + \angle C = 360°$$

$$\angle \alpha + \angle \beta = 360° - \angle ACB - (\angle \theta_1 + \angle \theta_2) = \phi$$

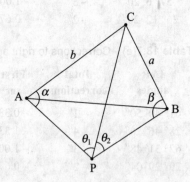

Fig. 18.18 Three-point problem

From triangle PAC

$$PC = b \sin \alpha / \sin \theta_1$$

From triangle PBC

$$PC = a \sin \beta / \sin \theta_2$$

Equating the two values of PC

$$\sin \beta = b \sin\alpha \sin \theta_2 / a \sin \theta_1$$

As

$$\beta = \phi - \alpha$$

$$\sin(\phi - \alpha) = b \sin\alpha \sin\theta_2 / a \sin\theta_1$$

$$\sin \phi \cos \alpha - \cos \phi \sin \alpha = b \sin \alpha \sin \theta_2 / a \sin \theta_1$$

Dividing by $\sin\phi \sin\alpha$, we get

$$\cot \alpha - \cot \phi = b \sin \theta_2 / a \sin \theta_1 \cos \phi$$

$$\cot \alpha = \cot \phi + b \sin \theta_2 / a \sin \theta_1 \sin \phi$$

$$\cot \alpha = \cot \phi (1 + b \sin \theta_2 \sec \phi / a + \sin \theta_1)$$

$$\beta = \phi - \alpha$$

PA, PC, and PB are calculated using the sine rule.

$$PA = b \sin(\angle ACP)/\sin \theta_1 = b \sin (180° - \theta_1 - \alpha)/\sin \theta_1 = b\sin(\theta_1+\alpha)/\sin \theta_1$$

$$PB = a \sin (\angle BCP)/\sin \theta_2 = b \sin(\theta_2 + \beta)/\sin \theta_2$$

$$PC = b \sin \alpha /\sin \theta_1$$

Example 18.32 In triangle ABC (Fig. 18.19), A, B, and C are the triangulation stations of a known position. To find the position of P, the following measurements are available: $\angle APC = 36° 12' 40''$, $\angle BPC = 54° 16' 40''$, $\angle ACB = 100° 10' 40''$, AC $= 2197.5$ m, and BC $= 2848.5$ m. Find the lengths PA, PB, and PC.

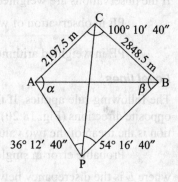

Fig. 18.19 Triangle of Example 18.32

Solution Let $\angle PAC = \alpha$ and $\angle PBC = \beta$. We find PC from triangles PAC and PBC and equate the two results.

$$2197.5 \sin\alpha/\sin(36° 12' 40'') = 2848.5 \sin\beta/\sin(54° 16' 40'')$$

Putting

$$\alpha + \beta = \phi = 360° - 36° 12' 40'' - 54° 16' 40'' - 100° 10' 40'' = 169° 20'$$

$$\sin (\phi - \alpha) = \sin \beta = \frac{2197.5\sin\alpha \sin 54°16'40''}{2848.5\sin 36°12'40''}$$

or

$$\sin \phi \cos \alpha - \cos \phi \sin\alpha = \frac{2197.5\sin\alpha \sin 54°16'40''}{2848.5\sin 36°12'40''}$$

Dividing by $\sin \phi \sin \alpha$,

$$\cot \alpha - \cot \phi = \frac{2197.5\sin 54°16'40''}{2848.5\sin 36°12'40'' \sin\phi}$$

$$\cot \alpha = \cot \phi + \frac{2197.5\sin 54°16'40''}{2848.5\sin 36°12'40'' \sin\phi}, \quad \phi = 169° 20'$$

$$\cot \alpha = 0.4185, \quad \alpha = 67° 17' 26'', \quad \beta = 102° 02' 34''$$

$$\angle ACP = 180° - 67° 17' 26'' - 36° 12' 40'' = 76° 29' 54''$$

$$\angle BCP = 180° - 102° 02' 34'' - 54° 16' 40'' = 23° 40' 46''$$

$$PA = \frac{2197.5\sin 76°29'54''}{\sin 36°12'40''} = 3616.97 \text{ m}$$

$$PB = \frac{2848.5\sin 23°40'46''}{\sin 54°16'40''} = 1409.31 \text{ m}$$

$$PC = \frac{2197.5\sin 67°17'26''}{\sin 36°12'40''} = 3431.39 \text{ m}$$

18.16 Adjustment of Levels

As in the case of angles and distances, levels running from benchmarks need to be checked and corrected. Levels between two points may be run on different routes and by different parties at different periods of time. Even with precise levelling, the difference in elevation measured by two different routes may differ. Probable errors (PEs) are calculated as follows:

$$\text{PE in observation of unit weight} = \pm 0.6745 \sqrt{\Sigma e^2 / (n-1)}$$

$$\text{PE in mean} = \pm 0.6745 \sqrt{\Sigma e^2 / [n(n-1)]}$$

If the observations are weighted

$$\text{PE in observation of weight } w = \pm 0.6745 \sqrt{\Sigma we^2 / w(n-1)}$$

$$\text{PE in weighted arithmetic mean} = \pm 0.6745 \sqrt{\Sigma we^2 / \Sigma w(n-1)}$$

Level lines

The following rule applies. If the lines of levels between two benchmarks run in opposite directions (Fig. 18.20), then the probable value of difference in the elevation is the mean of the two values obtained.

$$\text{Probable error in single observation} = \pm 0.4769E$$

where E is the discrepancy between the measurements.

$$\text{Probable error in the mean} = 0.3373E$$

General laws of weights   The following two rules are observed.

(a) The probable error in a line of levels varies inversely as the square root of the length of the line when the measurements are taken under similar conditions.

(b) The weight of the result due to any line of levels varies inversely as the length of the line when the measurements are taken under similar conditions.

In the case shown in Fig. 18.20, each line of levels should be weighted inversely as the length.

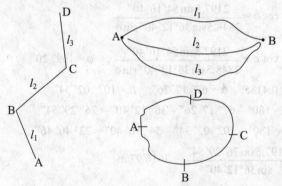

Fig. 18.20   Level lines

Closed lines

When the line of levels returns to the starting point, the level circuit is said to be closed. In such a case, the error of closure is distributed in direct proportion to the distances of the staff stations from the starting point. The following examples illustrate the above principles.

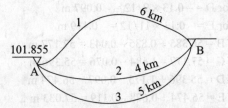

Fig. 18.21   Level net of Example 18.33

Example 18.33   From the following data, determine the probable value of elevation of *B* if the elevation of *A* is 101.855 m (Fig. 18.21).

Table 18.8   Data for Example 18.33

Point A (m)	Point B (m)	Route	Distance (km)
101.855	123.345	A-1-B	6
	123.352	A-2-B	4
	123.348	A-3-B	5

    Solution   The differences in elevation and their weights are calculated first.

    Difference in elevation, $V_1 = 123.345 - 101.855 = 21.49$ m (weight $\propto 1/6$)

    Difference in elevation, $V_2 = 123.352 - 101.855 = 21.497$ m (weight $\propto 1/4$)

    Difference in elevation, $V_3 = 123.348 - 101.855 = 21.493$ m (weight $\propto 1/5$)

    Weight ratio = $1/6 : 1/4 : 1/5 = 10 : 15 : 12$

    Weighted arithmetic mean = $(21.49 \times 10 + 21.497 \times 15 + 21.493 \times 12)/37$

$$= 21.494 \text{ m}$$

    Probable value of elevation of B = $101.855 + 21.494 = 123.349$ m

Example 18.34   Find the probable values of the reduced levels of B, C, D, and E from the following data given that the reduced level of A = 56.385.

Table 18.9   Data for Example 18.34

Line	Distance (km)	Difference in elevation (m)
AB	4	0.835
BC	3	− 1.684
CD	2	1.175
DE	2	0.678
EA	1	− 0.824

Solution   In a closed line of levels, the corrections are proportional to the distance from the starting point.

    Total distance = 12 km

    Discrepancy in elevation = $0.835 - 1.684 + 1.175 + 0.628 - 0.824 = -0.13$ m

    Correction for B = $-0.13 \times 4/12 = -0.043$ m

    Correction for C = $-0.13 \times 7/12 = -0.076$ m

Correction for D $= -0.13 \times 9/12 = -0.097$ m

Correction for E $= -0.13 \times 11/12 = -0.119$ m

Elevation of B $= 56.385 + 0.835 - 0.043 = 57.177$ m

Elevation of C $= 57.177 - 1.684 - 0.076 = 55.396$ m

Elevation of D $= 55.396 + 1.175 - 0.097 = 56.474$ m

Elevation of E $= 56.474 + 0.678 - 0.119 = 57.033$ m

Adjustment of a level net

A level net is an interconnecting line of levels connecting a number of benchmarks. The difference in elevation found using different routes is adjusted by the method of normal equations or the method of correlates. The probable differences in elevation or probable errors are found using these methods and the necessary corrections applied. The following example illustrates the principle.

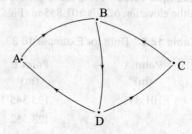

Fig. 18.22 Network for Example 18.35

Example 18.35 In the level net shown in Fig. 18.22, the difference in elevations in the different sections is as follows:

A to B $= +2.31$ m (weight $= 3$)

B to D $= 7.512$ m (weight $= 2$)

D to A $= -10.02$ m (weight $= 1$)

D to C $= -2.892$ m (weight $= 2$)

C to B $= -4.572$ m (weight $= 1$)

Find the probable elevations of points given that the elevation of point A $= 1050.805$ m.

Solution There are two level circuits here, ABD and BDC. In circuit ABD,

Error $= 2.31 + 7.512 - 10.02 = -0.198$ m

In circuit BDC,

Error $= 7.512 - 2.892 - 4.572 = 0.048$ m

If c_1, c_2, c_3, c_4, and c_5 are the corrections to the listed levels, respectively, then

$c_1 + c_2 + c_3 = 0.198$

$c_2 + c_4 + c_5 = -0.048$

Also, from the principle of least squares, Σwc^2 is minimum. Also $3c_1^2 + 2c_2^2 + c_3^2 + 2\,c_4^2 + c_5^2$ is minimum. Differentiating the three equations partially, we get

$$\delta c_1 + \delta c_2 + \delta c_3 = 0 \tag{i}$$

$$\delta c_2 + \delta c_4 + \delta c_5 = 0 \tag{ii}$$

$$3c_1\delta c_1 + 2c_2\delta c_2 + c_3\delta c_3 + 2c_4\delta c_4 + c_5\delta c_5 = 0 \tag{iii}$$

Multiplying Eqns (i) and (ii) by $-\lambda_1$ and $-\lambda_2$ and adding the three equations, we get

$$\delta c_1(3c_1 - \lambda_1) + \delta c_2(2c_2 - \lambda_1 - \lambda_2) + \delta c_3(c_3 - \lambda_1) + \delta c_4(2c_4 - \lambda_2) + \delta c_5(c_5 - \lambda_2) = 0$$

As the coefficients should vanish independently, we get

$$c_1 = \lambda_1/3, \quad c_2 = (\lambda_1 + \lambda_2)/2, \quad c_3 = \lambda_1, \quad c_4 = \lambda_2/2, \quad c_5 = \lambda_2$$

Substituting these values in the conditional equations, we get

$$11\lambda_1 + 3\lambda_2 = 1.188, \quad \lambda_1 + 4\lambda_2 = -0.096$$

Solving these equations

$$\lambda_1 = +0.123, \quad \lambda_2 = -0.055$$

Substituting these values, we get

$$c_1 = 0.041, \quad c_2 = 0.034, \quad c_3 = 0.123, \quad c_4 = -0.028, \quad c_5 = -0.055$$

The corrected differences in elevation are

$$A - B = 2.351$$
$$B \text{ to } D = 7.546$$
$$D \text{ to } A = -9.897$$
$$D \text{ to } C = -2.92$$
$$C \text{ to } B = -4.627$$

18.17 Adjustment of a Closed Traverse

For precise adjustment of a closed traverse, *Crandall's method* is used. In this method it is assumed that the error of closure is due to errors in the linear measurements only. The adjustment, thus, does not affect the direction or bearings of lines. The angular error, if any, is distributed before the adjustment is done. The following procedure is adopted.

1. Compute the angular error of the traverse and adjust the same by distributing it among the angles or bearings.
2. If c is the correction in latitude and k is the correction in departure of any side, then

$$\text{Total correction in latitude} = \Sigma cL \tag{18.19}$$
$$\text{Total correction in departure} = \Sigma cD \tag{18.20}$$

The values of c (c_1, c_2, c_3, etc.) depend on the bearings of the lines. By the theory of least squares,

$$\Sigma(c^2 L^2 / l) \text{ is a minimum} \tag{18.21}$$

Differentiating the three equations, we get

$$\Sigma L \delta c = 0, \quad \Sigma D \delta c = 0, \quad \Sigma cL \delta c = 0$$

Using correlates $- \lambda_1$ and $- \lambda_2$,

$$\Sigma(-\lambda_1 L \delta c) = 0, \quad \Sigma(-\lambda_2 D \delta c) = 0$$

Adding the three equations

$$\Sigma \delta c(cL - \lambda_1 L - \lambda_2 D) = 0$$

The coefficients of the terms have to individually become zero. Thus,

$$c_1 = (\lambda_1 L_1 + \lambda_2 D_1)/l_1, \quad c_2 = (\lambda_1 L_2 + \lambda_2 D_2)/l_2, \quad \text{etc.}$$

Substituting these values back into the equations, we get

$$\text{Total correction in latitude} = \lambda_1 \Sigma(L^2 / l)$$
$$\text{Total correction in departure} = \lambda_2 \Sigma(LD / l)$$

These two equations can be solved to get the values of λ_1 and λ_2.

Correction to latitude = $(\lambda_1 L_1^2 + \lambda_2 L_1 D_1)/l_1$, etc.

Correction to departure = $(\lambda_1 L_1 D_1 + \lambda_2 D_1^2)/l_1$, etc.

The following example illustrates the procedure.

Example 18.36   Adjust the traverse shown in Fig. 18.23. The following is the data of the traverse.

Line	PQ	QR	RS	ST	TP
Length	402.2	398.3	430.2	490.4	274.5
Bearing	N60° 45′ 10″E	S70° 30′ 12″E	S22° 30′ 10″W	N81° 45′ 14″W	N16° 22′ 18″W

Solution   We first calculate the latitudes and departures as shown in Tables 18.10–18.13. In these tables L is the latitude, D is the departure, and l is the length of line.

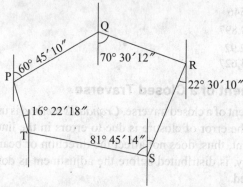

Fig. 18.23   Traverse for Example 18.36

Table 18.10   Calculation of latitudes and departures (Example 18.36)

Line	Length l	Bearing	Latitude L	Departure D
PQ	402.2	N60° 45′ 10″E	196.5064	350.9274
QR	398.3	S70° 30′ 12″E	– 132.9334	375.4618
RS	430.2	S22° 30′ 10″W	– 397.445	– 164.6497
ST	490.4	N81° 45′ 14″W	70.3358	– 485.3298
TP	274.5	N16° 22′ 18″W	263.3699	– 77.3725
Total			– 0.1663	– 0.9628

We now calculate L^2/l, LD/l, and D^2/l for the lines as shown in Table 18.11.

Table 18.11   Calculation of values (Example 18.36)

Line	Length	L^2/l	LD/l	D^2/l
PQ	402.2	96.0088	174.12	306.19
QR	398.3	44.366	– 125.311	353.933
RS	430.2	367.184	152.113	63.016
ST	490.4	10.087	– 69.63	480.312
TP	274.5	252.69	– 74.235	21.801
Total		770.3358	57.057	1225.252

The corrections are now calculated as shown in Tables 18.12 and 18.13.

Table 18.12 Calculation of corrections (Example 18.36)

Line	Correction to latitude			Correction to departure		
	$\lambda_1 L^2/l$	$\lambda_2 LD/l$	Total	$\lambda_1 LD/l$	$\lambda_2 D^2/l$	Total
PQ	0.0157	0.1329	0.1486	0.0288	0.2382	0.267
QR	0.0068	− 0.0978	− 0.0910	− 0.021	0.2753	0.2543
RS	0.0617	0.1175	0.1793	0.0254	0.0488	0.0745
ST	0.0001	− 0.0544	− 0.0545	− 0.0117	0.3737	0.3628
TP	0.042	− 0.0579	− 0.0159	− 0.0124	0.0169	0.0045

Table 18.13 Corrected latitudes and departures (Example 18.36)

Line	Correction to latitude	Corrected latitude	Correction to departure	Corrected departure
PQ	0.1486	196.6550	0.267	351.1948
QR	− 0.0910	− 133.0244	0.2543	375.7164
RS	0.1793	− 397.2654	0.0745	− 164.5755
ST	− 0.0545	70.2813	0.3628	− 484.9678
TP	− 0.0159	263.3540	0.0045	− 77.368

1. Calculate latitude L and departure D from the lengths and bearings as $l \cos\theta$ and $l \sin\theta$.
2. Find the sum of the latitudes and departures separately as − 0.1663 and − 0.9628.
3. The corrections are + 0.1663 and + 0.9628.
4. Calculate $\Sigma L^2/l$, $\Sigma LD/l$, and $\Sigma D^2/l$ by finding the values for each line separately and finding the sum.
5. $\lambda_1(\Sigma L^2/l) + \lambda_2(\Sigma LD/l) = 0.1663$ and $\lambda_1(\Sigma LD/l) + \lambda_2(\Sigma D^2/l) = 0.9628$ are the two equations obtained. Solve these by substituting values for the quantities. We get $\lambda_1 = 0.000168$ and $\lambda_2 = 0.000778$.
6. Calculate the corrections as follows:

    For latitude: $\lambda_1 L^2/l + \lambda_2 LD/l$

    For departure: $\lambda_1 LD/l + \lambda_2 D^2/\lambda$

Summary

Geodetic surveying is performed to locate points for horizontal control. Despite using precise equipment and methods, errors still remain. Most of the systematic errors (errors due to equipment, personnel, and natural causes) are eliminated by the precision with which the survey is conducted. However, errors still remain in the observations. Such errors are known as accidental errors. They follow the law of probability and the angles and distances are corrected using different methods.

Observations are assigned weights depending upon the precision with which the measurements are taken. The most probable value of a single quantity through a number of observations is the arithmetic mean or the weighted arithmetic mean.

The principle of least squares is the most valuable tool for correcting the values of measured quantities. The principle states that the sum of the squares of the errors from the mean is the least.

Quantities are corrected using the conditions governing them, such as the sum of the angles of a triangle is 180°, the sum of the angles measured around a point closing the horizon is 360°, etc. The method of normal equations, where the equations are multiplied by the coefficient of the unknown quantity and the weights and then added, is used to adjust the values. Another method is that of correlates, in which the quantities are multiplied by a constant $(-\lambda)$ and, along with the principle of least squares, the errors and constants are related by the partial differentiation of equations. The number of correlates required is equal to the number of conditions imposed.

The adjustment of triangulation networks is done using a stepwise process of station adjustment and figure adjustment. In station adjustment, the angles measured around a station are adjusted. In figure adjustment, the condition of the sum of angles in a geometrical figure is used.

Triangles are the basic figures in networks. Triangles are adjusted satisfying the condition that the sum of the angles is 180°. A quadrilateral has two triangles with eight angles measured. It has to satisfy the condition of triangles and quadrilaterals.

Polygons with central stations are adjusted in a similar way. The sum of the interior angles of a polygon is $(2n - 4)$ right angles. In addition, every triangle formed has to satisfy the condition of triangles.

The three-point problem is the problem of locating a station with the positions of three known points. This problem is solved analytically or graphically. Levels are adjusted similarly. The weights of measurements are inversely proportional to the length of the line of levels. Level nets are adjusted using principles similar to that of angles.

Exercises

Multiple-Choice Questions

1. The weight of an observation is
    (a) its probable value
    (b) a number indicating the trustworthiness of measurement
    (c) the average value from a number of observations
    (d) the value indicating least error in the value
2. Most probable value of a quantity is the value
    (a) having greater weight
    (b) having average weight
    (c) nearer to the true value than any other value
    (d) satisfying all conditional equations
3. The weight of the arithmetic mean of a number of weighted observations is
    (a) the product of the weights of the observations
    (b) the reciprocal of the sum of reciprocals of the weights
    (c) the sum of the weights of the observations
    (d) the average weight of the observations
4. The weight of the arithmetic mean of a number of observations of equal weight is
    (a) the weight of any observation
    (b) the sum of weights of observations
    (c) the product of the weights of observations
    (d) the number of observations
5. The weight of the sum of, or difference between, two weighted quantities is
    (a) the sum of or difference between the weights
    (b) the product of the weights

(c) the ratio of the weights

(d) the reciprocal of the sum of reciprocals of individual weights

6. If a weighted observation is multiplied by a factor, then the weight of the product is the weight

(a) divided by the factor

(b) multiplied by the factor

(c) divided by the square of the factor

(d) divided by the square root of the factor

7. If a measured angle α has a weight of 3, then the weight of 4α is

(a) ¾

(b) 12

(c) 1.5

(d) 3/16

8. Angle α has a weight of 3 and angle β has a weight of 4. The weight of $(\alpha + \beta)$ is

(a) 7

(b) 12

(c) ¾

(d) 12/7

9. If a number of observations of equal weight are taken of a quantity, then the most probable value of the quantity is the

(a) arithmetic mean

(b) least value among the observations

(c) maximum value among the observations

(d) median value of the observations

10. If an equation $4x + 5y = 32$ has a weight of 3, then the normal equation in x is

(a) $12x + 15y = 96$

(b) $(4/3)x + (5/3)y = 32/3$

(c) $48x + 60y = 384$

(d) $16x + 20y = 128$

11. If an equation $4x + 3y + 5z = 72$ has a weight of 2, then the normal equation in z is

(a) $8x + 6y + 10z = 144$

(b) $40x + 30y + 50z = 720$

(c) $32x + 24y + 40z = 576$

(d) $24x + 18y + 30z = 432$

12. The principle of least square states that the

(a) sum of the errors from the mean is the minimum

(b) sum of the squares of the errors from the mean is the minimum

(c) square root of the sum of the squares of the observations is the most probable value

(d) sum of squares of observations divided by the square of the number of observations has the least error

13. In the case of observed angles of given weights, any error is distributed amongst them

(a) in proportion to their magnitude

(b) equally

(c) in proportion to their log sine values

(d) in inverse proportion to the weights of angles

14. When the value of an angle is given as $\alpha \pm e$, then e is

(a) the index error of the instrument

(b) the error due to inclination of horizontal axis

(c) the error due to level not being at the centre while reading

(d) the probable error in the value of the angle

15. If 3 angles of equal weight are measured at a point closing the horizon, then the error e in their sum is distributed

(a) in proportion to their magnitude

(b) equally among the three angles

(c) in proportion to the square of the angle values

(d) in proportion to the square root of the angle values

16. In a geodetic triangle, the sides are

(a) straight lines joining the station points

(b) arcs of a sphere

(c) arcs on the surface of an ellipsoid

(d) arcs on the surface of a spheroid
17. Spherical excess is the amount by which
    (a) arc length of a spherical triangle exceeds the chord length between the points
    (b) angles measured exceeds the true value of the angles
    (c) probable error in spherical angles
    (d) the sum of angles of spherical triangle exceeds 180 degrees
18. Spherical excess depends upon
    a) the sides of the spherical triangle   (c) radius of the sphere
    (b) the angles of the spherical triangle (d) area of the spherical triangle
19. In DeLambre's method of solving spherical triangles, the term plane angles refers to
    (a) the angle subtended at the centre by the sides
    (b) the spherical angles reduced by spherical excess
    (c) the angles between the chords joining the station points
    (d) the angles obtained when the sides are reduced to mean sea level
20. Legendre's theorem can be stated as
    (a) the spherical triangle can be treated as a plane triangle if the sides are small
    (b) the spherical triangle can be treated as a plane triangle if the angles are not less than 30 and not more than 120
    (c) the spherical triangle can be treated as a plane triangle if the spherical excess is small
    (d) by reducing each angle by $1/3$ of spherical excess, the sine of these sides is proportional to the length of the opposite sides
21. In adjusting a triangulation system, station adjustment is adjusting the
    (a) closing error in the triangles
    (b) length of the sides for measurement errors
    (c) angles around a station so that their sum is 360
    (d) angles of a triangle so that their sum is 180
22. In adjusting a triangulation system, figure adjustment refers to adjusting the
    (a) closing error in the triangles
    (b) length of the sides for measurement errors
    (c) angles around a station so that their sum is 360
    (d) angles of a triangle such that their sum is 180
23. In adjusting a triangle with a central station, side equation can be stated as
    (a) sum of log sines of left angles = sum of log sines of right angles
    (b) sum of log cosines of left angles = sum of log cosines of right angles
    (c) sum of log tangents of left angles = sum of log tangents of right angles
    (d) sum of left angles = sum of right angles
24. Method of equal shifts in adjusting a polygon with a central station assumes that
    (a) shift required in the sides of the polygon is the same for each line
    (b) shift required in the vertices of the polygon is the same
    (c) correction necessary to the angles to satisfy the conditions of adjustment is the same for each triangle
    (d) the corrections are equally distributed as per their weights in each triangle

Review Questions

1. Explain the terms systematic error and accidental error.
2. Explain the concept of weight of a measurement.
3. State the laws of weights.
4. State and prove the principle of least squares.
5. Define the terms probable value and probable error.

6. What are normal equations? Explain the method of forming normal equations.
7. What is a correlate? Explain the method of finding probable values using correlates.
8. Explain the terms station adjustment and figure adjustment.
9. Explain how a geodetic triangle is adjusted.
10. Explain Legendre's and Delambre's method of adjusting a spherical triangle.
11. State and prove the side or log sine equation.
12. Describe the method of adjusting a triangle with a central station.
13. Explain the method of approximate adjustment of a quadrilateral.
14. Describe the conditions and the method of adjusting a polygon with a central station.
15. State the three-point problem and show how it is solved analytically.
16. Explain the method used to adjust a line of levels.
17. What is a level net? Explain how a level net is adjusted.

Problems

1. If $\angle A = 32° 24' 50''$ (weight = 3) and $\angle B = 26° 18' 40''$ (weight = 2), find the weight of $\angle A + \angle B$.
2. If $\angle A = 61° 30' 40''$ (weight = 3), find the weight of $2A$ and $A/2$.
3. Five values of $\angle P$ measured are given below. Find the most probable value of $\angle P$.

   $\angle A = 42° 32'$ 16.82''   17.3''   15.82''   14.32''   18.1''

4. Find the most probable value of $\angle A$ and its weight from the following measurements:
   $\angle A = 42° 32' 18''$ (weight = 3), $17''$ (weight = 2), $16''$ (weight = 1), $19''$ (weight = 4).
5. Find the probable value of $\angle P$, given that $P = 24° 42' 32''$, $2P = 49° 25' 02''$, and $3P = 74° 07' 40''$.
6. If $\angle A = 40° 20' 10.2''$, $\angle B = 25° 32' 18.6''$, and $\angle C = 65° 52' 29.6''$, find the probable values of $\angle A$ and $\angle B$ given that $A + B = C$. Use the method of normal equations.
7. If the angles of a triangle ABC are $\angle A = 60° 12' 38''$ (weight = 2), $\angle B = 56° 32' 44''$ (weight = 3), and $\angle C = 63° 14' 46''$ (weight = 1), find the probable values of $\angle A$, $\angle B$, and $\angle C$ by the method of normal equations.
8. Four angles observed around a station closing the horizon are $\angle 1 = 92° 30' 23''$ (weight = 3), $\angle 2 = 87° 25' 44''$ (weight = 2), $\angle 3 = 96° 37' 52''$ (weight = 1), and $\angle 4 = 83° 26' 12''$ (weight = 2). Find the probable values of angles 1, 2, 3, and 4 by the method of normal equations.
9. Solve Problem 6 by the method of correlates.
10. Solve Problem 7 by the method of correlates.
11. Solve Problem 8 by the method of correlates.
12. From the eight values of P measured, find the probable value of P. The weights are given in the brackets.

48° 24'	33.6''	34.8''	32.8''	35.2''	36''	34.8''	36.2''	35.3''
	(2)	(2)	(3)	(3)	(2)	(1)	(2)	(1)

13. If $\angle A = 62° 30' 42''$, $\angle B = 58° 40' 52''$, $\angle C = 71° 42' 36''$, $\angle A + \angle B = 121° 10' 28''$, and $\angle B + \angle C = 130° 23' 32''$, find the probable values of A, B, and C.
14. If A, B, C, and D are angles measured around a point closing the horizon, find the most probable values of the angles by (a) correcting the angles, the corrections being inversely proportional to the weights, (b) the method of normal equations, and (c) the method of correlates. $\angle A = 90° 32' 36''$ (weight = 3), $\angle B = 89° 28' 42''$ (weight = 4), $\angle C = 98° 44' 52''$ (weight = 2), and $\angle D = 81° 13' 44''$ (weight = 2).

15. From the following data, find the most probable values of the angles.

    ∠A = 57° 12′   34″   33″   32″   35″   36″   34″
    ∠B = 68° 42′   13″   12″   11″   14″   15″   10″   12″   14″
    ∠C = 54° 5′   16″   15″   14″   17″   18″

16. The angles in a geodetic triangle are ∠A = 56° 24′ 32.18″ (weight = 2), ∠B = 62° 34′ 18.32″ (weight = 3), ∠C = 61° 01′ 19.15″ (weight = 4). If the length of side *a* (opposite ∠A) = 6835.5 m, find the lengths of the other two sides using (a) spherical trigonometry, (b) Legendre's method, and (c) Delambre's method.

17. Two triangles ACD and BCD have a common side CD. The angles measured are ∠CAD = 94° 13′ 24″, ∠CBD = 92° 15′ 38″, ∠BCA = 78° 32′37″, ∠BDA = 94° 58′ 16″, ∠ACD = 38° 18′ 48″, ∠BCD = 40° 13′ 45″, ∠ADC = 47° 27′ 42″, ∠BDC = 47° 30′ 30″. Find the corrected values of the angles (see Fig. 18.24).

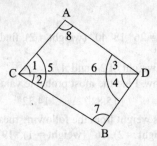

Fig. 18.24   Figure for Problem 17   **Fig. 18.25**   Figure for Problem 18

18. Adjust the angles of the triangle with a central station shown in Fig. 18.25. The angles are ∠1 = 26° 13′ 18″, ∠2 = 30° 15′ 37″, ∠3 = 28° 12′ 10″, ∠4 = 32° 18′ 42″, ∠5 = 30° 08′ 16″, ∠6 = 32° 51′ 56″.

19. Adjust the quadrilateral shown in Fig. 18.26. The angles are ∠1 = 58° 37′ 16.8″, ∠2 = 41° 07′ 46.2″, ∠3 = 36° 18′ 27.2″, ∠4 = 43° 56′ 31.2″, ∠5 = 46° 13′ 17″, ∠6 = 56° 31′ 52″, ∠7 = 42° 20′ 12.8″, ∠8 = 37° 54′ 49.2.

20. Adjust the quadrilateral with a central station shown in Fig. 18.27. Values of the angles are:

    ∠1 = 42° 10′57″, ∠2 = 45° 5′20″, ∠3 = 45° 37′25″, ∠4 = 40° 16′51″,
    ∠5 = 52′51″;
    ∠6 = 47° 52′29″, ∠7 = 44° 9′25″, ∠8 = 46° 57′ 51″, ∠10 = 92° 43′45″,
    ∠10 = 94°5′46″,
    ∠11 = 84° 14′46″, ∠12 = 88° 55′47″

21. Find the probable elevation of q from the data given below:

A	B	Route	Distance
203.135	217.845	A-1-B	8 km
	216.915	A-2-B	6 km
	218.215	A-3-B	5 km

22. Find the probable values of the elevations of Q, R, S, and T given that the elevation of P = 138.965 m.

Line	Distance (km)	Difference in elevation
PQ	3	0.915
QR	4	− 1.215
RS	3	2.315
ST	2	− 1.875
TP	5	− 0.465

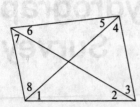 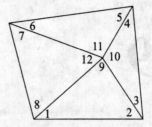

Fig. 18.26   Figure for Problem 19   **Fig. 18.27**   Figure for Problem 20

23. Analyse the level net shown in Fig. 18.28 and find the probable values of the elevations of B, C, and D given that elevation of A = 100.000 m.

Line	Difference in elevation	Weight
A-B	1.575	3
B-D	2.105	3
D-A	– 3.335	2
D-C	– 1.825	2
C-B	– 2.93	4

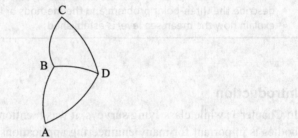

Fig. 18.28

CHAPTER

19

Hydrographic Surveying

Learning Objectives

After going through this chapter, the reader will be able to
- state the objectives of hydrographic surveying
- discuss the establishment of horizontal and vertical controls for hydrographic surveying
- describe the phenomenon of tides and the methods used to measure tides
- describe the equipment required for hydrographic surveying
- describe the methods used for sounding and recording sounding data
- describe the different methods used for locating soundings
- describe the three-point problem and the methods of its solution
- explain how the mean sea level is established

Introduction

In Chapter 1, while classifying surveys, it was mentioned that the survey of water bodies is important for many engineering applications. The survey conducted on water bodies is known as hydrographic surveying. The water body can be a stream, river, lake, or an ocean.

The objectives of hydrographic surveying can be stated as follows.

(a) To determine navigation routes or prepare nautical charts using surface data to construct navigation channels in rivers and ocean stretches
(b) To undertake underwater investigations to collect data for designing port and harbour facilities
(c) To determine the extent of scouring and silting in subaqueous floors to calculate areas and volumes
(d) To plan engineering projects such as bridges, dams, and reservoirs
(e) To determine the shorelines of water bodies
(f) To collect discharge data by measuring flows in rivers, etc.
(g) To collect data about tides and determine the mean sea level

The essential principles of surveying remaining the same, hydrographic surveying provides a different kind of challenge. In this chapter, we will study the instruments and methods used in hydrographic surveying.

19.1 Controls

As in land surveying, it is necessary to have horizontal and vertical control for hydrographic surveying. Hydrographic surveying essentially involves measuring the depth of water at points on the floor and establishing the locations of the points at which the depth is measured. Finding the depth of water at a point is known as *sounding*. For sounding data to be meaningful, vertical control is necessary. Benchmarks may be established on the shore to relate the sounding to surface levels. The water level also continuously varies; therefore, it is necessary to record the depth of the water when the sounding is taken. The reduced levels of points on the floor have to be related to the depth of water as well.

To locate the points where soundings are taken, horizontal control is required. Triangulation points or shoreline traverse points are used for this purpose. Wherever precision is required, the location is determined with reference to the triangulation points. Otherwise, traverses using a tape and compass or a theodolite are conducted near the shore. In the case of rivers, such points can be located on both shores.

19.2 Tides

Tides are important considerations in ocean surveys. The surface level of sea water continuously changes, with high water levels during high tides and low water levels during low tides.

The theory of tides is based upon Newton's equilibrium theory. All celestial bodies exert a force of attraction known as the force of gravitation, which is directly proportional to the masses of the bodies and inversely proportional to the square of the distance between them. In the case of the earth, significant forces are exerted by the sun and the moon, the moon having a greater influence due to its proximity to the earth. The generally accepted tidal theory essentially makes two assumptions: (a) the earth is covered all around by an ocean of uniform depth and (b) the sea is capable of taking, instantaneously, any new position as per the forces exerted on it. These assumptions neglect the inertia of the mass of water, its viscosity, and intrawater forces.

19.2.1 Lunar Tides

Lunar tides are variations in the ocean surface level due the moon. These can be explained with reference to the positions of the moon with respect to the earth. The moon has two types of motion—rotating around its own axis and revolving around the earth. Lunar tides can be explained with reference to Fig. 19.1.

In Fig. 19.1, let M_e be the centre of mass of the earth and M_m be the centre of mass of the moon. Let M_c be the common centre of gravity. If, at first, we do not consider their motions, the attractive gravitational forces between them will be acting on the earth as shown. This is not uniform as the distance of the mass particles from the moon varies. Now, due to the rotation of the earth, centrifugal forces will be acting on the mass particles. These forces will be fairly uniform. The equilibrium of the earth is maintained due to these forces. However, as one set of forces is not uniform, the net force will be acting on the mass particles of the earth. These forces will be the most effective on the face nearer to the moon and the least effective on the surface away from the moon.

The water rises to the maximum level on the side facing the moon and this is known as *superior tide*. On the opposite face, the water rises to the minimum level and this is known as *inferior lunar tide*. On the other two surfaces, the water level will be low and this is the phase of low tide.

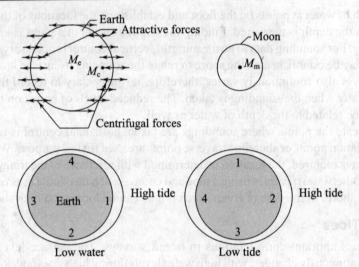

Fig. 19.1 Lunar tides

The rotation of the earth about its own axis brings different parts of the earth's surface to face the moon. Thus, if you consider points 1, 2, 3, and 4 on the surface of the earth, 1 will face the moon at some point in time, six hours later point 2 will face the moon, 12 hours later point 3 will face the moon, 18 hours later point 4 will face the moon, and point 1 will again face the moon after 24 hours. The tidal positions thus change and all the points will experience high and low tides at some point in time.

Further, due to the motion of the moon relative to the sun in about 29.5 days, there will be more changes in the tidal positions. It is sufficient for our purpose to know that the moon has a significant influence on the water surface, and in a complicated pattern, the tidal forces change the surface level during low and high tides.

19.2.2 Solar Tides

The tidal phenomenon due to the sun is of the same pattern as that due to the moon. In Fig. 19.2(a), if M_e is the mass of the earth, M_m the mass of the moon, and D_m the distance between the earth and the moon, then the net force at a point facing the moon is

$$F_m = KM_m[1/(D_m - R)^2 - 1/D_m^2]$$

As $D_m \gg R$,

$$F_m = KM_m(2R/D_m^2)$$

Similarly, the tide producing force due to the sun is

$$F_s = KM_s(2R/D_s^2)$$

where M_s is the mass of the sun and D_s is the distance between the sun and the earth.

$$F_s/F_m = (M \times D_m^2)/(M_m \times D_s^2)$$

The mass of the sun is 331,000 times the mass of the earth, the mass of the moon is $1/18$ times the mass of the earth, $D_m = 384,630$ km, and $D_s = 449 \times 10^6$ km.

When these values are substituted, it can be seen that the tide producing force due to the sun is about 0.46 times that due to the moon.

19.2.3 Spring and Neap Tides

The combined effects of the lunar and solar tides result in spring and neap tides (Fig. 19.2).

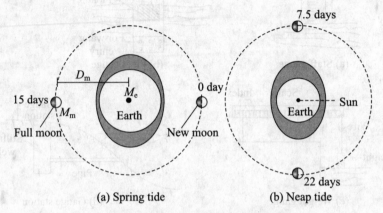

(a) Spring tide   (b) Neap tide

Fig. 19.2 Spring and neap tides

Spring tides occur during full moon when the sun and the moon have the same celestial longitude. Assuming that the sun and the moon lie along the same horizontal with the equator, the effect of the tidal forces of the sun and the moon are additive giving a maximum tide known as spring tide. After about 7.5 days, the longitudes of the sun and the moon are at 90° and the crest of the moon tide coincides with the trough of the sun tide and we have what is known as a neap tide. At this point the high water level is below the average and the low water level is above the average. The cycle of spring and neap tides repeats after about 29.5 days.

The equilibrium theory, on which these observations are based, does not give a true picture due to many factors such as the following:
(a) The orbits of the masses not being circular (they are elliptical)
(b) The varying relative positions of the sun and the moon
(c) The relative attraction between the sun and the moon
(d) The deviation from the equator known as the declination of the sun and the moon
(e) The distribution of masses
(f) The effect of land masses replacing water on the surface

The prediction of tides is thus very difficult and should be mainly based on the observational data of actual occurrence of tides.

19.2.4 Measurement of Tides

The elevations of high and low waters or tidal positions are measured using various types of gauges. There are self-registering and non-registering type of gauges. For non-registering type of gauges, an attendant is required to take measurements. The following types of gauges (Fig. 19.3) are commonly used.

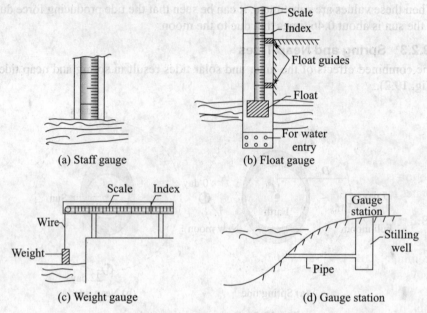

Fig. 19.3 Measurement of tides using various types of gauges

Staff gauge A staff gauge is a simple scale, graduated to 5 or 10 cm [Fig. 19.3(a)]. The scale is fitted vertically. The zero of the scale is fixed arbitrarily and its elevation is determined by levelling. The scale is read at intervals and the readings recorded. The scale should be sufficiently long to record the low and high water levels.

Float gauge Figure 19.3(b) shows a float gauge. It is a simple device consisting of a container in which a float is hung with a wire. Water enters the gauge through the openings at the bottom and lifts the gauge up to the level of the water. A graduated vertical rod with an index mark is used to read the position of the float. As the float rises with the rising water level, the reading against the index mark is taken to determine the water level.

Weight gauge The weight gauge shown in Fig. 19.3(c) is another device to determine water level. It consists of a wire or chain to which a weight is attached. The weight or chain is passed around a pulley to maintain a vertical position. A graduated scale with an index mark is placed near the wire for taking the reading. During measurement, the weight is lowered to touch the water level and a reading is taken against the index mark. The reduced level of the zero of the graduated scale is determined using a level and staff. The staff is held touchig the bottom of the weight when it touches the water level, at the same time placing the index over the zero mark of the scale. The reduced level of the zero mark is thus established.

Self-registering gauges These designed to automatically register water levels either on paper or store them electronically along with the time of the day. They essentially consist of a float that is attached to a float wheel by wires and kept under constant tension. The float moves with the water level and the motion is transferred to the wheel. Through an appropriate gear system, such

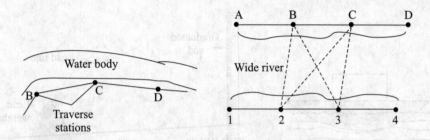

Fig. 19.4 Shoreline survey

motion is recorded on paper on a drum. The drum maintains a constant speed as it rotates and this establishes the time interval between readings.

19.3 Shoreline Survey

The objective of shoreline survey is to determine the location of shorelines, locate prominent features on the shoreline, and to determine high and low tides.

As shown in Fig. 19.4, the shoreline is located by running a traverse and taking offsets from the traverse lines to the shoreline points. An open traverse on one bank of a narrow river is sufficient to locate the shorelines on both the banks. In the case of wide rivers, traverses are run on both the banks to determine shorelines. Any important feature is also surveyed for use as a reference point later. The two traverses on the banks are interconnected by observations to points on the other traverse. This also acts as a check on the work being done. It is also possible to use a triangulation network along the banks of a wide river.

In the case of tidal water surface, the low and high water lines are observed from marks on the shore or from data collected earlier. Shore deposits and marks on the rocks can be used for determining high and low water lines. Contouring can be done to locate the points of high water line. Generally, interpolation from soundings is done to save time and effort.

19.4 Soundings

Sounding is the determination of the depth of water at different points. This is similar to determining the reduced levels of points in topography. As in topographic surveying, there are two measurements required—vertical measurements giving elevations and horizontal measurements to locate the points whose reduced levels are being determined. In flowing and turbulent water as in the sea or a river, sounding has to be done carefully, as the data collection of soundings may be disturbed by silting and scouring that changes the bed level.

19.4.1 Sounding Equipment

The types of equipment required for sounding (Fig. 19.5) are as follows:

Sounding boat It is a simple boat with a sounding platform. A rowing canoe or a flat-bottomed boat may be used in placid water. Sometimes, a special sounding boat with a central well is used. In flowing waters and difficult conditions, a motor launch is preferred.

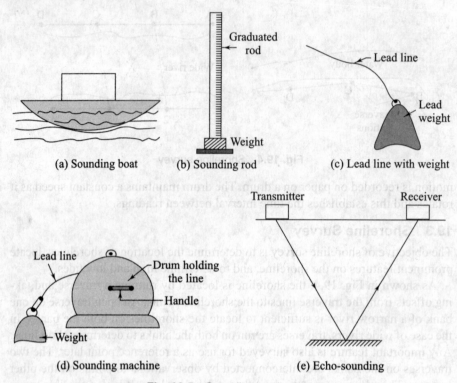

(a) Sounding boat   (b) Sounding rod   (c) Lead line with weight

(d) Sounding machine   (e) Echo-sounding

Fig. 19.5   Sounding equipment

Sounding rod or pole   It is a thick wooden pole 5–8 m long and about 80 mm in diameter. Sounding rods are suitable for shallow water sounding only. A weight of lead is attached at the end for stability and for holding it vertical.

Lead lines   There are made of cord, rope, or a brass chain with a sounding lead attached to it. The line is graduated in a stretched position after wetting it. It should be dried and stored. It is wetted for about one hour before it is used for sounding. The sounding lead is a weight attached to the line. It may be hollow at the bottom for sampling the material. It is bell shaped with a ring at the top to attach the lead line. The weight should be sufficient for the line to be stable in flowing water.

Sounding machine   It is used where extensive sounding work is expected. The sounding machine is a drum with a lead line and may be operated manually or by electric power. The lead weight is attached at the end and is lowered slowly in water. The depth of the sounding can be read from the dials. The sounding machine can be used for depths up to 30 m.

Fathometer   It works on the principle of echo-sounding. It can be used for greater depths and is hence used more often in ocean sounding. A fathometer determines water depths by measuring the time taken by sound waves to travel through water and back. The instrument may directly give the depth or record it on paper giving a profile of the ground. As the velocity of sound waves in water is dependent on many factors, the instrument can be adjusted to the velocity in a particular stretch of water based on its properties. A fathometer can be easily carried in a boat for sounding.

A fathometer has a transmitter for generating and transmitting sound waves and a receiver unit for receiving the echo sound waves. It also has recording and power units. The signal is transmitted by the transmitter and travels through the depth of the water and is reflected back and received by the receiving unit. The time of travel is recorded. As the distance travelled is two times the depth of the water, if V is the velocity of sound and t is the time taken, then

$$D = Vt/2$$

where D is the depth of the water. A correction can be applied if the boat is in motion, as the position of transmission and reception of signals is not the same (Fig. 19.5). The circuitry is so designed that when the sounding is taken, the depth is automatically calculated and displayed or recorded as a graph.

Advantages of Fathometer

The advantages of a fathometer are as follows.
(a) It is more accurate than other methods.
(b) It can be used in strong currents where reliable soundings from other methods are difficult to obtain.
(c) A true vertical sounding is obtained with a boat in motion and a correction can be easily applied.
(d) It is a very fast method of sounding.
(e) It can be used in conditions where other sounding methods may not be applicable.
(f) It can be used in all kinds of weather conditions.
(g) As it provides a continuous record of the soundings, the profile of the ground can be seen later after the fieldwork is over.

Modern systems, particularly those used for oceanographic studies, large-scale harbour projects, etc., use echo-sounding systems employing side-scan/single- or multiple-beam scanning systems. Multiple-beam scanning systems use a number of transducers and scan the floor in parallel lines in two perpendicular directions. These are known as *sonar*, an acronym for *sound detection and ranging*. Another system that finds application in hydrographic studies is *lidar* or *light detection* and *ranging*. These airborne systems can be used for a variety of purposes, such as to find the range or depth of water, i.e., sounding.

19.4.2 Fieldwork

Fieldwork in sounding involves the following steps.
1. As the sounding party moves along in a boat, the sounding man, standing at the end of the boat, is to keep the rod at a vertical angle and dip it at a forward position. Take the reading when the rod becomes vertical. Take a number of such readings at regular intervals.
2. If a lead line is used, throw the lead line at a forward position. Take the reading when the line becomes vertical at the time the boat is in position. This action requires experience, as the lead line should reach the bottom and become vertical when the boat reaches that point.

3. In placid waters, use this method directly. In turbulent waters, where the water level changes continuously, read the tide gauge and note the time. The sounding party will also note the time of reading the sounding. It will then be possible to reduce the readings to a common datum.

19.5 Locating Soundings

In order to determine the topography of the floor of a river or an ocean, it is necessary to have the location of soundings in addition to the vertical measurement. There are many ways to locate the soundings. The following terminology is relevant in this regard.

Range line   The soundings are generally taken along straight lines laid at intervals, at nearly right angles to the shore or banks of a river. Figure 19.6(a) shows range lines. This is possible at places where the shoreline is straight. In case the shoreline is not straight, the range lines are laid out radiating from a prominent, conspicuous object on the shore [Fig. 19.6(b)].

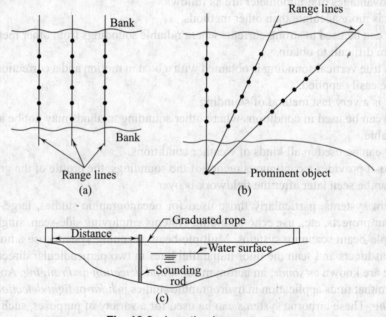

Fig. 19.6   Location by cross rope

Signals   In order to range the line, shore signals are placed on both the banks or at two different points far apart. The signals are rods or tripods fixed to the ground with a coloured flag on top. The signals must be clearly visible from a considerable distance. They are accurately located with respect to the traverse or triangulation points. The location of soundings can then be related to shore points on the plan through the signals.

19.5.1 Location by Cross Rope

One of the commonly used methods is to locate soundings by a cross rope [Fig. 19.6(c)]. This method consists of stretching a rope or wire across the width of a river, lake, or harbour area. The wire is fixed to points on the shore at both ends. The stretched line has tag marks indicating distances with respect to a zero position on the shore. As the soundings are taken, their positions with respect to these tags are noted and recorded along with the depth and time. This is a simple technique and is commonly employed. A variation of this method is to use two boats, one for stretching the rope and the other for taking soundings. One of the boats has a wire wound around a drum. One end of the wire is fixed to an anchor point on the shore. As the boat moves along the river, the wire is stretched and held taut by suitable means. The sounding boat follows the first boat and soundings are taken with respect to the tag marks along the stretched wire. Once a range line is completed, the boat with the wire returns to the other shore, winding the wire on the drum. The second boat also returns to the other shore. The position of the wire along the shore is changed and the process is repeated with a second range line.

19.5.2 Range and Time Intervals

In range and time intervals method, the sounding boat is rowed along the river at a constant speed and the soundings taken at time intervals, are noted. The range line is first fixed by shore signals and the sounding line is kept as the line joining the signals. The boat is moved at a constant speed along the line joining the signals. The sounding is taken and the time is noted. From the speed of the boat and the time intervals, the location of the sounding can be ascertained. This method does not give accurate results and hence should not be used when precision is necessary.

19.5.3 Range Line and One Angle from Shore

As shown in Fig. 19.7, range lines are laid out first. The method essentially involves observing the angle to a sounding point from the instrument station. The instrument station is so fixed in the original plan that very acute angles are not observed. With very acute angles, the observed point cannot be fixed accurately. The procedure is as follows.

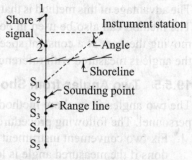

Fig. 19.7   Locating by range and one angle from shore

1. Fix two shore signals in line with the range line. Set the instrument to zero and observe the shore signal with the angle set to zero.
2. The sounding operation and angle observation will require great coordination. Move the boat along the range line. Fix the line of sight from the instrument to the bow of the boat or the sounding man.
3. Mutually arrange appropriate signals to inform the instrument man of the time of sounding. A flag may be held upright by an assistant on the boat. At the time of sounding, drop the flag, signalling the instrument man to take the reading.
4. Note the time using the parties on the boat and the instrument. As the whole operation has to be done very fast, at least two people are required at the instrument and two or more people at the boat.

5. Observe the sounding and the angle simultaneously. Measure the angles to an accuracy of $1' - 5'$.
6. Sometimes, the angles are observed only for some of the points. Locate the remaining points by the speed-and-time method by moving the boat at a constant speed.
7. Row the boat for sounding in both the directions on alternate range lines to make the operation faster.

19.5.4 Range Line and One Angle from the Boat

The range line and one angle from the boat method is very similar to the previous method except that the angle is measured from the boat itself. The following procedure is adopted.

1. Fix convenient range lines on the plan. Fix any prominent object on the shore for reference to the angular measurement. If no prominent feature is available, set up a station using tripods and flags.
2. The boat also carries the angle-measuring instrument. Measure the angles with reference to the shore station (Fig. 19.8).
3. Move the boat along the range line and take the sounding. Simultaneously, note the angle between the range line and the shore signal.

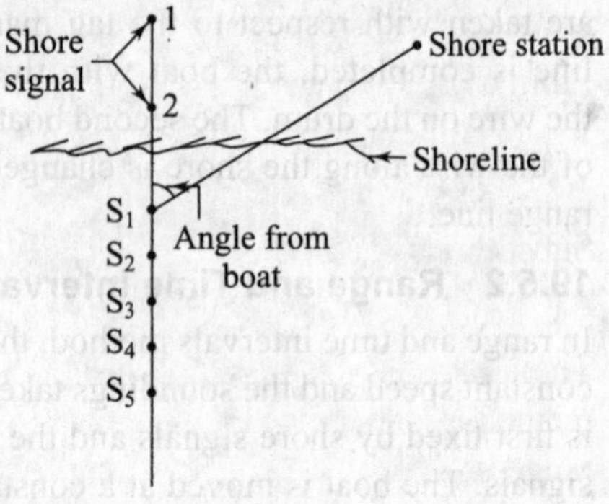

Fig. 19.8 Locating soundings by range line and one angle from boat

The advantage of this method is that there is a better control over the measurements. The method can also be used in conjunction with the time-and-speed method by moving the boat at a constant speed. Sometimes, two shore signals are kept and the angle is measured with reference to the second signal also.

19.5.5 Two Angles from Shore

The two angles from shore method requires two instruments and two instrument personnel. The following procedure is adopted.

1. Fix two convenient instrument stations on the shore. Shift the instrument stations if the measured angle is less than $30°$.
2. There is no need to accurately fix a range line and range signals. It is, however, preferable to run the boat approximately along a range line.
3. At the time of sounding, note the time with the help of the boatman and the two people handling the instruments. Evolve appropriate signals to alert the instrument men of the sounding.
4. If 1 and 2 are the instrument stations and α and β are the angles measured (Fig. 19.9), the coordinates of the sounding station S can be obtained as

$$x = D \tan\beta / (\tan\alpha + \tan\beta)$$

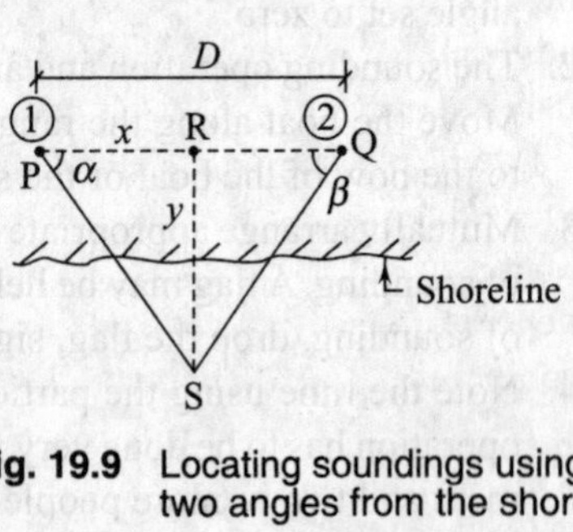

Fig. 19.9 Locating soundings using two angles from the shore

$$y = D \tan\alpha \, \tan\beta / (\tan\alpha + \tan\beta)$$

These formulae can be easily derived. From triangle PSR,

$$y = x \tan\alpha$$

From triangle QSR,

$$y = (D - x)\tan\beta$$

Equating the two,

$$x \tan\alpha = (D - x)\tan\beta$$

which gives

$$x = D \tan\beta / (\tan\alpha + \tan\beta)$$

From triangle PSR,

$$y/x = \tan\alpha$$

$$y = x \tan\alpha$$

Substituting for x,

$$y = D \tan\beta \tan\alpha / (\tan\alpha + \tan\beta)$$

The advantage of this method is that the exact range line is not required. This method is suitable when there is a difficulty in rowing the boat accurately along a range line.

19.5.6 Two Angles from the Boat

As in the previous method, two angles can be measured from the boat to locate soundings. In this method, as shown in Fig. 19.10, angles α and β are measured from the boat. The two angles are measured as the angles subtended by three prominent stations at the sounding point. The two angles are measured at the time of sounding.

The three prominent stations are either permanent features available on the shore or three points established by traverse or triangulation. The positions

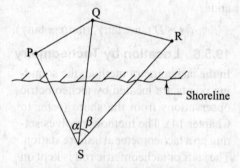

Fig. 19.10 Locating soundings using two angles from the boat

of the three points are known in the plan. With two angles measured, it is possible to locate the plan position of the sounding point by solving the triangles. This is known as the three-point problem. In Fig. 19.10, P, Q, and R are the three points and S is the position of the sounding. The angles α and β are observed simultaneously with the help of two sextants when the sounding is taken.

Two people are required to measure the angles while a third person is required to take the soundings. Range lines and range signals are not required. The method is best suited for measuring soundings at isolated points. There is better control over the operation as the whole party is in one boat. The solution of the three-point problem is discussed in a later section.

19.5.7 One Angle from Shore and One Angle from Boat

As measuring two angles from the boat simultaneously is difficult and there is no

great control over the time at which both
the angles are measured from the shore, a
combination of such measurements can be
used. As shown in Fig. 19.11, one angle is
measured from the shore and one angle is
measured from the boat. Two shore signals
are required in this case. P and Q are shore
signals and S is the location of the sounding. One instrument is kept at P and Q is a
prominent signal. When the sounding man
is about to take a sounding, a flag is raised

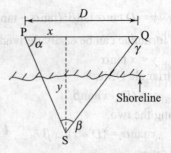

Fig. 19.11   One angle each from the shore and from the boat

from the boat to alert the instrument men. As the flag falls, both the instruments
are read.

α is the angle measured at P with a theodolite between the signal Q and the
sounding point S. β is the angle measured at the boat between the stations P and Q
with a sextant. Once these two angles are known, the position of S can be located
by calculating the coordinates x and y. Knowing the angles α and β, the third angle
at Q can be calculated as $180° - (\alpha + \beta)$. If this angle is γ, the problem reduces
to the same as the measurement of two angles from the shore. x and y are
calculated as

$$x = D \tan\gamma / (\tan\alpha + \tan\gamma)$$

and

$$y = D \tan\alpha \tan\gamma / (\tan\alpha + \tan\gamma)$$

19.5.8   Location by Tacheometry

In the tacheometry method, the sounding points are located by tacheometric
observations from the shore (refer to
Chapter 14). The method involves setting up a tacheometer at a shore station.
The staff or tacheometric rod is kept on
the boat. The staff is read at the time the
sounding is taken (Fig. 19.12).

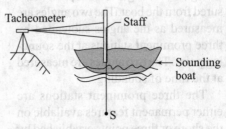

Fig. 19.12   Tacheometric method

All the three stadia hairs are read.
The line of sight may be inclined or horizontal. The distance from the instrument
to the staff is calculated as follows.
For horizontal sight,

$$D = Ks + C \quad (C \text{ is generally zero})$$

For inclined line of sight,

$$D = Ks \cos^2\theta + C \cos\theta \quad (\text{staff held vertical})$$

19.5.9   Intersecting Range Lines

If soundings are to be taken a number of times at the same point (to determine
scouring or silting), two sets of range lines can be set out by suitable signals
from the shore. Sounding points are fixed on the intersection of the range lines
(Fig. 19.13). The advantage of this method is that no angles are to be measured.

The boat is moved along a range line; the sounding man takes a sounding when his position comes in line with the other range line, which is inclined. These points are fixed in the plan by the intersecting lines, which are located by the traverse points or triangulation points on the shore.

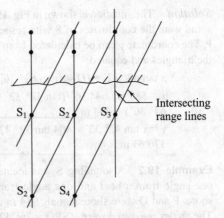

Fig. 19.13 Intersecting range lines

19.6 Reduction and Plotting

As mentioned earlier, sounding is a process of levelling. The soundings are measured as the depth of water at the time of taking the sounding. The water level changes over time due to tidal changes. Therefore, tidal gauge readings are normally taken at the time of taking the soundings as well. The sounding readings can be useful if these can be reduced to a common datum. The most commonly used datum is the mean sea level.

The datum used can be the mean level of low water of spring tides (LWOST) or the mean low water springs (MLWS). A correction is applied to the soundings according to the actual water level read on the tidal gauges at the time of taking the soundings. Table 19.1 and examples 19.1–19.3 illustrate this reduction.

Mean datum reading = 4.65 m

Gauge reading at 10 a.m. = 3.25 m

Correction to sounding = 4.65 − 3.25 = 1.4 m

Date of observation = 11 June 2005

Table 19.1 Sample sounding data

Time	Gauge	Distance (m)	Sounding	Corrected sounding
10 a.m.	3.25 m	20	1.85	3.25
		40	2.15	3.55
		60	2.55	3.95
		80	2.80	4.20
		100	3.10	4.50

Plotting

Soundings are plotted on the plan. If the range lines are previously fixed on the plan, the soundings are plotted on the range lines depending upon the distances measured on the cross rope or by any other means. If angles are used, either the distances are calculated or angles are laid out to locate the soundings. If the angles are measured from a boat, as in the three-point problem, then the distance is computed or the problem is solved by graphical means. The three-point problem is discussed in the next section.

Example 19.1 During a sounding fieldwork, P and Q were two stations on the shore. S was a sounding station. The angles measured were ∠SPQ = 42° 32′ and ∠SQP = 64° 36′. Find the coordinates of S with respect to P if the distance PQ is 1580 m.

Solution The situation is shown in Fig. 19.14. x and y are the coordinates of S with respect to P. The coordinate y can be calculated from both the triangles and equated:

$$x \tan 42° 32' = (D - x)\tan 64° 36'$$
$$x = 580 \tan 64° 36'/(\tan 42° 32' + \tan 64° 36') = 404 \text{ m}$$
$$y = x \tan 42° 32' = 404 \tan 32° 42' = 370.63 \text{ m}$$

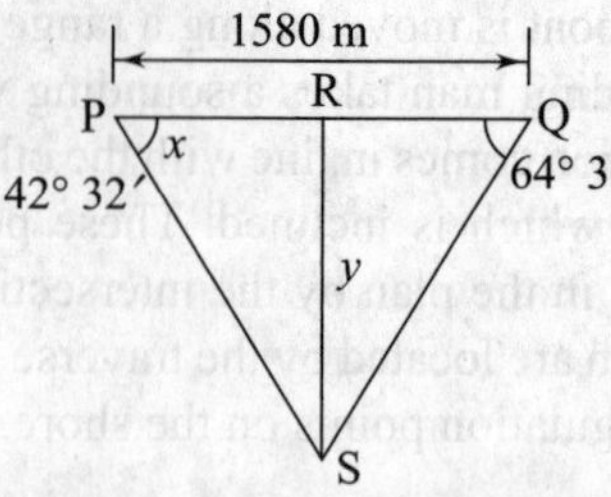

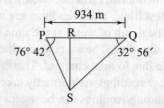

Fig. 19.14 Figure for Example 19.1

Example 19.2 A sounding S was located by one angle from a boat and one angle from the shore. P and Q were shore signals 934 m apart. The angles measured were $\angle SPQ = 76° 42'$ and $\angle PSQ = 32° 56'$. Find the coordinates of S with respect to P.

Solution Figure 19.15 illustrates the situation. x and y are the coordinates of S with respect to P. $\angle SPQ$ and $\angle PSQ$ are known. $\angle PQS$ can be calculated from the triangle.

Fig. 19.15 Figure for Example 19.2

$$\angle PQS = 180° - (76° 42' + 32° 56') = 70° 22'$$

x and y can now be calculated from $\angle SPQ$ and $\angle PQS$ as

$$x = 934 \tan 70° 22'/(\tan 70° 22' + 76° 42') = 372.25 \text{ m}$$
$$y = 372.253 \times \tan 76° 42' = 574.74 \text{ m}$$

Example 19.3 The following observations refer to tidal gauge readings and soundings. Find the corrected soundings referred to the datum. At 10.00 a.m., the gauge reading is 6.85 m. After 10 min, the gauge reading is 6.95 m. The datum gauge reading is 1.0 m. The soundings taken at 10.05 a.m. were – 2.35 m and 7.65 m.

Solution Mean gauge reading at 10.05 a.m. = (6.85 + 6.95)/2 = 6.9 m

$$\text{Correction} = - (6.9 - 1.0) = - 5.9 \text{ m}$$

The corrected soundings are as follows.
For 2.35 m,

$$- 5.9 + 2.35 = - 3.55 \text{ m}$$

For 7.65 m,

$$- 5.9 + 7.65 = +1.65 \text{ m}$$

The negative reading shows that the point is above the datum.

19.7 Three-point Problem

The three-point problem as mentioned earlier is a case of locating a sounding position knowing the plan positions of three points on the shore and the angles subtended at the sounding point. As shown in Fig. 19.16, P, Q, and R are the three points on shore. S is the sounding point. Angles α and β are measured in the field. The positions of the three shore points (P, Q, and R) are known in plan. The position of S is to be located. The three-point problem can be stated as 'Knowing the positions in plan of three well-defined points and knowing the angles subtended at a point S by the lines joining these points, locate the position of S with respect to the three given points.'

The three-point problem can be solved by mechanical or graphical methods as well as analytically. The methods commonly used are outlined below.

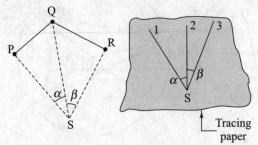

Fig. 19.16 Tracing paper method

19.7.1 Tracing Paper Method

The tracing paper method is very simple to use. As shown in Fig. 19.16, the three given points, P, Q, and R, are available in the plan. From the sounding station S, angles subtended by the two lines PQ and QR at S have been measured. To locate P, mark angles at a point on the tracing paper as shown. When the tracing paper is kept on the plan, the three lines S-1, S-2, and S-3 will not pass through P, Q, and R simultaneously. Keep the tracing paper on the plan containing P, Q, and R and adjust the position of the paper such that the line segments S-1, S-2, and S-3 pass through the points P, Q, and R simultaneously. The position of S can be marked on the plan when the line segments satisfy this condition.

19.7.2 Using Station Pointer

A station pointer (Fig. 19.17) is a special protractor with three arms. The central arm is fixed to the hold-ing ring while the other two arms are movable. The central ring is graduated in degrees and minutes and the angle between the arms can be set to given values. The arms of the station pointer have bevelled edges which indicate the lines be-tween which the angles are set.

As in the tracing paper method, the angles are set in the station

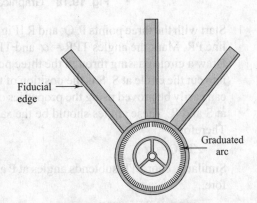

Fig. 19.17 Station pointer

pointer between the arms. These angles are the angles subtended at the sounding point S (angles measured from the boat) by the lines joining P, Q, and R. Once these angles are set, the station pointer is moved and rotated on the plan such that the bevelled edges of the arms simultaneously pass through the points P, Q, and R. The station S can then be located as the centre of the circular ring holding the arms.

19.7.3 Graphical Method

There are many graphical methods (Fig. 19.18). Some of these are outlined below.

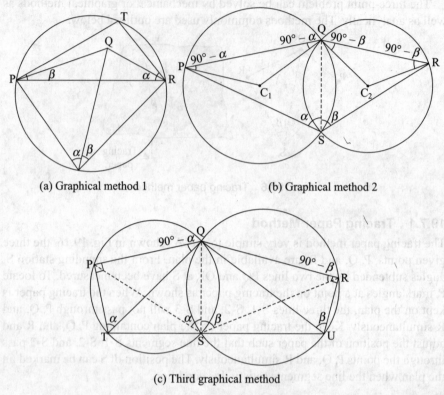

(a) Graphical method 1   (b) Graphical method 2

(c) Third graphical method

Fig. 19.18   Graphical methods

1. Start with the three points P, Q, and R [Fig. 19.18(a)] available in plan. Join the line PR. Mark the angles TPR = α and TRP = β at P and R to get the point T. Draw a circle passing through the three points P, R, and T. Join TQ and extend it to cut the circle at S. S is the position of the sounding station in the plan. This can easily be proved using the properties of a circle. Chord PT subtends angles at S and R. These angles should be the same (using the properties of circles). Therefore,

   $$\angle PSQ = \alpha$$

   Similarly, chord TR subtends angles at P and S. These angles are equal. Therefore,

   $$\angle TPR = \beta = \angle TSR$$

   Therefore, the angles at S are α and β and hence the position of S is located as per angles measured from the boat.

2. A second graphical method is illustrated in Fig. 19.18(b). In this method, we proceed as follows.
   (i) Join PQ and QR. Set off angle $(90° - \alpha)$ at P and Q as shown. The lines intersect at C_1.

(ii) Set off angle $(90° - \beta)$ at Q and R as shown. The two lines intersect at C_2.

(iii) Draw a circle with C_1 as the centre and C_1P or C_1Q as the radius. The circle passes through P and Q.

(iv) With C_2 as the centre, draw a circle with C_2Q or C_2R as the radius. The circle passes through Q and R.

(v) The two circles intersect at a second point S. S is the position of the sounding boat from which angles α and β were previously measured.

(vi) From triangle PC_1Q,

$$\angle PC_1Q = 2\alpha \,[180° - (90° - \alpha) - (90° - \alpha)]$$

This is the angle subtended at the centre by chord PQ. The same chord will subtend angle α at S as per the properties of circles:

$$\angle PSQ = \alpha$$

(vii) The chord QR subtends angle 2β at the centre:

$$\angle QC_2R = 2\beta$$

The same chord will subtend an angle β at a point on the circle:

$$\angle QSR = \beta$$

(viii) These are the two conditions to be satisfied by the location of S. Thus, S is the point at which these angles were measured.

3. A third graphical method is shown in Fig. 19.18(c). We proceed as follows.

(i) Join PQ and QR. Erect perpendiculars at P and R.

(ii) Draw lines making angles $(90° - \alpha)$ and $(90° - \beta)$ at Q as shown. These intersect the perpendiculars at P and R at T and U.

(iii) Join TU. Drop a perpendicular from Q to TU. The foot of the perpendicular is at S, the required boat station. The proof is as follows: In the quadrilateral PQST, the opposite angles are 90° each. PQST is a cyclic quadrilateral. Similarly, QRUS is also a cyclic quadrilateral. Point S is a common point of the two circles.

From triangle PTQ,

$$\angle PTQ = 180° - (90° - \alpha) - 90° = \beta$$

Similarly, from triangle QRU,

$$\angle QUR = 180° - (90° - \beta) - 90° = \beta$$

Chord PQ subtends an angle α at T. The angle subtended by the same chord at S will also be α from the properties of circles. Thus

$$\angle PSQ = \alpha$$

On similar lines,

$$\angle QSR = \beta$$

Thus point S satisfies the two angle conditions and is the station required. This method will not work if the four stations, P, Q, R, and S are concyclic.

19.7.4 Analytical Method

The analytical method has already been outlined in Chapter 18 with reference to triangulation surveys. The same solution is applicable in this case also. A brief account of the solution is given in Fig.19.19(a).

S is the sounding station from which angles α and β have been measured. P, Q, R are the three shore stations, which are located through traversing or triangulation. We have to calculate any of the two distances PS, QS, and RS, or the angles x and y as shown. The known angles are $\angle PQR = \gamma$, $\angle PSQ = \alpha$, and $\angle QSR = \beta$. Let the sum of angles x and y be θ:

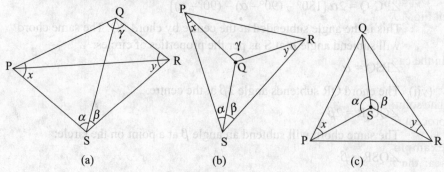

(a)   (b)   (c)

Fig. 19.19 Analytical solution

$$x + y = \theta$$

$$\theta = 360° - (\alpha + \beta + \gamma)$$

θ can be calculated from the known values. We can find the values of SQ from the two triangles SPQ and SRQ using the sine rule and equate them. Thus, from triangle SPQ,

$$SQ = PQ \sin x / \sin \alpha$$

and from triangle SQR,

$$SQ = QR \sin y / \sin \beta$$

Equating,

$$PQ \sin x / \sin \alpha = QR \sin y / \sin \beta$$

$$\sin y = \frac{PQ \sin x \sin \beta}{QR \sin \alpha}$$

As $y = \theta - x$,

$$\sin y = \sin(\theta - x) = \sin \theta \cos x - \cos \theta \sin x$$

Therefore,

$$\sin \theta \cos x - \cos \theta \sin x = \frac{PQ \sin x \sin \beta}{OR \sin \alpha}$$

Dividing both sides by $\sin \theta \sin x$, we get

$$\cot x - \cot \theta = \frac{PQ \sin \beta}{QR \sin \alpha \sin \theta}$$

$$\cot x = \cot \theta + \frac{PQ \sin \beta}{QR \sin \alpha \sin \theta}$$

$$= \cot \theta \left(\frac{1 + PQ \sin \beta \sec \theta}{QR \sin \alpha} \right)$$

As the right-hand side contains all the known values, x can be calculated. y can be calculated from the relation $y = \theta - x$. The lengths PS, QS, and RS can now be calculated by applying the sine rule to the triangles, thus locating S.

Depending upon the location of P, Q, R, and S, another case may arise, as shown in Fig. 19.19(c). In the case shown in Fig. 19.19(b),

$$x + y = \gamma - (\alpha - \beta) = \theta$$

In the case shown in Fig. 19.19(c),

$$x + y = 360° - (\alpha + \beta + \gamma) = \theta$$

The solution is similar to the case described. The following example illustrates the procedure.

Example 19.4 From a sounding boat at sea, the following angles were measured (Fig. 19.20): $\alpha = 32° \ 46'$ and $\beta = 41° \ 24'$. The three shore stations P, Q, and R are located by traversing. PQ = 596 m, QR = 678 m, and $\angle PQR = 132° \ 52'$. Find the location of S by calculating the distances PS, QS, and RS.

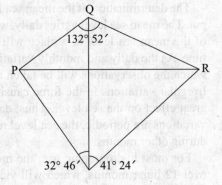

Fig. 19.20 Figure for Example 19.4

Solution If $\angle QPS = x$ and $\angle QRS = y$, then

$$x + y = 360° - 32° \ 46' - 41° \ 24'$$
$$- 132° \ 52' = 152° \ 58'$$

From triangle QPS, using the sine rule,

$$QS = 596 \sin x / \sin 32° \ 46' = 101.22 \sin x$$

From triangle QRS, using the sine rule,

$$QS = 678 \sin y / \sin 41° \ 24'$$
$$= 1025.23 \sin y$$

Equating these two values, we get

$$\sin y = 1101.22 \sin x / 1025.23$$
$$= 1.0741 \sin x$$
$$y = 152° \ 58' - x$$

Thus,

$$\sin (152° \ 58' - x) = 1.0741 \sin x$$
$$\sin 152° \ 58' \cos x - \cos 152° \ 58' \sin x = 1.0741 \sin x$$

Dividing by $\sin 152° \ 58' \sin x$, we have

$$\cot x - \cot 152° \ 58' = 1.0741 / \sin 152° \ 58'$$

This gives

$$x = 70°$$

$$y = 152° 58' - 70 = 82° 58'$$

Also

$$\angle PQS = 180° - 70° - 32° 46' = 77° 14'$$
$$\angle RQS = 180° - 82° 58' - 41° 24' = 55° 38'$$

The distances can now be calculated from the sine rule as

$$PS = 596 \sin 77° 14'/\sin 32° 46' = 243.35 \text{ m}$$
$$QS = 596 \sin 70°/\sin 32° 46' = 1034.8 \text{ m}$$
$$RS = 678 \sin 55° 38'/\sin 41° 24' = 846.27 \text{ m}$$

19.8   Mean Sea Level

We have seen the necessity for a datum plane when levelling is done. The observations are reduced with reference to such a datum. While it is sufficient to have an arbitrary datum in the case of small surveys, it is necessary to have a fixed datum in the case of large-scale surveys. It is also necessary to have such a datum if surveys conducted by different parties on different areas have to be related. The common datum used in such cases is the mean sea level (MSL).

The determination of the mean sea level depends upon the purpose to which it is put. The mean sea level varies daily, weekly, monthly, and yearly. A good estimate of the mean sea level at a place will require observations over a long period of time. As the daily and monthly variations are irregular, one has to fix a period for which the observations will be taken and determine the mean water surface level. Irregular variations in the force causing the tides and the wind directions have a great effect on the sea level. Thus, daily variations are highly irregular. Monthly variations are periodic, the sea level remaining low during some months and high during other months.

For most hydrographic use, the mean sea level may be found by observation over 12 lunar months, which will yield sufficiently accurate results. However, a very precise determination of the mean sea level is obtained by observations over a period of 19 years. This is the period over which the lunar nodes will have completed one complete cycle.

Referring survey data to the mean sea level is necessary in order to have a common benchmark at different places and relate the levels over a wider area. A relation between the sea level and the terrestrial level is also necessary for executing many engineering projects. The mean sea level is determined at a place by observations over a period of time and by relating all the vertical measurements to a common datum.

PSMSL data   PSMSL stands for Permanent Service for Mean Sea Level. PSMSL is a global data bank for mean sea levels from across the world. The data pertaining to India as listed in their website is given in the table below.

PSMSL keeps records of tide gauge readings and bottom level pressure recorders obtained from various countries. These data are continuously updated based upon the fresh readings and data obtained. Started in 1933, PSMSL is located at the National Oceonographic centre, Liverpool, UK. The centre obtains data from nearly 2000 tide gauge stations from around the world. The revised Local Reference (RLR) data is based on a datum which is 7 m below mean sea level at a place.

Table 19.2 Tide gauges in India

Station Name	ID	Lat	Long		Date	C/L code	Station Code
KANDLA	596	23.017	70.217		15/09/2010	500	011
VADINAR	1943	22.450	69.683		15/04/2009	500	013
OKHA	1395	22.467	69.083		14/09/2010	500	014
VERAVAL	770	20.900	70.367	31	07/11/1990	500	021
BHAUNAGAR I	420	21.800	72.300		01/01/1980	500	030
BHAUNAGAR II	804	21.750	72.233		01/01/1980	500	031
MUMBAI / BOMBAY (APOLLO BANDAR)	43	18.917	72.833		15/04/2009	500	041
BOMBAY (PRINCES DOCK)	212	18.950	72.833		01/01/1980	500	051
MARMAGAO	1249	15.417	73.800	281	14/09/2010	500	065
KARWAR	1273	14.800	74.117		25/07/2007	500	067
MANGALORE (PANAMBURU)	1423	12.917	74.800		06/10/2003	500	070
MANGALORE	696	12.850	74.833		07/11/1990	500	071
COCHIN (WILLINGDON IS.)	438	9.967	76.267	32	14/09/2010	500	081
TUTICORIN	1072	8.750	78.200		14/09/2010	500	083
TANGACHCHIMADAM	1258	9.283	79.250		21/03/1990	500	085
NAGAPPATTINAM	1308	10.767	79.850		07/09/1994	500	087
CHENNAI / MADRAS	205	13.100	80.300	34	20/10/2010	500	091
VISHAKHAPATNAM	414	17.683	83.283	35	14/09/2010	500	101
PARADIP	1161	20.267	86.700		14/09/2010	500	106
GANGRA	1369	21.950	88.017		15/04/2009	500	109
HALDIA	1270	22.033	88.100		14/09/2010	500	110
SAUGOR/SAGAR	417	21.650	88.050		09/09/1996	500	111
DUBLAT (SAUGOR IS.)	49	21.633	88.133		01/01/1980	500	121
DIAMOND HARBOUR	543	22.200	88.167		15/09/2010	500	131
CALCUTTA (GARDEN REACH)	369	22.550	88.300		14/09/2010	500	141
KIDDERPORE	48	22.533	88.333		01/01/1980	500	151
TRIBENI	1002	22.983	88.400		16/12/2008	500	161

The data pertaining to India is supplied by the Geodetic and Research branch of the Survey of India and is periodically updated. Some details of the station can also be obtained from its website. Some sample details are given below.

1. Tuticorin (Station ID 1072)
   Latitude : 8.75; Longitude: 78.2
   Coastline code : 500

Station code: 83
MSL is 8.2 m below benchmark 136;
BM 136 is 1.76 m above chart datum.
2. Chennai (Station ID 205)
Latitude 13.1; Longitude: 80.3
Coast line code 500; Station code: 91
Present BM 160/23 is 5.04 m above chart datum.
3. Mumbai (Apollo Bandar)
Station ID: 43
Latitude: 18.916667; Longitude: 72.83333
Coastline code: 500; Station code: 41
Time span of data: 1878–2006
Local benchmark 5/88 is 7.44 m above chart datum.

Summary

Hydrography or hydrographic surveying is the survey of water bodies such as lakes, rivers, seas, and so on. It is essentially the profiling of the underwater ground surface by measuring levels. This is done by determining the depth of water at different points. Due to tidal effects, particularly in the sea, the water level does not remain constant. Tides are caused due to the forces exerted by the sun and the moon. The moon has a greater effect on tides due to its proximity to the earth. The net forces arising due to the gravitational forces of attraction and the centrifugal forces due to rotation of bodies around the sun cause tides or variations in water levels at different points. Due to the rotation of the earth about its own axis, the maximum water level is reached at various points at different periods of time. The complex phenomenon of tides is studied over a long period of time and there is no exact method of predicting tides. Tides are measured at frequent intervals using tide gauges.

A boat or motor launch is used to conduct hydrographic surveys. Two operations are involved in hydrography—measuring the depth of the water and locating the point at which the depth is measured. Measuring the depth of the water is known as *sounding*. This is done using measuring rods, flexible ropes or wires, or an echometer, depending upon the depth of the water. Soundings can be located and related to shore points using many methods that involve measuring angles and distances. A series of soundings are taken along a line known as the *range line*, which is related to the shore points for horizontal control.

When angles are measured from a boat with reference to three points on the shore, we have a three-point problem of boat location using the measurements made. This is solved graphically or analytically.

The mean sea level is the common datum used to relate surface and underwater levels. The mean sea level at a point is determined over a long period of time, usually 19 years. The reduced levels of all the points over a wider area can be correlated and used for all applications by such data.

Exercises

Multiple-Choice Questions

1. The term sounding in survey of water bodies refers to
   (a) the use of a sound meter to find water levels
   (b) the determination of depth of water at different points
   (c) getting horizontal control points in water
   (d) getting vertical control points in the water body

2. The term tides in the sea refers to
   (a) waves in the sea water
   (b) change in the sea level, between a maximum and minimum
   (c) constant rise in the sea level due to various reasons
   (d) constant decrease in the sea level due to evaporation of sea water
3. The effect of moon on tides in the ocean is
   (a) less than that due to the sun
   (b) equal to that due to the sun
   (c) more than due to the sun
   (d) due to its rotation around the sun along with the earth
4. The sun has less effect on the tides than moon because
   (a) of its large mass   (c) it is not a solid mass
   (b) of its larger distance from the earth (d) it is a star and not a planet
5. Spring tide occurs due to the
   (a) combined additive effect of the sun and the moon
   (b) opposing effect of the Sun and the Moon
   (c) rotation of the earth about its own axis
   (d) rotation of the moon about its own axis
6. Neap tide refers to the
   (a) combined additive effect of the sun and the moon
   (b) opposing effect of the sun and the moon
   (c) rotation of the earth about its own axis
   (d) rotation of the moon about its own axis
7. Echo sounding is a method where
   (a) a metal rod is used to impact the floor of the water body
   (b) water depths are determined by the time taken by sound waves
   (c) a sounding machine is used to determine water depths
   (d) depths are determined by two different methods
8. A fathometer uses the principle of
   (a) direct levelling   (c) echo sounding
   (b) barometric levelling   (d) hypsometry

Review Questions

1. Explain the objectives of hydrographic surveying.
2. Explain how horizontal and vertical control is achieved during hydrographic surveying.
3. Define the term sounding. Enlist the equipment required for sounding.
4. Briefly explain the phenomenon of tides.
5. Explain the different methods of measuring tide levels.
6. Explain the different methods of locating sounding positions.
7. Explain the three-point problem. Explain the method of solution of the three-point problem using (a) tracing paper and (b) station pointer.
8. Explain the graphical methods of solution of the three-point problem.
9. Derive the analytical solution of the three-point problem.
10. Explain the term mean sea level. Explain the method used to arrive at the mean sea level at a place. Explain the importance of mean sea level.

Problems

1. Two shore stations, P and Q, are 600 m apart. S is a point of sounding. The angles measured at P and Q are $\angle QPS = 44° \ 16'$ and $\angle SQP = 52° \ 18'$. Find the coordinates of S with respect to P.

2. In a sounding operation, angles were measured from the shore and the boat. A and B were the two shore points 1100 m apart. The angles measured were ∠SAB = 48° 15′ and ∠ASB = 62° 24′, where S is the sounding point. Find the coordinates of S with respect to A.

3. From the following tidal gauge readings, find the corrected value of the sounding referred to the datum gauge reading.

Time	Gauge reading
10.00 a.m.	3.25 m
10.20 a.m.	3.15 m

Datum gauge reading = 1.85 m. The soundings taken at 10.10 a.m. are 1.15, 2.45, 3.85, and 6.54 m at four points.

4. P, Q, and R are three stations on shore. PQ is 985.6 m and QR is 1020.8 m. ∠PQR = 108° 16′. S is the location of the sounding boat. The angles measured from the sounding boat were ∠PSQ = 40° 6′ and ∠QSR = 55° 45′. Find the location of the sounding boat with respect to P, Q, and R.

20 Engineering Surveys

Learning Objectives

After going through this chapter, the reader will be able to

- describe the purpose, equipment, and methods of route surveys
- define the term profile levelling and explain the method of longitudinal sectioning and plotting of profiles
- describe the method used to conduct cross-sectioning and the method used to plot cross sections
- describe the equipment and methods used for topographic surveying
- describe the equipment and methods used for city surveying
- describe the methods used to conduct underground surveys and special features of underground surveys

Introduction

In the preceding chapters, we have outlined various methods of surveying using different instruments. Some of the instruments have become obsolete with the availability of better equipment and methods. Surveying is the first step undertaken in most engineering projects. Survey data is required for designing projects and locating the position of the project accurately. Engineering projects cannot take off without survey data. Some of the main applications of surveys in engineering are outlined in this chapter.

20.1 Route Surveys

The term route survey is generally applied to surveys of narrow strips of land that stretches to long distances. Such surveys are commonly undertaken for transportation system design such as roads, railway lines, water supply, and sewerage lines, aqueducts, oil and gas pipelines, transmission lines, and so on. In general, a route survey includes the surveying of terrain for horizontal distances and elevations, calculation of earthwork, laying out the project in position, and preparing drawings for required profiles. Route surveys generally include reconnaissance, preliminary survey, location survey, and construction survey.

20.1.1 Reconnaissance

The objective of reconnaissance is to establish the best location for a detailed survey. The reconnaissance party surveys the entire area and makes sketches of the important land features such as existing infrastructure, rivers, or other obstacles. This helps narrow down the choice of location for a detailed survey.

While reconnaissance is a preliminary survey of the project using rudimentary instruments, it should not be considered insignificant. In fact, reconnaissance has to be conducted by experienced surveyors who have the technical knowledge and experience to make the best use of it. For a road or railway project, for example, a very thorough examination of a vast tract of land is necessary to determine the most economical and useful route from the many choices available. The time, effort, and money spent on reconnaissance is not wasted, as it helps refine further survey work. Only one or two of the most probable routes will be chosen for a detailed survey after reconnaissance.

In order to cover a large area, surveyors use various transportation means such as jeeps, horses, or any other means of effective movement. Not many instruments are carried during this survey. Distances may be measured from available maps or by rough judgement, pacing, etc. Minor instruments may be carried for ascertaining levels, slopes, etc. In order to make the survey data useful, the following points have to be kept in mind.

(a) If a topographic map is available, a lot of information is recorded on the map itself. In the absence of maps, rough sketches detailing the prominent features must be prepared.

(b) Extensive notes of all the observations of the terrain must be maintained with sketches. Approximate distances and differences in levels must be recorded.

(c) Route gradients are worked out from the data available. The crossing of streams and rivers by the route have to be recorded, waterway requirements have to be calculated from the flow data, and the preliminary requirements of culverts and bridges have to be worked out.

(d) The route cannot be designed on the shortest length; it has to be based on the requirements of the population living in the vicinity. The final location of the route will have to take into account this social aspect and the utility of the route.

(e) A major consideration in route design is the drainage of water from the route. The requirements of the drainage system must be kept in mind while collecting data.

20.1.2 Preliminary Survey

An analysis of the reconnaissance data will give a good idea of the location of the route. One or two of the most possible routes are selected for further survey. The survey done on such selected routes is known as preliminary survey. The objectives of the preliminary survey are as follows: (a) preparation of an accurate topographic map of the tract of land containing the route, (b) preparation of an estimate and working out the cost of the project, (c) selection of the most economical and viable route, if more than one route is surveyed.

Preliminary survey is done with accurate instruments and the data collected is used for the preparation of a topographic plan. One or two selected routes will be surveyed with equal precision. A much wider width of land than actually required for the construction of the route is surveyed. A width of about 120–150 m for roads and 400–500 m for railways is commonly reserved for preliminary survey.

Fieldwork Fieldwork consists of running an open traverse within the tract of land selected, along the proposed route, contouring along the centre line and cross-sectioning. In general, a minimum of three survey parties work simultaneously on large projects.

Traversing   Horizontal control points should be established near the route by running a traverse from the nearby triangulation points. As such routes are likely to be very long, intermediate control points are also established. An open traverse is ideal for route surveying. The traverse may be done using deflection angles or included angles. As a check, the bearings of the lines should be observed at various intervals. A theodolite is used for traversing. From the reconnaissance data, the surveyor locates the stations for open traversing on paper. These are then marked on the ground with pegs and stations. Traverse lines can be as long as practicable from the point of view of visibility and change in directions of the route. The route will thus be a series of straight lines. Curves can then be designed according to the requirements.

Levels along the centre line   Levels are taken along the centre line of the proposed route. The centre line is laid out by the traversing party using pegs driven at suitable intervals. The levelling party will take levels at points along these lines. A dumpy level may be used for this purpose. Depending upon the terrain, levels are taken along suitable intervals. The interval will be larger in the case of flat terrain and much smaller in the case of hilly terrain. This survey will help to work out a profile of the proposed route as well as design the earthwork requirements. Profile levelling is explained in detail in Section 20.2.

Cross-sectioning   In the case of large projects, a third survey party may be required to do the cross-sectioning. Cross-sectioning is the process of finding levels along the lines perpendicular to the route. Equal lengths on either side of the proposed route are surveyed for this purpose. Lines are laid out on either side by driving pegs at the end of cross lines. Points are also marked in between the end pegs. Levels are taken at these points. A number of sections at suitable intervals, depending upon the terrain, are taken. This data together with the profile data will help in designing the final profile of the proposed route, earthwork, and drainage system.

20.1.3   Location Survey

The data collected from the preliminary survey of one or more routes is analysed to find the most suitable location for the route. The route is first finalized on the plan, or a paper location is made. The paper location is necessary to arrive at the optimum project location.

A complete map is prepared from the survey data to a scale of 1:4000. A route anywhere within the strip surveyed can be selected. The major considerations include the minimum gradients, the number and radius of the curves for change in direction, volume of the earthwork, balancing of the earthwork, drainage along the route, crossing of streams and rivers, and satisfying obligatory requirements such as passing through habited areas. A number of trials are conducted before finalizing the location on the plan. This calls for a very detailed analysis of data from all the survey teams, from which plans and plots are made.

Field location   Once the location is finalized on the plan, it has to be transferred to the ground. This is known as location survey. The final location of the route is established on the ground by transit-tape survey. Improvements are made on the location depending upon the ground situation by making minor alterations wherever required. Profile levelling and cross sections are again taken on the route that has been fixed. Curves are laid out as per the requirements. The earthwork calculations are repeated to estimate the fill requirements, source, and haulage. The complete route is now available on paper as well as on ground with stakes driven at intervals.

20.1.4 Construction Survey

Construction survey is the name given to the survey carried out as the construction proceeds to realize the paper plan. Levels, gradients, superelevations on curves, locations of approaches, abutments, piers for bridges, etc., will have to be checked as the construction work progresses. The depth of cutting and the filling in of earthwork is checked at sections at various intervals using a level. If there is a private property on the proposed route, its acquisition as per the proposed location and boundary lines must be undertaken. The survey work has to continue until the construction is complete. A theodolite and level are essential parts of the equipment needed at a construction site.

20.2 Longitudinal and Cross-section Levelling

In the discussion on route surveying, profile levelling and cross-sectioning were mentioned. In the following sections, we will discuss longitudinal or profile level-ling and cross-section levelling.

20.2.1 Profile Levelling

Profile levelling is also known as longitudinal sectioning or sectioning. It is done along the centre line of a proposed route, road, or railway at suitable intervals. Points on the centre line are located by distances measured from the starting point. The elevations of these points are obtained by levelling. A plot of the horizontal distances on the horizontal axis with corresponding levels on the vertical axis is known as a profile. This plot shows the undulations of the ground and is used as mentioned earlier for many purposes such as calculating volumes and intervis-ibility of stations.

Profile levelling is done on points lying at fixed intervals along the lines previ-ously established by a traverse. The intervals can vary depending upon the condi-tions of the terrain. On flat terrain, the interval is larger, and on hilly terrain, points are placed at smaller intervals. The interval can vary from about 10–30 m or more. The points surveyed are already located by relating them to the horizontal control points. For vertical control, benchmarks are established by fly levelling from the known benchmarks nearby.

To get a true picture of the ground, the distance between the points should be suitably fixed. It will be necessary to take additional points at sites of sudden changes of slope or where the characteristics of the terrain change. The profile obtained will be distorted if this factor is not taken into account as shown in Fig. 20.1(a).

Figure 20.1(b) shows a long line of levels with benchmarks and points where the levels are taken. Many points are covered from one instrument location. When the line of sight becomes long, about 100 m or more, the instrument is shifted. The instrument is shifted to a position midway along the next stretch of the route that can be covered from the new instrument position. Thus the first instrument station is A, the second station is B, the third instrument position is C, and so on. The first sight is taken to a benchmark established for the survey. The reduction is generally done by the height of instrument method (refer Chapter 7), as the levels consist of a number of intermediate sights. Such levels can be easily handled by this method. A sample of the level book is shown in Table 20.1.

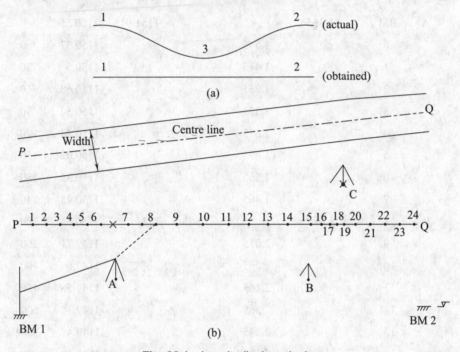

Fig. 20.1 Longitudinal sectioning

Profiles are plotted on a drawing sheet with distances measured along the x-axis and elevations along the y-axis. As the profile has large distances along the x-axis and smaller distances along the y-axis, it is usual to plot the two values to different scales to emphasize the differences of elevation in the profile. If the horizontal scale is 1:4000, the vertical scale for plotting may be 1:100 to 1:1000. Data from Table 20.1 are used to plot the profile shown in Fig. 20.2.

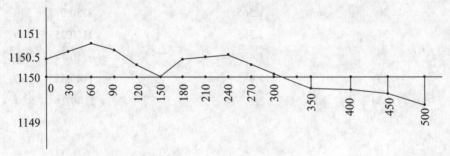

Fig. 20.2 Plotting the profile

Table 20.1 Profile levelling data

Traverse line from P to Q; length = 812.45 m, bearing = 125° 35′ 40″

Instrument at	Staff at	Backsight	Intermediate sight	Foresight	Height of instrument	Reduced level	Distance from start
A	BM 1	1.365			1151.915	1150.55	
	0		1.545			1150.37	0
	1		1.415			1150.5	30
	2		1.235			1150.68	60
	3		1.375			1150.54	90
	4		1.675			1150.24	120
	5		1.895			1150.02	150
	6		1.565			1150.35	180
	7		1.465			1150.45	240
B	8	1.685		1.875	1151.725	1150.04	300
	9		2.015			1149.71	350
	10		2.035			1149.69	400
	11		2.165			1149.56	450
	12		2.365			1149.36	500
	13		2.245			1149.48	550
	14		2.145			1149.58	600
	15		2.065			1149.66	650
C	16	1.985		2.075	1151.635	1149.65	670
	17		1.235			1150.4	690
	18		1.135			1150.5	700
	19		1.655			1149.98	710
	20		1.565			1150.07	730
	21		1.385			1150.25	740
	22		1.495			1150.14	760
	23		1.985			1149.65	780
	BM 2		2.375			1149.26	800

20.2.2 Cross-sectioning

Cross-sectioning is the process of levelling across the centre line. Depending upon the width chosen for the survey, lines perpendicular to the centre line are laid out at fixed intervals. The interval depends upon the nature of the terrain. If the ground is reasonably level, cross sections may be taken at 20-m or 30-m intervals along the centre line. The sections may be taken at shorter intervals if the nature of the terrain demands it (Fig. 20.3).

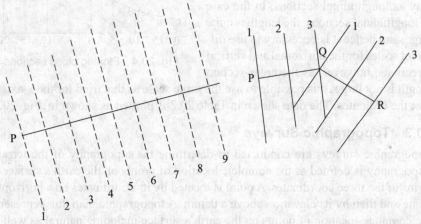

Fig. 20.3 Cross-sectioning

Table 20.2 Level data for cross sections

Line PQ; length = 345 m, bearing = 127° 45′ 34″

Chainage	Cross section number	Distance from centre L R	Backsight	Intermediate sight	Foresight	Height of instrument	RL
–	BM 1		1.935			570.435	568.5
0		0					568.48
		5			2.015		568.42
		10			2.455		567.98
		15			2.675		567.76
		20			2.875		567.56
		5			2.345		568.09
		10			2.015		568.42
		15			1.985		568.45
		20			1.565		568.87
30		0					570.24

To take cross-section levels, starting from the beginning of the line, lines perpendicular to the centre line are laid out using an optical square or theodolite. The level can be kept along the centre line and the staff at suitable intervals along the cross line. Depending upon the width of the tract and the nature of the terrain, three to five points on each side of the centre station may be chosen. The data may be recorded along with the longitudinal section data or separately. It is preferable to record the data of cross sections separately to avoid confusion. The cross sections should be numbered sequentially along the line. The data may be recorded as shown in Table 20.2.

When a line changes direction, two cross sections are taken at that point, one perpendicular to each line. Data from cross sections along with the profile data is used to calculate volumes.

Cross sections are plotted in the same way as longitudinal sections. In the case of longitudinal sections, the length is quite large, and hence it is necessary to use different scales for the horizontal and vertical directions; however, as the cross-section

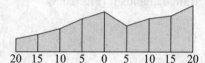

Fig. 20.4 Plotting cross sections

length is not large, it is possible to use the same scale as that used for the *x*-axis to plot the ordinates. The data shown in Table 20.2 is plotted as shown in Fig. 20.4.

20.3 Topographic Surveys

Topographic surveys are conducted to determine the topography of the terrain. Topography is defined as the complete location of points on the earth's surface in terms of the three coordinates. A point is located by its coordinates on a horizontal plane and then by its elevation above a datum. A topographic map thus represents the complete location of points on the earth's surface including natural as well as man-made features.

Topographic maps are important tools for the design of various engineering projects. The maps are prepared to a suitable scale depending upon the purpose. In order to represent the three-dimensional location of points on the surface, hachures and colour schemes are used in topographic maps. Hachures are lines drawn on a map to depict changes in elevation. The lines are thicker and closer if the slope is steep, and thinner and far apart if the slope is gentle. Colour schemes also serve the same purpose. Contour lines drawn on maps also give an idea about the relative elevation of points.

Controls Topographic surveys involve measuring distances, directions, and elevations. To prepare a topographic map, therefore, one will need equipment to measure angles, distances, and elevations. To correlate topographic maps, control points have to be established for horizontal and vertical control. Horizontal control is established by determining points that are related to triangulation stations through traverse surveys. A number of control points are established near the location of the survey. For vertical control, benchmarks are established at the location by fly levelling from the nearest GTS (Great Trigonometrical Survey) benchmark. Control points are established using precise equipment and methods with sufficient checks on their locations.

Fieldwork Fieldwork in topographic surveying consists of establishing control points, local traverse surveys, and contouring. All the methods and equipment required for topographic surveys have already been discussed in the earlier chapters.

Establishing controls: Controls are established by transferring points by triangulation or traversing to local areas. For important works, this may be done by primary and secondary triangulation. Here control points are established using precise equipment and methods. Once a primary set of control points is established, further points are established by traversing with less precise equipment and methods. Benchmarks are established near the site of work using precise levels. Precise levelling methods are employed to transfer benchmarks to the local area. The control points are permanently marked and also located with respect to some permanent ground features.

Locating points: A transit-tape survey is used to locate points at the site. Using the data collected from reconnaissance, the points to be located are determined. A traverse survey is the best method to establish a skeleton for a more detailed survey or for locating details. A vernier theodolite or a tacheometer may be used for this purpose, as the reduced levels of the points surveyed are also determined.

Contouring: This is an essential part of topographic surveying. Methods of contouring have already been discussed in Chapter 9. Longitudinal sectioning, cross-sectioning, and spot levels on grids are commonly employed. The spot levels are interpolated to obtain the desired contours. A tacheometer may be beneficial, as the distances and elevations are recorded simultaneously during the survey.

Locating details To prepare an exhaustive topographic map showing all the ground features, the details are located by establishing points on the surface. This is done with reference to the traverse lines by taking offsets or through linear and angular measurements. All ground features such as buildings, rivers, streams, roads, railways, properties, agricultural lands are located by establishing points on their boundaries or outlines.

Topographic or relief maps The data collected from the fieldwork is transferred to paper by plotting. Topographic maps may be made to a number of scales varying from 1:250,000 to 1:25,000 or 1 cm = 2.5 km to 1 cm = 0.25 km (250 m). The scale chosen depends upon the purpose of the map.

Relief maps are topographic maps in which the distances and elevations are shown using different schemes. The three coordinates can be represented in a relief map. This is done by hachures, shading, colours, contour lines, or other means. As many of the schemes only provide an idea of the elevations but not the actual height, contour lines is the most preferred method for engineering use.

Surveys for engineering projects Details of the surveys required for engineering projects vary with the nature of the project. Route surveying for roads and railways was discussed in the earlier sections. The essential principles remaining the same, similar surveys have to be conducted for other engineering projects.

In the case of a large building project, similar surveys have to be conducted for preparing topographic maps of the area. From a contour map, earthwork in excavation and filling can be worked out. The design of basements, plinth levels, storey heights, water supplies, and sewage lines are all based upon such survey data.

As another example, consider the design of a dam. Based upon the contour map prepared for the site, the dam centre line is located on the map. From the contour map, the capacity of the reservoir can be ascertained for different heights of the dam. The catchment area will vary with the height of the dam. The rainfall data will have to be considered for determining the height of the dam.

Surveys necessary for the design of a dam include triangulation survey to locate control points, precise levelling to establish benchmarks, topographic survey of the area, and hydrographic survey to establish the underwater surface. Extensive surveys of the surrounding areas are required to identify all the details required for the design. The construction of the dam will create a large lake, which is likely to submerge surrounding localities. A survey of the state and private properties is required to be conducted for acquiring land and rehabilitating the affected population. Similarly, route surveys are required for establishing communication links required during and after the construction. The different methods of surveying are thus an integral part of the design of all the engineering structures.

20.4 City Surveys

City surveys are done within a municipality or corporation area for various purposes. Such surveys have to be accurate, as the land is costly and a precise determination of the various boundaries is required to avoid disputes. City surveys are different from surveys of virgin land or sparsely occupied land in that these surveys are done in inhabited areas. Many obstacles have to be overcome while conducting city surveys. City surveys are done for two purposes—to prepare the existing topographic profile and for the design of development projects.

The existing topographic profile is required to delimit property lines, land holdings, buildings, roads, railway lines, water supply and sewerage systems, streams, rivers passing within the city limits, etc. Both surface and underground maps have to be prepared for various uses, such as to keep records of property distribution in private and state holdings within the limits of the city. All property disputes are settled based upon the maps prepared after a survey of the area. As the development work continues, such maps have to be continuously updated by surveys. Any development work such as new buildings, new roads, and augmenting water supplies and sewerage systems are based upon the existing maps and the surveys conducted for the new structures. As development work is continuous, survey work is undertaken frequently.

City surveys for development work include surveys required for specific projects. City surveys are challenging because of the existing profile and the many obstructions that have to be overcome. Plans and maps are prepared to small scales, such as 1:2000 to 1:500. Such detailing requires very precise measurements. Both linear and angular measurements have to be very accurate.

Horizontal and vertical controls are established permanently, as these are required frequently. Triangulation points are used as control points. Chapter 17 outlined the procedure used for establishing control points. The base line and angles are measured accurately with invar or steel tapes and a precision theodolite. Check bases, as may be required, are established in the triangulation system. Very stringent requirements of precision have to be established.

Vertical control points are established by precise levelling. A series of benchmarks is established in the area for future use. These are permanently marked.

Monuments are set up at the intersection points of streets away from the traffic. Monuments are permanent marks for stations established by precise traversing. They are permanently marked with concrete posts, pipes, or other metal marks, with the stations marked accurately with a cross.

City surveys require the preparation of many different types of maps such as topographic maps, property maps, wall maps, and so on. Maps are prepared for different purposes and to different scales. While wall maps are prepared for the whole area to a large scale (1:20,000), other maps are prepared to much smaller scales (1:500). A large number of sheets are prepared to cover larger areas. The following are the general features of a city survey.

(a) Precise equipment and methods are required for the survey. This is particularly true as properties are costly and need to be accurately demarcated.

(b) A high degree of planning is involved as the survey is conducted in built-up and inhabited areas. Many obstacles have to be overcome by different techniques.

(c) A large number of maps and plans have to be prepared to cover the area in small lots.

(d) City survey work is more or less continuous as developmental work takes place all the time. Maps and plans have to be modified frequently. Modern methods of electronic drafting are useful for making such changes.

(e) Location and construction surveys have to be very precise, at least to the second or third order of triangulation.

(f) City surveys for property lines require precise location of details, which is a very laborious process. Plans are drawn to small scales and in sections to cover the required area.

20.5 Underground Surveys

Underground surveys are undertaken to explore the ground underneath for minerals and ores as also to construct tunnels. Underground surveys are also required to construct transportation lines underground as for metro railways. Two major forms of underground surveys are mine and tunnel surveys.

The basic principles of underground surveys remain the same as for surface surveys. Some special features are as follows:

(a) The space within a tunnel or mine is limited, hence special equipment and methods are required.

(b) The surface measurements have to be related to underground measurements. Special methods are used for this.

(c) Since natural lighting is not available or is very limited, special lighting arrangements have to be made.

(d) The safety of equipment and personnel should be given due consideration while planning the work.

(e) Stations, in general, cannot be established on the base of the mine or tunnel, as these will be disturbed during work. They may be situated on the roof and hence special procedures are required.

(e) A high degree of precision is required in the alignment, say, of a tunnel, as work will proceed from at least two ends.

(f) As measurements will possibly be required on steep slopes, both linear and angular measurements have to be precise.

20.5.1 Mine Surveys

The main objective of mine surveying is to explore the mineral wealth of an area. Mine survey is a combination of geological and engineering surveys. The geology of mines affects their stability. All engineering aspects are required to be taken into account to ensure the safety of structures constructed underground.

Mine survey equipment

Theodolites and precise levels are used for mine surveys. Surface surveys are conducted as usual. Horizontal and vertical controls are established prior to undertaking the survey of the area. A major problem is the transfer of coordinates and levels from the surface to underground. These factors have been discussed in Chapter 13.

In earlier surveys, special theodolites were used. They were small in size. As the line of sight may generally be inclined steeply, special theodolites with auxiliary telescopes (either on the side or above the regular telescope) were used. The lines of sight being parallel through both the telescopes, corrections were made to the angles measured for eccentricity of the centre. However, with modern theodolites and total stations, surveys can be done precisely.

20.5.2 Tunnel Surveying

Tunnel surveying is similar to mine surveying. Fieldwork in tunnel surveying involves the following steps.

1. Establish horizontal and vertical control points by triangulation from nearby points, which have been established earlier.
2. Lay out the centre line of the tunnel as designed over the surface. In the case of short tunnels, start the tunnelling work from both the ends. In the case of long tunnels, open an intermediate shaft in order to have secondary access and start the work from several fronts to speed up the work.
3. Lay out the centre line of the tunnel on the surface and mark it as a permanent station. The alignment of the tunnel will require great precision. As the work proceeds from both ends, they should meet exactly over the alignment when they come close together. Use total stations and distomats to align the tunnel accurately.
4. Transfer the alignment and levels underground using the methods discussed in Chapter 13. From the ends, where the tunnelling work starts, transferring alignment and levels is a difficult task.
5. Preferably establish the survey stations on the roof to avoid them being disturbed during work. Similarly, establish benchmarks along the tunnel on protruding rock surfaces or on specially prepared platforms. Check the stations frequently for any disturbance due to the settlement of the surfaces caused by the tunnelling work.
6. With modern instruments, survey work can be done with great precision and ease. If there is any problem of alignment, widen the tunnel or introduce curves to correct the same, which, however, will add to the cost.

Control in underground surveys Both horizontal and vertical controls are critical in underground surveys. In open cast mining, the problem is not there because the controls can be established with modern instruments like GPS or the controls can be transferred by optical means.

Surface surveys are carried out accurately using triangulation or precise traversing. These can be checked and adjusted by the methods outlined in chapter 18. Tunnel alignment can be accurately done on the surface and control points established along the alignment close to access shafts. The control points must be permanent, preferably made on concrete blocks. Benchmarks for vertical control can be similarly established. Both horizontal control points and vertical control points have to be transferred underground. Mechanical methods using wires or optical transfer where possible can be done.

Mechanical method

One method of transferring horizontal control was discussed in Chapter 13. The method was to transfer the alignment from surface survey to underground through wires. The technique used is shown in Fig. 20.5.

This has already been discussed in Chapter 13. The two wires hung from any system from the surface are kept in drums containing oil or water to reduce oscillations of the wire.

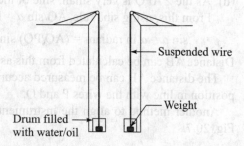

Fig. 20.5 Transferring control to underground points

The general precautions in using this method is to ensure the following.

(a) The wires should be as fine as possible. The wires are handled with care so that there are no kinks in the wire.
(b) The wires must be loaded so that they remain straight. The loads should be such that they are loaded to 50 per cent of their ultimate strength.
(c) The weights are placed in a drum of water or oil to keep them free of vibrations.
(d) The wires may be coloured or colour coded to identify them.

The two wires, when they are still, give the alignment of the tunnel as transferred from surface survey. This is very critical in tunnel surveys as any small change in alignment can cause considerable deviation in the final positions and lead to extra cost and labour.

Assuming that the wires give the alignment accurately, the idea is to place a theodolite or total station in line with the orientation given by the wires. The placement of the station in line with the one given by the wires is a very tedious process. Many trials have to be made to achieve this. To help in this operation, the Weisbach triangle method is used.

The Weisback Triangle The Weisbach triangle is a method to accurately find the position of the theodolite so that it is in line with the two positions given by the wires. Let P and Q be the two positions given by the wires. The theodolite is set up at A in line with the wires as close as possible to the correct alignment and as judged by the eye. The procedure is as follows:

(a) Measure the $\angle PAQ$ (which will be a very small angle) using a theodolite or total station very accurately. Both face observations are made and repeated if required.

(b) Measure the distances AP and AQ accurately. Also, measure the distance between the wires as a check. Use a steel tape

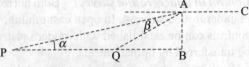

P, Q, – Positions of suspended wires
A – Theodolite position
B – Aligned position of instrument
AC – Line parallel to PQ

Fig. 20.6 The Weisbach triangle

and repeat the measurements to accurately measure the distances.

(c) A line AC can be set out parallel to CD using any of the methods outlined earlier in Chapter 4.

(d) As the $\angle APQ$ is very small, sine of the angle is equal to the angle in radians. From this, using sine rule, $AQ/\sin \alpha = PQ/\sin \beta$. From this

$$\sin \alpha = \alpha \text{ in radians} = (AQ/PQ) \sin \beta$$

Distance AB can be calculated from this as $AB = AP \sin \alpha = AP \times \alpha$ in radians.

The distance AB can be measured accurately using steel tape to get the station position in line with the wires P and Q.

Another method to align the instrument along the wires may be as shown in Fig. 20.7.

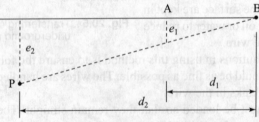

Fig. 20.7 Aligning underground

The method consists of measuring distances d_1, d_2, and e_1. These distances are very accurately measured using a steel tape. Point 1 can be established perpendicular to the alignment and in line with line PB. Then from similar triangles,

$$\text{Distance } e_2 = e_1 \times \frac{d_2}{d}$$

The instrument can be shifted by a distance e_2 to be in alignment with the wires.

Optical transfer of control   In the case of shallow tunnels, an optical transfer of control may be possible as shown in Fig. 20.8.

In Fig. 20.7, A and B are surface control points established by triangulation survey in the alignment of the tunnel. C and D are positions established close to the shaft in line with observations and repeated measurements are done to ensure that C and D are in line with AB.

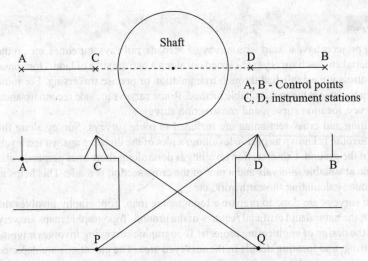

Shaft

A, B - Control points
C, D, instrument stations

Fig. 20.8 Optical transfer of control

Keeping the theodolite at C and sighting the control point B, the sight direction can be transferred as shown. A second point can be established by keeping the theodolite at D and sighting A. Both face observations can be done as a check.

In the case of large projects, like the tunnel under the English channel, where much greater accuracy is required, the alignment operations are done using a gyrotheodilite. Such instruments give automated measurements but are very expensive to use in normal works. The instrument uses a gyroscope to accurately determine azimuthal directions. The instrument ensures a fully automated system of azimuth measurements using a gyroscope. The measurements are fully automatic and controlled by a PC. The instrument is particularly useful in tunnel alignment where greater accuracy is required and is not achievable by other methods or instruments.

Fig. 20.9 Gyromat 2000

Summary

All engineering projects have to start with surveys. For roads, railways, pipelines, etc., route surveys are undertaken. Such surveys are carried out along a narrow strip of land. Horizontal and vertical controls are established through triangulation or precise traversing. For route surveys, an open traverse is the most suitable method. Route surveys include reconnaissance, preliminary survey, location survey, and construction survey.

Profile levelling and cross-sectioning are included in route surveys. Survey along the centre line of the route is known as profile levelling; a plot of the distance against the levels gives a profile of the ground. Cross-section levelling is done along short lines perpendicular to the centre line at suitable intervals and a plot of the cross section is made. This helps in designing the route, calculating the earthwork, etc.

Topographic surveys are done to prepare a topographic map. Topography involves the determination of the natural and artificial features of the ground. Topographic maps are very useful tools for the design of engineering projects. Topographic surveying involves traverse surveys, contouring, and locating details in the surveyed area. The elevations are depicted in a topographic map by hachures, colour schemes, or contours. For engineering purposes, contours are more commonly used.

City surveying also involves topographic surveys within city limits. Essentially, property delimitation and infrastructure details are depicted by city survey maps. These are vital for executing developmental works such as water supply and sewerage systems and transportation systems.

Underground surveys, such as mine surveys, are undertaken to prospect mineral wealth or, as in the case of tunnels, to explore the viability of underground transportation systems. Underground surveys pose special problems due to limited lighting, cramped space, and sometimes steeply inclined sights. Special techniques are required for transferring surface coordinates to underground points.

Exercises

Multiple-Choice Questions

1. Route survey refers to
   - (a) preparation of contour maps
   - (b) surveys of transportation networks
   - (c) traversing with tape
   - (d) survey of city areas
2. Objective of profile levelling is to obtain
   - (a) cross section plots of railways
   - (b) cross section plots of road works
   - (c) longitudinal section of road or railway
   - (d) topographic details of an area
3. Objective of cross sectioning is to find the
   - (a) gradient of a pipe line
   - (b) levels of points perpendicular to the route
   - (c) longitudinal section of a road route
   - (d) topographic details of an area
4. A topographic survey involves measuring
   - (a) elevation of points
   - (b) horizontal distances of points from a reference station or line
   - (c) distances, directions, and elevations
   - (d) vertical angles

Review Questions

1. Explain the term route survey. What is the objective of such a survey?
2. Explain the steps involved in route surveys.
3. Explain the term longitudinal sectioning. How is a profile map plotted and what are the considerations while deciding the scales for the plot of a section?
4. Explain the term cross-sectioning.
5. Explain the term topographic survey. Explain the stepwise procedure for such a survey.
6. Explain the objectives of city surveying. What are the special features of such a survey?
7. Explain the special problems encountered in underground surveying.
8. Explain the objective of mine surveying and the procedure for undertaking such a survey.
9. Explain the objectives and special features of tunnel surveying.
10. Explain the methods common to surface and underground surveys.
11. Briefly explain the precautions to be undertaken in tunnel surveys.

Astronomical Surveying

Learning Objectives

After going through this chapter, the reader will be able to

- state the important properties of a sphere and apply formulae from spherical trigonometry
- define the terms latitude and longitude
- draw a neat sketch to depict important astronomical terms
- define different types of times and their interrelationships
- explain the methods used to determine the azimuth, true meridian, latitude, and longitude
- explain the corrections required to be made to astronomical observations

Introduction

In the preceding chapters, we have studied several methods of surveying. Many different types of equipment have been dealt with and many different methods have been detailed. We know from our daily experiences that nothing in this universe is stationary; when we determine the bearings of lines or the coordinates of a point, they are not absolute. The reference lines on which these are based are not fixed, as a result of the motion of the earth and other heavenly bodies. For example, the magnetic meridian at a place does not remain constant. We also know that the true meridian at a place is a more reliable direction, as it maintains a constant direction and is not subject to variations. The true meridian at a place is determined with reference to celestial bodies such as the sun. Similarly, the location of a point on the surface of the earth is determined with respect to its latitude and longitude. Time is another important dimension when locating or specifying many other parameters. All these factors are taken into account in a branch of surveying known as *field astronomy*. Astronomy is a very vast field covering the motion of all heavenly bodies. For the purpose of surveying, we have to consider only a few of the millions of heavenly bodies around us. This chapter briefly describes surveys based on the principles of astronomy.

The earth is an oblate spheroid or ellipsoid. However, for practical purposes, it is considered a sphere. The properties of a sphere and of the lines (for example, survey lines) lying on it thus need to be studied, which are covered by a branch of mathematics called spherical trigonometry. We thus start with a brief outline of relevant principles from spherical trigonometry.

21.1 Spherical Trigonometry

Consider the spheres shown in Fig. 21.1. The following points may be noted.

(a) A sphere is a solid obtained by the revolution of a semicircle or a circle about its diameter [Fig. 21.1(b)]. All points on the surface of a sphere are equidistant from its centre O.

(b) When a sphere is cut by any plane, the resulting section is a circle. When the cutting plane passes through the centre of the sphere, the circular section is known as a *great circle*. When the cutting plane does not pass through the centre of the sphere, the circular section is known as a *small circle* [Fig. 21.1(c)].

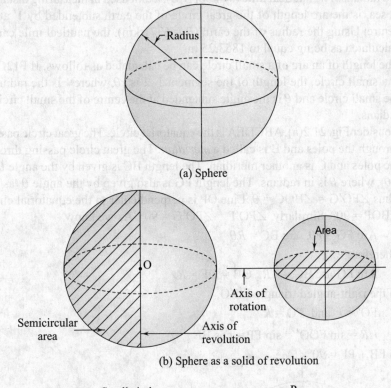

(a) Sphere

(b) Sphere as a solid of revolution

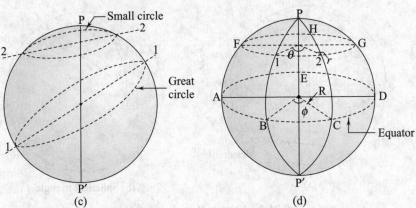

(c)

(d)

Fig. 21.1 Sphere and its properties

(c) The great circle ABCDEA is also called the *equatorial circle* and divides the circle into two equal parts. In the case of the earth, such a circle is called the *equator*.

(d) A line perpendicular to the equatorial circle and passing through the centre passes through two points on the surface of the sphere known as the *poles*. P and P′ are the poles of the great circle ABCDEA [Fig. 21.1(d)].

(e) The length of an arc of a great circle is given by $R\phi$, where R is the radius of the circle and ϕ is the angle subtended by the segment BC at the centre O in circular measure (radians). If the radius is equal to unity, the length of BC $= \phi$ radians. The nautical mile (symbol NM), a unit used in measuring distances at sea, is the arc length of the great circle of the earth subtended by 1′ at the centre. Using the radius of the earth ($R = 6371$ km), the nautical mile can be calculated as being equal to 1853.25 m.

(f) The length of an arc of a small circle can be calculated as follows. If F12GHF is a small circle, the length of the segment 1–2 is $r\theta$, where r is the radius of the small circle and θ is the angle subtended at the centre of the small circle in radians.

(g) Consider Fig. 21.2(a). ABCDEA is the equatorial circle. The great circle passing through the poles and B is called a *meridian*. The great circle passing through the poles and C is another meridian. The length BC is given by the angle θ (as $R\theta$), where θ is in radians. The length FG is also given by the angle θ (as $r\theta$). Thus $\angle FO'G = \angle BOC = \theta$. Line OP is perpendicular to the equatorial circle. $\angle BOP = 90°$. Similarly, $\angle PO'F = \angle PO'G = 90°$. As we know,

Arc FG $= r\theta$, arc BC $= R\theta$

Therefore,

Arc FG/arc BC $= r/R$, OB $=$ OF $= R$

In the right-angled triangle FO′O,

FO′ $= r$ and FO $= R$

$r/R = \sin\text{FOO}' = \sin\text{FP} = \cos\text{BF}$

as FB + PF $= 90°$.

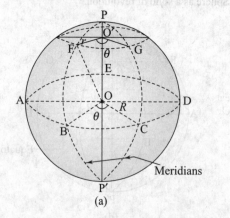

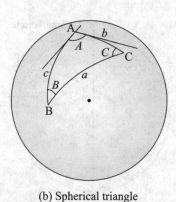

(a)

(b) Spherical triangle

Fig. 21.2 Length of circle segments

Arc FG/arc BC = sin FP = cos BF

Length of small circle arc FG = (length of great circle arc BC) × sin FB

= BC cos PF

Note that arcs of great circles are represented by the angles subtended by them at the centre of the sphere.

(h) The shortest distance between any two points on the surface of a sphere is along the great circle passing through the two points.

(i) A spherical triangle is formed by three intersecting arcs of the great circles of a sphere [Fig. 21.2(b)]. Thus, ABC is a spherical triangle. The plane angles between the tangents to these arcs give the spherical angles. Thus, the spherical angle *A* is the plane angle between the tangents to the arcs meeting at A. The following are some important properties of spherical triangles.

   (i) The sum of the angles of a spherical triangle is more than 180°. Therefore, $A + B + C = 180° + E$. *E* is known as *spherical excess*.

  (ii) Any angle of a spherical triangle is smaller than 180° or π rad.

 (iii) The sum of the three spherical angles is less than six right angles or 3π rad and more than two right angles or π rad.

 (iv) The sum of the lengths of any two sides is greater than the length of the third side (same as in a plane triangle).

  (v) If the sum of any two angles is equal to two right angles or π rad, then the third angle is equal to two right angles or π rad.

 (vi) The smaller angle lies opposite the smaller side and vice versa (same as in a plane triangle).

 (vii) The area of a spherical triangle is obtained from the following formula:

$$\text{Area of spherical triangle} = \pi R^2 E$$

where $E = A + B + C - 180°$ is the spherical excess.

(viii) The following formulae relate the angles and sides of a spherical triangle. Let *a*, *b*, and *c* be the sides and *A*, *B*, and *C* be the corresponding angles, as shown in Fig. 21.3.

Sine formula

$$\sin a/\sin A = \sin b/\sin B = \sin c/\sin C$$

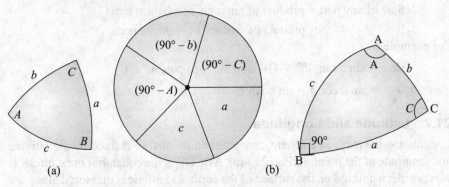

Fig. 21.3 Relation between sides and angles

Cosine formulae

$$\cos a = \cos b \cos c + \sin b \sin c \cos A$$
$$\cos A = (\cos a - \cos b \cos c)/\sin b \sin c$$

Similar expressions can be derived for the other sides and angles.

Half-angle formulae

$$\sin(A/2) = \sqrt{[\sin(s-b)\sin(s-c)]/\sin b \sin c}$$
$$\cos(A/2) = \sqrt{[\sin s \sin(s-c)]/\sin b \sin c}$$
$$\tan(A/2) = \sqrt{[\sin(s-b)\sin(s-c)]/\sin s \sin(s-c)}$$

where $s = (1/2)(a + b + c)$. Similar expressions can be derived for the other angles.

Sum and difference formulae

$$\tan[(a+b)/2] = \frac{\cos[(A-B)/2]\tan(c/2)}{\cos[(A+B)/2]}$$

$$\tan[(a-b)/2] = \frac{\sin[(A-B)/2]\tan(c/2)}{\sin[(A+B)/2]}$$

$$\tan[(A+B)/2] = \frac{\cos[(a-b)/2]\cot(C/2)}{\cos[(a+b)/2]}$$

$$\tan[(A-B)/2] = \frac{\sin[(a-b)/2]\cot(C/2)}{\sin[(a+b)/2]}$$

Napier's rules of circular parts   These rules are used in the case of a right-angled spherical triangle. Considering the spherical triangle ABC right-angled at B, we write the sides containing the right angle and the complement of the remaining parts in a circle as shown in Fig. 21.3(b). These parts are b, c, B, a, and C. The circle is thus divided into five parts. The quantities to be filled in the parts are b, c, $90° - B$, $90° - a$, and $90° - C$. Considering any part, it has adjacent and opposite parts. For any part, two adjacent and two opposite parts exist. The following rules apply:

Sine of any part = product of tangents of adjacent parts

= product of cosines of opposite parts

For example,

$$\sin c = \tan a \tan(90° - A) = \cos(90° - b) \cos(90° - C)$$

$$= \tan a \cot A = \sin b \sin C$$

21.2   Latitude and Longitude

Considering the earth as a sphere, any point on its surface is fixed by the latitude and longitude of the point. In Fig. 21.4(a), ABCDE is the equatorial great circle. If we consider a point M on the surface of the earth, its latitude is measured along the great circle passing through the poles and the point M. Such a circle is known as

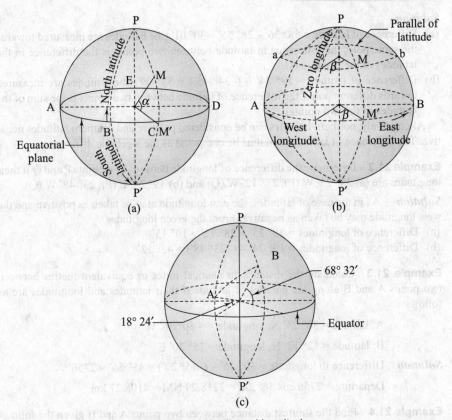

Fig. 21.4 Latitude and longitude

the *meridian circle*. The latitude is the angle subtended by the arc MM′, where M′ is the point where the meridian circle through M intersects the equatorial circle. α is the latitude of M.

The equatorial circle divides the earth into two halves. The part above the equatorial circle is known as the northern hemisphere and the part below is known as the southern hemisphere. If the angle to a point is measured above the equator, the point has a northern latitude; if the angle to a point is measured below the equator, it has southern latitude.

The longitude of a point is the angle between the standard meridian and the meridian through that point. The meridian through the Greenwich is internationally accepted as the standard zero meridian. A point may lie to the east or west of the standard meridian. A longitude can thus be an east longitude or a west longitude.

Point M lies to the east of the standard meridian and thus has an east longitude. The longitude is the angle between the two meridianal planes. Thus β is the longitude of M.

A parallel of a latitude is a small circle having P as its pole (axis). All points on the parallel of a latitude have the same latitude. In Fig. 21.4(b), aMba is a parallel of latitude. Points a, M, and b have the same latitude.

Example 21.1 Determine the difference of latitude between two points 1 and 2 if their latitudes are (a) 28° 35′ N, 58° 36′ N and (b) 18° 24′ N, 34° 45′ S.

Solution

(a) Difference of latitude = 58° 36′ – 28° 35′ = 30° 01′ (the latitudes are measured towards the same pole; the difference in latitude between two points is the difference in the latitudes of the two points).

(b) Difference of latitude = 18° 24′ – (–34° 45′) = 53° 09′ (the latitudes are measured towards different poles; the difference of latitude between two points is the sum of the latitudes of the two points).

Alternatively, northern latitudes can be considered positive and southern latitudes negative. The difference of latitudes can thus be computed as the algebraic difference.

Example 21.2 Determine the difference of longitude between two points P and Q if their longitudes are (a) 38° 25′ W (P), 28° 12′ W (Q) and (b) 19° 24′ E (P), 23° 48′ W (Q).

Solution As in the case of latitudes, the east longitude may be taken as positive and the west longitude may be taken as negative. From the given longitudes

(a) Difference of longitudes = 38° 25′ – 28° 12′ = 10° 13′
(b) Difference of longitudes = 19° 24′ – (–23° 48′) = 43° 12′

Example 21.3 Calculate the distance in nautical miles or equivalent metres between two points A and B along the parallel of latitude if their latitudes and longitudes are as follows:

A: latitude = 36° 24′ N, longitude = – 30° 36′ W

B: latitude = 12° 12′ N, longitude = 15° 20′ E

Solution Difference in longitude = 30° 36′ – (–15° 20′) = 45° 56′ = 2756′

Departure = 2756 cos 36° 24′ = 2218.29 NM = 4108.27 km

Example 21.4 Find the shortest distance between two points A and B given the following:

A: latitude = 18° 24′ N, longitude = 36° 18′ E

B: latitude = 68° 32′ N, longitude = 126° 34′ E

Solution Refer to Fig. 21.4(c). Considering the triangle PAB with the pole as one of the vertices, it is clear that

$$PA = 90° – 18° 24′ = 71° 36′, \quad PB = 90° – 68° 32′ = 21° 28′$$

∠APB of the triangle is the difference of longitudes:

$$∠APB = 126° 34′ – 36° 18′ = 90° 16′$$

We apply the cosine formula to calculate AB:

$$\cos AB = \cos PA \cos PB + \sin PA \sin PB \cos APB$$

$$= \cos 71° 36′ \cos 21° 28′ + \sin 71° 36′ \sin 21° 28′ \cos 90° 16′$$

$$= 0.2937 – 0.0016 = 0.2921$$

Therefore,

$$\text{Angle } AB = 73° 1′$$

Distance AB = R × angle AB. Taking R = 6370 km and converting the angle into radians, we get

$$AB = 6370 × 73° 1′ × π/180 = 8117.76 \text{ km}$$

21.3 Celestial Sphere

On a clear night, innumerable stars and planets, in fact millions of them, are visible to the naked eye; many more can be sighted with a telescope. As the earth itself is in motion, these celestial bodies appear to be changing their positions daily. It has been found convenient to consider these celestial bodies as lying on the surface of a sphere of infinite radius. Such a sphere is known as the celestial sphere. The earth is at the centre of this sphere (Fig. 21.5).

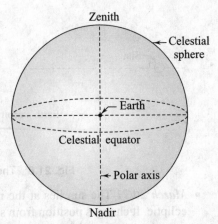

Fig. 21.5 Celestial sphere

It has been found convenient to consider the earth as fixed at the centre of the celestial sphere. As the earth rotates from west to east, it is convenient to consider instead the celestial sphere as moving from east to west with the earth's motion halted. The positions of the celestial bodies are then fixed with respect to their positions on the celestial sphere. From the point of view of astronomical surveys, the distances to celestial bodies are not significant but the angles at which they are seen are significant. Our interest is also limited to the sun and a few fixed stars.

21.4 Solar System

The solar system comprises the sun at the centre and a number of planets revolving around it, the earth being one of them.

Sun

The sun is a small star belonging to the *Milky Way* galaxy. It is about 150 million km away from the earth. It is very massive compared to the earth with a diameter 100 times that of the earth. The sun is small compared to other stars. It is assumed that it moves about its own axis. In addition, the sun appears to move with respect to the earth and the other stars.

Earth

The earth is the third planet from the sun after mercury and venus. The earth rotates about its own axis as well as about the sun. The orbit of the earth around the sun is shown in Fig. 21.6(a). The sun appears to trace an elliptical path in the sky known as the ecliptic, which is inclined to the equatorial plane at an angle of 23° 27′ [Fig. 21.6(b)]. While the earth experiences day and night alternately due to the rotation of the earth about its own axis, the seasons are experienced due to the earth's rotation around the sun. The sun lies at one of the foci of the earth's elliptical orbit. The four seasons correspond to four different positions of the sun with respect to the earth in the northern hemisphere (Table 21.1).

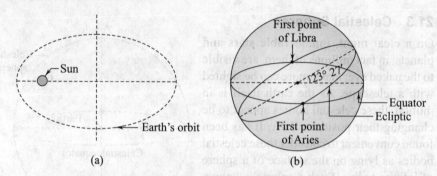

Fig. 21.6 The earth and the sun

- *March 20/21* The sun lies at the intersection of the equatorial plane and the ecliptic. It changes position from south to north. This position is known as the *vernal equinox or the first point of Aries*.
- *June 20/21* The sun lies at its northernmost point called the *summer solstice*.
- *September 22/23* The sun lies again at the intersection of the equator and the ecliptic. This position is called the *autumnal equinox or the first point of Libra*.
- *December 21/22* The sun is at its southernmost position called the *winter solstice*.

The sun is said to have a north declination for six months from March to September and a south declination from September to March. In the southern hemisphere, June 21 will be known as winter solstice and December 21 is known as the summer solstice.

Table 21.1 Positions of the sun with respect to the earth in the northern hemisphere

Dates	Position	Declination	Horizontal	Remarks
March 20/21	Vernal equinox	Zero	Zero	First point of Aries
June 20/21	Summer solstice	23° 27′	90°	Winter solstice in the southern hemisphere
September 22/23	Autumnal equinox	Zero	180°	First point of Libra
December 21/22	Winter solstice	23° 27′	90°	Summer solstice in the southern hemisphere

Moon

The moon is the earth's only natural satellite—it revolves around the earth, completing one revolution in 29.5 days. The plane of the moon's orbit is inclined to the ecliptic at about 5° 8′. The moon takes the same amount of time to rotate about its own axis as it does to revolve around the earth. Thus, about 60 per cent of the moon is always seen by an observer on earth. The radius of the moon is 1738 km and its mass is 7.349×10^{22} kg. The mean orbital distance is 3.844×10^5 km.

21.5 Astronomical Terms

Based on the definition of the celestial sphere, the following astronomical terms must be understood (Fig. 21.7).

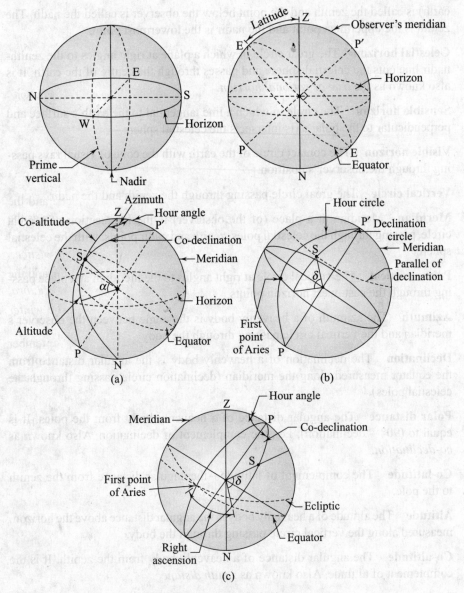

Fig. 21.7 Astronomical terms

Polar axis   The diameter joining the poles about which the earth rotates.

Terrestrial equator   The great circle on the earth's surface cut by a plane perpendicular to the polar axis.

Celestial equator   The great circle of the celestial sphere in which the plane of the equator cuts the celestial sphere.

Celestial poles   Points at which the polar axis intersects the celestial sphere.

Zenith and nadir points   A vertical line drawn through the observer's station intersects the celestial sphere at two points. The point above the observer (on the earth) is called the zenith and the point below the observer is called the nadir. The zenith is the uppermost point and the nadir is the lowermost point.

Celestial horizon The great circle in which a plane at right angles to the zenith–nadir line cuts the celestial sphere and passes through the centre of the earth. It is also known as the *true or rational horizon*.

Sensible horizon The circle in which a line tangential to the earth's surface and perpendicular to the polar axis intersects the celestial sphere.

Visible horizon The contact circle of the earth with the cone of visual rays passing through the observer's position.

Vertical circle The great circle passing through the zenith and the nadir.

Meridian Meridian at a place (of the observer) is the intersection of a great circle, passing through the celestial poles, zenith and nadir points, with the celestial sphere.

Prime vertical The vertical circle at right angles to the meridian at a place passing through the east–west horizon points.

Azimuth The azimuth of a heavenly body is the angle between the observer's meridian and the vertical circle passing through the body.

Declination The declination of a heavenly body is the angular distance from the equator measured along the meridian (declination circle passing through the celestial poles).

Polar distance The angular distance of a heavenly body from the poles. It is equal to (90° – declination), i.e., the complement of declination. Also known as *co-declination*.

Co-latitude The complement of latitude—the angular distance from the zenith to the pole.

Altitude The altitude of a heavenly body is the angular distance above the horizon, measured along the vertical circle passing through the body.

Co-altitude The angular distance of a heavenly body from the zenith. It is the complement of altitude. Also known as *zenith distance*.

Right ascension The right ascension of a heavenly body is the angular distance along the equator measured towards the east from the vernal equinox.

21.6 Coordinate Systems

To specify the position of a celestial body on the celestial sphere, three coordinate systems are used. These are briefly described here.

21.6.1 Altitude–Azimuth System

Referring to Fig. 21.8(a), the celestial body X can be identified by the altitude α and the azimuth A of the celestial body on the celestial sphere. O is the centre of the celestial sphere (earth) and P and Z are, respectively, the pole and zenith of the celestial sphere. NWS is the horizon.

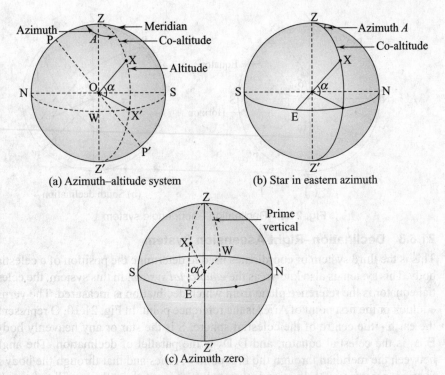

Fig. 21.8 Horizon system of coordinates

A great circle is drawn through the body and the Z–N points cut the horizon circle at X'. ∠XOX' = α, the altitude of the body. The arc NX' along the horizon is called the azimuth. This is the same as ∠PZX. Thus, the position of the celestial body X can be specified by the coordinates (α, A), i.e., the altitude and the azimuth. The azimuth is west in this case. The great circle arc ZX is the co-latitude or zenith distance, denoted by z.

A celestial body may have an east azimuth, as shown in Fig. 21.8(b). When the azimuth of a body is zero, it lies on the prime vertical through the points E and W [Fig. 21.8(c)].

21.6.2 Declination–Hour Angle System

Referring to Fig. 21.9(a), the celestial sphere has its centre at O (earth) and X is the celestial body. The great circle through the celestial poles and the celestial body intersects the celestial equator at X'. NWS is the horizon. Z denotes the zenith and Z' denotes the nadir. X_1X_2 is the parallel of declination, a small circle parallel to the equatorial circle. All points on this circle have the same declination. ∠XOX' = δ is the declination of the body X. This is the length of the arc XX' along the great circle PXX'P' and is measured from the equator. The declination can be towards the north or the south of the equator [Fig. 21.9(b)]. ∠ZPX is the *hour angle*. It is the angle between two meridians—the observer's meridian PZP'Z' and the meridian PXX'P' through the celestial body. If this angle is denoted by H, then the coordinates of the body are (δ, H).

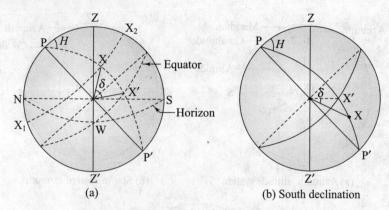

Fig. 21.9 Declination—hour angle system

21.6.3 Declination–Right Ascension System

This is the third system of coordinates used to determine the position of a celestial body. This system is also known as the *equatorial system*. In this system, the celestial equator is the reference plane from which declination is measured. The vernal equinox or the first point of Aries is the reference point. In Fig. 21.10, O represents the earth—the centre of the celestial sphere, S is the star or any heavenly body, E_1E_2 is the celestial equator, and D_1D_2 is the parallel of declination. The angle between the meridian through the first point of Aries and that through the body is the right ascension. Declination is the angle measured along the meridian from the equatorial plane.

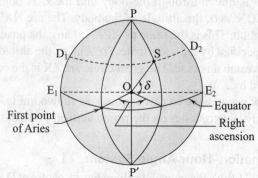

Fig. 21.10 Declination—right ascension system

21.6.4 Relation Between Coordinate Systems

It can be easily shown that the altitude of the celestial pole is equal to the latitude of the observer. Referring to Fig. 21.11(a), let M be the position of the observer. NMS is the observer's horizon, and the vertical through M passes through the centre of the earth, O, and the zenith. PP' is the polar axis of the earth. When extended, the polar axis intersects the surface of the celestial sphere at P_1. As the radius of the celestial sphere is infinity, the observer will be looking at the pole along a parallel line MP_2.

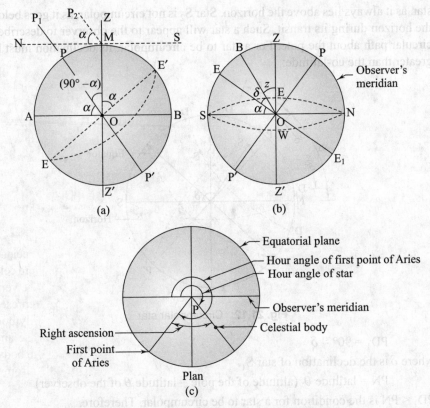

Fig. 21.11　Altitude of celestial pole and latitude

From the observer's position, the altitude of the celestial pole is given by $\angle P_2MN$. If the altitude of P_2 is α, then $\angle MOE' = \alpha$. Now, $\angle POM = 90° – \alpha$. MP_2 and OP are parallel; thus, $\angle POM = \angle P_2MZ = 90° – \alpha$. As $\angle P_2MZ + \angle P_2MN = 90°$, $\angle P_2MN = \alpha$, the latitude of M. Thus the altitude of the celestial pole is the same as the latitude of the observer.

We can also derive the relationship between latitude, declination, and altitude as follows [Fig. 21.11(b)]. Let $EE1$ be the equator and $NESW$ be the horizon. Let B be the celestial body. $\angle EOB$ is the declination δ, $\angle SOB = \alpha$, which is the altitude of the star, BZ is the zenith distance z, and EZ is the latitude of the place of observation. Then, $\theta = \delta + z$. If δ is in the south of the equator, it will be negative. If B is to the south of equator, z will be negative.

Referring to Fig. 21.11(c), it is easy to see that the hour angle of the celestial body B is equal to the difference between the hour angle of the first point of Aries and the right ascension of the celestial body.

Circumpolar stars

In Fig. 21.12, let O be the centre of the celestial sphere (earth). The observer has a north latitude of θ. The circle $PZP'Z'P$ is the meridian circle. The horizon circle and equatorial circle are also shown. The star S_1 has the parallel of declination D_1D_1' and the star S_2 has the parallel of declination D_2D_2'. Star S_1 is known as a circumpolar star, as it always lies above the horizon. Star S_2 is not circumpolar, as it goes below the horizon during its transit. Such a star will appear to the observer to describe a circular path about the pole. For a star to be circumpolar, its declination must be greater than the co-latitude:

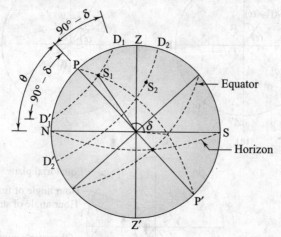

Fig. 21.12 Circumpolar star

$$PD_1 = 90° - \delta$$

where δ is the declination of star S_1.

PN = latitude θ (altitude of the pole = latitude θ of the observer)

$PD_1 < PN$ is the condition for a star to be circumpolar. Therefore,

$$90° - \delta < \theta \quad \text{or} \quad \delta > 90° - \theta$$

which is the co-latitude.

Stars at culmination

When a heavenly body crosses the observer's meridian, it is said to be in culmination. In one revolution around the pole, a star crosses the observer's meridian twice. One position is known as the *upper culmination* or *transit* and the other position is known as the *lower culmination* or *transit*. In Fig. 21.13(a), the star S_1 is at upper culmination at D_1 and its lower culmination is at D_1'. For the star S_2, the upper transit point is D_2 and the lower transit point is D_2'. The upper culmination may occur on the north of the zenith towards the north pole P or on the south of the zenith as for star S_2. Let θ be the latitude of the observer, δ the declination of the star, ZP = co-latitude = $90° - \theta$, PD_1 = co-declination = $90° - \delta$.

For the star S_1, the zenith distance at upper culmination is given by

$$ZD_1 = ZP - PD_1 = (90° - \theta) - (90° - \delta) = \delta - \theta$$

The declination is greater than the latitude. For the star S_2, the zenith distance at upper culmination is given by

$$ZD_2 = PD_2 - PZ = (90° - \delta) - (90° - \theta) = \theta - \delta$$

The latitude is greater than the declination of the star. For the upper culmination to occur on the north of the zenith, the declination of the star must be greater than the latitude of the observer.

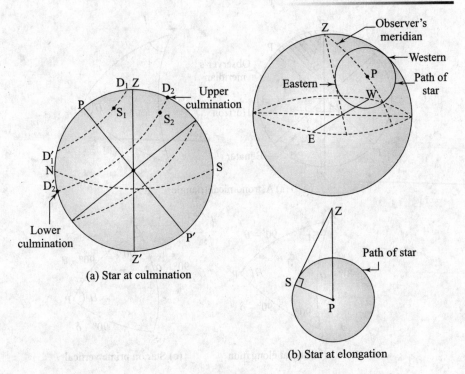

(a) Star at culmination

(b) Star at elongation

Fig. 21.13

Stars at elongation

A star is said to be at elongation when it is at its maximum distance from the meridian. Figure 21.13(b) shows the position of a star at elongation. The elongation can be eastern or western. At elongation, the angle at S (of the astronomical triangle) is a right angle.

Astronomical triangle

The pole P, the zenith Z, and any heavenly body S form a triangle known as the astronomical triangle. PZ lies on the observer's meridian great circle, PS on the great circle of the star's meridian, and ZS lies on a great circle ZSNZ (Fig. 21.14). The three sides and angles of this spherical triangle are the following.

Zenith distance: $ZP = 90° - \theta$ (θ is the latitude of the observer)

Co-altitude: $ZS = 90° - \alpha$ (α is the altitude)

Co-declination: $PS = 90° - \delta$ (δ is the declination of the star)

If the latitude, altitude, and declination are known, the astronomical triangle ZPS can be solved for the azimuth A and the hour angle H as follows:

$$\cos A = \frac{\sin \alpha}{\cos \alpha \cos \theta} - \tan \alpha \tan \theta$$

A can also be calculated from the tangent half-angle formula:

$$\tan(A/2) = \sqrt{\sin(s - ZS)\sin(s - ZP)/[\sin s - \sin(s - PS)]}$$

where s is the semi-perimeter,

$$s = (1/2)[PS + SZ + ZP]$$

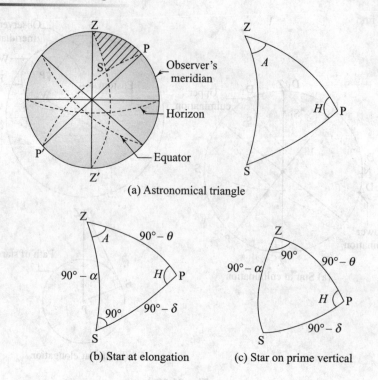

(a) Astronomical triangle

(b) Star at elongation (c) Star on prime vertical

Fig. 21.14

H can be calculated as follows:

$$\cos H = \frac{\sin \delta}{\cos \alpha \cos \theta} - \tan \delta \tan \theta$$

or from the tangent half-angle formula

$$\tan(H/2) = \sqrt{\sin(s - \mathrm{ZP})\sin(s - \mathrm{PS})/[\sin s - \sin(s - \mathrm{ZP})]}$$

At elongation, the astronomical triangle is right-angled at S [Fig. 21.14(b)]. When the star lies on the prime vertical (the great circle passing through the east-west points of the horizon, at right angles to the meridian), the astronomical triangle is right-angled at the zenith point Z [Fig. 21.14(c)]. At elongation, the angle at S is a right angle. In such a case,

$$\sin \alpha = \sin \theta/\sin \delta, \quad \sin A = \cos \delta/\cos \theta, \quad \cos H = \tan \theta/\tan \delta$$

When the star lies on the prime vertical, A is a right angle. Then,

$$\sin \alpha = \sin \delta/\sin \theta, \quad \cos H = \tan \delta/\tan \theta$$

Example 21.5 Find the coordinates in terms of the azimuth and altitude of a star if the latitude of the observer is 42° north, the declination is 26° 30′, and the hour angle is 42°.

Solution Refer to Fig. 21.15(a). In the astronomical triangle ZPS, the co-latitude

PZ = 90° − 42° = 48°

PS = 90° − δ = 90° − 26° 30′ = 63° 30′

Hour angle, ∠ZPS = 42°

Side ZS = 90° − α

If we find side ZS, we can find the altitude:

Azimuth $A = \angle PZS$

cos ZS = cos PS cos PZ + sin PS sin PZ cos SPZ

= cos 63° 30′ cos 48° + sin 63° 30′ sin 48° cos 42°

= 0.4462 × 0.4461 + 0.8949 × 0.7431 × 0.7431

= 0.7926

ZS = 37° 34′ 15″

Altitude α = 90° − 37° 34′ 15″ = 52° 25′ 45″

cos A = sinδ/cosα cosθ − tanθ tanα

= sin 26° 30′/cos 52° 25′ 45″ cos 42° − tan 52° 25′ 45″ tan 42°

= − 0.1857

= 100° 42′ 07″ W

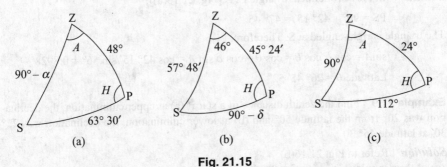

Fig. 21.15

Example 21.6 Determine the coordinates in terms of the hour angle and declination from the following observations: latitude of the observer = 44° 36′, azimuth of the star = 46° W, and altitude = 32° 12′.

Solution In the astronomical triangle ZPS [Fig. 21.15(b)],

ZP = 90° − 44° 36′ = 45° 24′

ZS = 90° − 32° 12′ = 57° 48′

A = 46° W

By the cosine rule,

cos SP = cos ZS cos ZP + sin ZS sin ZP cosA

= cos 57° 48′ cos 45° 24′ + sin 57° 48′ sin 45° 24′ cos 46°

= 0.7925

SP = 37° 34′ 49″, declination δ = 90° − 37° 34′ 49″ = 22° 25′ 11″

cos H = [cos ZS − cos ZP cos PS]/(sin ZP sin PS)

= [cos57° 48′ − cos45° 24′ cos37° 34′ 49″]/(sin45° 24′ sin37° 34′ 49″)

= − 0.0541

Hour angle H = 93° 06′ 09″

Example 21.7 Calculate the hour angle and azimuth of the sun from the following data: (a) latitude of the place is 46° N, declination is 22° S at sunrise; (b) latitude is 46° S declination is 20° N at sunset.

Solution In the astronomical triangle ZPS [Fig. 21.15(c)],

$$ZP = 90° - 46° = 24°, \quad \delta = 22°S = 90° + 22° = 112°$$

At sunrise, the sun is on the horizon. Hence ZS = 90°. Applying the cosine rule,

$$\cos H = [\cos ZS - \cos ZP \cos PS]/(\sin ZP \sin PS)$$

$$= [0 - \cos 44° \cos 112°]/(\sin 44° \sin 112°) = 0.4183$$

Hour angle $H = 65° \, 16' \, 22''$

When ZS = 90°,

$$\cos A = \cos PS/\cos ZP = \cos 112°/\cos 44° = -0.5207$$

Azimuth $A = 121° \, 23'$

Example 21.8 Find the latitude of the place of observation from the following data. Star at elongation: azimuth of the star = 56° E and declination = 42° 15'.

Solution In the astronomical triangle ZPS [Fig. 21.16(a)],

$$PS = 90° - 42° \, 15' = 47° \, 45'$$

The triangle is right-angled at S. Therefore,

$$\sin A = \cos \delta/\cos \theta = \cos \theta = \cos \delta/\sin A = \cos 42° \, 15'/\sin 56° = 0.8928$$

Latitude $\theta = 26° \, 45' \, 52''$

Example 21.9 Find the zenith distance of a star if (a) at upper culmination, the declination was 20° from the latitude 50° and (b) at lower culmination, the declination was 22° 30' at latitude 45° 30'.

Solution Refer to Fig. 21.16(b).
(a) At upper culmination, as $\delta < \theta$, the star is on the south side of the zenith.
  Zenith distance = 50° − 20° = 30°
(b) At lower culmination,
  Zenith distance = 180° − θ − δ = 180° − 22° 30' − 45° 30' = 112°

Example 21.10 If the altitudes of a star at upper and lower culmination are 72° 18' and 21° 30', respectively, both on the north side of the zenith, find the declination of the star and latitude of the place.

Solution Refer to Fig. 21.16(c). At upper culmination,

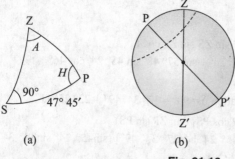
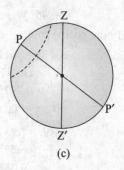

(a)   (b)   (c)

Fig. 21.16

Zenith distance = $\delta − \theta = 90° −$ altitude = 90° − α

At lower culmination,

Zenith distance = 180° − δ − θ = 90° − α

Therefore,

$$\Delta - \theta = 90° - 72° \; 18' = 17° \; 42'$$

$$180° - \delta - \theta = 90° - 21° \; 30' = 68° \; 30'$$

Solving the two equations,

$$\theta = 46° \; 54', \quad \delta = 64° \; 36'$$

21.7 Time and Astronomical Work

The earth has two motions—one about its polar axis, completing one rotation in one day, and the second around the sun, completing one revolution in one year. The sun lies at one of the foci of the elliptical orbit of the earth. This makes the sun appear to move with variable speed along the ecliptic. The ecliptic is inclined to the equatorial plane at an angle of 23° 27′. It cuts the equatorial plane at the first point of Aries and the first point of Libra, as explained earlier. The motion of the earth about its own axis from west to east and this makes the celestial bodies like the sun appear to move from east to west.

Astronomical measurements with reference to celestial bodies have to take into account the time element at which the required observations are made. Different time scales are used in astronomy, as compared to day-to-day work. *Sidereal time* and *apparent solar time* are used by astronomers while mean solar time and standard time are used for day-to-day work.

21.7.1 Sidereal Time

Sidereal time depends upon the movement of the first point of Aries. *The time interval between the movement of the first point of Aries over the same meridian twice is called a sidereal day.* The first point of Aries moves backwards opposite the motion of the sun along the ecliptic at the rate of 50.2 sec in one year. A mean first point of Aries can be defined, which moves with uniform angular velocity equal to the average angular velocity of the true first point of Aries. A sidereal day is divided into 24 hours (hr), each hour into 60 minutes (min), and each minute into 60 seconds (sec).

The sidereal year is the time taken by the sun to complete a complete revolution along the ecliptic with reference to a fixed star. The solar or tropical year is the time taken by the sun for a complete circuit about the first point of Aries. The backward movement of the first point of Aries makes the solar year a little shorter than the sidereal year. A sidereal year consists of 365.2564 mean solar days and a tropical year consists of 365.2422 mean solar days.

Sidereal time is equal to the hour angle of the first point of Aries. Local sidereal time is defined as the right ascension of the meridian of a place. In other words, local sidereal time is the time that has elapsed after the transit of the first point of Aries over the meridian. The hour angle of a star is also the sidereal time that has elapsed after its transit: local sidereal time = right ascension of a star + westerly hour angle of the star. If, by this calculation, the time comes out to be more than 24 hr, we have to subtract 24 hr from it, and if the sum is negative, we have to add 24 hr to it. When the star is at upper culmination, its hour angle is zero and the sidereal time is equal to the right ascension.

21.7.2 Apparent Solar Time

The apparent solar time is based upon the daily motion of the sun along the ecliptic. The motion of the sun along the ecliptic is not uniform, and hence cannot be kept by a clock that keeps time based upon regular, uniform motion. *The apparent solar day is defined as the time interval between two consecutive lower transits of the sun over the meridian.* The lower transit is taken so that the time at midnight starts at the zeroth hour. The apparent solar day is divided into 24 hours, each hour into 60 minutes, and each minute into 60 seconds. The movement of the sun is not uniform and hence cannot be used for day-to-day use. Only a sundial can show the apparent solar time.

21.7.3 Mean Solar Time

As the apparent solar time cannot be used as civil time, astronomers have devised a mean solar time. The mean solar time is based upon the diurnal motion of a *mean sun*, which is a point that moves uniformly along the equator. The time interval between two consecutive passages of the mean sun over the meridian is the *mean solar day*. It is the average time interval between the apparent solar days of the year. The mean sun (an assumed point) starts from the first point of Aries at the same time as the true sun and returns to the vernal equinox at the same time as the true sun.

The mean solar day is divided into 24 hours, each hour into 60 minutes, and each minute into 60 seconds. Starting at 0 hr at midnight, time can be measured as 24 hr or, as is most commonly used, 12 hr up to noon and then another 12 hr up to midnight. When time is measured in 12-hr spans, it is suffixed with the letters a.m. (ante meridiem) for the first 12 hours (0 to 12 hr) to indicate that it refers to time before noon. It is suffixed with the letters p.m. (post meridiem) for the next 12 hr, from noon to midnight (12 to 24 hr). Thus, 10 a.m. means 10 hr, while 4 p.m. means 16 hr.

The time at which the sun crosses the meridian at a place is known as *local apparent noon* and the time at which the mean sun crosses the meridian is known as *local mean noon. Local apparent midnight and local mean midnight* are similarly defined. The term 'local time' refers to the time corresponding to the meridian of the place. All locations on the same meridian have the same local time. However, if we go by this definition alone, there will be so many local times corresponding to the meridian of the place in a country. To avoid such confusion, for civilian use, standard time is defined based on some meridian for a country.

21.7.4 Standard Time

As mentioned in the preceding section, the standard time is based upon a standard meridian lying within the country. For India, the standard meridian selected is at 82° 30′ E longitude and is 5 hr and 30 min ahead of the Greenwich meridian time. All clocks in India maintain this standard time for daily use. The local time at any place can be determined if the longitude of the place and the standard longitude and time are known.

Equation of time

The difference between the apparent time and the mean time is known as the equation of time (ET). These differences are listed in the nautical almanac for every day of the year for Greenwich 0 hr. The equation of time, arising from the above-mentioned difference, can be attributed to two factors: (i) the orbit of the earth is an ellipse and hence the motion of the earth is not uniform in its orbit and depends upon its distance from the sun; (ii) the mean sun moves uniformly along the equator while the sun moves along the ecliptic, which lies on a plane inclined to the equator.

Equation of time = right ascension of the mean sun – right ascension of the sun

This time does not remain constant and varies over the year in a complex pattern. The equation of time is zero on four days in a year—April 15, July 14, September 1, and December 25—as the position of the mean sun and that of the sun coincide on the same meridian. The variation of the equation of time can be represented as shown in Fig. 21.17. The value of the equation of time varies between 0 and 16 min.

Conversion of time

This section discusses the conversion of time from one system to another. The following abbreviations are used:

GMT Greenwich mean time

GAT Greenwich apparent time

GMN Greenwich mean noon

GMM Greenwich mean midnight

GST Greenwich sidereal time

LMT Local mean time

LAT Local apparent time

LMN Local mean noon

LMM Local mean midnight

LST Local sidereal time

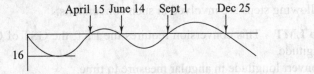

Fig. 21.17 Variation of equation of time

Conversion of degrees into time and vice versa   As 360° is equal to 24 hours, one hour is equal to 15°. Further, as the conversion of degrees to minutes and seconds is by the same factor as that of hour into minutes and seconds,

15° = 1 hr, 15′ = 1 min, 15″ = 1 sec

1° = 4 min, 1′ = 4 sec

Minutes and seconds are common in the hour system of time and the degree system used for angles. Therefore, to distinguish between the two measures, h, m, s will be used for time while the symbols °, ′, ″ will be used for angles.

Mean solar time and sidereal time are related by

365.2422 mean solar days = 366.2422 sidereal days

1 mean solar day = 1 + 1/365.2422 sidereal days = 24h 3m 56.56s

One mean solar day is 3m 56.56s longer than a sidereal day.

1 mean solar hour = 1 hr + 9.8565 sec sidereal time

1 mean solar minute = 1 min + 0.1643 sec sidereal time

1 mean solar second = 1 sec + 0.0027 sec sidereal time

The time of 9.8565 sec, which is added to the mean hour to obtain the sidereal hour, is known as *acceleration*. The inverse relationship can be similarly derived:

1 sidereal day = 23h 56m 4.09s mean solar time

From this, one can derive

1 hr sidereal time = 1 hr mean solar time − 9.8296 sec

1 min sidereal time = 1min mean time − 0.1638 sec

1 sec sidereal time = 1 sec mean time − 0.0027 sec

The time deducted from the mean time to obtain the sidereal time is known as *retardation*.

Difference in longitude   When the difference in longitude between two places is converted into time, it will be the difference in time between two places lying on these longitudes for all types of time—sidereal, apparent, or mean. Thus, one can write

Local time = Greenwich time ± longitude of the place   (+ for east longitude)

Greenwich time = local time ± longitude of the place   (+ for west longitude)

Local sidereal time to local mean time and vice versa   The following formulae are used:

LST at LMN/LMM = GST at GMN/GMM ± 9.96 sec/hr of longitude

LST = LST at LMN/LMM + sidereal interval from LMN/LMM

The following steps are involved in such conversions.

LST to LMT   This conversion requires the LST, the GST of GMN/GMM, and the longitude.

(a) Convert longitude in angular measure to time.

(b) Applying the rule given above, find LST of LMN.

(c) Obtain the sidereal interval from LMN/LMM.

(d) Convert sidereal interval into mean time by deducting 9.8296 sec/hr to obtain LMT.

LMT to LST   This conversion requires the LMT, the GST of GMN/GMM, and the longitude.

1. Convert longitude in angular measure to time.

2. Determine the LST of LMN/LMM using the given rule.

3. Convert LMT into sidereal units by adding 9.8565 sec/hr of LMT. This gives the sidereal interval from LMN/LMM.

4. Calculate the LST by adding the sidereal interval to the LST of LMN/LMM.

An alternative procedure is available if the GMT of the transit of the first point of Aries (TFA) (0 hr sidereal time at Greenwich) is given. The rule is as follows:

LMT of TFA = GMT of TFA ± 9.8256 sec/hr of longitude (+ sign for east longitude)

LST to LMT   This conversion requires the LST, the GMT of the TFA, and the longitude.

1. Convert the longitude in angular measure to time.
2. Find the LMT of the TFA using the given rule.
3. Deduct 9.8256 sec of the given sidereal time to obtain the corresponding mean time interval.
4. Find the LMT by adding this mean time interval to the LMT of the TFA.

LMT to LST   This conversion requires the LMT, the GMT of the TFA, and the longitude.

1. Convert the given longitude in angular measure to time.
2. Find the LMT of transit from the given GMT of TFA.
3. Determine the mean time interval as the difference between the mean time of transit and the given local mean time.
4. Convert this mean time interval to sidereal time by adding 9.8565 sec/hr of the mean time interval. This gives the LST corresponding to the given LMT.

Example 21.11   Carry out the following conversions. (i) Express the following times in degree measure: (a) 4h 12 m 48s and (b) 16h 32m 24s. (ii) Express the following angles in hours, minutes, and seconds: (a) 12° 32′ 48″ and (b) 124° 12′ 36″.

Solution
(i) (a)   4 hr = 60°, 12 min = 12 × 15 = 180′ = 3°, 48 sec = 48 × 15″ = 12′
         Therefore, 4h 12m 48s = 63° 12′.
    (b)   16 hr = 240°, 32 min = 32 × 15 = 480′ = 8°, 24 sec = 24 × 15 = 360″ = 6′
         Therefore, 16h 32′ m 24s = 248° 06′.

(ii) (a)   12° 32′ 48″ − 1 = 4 m; 1′ = 4s
         Therefore,

$$12° = 12 \times 4 = 48 \text{ m}$$

$$32m = 32 \times 4 = 128s = 2m \, 8s$$

$$48″ = 48 \times 4/60s = 3.2s$$

    Hence,

$$12° \, 32′ \, 48″ = 50m \, 11.2s$$

    (b)   124° 12′ 36″ − 124° × 4 = 496m = 8h 16m

$$12′ = 12 \times 4s = 48s$$

$$36″ = 36 \times 4/60 = 2.4s$$

    Hence

$$124° \, 12′ \, 36″ = 8h \, 16m \, 50.4s$$

Example 21.12   Places P and Q have longitudes of 88° 10′ E and 64° 36′ W. Find the local mean time at P and Q if the standard meridian is 72° E and the standard time is 9h 36m 48s.

Solution   Longitude of P = 88° 10′ E, standard meridian = 72° E
Difference in longitude = 88° 10′ – 72° = 16° 10′ = 1h 4m 40s
Standard time = 9h 36m 48s
Local mean time at P = 9h 36m 48s + 1h 4m 40s = 10h 41m 28s
Longitude of Q = 64° 36′ W
Difference in longitude = 64° 36′ + 72° = 136° 36′ = 9h 6m 24s
Standard time = 9h 36m 48s
Local time at Q = 9h 36m 48s – 9h 6m 24s = 0h 30m 24s

Example 21.13   Find the Greenwich mean time corresponding to the local times at P and Q given that (a) P has a longitude of 62° 38′ E and the local mean time is 8h 10m and (b) Q has a longitude of 50° 48′ W and the local mean time is 6h 12m 30s.

Solution   (a) Longitude of P = 62° 38′ E = 4h 10m 32s
Greenwich mean time = local mean time at P – longitude in time
= 8h 10m 0s – 4h 10m 32s = 3h 59m 28s
(b) Longitude of Q = 50° 48′ W = 3h 23m 12s
Greenwich mean time = 6h 12m 30s + 3h 23m 12s = 9h 35m 42s

Example 21.14   If the time at a place in India as per a clock is (a) 9 a.m. and (b) 10 p.m. and the latitude and longitude of the place are 32° 30′ N and 72° 30′ E, respectively, find the local mean time and Greenwich mean time. The standard meridian for India is 82° 30′ E.

Solution   (a) Standard time = 9 a.m.
Longitude = 72° 30′ E
Difference in longitude = 82° 30′ – 72° 30′ = 10° = 0h 40m
LMT at the place = 9h – 0h 40m = 8h 20m
GMT = 9h – 82° 30′ in time = 9h – 5h 30m = 3h 30m
(b) Local time = 10 p.m. = 22h
82° 30′ = 5h 30m, 72° 30′ = 4h 50m
Difference in longitude = 82° 30′E – 72° 30′E = 10s = 0h 40m in time
Local mean time at the place = 22h 0m – 0h 40m = 21h 20m
Corresponding Greenwich mean time = 21h 20m – 4h 50m = 16h 30m
or
22h 00m – 5h 30m = 16h 30m

Example 21.15   Find the local mean time for a place having a longitude of 70° 30′ E if the Greenwich mean time is (a) 6 a.m. and (b) 8 p.m.

Solution   A longitude of 70° 30′ E is equivalent to 4h 46m in time. As the longitude is east, the local time will be ahead of the GMT.
(a) 6 a.m.: LMT = 6h + 4h 46m = 10h 46m a.m.
(b) 8 p.m.: LMT = 8h + 4h 46m = 12h 46m = 0h 46m a.m. of the next day.

Example 21.16   Find the local mean time at a place of longitude 78° 30′ W if the local sidereal time is 13h 30m and the GST of GMN is 6h 20m.

Solution   Longitude 78° 30′ W = 5h 14m in time = 5.2333 hr
Acceleration = 5.2333 × 9.856 = 51.6 sec
GST of GMN = 6h 20m
LST of LMN = GST of GMN + acceleration = 6h 20m + 51.6s
= 6h 20m 51.6s

Local sidereal time = 13h 30m 0s

Sidereal interval from LMN = LST – LST of LMN

= 13h 30m 0s – 6h 20m 51.4s = 7h 9m 8.4s

Retardation = [7 × 9.8296 + 9 × 0.1638 + 8.4 × 0.0027] sec = 1m 10.3s

Local mean time = 7h 9m 8.4s – 1m 10.3s = 7h 7m 58.1s

This is the interval from LMN. Therefore, the time will be designated as p.m.

Example 21.17 Find the local sidereal time at a place of longitude 68° 40′ E if the local mean time is 22h 40m and the GST of GMN is 5h 38m.

Solution   Longitude = 68° 40′ = 4h 34m 40s

Retardation = 4.345 × 9.856s = 45s

GST of GMN = 5h 38m

LST of LMN = 5h 38m – 45s = 5h 37m 15s

Local mean time = 22h 40m

Mean time interval from LMN = 10h 40m = 10.6667 hr

Acceleration for this interval = 10.6667 × 9.856s = 1m 45s

LST interval = LMT interval + acceleration = 10h 40m + 1m 45s

= 10h 41m 45s

Local sidereal time = 10h 41m 45s + 5h 37m 15s = 16h 19m

21.8 Application of Astronomy in Surveying

So far, we have studied the general principles of astronomy. The applications of astronomy in surveying include determining the true meridian at a place, the azimuth of a survey line, the latitude and longitude of a place, and the time that is useful to astronomers and surveyors.

For astronomical observations, fieldwork is generally done during the early or late part of the day for solar observations and at night for stellar observations. The meridian observation of the sun is done around noon. Most of the instruments used for astronomical surveys, such as theodolites, have illuminated cross hairs. Reference marks and triangulation stations also need to have illuminated signals for night observations.

Generally, stellar observations are done with respect to the sun or the star Polaris. Polaris is a bright star near the celestial north pole and is easily visible to the naked eye. Polaris observations can be taken during late evening or at night. This star can be recognized easily by locating the star constellation Ursa Major as shown in Fig. 21.18. To observe any other star, its location and other data can be obtained from star charts. An important document to be kept in hand for astronomical surveying is the nautical almanac, which has important data about stars and planets.

Astronomical surveying equipment essentially includes an accurate theodolite with clear cross hairs. Solar observations are taken when the sun becomes tangential to the vertical and horizontal cross hairs, as shown in Fig. 21.18. Solar attachments, such as the Burt attachment or the Saegmuller solar attachment, are also sometimes used. These accessories were mounted on a transit and used for solar observations in earlier times.

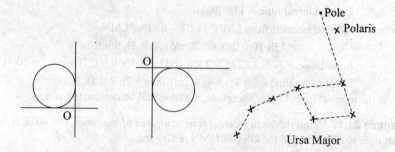

Fig. 21.18 Solar observations

21.8.1 Determination of True Meridian

The true meridian at a place is generally determined by two methods—(i) observation of Polaris at culmination and (ii) observation of Polaris at elongation. The second method is preferred and is more accurate. To use either of the methods, the latitude of the place of observation must be known. The latitude can be obtained from a map or table or can be determined by measuring the altitude of Polaris, which is very nearly equal to the latitude of the place.

Observation of polaris at culmination

Polaris can be observed during the early part of the evening and at night. A suitable day is chosen and the time of culmination of Polaris is computed as follows. At the time of upper culmination, the local sidereal time is equal to the right ascension of Polaris. Lower culmination will occur 12 hours after the upper culmination. So, the time for lower culmination can be found by adding or subtracting 11h 58min to or from the time of upper culmination. The time of culmination can also be obtained from special tables available. Once these times are determined, the following procedure is adopted.

1. Select a suitable spot with about 100 m of clear view in the approximate direction of the meridian. Set up the theodolite at the station about 15 min before the time of culmination is obtained from tables or computed.
2. Accurately centre and level the theodolite. Mark the station point accurately. The vertical circle vernier can be set to read the latitude of the place. This will facilitate locating the star, as the star is very close to the pole.
3. Just a few minutes before the culmination, direct the telescope towards the star. The in-built illumination facilitates clear sighting of the cross hairs.
4. Clamp the lower and upper plates. Follow the star continuously using the tangent screw. If upper culmination is being followed, the star will appear to move leftwards. During lower culmination, the star will appear to move right.
5. At the time of culmination, depress the telescope and obtain the line of sight along the ground. Mark a point accurately along the line of sight in line with the vertical cross hair. The line joining the station point and the point marked gives the direction of the meridian at the place.

Observation of polaris at elongation

A star is said to be at elongation when it is at the maximum distance from the poles. The time of elongation can be obtained from astronomical tables. The star will be at elongation approximately six hours after culmination. The star can be observed at western elongation or eastern elongation. The procedure is as follows.

1. Select an instrument station such that as the star reaches elongation along the line of sight of the telescope, about 100 m of clear distance is available for marking the meridian.
2. Set up the theodolite over the station and centre and level the instrument accurately. Set the vertical circle to read the latitude of the place for ease of locating the star. Adjust the horizontal circle to read zero.
3. Just a few minutes before the time of elongation, clamp both the motions and follow the star accurately with the vertical cross hair, using the tangent screw. The star that appears to move leftwards while approaching the western elongation appears to move vertically at the time of elongation. Depress the telescope and mark the line of sight with a stake at about 100 m from the instrument station.
4. To be more accurate, change the face, observe the star again, and mark another point in line with the line of sight of the telescope. The point midway between the two marks gives the direction of the line of sight of the star at elongation.
5. Calculate the azimuth of the star at elongation using the relation $\cos A = \cos \delta / \cos \theta$, where A is the azimuth at elongation, δ is the declination, and θ is the latitude of the place.
6. Lay off the azimuth angle to the east or west of the line marked from the instrument station as the direction of the star at elongation. If the star is observed at western elongation, mark the angle to the east of the line. Mark a stake in line with the line of sight, which gives the direction of the true meridian at the place.

21.8.2 Determination of Azimuth

The azimuth of a survey line is the angle made by the survey line with the direction of the true meridian at a place. It is determined by observing the azimuth of a celestial body, the horizontal angle between any reference mark and this celestial body, and the angle between the reference mark and the survey line. The reference mark itself may be at the other end of the survey line.

There are many methods to obtain the azimuth of a line. A method that permits the measurement of the angle with both faces is preferable, in order to eliminate instrumental errors. Circumpolar stars are the best choice of celestial body, as they move slowly in azimuth. Solar observations are also used to find the azimuth of a survey line. The following methods are commonly used.

Extra-meridian observation of the sun or circumpolar star

Observation of the sun   This method involves the observation of the altitude of the sun and the measurement of the horizontal angle between the sun and the survey line. The procedure is as follows.

1. Set up a theodolite at one end of the line for observations, which are taken with both faces to eliminate instrumental errors.
2. Note the local mean time of observation of the altitude.
3. Obtain or measure the latitude of the place.

4. Knowing the altitude of the sun, the horizontal angle between the sun and the line, and the latitude of the place, compute the azimuth of the line.

The azimuth of the sun is computed by solving the *astronomical triangle* (Fig. 21.19). The observed altitude of the sun is corrected for various factors and the correct altitude is obtained. For better results, the sun is observed between 8 and 10 a.m. or from 2 to 4 p.m. If the observation is made when the sun is near the prime vertical, the results will be better, as the movement of the sun is slow in azimuth.

In Fig. 21.19, let O be the position of the observer and OL be the survey line. S is the position of the sun. ZPS is the astronomical triangle. Angle PZS or EOR is the azimuth of the sun. Let α be the corrected altitude of the sun, θ be the latitude of the place, and δ be the declination of the sun at the time of observation. The declination of the sun can be computed from the local mean time and the declination at Greenwich is given in the nautical almanac.

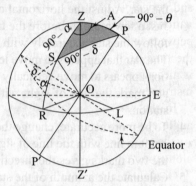

Fig. 21.19 Determining the azimuth

In the astronomical triangle ZPS, the three sides are known. The azimuth angle PZS can be computed from the formula

$$\cos A = (\sin\delta - \sin\alpha \sin\theta)/(\cos\theta \cos\alpha)$$

or from

$$\tan(A/2) = \sqrt{\sin(s - ZP)\sin(s - ZS)/[\sin s \sin(s - PS)]}$$

where s is the semi-perimeter. Angle LOR is the angle measured between the sun and the survey line. Then the azimuth of the line is

$$\angle EOL = A - \angle LOR$$

Observation of circumpolar star A circumpolar star (or any other star) is observed similarly, following the same procedure. The star should preferably be near the prime vertical so that there is enough time for making face-left and face-right observations.

Observation of a circumpolar star at elongation

At elongation, a circumpolar star is farthest from the pole. The elongation can be east or west (Fig. 21.20). In this position, $\angle ZSP$ of the astronomical triangle is a right angle. The vertical circle in which the telescope moves (through the observer's zenith) is tangential to the path of the star at elongation. The star moves vertically as seen by the observer. This is found to be a very favourable position for observing the star.

The time of elongation of the star can be computed if the declination of the star and the latitude of the place from the astronomical triangle ZPS are known. As $\angle ZSP$ is a right angle, the hour angle of the star can be computed as

$$\cos H = \tan\theta / \tan\delta$$

where H is the hour angle, δ the declination, and θ is the latitude of the place. The hour angle is then converted to time and added to (for western elongation) or subtracted from (for eastern elongation) the right ascension of the star. The time so obtained is the local sidereal time, which can be converted to local mean time to know the time of elongation.

Polaris is the most favoured star for such observations, as it is very bright and very close to the pole. The observation involves measuring the angle between the survey line and the star at the time of its elongation. The angle is measured with both faces to reduce instrumental errors.

In the astronomical triangle, the angle at s is the azimuth of the star. As the angle at S is a right angle at the elongation, A can be calculated as

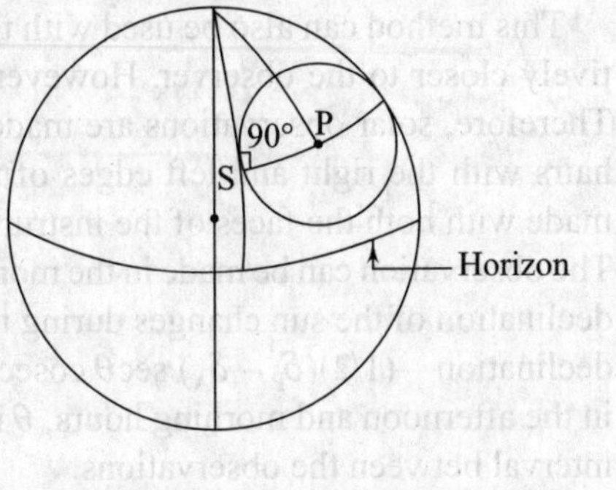

Fig. 21.20 Circumpolar star at elongation

$$\sin A = \cos(\text{declination})/\cos(\text{latitude}) = \cos\delta/\cos\theta$$

If the azimuth of the star is known, the azimuth of the line can be calculated.

Equal altitudes of the sun or circumpolar star

In this method, the horizontal angle between the star and the survey line is measured when the star reaches the same altitude. In Fig. 21.21, O is the observer's position and OL is the line whose azimuth is to be observed. The procedure is as follows.

1. Set up the instrument at O, observe the star at a convenient position, and clamp the vertical circle.
2. Read the horizontal angle between the line and the star from the horizontal circle.
3. Follow the star with the telescope, with the vertical circle reading remaining the same.
4. When the star reaches the same altitude again, note the horizontal angle between the line and the star.

The two positions of the star will be on either side of the meridian. However, the same-altitude positions of the star may be on the same side of the line OL [Fig. 21.21(a)] or on different sides [Fig. 21.21(b)]. Let A_1 and A_2 be the angle measured between the line OL and the star in the two positions. If A is the azimuth of the line and the two positions of the star are on the same side of OL, then

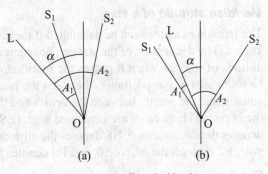

Fig. 21.21 Equal altitudes

Azimuth $A = A_1 + (A_2 - A_1)/2 = (A_1 + A_2)/2$

The azimuth in this case is the average of the observed angles. When the two positions of the star are on different sides of OL,

Azimuth $A = (A_1 + A_2)/2 - A_1 = (A_2 - A_1)/2$

The azimuth here is half the diference between the two angles.

This method can also be used with the observation of the sun, as the sun is relatively closer to the observer. However, the centre of the sun cannot be bisected. Therefore, solar observations are made by making the sun tangential to the cross hairs with the right and left edges of the circle. The observations are preferably made with both the faces of the instrument. The time of observation is also noted. The observation can be made in the morning and afternoon hours and averaged. The declination of the sun changes during these observations. A correction is made for declination—$(1/2)(\delta_P - \delta_A) \sec\theta \csc(T/2)$, where δ_P and δ_A are the declinations in the afternoon and morning hours, θ is the latitude of the place, and T is the time interval between the observations.

21.8.3   Determination of Latitude

The latitude of a place is the distance along the meridian from the equatorial plane. Latitude and longitude are the coordinates that fix the position of a point on the surface of the earth. If the elevation of the point is known, the point is completely determined. There are many methods to determine the latitude of a place. Some of these methods are described in this section.

Observation of a circumpolar star at upper and lower culmination

A circumpolar star such as Polaris is commonly used in this method. At culmination, a star is on the meridian. The procedure consists of observing the altitudes of a star at upper and lower culminations. The altitude of the pole is the mean of these two altitudes. The latitude of the pole is equal to the altitude of the pole. If the altitudes of the circumpolar star are α_L and α_U at lower and upper culminations, respectively, then the altitude of the pole is $(\alpha_L + \alpha_U)/2$. This is equal to the latitude of the place. The disadvantage of the method is that the altitudes are observed at an interval of nearly 12 hours. If one observation is made at night, the other observation has to be made during the day, which requires a high-power telescope. This is difficult or impossible with a small instrument. This method, therefore, finds little use.

Meridian altitude of a star

The latitude of a place can be calculated if the altitude and declination of a star are known. The declination of the star can be obtained from the nautical almanac. The altitude of the star when it passes the meridian is observed. There can, however, be four cases of star positions—between the horizon and the equator, between the equator and the zenith, between the zenith and the pole, and between the pole and the horizon. These cases are explained here. (ZS denotes the zenith distance z, ES denotes the declination δ, NS denotes the altitude α, NP denotes the altitude of the pole, EZ denotes the latitude θ, and PS denotes the co-declination.)

Star between horizon and equator   In this case [Fig. 21.22(a)],

$$\theta = ZS - ES = z - \delta$$

Star between equator and zenith In this case [Fig. 12.22(b)],

$$EZ = \theta = ES + SZ = z + \delta$$

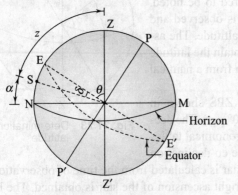

(a) Between horizon and equator

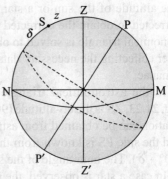

(b) Between equator and zenith

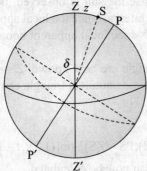

(c) Between zenith and pole

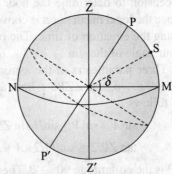

(d) Between pole and equator

Fig. 21.22 Meridian altitude of a star

Star between zenith and pole In this case [Fig. 21.22(c)],

$$\theta = \delta - z$$

Star between pole and horizon In this case [Fig. 21.22(d)],

$$\theta = NP$$
$$z = 90° - \alpha$$
$$NS = 90° - z$$
$$PS = 90° - \delta$$
$$\theta = NP = NS + PS = 90° - z + 90° - \delta = 180° - (z + \delta)$$

The observation can also be that of the sun. The altitude of the sun is observed as it crosses the meridian. The observed altitude is corrected for all errors, instrumental, refraction, parallax, and semidiameter. The corrected altitude is used to calculate the latitude as explained above. The declination of the sun can be obtained if the longitude of the place is known.

Extra-meridian altitude of the sun or a star

The latitude can also be calculated by measuring the altitude of a star or the sun at any position. The time of observation is required to be noted. The altitude of the sun or a star is observed and corrected to obtain the corrected altitude. The astronomical triangle is solved to obtain the latitude after collecting the necessary data from a nautical almanac.

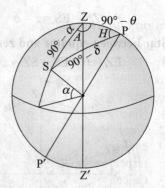

Fig. 21.23 Determination of latitude

In the astronomical triangle ZPS shown in Fig. 21.23, ZS is the co-altitude ($90° - \alpha$). The declination can be obtained from astronomical tables and the side PS is known from the co-declination ($90° - \delta$). The hour angle of the star is calculated from the time of observation.

In case a star is observed, the right ascension of the star is obtained. The local mean time of this observation is converted to sidereal time and added to the right ascension to determine the hour angle. In case the sun is observed, the interval since the local mean noon is converted to the interval since the local apparent time using the equation of time. The time interval since local apparent noon gives the hour angle of the sun.

Three parameters of the astronomical triangle are now known. ZS = $90° - \alpha$, PS = $90° - \delta$, and $\angle$ZPS = hour angle H. The astronomical triangle can be solved as follows:

$$\sin A = \sin \text{PS} \sin H / \sin \text{ZS}$$

$$\tan \text{ZP}/2 = \sin[(1/2)(A + H)] \tan[(1/2)(\text{PS} - \text{ZS})]/\sin[(1/2)(A - H)]$$

ZP is the co-latitude $90° - \theta$. The latitude θ can now be calculated.

Extra-meridian observation of Polaris

The altitude of Polaris is observed when it is out of meridian at any convenient time. Face-left and face-right observations must be taken. The time of observation is also noted. As Polaris is very close to the pole, the latitude is obtained by a special formula:

$$\text{Latitude } \theta = \alpha - \text{PS} \cos H + (1/2)\sin 1'' (\text{PS} - H)^2 \tan\alpha$$

where α is the corrected altitude, PS = $90° - \delta$ is the polar distance, and H is the hour angle.

21.8.4 Determination of Longitude

The longitude of a place, as we have seen, can be east or west of the standard zero longitude. Longitude is an important parameter with reference to time, as the difference in longitude gives the difference in time, be it mean, apparent, or sidereal times. Longitude can be determined very accurately by triangulation but it is an expensive and complex process. Two other methods of determining the longitude are explained below, though both are not very accurate.

Using chronometers

A chronometer is like a clock but is heavier and more accurate. The chronometer time is affected by temperature and other inherent factors. The amount of time the chronometer gains or loses with time is known as the *rate*. The procedure for determining the longitude with a chronometer is as follows.

1. Determine the local mean time.
2. Simultaneously, obtain a chronometer reading. The chronometer should be compared with the Greenwich mean time by noting its error and rate with respect to the GMT.
3. Calculate the GMT. The difference in time gives the longitude of the place.

The local times of two places can also be compared if chronometers with known rates are available. Transporting the chronometer from place to place is a source of error for comparison, as the rate during transportation is unknown.

Using wireless signals

Wireless signalling is a very accurate method. The local mean time at the place is determined by astronomical observations. The time signals can be obtained for the Greenwich mean time by wireless from many stations. These can be compared and the longitude determined from the difference of times. These days, with satellite signals available, the time method of determining the longitude has become very accurate.

21.8.5 Determination of Time

While making observations for the altitude or meridian, it is necessary to have accurate knowledge of time. While time can be maintained and recorded from a clock or chronometer, any error in the time shown by the clock can be detected from astrological observations. Time can be determined by many methods. Some of the methods are described below.

Extra-meridian observation of a star or the sun

In this method, the altitude of a star is obtained very accurately. As time has to be determined, the observations are taken using both faces, and the time of observation is also noted from a chronometer. For greater accuracy, several altitude values are obtained in a short span of time using both faces.
The mean of these observed values is taken as the correct altitude. Appropriate corrections are also applied to the observations. Better results are obtained when the star is on the prime vertical.

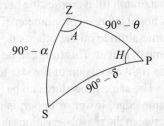

Knowing the corrected altitude α, declination δ, and latitude θ of the place, the astronomical triangle (Fig. 21.24) can be solved to get the hour angle of the star. The hour angle is calculaed from

Fig. 21.24 Determination of time

$$\sin\alpha = \sin\theta \sin\delta + \cos\theta \cos\delta \cos H$$

or from

$$\tan(H/2) = \sqrt{\sin(s - \text{ZP})\sin(s - \text{PS})/\sin s \sin(s - \text{ZS})}$$

If the observations are taken when the star is on the prime vertical, then the hour angle is calculated from $\cos H = \tan\delta/\tan\theta$. The hour angle is converted to time and added to (for a western star) or subtracted from (for an eastern position) the right ascension of the star to get the local sidereal time. The local mean time can then be obtained.

A similar procedure is adopted for solar observations. As the centre of the sun cannot be bisected, the lower or upper edge of the sun is made tangential to the cross hair. Both face observations are taken, along with the time of observation. For greater accuracy, several observations are taken, half with one face and half with the other face. All corrections including the correction for semidiameter are applied. The corrected altitude is then obtained as the mean of all the observations.

Meridian transit of a star or the sun

This method can be used if the direction of the meridian at the point of observation is known. The transit of a star is followed with a theodolite and the exact time at which it crosses the meridian is noted with an accurate chronometer. The right ascension of the star gives the local sidereal time at transit, as the hour angle of the star is zero at that position. The local mean time can be calculated from the sidereal time.

In the case of solar observations, the times at which the sun's edge crosses the vertical cross hair (as it is not possible for the centre to cross) are noted. The mean of the two times gives the mean time of crossing. This time corresponds to the local apparent time as per the clock. The local mean time can then be calculated.

Equal altitudes of a star or the sun

Another method is to observe a star or the sun when it has equal altitudes on either side of the meridian. The essential principle is to observe a star when it achieves the same altitude on either side of the meridian and note the time required for this transit. The method is simple, accurate, and particularly useful when the meridian direction is not known.

A theodolite is used to make the observations. The instrument is set up and accurately levelled at a station. Knowing the altitude of the star, the vertical circle of the theodolite is set to read the altitude. The star is tracked using the tangent screw in azimuth till it crosses the horizontal cross hair (or reaches the altitude). The time is noted using a chronometer. The procedure is repeated when the star reaches the same altitude on the other side of the meridian. The mean of the two times is the time of transit. This time can be compared with the local mean time computed from the right ascension of the star to find the error in the chronometer.

The same procedure can be adopted in the case of observations of the sun. The same edge, lower or upper, is observed in the case of the sun. The time interval between the two observations is large and this changes the condition of observation. It must be noted that the exact altitude is not being measured; only the time for the star to achieve the same altitude in two positions on either side of the meridian is noted. One does not require any other data such as declination of the star or latitude of the place.

21.9   Corrections to Astronomical Observations

Astronomical observations need to be corrected for various factors—instrumental and observational errors.

Correction for refraction

Refraction of light rays due to varying air densities along the depth makes a celestial body appear higher than it actually is. Thus, in Fig. 21.25(a), to an observer at O, light rays from the star S appear to come from S'. The observer thus estimates a higher altitude than what it actually is. The correction for refraction is thus subtracted from the observed altitude. This correction does not depend upon distance. It can be determined from refraction tables or can be calculated from the following formulae:

$$\text{Correction for refraction} = 58'' \cot \alpha$$

where a is the apparent altitude or

$$\text{Correction for refraction} = 58'' \cot z$$

where z is the zenith distance.

Correction for parallax

The correction for parallax has to be made in the case of observations made to the sun only, as it is the nearest star. When observations are made to other stellar bodies, this correction can be ignored as being small. This correction is required because the observations are made from a point on the surface of the earth, whereas they are supposed to be made from the centre of the earth. The sun's parallax in altitude is the angle subtended at the centre of the sun by the line joining the centre of the earth and the position of the observer.

Referring to Fig. 21.25(b), S is the position of the sun observed from O. O' is the centre of the earth. S' is the position of the sun on the horizon. The angle measured as the altitude is $\angle SOS'$. O'H is the true horizon and the true altitude of the sun is given by $\angle SO'H$. $\angle OS'O'$ is the *horizontal parallax* P_h. P_a is the *parallax in altitude*. P_h is the angle subtended by the line OO'' at the centre of the sun when the sun is on the horizon. Horizontal parallax varies inversely as the distance of the sun from the earth and is maximum in December and minimum in July. The values of horizontal parallax are listed in the nautical almanac. The average value of P_h is 8.8″. Applying the sine rule to triangle O'OS,

$$OO'/\sin OSO' = O'S/\sin O'OS$$

or

$$\sin O'SO = [O'O/O'S] \times \sin O'OS$$

$$\sin O'OS = \cos\alpha$$

where α is the observed altitude. Also,

$$O'O/O'S = O'O/O'S' = \sin OS'O' = \sin P_h$$

Since the angles involved are very small, we can write

$$P_a = P_h \cos\alpha = 8.8'' \cos\alpha$$

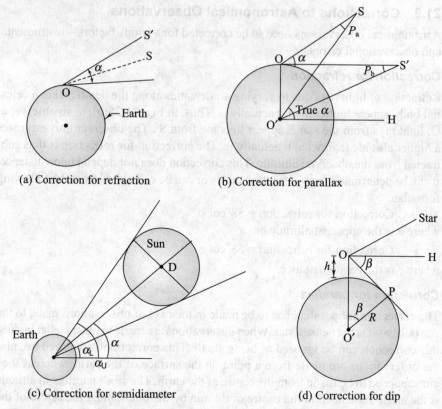

(a) Correction for refraction   (b) Correction for parallax

(c) Correction for semidiameter   (d) Correction for dip

Fig. 21.25 Observational corrections

Correction for semidiameter

While observing the sun, it is not possible to bisect the centre of the body. In Fig. 21.25(c), the semidiameter is one-half of the angle subtended at the centre of the earth by the diameter of the sun. As the distance of the earth from the sun varies over the year, the semidiameter does not remain constant and varies from 15′ 45″ to 16′ 18″. The values are listed in the nautical almanac.

The sun is observed by touching the upper or lower end of its diameter with the cross hairs, as the centre of the sun cannot be observed. The reading so obtained is corrected for semidiameter by subtracting the semidiameter if the upper end is made tangential and adding the semidiameter if the lower end is made tangential.

Correction for dip

Correction for dip is applied when the observations are taken from sea using a sextant. As shown in Fig. 21.25(d), such observations are made from the visible horizon and not from the sensible horizon. The angle between the visible horizon and the sensible horizon is known as dip. From the figure, if β is the dip angle, then

$$\tan\beta = \text{OP/O}'\text{P} = \frac{\sqrt{(R+h)^2 - R^2}}{R} = \frac{\sqrt{h^2 + 2hR}}{R}$$

If h is taken to be small compared to R, then

$$\tan\beta = \sqrt{\frac{2h}{R}}$$

Correction for dip is always negative. It varies with the height of the observer above the sea. The values for these corrections can be found in mathematical tables.

Example 21.18 Determine the azimuth of the line AB (Fig. 21.26), the altitude of the star, and the local mean time if the star is observed at western elongation from a place of latitude 48° 40′ N and longitude 51° 30′ W. The declination of the star was 59° 42′ and right ascension of the star was found to be 9h 42m. The GST of GMN was 7h 32m 10s and the angle between the line and the star measured to the right was 102° 35′ 51″.

Solution As the star was observed during elongation, ∠ZSP of the astronomical triangle (Fig. 21.26) is a right angle. Then,

$$\sin\alpha = \sin\theta/\sin\delta, \quad \sin A = \cos\delta/\cos\theta, \quad \cos H = \tan\theta/\tan\delta$$

In these expressions, α, A, and H represent the altitude, azimuth, and hour angle of the star, respectively. From these equations,

$$\sin\alpha = \sin 48° 40′/\sin 59° 42′, \quad \alpha = 60° 42′ 17″$$

Therefore, the altitude of the star is 60° 42′ 17″.

$$\sin A = \cos 59° 42′/\cos 48° 40′, \quad A = 49° 48′ 42″$$
$$\cos H = \tan 48° 40′/\tan 59° 42′ = 48° 21′ 56″$$

Azimuth of the line AB = azimuth of the star + mean horizontal angle
between the star and the line

$$= 49° 48′ 42″ + 102° 35′ 51″ = 152° 24′ 33″$$

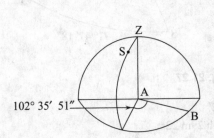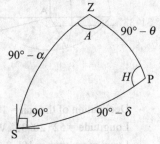

Fig. 21.26

We find the local mean time of observation as follows:

Longitude = 51° 40′ W = 3h 14m

Acceleration @ 9.8596 sec/hr = 31.8 sec

Now,

GST of GMN = 7h 32m 10s

LST of LMN = GST of GMN + acceleration
= 7h 32m 10s + 31.8s = 7h 32m 41.8s

Hour angle H = 44° 20′ 22″ = 2h 57m 21s

LST = right ascension + hour angle = 9h 42m + 2h 57m 21s
= 12h 39m 21s

Sidereal interval = LST − LST of LMN
= 12h 39m 21s − 7h 32m 41.8s = 5h 6m 39.2s

Retardation @ 9.8296 sec/hr = 50.2 sec

Mean time interval = sidereal interval – retardation

= 5h 6m 39.2s – 50.2s = 5h 5m 49s

As this is the mean time interval from LMN,

Local mean time = 17h 5m 49s or 5h 5m 49s p.m.

Example 21.19 Solar observations were made from a place of latitude 55° 36′ 15″ N and longitude 52° 20′ 35″ W. The horizontal angle between the reference line AB and the sun was 68° 31′ 33″. The mean altitude (of the upper edge) was observed to be 44° 28′ at 10 a.m. The sun's declination was found to be 18° 12′ at noon, increasing at 28.2″ per hour. The corrections for semidiameter, parallax, and refraction were found to be 15′ 32.8″, 8.8″cos(altitude), and 57″ cot(altitude), respectively. Find the azimuth of line AB (Fig. 21.27).

Solution We first find the corrected altitude:

Correction for refraction = 57″cot 44° 28′ = 58.07″ (–)

Correction for parallax = 8.8″cos 44° 28′ = 6.28 (+)

Altitude corrected for refraction and parallax = 44° 28′ – 58.07″ + 6.28″

= 44° 27′ 8.21″

Altitude corrected for semidiameter = 44° 27′ 8.21″ – 15′ 32.8″

= 44° 11′ 35.41″

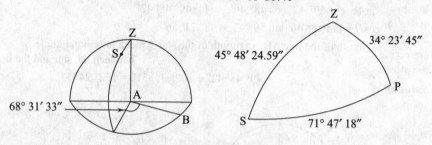

Fig. 21.27

Declination of the sun = 18° 12′

Longitude = 52° 20′ 35″ W = 3h 29m 22s

Local mean time = 10h 0m 0s

GMT = LMT + west longitude = 10h + 3h 29m 22s = 13h 29m 22s

Increase for 1h 29m 22s @ 28.2 sec/hr = 42 sec

Sun's declination at noon = 18° 12′ N

Sun's declination at GMN = 18° 12′ + 42″ = 18° 12′ 42″ N

In the astronomical triangle ZPS,

$$ZS = 90° – \alpha = 90° – 44° 11′ 35.41″ = 45° 48′ 24.59″$$

$$PS = 90° – \delta = 90° – 18° 12′ 42″ = 71° 47′ 18″$$

$$ZP = 90° – \theta = 90° – 55° 36′ 15″ = 34° 23′ 45″$$

Azimuth A can be obtained as follows:

$$\cos PS = \cos ZS \cos ZP + \sin ZS \sin ZP \cos A$$

$$\cos A = \sin\delta/\cos\alpha \cos\theta – \tan\alpha \tan\theta$$

$$= \sin 18° 12′ 42″/\cos 44° 11′ 35.42″ \cos 55° 36′ 15″$$

$$- \tan 55° \ 36' \ 15'' \ \tan 44° \ 11' \ 35.41''$$

Azimuth $A = 130° \ 25' \ 15.8''$ E

Therefore,

Azimuth of line AB $= 130° \ 25' \ 15.8'' - 68° \ 31' \ 33'' = 61° \ 53' \ 42.8''$

Example 21.20 Find the latitude of the place of observation if the meridian altitude of a star was found to be $69° \ 32' \ 31.8''$ and its declination was found to be $48° \ 26' \ 38.31''$. The position of the star was between the zenith and the equator.

Solution Observed altitude $\alpha = 69° \ 32' \ 31.8''$

Correction for refraction $= 57'' \cot 69° \ 32' \ 31.8'' = 21.26''$

Corrected altitude $\alpha = 69° \ 32' \ 31.8'' - 21.26'' = 69° \ 32' \ 10.54''$

Zenith distance $= 90° - \alpha = 90° - 69° \ 32' \ 10.54'' = 20° \ 27' \ 49.46''$

As the star lies between the zenith and the equator,

Latitude θ = zenith distance $+ \delta$

$$= 20° \ 27' \ 49.46'' + 48° \ 26' \ 38.31'' = 68° \ 54' \ 27.77''$$

Example 21.21 Find the latitude of a place from the following observations with respect to the sun. Mean altitude of the lower limb was $48° \ 12' \ 36''$. Declination was $26° \ 36' \ 18.4''$ S at 0 hr, increasing at the rate of $11.2''$ per hour. The equation of time was 8m 31.2s, decreasing at the rate of $1.3''$ per hour. The semidiameter correction for the sun was $15' \ 21.2''$. The correction for refraction was $57'' \cot \alpha$ and that for parallax was $8.8'' \cos \alpha$. The longitude of the place of observation was $106° \ 10' \ 40''$ and the time of observation was 14h 45m 30s.

Solution Longitude $= 106° \ 10' \ 40'' = $ 7h 4m 43s in terms of time

We first correct the observed altitude:

Refraction correction $= 57'' \cot 48° \ 12' \ 36'' = 50.94''$

Parallax correction $= 8.8'' \cos 48° \ 12' \ 36'' = 5.86''$

Net correction $= 50.94'' - 5.86'' = 45.06'' (-)$

Corrected altitude $= 48° \ 12' \ 36'' - 45.06'' = 48° \ 11' \ 50.92''$

The semidiameter is added, as the lower edge was observed:

Corrected altitude $= 48° \ 11' \ 50.92'' + 15' \ 21.2'' = 48° \ 27' \ 12.12'$

Declination at 0 hr $= 26° \ 36' \ 18.4''$

Increase @ 11.2 sec for 7h 4m 43s $= 4' \ 4.57''$

Declination at the time of observation $= 26° \ 36' \ 18.4'' + 4' \ 4.57''$

$$= 26° \ 40' \ 22.97''$$

To find the hour angle,

Local mean time $=$ 14h 45m 30s

GMT = LMT + longitude $=$ 14h 45m 30s + 7h 4m 43s $=$ 21h 50m 13s

Equation of time $=$ 8m 31.2s

Variation for 21h 50m 13s $=$ 28.38s

Interval from local mean noon $=$ 2h 45m 30s

Adding the equation of time,

Hour angle $=$ 2h 45m 30s + 8m 31.2s

$$- \ 28.38s$$

$$= \text{2h 53m 32.82s} \quad \text{or} \quad 43° \ 23' \ 12'' \text{ in angular measure}$$

In the astronomical triangle ZPS (Fig. 21.28),

$$ZS = 90° - 48° 27' 12.12'' = 41° 32' 47.88''$$

$$PS = 90° - 26° 40, 22.97'' = 63° 19' 37''$$

$$\angle ZPS = H = 43° 23' 12.3''$$

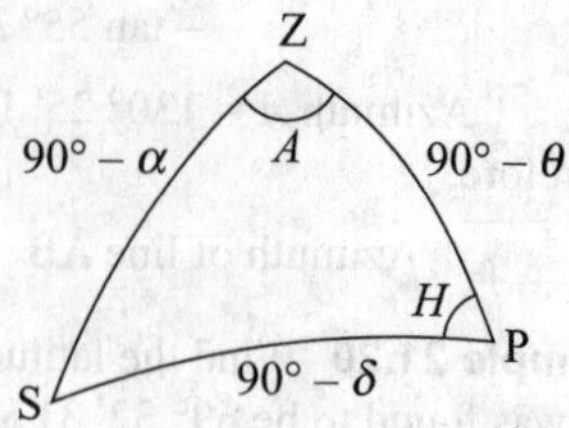

Angle PZS = A, which can be calculated from the sine rule:

$$\sin A = \sin 43° 23' 12.3'' \sin 63° 19' 37''/\sin 41° 32' 48'' = 0.9255$$

$$A = 67° 44' 38''$$

Fig. 21.28 Astronomical triangle (Example 21.21)

Side ZP is computed from the following formula:

$$\tan = \left(\frac{ZP}{2}\right) = \frac{\sin\frac{1}{2}(A+H)}{\sin\frac{1}{2}(PS-ZS)} \times \tan\frac{1}{2}(A-H)$$

$$A + H = 67° 44' 38'' + 43° 23' 12.3'' = 110° 7' 50.3''$$

$$A - H = 67° 44' 38'' - 43° 23' 12.3'' = 24° 21' 25.7''$$

$$PS - ZS = 63° 19' 37'' - 41° 32' 48'' = 21° 46' 49''$$

$$\tan(ZP/2) = \sin 55° 3' 55'' \tan 10° 53' 24.5''/\sin 12° 10' 42.85'' = 0.7476$$

$$ZP = 73° 34' 1.7''$$

$$\text{Latitude} = 90° - 73° 34' 1.7'' = 16° 25' 58.3'' \text{ N}$$

Summary

Any point on the surface of the earth can be located by its *latitude* and *longitude*. The earth is considered a sphere, and a great circle passing through the poles and a particular location is known as the *meridian* of that location. The great circle perpendicular to the polar axis is the *equator*. The latitude is measured from the equatorial plane along the meridian and the longitude is measured with reference to some standard meridian.

The earth has two types of motion—one about its own axis and one around the sun. The first motion causes day and night while the second causes the seasons. The earth revolves around the sun in an elliptical orbit with the sun at one of its foci. The actual motion of the earth from west to east makes the sun and other celestial bodies appear to move from east to west. The sun appears to make a complete revolution in one year along a path called the *ecliptic*, which is inclined to the equatorial plane at 23° 27'.

The *celestial sphere* is an imaginary sphere of infinite radius with the earth at its centre and all the stars lying on its surface. Angles made by the stars, rather than distances, are important in the celestial sphere. Considering an observer on the surface of the earth at some latitude and longitude, a vertical line drawn from the observer's position cuts the celestial sphere at two points known as the *zenith* and the *nadir*. The earth's equatorial plane when extended cuts the *celestial sphere* at the celestial equator. When the polar axis of the earth is extended, it cuts the celestial sphere at two points known as *celestial poles*. Azimuth, altitude, declination, and hour angle are the parameters of a celestial body. The position of a star is given by coordinates such as azimuth–altitude, declination–hour angle, or declination–right ascension. The astronomical triangle is a spherical triangle and can be solved using formulae from spherical trigonometry. It is formed by three points—the zenith (Z), a pole (P), and a celestial body (S).

Time is an important parameter in astronomy. Four types of times are used—sidereal time, apparent solar time, mean solar time, and standard time. The meridian at Greenwich is the standard meridian used as a reference for meridians of other places. The application of astronomy in surveying is to find the true meridian at a place, the azimuth of a survey line, and the latitude and longitude of a place. These are done by stellar observations either during the day by observing the sun or during the night by observing some star such as Polaris.

Exercises

Multiple-Choice Questions

1. A great circle considering the earth as a true sphere is
   (a) a circle obtained by a cutting plane passing through the centre
   (b) a circle obtained by a cutting plane passing through one of the poles
   (c) a circle obtained by any cutting plane
   (d) the equatorial circle
2. A small circle considering earth as a sphere is
   (a) obtained by a cutting plane passing through the centre
   (b) obtained by a cutting plane not passing through the centre
   (c) any circle other than the equatorial circle
   (d) a circle obtained by a cutting plane passing through both the poles
3. Poles are the points obtained by the intersection of a line
   (a) perpendicular to the equatorial plane
   (b) perpendicular to any great circle
   (c) perpendicular to a small circle
   (d) in the equatorial plane with the surface
4. The latitude of a point on the surface of the earth is
   (a) the angle between the standard meridian and the meridian through the point
   (b) the elevation of the point above a selected datum
   (c) the distance to the point along the meridian from the equatorial circle
   (d) the angle formed at the centre by the arc joining the point and a point on the equatorial circle along the meridian
5. A meridian circle is
   (a) a small circle between any point and the standard meridian
   (b) a circle formed by a cutting plane perpendicular to the line joining the poles
   (c) the great circle passing through the poles and any point
   (d) the equatorial circle or circle parallel to that
6. The longitude of a point on the surface of the earth is
   (a) the distance along the equator to a meridian through the point
   (b) the angle between the standard meridian and the meridian through the point
   (c) the radial distance to the point
   (d) the distance from the pole to the point along the meridian
7. A parallel of latitude is
   (a) a great circle through a given point
   (b) a circle passing through the point and the centre
   (c) a meridian circle through the point
   (d) a small circle the points on which have the same latitude
8. A celestial sphere has
   (a) the sun as the centre
   (b) the moon as the centre
   (c) the earth as the centre
   (d) the centre at the earth's centre and having a finite diameter

9. Zenith is a point on the
   (a) celestial sphere
   (b) moon's surface
   (c) sun's surface
   (d) earth's surface

10. In the altitude–azimuth system of coordinates of a celestial body, azimuth is
    (a) longitude of the body
    (b) latitude of the body
    (c) the angle between the observer's meridian and the vertical circle through the body
    (d) the angle between the horizon and the line joining the centre to the body

11. The co-latitude of a celestial body is
    (a) angular distance of a heavenly body from the zenith
    (b) angular distance of a heavenly body from the observer's meridian
    (c) remaining angle when latitude is subtracted from 180 degrees
    (d) the azimuth angle of the heavenly body

12. First point of Aries refers to the position of the sun in the month of
    (a) March   (b) June   (c) September   (d) December

13. The astronomical triangle is formed by
    (a) the arcs of any three great circles
    (b) the meridian circle, equatorial circle and any vertical circle
    (c) the pole, zenith and a celestial body on the celestial sphere
    (d) the pole, the observer and parallel of latitude on the earth's surface

14. A sidereal day is the time taken by
    (a) the earth to move around the sun once
    (b) the moon to move around the earth once
    (c) the first point of Aries to cross the same meridian twice
    (d) the earth to move around its own axis once

15. An angular measure of 36° 42′ 30″ converted into time measure is equal to
    (a) 3 hr 20 min 40 sec   (c) 2 hr 10 min 30 sec
    (b) 3 hr 10 min 10 sec   (d) 2 hr 26 min 50 sec

16. A time measure of 12 hrs 30 mins 20 secs when converted to angular measure will be
    (a) 187° 35′ 00″   (b) 165° 25′ 30″   (c) 135° 35′ 20″   (d) 120° 25′ 30″

17. If the local time at a place of longitude 60° 40′ E is 8 hours 30 minutes, then the Greenwich Mean Time is
    (a) 12 h 32 m 40 s   (b) 7 h 27 m 20 s   (c) 4 h 27 m 40 s   (d) 3 h 20 m 40 s

18. When the Greenwich Mean Time is 10 a.m., the local time at a place of longitude 63° 45′ W is
    (a) 4 h 15 m (p.m.) (b) 2 h 15 m (p.m.) (c) 4 h 45 m (a.m.) (d) 5h 45 m (a.m.)

19. If the Greenwich Mean Time is 8 a.m., the local time at a place of longitude 45° 30′ E is
    (a) 11 h 2 m   (b) 4 h 58 m   (c) 3 h 12 m   (d) 14 h 12 m

20. If the GMT is 10 p.m., the local time at Nagpur is, given that the standard meridian in India is 82° 30′ E,
    (a) 16 h 30 m   (b) 18 h 30 m
    (c) 3 h 30 m (next day)   (d) 1 h 30 m (next day)

21. If the longitude of a place is 80° 30′ W, the local time at that place when the GMT is 11 a.m. is
    (a) 4: 22 p.m.   (b) 2: 32 p.m.   (c) 6:38 a.m.   (d) 5:38 a.m.

22. Correction for refraction in astronomical observation is taken to be equal to, where α is the apparent altitude,
    (a) 58″ sin α (b) 58″ cot α (c) 58″ tan α (d) 58″ cos α

23. While observing the sun, semidiameter refers to
    (a) one-half of the angle at the centre of the sun by the diameter of the earth
    (b) one-half of the angle subtended at the centre of the earth by the diameter of the sun
    (c) angle subtended at the centre of the earth by the diameter of the sun
    (d) angle subtended at the centre of the sun by the diameter of the earth

Review Questions

1. Draw a neat sketch to show the latitude and longitude of a place on the surface of the earth.
2. Draw a sketch of the celestial sphere and show the following—zenith, nadir, poles, zenith distance.
3. Define and show the following in a sketch-horizon, declination, hour angle, meridian, altitude, right ascension.
4. Briefly describe the coordinate systems used to locate celestial bodies with neat sketches.
5. Explain the time systems employed in astronomy.
6. Explain the method used to convert sidereal time to mean time.
7. Explain any three methods used to determine the azimuth of a celestial body. Explain how the azimuth of a survey line can be determined from the azimuth of a celestial body.
8. Explain any three methods used to determine the true meridian at a place.
9. Explain the methods used to determine the latitude of a place.
10. Explain the methods used to determine the longitude of a place.
11. Briefly explain, with neat sketches, the corrections to be made to astronomical observations.

Problems

1. Calculate the following. (a) Difference in latitude between two points A and B if the latitudes are as follows: (i) 72° 36′ N at A, 38° 24′ N at B; (ii) 22° 48′N at A, 32° 32′ S at B. (b) Difference in longitude between two points P and Q if the longitudes are as follows: (i) 82° 30′ E at P, 64° 44′ at Q; (ii) 32° 48′ E at P, 22° 36′ W at Q.
2. Find the shortest distance between two points P and Q if their latitudes are 22° 10′ N and 34° N and longitudes are 32° 30′ E and 64° 48′ E, respectively.
3. Find the shortest distance between points A and B if their latitudes are 34° 40′ N and 48° 30′ N and longitudes are 10° 15′ E and 30° 30′ W, respectively.
4. Find the zenith distance at upper culmination of a star if (a) latitude = 48° 15′ N, declination = 26° 15′ N and (b) latitude = 24° 30′ N, declination = 10° 15′ S.
5. Find the zenith distance at the lower culmination of stars if (a) latitude = 48° 15′ N, declination = 54° 30′ N and (b) latitude = 39° 30′ S, declination 51° 15′ N.
6. If the declination of a star is 36° 30′ N and the upper culmination is at the zenith, find the altitude of the star at lower culmination.
7. If the altitudes of a star at the culminations are 68° 30′ and 22° 40′, both on the north of the zenith, find the declination of the star and the latitude of the place.
8. Find the azimuth and altitude of a star if the latitude of the place = 42° N, hour angle = 48°, and declination = 26°.
9. Find the azimuth and altitude of a star if latitude = 45° N, hour angle = 19h 30m, and declination = 22° 15′ S.
10. Find the hour angle and declination of a star if the latitude of the place = 45° N, azimuth angle = 35° W, and altitude = 32° 15′ N.
11. Find the hour angle and declination of a star if latitude = 48° N, azimuth = 98° 30′ E, and zenith distance = 58° 30′.

12. Calculate the hour angle and altitude of the sun at sunset if the latitude = 52° N and declination = 28° 30′ N.

13. Calculate the hour angle and altitude of the sun at sunrise if the latitude = 50° N and declination = 14° 20′ S.

14. (a) Find the local mean time at a place of longitude if the standard time is 6h 10m and the standard meridian is 82° 30′. (b) Find the standard time corresponding to the LMT of 10h 12m 30s at longitude 60° 15′ E.

15. Calculate the following.
    (a) Sidereal time interval corresponding to mean time interval of 6h 36m.
    (b) Mean time interval corresponding to sidereal time interval of 10h 12m 30s.
    (c) Angles corresponding to times of 10h 12m and 23h 36m.
    (d) Time interval corresponding to angles of 78° 30′ and 240° 10′.

16. (a) Find the GMT corresponding to the LMT of 10h 12m at longitude 42° 36′ E. (b) Find the LMT at longitude 72° 10′ if the GMT at an instant is 8h 12m.

17. Find the local apparent time of an observation taken at local mean time 10h 30m at longitude 78° 30′ E. The equation of time at GMN is 3m 4.52s subtractive from the apparent time and decreasing at the rate of 0.3 sec/hr.

18. Find the LMT if the LAT is 16h 30m. The equation of time at GMN is 4m 3.8s and increasing at the rate of 0.2 sec/hr. The longitude is 46° 12′ E.

19. Find the LMT at longitude 82° 42′ W corresponding to the LST 16h 32m if the GST of GMN is 6h 30m 32s.

20. Find the local sidereal time corresponding to the LMT of 19h 30m at longitude 68° 12′ E if the GST of GSN is 4h 30m.

21. Find the azimuth of a line AB from the following observations of a star at western elongation: latitude of A = 48° 30′, longitude of A = 52° 20′, declination = 59° 35′, right ascension = 9h 39m, mean horizontal angle between the line and the star = 102° 30′ 15″.

22. Find the latitude of a particular place of observation from the following data: mean meridian altitude of the star = 69° 32′ 30″, declination = 43° 32′ 15.8″. The star lies between the zenith and the equator.

23. The following data refers to the observation of the sun lying south of the zenith. Longitude of the place = 72° 30′, mean meridian altitude = 43° 36′ 10.8″, declination at Greenwich apparent noon = 19° 25′ 18.6″ increasing @ 6.98 sec/hr, semidiameter of the sun = 15′ 56.8″. Find the latitude of the place.

CHAPTER 22

Aerial Surveying

Learning Objectives

After going through this chapter, the reader will be able to

- explain the terms photogrammetry and photo-interpretation
- explain the basic principle and procedure of terrestrial photography
- describe the equipment required for and principle of aerial photogrammetry
- explain the basic principle of stereoscopic vision and how it is employed in stereoscopes to study aerial photographs
- explain the basic principle of photo-interpretation

Introduction

Aerial surveying, sometimes also called *photogrammetry*, is a method of surveying generally used to cover large areas. The term photogrammetry is today a misnomer because the method does not necessarily depend upon photographs taken from the air or ground. Any type of imagery—photographs, digital or CCD images, images using thermal or other scanning systems from a satellite, etc. may be used in this method.

Terrestrial photography, in which photographs are taken from ground stations, was the beginning of this technique. With the advent of aviation, aerial photographs taken from cameras onboard aircraft started gaining popularity. Other scanning systems were developed later and are being used extensively today. The term *digital photogrammetry* is used for collecting images in electronic format and manipulating them with software for preparing maps.

Whether we take photographs or images through other scanning systems, all these methods can be classified under a general technique known as *remote sensing*. The term remote sensing, in simple terms, refers to accessing information about an object without being in physical contact with it. Sometimes, the objective is to obtain qualitative information about the area. At other times qualitative as well as quantitative information is sought. These methods come under a technique known as *photo-interpretation*. Both these techniques are covered under aerial surveying.

As can be intuitively observed, aerial surveying is an expensive technique. In earlier times, only state agencies undertook such surveys. Today, many private companies undertake such jobs. Such surveys have become an integral part of large projects, as the coverage of area by this method is comparatively larger and requires much less time, making the use of this technique viable. This chapter deals with some elementary aspects of aerial surveying.

22.1 Terrestrial Photogrammetry

Terrestrial photography is done with a *phototheodolite*. A phototheodolite consists of a camera mounted on a levelling base, such as a theodolite with a telescope on top of the camera box. The line of collimation of the telescope and the optical axis of the camera must lie in the same vertical plane. The phototheodolite also has a horizontal plate for measuring horizontal angles. The camera has a frame with cross hairs inside. The line joining the intersection of the cross hairs and the optical centre of the camera lens defines the line of collimation.

22.1.1 Fieldwork

Fieldwork in terrestrial photogrammetry includes reconnaissance, establishing control points, and taking photographs. Reconnaissance is done to locate triangulation stations and stations for placing the camera. The area is thoroughly surveyed so that a minimum number of photographs is required to cover it. Triangulation stations are located or established from existing locations. Triangulation stations can also be used as camera stations. The camera stations are so selected that the features to be located are clearly visible in at least two photographs. In addition, the base line joining the camera stations, which is carefully measured, and the lines of collimation from the cameras must make well-formed triangles. The photographs should also contain a triangulation station or any other station whose location is already available on a map. A point or feature to be located can be plotted graphically or its location determined analytically.

The basic principle of terrestrial photogrammetry is similar to that of plane tabling. The principle can be explained in terms of the situation shown in Fig. 22.1. In Fig. 22.1(a), C_1 and C_2 are two camera stations. C_1L and C_2L are the lines of collimation of the cameras at the two stations intersecting at L. Let P be the point to be established. C_1C_2 is the base line and is measured accurately. Let α and β be the angles made by the lines of collimation with the base line. The coordinates of point P measured from the photographs are (x_1, y_1) from the camera at C_1 and (x_2, y_2) from the camera at C_2. f is the focal length of the camera. To locate the point P graphically, the following procedure is adopted.

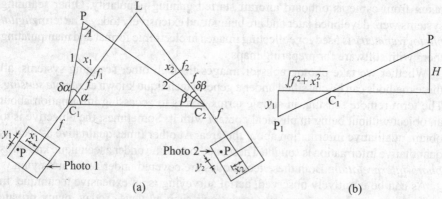

Fig. 22.1 Terrestrial photogrammetry

1. On the given plan, locate the camera stations C_1 and C_2. If one of the points is known, locate the other point by plotting the base line distance to the same

scale as that of the plan. Measure angles α and β to locate the lines of collimation of the cameras and their intersection point L.

2. Mark points f_1 and f_2 by measuring the focal length of the camera in the direction of the line of collimation.

3. From the photograph, measure the coordinate distances. Mark the distances x_1 and x_2 at right angles to the lines of collimation from f_1 and f_2.

4. Join $C_1 1$ and $C_2 2$. Extend these lines to intersect at P. P is the location of the point in the plan.

Analytically, if P is the location of the point in plan, let $\angle PC_1L$ be $\delta\alpha$ and $\angle PC_2L$ be $\delta\beta$. It is clear from the figure that $\tan\delta\alpha = x_1/f$ and $\tan\delta\beta = x_2/f$. Angles PC_1C_2 and C_1PC_2 can be calculated as $\alpha + \delta\alpha$ and $A = 180° - (\alpha + \delta\alpha) - (\beta + \delta\beta)$, respectively. Knowing the length C_1C_2, the lengths C_1P and C_2P can be calculated using the sine rule:

$$C_1P = C_1C_2\sin(\beta - \delta\beta)/\sin A \quad \text{and} \quad C_2P = C_1C_2\sin(\alpha + \delta\alpha)/\sin A$$

To determine the elevation of point P, we find the elevation of P above the line of collimation at C_1. Figure 22.1(b) shows the elevation above the line of collimation. From the similarity of triangles, it can be seen that $H = C_1P \times y_1/\sqrt{f^2 + x_1^2} \cdot C_1P$ can be measured from the plan to scale. The reduced level of P can be found if the RL of the line of collimation at the camera station C_1 is known from observations to a benchmark.

22.1.2 Terrestrial Stereophotogrammetry

An advanced method of terrestrial photogrammetry is to take photographs in pairs from the ends of a base line. Photographs are generally taken with the optical axis at right angles to the base line. This is achieved by taking a backsight to the first station from the second station. The camera stations are located by triangulation. The base line joining the camera stations is measured very accurately. The camera stations C_1 and C_2 (Fig. 22.2) are selected such that the required feature appears in both the photographs. When the two photographs are properly adjusted for viewing and seen through a stereoscope, the ground features are seen in three dimensions.

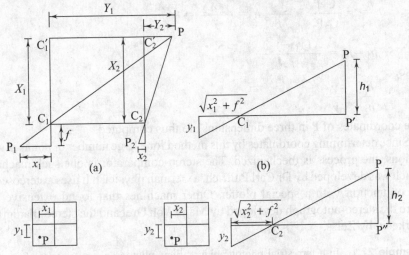

Fig. 22.2 Stereophotogrammetry

Principle of plotting

Plotting is generally done mechanically using plotting machines. We explain the principle with reference to Fig. 22.2. Let the coordinates of P with reference to points C_1 and C_2 be (X_1, Y_1) and (X_2, Y_2), respectively. Also, it is clear that $X_1 = X_2$. $C_1'P = Y_1$ and $C_2'P = Y_2$. Length of the camera base line = $C_1C_2 = Y_1 - Y_2$. Let (x_1, y_1) be the coordinates of P in the photograph taken from the first station and (x_2, y_2) be the coordinates of P in the photograph taken from station C_2. At station C_1, from similar triangles,

$$Y_1/X_1 = x_1/f$$
$$Y_1 = X_1 x_1/f$$

Similarly, at station C_2,

$$Y_2/X_2 = x_2/f$$
$$Y_2 = X_2 x_2/f = X_1 x_2/f \text{ (as } X_1 = X_2)$$

From this, the base line length

$$C_1C_2 = X_1(x_1 - x_2)/f$$

or

$$X_1 = Bf/(x_1 - x_2)$$

where $B = C_1C_2$ is the camera base line. Now,

$$Y_1 = X_1 x_1/f = Bx_1/(x_1 - x_2) \text{ (substituting for } X_1)$$

Similarly,

$$Y_2 = X_1 x_2/f = Bx_2/(x_1 - x_2)$$

To determine the elevation of P, let h_1 be the height of P above the optical axis at C_1. From Fig. 22.2(b), from similar triangles,

$$\frac{h_1}{y_1} = \frac{C_1P}{C_1P_1} = \frac{X_1}{f}, h_1 = \frac{y_1 X_1}{f} = \frac{y_1 X}{f} \quad (X = X_1 = X_2)$$

$$\frac{h_2}{y_2} = \frac{C_2P}{C_2P_2} = \frac{X_2}{f}, h_2 = \frac{y_2 X_2}{f} = \frac{y_2 X}{f}$$

$$h_1 - h_2 = \frac{X(y_1 - y_2)}{f}$$

Also,

$$\frac{h_1}{h_2} = \frac{y_1}{y_2}$$

The coordinates of P in three dimensions are thus computed.

Since determining coordinates by this method for a large number of points is very tedious, the process is mechanized. The stereo-comparator is one such machine, which was developed by Dr Carl Pulfrich, a German physicist. It uses a stereoscope in conjunction with a special plotter. Other machines that found extensive use were the stereo-autograph developed by Major Von Orel and the stereo-planigraph marketed by Zeiss.

Example 22.1 In a terrestrial photographic survey, photographs were taken from two stations 220 m apart. The focal length of the camera was 180 mm. The lines of collimation from the camera stations made angles of 36° 30′ and 66° 45′ from the left and right stations, respectively, while observing station S. A point P appeared on the photographs from both stations. P was 10 mm to the left and 8 mm above the cross lines in the first photograph and 14 mm to the left and 3 mm above the cross lines in the second photograph. The reduced level of the camera axis at the first station was found to be 182.5 m by observations to a benchmark. Find the coordinates and RL of point P.

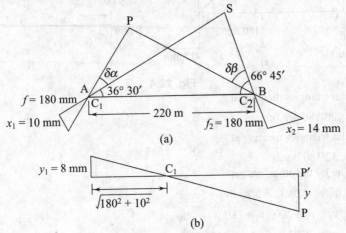

Fig. 22.3

Solution   Refer to Fig. 22.3(a):

$\tan \delta\alpha = x_1/f = 10/180$, $\delta\alpha = 3° \, 10′ \, 47″$

$\tan \delta\beta = x_2/f = 14/180$, $\delta\beta = 4° \, 26′ \, 51″$

$\angle PAB = 36° \, 30′ + 3° \, 10′ \, 47″ = 39° \, 40′ \, 47″$

$\angle PBA = 66° \, 45′ - 4° \, 26′ \, 51″ = 62° \, 18′ \, 9″$

$\angle APB = 180° - 39° \, 40′ \, 47″ - 62° \, 18′ \, 9″ = 78° \, 1′ \, 4″$

The distance PA can be calculated by applying the sine rule to triangle ABP:

AB/sin APB = PA/sin PBA

PA = AB sin PBA/sin APB = 220 sin 62° 18′ 9″/sin 78° 1′ 4″ = 202.08 m

The coordinates of P are (AP cos PAB, AP sin PAB), i.e., $x = (202.08 \text{ m}) \times \cos 39° \, 40′ \, 47″$ = 155.52 m and $y = (202.08 \text{ m}) \times \sin 39° \, 40′ \, 47″ = 129.03$ m. Thus, the coordinates of P are (155.52, 129.03). To find the elevation of P, refer to Fig. 22.3(b):

$8/180 = y/202.08$

$y = 8 \times 202.08/180 = 8.98$ m

Elevation of P = 182.5 + 8.98 = 191.48 m

Example 22.2   A and B are two camera stations 200 m apart. Stereo-pairs were taken with the optical axis at right angles to the camera base line. In the photograph exposed at A, a point P was found to be 20 mm to the right and 8 mm above the cross lines. The same point was 32 mm to the left and 12 mm above the cross lines in the photograph taken from B. If the focal length of the camera lens was 180 mm, find the coordinates of P with the origin at A.

Solution   The situation is represented in Fig. 22.4(a). $AA' = BB' = x$, $A'P = y_1$, and $B'P = y_2$. We can write

$y_1 + y_2 = 200$

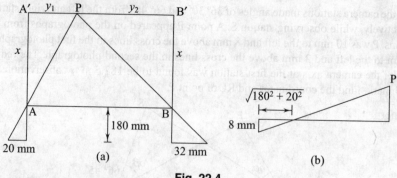

Fig. 22.4

From similar triangles meeting at A,

$$20/180 = y_1/x$$

From similar triangles meeting at B,

$$32/180 = y_2/x$$

Dividing one by the other, we get

$$y_1/y_2 = 20/32 = 5/8, \quad 8y_1 = 5y_2$$

$$y_1 + y_2 = 200, \quad y_1 + (8/5)y_1 = 200$$

$$y_1 = 76.92 \text{ m} \quad \text{and} \quad y_2 = 123.08 \text{ m}$$

$$x = 180 \times 76.92/20 = 692.28 \text{ m}$$

Thus, the coordinates of P are (692.28 m, 76.92 m). To find the height of P, refer to Fig. 22.4(b). $P = 8 \times 692.28/180 = 30.77$ m above the optical axis at A.

22.2 Aerial Photogrammetry

Terrestrial photogrammetry virtually went out of use with the advent of aerial surveying techniques. Aerial photogrammetry makes use of cameras fitted in an aircraft to photograph an area from an overhead position. The principle of stereo-scopic vision is used in studying and interpreting aerial photographs. Therefore, overlapping photographs are taken in the direction of flight as well as in the lateral direction as the aircraft flies along a parallel path. It must be understood that while a map is an orthographic projection by projecting points perpendicular to the plane, a photograph is a perspective projection, as all the light rays for forming the image pass through a point.

22.2.1 Basic Terminology

An aerial photograph is a record of the ground features at a point in time. Aircraft fitted with cameras moves along predetermined paths and takes photographs at planned intervals. The following are the basic terminology used to describe aerial photography (Fig. 22.5).

Altitude Height of the aircraft above the ground.

Flying height Height of the aircraft above a chosen datum.

Exposure station Position of the aircraft at the time of exposure of the film. It is essentially the position of the optical centre of the camera lens when the film is exposed.

Air base   Distance between two consecutive exposure stations.

Tilt and tip   Tilt is the inclination of the optical axis of the camera about the line of flight. In Fig. 22.5, ϕ is the tilt. Tip is the inclination of the camera axis about a line perpendicular to the line of flight.

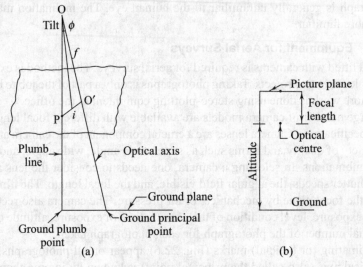

Fig. 22.5   Terminology

Picture plane   Plane that contains the image at the time of camera exposure.

Ground plane   Horizontal surface from which heights can be measured and which can be used as a datum surface.

Perspective centre   Optical centre of the camera through which all rays of light pass.

Principal point   Point of intersection of the optical axis of the camera with the photographic plane. O is the optical centre and O′ is the principal point. When the optical axis is extended downwards, the point of intersection with the surface is known as the *principal ground point*.

Isocentre   Point on the photograph at which the bisector of the angle of tilt meets the photographic plane. 'i' is the isocentre, at a distance of $f \cos\phi$ along the principal line, where f is the focal length of the camera.

Plumb points   The points at which the vertical line through the optical centre meets the photographic plane and the ground surface. The plumb point on the ground surface is also known as ground nadir point. The plumb point on the photograph is known as nadir point.

Homologous points   Points on the ground and their representations in the photograph in perspective projection.

Vertical photograph Obtained with an aerial camera when the ground is perfectly flat and the optical axis is vertical. This can rarely be obtained.

Tilted photograph Obtained when the optical axis is inclined unintentionally to the vertical at an angle not greater than 3°.

Oblique photograph Photograph taken with the optical axis deliberately inclined to the vertical at a large angle. This is done because the view as seen in a vertical photograph is generally unfamiliar to the human eye. The inclination makes the view more familiar.

22.2.2 Equipment for Aerial Surveys

Aircraft fitted with cameras is required for aerial surveys. The method is expensive but viable for large projects. Taking photographs is only a part of the job to be done. A lot more work is done using stereo-plotting equipment in the office.

A large number of camera models are available with different focal lengths and other specifications. Camera lenses are a crucial component of the camera and come in a variety of quality and forms such as super-wide angle, wide angle, and normal lens combinations. In selecting a camera, one needs to consider the lens aperture sizes, shutter speeds, the angular field visible, and the focal length. The film is kept flat on the focal plane by mechanical or air pressure. The camera also records the date of exposure, level condition of the camera, time of exposure, altitude of flight, and serial number of the photograph for each photograph taken.

Collimating (or fiducial) marks (Fig. 22.6) appear on all photographs, so that the optical axis (also called the principal point) is known by joining them on the photograph.

Fig. 22.6 Collimating marks

Terminology

The following terms are important in aerial photography and illustrated in Fig 22.7.

1. **Aerial cameras** These are no different in their working principle than any other camera. There are several types of cameras in use in aerial photography including digital cameras. The basic principle of working is illustrated in Fig. 22.7(a).
2. **Aperture** It is the opening of the shutter at the time of exposure, which can be adjusted, also called f-stop. The f-stop numbers are given as f/8, f/11, etc. which is the focal length divided by the effective diameter of the lens opening.
3. **Shutter speed** It is the time the shutter remains open to allow the light to fall on the film. This can vary from 1/2000 to 1 or more seconds.
4. **Film speed** It is indicated by a number as 100, 200, 400, etc. This indicates the amount of radiation required for the film response. Larger the number,

larger is the speed and requires less aperture and small shutter speeds for a given response.

5. **Focus** It is the operation bringing the image clear and is done by varying the distance between the lens and the image plane. When the camera is focused, a clear image of the object will be formed on the plane of the film.

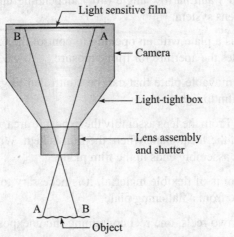

Fig. 22.7 Basic principle of aerial camera

22.2.3 Types of Aerial Camera

Numerous types of cameras are in use for aerial photography from conventional cameras to digital cameras. The cameras in use can be classified into metric camera, multi spectral camera, and other special purpose cameras.

Metric camera

Metric cameras are commonly used for aerial photography. A typical diagrammatic sketch of a metric camera is shown in Fig. 22.8.

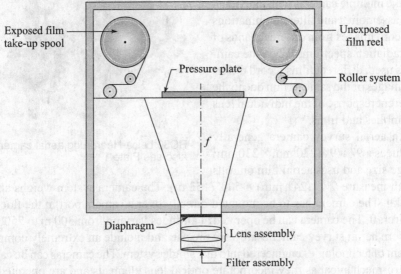

Fig. 22.8 Parts of an aerial camera

The metric camera consists of the following components:

Lens assembly   It comprises lenses and filters as required. In practice, the lens assembly is not a single lens but a complex system of lenses to give clear images. The lens assembly can also be classified into superwide angle (125°), wide angle (95°), normal (60°) and narrow angle (40°), depending upon the angle of view provided by the lens system.

Diaphragm   It is a plate with an opening to control the exposure of the film to light. This provides the aperture for film exposure.

Shutter   It is a movable plate that can be controlled to give a specific time of exposure of the film to light.

Exposure area   From the lens assembly the exposure area widens through a conical structure to fit the focal plane where the film is kept. When in focus, the focal length of the lens assembly ends in the film plane.

Film   As the film is of flexible material, it is necessary to keep it in a plane by using the pressure from a flattening plate.

Reel   There are two reels, one reel containing the unexposed film and the other reel to take up the exposed film.

Aerial cameras also have associated instrument panel to record data such as altitude, time of exposure, and so on. The camera is mounted with a bracket mount and will have many other associated items such as power supply, pressure system, and other units to control the filming operation.

Multi-spectral camera

Multi-spectral cameras use a multi-lens assembly to have exposure to multiple bands of the radiation spectrum. These act like multiple cameras with individual lens assemblies and filter combinations to record images from different bands of the radiation spectrum. When the camera is used, different films record different images of the same terrain due to the different response of the individual lens assemblies and films.

An aerial survey camera generally produces a 9″ × 9″ (230 mm × 230mm) image size and uses aerial film on rolls

RC30 Leica-Heerbrugg aerial camera; also see Plate 1

which measure 9½″ (241 mm) × 500′ (152 m). The camera system weighs about 150 kg. The camera has to be mounted directly over a camera port in the floor of an aircraft. The camera can be operated at altitudes varying from 300 m to 7500 m. Modern aerial survey cameras are very precise and include an extremely complex system of technologies engineered into one single system. The cameras can be called electro-mechanical as they incorporate optical lens elements and are operated by multiple computer chips and may also be linked to GPS satellites.

Most camera systems incorporate mechanisms for Forward Motion Compensation. The technology enables the apparent ground image motion to freeze on the film emulsion during the exposure cycle. The image quality with respect to sharpness and resolution is substantially increased by use of this technology. This is invariably incorporated in most camera systems today.

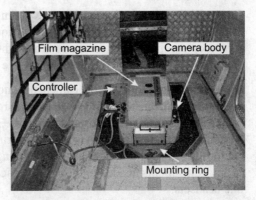

Zeiss RMK camera system and mounting

Global Positioning System

Global Positioning systim (GPS) is utilized extensively throughout the flight planning and project acquisition phases of all aerial survey projects. Many survey camera systems in use today are linked to survey grade GPS systems to ensure extremely accurate navigation of flight lines and to precisely expose predetermined photo centre positions throughout an aerial survey mission.

The Zeiss RMK camera system has five parts. The system weighs well over 100 kg.

Mounting   The camera is fitted on the aircraft floor on a motorized mounting ring. The camera looks vertically downwards through a port in the aircraft belly. The mounting ring is self-levelling to keep the camera level during flight. It can also be rotated to compensate for crabbing in the aircraft flight. Crab occurs when the aircraft is slightly rotated due to cross winds during flight. Tilt and crab of the camera affect the coverage area of the photographs and must be compensated for by the camera system.

Camera body   The camera body contains the shutter system and the lens cone as described earlier. The focal length of the lens is 153 mm which gives a photograph scale of 1:10,000 at 1524-m flying height and 1:20,000 at 3048 m. The camera lens is not a single lens but a complex lens system made up of several elements to give very high image resolution with minimal distortion.

The Zeiss camera also has a special shutter which opens and closes radially. This minimizes time lag in the operation of the shutter and reduces blurring of the image due to motion of the aircraft during the time that the shutter is open.

The camera has a maximum shutter speed of 1/1000 second and a large aperture range. The shutter speed and aperture can be manually set or calculated automatically by the camera system using a light sensor in the lens housing.

The lens also can be used with a variety of filters for different atmospheric and light conditions to equalize the amount of light passing across the large lens.

The film magazine   The camera uses roll film that is 24 cm wide and generally 75 m or 80 m long. This gives about 300 photographs per roll. The magazine contains a complex series of rollers to move the film quickly and smoothly across the camera back between exposures. There is also a vacuum system to ensure each negative is completely flat at the time of exposure.

Controller   It is a device used for calculating the time interval between exposures to ensure full coverage of the area with adequate overlaps. The controller uses information received from the navigation sight and the camera settings to calculate the time interval needed between successive photographs.

Navigation sight   It helps the camera operator to navigate along the planned survey flight track. This instrument is fitted at the front of the aircraft and has a periscope head which looks downwards and forwards. This is retractable during take-off and landing. The navigation sight is lowered during flight. The camera operator can see the terrain ahead and match it to the planning map. The operator can tell the pilot of any adjustments needed to the aircraft track through communication equipment like headsets.

The navigation sight is also used to compensate for crab of the aircraft. The operator can manipulate the navigation sight to point in the direction of flight and the camera mount follows the sight direction.

The navigation sight also allows the time interval between photographs to be varied to keep coverage without gaps or a specific overlap between photographs. As the aircraft flies over a mountain, the height of the aircraft is less from the ground. The photographs have to be taken closer together to avoid gaps. When the aircraft flies over a valley, the time interval can be adjusted to suit the flight height. The navigation sight has an ingenious method for doing this. Illuminated lines move through the field-of-view. The operator adjusts the speed of travel of the lines to match the speed of aircraft through the field-of-view of objects on the ground such as rock outcrops or crevices. The controller unit uses this information to trigger the camera at the correct time intervals to keep the desired coverage.

22.2.4   Procedure for Aerial Surveying

The procedure for aerial surveying includes reconnaissance of the area, establishing ground controls, flight planning, photography, and then paperwork including computation and plotting.

Reconnaissance is undertaken to study the important features of the ground for reference purposes. Ground control is required in order to obtain a set of points of known position based on which other points are located and plotted. The number of ground control points depends upon the extent of area covered, scale of the map to be prepared, flight plan, and the process of preparing the maps. A minimum of three control points must appear in each photograph. These points are established by triangulation or precise traversing.

Flight control is achieved by flight planning, which takes into account the extent of the area, type of camera and its focal length, scale of the photograph, altitude, speed of aircraft, and the overlaps of the photograph. The area covered by each photograph, time interval between exposures, and the number of photographs required are decided based upon such flight planning.

Ground control   Aerial photographs cannot be used effectively if ground control points are not established and available in photographs. In the earlier days of aerial photography, it formed an important step. Here, one conducts a triangulation or a precise traverse survey to establish control points. These are carefully selected as points seen prominently and identifiable in the photographs to be taken. Both horizontal and vertical control points must be established. The ground coordinates and elevations of these points are thus known. Well-selected features or specially -built stations can be used as ground control points. A photograph must have three or more control points for effective use of the photograph for plotting and map preparation.

In modern methods of aerial photography, a number of new technologies are available with total stations and satellite GPS navigators. The ground control points can be established to great accuracy using these new devices. Coordinates in terms of latitude, longitude, and elevations can be easily established by these instruments.
Flight plan: After the reconnaissance and ground survey, a flight plan is prepared. The flight plan will depend upon the extent of the area covered and the equipment (camera) characteristics. The flight plan will include the speed of the aircraft, altitude, area covered by each exposure and the overlaps (see below) in the flight and transverse directions, number of flights, number of exposures, and scale of photographs.

Overlaps

Overlaps are required to ensure complete coverage of the area. Stereoscopic vision is possible only with overlapped parts of the photographs (parts common to two photographs). Overlaps also ensure that the central part of the photograph is less distorted than the outer edges. The photographs are made to overlap longitudinally as well as laterally. Lateral overlap ensures that no area is left without being photographed. Longitudinal overlap is generally 60 per cent while lateral overlap is about 30 per cent.

To obtain overlaps, the aircraft flies in a straight line to the extent possible, and the camera provides exposures as per desired time intervals. The interval between exposures must be such that the desired longitudinal overlap of the area covered is attained. The next flight is undertaken in a parallel line such that the required lateral overlap is obtained. Such parallel flights can in general be achieved automatically with modern aircraft. Figure 22.9 shows how longitudinal and lateral overlaps are obtained. There can be problems in controlling the flight path of the aircraft. The aircraft may not fly in a straight line; it may be thrown off track due to atmospheric conditions such as air currents.

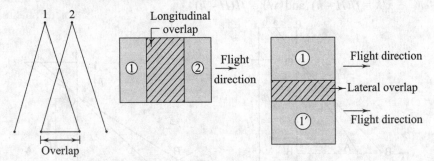

Fig. 22.9 Overlaps

Scale of photographs

As the camera film is exposed to take a photograph, it is clear that the rays of light pass through the optical centre of the camera lens. The representation of any object in the photograph depends upon the distance of the object from the camera. In aerial surveying, this distance is represented by the flying height. As all the ground features do not have the same elevation, it is clear that the scale representation will be different for features lying at different elevations. A photograph, unlike a map based on orthographic projection, does not have a uniform scale. The scale can be varied by changing the flying height. To account for changes in scale due to ground relief, an average elevation and average scale for an area can be worked out.

Vertical photographs

In a vertical photograph, if the ground was perfectly plane, we will have a photograph similar to a map having uniform scale. Referring to Fig. 22.10(a), the length AB (line AB having uniform elevation) on the ground is represented by ab in the photograph. The scale of the photograph is ab/AB. From similar triangles abO and ABO, the scale of the photograph S is as follows:

$$S = ab/AB = f/(H - h)$$

where f is the focal length of the camera, H is the flying height, and h is the elevation of line AB. H and h are measured from the same datum. If A and B were at different levels, as in Fig. 22.10(b), with elevations h_a and h_b, an average elevation for the line can be used: $h_{av} = (h_a + h_b)/2$. The scale for line AB can be taken as $f/(H - h_{av})$. In the case of an area, the elevations of a large number of points can be averaged and an average scale factor can be calculated. Thus, if $h_1, h_2, h_3, ..., h_n$ are the elevations of n points in an area, $h_{av} = (h_1 + h_2 + h_3 + \cdots + h_n)/n$. The scale for the area in the photograph can be found using this average elevation. If the flying height also changes during the coverage of the area, an average flying height can also be worked out and used in the above formula.

A vertical photograph, thus, does not have a uniform real scale owing to ground relief. The lengths of lines, however, can be determined using the coordinates in the photographs and on the ground. The ground coordinates are determined with reference to the ground plumb point. The photograph coordinates are determined from the principal point of the photograph. If (X, Y) are the ground coordinates and (x, y) are the photographic coordinates, then

$$x/X = f/(H - h) \quad \text{and} \quad y/Y = f/(H - h)$$

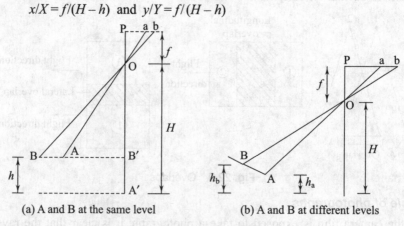

(a) A and B at the same level   (b) A and B at different levels

Fig. 22.10 Scale of photograph

The length of a line can be determined from the coordinates of its end points. If PQ is a line, (X_p, Y_p) are the ground coordinates of P, and (x_p, y_p) are the photographic coordinates of P $(X_q, Y_q, x_q,$ and y_q representing similar terms for Q),

$$X_p = (H - h_p)x_p/f, \; Y_p = (H - h_p)y_p/f$$
$$X_q = (H - h_q)x_q/f, \; Y_q = (H - h_p)y_p/f$$
$$\text{Legth PQ} = \sqrt{(X_p - X_q)^2 + (Y_p - Y_q)^2}$$

Tilted photographs

In the case of tilted photographs, the scale factor is more difficult to derive. If ground relief is also to be taken into account, the determination of scale becomes very complex. While a formula can be derived for the scale, the tilt effect is generally removed by a camera process called *rectification*; the vertical scale factor can then be used.

Scale of tilted photographs can be derived as follows: In Fig. 22.11 the picture plane or the plane of the film is tilted by an angle θ to the ground. Point N is the ground nadir point, the point of intersection of the plumb line through the optical centre O and the ground. The corresponding point N′ is the nadir point. P′OP is the line perpendicular to the picture plane. P′ is the principal point in the photograph and P is the corresponding point on the ground. I′ is the isocentre or the point at which the bisector of the angle of tilt meets the picture plane. I is the corresponding ground point.

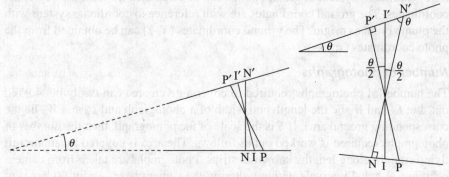

Fig 22.11 Scale of tilted photograph

From the figure, it is clear that from $\triangle$ N′MN

$$\angle \text{MN'N} = 90 - \theta$$

In $\triangle$ P′OI′

$$\angle \text{I'P'O} = 90; \qquad \angle \text{P'OI'} = \theta/2; \qquad \angle \text{P'I'O} = 90 - \theta/2$$

In $\triangle$ NOI

$$\angle \text{ONI} = 90; \qquad \angle \text{NOI} = \theta/2; \qquad \angle \text{OIN} = 90 - \theta/2$$

The triangles P′OI′ and NOI are similar.

Also, OP, the perpendicular distance of the picture plane from the optical centre is the focal length f.

$$\text{OI'} = f \sec \theta/2 \qquad \text{and} \qquad \text{ON} = f \sec \theta$$
$$\text{ON} = h, \text{ the flying height at the time of exposure}$$

Therefore, OI = $h \sec \theta/2$.
The scale at the isocentre

$$I' = \frac{\text{OI}}{\text{OI}} = \frac{f \sec\left(\theta/2\right)}{h \sec\left(\theta/2\right)} = \frac{f}{h}$$

as in a vertical photograph.
The scale however will change at other points. For example,

$$\text{scale at N}' = \frac{\text{ON}'}{\text{ON}} = \frac{f \sec\theta}{h}$$

$$\text{and scale at P} = \frac{\text{OP}'}{\text{OP}} = \frac{f}{h \sec\theta}$$

As mentioned earlier, the scale factor is rarely used for tilted photographs. The photographs are first rectified for the effect of tilt and then the scale of vertical photograph is used.

It can also be shown that the distortions due to tilt are radial.

Ground and photo coordinates   In a photographic image, the coordinate system is established with the principal point as the origin. In the 2D photograph, the x-axis is generally chosen along the flight direction and y-axis is perpendicular to it. This (x, y) coordinate system has its origin at the principal point in the photograph.

Points on the ground have ground coordinates in three-dimensional space. The (X, Y) coordinate system of ground points can be calculated from the photographic coordinates. The ground coordinates are with reference to coordinate system with the plumb point as origin. The ground coordinates (X, Y) can be obtained from the photo coordinates (x, y).

Number of photographs

The number of photographs required to cover a given area can easily be worked out. Let L_p and W_p be the length and width of a photograph and L_g and W_g be the corresponding ground area. If S is the scale of the photograph, then the number of photographs required is worked out as follows. The area is covered by an aircraft flying straight along lengths known as strips. Photographs are taken from camera positions at equal intervals without altering the camera base. About 60 per cent overlap is provided in the longitudinal direction. Let the longitudinal overlap be represented by O_l. The aircraft then moves along parallel strips to cover the area and there is also about 30 per cent overlap between strips known as side overlap. Let the side overlap be represented by O_s.

Effective ground length covered by one photograph $L_g = SL_p(1 - O_l)$

Effective ground width covered by one photograph $W_g = SW_p(1 - O_s)$

If L and W are the length and width of the area to be covered, then

Number of photographs/strip, $N_L = L/L_g + 1$

1 is added to cover the end areas.

Number of strips, $N_S = W/W_g + 1$

Number of photographs, $N = N_L N_S$

The number of photographs can also be calculated as

$N = LW/L_g W_g$

Interval between exposures

The time interval between exposures is decided by the speed of the aircraft and the distance it has to cover between exposures. The time interval is given by T (in seconds) $= 3600L/V$, where L is the distance travelled by the aircraft (in kilometres or metres) and V is the speed of the aircraft (in km/hr or m/hr).

22.3 Distortions in Aerial Photographs

The photographic positions of objects as taken from air cameras differ from their orthographic positions, due to the tilt of the camera axis as well as ground relief. These displacements are corrected to some extent before maps are prepared from photographs.

Displacement due to tilt

Figure 22.12 shows the displacement due to tilt, where O is the optical centre , points A and B are seen as a' and b' in the tilted photograph. a and b are the points which would have been obtained in a vertical photograph. i is the isocentre bisecting the angle of tilt and passing through O. P is the principal point, a line perpendicular to the picture plane and passing through O. θ is the angle of tilt. With i as centre draw two arcs with ia' and ib' as radii. These intersect the vertical photograph plane at a'' and b''.

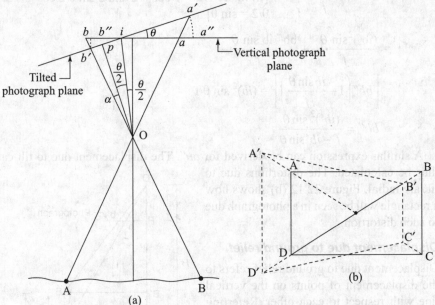

Fig. 22.12 Displacement due to tilt

Since $ib' = ib''$ by construction, $\angle ib' b''$ is isosceles.

In $\angle ib'b''$, angle at I $= \theta$, the tilt angle. The remaining two angles at b' and b'' will be equal, i.e., $90 - \theta/2$. Thus

$$\angle ib'b'' = \angle ib''b' = 90 - \theta/2$$

Also $\angle PiO = 90 - \theta/2$ from $\angle PIO$ as angle at P is 90°.

If you consider the lines $b'b''$ and Oi cut by line $b'a'$, then $\angle PiO$ and $\angle b''b'i$ are alternate angles. This shows that lines $b'b''$ and Oi are parallel. From this we get that triangles iOb and $b''b'b$ are similar. Therefore,

$$\frac{bb'''}{b'b''} = \frac{ib}{Oi} \Rightarrow Oi = \frac{f}{\cos(\theta/2)} = f\sec(\theta/2)$$

From triangle $ib'b''$,

$$\angle b'ib'' = \theta; \qquad \angle ib'b'' = \angle ib''b' = 90 - \theta/2$$

Applying sine rule,

$$\frac{Ib'}{\sin(90 - \theta/2)} = \frac{b'b''}{\sin\theta}$$

$$b'b'' \frac{ib'\sin\theta}{\sin\left(90 - \dfrac{\theta}{2}\right)} = 2\,ib'\sin\left(\theta/2\right)$$

[from $\sin(90 = \theta/2) = \cos\theta/2$ and $\sin\theta = 2\sin\left(\dfrac{\theta}{2}\right)\cos\left(\dfrac{\theta}{2}\right)$]

From the earlier equation for similar triangles,
$bb'' = b'b'' \times ib / Oi = 2\,ib'\sin(\theta/2)\,(ib'' + b''b)\,/f\sec\theta/2$

$$= \frac{ib'\sin\theta\,(ib' + b''b)}{f} \quad [\text{from } \sec\theta/2 = 1/\cos\theta/2 \text{ and } 2\sin\theta/2\cos$$
$$\theta/2 = \sin\theta]$$

$$= \frac{(ib')^2\sin\theta}{f} + \frac{bb''\,ib\sin\theta}{f}$$

$$= \left[bb''\left(1 - \frac{ib\sin\theta}{f}\right)\right] = (ib)^2\sin\theta/f$$

$$= bb'' = \frac{(ib')^2\sin\theta}{f - ib'\sin\theta}$$

A similar expression can be derived for aa''. The displacement due to tilt can thus be calculated. The distortions due to tilt are radial. Figure 22.12 (b) shows how a rectangle will be seen in a photograph due to such distortion.

Displacement due to ground relief

Displacement due to ground relief refers to the displacement of points on the vertical line with respect to each other. Referring to Fig. 22.13, the vertical line AB, say, the vertical edge of a wall, will be seen in the photograph as ab, displaced radially instead of being superposed as in a map. The radial

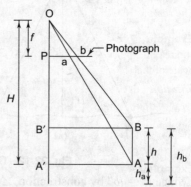

Fig. 22.13 Displacement due to ground relief

shift may be worked out as follows, with H denoting the flying height, f the focal length, and h the height of B over A. From triangles PbO and OBB′,

$$Pb/f = BB'/(H - h)$$

where $h = h_b - h_a$. Similarly, from triangles PaO and OAA′,

$$Pa/f = BB'/H$$

Dividing one by the other,

$$Pb/Pa = H/(H - h)$$

or

$$(Pb - Pa)/Pa = h/(H - h)$$

The displacement is ab = Pb – Pa. Therefore,

$$ab = h \times Pa/(H - h)$$

Thus the displacement due to ground relief is proportional to the radial distance from the plumb line. It is assumed that the photograph is a vertical one.

Example 22.3 Aerial photographs were taken with a camera having a focal length of 180 mm. The average elevation of the ground in the photograph was 160 m. Find (a) the scale of the map if the flying height was 2500 m and (b) the flying height required to have a photo scale of 1 in 6000.

Solution
(a) The photo scale can be found from $S = f/(H - h_{av}) = 0.180/(2500 - 160) = 1/13,000$.
(b) The flying height can be found from $1/6000 = 0.180/(H - 160)$; $H = 1080$ m.

Example 22.4 On an aerial photograph taken with a camera having a focal length of 150 mm, an 1800-m-long line PQ had a length of 125 mm. The average elevation of the line, PQ, was 290 m. Find the scale of another area in the same photograph with an average elevation of 950 m.

Solution The scale of the photograph, S, is given by 12.5 cm = 1800 m or 1 cm = 144 m. Taking the units into account, we get

$$1/144 = 15/(H - 290)$$

$$H = 2450 \text{ m}$$

If the average elevation is 950 m, then the scale of the photograph is given by

$$S = 0.15/(2450 - 950) = 1/10,000$$

The scale is given by the representative fraction 1/10,000.

Example 22.5 Find the number of photographs (size 250×250 mm) required to cover an area of 20 km $\times$ 16 km if the longitudinal overlap is 60% and the side overlap is 30%. Scale of the photograph is 1 cm = 150 m.

Solution We take the length of the area as 20 km and width as 16 km.
Number of photographs per strip = $1 + 20,000/[150 \times 25(1 - 0.6)] = 15$
Number of strips required = $1 + 16,000/[150 \times 25 \times (1 - 0.3)] = 8$
Number of photographs required = $15 \times 8 = 120$

Example 22.6 The scale of a photograph of size 250×250 mm is 1/12,000. Determine the number of photographs required to cover an area of 250 sq. km if the longitudinal and side overlaps have to be 60% and 30%, respectively.

Solution The ground area covered by one photograph can be determined by $L_g = SL_p$ $(1 - O_l)$ and $W_g = SW_p(1 - O_w)$:

$$L_g = 12 \times 250(1 - 0.6) = 1200 \text{ m}$$

$$W_g = 12 \times 250(1 - 0.3) = 2100 \text{ m}$$

Area covered by one photograph $= 1.2 \times 2.1 = 2.52$ sq. km

Number of photographs required $= 250/2.52 = 99.2 = 100$

Example 22.7 In a photograph taken with an aerial camera having a focal length of 200 mm, a line PQ measures 108 mm. If the same line has a length of 40 mm on a map drawn to a scale of 1:40000, find the flying height of the aircraft if the average elevation is 400 m.

Solution The scale ratio will be the same as the length ratio. Therefore,

$$\frac{\text{Scale in photo}}{\text{Map scale}} = \frac{\text{Length in photo}}{\text{Distance in map}}$$

$$\text{Scale in photo} = \frac{\text{distance in photo} \times \text{map scale}}{\text{distance in map}}$$

$$= 108 \times \left(\frac{1}{40}\right) \times \left(\frac{1}{40000}\right) = \frac{1}{14814}$$

Photoscale is also given by

$$\frac{f}{(H - h_a)} = \frac{0.2}{(H - 400)}$$

Therefore, $\dfrac{0.2}{(H - 400)} = \dfrac{1}{14814}$; $H = 3362$ m

Therefore the flying height is 3362 m.

Example 22.8 In a vertical photograph, a line, 2100 m long, measures 110 mm. If the average elevation of the line is 480 m, find the scale of an area in the same photograph with an average elevation of 1200 m. The focal length of the camera used was 210 mm.

Solution The scale of the photo $= \dfrac{0.11}{2100}$

The scale is also given by

$$\frac{f}{(H - h_a)} = \frac{0.21}{(H - 480)}$$

$$\frac{0.11}{2100} = \frac{0.21}{(H - 480)}$$

$$H = 4489 \text{ m}$$

The average elevation of the area is 1000 m. The scale at 1200 m elevation is

$$\text{Scale} = \frac{0.21}{(4489 - 1000)} = \frac{1}{16614}$$

Example 22.9 Points P and Q have elevations of 600 m and 300-m respectively. The photographic coordinates of points A and Q were measured as P (35, 25) and Q(20, 50) in millimetres. The photograph was taken with a camera having a focal length of 210 mm and from an altitude of 2500 m. Find the length of line PQ.

Solution The ground coordinates can be found as follows:

For P, X-coordinate = $\dfrac{(H-h_a)}{f}$ = $(2500-600) \times \dfrac{35}{210} = 316.7\,\text{m}$

Y-coordinate = $(2500-600) \times \dfrac{25}{210} = 226.2\,\text{m}$ $(2500-600) \times 25/210 = 226.2\,\text{m}$

Ground coordinates of P are (316.7, 226.2).

For Q, X-coordinate = $(2500-600) \times \dfrac{20}{210} = 209.5\,\text{m}$

Y-coordinate = $(2500-300) \times \dfrac{50}{210} = 523.8\,\text{m}$

Coordinates of Q are (209.5, 523.8)

The length PQ = $\sqrt{\left[(316.7-209.5)^2 + (226.2-523.8)^2\right]} = 316.3\,\text{m}$

Example 22.10 The height of a chimney is 110 m. In a vertical photograph taken from an altitude of 2800 m, the chimney appears at a distance of 70 mm from the principal point. Determine the displacement of the top of the chimney from the bottom if the elevation of the base is 1200 m.

Solution The displacement is proportional to the radial distance.

Displacement = radial distance $\times \dfrac{h}{(H-h_a)} = \dfrac{70 \times 110}{(2800-1200)} = 4.8\,\text{mm}$

Example 22.11 The image of the base and top of a 200 m tower is seen in a photograph of scale 1:15000. If the base point is 65 mm from the principal point, find coordinate of the top from the principal point. The focal length of the camera was 200 mm and assume that the base of the tower is the datum plane.

Solution Scale = $\dfrac{f}{(H-h_a)} = \dfrac{1}{15000}$

Therefore, $(H-h_a) = 15000 \times 0.2 = 3000\,\text{m}$

Radial ground distance = $\dfrac{65(3000-200)}{200} = 910\,\text{m}$

Distance to the base = $910 \times \dfrac{200}{3000} = 60.6\,\text{mm}$

Example 22.12 An aerial photo 250 × 250 mm was taken with a camera having a focal length of 200 mm. At the time of exposure, the tilt was assumed to be 1.5°. Find the displacement of a point lying at 70 mm on a line passing through the principal point.

Solution Distance to the point from isocentre = $70 + 200 \tan(1.5) = 75.24\,\text{mm}$

Displacement = $\dfrac{(75.24)^2 \times \sin 1.5}{(200 - 75.24 \times \sin 1.5)} = 0.74\,\text{mm}$

Mosaics

One way to use the photographs taken from an aerial platfrom is to arrange them as stereo pairs. This give a better idea of the ground relief, thereby enabling to study and interpret the photographs. Another method of using them is to get a photographic view of the area by cutting and positioning them to form a total image of the area. The photograph that results after the tilt and ground relief corrections have been made is known as a controlled mosaic. If corrections are not made beforehand, then the built-up image is called an uncontrolled mosaic.

A mosaic is built by first obtaining the contact prints in the serial order of taking them and removing the overlapping portions. Generally the mosaic contains the centre portion of the photograph which has generally an average scale. The first photograph is suitably cut with a fine cutter to give a slanted edge. The next photograph is cut similarly so that when pasted to the first it forms a complete image of the area. A number of photographs are used in building up a mosaic in this way to give a complete photographic image of the area surveyed.

Photographs and maps An aerial photograph gives a different view from what we see on the ground as the view is from the top which is unfamiliar to the eye. Human eye is more accustomed to seeing maps. The following are the differences between photographs and maps:

(a) An aerial photograph is a two-dimensional view of an area at an instant of time. Maps, on the other hand, are made with data collected at some point in time. As maps take time to prepare, some of the measured data may be outdated by the time maps get prepared.

(b) Aerial photographs do not have uniform scale over the whole area. The image is also distorted due to tilt and ground relief. Maps are drawn to a uniform scale.

(c) A photograph is a perspective projection whereas a map is an orthogonal projection.

(d) A photograph, gives full details of the landscape, including details which may not be significant. Maps, on the other hand, can be made to exact details the customer wants – showing only those features which are important for the project.

(e) Photographs show much more details compared to maps. But clarity is less in photographs compared to maps.

(f) An air photograph shows details as seen by the camera. A map, on the other hand, shows details using conventional symbols.

(g) Air photographs can show changes in the landscape or land use by pictures taken over time periods. A map is the representation of landscape at a fixed point in time.

22.4 Stereoscopic Vision

An aerial photograph is a two-dimensional view of an object as it lies on a plane. However, if photographs are available in stereo-pairs, it is possible to get a three-dimensional view of the object. Stereoscopic vision essentially lends height or depth to the object in the photograph as it would have been seen by the human eye.

Stereoscopic or binocular vision

The human brain receives two images of the same object through the two eyes as seen from slightly different angles and fuses them into a three-dimensional view. The base length for vision by human eyes is about 65 mm. As shown in Fig. 22.14, PQ is the object. Each eye sees the object but from a slightly different angle. β_1 and β_2 are known as *parallax angles*. The two images received by the retina

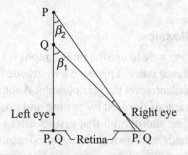

Fig. 22.14 Binocular vision

are fused into a single image giving depth to vision. When the distance between the eye and the object is large, the parallax angle becomes smaller and at about 600 m the relative depth is no longer perceived by the eye.

Stereoscopic pairs

To enable stereoscopic vision, it is essential to have photographs in pairs covering common areas. In aerial photography, photographs are taken from two camera positions with sufficient overlap in the photographs. Only the areas seen in both the photographs are amenable to stereoscopic vision. The common part of the photograph gives us two views of the same object from slightly different angles, as seen by the human eye. When the photographs are suitably placed and the left eye views one and the right eye the other, the common area comes in relief or in three dimensions due to stereoscopic fusion. Stereoscopic vision enables one to view the depth or height of points on the photograph.

Stereoscopes

There are many types of stereoscopes—mirror stereoscopes, lens stereoscopes, scanning mirror and zoom stereoscopes. Lens and mirror stereoscopes are handy and commonly used.

Mirror stereoscope   The schematic diagram of the mirror stereoscope is shown in Fig. 22.15(a). The mirror stereoscope consists of two viewing eyepieces. A stereoscopic pair of photographs is placed at a distance from the stereoscope. The photographs are adjusted so that one photograph is seen through one eyepiece. The instrument has four mirrors, two mirrors attached to each eyepiece. As the viewer looks through the stereoscope, he/she sees the image of the same object (the overlapping part) on the two photographs and this gives a stereoscopic view by fusion. The terrain is seen in relief due to this.

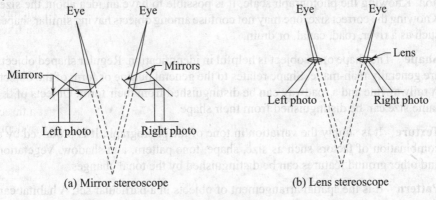

(a) Mirror stereoscope   (b) Lens stereoscope

Fig. 22.15

Lens stereoscope   A lens stereoscope has two eyepieces through which the observer sees the photographs, providing the experience of stereoscopic or spatial view [Fig. 22.15(b)]. The lenses help to magnify the image as seen by each eye. The distance between the eyepieces is adjustable and can be set by the observer as per requirement. This distance is approximately equal to the distance between human eyes. The lenses tend to magnify the object and its height. Lens stereoscopes are more compact than mirror stereoscopes.

22.5 Photo-interpretation

Photo-interpretation is the key to effective use of photographs. It refers to the accurate identification of the features seen in photographs. Objects seen in photographs are often not easy to recognize, and it takes some amount of skill on the part of the interpreter to correctly identify the objects and judge their significance. It is more difficult to identify objects in vertical photographs than in tilted photographs owing to the familiarity of view in oblique photographs. Colour photographs are easier to interpret than black and white photographs due to tonal variations. A stereoscopic pair is easier to interpret due to the depth available in the photographs when seen through a stereoscope. Considerable amount of practice and experience is required to correctly interpret photographs.

Interpretation of aerial photographs is required extensively in developmental project design and execution. It has been successfully applied in a variety of fields. The success of project planning depends on the effective and efficient interpretation of photographs by engineers and others. A good deal of patience and ingenuity is required to interpret photographs.

22.5.1 General Features of Photographic Images

The knowledge of some of the basic characteristics of the images in aerial photographs helps one to interpret these images. Photo-interpretation requires large-scale photographs. The success of the interpretation depends upon the experience of the person in addition to the conditions under which the photographs were taken and the quality of photographic material. Additionally, the photographs should be studied in the correct orientation with respect to the light conditions at the time of photography. Some of the basic features of photographs that help in identification are discussed here.

Size   The size of an object in the photograph is sometimes helpful in interpretation. Knowing the photograph scale, it is possible to have an idea about the size. Knowing the correct size, one may not confuse among objects having similar shapes such as a river, road, canal, or drain.

Shape   The shape of an object is helpful in identification. Regular shaped objects are generally man-made. Shape relates to the general outline or form of the object. A railway line and a roadway can be distinguished from their form. Objects of the same size can be distinguished from their shape.

Texture   It is simply the variation in tone of the photograph. It is produced by a combination of factors such as size, shape, tone pattern, and shadow. Vegetation and other ground features can be distinguished by the tonal changes.

Pattern   It is the spatial arrangement of objects in a particular set. A habitat can be easily distinguished by the arrangement of roads, houses, etc. because of the pattern.

Shadow   The shadow of an object formed during photography is sometimes helpful in identification, as it shows the outline of the object.

Tone   It is produced by the amount of light reflected back by the object to the camera. If the particular tones associated with specific objects are known, it is easy to identify them.

Location The location of an object in the photograph helps in identifying the object itself. Knowing the objects or areas surrounding the object, one can identify the main object. Refer Chapter 24 for more on visual image processing.

22.6 Parallax

Parallax is caused by the motion of the point of observation and the difference in elevation between the points observed. The apparent movement of the point under observation with respect to the reference system caused by the movement of the camera position is known as parallax. The parallax of a point can be measured from photographs, as the total movement of the image of the point between two exposure positions. For this, two consecutive photographs are taken and the principal points of both exposures are

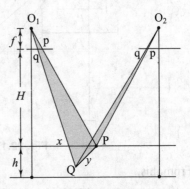

Fg. 22.16 Parallax

marked in the two photographs. When joined, this gives the line of flight between the two exposures. A line perpendicular to this line of flight is now drawn and the two sets of lines provide a reference coordinate system. If (x_1, y_1) and (x_2, y_2) are the coordinates in the two photographs, then the parallax p is given by $p = x_1 - x_2$. Refer Fig. 22.16. Also, if X and Y are the coordinates from the plumb point,

$$X = Bx_1/p, \quad Y = By_1/p, \quad H - h = Bf/p$$

where B is the camera base line length.

22.6.1 Parallax Equation

The parallax equation helps one to find the difference in elevation between points from the parallax in two photographs. Referring to Fig. 22.17, O_1 and O_2 are the two camera or exposure stations, H is the flying height, D is the camera base line length on the ground and d is the same in the photographs, f is the focal length of the camera, A and B are two points seen in the photographs at heights h_a and h_b, a_1 and b_1 are the positions of A and B in photograph 1, and a_2 and b_2 are the positions of the same points in photograph 2.

Two photographs are taken from positions O_1 and O_2. The total movement of the points A and B between the two photographic positions is $P_a = P_1a_1 + P_2a_2$ and $P_b = P_1b_1 + P_2b_2$. Considering the two similar triangles $O_1a_1a_2$ and O_1AO_2, we have

$$D/P_a = (H - h_a)/f$$
$$P_a = Df/(H - h_a)$$

Similarly,

$$P_b = Df/(H - h_b)$$

The difference in height between A and B can be found from the parallax between them as

$$(P_a - P_b) = \delta P = Df/(H - h_a) - Df/(H - h_b)$$

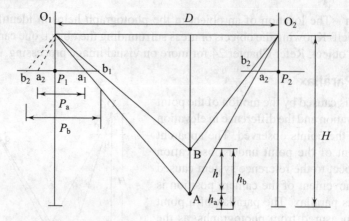

Fig. 22.17 Parallax equation

From this,

$$h_b - h_a = \delta P(H - h_a)(H - h_b)/Df$$

If h_a and h_b are small compared to H, we have

$$h_b - h_a = \delta P H^2/Df$$

This is the parallax equation, which can be used to calculate the difference in height. This equation can be put in many forms as follows:

$$h_b - h_a = \delta P(H - h_a)/(P_a + \delta P)$$
$$= \delta PbH/[P_a(P_a + \delta P)]$$
$$= H\, \delta P/(b + \delta P)$$

22.6.2 Measuring Parallax

Parallax, or strictly the difference in parallax, between points is measured using a stereoscopic viewer or a parallax bar, also called a stereometer.

A *stereoscopic viewer* is a stereoscope with a floating point mark in each viewing eyepiece. When stereoscopically viewed, the two floating points float and fuse, giving an apparent impression of height. The floating points are moved over the photographs using a micrometer screw. To determine the parallax between two points as the photographs are viewed stereoscopically, the floating point is moved and kept over, say, point 1, and the micrometer reading recorded. The floating is now moved and kept over point 2 and the reading recorded. The difference in the two micrometer readings is a measure of the parallax between the two points.

A *parallax bar* (Fig. 22.18) consists of a rigid bar with two glass reticules, each having an index point. One of them can be adjusted by moving it along the rigid bar and can be clamped in any position. The other reticule can be moved very little using a micrometer screw. When two photographs have been adjusted for stereoscopic viewing, the photographs are fixed in position and the parallax bar is placed over them, the glass reticules over the photographs. The index points are fused during viewing. These dots when fused for viewing can be placed over any point at its elevation by moving the micrometer screw. A similar procedure is followed with a different point, recording the micrometer reading in each case. The difference in the micrometer reading gives the parallax difference between the points.

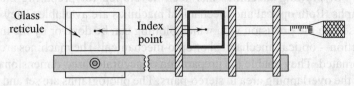

Fig. 22.18 Parallax bar

Example 22.12 Find the parallax difference between the top and bottom of a building given that the mean base length in the photographs is 102.5 mm, the flying height during exposure is 1600 m, and the focal length of the camera lens is 200 mm. The height of the building is 30 m.

Solution Scale of photograph $= f/H = 0.200/1600 = 1/8000$

Mean base length in the photograph, $b = 102.5$ mm

Actual base length $B = bH/f = 102.5 \times 1600/200 = 820$ m

Parallax is given by $Bf/(H-h)$.

Parallax at bottom $= 820 \times 200/1600 = 102.5$

Parallax at top $= 820 \times 200/(1600 - 30) = 1044.45$ mm

Difference in parallax $= 1.95$ mm

22.7 Plotting

Plotting is the process of adding details to maps having control points. The *radial line method*, which is a graphical method, is commonly used. Other methods include the *mechanical method* and *automated machine method*.

Radial line or photo triangulation method

The radial line method is used when the plan positions of at least three points are known in the photograph. The scale of the photograph is either known or can be worked out. The three points available in the photographs and already plotted on the map are identified. The first step is to locate the *principal point* of the photograph on the map. This is done using a tracing paper, which is kept on the photograph, and drawing radial lines from the three points to the principal point. The tracing paper is then kept on the map and adjusted by trial and error such that the radial lines pass through the three points. Their intersection can now be plotted on the map as the principal point. The principal point on the other photograph can be plotted similarly.

Using two consecutive photographs in which the required points to be plotted are available, a similar procedure that uses tracing paper is used to plot such points as the intersection of lines from the two principal points.

Slotted template method

The slotted template method, which is mechanical, uses templates. A template is a transparent plastic sheet with an aperture for the principal point and short radial slots that passes through various points. One template is prepared for each photograph. After the templates are prepared, they are assembled on a base grid. Once the templates are properly set, points can be plotted on the map.

Plotting machines

A variety of plotting machines has been developed for preparing maps from photographs. Both optical and mechanical machines are available. They are classified based on the precision with which plotting can be done or by the method of reproduction—optical, mechanical or opto-mechanical. The machines are more or less automated. They enable the preparation of accurate three-dimensional optical views of the overlapping area in stereo-pairs. The photographs are set and adjusted to remove tilt, etc. and obtain a stereoscopic view. The machines use a floating mark, which can be moved over any point on the photograph. The movement can also be transferred to a tracing point, which can trace the movements on paper. Vertical control points of known elevations are used for obtaining the elevations of other points. Contours can also be traced on the map. The multiplex and autograph are two such instruments used by large organizations for plotting from aerial photographs.

22.8 Applications and Advantages of Aerial Surveying

As has been discussed in the preceding sections, aerial surveying finds many applications in map preparation and map revision for large areas. Modern plotting machines and mostly automated operations have simplified the process of preparing maps from aerial photographs. Aerial surveying also finds extensive application in urban planning and development, transportation network design and calculations, disaster management, forestry, mining operations, reservoirs, agriculture, etc.

Aerial surveying, though expensive (as the operation involves using an aircraft fitted with a camera for taking photographs), is most suitable for covering large areas efficiently and is an indispensable method in modern surveying. It finds extensive application in covering remote and inaccessible areas, which are otherwise unsuitable for other methods of surveying. With better cameras and lenses and films having better resolution and picture quality, aerial surveying can now provide a wealth of detail.

With advancements in technology, aerial photography has given way to aerial image processing. High-resolution digital images (soft images) can be made and processed using software to prepare excellent maps. All forms of rectification and corrections can be done automatically before converting the data into a map. Digital photogrammetric equipment and software have developed sufficiently to facilitate the preparation of very accurate maps.

Summary

Aerial surveying or photogrammetry is a method of surveying in which cameras fitted in aircraft are used to take aerial photographs of the ground features. Digital images and images from other sensor systems (thermal, infrared, and radio) can also be used. Before aircraft became available for surveying, photo-theodolites were used to take terrestrial photographs.

During aerial photography, an effort is made to keep the camera axis vertical. However, the axis often gets tilted due to air currents, etc. The camera axis may also be intentionally tilted to obtain an oblique view of the area. Longitudinal and side overlaps are made as photographs are taken. These overlaps are required to ensure that the whole area is covered as also to obtain common areas in photographs, which can then be viewed stereoscopically— photographs of the same area taken from two stations can be viewed stereoscopically to get a relief view of the ground surface.

The *scale of photographs* depends upon the focal length and flying height of the aircraft. It also depends upon the elevation of the point being photographed. Distortions in photographs occur due to ground relief as well as tilt in the photographic plane. Parallax is the displacement of points in photographs due to differences in height. The parallax equation is used to determine the elevation of points.

Stereoscopes are handy instruments to view two photographs simultaneously to get a three-dimensional view of the surface. This three-dimensional view is used in *photo-interpretation*, which is a method of identifying and classifying objects seen in the photograph.

Most of the work with aerial photographs is automated using plotting machines such as multiplexes and autographs. These machines automatically plot a relief map from aerial photographs and provide the details of coordinates of points, lengths, areas, etc. Aerial surveying finds extensive application in areas such as engineering, mining, urban planning, and forestry, to name only a few.

Exercises

Multiple-Choice Questions

1. Terrestrial photogrammetry is taking photographs of
   (a) the terrain of the earth from an aircraft
   (b) the terrain of the earth from cameras on ground
   (c) celestial bodies from the earth
   (d) celestial bodies from an aircraft
2. In terrestrial stereophotogrammetry, photographs are taken
   (a) from two cameras in two aircrafts
   (b) twice of the same features from an aircraft
   (c) in pairs using two cameras on the ground
   (d) twice of the same features with the same camera
3. The scale of photograph taken from an aircraft essentially depends upon
   (a) the distance of the object from the camera
   (b) the focal length of the lens in the camera used
   (c) the aperture of the camera lens
   (d) the speed of the film used
4. Overlap in aerial photography refers to
   (a) the overlap of ground features due to inclination of camera
   (b) the overlap of ground features due to difference in elevation of objects
   (c) the same ground features taken from two camera positions
   (d) the blurring due to lack of focus in photographs
5. Distortions in aerial photographs is caused by
   (a) tilt and ground relief
   (b) faulty camera lens
   (c) lack of focus during photography
   (d) large distance between camera and the object
6. Displacement due to ground relief in aerial photography is
   (a) the radial distortion of ground features like a vertical line
   (b) the reduction in size of line or area
   (c) the lack of depth of field in the photographs
   (d) the lack of focus of different objects due difference in distance
7. Stereoscopic vision is
   (a) the three-dimensional view from stereoscopic pairs
   (b) the view seen by the camera lens

(c) the view obtained by special lens

(d) the view obtained in a photograph

8. Parallax in aerial photographs is an error due to

(a) movement of camera and ground relief

(b) overlap in photographs

(c) distortion caused by camera lens

(d) distortion due to lack of focus

Review Questions

1. Define the term photogrammetry.
2. Briefly explain the equipment and procedure used for terrestrial photogrammetry.
3. Explain how the coordinates of a point are worked out from terrestrial photographs.
4. Briefly describe the equipment required for aerial photogrammetry and explain the procedure followed to take aerial photographs.
5. Derive an expression for the scale of aerial photographs.
6. Derive an expression for distortion due to ground relief in aerial photographs.
7. Given the length and width of an area, explain how the number of photographs required is worked out.
8. Define the term parallax. Derive the parallax equation.
9. Define the term photo-interpretation. What are the features that help to identify objects in photographs.
10. Explain briefly the radial method of plotting details from photographs onto maps.
11. Briefly explain the advantages of aerial surveying.

Problems

1. Photographs were taken from two ground stations 200 m apart with a camera having a focal length of 150 mm. The camera axis made an angle of 42° 36′ at the left station and 60° 48′ at the right station with the base line. On the photograph taken from the left station, a point M was 12 mm to the left and 8 mm above the cross lines. On the photograph taken from the right station, M was 16 mm to the left and 6 mm above the cross hairs. The reduced level of the camera axis at the left station was 106.85 m. Find the coordinates of M with reference to the left station and its reduced level.

2. Photographs were taken from camera stations 300 m apart, the camera axis being at right angles to the base line at both stations. The focal length of the camera was 200 mm. On the photograph taken from the left station, a point appears 32.5 mm to the right and 18.5 mm above the cross lines. On the photograph taken from the right station, the same point appears 68.5 mm to the left and 12 mm above the cross lines. Find the coordinates of the point with the left station as the origin.

3. The elevations of points in an area vary from 136 m to 184 m. Aerial photographs were taken with a camera having a focal length of 200 mm. Determine (a) the flying height required to have a photographic scale of 1 in 8000 and (b) the photographic scale if the flying height is 2500 m.

4. The size of photographs to be taken with a camera is 230×230 mm. The photographic scale required is 1 in 50,000. The longitudinal and side overlaps are 60% and 30%, respectively. Find the number of photographs required to cover an area of 2000 sq. km.

5. If the required photograph size is 250×250 mm^2, the scale required is 1 in 10,000, and the longitudinal and side overlaps are 60% and 30%, respectively, find the number of photographs required to cover an area of size 30 km × 40 km.

6. A 1000-m-long line on the ground with an average elevation of 685 m measures 11.35 m in a photograph. The focal length of the lens is 210 mm. Find the scale of the photograph for an area having an average elevation of 900 m.

7. In two photographs forming a stereo-pair, the mean distance between the principal points is 68.5 mm. The focal length of the camera lens is 200 mm. Using a parallax bar, the difference in parallax between the top and the bottom of a vertical structure was found to be 2.75 mm. Find the height of the structure.

8. The mean distance between the principal points in two photographs was 110.5 mm. The photographs were taken from an elevation of 3000 m above the datum. The difference in parallax between the top and the bottom points of a tower was found to be 3.5 mm. If the elevation of the lower point of the tower is 200 m, find the height of the tower.

23

Modern Surveying Instruments

Learning Objectives

After going through this chapter, the reader will be able to

- explain the term electromagnetic spectrum
- describe the basic principle of electromagnetic distance measurement
- name and briefly explain any two EDM instruments
- describe the basic features and capabilities of a total station

---

Introduction

In the preceding chapters we have discussed many surveying equipment and methods. Surveying equipment has developed phenomenally in the last few decades. We have also mentioned some modern equipment such as auto levels, micro-optic theodolites, and electronic or digital theodolites. Digital equipment has become very popular because of simplicity of operation, ease and speed of working, and the facility to record the observations and retrieve them at a later time. This does away with manual reading and recording which is always error-prone.

In this chapter, we will discuss some modern instruments that have become an indispensable part of surveying. Electromagnetic distance-measuring (EDM) equipment and total stations have become common in almost all important engineering projects. They greatly facilitate surveying operations.

23.1 Electromagnetic Spectrum

Many types of equipment use some component of electromagnetic waves in their operation. Electromagnetic spectrum is the name given to a cluster of radiations that carry energy. The radiations consist of both electric and magnetic effects. These include x-rays, ultraviolet rays, light, infrared rays, and radio waves.

Figure 23.1 shows the electromagnetic spectrum and Table 23.1(a) shows the different components and their properties.

Electromagnetic waves are energy carrying waves; they carry composite energy due to electrical and magnetic fields. These sinusoidal waves have the two components at right angles to each other and move in the same direction. At one end of the spectrum are the gamma rays and x-rays (high frequency, small wavelength) and at the other end are radio waves (low frequency and large wavelength). Visible light is a very small component of this spectrum, with each colour of the band having a different frequency [Refer Table 23.1(b)].

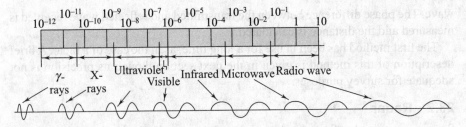

Fig. 23.1 Electromagnetic spectrum

Table 23.1

(a) Electromagnetic spectrum

Type of wave	Frequency range (Hz)	Wavelength range (m)	Remarks
Gamma rays	3×10^{22}– 3×10^{18}	10^{-14}–10^{-10}	Very high energy wave
X-rays	3×10^{19}– 3×10^{15}	10^{-11}–10^{-7}	Used in radiology
Ultraviolet	3×10^{16}– 3×10^{14}	10^{-8}– 4×10^{-7}	Causes skin burn
Visible light	7.5×10^{14}– 3.75×10^{14}	4×10^{-7}– 8×10^{-7}	Sensation of light and EDM
Infrared	3.75×10^{14}– 3×10^{11}	8×10^{-7}–10^{-8}	Used in TV remotes and EDM
Microwaves	30×10^{9}– 50×10^{6}	10×10-3^{-6}	Used for cooking and EDM
Short waves	30×10^{6}– 3×10^{6}	10–100	Radio broadcast
Medium waves	2×10^{6}– 400×10^{3}	150–750	Radio broadcast
Long radio waves	300×10^{3}– 10×10^{3}	1000–30,000	Radio broadcast

(b) Visible light spectrum

Colour band	Frequency (Hz) $\times 10^{14}$	Wavelength (m) $\times 10^{-9}$
Violet	6.7–7.5	450–400
Indigo	6.5–6.7	460–450
Blue	6.0–6.5	500–460
Green	5.3–6.0	580–500
Yellow	5.1–5.3	590–580
Orange	4.8–5.1	620–590
Red	4.0–4.8	750–620

The electromagnetic waves travel with the speed of light, which is approximately 300,000 km/s. As their speed is known, one method of distance measurement is the transmission and reception of these waves. The speed does not remain constant but depends upon many factors such as pressure, temperature, and humidity.

These waves can be used in two ways. One method is to generate the wave, transmit it, and receive it back after reflection from an object. The time of travel will be in microseconds and can be measured. Knowing the speed of the wave, the distance can be calculated. The second method, which is more common in surveying instruments, is to generate and transmit the wave and receive back the reflected wave. The phase difference between the transmitted and reflected wave received is measured and the distance is calculated.

The first method has been in use for a long time, as in the case of a radar. A brief description of this method is given in the next section though its precision is not adequate for survey purposes.

23.2 Radar

Radar is an acronym for radio detection and ranging. Radar works with radio frequencies. It uses short waves of 1–10 m (frequency 30–300 MHz). Radar helps to locate objects and determine their distances from the observer. This technique is most commonly employed in aviation, marine operations, defence operations, etc.

The principle of operation of the radar is simple. An intense burst of short-wave radio energy is transmitted from the radar station. The wave is reflected back from the object or another station and is received back at the transmitting station. The time (t) of travel is noted by the electronic circuitry and the distance between the two points can be calculated knowing the velocity (v) of propagation, which is the speed of light. The distance is $vt/2$—the distance travelled once during the transmission and reception.

There are two forms of radar operation. One is the cooperative radar, in which the transmitted wave is received at a receiving station and is transmitted back after a delay of known duration. In the non-cooperative radar operation, the waves are directly reflected from the object and received back. Cooperative radars are more accurate and can be used for survey purposes. The distance between two stations can be measured within the range of visual observation. While a distance of 100 km can be measured, the accuracy depends upon many factors. The likely errors are prohibitive for geodetic work. Errors in the range of 10–20 m can be expected.

While expensive forms of radar using an aeroplane can be employed to measure the distance between two points a number of times and averaged to get a fairly accurate value, it is still not precise enough for survey work. Radar thus finds greater application in fields other than surveying. Electromagnetic distance measurement using other forms and different techniques is more commonly used in surveying.

23.3 Electromagnetic Distance Measurement

Electromagnetic distance measurement has become very popular in recent times. The instruments have become very compact and their electronic components have become inexpensive. The accuracy has improved over the years and the instruments have become easy to handle and operate. Electromagnetic distance measurement is based upon the measurement of phase difference between the transmitted and the received signals. A measuring wave is generated and put on a carrier wave for transmission.

Electromagnetic waves of all frequencies travel nearly at the same speed in vacuum. This constant speed is taken to be (299792.5 ± 0.4) or 300,000 km/s. The velocity, wavelength, and frequency of the waves are related by $C = f\lambda$, where C is the velocity, f is the frequency, and λ is the wavelength.

The wavelengths used in EDM vary from 30 m to about 400 nm (1 nm = 10^{-9}m).

When the wavelength used is shorter, as in electro-optical instruments, they are accurate but their penetration is less. Longer wavelengths, as in microwave instruments, can travel through fog and haze and difficult climatic conditions. Both types of waves are employed in equipment made by different manufacturers. The basic principle of operation is explained below.

Basic principle of operation

As shown in Fig. 23.2, let the horizontal distance between the two stations be D m. Using two different waves of frequencies f_1 and f_2, the distance D can be calculated as

$$D = nu_1 + l_1 = nu_2 + l_2$$

where n is the number of measuring units u_1 and u_2 and l_1 and l_2 are the phase shifts. The two frequencies are so chosen that

$$ku_1 = (k + 1)u_2$$

Thus

$$u_1 = (k + 1)u_2/k$$

and

$$D = n(k + 1)u_2/k + l_1 = nu_2 + l_2$$

from which

$$D = (k + 1)(l_2 - l_1) + l_1$$

The distance is obtained independently of n, the number of whole units in the measurement. In practice, l_1 and l_2 are measured electronically by a delay-null method. The distance will be measured without ambiguity only up to a certain value. l_1 and l_2 can be converted into distance by a simple calculation. If $k = 10$, then $10u_1 = 11u_2$. The distance can be measured uniquely to a small value only by using the two frequencies given above.

Let us say $l_1 = 0.2$ m and $l_2 = 0.8$ m. Then, $D = 11(0.8 - 0.2) + 0.2 = 6.8$ m. $n = 6$. Distances up to 11 m can be calculated using the two given wave patterns. The same phase difference of $0.8 - 0.2 = 0.6$ will be repeated for distances of $11 + 6.8$, $11 + 2 \times 6.8$, etc. A third frequency can be added such that $100u_2 = 101u_3$, which will increase the range beyond 100 m.

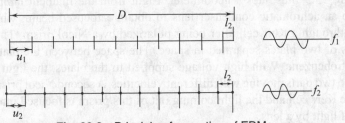

Fig. 23.2 Principle of operation of EDMs

The same value of the difference $(l_2 - l_1)$ will be true for other distances as well. In order to increase the range and remove ambiguity, a number of wave patterns are used for measurement. Thus we have instruments with short ranges of about 3–150 km. Thus an instrument may use a number of frequencies such as 10,

9.99, 9.9, 9 MHz. Three different ranges can be obtained using these frequencies.

23.4   EDM Equipment

EDM equipment can be classified based upon the type of wave used, into M (microwave) DM and EO (electro-optical) DM equipment. The first type uses low-frequency short radio waves while the second type uses high-frequency light waves. They can also be classified based upon the range as follows.

(a) Short-range equipment such as tellurometers and mekenometers with a range of up to 3 km.
(b) Medium-range equipment such as geodimeters with a range of up to 25 km. The range is about 5 km during the day and can go upto 25 km at night.
(c) High-range equipment with a range of up to 150 km. Tellurometers and distomats come under this category.

The accuracy varies with the range. Short-range equipment has an accuracy of ± (0.2 mm + 1 mm/km). Medium-range equipment has an accuracy of ± (5 mm + 1 mm/km) while high-range equipment has an accuracy of ± (10 mm + 3 mm/km). Distomats have replaced other forms of equipment due to their compact design, ease of operation, and precision.

All types of equipment using electromagnetic waves perform the following functions.

(a) Generation of two waveforms for carrier and measurement functions
(b) Modulation and demodulation of waves
(c) Measurement of phase difference
(d) Computation and display of distance or the results of measurement

A brief description of the various types of equipment is given below.

23.4.1   Geodimeter

The development of the geodimeter is credited to E. Bergestrand of Sweden. The instrument uses light waves and three to four frequencies. A tungsten filament lamp operated by a battery or a mercury vapour lamp lighted with a generator for night operation is used. Models using the helium–neon laser have also been made. The instrument has a transmitter and receiver system. The essential operation is as follows.

Figure 23.3 illustrates a geodimeter. Light from the filament lamp is passed through an achromatic condenser lens to obtain a focused beam, which is then passed through a Kerr cell after being polarized by a Nicol prism. The Kerr cell consists of two plates separated in space. The space between the plates is filled with nitrobenzene. With high voltage supplied to the plates, the light beam gets split into two parts moving with different velocities. A second Nicol prism is placed after the Kerr cell and the light coming out of this prism is focused into a parallel beam of light by a lens.

The geodimeter is kept at one end of the line to be measured. At the other end of the line is a reflex prism system or spherical reflector. The reflector, made up of single or multiple prisms, is retro-reflective, reflecting the incident light along the same direction. This eliminates the need for exact ranging of the reflector and the transmiter.

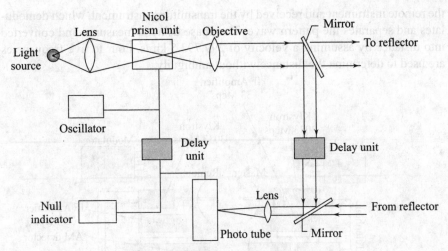

Fig. 23.3 Geodimeter

The receiver system of the geodimeter has a mirror–lens system to receive the signal reflected back from the target prisms. This is fed to a photo-detector tube. The signal is magnified many times. A comparator or delay-null unit compares the trans-mitted and received signals and the value obtained through such phase comparison is proportional to the distance. This value is directly sent to the display unit.

The range is not very high during the day. It is about 5 km during the day, and 15–25 km can be obtained with a mercury lamp at night. With a laser, the range and accuracy are improved during the day. The distance obtained may be the slant distance and has to be corrected for various atmospheric factors and reduced to the MSL distance.

Geodimeter

23.4.2 Tellurometer

Unlike the geodimeter, which uses light waves, a tellurometer uses high-frequency radio waves or a microwave-based system. A block diagram of tellurometer operation is shown in Fig. 23.4. The carrier signal has a frequency of 10 GHz ($\lambda = 30$ mm). This is modulated by a measuring frequency of 10 MHz. The development of this instrument is credited to the South African Trigonometric Survey and Wadley of the South African Telecommunications Society.

Two types of tellurometers have been developed—one giving a delay-line output in terms of time and the other directly in terms of distance. The instruments generally use five frequencies.

Two instruments are placed, one at each end of the line. The transmitting unit is known as the master unit and the other unit is the remote unit. The master unit transmits microwaves modulated by the pattern wave. This wave is reflected back by the remote instrument and received by the transmitting instrument, which demodulates and separates the pattern wave. The phase delay is measured and converted into distance by assuming a velocity of 299,792.5 km/s. Four to five frequencies are used to determine the distance without ambiguity.

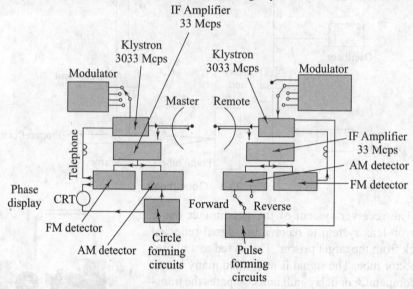

Fig. 23.4 Block diagram of the tellurometer system

All instruments have the facility for communicating with the other unit. The communication system provided in the instruments enables the two parties to communicate with each other during operation.

It will be necessary to use two identical instruments, one at each end of the station. This increases the cost, as two instruments and two sets of skilled personnel are required. The distance is more accurately determined this way, as each instrument can act as a transmitter as well as a receiver. The distance is thus measured twice and averaged. The instrument has a high range of 100 km. The accuracy obtained is ± (10 mm + 3 ppm). Instruments using higher frequencies (37.5 GHz and 75 MHz) have also been developed, giving a range of 50 km and an accuracy of ± (5 mm + 3ppm).

Instruments using infrared range waves (frequency of 300,000 GHz and measuring frequency of 75 MHz) have also been used. These eliminate the need for a second instrument, which is replaced by a reflector. The instrument has a range up to 3000 m and a precision of ± (5 mm + 2 ppm).

23.4.3 Mekenometer

A mekenometer is similar in principle to a geodimeter. It is an electro-optical distance-measuring equipment. The development of this instrument is credited to K.D. Froome of the National Physical Laboratory in England. It uses an intermittent signal of a high frequency of 500 MHz.

The light source is a tungsten lamp or a helium–neon laser. This can be generated using a battery. The instrument is kept at one station and a retro-reflector is kept at the other station. Accurate ranging of the reflector is not required. The reflected signal is received by the receiver system of the instrument. A phase comparison of the transmitted and received signals is made by a variable delay unit. The distance can then be calculated and digitally displayed by the instrument.

Tellurometer

The range of the mekenometer is 3000 m. It is very accurate in measurement with a precision of ± (1 mm + 3 ppm).

23.4.4   Distomat

Distomats are latest in the series of EDM instruments. The use of a distomat is similar to that of a tellurometer. Two identical instruments are used, one at each end of the line to be measured. A communication system is provided in the form of microphones and loudspeakers for contacting the remote or master unit.

The instrument uses 30-mm carrier waves. The frequency used in the master unit is 10,324.3 MHz and 10,335 MHz for the remote unit. The range of the instrument is 20–150,000 m. This is made possible by using five different frequencies in the 14-MHz band. The master unit sends the signals to the remote unit, which receives and reflects back the signals. The instrument can automatically send each of the signals and calculate the phase shift in each case. The distance is then automatically diplayed.

23.5   Corrections to Measurements

The distances calculated using EDM instruments have to be corrected for reduction to mean sea level, curvature and slope, and zero correction. The instrument manufacturers supply some of the values of the corrections due to the instrument and the zero correction. For ordinary work, only the corrections specified by the manufacturers may be applied. For precise work, the corrections have to be computed and applied by the surveyor.

Zero correction is done when the measuring centre of the instrument does not coincide with the station mark. Other corrections are made for the atmospheric conditions, such

Distomat; also see Plate 2

as pressure and temperature, which may be different from the standard calibrating conditions. Reduction to mean sea level can be done using formulae.

23.6   Total Stations

One of the recent developments in surveying equipment is the integration of distance- and angle-measuring components in one piece of equipment. A total station is the integration of an electronic theodolite with the EDM equipment. Many companies market total stations. Though the technology details used by different manufacturers may be different, they all have common features, which will be discussed below.

A digital theodolite is combined with one of the many forms of EDM equipment to obtain a very versatile instrument that can perform the required functions very easily.

Digital theodolite

The electronic or digital theodolite was discussed in Chapter 4. We will just recapitulate some salient points. These instruments have glass circles, which are encoded in the incremental or absolute mode. These are read by an optical scanning system and the reading is converted into angles and displayed or stored by the instrument. All the instruments are provided with an optical plummet for centring and a compensator system (single-axis or dual-axis) to take care of the tilt of the instrument axis. Any tilt is taken care of by the microprocessor in the theodolite and the displayed angles and distances are previously corrected for such minor errors. The user can choose the required accuracy of angular measurement. These theodolites are normally operated by a rechargeable battery pack. The charged batteries can work for 40–80 hours. Some instruments need a prism target. The prism targets can be of many types with single or multiple prisms. Even reflecting tapes are used. A digital theodolite comes with the following facilities.

(a) Zero-setting
(b) Bidirectional measurement
(c) Precision setting
(d) Horizontal and vertical angles
(e) Slant distance and horizontal distance
(f) Difference in elevations
(g) Entry and display of data
(h) Display and storage of results
(i) Data management system and data transfer facility

A total station has all the above facilities and in addition measures horizontal distance using a built-in EDM module. Total stations come with a lot more facilities of data storage and manipulation. The following are the salient features of a total station.

Angle measurement   Horizontal and vertical angles are measured to an accuracy of $1''$–$5''$. The angles are displayed on the display unit of the console. Many instruments have console units on both sides of the instrument.

Total station

Electronic theodolite

Distance measurement This is done with an EDM module functioning coaxially with the telescope tube. The distance measured is the slant distance if the stations are at different elevations. Reflecting multiple prisms are commonly used as targets, even though reflectorless distance measurement has also been made possible. The instrument uses the vertical angle measured by the theodolite and calculates the horizontal distance. All values are displayed as per the settings made by the user. Distance measurement can be done in different modes such as standard or coarse mode, precision mode, and fast mode. The precision and time taken vary depending upon the mode.

Microprocessor and software The onboard software in total stations can perform many functions. The processor is pre-programmed, and in some cases can be programmed by the user to perform many useful functions with the measured data. The details may vary with the manufacturers but some of the common features are as follows.

Automatic target recognition: Most of the modern total stations have the facility of automatic target recognition (ATR). In ATR, the telescope has to be roughly pointed towards the target while the measurement key is pressed. The instrument automatically points to the target before measurement. The instruments have motorized endless drives to facilitate ATR.

Reflectorless distance measurement: Until recently, total stations had to be used with special multiple prisms as targets for EDM. The new versions of total stations can measure distances without a prism target. This means that distances to points where a target cannot be erected can now be measured easily without any extra survey effort. This has been made possible by a red laser, which can direct to a point on any surface.

Computation of reduced levels: The reduced levels are measured from slope distance and vertical angle. Data input enables the user to input the height of instrument, height of target prism, and the RL of the station occupied. The instrument calculates the RL of the target station and displays the same.

Orientation: The instrument automatically orients to any direction specified by the user. If the coordinates of two points are input, the horizontal circle will be oriented to measure the bearing of the line automatically.

Automated processes: Automatic computation of coordinates of points, areas, offsets, etc. is possible with a total station. More and more on-board functions are being incorporated in total stations. Setting out points on the ground using coordinates or directions is possible.

Wireless keyboard and remote unit: Many new total stations come with a separate wireless keyboard. The input of data to the total station becomes very easy with a handheld keyboard. Another development is the availability of a remote unit so that the person at the prism can operate the total station for almost all the functions. As there is no need to bisect a target or read the angle, the system can be operated by one person positioned near the target.

Data management system: Total stations have a very efficient data management system. Data transfer to data recorders, computers, or flash cards is possible. The in-built memory can store up to 10,000 blocks of data.

Graphic display: Many new instruments have extremely powerful graphic display programmes. With large display panels, the data can be plotted and displayed.

Working with total station

Total stations are manufactured by many leading manufacturers of Survey equipment. Leica geosolutions, Topcon, Pentax, Nikon tripod data systems, Stonex are some of the major manufacturers of total stations. Figure 23.5 shows models of different manufacturers. While specific details may vary with the manufacturers, some features are common to all of them.

A total station, as mentioned earlier, is a versatile equipment for surveying operations. The equipment details and operations can be understood by referring to the user manual provided with the equipment.

Safety aspects Total stations incorporate some form of laser to measure distances. As lasers are harmful to the eyes, the surveyor should ensure not to look at the laser beam continuously. The precautions to be taken in this regard are given in the equipment manual.

The components, functions, and operation of the Stonex series of total station will be discussed here. The equipment details may vary, but will have invariably the same functions: Each equipment set-up comes with a user manual which gives details of its operation. In the following pages, we will highlight only the salient features of operation. The following is an abstract of the online user manual of the Stonex series.

(a)   Sokkia SR 50 series total station

(b)   NIKON total station

(c) Pentax total station; also see Plate 2

Fig. 23.5   Photographs of total stations; also see Plate 2

Salient Features

Software functions   The internal software installed in STS2/5R series total Station has a compact menu structure and complete and practical application programs to aid in accurate measurement.

Simplified operation   (STS2/STS5R) series Total Station has various functional keys, coupled with an input mode that combines characters and numbers precisely. It is simple, practical and convenient to use. Since it is easy to work with, even engineers with little experience in surveying can master the operation easily.

Absolute encoding circle   The pre-assembled absolute encoding circle enables the user to start measurement directly after switching on the instrument. Even if there is a disruption during operation (such as battery replacement), the azimuth data will not be deleted.

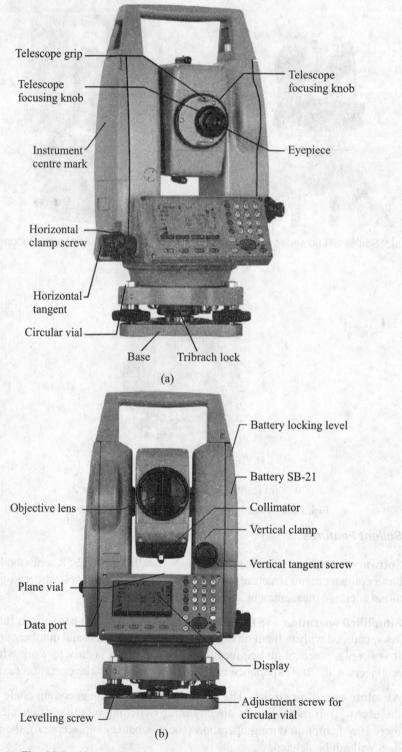

Fig. 23.6 Nomenclature and functions of Stonex total station

Reflectorless EDM The reflectorless laser EDM enables precise measurement from long distances on various objects such as the walls of buildings, telegraph poles, wires, cliffs and mountains, soil, stumps, and so on. Thus, it will be helpful in measuring a target that is inaccessible and of uneven texture/nature.

High precision and long measuring range The measuring range of STS-(R) Series Total Station is 2.4 km with single prism.

Reliable water/dust proof function STS-(R) Series Total Station supports water/ efficient dust proof function, which ansures the hardware performance of total station.

Auto power off An auto power off function switches off the instrument after a set period time to save battery. The display module of the total station is shown in Fig.23.7.

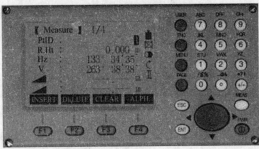

Fig. 23.7 Display module of total station

Table 23.2 Soft keys and their functions

Keys	Function	Keys	Function
[All]	Starts angle and distance measurements, and saves measured values.	[EDM]	Displays EDM settings.
[DIST]	Starts angle and distance measurements without saving measured values.	[Esc]	Returns to the previous mode or display.
[REC]	Saves displayed values.	[Con]	Continues to next mode or display.
[ENH]	Opens the coordinate input mode.	[ ]	Returns to highest soft key level.
[List]	Displays the list of available points.	[↓]	Returns to next soft key level.
[Search]	Starts the search for the input points.	[ENT]	Sets displayed message or dialog and quits dialog.

Icons

Many icons will be seen in the display panel. They refer to the following.

Measurement mode icons

- Infrared EDM (invisible) for measuring prisms and reflective targets.
- Reflectorless EDM (visible) for measuring all targets.
- Use reflective foils as reflective targets.

Battery capacity status icon The battery symbol indicates the level of the remaining battery capacity (80% full shown in the example).

Compensator status icons

- Compensator is ON
- Compensator is OFF

Character/Number inputting mode icons

- 01 Numeric Mode
- AB Alphanumeric Mode

Menu trees

The instrument works with many useful functions and can be accessed as a part of many pages displayed. The menu trees have the following details:

MENU (P1)

Programs -----------Surveying
- Stake Out
- Free Station
- COGO
- Tie Distance
Area (plan)
- Remote Height
- Reference Line/Arc
- Roads
- Construction
Settings --------Contrast, Trigger Key,
User Key, V- Setting,
Tilt Crn, Coll. Crn.
- SectorBeep, Beep, Hz<=>,
Facel Def., Data Output,
Auto-Off,
- MinReading, Angle Unit, Dist.Unit, Temp.
Unit, Press
Unit, Code Rec.
- GSI 8/16, Mask 1/2
EDM Settings ----------EDM Mode
- Prism
- Atmospheric Data
- Grid Factors
- Signal
- Multiply Constant
File Management --------Job
- Known points
- Measurements
- Codes
- Initialize Memory
- Memory Statistic

MENU (P2)

Adjustment ----------V-index
- Hz-collimation
- Horizontal Axis
- VO/Axis (Cons,list)
- lnst.Constant
- Tilt Parameter
- State
Comm Parameters ------Baud rate
- Data Bits
- Parity
- End Mark

- Stop Bit
Data Transfer -------Data Send ---- Job

- Data
- Format

System Information --------Battery
- Date
- Time
- Version
- Type
- Number
- End Mark
----Stop Bit
Data Transfer ---------Data Send ---- Job
- Data
- Format

Instrument set-up First the tripod is set up over a station point and ensured that the station point and the vertical axis are nearly in line. The tripod legs are extendable and are adjusted to facilitate comfortable viewing and stability. The instrument is placed over the tripod and levelling and centring are done. For rough levelling, a circular bubble is available and for precise levelling, a plate bubble is used as in conventional instruments. The instruments invariably have an optical plummet for centring. The centring and levelling operations are repeated many times to final setting of the instrument.

Reflector prisms For distance measurement, we need to have reflector prisms at the target station. The prism set-up can be of single or multiple prisms for more accuracy. Some of these are shown here.

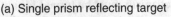

(a) Single prism reflecting target   (b) Triple prism reflecting target

Some versions of total stations can also measure distances in a reflectorless mode to any point. The accuracy of measurement depends upon many factors.

The instrument has many versatile features:
1. Input of characters and numbers can be done easily from the keypad.
2. There is a powerful search function for known points from memory.
3. Distance measurements can be done in different modes—fine single, fine two times to five times, fine repeat and tracking.
4. The total station is equipped with red laser and invisible laser EDM as well as reflector with prism(s), without prism, and reflective foil.
5. Appropriate prism constant can be input by the user for the type of prism used.
6. EDM measurement corrections: The instrument automatically corrects for curvature and refraction. The formula used for these corrections are as follows.
   Corrected horizontal distance $= S\,[\cos\alpha + S\sin\alpha\cos\alpha\,(k-2)/2R]$
   Corrected vertical distance $= S\,[\sin\alpha + S\cos^2\alpha\,(1-k)/2R]$
   where S is the oblique distance, α is the vertical angle, k is the atmospheric refraction constant, and R is the radius of the earth.

7. Atmospheric correction: Distance measurement is influenced directly by the atmospheric conditions of the place where distance measurements are taken. In order to take into consideration these influences distance measurements are corrected by using atmospheric correction parameters.

Temperature: Air temperature at instrument location.

Pressure: Air pressure at instrument location.

Atmospheric PPM: Calculated and indicated atmospheric PPM.

The formula for calculating atmospheric correction is as follows: (calculating unit:metre)

$$\text{PPM} = 273.8 - \frac{0.2900 \times \text{pressure value (hPa)}}{1 + 0.00366 \times \text{temperature value (°C)}}$$

If the pressure unit adopted is mmHg: make conversion with: 1hPa = 0.75 mmHg.

The following is the standard atmospheric condition of STS Total Station instrument (e.g. the atmospheric condition under which the atmospheric correction value of the instrument is zero):

Pressure–1013 hPa

Temperature–2°C

Major functions that can be performed with a total station

1. Routine survey operations like measurement of horizontal and vertical angles, distances and coordinates can be performed. These are supported by numerous menu-driven functions that are provided in the user manual. The instrument height can be accurately measured to the mark on the side of the instrument. The target height can be similarly measured. The heights can be input manually by the user. Several functions can be called and used independently or from different application programmes.

2. In levelling operations, the compensator mechanism must be turned for automatic adjustment for any tilt. The tilt mechanism compensates for small tilt of the axis. A warning sign will come if the compensator mechanism is out of range and then will require manual levelling of the instrument.

3. Free station is an application programme which can be used to locate the station occupied by the instrument by observations to a minimum of two known points and up to five points. This is similar to the two-point or three-point problem in plane tabling. This is made possible by measuring the horizontal and vertical angles and/or distances to the known point from the station occupied.

4. COGO is an application programme to perform coordinate geometry calculations. This enables calculation of coordinates, azimuth between points, and distances between points.

5. Area is another application programme to calculate the area enclosed by a number of points whose coordinates are measured, recalled from memory, or input by the user. Areas and perimeter are calculated and displayed. The area is the plane as projected on to a plane.

6. Stake out application is available for straight line curves and spirals. Curves can be set by many methods based on arc length, chord length, chainages, incremental stakeouts, and offsets.

7. A powerful file management function enables the user to input values, delete, edit, and recall values from memory. The measurement data is stored as part of a job file created by the user. New jobs can be created by inputting particulars such as job name, number, operators and so on. Memory statistics can be displayed to know the amount of data stored, available memory, and so on etc.

8. Communication and data transfer functions are used to transfer data to an external device such as a PC.

9. System information like battery status, date and time, instrument details, and so on can be accessed.

10. Instrument will need permanent adjustments as in conventional instruments after long usage. The test and adjustment are done as in conventional instruments. Plate level adjustment, circular bubble, reticle adjustment, vertical index error, horizontal axis, and optical plummet adjustment can be done.

Instrument Specifications

(a) Distance Measurement
- Type — visible red laser
- Carrier wave — $0.670\mu m$
- Measuring system — basis 60 MHz
- EDM Type — coaxial
- Display (least count) — 1 mm
- Laser dot size — approx.7 × 14 mm / 20 m (reflectorless mode only)
- Accuracy — approx. 10 × 20 mm / 50 m

With reflector
- Fine – 2 mm ± 2 ppm; Time < 1.8 sec
- Fast – 3 mm ± 2 ppm; Time < 1.2 sec
- Tracking – 5 mm ± 2 ppm; Time < 0.8 sec
- IR – tape – 5 mm ± 2 ppm; Time < 1.2 sec

Without reflector
- Fine – 5 mm ± 2 ppm; Time < 1.2 sec
- Tracking – 10 mm ± 2 ppm; Time < 0.8 sec.
- Range: 300 m to 4000 m with reflector and 100 m to 200 m without reflector.

(b) Angle measurement: 1″/5″/10″ selectable.

Smart stations

A more recent development is the integration of the total station with a GPS (global positioning system) module. The global positioning system works with satellite signals (Chapter 24).

The integration may be with a common data management system and inter-transferability of data. The surveyor, depending upon the circumstances, can use the total station or the GPS unit for locating his position. Each method has its advantages and works better in certain conditions. The integration enables one to make the best use of both. The recent instruments come with an antenna that can fit on top of the total station. The position data from the GPS reside in the total station and there is perfect integration of both.

Smart station

GPS receivers need open areas to receive satellite signals. Therefore, they cannot be kept at places where there are obstructions to a clear view of the sky. Total stations need to be placed at known stations or need local known stations for locating the position of the station. Using GPS and RTK (Real Time Kinematic), which is a technique used to obtain the position occupied, a smart station can be placed wherever convenient and its coordinates can be obtained, using GPS signals, in a very short time.

Total stations and smart stations can be used for a variety of surveying purposes ranging from first-order triangulation to setting out works. They have immense potential to speed up the work and are gaining popularity.

Summary

Phenomenal developments have taken place in the design features and technology used in survey instruments in the past few decades. Earlier, accurate measurement of distance directly was a time-consuming task. The radar technology used in aviation and other fields has been refined to make it viable for precision surveying.

Electromagnetic distance measurement using electromagnetic waves in the visible light, infrared, and radio waves spectrum has developed considerably over the years. The basic principle involves generating and transmitting a carrier wave modulated by a measurement wave and receiving it back after reflection from a suitable target. The phase difference between the transmitted and reflected waves is determined and the distance is calculated. Most of the instruments use a number of waves in order to obtain better range and accuracy.

In terms of equipment, geodimeters, tellurometers, and mekenometers were used earlier. Instruments using infrared lasers have become more popular now. The instruments may use visible light waves as in electro-optical instruments or microwaves as in microwave-based instruments. The range can vary from 1 km to 150 km.

Total station is a combination of an electronic theodolite with an EDM instrument. Total stations are useful in all types of survey work, from first-order triangulation to setting out of works with great precision. When set up at a point, the total station can accurately measure horizontal and vertical angles and the distance to the target. Modern instruments have very powerful on-board software that can calculate, store, and display coordinates, corrected angles for tilt of telescope, slant distance and horizontal distance, areas, etc. Total stations can be set up and used very easily due to facilities such as automatic target recognition, wireless input units, remote control units, huge memory capacity, reflectorless measurement, data management systems, and data transfer facilities.

A recent development is a smart station that combines a total station with a GPS rover unit. The system makes use of the best capabilities of a total station and the GPS. This makes the smart station a highly versatile instrument.

Exercises

Multiple-Choice Questions

1. Electromagnetic distance measuring instruments use
   (a) radiation frequencies from visible light to microwaves
   (b) radiation frequencies like x-rays
   (c) radiation frequencies like gamma rays
   (d) radio waves

2. Radar is an acronym for
   (a) radiation and ranging
   (b) radio detection and ranging
   (c) radiation and response
   (d) radiation and reporting
3. Electromagnetic spectrum is
   (a) a cluster of radiations carrying energy
   (b) the radiation of visible light
   (c) the colour band in the visible light
   (d) radiation emitted by the sun
4. Modern EDM instruments are based on the principle of
   (a) measurement of phase difference in the transmitted and received radiations
   (b) the cooperative radar
   (c) the speed of all radiations being constant or the same
   (d) high frequency radiations having the same speed in any medium
5. The range and accuracy of EDM instruments vary such that
   (a) short-range instruments have better accuracy
   (b) short-range instruments have less accuracy
   (c) high-range instruments have better accuracy
   (d) the range and accuracy are the same for all instruments
6. A total station is
   (a) a combination of an electronic theodolite and digital level
   (b) a combination of an electronic theodolite and a tacheometer
   (c) an electronic theodolite with levelling capabilities
   (d) a digital theodolite combined with an EDM module
7. A total station can measure
   (a) distances electronically
   (b) horizontal angles accurately
   (c) vertical angles and distances
   (d) horizontal and vertical angles and distances
8. A smart station is
   (a) a total station with software to calculate and display many quantities
   (b) a total station with an integrated GPS module
   (c) with display units on both sides
   (d) a total station attached to a computer

Review Questions

1. List the major components of the electromagnetic spectrum and the use of each type of radiation.
2. Explain the principle on which radar works.
3. Explain the basic principle of EDM.
4. Explain the functioning of a geodimeter and a distomat.
5. Explain the functioning of a tellurometer and a mekenometer.
6. Explain the basic features of a total station.
7. Explain the functioning and capabilities of a total station.

Modern Methods of Surveying

24

Learning Objectives

After going through this chapter, the reader will be able to

- explain the concept and methods of remote sensing
- describe the applications of remote sensing
- explain the basic concept of satellite-based positioning systems
- describe briefly the global positioning system (GPS)
- explain the application of GPS in surveying
- explain the purpose and functions of the geographic information system (GIS)
- briefly explain the applications of GIS

Introduction

In Chapter 23 we discussed some modern instruments. These instruments are undergoing continuous improvement in terms of precision, size and weight, and ease of operation. At the same time the technologies used in surveying are also undergoing change. In this chapter, we will discuss methods based on satellites. Satellite-based remote sensing and positioning systems are two popular methods in surveying.

24.1 Remote sensing

Remote sensing, as the name implies, refers to collecting data from a remote location without being in physical contact with the object. Remote sensing is not as uncommon as we may think. We have many remote sensing activities in day-to-day life. When we see an object and recognize its colour as red, we are using the concept of remote sensing. Similarly, our sense of smell also helps us to use remote sensing. Some of the common methods of remote sensing are described below.

Active and passive system of remote sensing

In an active system of remote sensing, the sensing equipment emits radiation, which is reflected back from the object. Radar is a typical example of such a system. Radar equipment transmits radiation and the reflected radiation is analysed to determine the distance and presence of any object in the ranging area.

In a passive system of remote sensing, the instrument does not generate and emit radiation. The radiation reflected from an external source is made available to the object. We use the passive system exhaustively in the form of the sun's radiation. Taking a photograph using light from the sun is an example. Photographic cameras, still or motion picture, and television cameras use the passive system of remote sensing.

Aerial photography

If we use an aircraft as a platform for photography, then we have a useful remote sensing system. Aerial photography was a very common method for a long time since the world wars. Improvements made in the optics and stability of the platform rendered it a very successful method. It is still very popular because of the following reasons:

- The comparative economy in covering large areas
- Easy availability
- The ability to detect small features and spatial location in difficult conditions that may not be evident even from the ground
- Time-freezing or recording of the features at a point in time
- Better resolution in the visible, ultraviolet, and near infrared bands of the spectrum
- The ability to provide a stereoscopic view of the area from which the horizontal and vertical positions of objects can be ascertained

Photo-interpretation is the study of aerial photographs for identifying ground features in all the three dimensions. Stereoscopes are used to study photographs. They serve many useful purposes in engineering, agriculture, forestry, etc.

Scanning systems

Scanning systems use radiations from many bands of the electromagnetic spectrum to collect data. While photographic methods use the visible spectrum and visible infrared spectrum, scanning systems use reflected radiations in the invisible infrared and radio waves spectrum too. Scanning systems can be based upon platforms in an aircraft or a satellite. Today, there are many satellites dedicated exclusively to remote sensing. The scanning systems are classified based upon the radiation spectrum as follows.

(a) Microwave systems that use microwaves in the wavelength range of 5–500 mm
(b) SLR (sideways looking radars) that works in the microwave length region of 5–500 mm
(c) Infrared thermal scanners that works in the infrared spectral band of wavelength 5–14 μm (10^{-6} m)
(d) Multi-spectral scanners that record in the visible and near infrared region of wavelength 0.3–1.4 μm

24.2 Basic Concepts in Remote Sensing

Remote sensing is essentially used for purposes such as land use and land cover analysis, mineral exploration, urban planning, disaster management, and water resources analysis. The two basic processes involved in this are data acquisition and data analysis.

Data acquisition involves the source of energy (electromagnetic waves) propagated through the atmosphere, the interaction of energy waves with the objects, the reflected energy waves sensed through a suitable sensing system, and the recording of data in a suitable form. Data analysis involves a suitable interpretation device to interpret the data, compiling the interpretation in graphical or other forms, and presenting the data for decision-making.

The electromagnetic spectrum has already been discussed in the previous chapter. These include gamma rays, x-rays, ultraviolet rays, visible light, infrared rays, microwaves, and radio waves. In remote sensing, the radiated waves from the object are scanned and images are obtained, which are interpreted to understand the surface and sub-surface features. Most of the techniques use the visible-to-microwave band of the spectrum.

While we considered electromagnetic energy in the form of waves in the previous discussions, particle theory offers another basic concept for electromagnetic energy. Electromagnetic radiation consists of particles known as photons, or, in general, quanta. This theory is thus known as the quantum theory. The energy of a quantum is directly proportional to its frequency. The proportionality constant is known as Planck's constant.

Energy $Q = hf$

where $h = 6.63 \times 10^{-34}$ J s and f is the frequency of the radiation. As $f = c/\lambda$, $Q = hc/\lambda$, where c is the velocity of propagation $(= 3 \times 10^5$ km/s) and λ is the wavelength. The energy is thus inversely proportional to the wavelength.

24.2.1 Basic Laws of Electromagnetic Radiation

Two basic laws cover the propagation of electromagnetic waves.

Stefan–Boltzmann law The Stefan–Boltzmann law states that the energy radiated by a body is equal to σT^4, where σ is the Stefan–Boltzmann constant and T is the absolute temperature of the body. All bodies above the absolute zero (in kelvin) emit energy. The law refers to an ideal black body that totally absorbs and radiates all energy. This is an ideal radiating body. Figure 24.1 illustrates the Stefan–Boltzmann law in a graphical form.

The graph shows radiation along the vertical axis and wavelength along the horizontal axis. The radiation intensity varies with temperature. The peaks of the curves shift towards shorter wavelengths as the temperature increases.

Wien's displacement law Wien's displacement law states that the maximum spectral wavelength at which radiation occurs from a black body is inversely proportional to its absolute temperature.

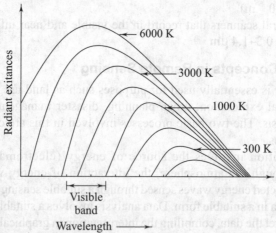

Fig. 24.1 Spectral energy distribution from a black body at various temperatures

$$\lambda_m = A/T$$

where A is a constant and λ_m is the dominant wavelength of the radiation.

24.2.2 Atmospheric Effects on Radiation

Due to the varying properties of the atmosphere through which radiation passes, the following effects take place.

Absorption Some of the radiation from the sun does not reach the earth's surface due to absorption. Examples of these are ultraviolet rays and others of wavelength less than 0.3 μm which are otherwise harmful to life on earth. These rays are absorbed by the ozone layer in the upper reaches of the atmosphere. Carbon dioxide and water vapour also absorb some radiation. Only microwaves and waves of longer wavelength are able to completely penetrate the atmosphere. Absorption takes place in specific wavelength bands known as the *absorption bands*.

Scatter Gas, dust, and smoke in the atmosphere cause diffusion of the radiations, known as scatter. Scatter may be selective or non-selective depending upon the size of the particles. Water particles cause non-selective scatter. Gas molecules cause selective scatter, resulting in haze in the images. Shorter wavelengths are scattered more, as the scatter effect is inversely proportional to the fourth power of the wavelength. Dust, water vapour, and smoke are predominant in the lower layers and affect the radiation in the longer wavelength band.

Absorption and scatter affect the selection of radiation for remote sensing. The following factors need to be considered.

(a) Wavelength (preferably regions with high transmission or low absorption).

(b) Spectral sensitivity of the sensors

(c) The magnitude and composition of the source of energy

The wavelengths for common sensors are shown in Table 24.1:

Table 24.1 Common sensors and their wavelength

Sensor	Wavelength (μm)
Human eye	0.3–0.7
Photography	0.3–2
Thermal scanners	3–11
Multi-spectral	0.3–10
Radar	1 mm–1 m

24.2.3 Interaction of EM Energy with a Surface

As the electromagnetic radiation falls on any type of surface, a part of the radiation gets scattered, a part gets absorbed, a part gets reflected, and some part gets transmitted through the surface. These effects depend upon the nature of the surface and the wavelength of the radiation. The surface itself can emit radiation. Due to these effects, the properties of the radiation change. The properties that vary include intensity, wavelength, direction, phase, and polarization.

Scattering, reflection, and emission are known as surface or area phenomena while absorption and transmission are known as volume phenomena. Both these phenomena are valuable in terms of interpreting the matter or surface feature with which the radiation interacts.

Surface phenomena such as scattering and absorption depend upon the nature of the surface feature. Thus they enable the interpreter to distinguish these features in the imagery formed by remote sensing. The distinguishing capability also depends upon the wavelength of the radiation. To deal with such variations and to enhance the ability to identify the nature of the surface, appropriate sensors are deployed for different radiations. Volume phenomena also depend upon the internal structure and characteristics of the matter. These can then be identified from the interaction of the radiation with the material.

Remote sensing essentially depends upon the reflectance properties of the matter. The wavelengths used are those which produce the maximum reflection from the surface. Flat and smooth surfaces reflect in one direction whereas rough surfaces reflect energy in all directions or in a diffused manner. These are essential characteristics on which remote sensing systems are operated and interpreted.

24.2.4 Resolution in Remote Sensing

Resolution is another important factor in remote sensing. Resolution is the ability to obtain the finer details of an image. Resolution also applies to photographic images. It determines the details of the surface features that can be distinguished. Four types of resolutions are generally considered—spatial resolution, spectral resolution, temporal resolution, and radiometric resolution.

Spatial resolution This refers to the ability of the remote sensing system to record spatial details. The concept of spatial resolution is different for photographs and digital images. In the case of aerial photographs, spatial resolution is defined as the sharpness of the image. In real terms, it is considered with reference to a standard chart such as the number of lines per mm that can be identified from the photograph under magnification. For photographs, this depends upon the camera lens and the film used. In the case of optical scanning systems, resolution depends upon the focal length of the optical system, the height of the imaging platform, and the size of the sensor element. Resolution is usually expressed in terms of the instant field of view—the solid angle through which the system is sensitive to the electromagnetic energy. For aerial photographs, the resolution is defined in terms of lines per millimetre; the equivalent definition for electro-optical systems is in terms of pixels.

Spectral resolution This refers to the wavelengths to which the remote sensing system is sensitive. Aerial photography uses only a narrow band of the spectrum (visible light). The spectral resolution of aerial photographs thus happens to be low. For other scanning systems, a number of wavelength bands are used. The resolution depends upon the number of wavelength bands as well as on the width of each band. For higher resolutions, a large number of bands of narrower bandwidth are used. The type and number of wavelength bands to be used depend upon the information to be extracted from the imagery. The spectral response of different features becomes unique with narrower bandwidths. This allows finer identification of objects from the images.

Temporal resolution This refers to the frequency with which images are obtained in terms of time. For satellite imagery, which is the most common form of remote sensing, this refers to the frequency with which the satellite covers a particular area.

Temporal resolution is important for predicting environmental changes, vegetation cover, and crop patterns.

Radiometric resolution   This refers to the sensitivity of the system to small changes in the radiation (wavelength). In photography, this is referred to as contrast or the number of grey tones in the image. In scanning systems giving digital images, this refers to the discrete levels into which the signal can be resolved during analog-to-digital conversion.

24.2.5   Satellite Remote Sensing

In satellite remote sensing, the remote sensing platform is a satellite. This form of remote sensing started in the 1960s. For remote sensing, unmanned satellites are used. The satellites use different types of sensors and cameras specific to the intended applications. There are two types of satellites based on the orbits they follow. Geostationary satellites orbit over the equatorial plane at very high altitudes (about 36,000 km) and always face a particular region/portion of the earth. This ensures that these satellites receive a continuous stream of data from that portion of the earth. Such satellites are commonly used as communication satellites. The orbital speed of the satellite is the same as that of the earth's rotation. On the other hand, remote sensing satellites use polar or near polar orbits going over the poles. Their orbits are at much lower heights and visit points on the surface at regular intervals.

24.2.6   Problems Confronting Remote Sensing Systems

In practice, remote sensing systems do not have ideal components and environment. Most remote sensing systems are passive and use an external energy source such as the sun's radiations reflected from the surface. The emitted and reflected radiations are not uniform and vary with time and place. This can create problems in the design of the sensing systems. As the platform for remote sensing systems is way up in the sky, the reflected radiation received by the sensors is weak and is modified by atmospheric interference. The sensitivity of practical sensors is limited in terms of the spectral distribution of wavelengths and resolution. To identify the surface features, the interaction of radiation with surface features must be uniquely known for different features. Spectral response patterns, also known as *signatures*, are subjected to many variations, making it difficult to analyse the images and identify a particular feature. Efficient data collection and interpretation is required to utilize the full potential of remote sensing techniques. Satisfying the specific needs of individual users may also create some problems in applying this technique.

24.3   Space Platforms for Remote Sensing

Many countries including India use space platforms provided by satellites for remote sensing. The orbital plane of a satellite varies depending upon the purpose for which it is used. The equatorial orbit of a satellite is one in which the satellite moves in an orbit containing the equatorial plane. Such orbits are said to have zero inclination. A polar orbit, on the other hand, has an inclination of 90 degrees as the orbital plane passes through the poles and is perpendicular to the equatorial plane. Any orbit in between these extremes has an inclination between zero and 90 degrees. Remote sensing satellites are placed in near-polar orbits. The satellites are in general placed in two types of orbits for different purposes. These can be of the following two types:

1. **Geostationary earth-synchronous satellites**   Such satellites are useful when the data or service has to be provided to a designated area only. Such satellites are commonly used for weather-related service or communication networks. The satellite in this case appears stationary with reference to the earth and hence the name geostationary. The orbital period of the satellite is the same as that of the earth and the satellite completes one orbit in a period of one year.

2. **Sun-synchronous orbit satellites**   The orbit of a satellite, when placed in orbit, does not remain fixed. This is due to the rotation of the earth around the sun and the magnetic field of the earth which influences the orbital plane. These influences cause the orbital plane of the satellite to precess with reference to a reference plane. This change in the relative position of the orbital plane causes problems particularly for satellite imaging of the earth. This is because the satellite footprint for a given area occurs at different periods of time and does not give good imaging illumination. This problem is solved by rotating the orbital plane to compensate for its precession. If the precession for rotation of the orbital plane is the same as earth's rotation, then the satellite crosses the equatorial plane at the same time every day giving uniform imaging environment. Such orbits are known as sun-synchronous orbits.

Table 24.2 shows the data for some sun-synchronous satellites.

Table. 24.2   Data for remote sensing satellites

Parameters	Landsat 1, 2, 3	Landsat 4, 5	Spot	IRS-1a	IRS-1c
Altitude	919	705	832	904	817
Orbital period	103.3	99	101	103.2	101.35
Inclination	99	98.2	98.7	99	98.69
Temporal resolution	18	16	26	22	24
Revolution	251	233	369	307	343
Equatorial crossing	9:30	9:30	10:30	10:00	10:30
Sensors	RBV, MSS	MSS, TM	HRV	LISS I, II	LISS III, PAN, WIFS

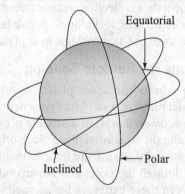

Fig. 24.2   Orbits of satellites

24.3.1 Imaging Sensors

We have mentioned some sensor characteristics earlier. Remote sensing platforms use a variety of sensors for different purposes. Indian remote sensing satellites use a number of sensors for optical and microwave remote sensing. Some of these are as follows:

(a) Linear imaging self-scanning sensors are multispectral sensors for collecting data. These are of four types, LISS I, LISS II, LISS III, and LISS IV.
(b) Multispectral optoelectronic scanner (MOS) used for oceanographic applications.
(c) Wide field sensor (WiFS) collects data in two bands in a wide swath.
(d) Advanced wide field sensors (AWiFS) collect data in a wide swath in four bands.
(e) Panchromatic sensors (PAN) collect data in a single band.
(f) Multifrequency scanning microwave radiometer (MSMR) collects data in four different frequencies.
(g) Ocean colour monitor (OCM) used for oceanographic application in four narrow bands.

Sensors and data collection

The sensors used in remote sensing on the earth can be classified into imaging and non-imaging sensors. The technique may involve scanning or non-scanning methods. Some of the major sensors for imaging and non-imaging purpose are shown in Table 24.3:

Table 24.3 Sensors and their wavelength

Wavelength	Imaging	Non-imaging
Visual (0.4 – 0.75 μm)	Photographic camera, TV camera, optical scanner, laser radar	Spectrometer, radio ranger
Near infrared (0.75 – 1.0 μm)	Same as above	Spectrometer
Middle IR (3 to 5 μm)	Optical scanner	Spectrometer, IR radiometer
Middle IR (8 – 14 μm)	Optical scanner	IR radiometer
Microwave (5 mm to 1 m)	Scanning MW radiometer, MW radar	MW radiometer, radar scattermeter

Sensors can also be classified into scanning and non-scanning types. A photographic camera which takes in the whole view seen by the lens in one go is a non-scanning system. On the other hand, a TV camera is based on scanning the scene across and down, line by line and building up the scene. Remote sensing of the earth uses mainly imaging sensors which provide a visual of the surface. Non-imaging sensors do not provide an image but provide data.

24.3.2 Imaging Techniques

Satellite platforms meant for remote sensing nowadays invariably use digital imaging techniques. The platform can be many like balloons, aircrafts, and satellites. We confine the discussion to satellite imaging. Most remote sensing images are obtained by scanning techniques where images are built up by a number of scan cycles.

Swath The concept of swath can be understood from the Fig 24.3.

The term swath refers to the width of the area scanned by the scanners in the satellite as it moves in orbit. The width, swath, depends upon the number of detectors, spatial resolution, and the orbital height of the satellite. Swath is the width perpendicular to the flight direction. For a satellite because of the height above the surface, swath can be in hundreds of kilometres.

There are two basic methods for getting images by scanning.

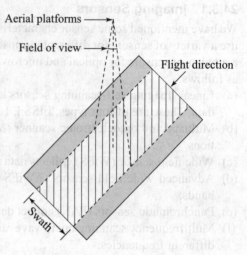

Fig. 24.3 Concept of swath

Across the track scanning As is clear from Fig. 24.4(a), across-the-track scanning by scans perpendicular to the direction of flight in a series of lines. This method of scanning uses a combination of motions—one by the satellite and the other by the opto-mechanical sensor. The scanning system consists of an optical system, spectrographic system along with scanning, detector, and reference systems.

The instantaneous field of view is the segment of the line being scanned and seen by the sensors. This determines the spatial resolution that depends upon the altitude of the space platform. A series of such segments are scanned and this line of segments is across the flight path. This is achieved by a rotating mirror which helps to scan a series of such cells along the line being scanned.

The total angular view of the sensors gives the magnitude of the swath. This angle can be small in the case of satellites because of the orbital height.

Along-the-track scanning The basic principle of along-the-track scanning can be seen from Fig.24.4(b). In this method, a series of line scans are obtained one by one along the flight path. The method uses a series of detectors kept in a line along the swath as shown. The field of view of a detector gives the spatial resolution of the scanning system. Each detector obtains an image of the cell viewed by it and forms the total image of the line scanned. As the platform moves, the next line along the flight path is scanned.

In scanning each line, a linear array is used to detect a spectral band. A series of such linear arrays is required for multi-spectral scanning. The electronic circuitry digitally records the information on CCDs (Charge-Coupled Devices). CCDs are light sensitive silicon chips. Because of their small size thousands of them can be stacked in a small linear space. The electric signal generated by the striking radiation is processed by the electronic circuitry and then made into a digital image of the space scanned.

Along-the-track scanners have the advantage of better spatial and spectral resolution because of the linear array sensors having a longer sensing time.

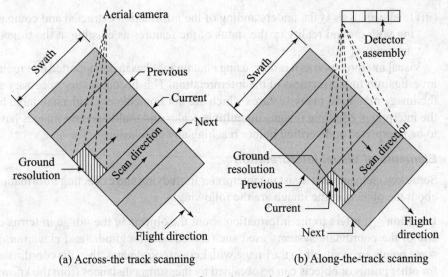

(a) Across-the track scanning   (b) Along-the-track scanning

Fig. 24.4 Scanning methods

24.4 Image Interpretation

Image interpretation is the process of extracting useful information from remote sensing data. Both qualitative and quantitative information can be extracted from maps. Earlier, the data was in analog form which is generally interpreted by humans. Today, the data is generally is digital form which can be interpreted by humans or processed by computers. The correct interpretation of remote sensing data is very important if it is to be useful for the various purposes for which it has been obtained.

24.4.1 Visual Image Processing

The remote sensing data can come in either of the two forms—raw data or processed after certain corrections. Visual images can be monochromatic or grey scale images or colour composites or colour photographs. The objective of visual interpretation is to obtain qualitative and quantitative information about objects seen in the image. This includes finding their size, location, and relationship with other objects. The way our eye perceives an object is different from the way remote sensing data is obtained. First, the image is taken from an aerial platform—an aircraft or a satellite. The view from above will be quite different from the view seen from the ground. Second, the sensors used for imaging record radiations from many parts of the electromagnetic spectrum including the visible band. This makes the imagery look different from what we see otherwise. Third, resolution obtained and scale of the image may be quite unfamiliar to the eye. Finally, the ground relief feature may not be evident in two-dimensional photograph or image. Stereoscopes are used to view photo pairs having common imagery to get a feeling of depth.

The following three processes are involved in image interpretation:

(i) Image reading is the first step in image interpretation and involves identifying objects in the image by their size, shape, pattern, etc.

(ii) Measurement from images is the extraction of information such as length, width, height, and other parameters like density or temperature from data keys as reference.

(iii) Image analysis is the understanding of the information extracted and comparing with ground reality or the status of the features as existing at the time of imaging.

Visual interpretation as it is done using photographs has to be supported by ground investigation for correctness of the interpretation. This becomes very necessary as the image may have many features which are not immediately understandable by the interpreter. Multiple images in multiple scales and multi-spectral images have to be interpreted and verified before reaching any conclusion.

Elements of Visual Interpretation

Some key elements that assist the interpreter in studying and extracting information about the objects in the image are the following:

Location It refers to the information about the objects in the image in terms of any of the coordinate systems used such as latitude, longitude, and elevation. If some points are available in the image with known coordinates, then the coordinates for other points or objects can be obtained by measuring distances from the known points. Actual ground surveys can also be performed using easier methods that use GPS or by traditional methods of surveying that use total station to get coordinates. Computer processing of the image after rectification can also be employed to get information about coordinates.

Size The size of an object seen in an image depends upon the scale of the image. Knowing the scale of the image, the length, width, perimeter or area can be used to extract information about the object. The absolute size of an object along with its relative size with reference to other objects is helpful in interpreting the image. Relative size is also important in distinguishing between features having the same shape. The size can help distinguish between objects of the same shape such as a building or a football field.

Shape The shape of an object is distinguishable in the image and can help the interpreter to identify the object. Objects of regular shapes such as rectangles square, circle or oval are generally man-made structures. Irregular boundaries of an object generally mean that the object is of natural origin such as forest area or a lake. Since the imaging is done from above, it is necessary to know how an object looks from the top.

Shadow Shadows are generally not desirable in images as they change the nature of the image that would have been seen otherwise. However, shadows help in finding the heights of tall structures like towers and multi-storey buildings. Shadows are created due to low sun angles. In addition to aiding in ascertaining the height of objects, shadows also provide a profile view of objects which is helpful in identification.

Tone It is the relative brightness or colour intensity of the image. A black and white photograph is a grey tone image with brightness ranging from black to white. The remote sensing sensor receives and displays a band of the spectrum of electromagnetic radiation and this is displayed as continuous shades of grey which gives different tones in the image. Tones are useful features of interpretation because different objects give unique tonal qualities due to their reflectance. Tonal differences can occur due to different bands in multi-spectral images. Experience and a clear eye help to distinguish the tonal variation.

Colour Colour images are obtained from colour films. Colour photographs or images hold a lot more information than black and white grey tone images. From the natural colour of the image in the film many features like vegetation can be identified. Colour can change depending upon the type of film and filters used. Colour corrections can be done to images to give true colours of the objects.

Texture It can be defined as the characteristic placement and variations in definite patterns for objects in the grey tone image. Textures are classified as smooth or coarse. This is due to the visual impression created by the tonal changes. Coarse textures are due to sudden changes due to abrupt changes in tone in small patches giving a mottled appearance. Smooth texture comes from very little changes in tone. Texture helps to identify objects in an image due to characteristic textures of objectse, especially vegetation and forest trees.

Pattern It refers to the randomness or regularity of similar objects in the image. The pattern seen in the image is helpful in identification. Arrangement of trees in a forest is random, while trees in an orchard are placed in an orderly way. Same is true of houses in a neighbourhood or buildings in a developed area. Such patterns can be identified and the objects recognized from the pattern.

Elevation As mentioned in Chapter 22, stereoscopes are used in association with photo pairs to have a view of the difference in elevation of objects. The overlapping areas of the images in photo pairs are useful in finding the elevations of points and also to have an idea about the relative heights of different objects seen in the image.

Interpretation keys These are used to help in visual interpretation of images. The keys are prepared by experienced interpreters who from past experience and ground verification prepare keys based on major elements of identification. Keys can be prepared for specific uses such as forestry, urban studies, network studies, and so on.

24.4.2 Digital Image Processing

Much of the present-day satellite data is in the digital form. Digital image processing can be done by computer software. The following is a brief description of digital image processing.

A digital image is a three-dimensional array of pixels having values corresponding to the radiation level received at that position. (See Fig 24.5). A pixel is a discrete picture element and positioned in a three-dimensional matrix of such elements. A two-dimensional array of pixels forms the basis for building up a three-dimensional array. A digital image consists of pixels having some parameter value assigned to it in terms of the radiation received by the point. A pixel has coordinates (i, j, k), i and j determining the position of the point in a plane which corresponds to a particular wavelength of radiation or band. The k coordinate comes from the plane in the third dimension corresponding to the band of radiation. The digital image data

line-by-line scanning and the glow of the pixel unit (three phosphors for R,G and B) depends upon the intensity of the driving signal. The colour monitor has three scan beams and compiles the image by line scans starting from the top left corner to the bottom right corner.

The number of colours that can be displayed by the monitor is called colour resolution. It depends upon the memory available in the graphics display card. Video cards can be of 16-, 24- or 32-bit capacity. The 24-bit video card is commonly used for display purposes in video monitors.

Processing digital images

The digital images undergo many processing steps before they become useful. The following procedures are commonly performed with different types of software.

Image rectification   It is the pre-processing technique used to correct the image data for various deficiencies. The two major corrections to be done on the image are radiometric corrections and geometric corrections. Radiometric corrections eliminate errors due to the source (sun's azimuth and elevation), atmospheric conditions, sensor responses, etc. They are also done to improve the fidelity of the brightness value. Different software tools are used to make such corrections. Geometric corrections are due to earth's curvature, motion of the air-borne platform, ground relief, instrumental errors and earth's rotation. Some corrections are also applied based on data collected from ground and maps available.

Image enhancement   It is the process of improving the image quality by processes to assist in image interpretation. These include image reduction processes whereby the image size is reduced in terms of number of pixels in columns and rows to give a suitable size for viewing, colour compositing, where different types of images can be obtained by combining and selecting the three primary colour bands, contrast enhancement for bettering image features as seen by the user, and many filtering techniques.

Image transformation   It includes processes that are intended to produce an image by manipulating and combining image data from many sources, both in terms of multi-spectral, multi-temporal, and multi-resolution images. The output will be an image better than the different image sources from which it built.

Digital image processing is performed generally with computer software available for various uses. This enables the users to get digital images from raw data obtained from satellite imagery.

24.5   Remote Sensing in India

In India, remote sensing using aerial photography was used during and after the British regime. In the 1970s and 1980s, remote sensing data was collected from other countries. Data from many satellites, such as Landsat, NOAA, and SPOT, was used. In 1979, an earth station was set up at Hyderabad, making a landmark in satellite reception in the country. These activities were undertaken by a state agency named National Remote Sensing Agency, NRSA.

The National Remote Sensing Agency (NRSA) is an autonomous organization under the Department of Space, Government of India, engaged in operational remote sensing activities. The operational uses of remote sensing data span a wide spectrum of areas including water resources, agriculture, soil and land degradation, mineral exploration, groundwater targeting, geomorphologic mapping, coastal and ocean resources monitoring, environment, ecology and forest mapping, land use and land-cover mapping, urban area studies, and large-scale mapping.

The activities of NRSA include satellite and aerial data reception, data processing, data dissemination, applications for providing services, and training and distribution of data from foreign satellites such as RADARSAT, IKONOS, Quickbird, and OrbView. NRSA established its own ground station at Shadnagar, 60 km south of Hyderabad, to acquire remote sensing satellite data from Indian remote sensing satellites. Data is also collected from foreign satellites such as Landsat, NOAA, ERS, Terra, and Aqua. NRSA also has facilities for aerial remote sensing using two modern aircraft (Beachcraft 200) having INS, K-GPS fitted with multi-spectral scanners, photogrammetric cameras, SAR, and electromagnetic sensors. The agency is also engaged in designing, developing, and deploying multi-sensor satellite-based ground systems comprising ground and application segments to meet domestic and international requirements. The acquired data is processed and a variety of data products are made available to users. The Indian Institute of Remote Sensing, located at Dehradun, is an organization established by NRSA for training professionals in remote sensing and GIS. NRSA has its headquarters at Hyderabad, which also organizes training programmes.

India has launched a number of remote sensing satellites since 1988. The first satellite IRS-1A was launched in March 1988, which heralded an era of remote sensing in India. The satellite had two LISS sensors which supplied valuable data that was used in large-scale mapping applications. The second satellite, IRS-1B, having similar sensors, was launched in August 1991. The two satellites together provided better receptivity. In December 1995, IRS-1C was launched with LISS-III, PAN, and WiFS sensors. IRS-1D was launched in September 1997; this further strengthened the scope of remote sensing with increased coverage and was used in applications such as resources survey and management, urban planning, forest studies, disaster monitoring, and environmental studies. To test the launch vehicle programme, IRS-P3 and IRS-P4 satellites were launched. IRS-P3 had an x-ray astronomy payload for space science studies, it also had WiFS (wide field sensor) and MOS (modular optoelectronics scanner) sensors. IRS-P4, a satellite dedicated for ocean applications, was launched in May 1999. Its OCM (ocean colour monitor) and MSMR (multi-frequency scanning microwave radiometer) sensors have opened up new vistas in ocean studies. The launch of IRS-P6 (RESOURCESAT-1) in October 2003 provided an excellent opportunity to obtain high-resolution, multi-spectral and moderate-resolution continuous data. IRS-P5 (CARTOSAT-1), launched on 5 May 2005, provided much greater capabilities to the Indian remote sensing programme.

The following are the Indian remote sensing satellites currently in service:

(a) Cartosat 2 launched on 10 Jan 2007
(b) Cartosat 2A launched on 28 April 2008
(c) Oceansat 2 launched on 23 September 2009
(d) Cartosat 2B launched on 12 July 2010
(e) Resourcesat 2 launched on 20 April 2010

Future launches

Following are the remote sensing satellites planned by Indian Space Research Organization to be launched for strengthening the fleet of IRS satellites and widening their applications:

1. **Risat (Radar Imaging Satellite)** A microwave remote sensing mission with Synthetic Aperture Radar (SAR) that operates in C-band and has a 6 × 2 metre planar active array antenna based on trans-receiver module architecture. SAR is an all weather-imaging sensor capable of taking images in cloudy and snow covered regions and also both during day and night. RISAT weighs 1,750 kg.

2. **RESEOURCESAT-3** A follow on to Resourcesat-2, it will carry more advanced LISS-III-WS (Wide Swath) sensor having similar swath and revisit capability as Advanced Wide Field Sensor (AWiFS), thus overcoming any spatial resolution limitation of AWiFS. The satellite would also carry Atmospheric Correction Sensor (ACS) for quantitative interpretation and geophysical parameter retrieval. It is slated to be launched in 2012.

3. **CARTOSAT 3** A continuation of Cartosat series , it will have a resolution 30 cm and 6 km swath suitable for cadastre and infrastructure mapping and analysis. It would also enhance disaster monitoring and damage assessment. It is slated to be launched in 2012.

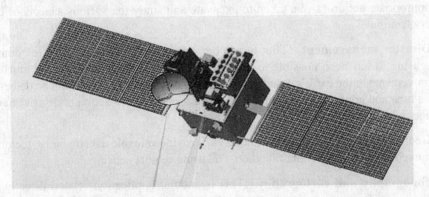

OCEANSAT 2 Satellite

4. **OCEANSAT-3**: Oceansat-3 would carry thermal IR sensor, 12-channel ocean colour monitor, scatterometer, and passive microwave radiometer. IR sensor and ocean colour monitor would be used in the analysis of operational potential fishing zones. The satellite is mainly for ocean biology and sea state applications. It is slated to be launched aboard PSLV in 2012–13.

24.6 Applications of Remote Sensing

Remote sensing has applications in a wide spectrum of areas. Remote sensing data can be used for taking sound decisions for planning many human developmental activities. It is also possible to take preventive action as in the case of forest fires and natural disasters. Weather forecasting is another important application. Some of the application areas are given below.

Land use and land cover analysis Perhaps one of the prime uses of satellite remote sensing is in the study of land use and land cover. Land cover through vegetation and specific crop areas can be studied using remote sensing data. Forest cover is an important aspect, which has been studied; the depletion of forest areas has been identified with the help of remote sensing. It is also possible to study crop diseases over large areas.

Mineral exploration It will be possible to use satellite data and discover the presence of valuable minerals and ores that are vital to economic development. Non-renewable energy resources, such as fossil fuels, can be identified using remote sensing data.

Environmental studies Global weather phenomena are a major area for study using remote sensing data. Global warming and ozone-layer depletion can be continuously monitored using remote sensing. Similarly, oceanographic studies also provide valuable information about the various characteristics of oceans around the world. Assessing water resources, their extent and depletion, snow cover studies, etc. have proved to be very valuable.

Archaeology Archaeological studies can make use of remote sensing data. The underlying old settlements can be recognized from remote sensing data and appropriate action can be taken to excavate and study the various aspects of old civilizations.

Disaster management This is another important application area of remote sensing. It has been possible to predict earthquake hazards by detecting unusual movements in the earth's crust. Floods, landslides, forest fires, etc. can be detected on time and appropriate action can be taken for preventive action in disaster management.

Geomorphology Geological studies can provide valuable data on faults, tectonic movements, rock-type identification, etc. using remote sensing data.

Topography and cartography This is another application related directly to surveying. Remote sensing can be used to accurately locate points with reference to ground coordinate systems when ground surveys are difficult or time-consuming. This data can be used to prepare maps or revise existing maps.

Other applications Remote sensing data is now being used to study troop movements, etc. for defence purposes. Other applications include urban planning studies, traffic studies, assessment of earth's resources for various purposes, and so on.

24.7 Satellite-based Positioning Systems

Satellite-based positioning systems are becoming more and more popular. The global positioning system (GPS) is operated by the US Department of Defence. GLONASS (Global Navigation Satellite System) is run by the Russians and Galileo is run by the European Union. These systems work generally on the same principles. Such positioning systems are generally used for navigation purposes and also find applications in surveying. GPS enables the user to locate his/her position in three dimensions as well as with respect to time. The system is run with the help of satellites launched for this purpose.

The systems used earlier were not very precise. The TRANSIT system run by the United States and the TSIKADA system by Russia are two such systems. Six satellites were used by the TRANSIT system primarily to locate the positions of aircraft and sea vessels. The system worked on the principle of Doppler effect, which is the change in frequency of signals transmitted when the source is moving. It is possible to compute the location of a ground receiver by receiving signals from the satellite as it passes over by moving in and out. The accuracy of positioning was not high and the frequency of obtaining position coordinates was unsatisfactory. This led to the development of modern systems.

24.7.1 Basic Principle of GPS

For the purpose of discussion, we describe the GPS run by the US Department of Defence. The system has a minimum of 24 satellites. The actual number may be more as new satellites are launched to replace old ones. The Galileo system has 30 satellites.

Satellite configuration

The satellites are placed in orbits such that there are six orbits having four satellites each. One needs to receive signals from at least four satellites to uniquely determine the position of the user. Figure 24.6 shows the orbits of the satellites and their planar projection. GPS receivers, hand-held or otherwise, are used for navigational purposes.

The satellites are placed in nearly circular orbits at a height of nearly 26,500 km. Orbital time is nearly 12 hours. Orbital planes are at 60° and are inclined at nearly 55° to the equatorial plane. Each satellite can be uniquely identified by a pseudo-random number (PRN). Each

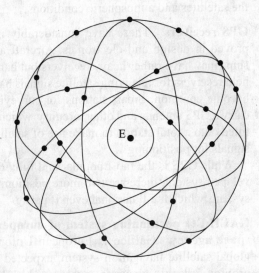

Fig. 24.6 GPS satellite configuration

satellite also produces two unique codes, the coarse acquisition code (C/A code) and the precise code (P-code). The C/A code is a 1-MHz signal repeating continuously. The P-code is a 10-MHz signal repeating with a seven-day frequency.

The satellite vehicle system of 24 satellites is the main platform for the GPS. The control segments consist of a master-control station, monitoring stations around the world and ground antennae. The satellites send signals on two frequencies identified as L_1 and L_2. The L_1 frequency is 1575.42 MHz and L_2 frequency is 1227.6 MHz.

Both these signals are modulated using the C/A code and P-code. Only the C/A code is available to civilian users. The L_2 signal carries the P-code only.

The C/A code is a PRN code that repeats continuously every millisecond. The code of each satellite is unique and the satellite can be identified by its PRN. This code modulates the L_1 and L_2 carrier signals and is the basis of civilian positioning systems.

The P-code modulates only the L_1 carrier frequency. The P-code is a long duration code having a seven-day cycle. This is the basis for precise positioning systems and is available to the military and other authorized or encrypted users.

Positioning using satellite signals

The distance between the satellite and the receiver can be computed using the satellite signals. This puts the receiver on the surface of a sphere with the satellite at the centre. With two more signals received from two other satellites, the position of the receiver can be located at any two points with three coordinates. Any ambiguity in position is resolved with the position of the receiver. The signal from a fourth satellite can be used to locate time. Thus with signals from four satellites, the position and time can be determined. The satellite configuration is such that between five and eight satellites are in the range of a receiver.

The satellites have atomic clocks to give accurate times, which are also monitored by monitoring stations. The positioning accuracy depends upon many factors of the satellites and atmospheric conditions.

GPS receivers   These have considerably improved in design and electronics. Aircraft and ships may have rather heavy receivers but hand-held receivers for receiving satellite signals have become common. Smart stations for surveying carry a GPS antenna and other circuitry, which is fitted onto a total station to make use of satellite signals for positioning.

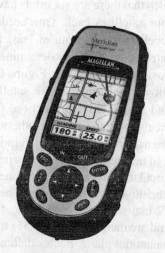

While GPS is the most popular and fully developed system, there are two more positioning systems which are functional even if partly.

GALILEO positioning system of european space agency   Galileo is European Union's global satellite navigation system, expected to provide a highly accurate, guaranteed global positioning service under civilian control. When fully operational, it will be compatible with GPS

Hand-held GPS receiver

and GLONASS, the two other global satellite navigation systems.

Galileo offers dual frequencies as standard. It delivers real-time positioning to an accuracy of one metre. The system will guarantee availability of the service under all but the most extreme conditions. The users will be informed of any breakdown in service due to satellite failure within seconds. The service will be useful for safety-critical applications such as guidance to vehicle users, train service, and aircraft landing and take-off.

European Space Agency's first two navigation satellites, GIOVE-A and-B, were launched in 2005 and 2008 respectively. Radio frequencies were set aside for Galileo by the International Telecommunications Union for testing key Galileo technologies.

On 21 October 2011, the first two of four operational satellites were launched. These were designed to validate the Galileo concept in both space and on earth. These were expected to verify the validity of the space, earth and user segments of the system. Two more will be launched in 2012. Once this in-orbit validation (IOV) phase has been completed, additional satellites will be launched to reach Initial Operational Capability around mid-decade.

Galileo services will be the same as provided by GPS and GLONASS. The services will guarantee quality and integrity. The first global positioning system provided by Galileo is expected to herald the success of this first complete civil positioning system from the basically military systems that have been put in place earlier.

A wide spectrum of services will be available once the system is built up from the initial validation service position and when the system becomes fully operational.

The fully deployed Galileo system will consist of 30 satellites. Of these 27 satellites will be fully operational and three satellites will be available as active spares. These will be positioned in three circular medium earth orbit (MEO) planes at 23,222 km altitude above the earth. The orbital planes will be inclined at 56° to the equatorial plane.

GLONASS   Russia's *Global Orbiting Navigation Satellite System* (GLONASS) is similar to GPS and Galileo as a satellite navigation system. The network of navigation satellites in orbit will function as a global positioning system similar to the American and European GPS networks.

GLONASS is operated by the Russian Space Forces. The system uses radio time signals to locate people and vehicles on and above the surface of earth and also will provide a variety services for navigation and other applications.

GLONASS is a system similar to the US GPS network. GLONASS satellites function in a very similar manner as the US NAVSTARs. The satellites of GLONASS orbit at about 12000 miles (19200 km) above, the earth and transmit signals on two frequencies in the 1200–1600 MHz range. These orbits are a little lower than the GPS satellite orbits.

The orbital period of the satellites is 11 hours 15 minutes. They are positioned in space in orbits such that a user on the ground can see at least five satellites at any time. GLONASS satellites transmit standard precision (SP) and high precision (HP) signals at a frequency around 1.6 GHz.

The GLONASS system offers a standard C/A positioning and timing service giving horizontal position accuracy within 180 feet (55 metres) and vertical position within 230 feet (70 metres) based on measurements from four satellite signals. P is a more accurate signal for Russian military use.

There are very few inexpensive GLONASS-only receivers for consumers on the market. However, commercial GPS receivers often are capable of receiving both NAVSTAR and GLONASS data.

Earlier, GLONASS satellites had very short space life. Today, GLONASS-K satellites have a 12-year life and offers greater accuracy and precision in determining locational coordinates.

Russia initiated GLONASS by sending the first operational satellites into space in December 1983. It continued building the GLONASS system by launching more satellites. The system was in full operation in December 1995.

Russia is improving the navigation satellite programme by launching improved GLONASS–M satellites with a seven-year working life. Six satellites were launched by 2005. Russia has also developed a next generation GLONASS–K satellite with improved capabilities and a 12-year life.

Russia also took help from India to launch two GLONASS–M satellites by ISRO. The complete GLONASS system is designed to have 24 satellites, with 21 operational satellites for services and three satellites as back-up active spares.

The GPS technology basically can perform the following functions:

(a) **Determining position** GPS helps us to locate the receiver, in terms of geo-coordinates of the station, on GPS was the first positioning to help people to locate the position of something or someone in three-dimensional space on the planet. It has immense applications in surveying which has the same basic objective.

(b) **Navigation** Another basic objective of GPS is to provide information about how to navigate from one place to another and to get route details. This aspect of GPS finds wide application in the movement of all types of vehicles on the road, boats and ships and aircraft. Today, many vehicles are provided with simple GPS systems for route determination to navigate from one place to another.

(c) **Tracking** GPS is used to track vehicles in motion like taxis, transport vehicles, ambulances, ships and aircraft. GPS used in conjunction with communication equipment and computer systems provide an answer for automatic vehicle location. In the case of taxis, fire engines and ambulances, this helps to locate the nearest available vehicle for activation in case of emergency reducing the reaction time.

(d) **Mapping** A major application of GPS as a part of surveying is preparation of maps. The accurate surveying of a vast area using conventional equipment was never an easy task. The Great Trigonometrical Survey of India and the Great Survey of Britain were done over many years with rudimentary equipment facing many hardships. The GPS has made it very easy to conduct such surveys very accurately and prepare maps of all geographic features.

(e) **Precise time determination** The atomic clocks provided in the space and ground segments of the GPS system make it very convenrient for any user to accurately determine the time and synchronize time for many uses.

24.5.2 Application in Surveying

The GPS is basically used for navigation control. Sea and aerial navigation use the GPS extensively. Since GPS provides a means to locate positions, it is evident that it can be used for surveys, as surveying is the process of locating points on the surface of the earth. Since GPS receivers need to receive signals from satellites, it is clear that the ground equipment should be placed such that it is possible to receive signals from satellites. There are many ways in which this technology can be used.

A GPS receiver can determine positions when it receives signals from at least four satellites in its range. Due to many problems in receiving signals, the position determined might not be accurate. A differential GPS technique is used in such a case. In this technique, the GPS can be used for base line measurement. With two receivers placed at either ends of the line, their coordinates can be worked out and the distance can be calculated. One receiver, known as the reference receiver, is placed on a point of known coordinates. A second receiver is placed at other positions and the distance between the points can be calculated. The signals are sent from the reference receiver through any communication mode such as mobile, radio, or the internet. The rover receiver receives these signals and, using the signals received from the satellites, computes its own position. The coordinates computed are the latitude, longitude, and the height above a chosen datum. Accuracies to the extent of 10 mm + 2 ppm of the length are achievable.

Recently, a technique of Real Time Kinematic (RTK) was developed for precise positioning. In RTK, a reference receiver is placed at a point of known coordinates. The reference receiver receives phase and code measurements from satellites and transmits them to the rover receiver. The rover receiver receives its own data of phase and code measurements from the satellites. The differential phase requires special dual frequency receivers which can receive measurements from L_1 and L_2 frequencies. A high level of accuracy can be achieved by such measurements.

GPS can be used in base line measurements for geodetic surveying, staking out, and all other forms of surveys.

Surveying with GPS   Within a few years of its launch, GPS has revolutionized the functioning of many operational systems with precise location and communication. The technology developed very fast so that it is available to many civilian users for a variety of purposes. The technology has found its way into transportation vehicles like cars, buses and trucks, marine navigation vehicles like boats and ships, aircraft, construction equipment, farm machinery, and even laptop computers. It is then not surprising that surveying which requires accurate measurements for positioning of points on the earth's surface is one segment where this technology has a great impact. We have mentioned about smart stations which have an integrated GPS module for use with total stations. There are also GPS instruments developed exclusively for surveying.

Leica GS12 GPS Smart Rover

Surveying with hand held GPS   With hand-held GPS instruments, it is possible to undertake survey work accurately for many purposes.

Two of the recent GPS survey equipment and their brief descriptions are given below:

Leica Geosystems has launched the Leica GS12 GPS smart rover for surveying purpose. This is one among the many GPS survey equipment in their supply line. The equipment boasts of the following features:

Proven GNSS technology

Built on years of knowledge and experience, the Leica GS12 delivers the hallmarks of Leica GNSS—reliability and accuracy.

- SmartCheck – RTK data-processing to guarantee correct results
- SmartTrack – advanced four-constellation tracking of all GNSS satellites
- SmartRTK—delivers consistent results in all networks

Light weight and full functionality

The Leica GS12 delivers ultimate ergonomics through extreme light weight.

- It weighs only 1 kg, ideal for ergonomic handling with ideal balance.
- It supports full GNSS compatibility: GPS, GLONASS, Galileo, and compass.
- It has fully scalable sensor that allows you to buy only what you need today and upgrade with additional functionality as you need it.

Rugged

The Leica GS12 is built for the most demanding environments.

- IP67 protection against dust and immersion to 1m
- Built for extreme temperatures of –40°C to +65°C
- Withstands 2 m pole topple over test
- Complete cable-free operation

Sokkia has launched its latest advanced system for surveying professionals around the world. The new GSR2700 ISX features a fully integrated, triple-frequency, high-performance receiver with GPS plus Russian-based GLONASS satellite tracking capability. The receiver offers seventy-two universal channels for increased satellite coverage and improved performance on the job.

Sokkia GSR 2700 ISX; also see Plate 2

The GSR2700 ISX also features numerous additional enhancements, including improved RTK performance, seamless virtual reference station support, multiple bluetooth connection options and support for GPS L2C and L5 signals. The system is completely cable free and extremely easy to set up and operate in base and rover modes. The receiver is also the only one of its kind to offer audible status notification in the field, now available in multiple languages and generic.

Equipped with rugged magnesium alloy housing, the GSR2700 ISX provides complete protection against dust ingress and is water immersible till 1.0 m (3.3 ft).

Application of GPS

GPS applications for survey work include the following:

(a) Geodetic work – High-precision zero-order national network of geodetic survey control network can be established with GPS stations.
(b) Existing primary control network points can be strengthened and adjusted and densification for finer mesh can be achieved.
(c) Offshore points such as in islands can be connected to mainland network of control points easily.
(d) Hydrographic surveying is greatly facilitated by using GPS stations to establish control points in water bodies.
(e) Highly accurate vertical control network can be established from the three -dimensional coordinate system provided by GPS stations.

24.8 Introduction to GIS

Geographic Information System (GIS) is a data management system that provides many facilities for surveyors and planners. The difference between GIS and other data management systems is the need to have special database management wherein the locational coordinates of points need to be maintained.

GIS is a computer-based information management system which collects and stores spatially referenced data with other relevant attributes and enables us to manipulate, analyse, and display, in suitable formats, such data for various planning and design purposes. The name implies that this system deals with the geographic space on the earth and its features. The data is organized in such a way that it provides useful information to the users. It is a system because it is composed of many interrelated components. One of the earliest GISs was developed in Canada and is credited to Tomlinson. This was a true operational GIS for land records and many other uses.

The data for the GIS can come from many sources. The data may come from existing maps, surveys, aerial photography, or satellites. The data is digitized using scanners to make it compatible for computer processing. The basic objectives of the GIS are as follows.
(a) To collect, analyse, and manipulate spatial data
(b) To produce maps and other products in standardized formats for use by different agencies
(c) To supply information in useful formats for logical decision-making
(d) To support research activities using spatial data

Definition of GIS   GIS has been defined in many ways by many people. We define GIS as follows:

Geographic Information System is a special purpose information system designed to acquire, store, retrieve, manipulate, analyse and output in desired forms geo-referenced data for the purpose of planning and management of land, water and other natural resources, transportation, environment, urban facilities, socio-economic development projects and other administrative records.

Components of GIS

GIS is considered to have four key components—hardware, software, data, and users.

Hardware for GIS   It is essentially computer hardware and peripherals. GIS can run on many types of systems from personal computers to mainframe systems to networked supercomputers. The developments in computer hardware have been tremendous in the past few decades with costs coming down and system capabilities increasing manifold. These have made it possible to design and implement GIS systems at a comparatively cheaper cost.

The computer systems have to be supported by peripheral systems for both input and output. Input devices include scanners, digitizers, GPS receivers, and many forms of storage devices such as disks and tapes, CD ROMS, and optical discs. These have to be supported for the user interface with output devices such as display units, printers, and plotters.

Software   Software refers to the programmes that run in computers. There are basic software that manages the computer system and special software for specific applications. The GIS software is a special application software that provides the functions and tools necessary to acquire, store, retrieve, analyse, and manipulate input and output data in a desired form desired by the user.

Some common types of software for GIS applications are a group of software by the brand name arc by ESRI like Arc-GIS, Arcview, ArcSDE, ArcIMS, and ArcINFO. and. Microsoft software packages like Geographics, Imaging solutions, Geowater and Geowaste water, and GEOPAK are also useful for many GIS applications. GIS solutions from PCI form another set of software packages used for GIS applications. The software packages can be customized to perform in stand-alone systems or networked environment.

Data   Data for GIS applications comes from many sources. Data for GIS is georeferenced, that is, they have spatial reference in addition to attribute data. The special nature of GIS is the integration of spatial data and attribute data. Spatial data that is obtained from digital maps, satellite images, and aerial photography. Attribute data comes from many documents and can be integrated with the spatial data in GIS. Digital map forms are the basic data input for GIS. Attribute data related to the maps can be in any of the document forms like tables.

Users   These are the people for whom the GIS is developed. There are different users looking for different kind of information and analysis and output in desired formats. Customized applications can thus be developed for classes of users. Geospatial data analysis is the core of GIS. This finds wide applications in a variety of areas by users from different walks of life. Their demands have to be met by the system.

Features of GIS

As mentioned earlier, GIS is specifically intended to integrate geospatial data and attribute data provided in many forms as documents like tables which are related. GIS uses themes, or layers which contain different types of information and which can be integrated to provide a composite geographical information. The essential objective is to provide information about features, objects, and classes in a geo-referenced format. GIS can be considered as information provided in three basic formats:

Data or geographic information is provided in a variety of forms about features, objects and classes. The common underlying feature is that the information is based on a structured database that describes the world around with geographic reference.

Maps are one of the basic features of GIS; intelligent, digital maps can be obtained that give complete description of required features or objects. This helps the users to query for a variety of information, edit or update the information, and analyse the data for planning and decision-making.

Modelling can be done with GIS to represent the abstractness of geographic reality into an analysable form. Modelling is used for a variety of purposes for analysis and decision-making about geography-related systems.

GIS is knowledge base for a variety of applications. GIS is a multi-disciplinary tool for planning and decision-making. It has inherent relationships and applications in a variety of disciplines such as geography, photogrammetry, remote sensing, computer science, information technology, environment, mathematics and statistics, civil engineering, social sciences, surveying and geodesy, and operations research.

GIS subsystems

GIS has a number of subsystems intended for specific uses and applications. These include the following:

Data acquisition   The GIS acquires data from maps and a variety of image formats. Such data can be input to the system for storing in the digital database.

Data processing and analysis   The database created can be accessed by users and analysed in a variety of ways. The resultant output can be given in a format desired by the user as maps, tables or other types of information as displays, hard copies or transferred to other storage devices.

Communication   It is another important aspect of GIS. The database of the GIS has to acquire information in a fast and real-time environment. This is a very difficult stage to design and implement. The system has to have a variety of communication channels to acquire data available in different formats. Communication channels are also important for the users to have the required information output in a variety of ways through display units, printers, and plotters.

Management of GIS   The system has to be planned and implemented effectively. This requires extensive planning and interactions with users. The organization, implementation and use of GIS require specially trained managers and staff who can design, operate, look to the needs of users, and maintain the system.

GIS Capabilities

GIS, in general terms, is the acquisition of data from different sources and integrating them with locational parameter as common identifying factor. GIS must be able to answer the following questions of the user.
(a) What is existing at a particular location? The user can specify the location using coordinates, geographic reference, or name of the place.
(b) Where is the location of something that satisfies certain conditions? This is the inverse of the first question and the user can identify the location of something he wants to find. This could be a hospital or a school or a police station.
(c) What has changed in a location over a period of time? As developments take place in a geographic area, there will be changes taking place in many uses and features in the area. The user can find the changes that have taken place over time in the given location.

(d) What spatial patterns exist and what anomalies are found in the patterns? Often, one can find patterns of habitats developing near water sources or near transportation facilities.

(e) What will happen if we do something? Such a question can be answered by modelling the feature in the geographic area and determining locational factors, risks involved, etc.

Operating GIS

The following are the steps involved in operating a GIS system:

Acquiring and relating information from different sources   A GIS can acquire and store information from many sources with integrated location data. The information like maps if they are not in machine readable form can be converted to digital format in many ways. GIS can not only acquire and store such information in various forms, but can also relate them based on criteria specified by the user.

Data storage and management   The computer system must provide facilities of storing the large amount of information that forms the vital part of the system. The major points to be considered are data security, data integrity, data storage and retrieval, and data updating and maintenance.

Data manipulation and registration   The data available in different forms may be required to be converted for integration and collation. Maps drawn in different scales can be adjusted for integrated use. Similarly, many projection systems are used to prepare two-dimensional maps of three-dimensional surfaces. Each projection is appropriate for some use. GIS has the capability to make projection in required form from any given geographic data. Data acquired in different ways and formats may not be compatible for integrated usage. GIS has the capability to convert geographic data in different data structures to any of the useful data structure forms for use by different users.

Spatial data modelling   There are many ways in which data is modelled in GIS. The raster approach of cell attributes or the vector approach of points, lines and polylines are modelling tools for different purposes. Spatial data modelling allows the user to spatial characteristics of a set of real-world phenomena and their interrelationships. This concept of spatial data modelling is implemented in GIS using spatial data structures and manipulated by various algorithms to understand the structure and behaviour of the real world. It is also possible to model different scenarios anticipated and analyse them for the likely outcomes.

Spatial analysis   The analysis of the collected information for various uses qualitatively and quantitatively is a powerful feature of GIS. This is done by functions available in GIS such as spatial interpolation, buffering, and overlay operations.

Data Types for GIS

There are two basic data types that are used in GIS in addition to other data models.

Raster data   This is a series of cells or dots or pixels that represent an image (see Fig 24.7) Raster data is obtained when you scan a paper image, blueprint, or photograph. The raster data is a series of cells to which values are assigned. This column–row format is used to represent geographic data and other images. Quality of raster data depends upon the cell size. Small size cells increase the reliability because there are more values to represent the image details. Raster models are simple to create and use.

Vector data This consists of points, lines and polylines, and arcs. This is similar to traditional approach of representing objects in a map. They form a group of mathematical equations that generate points, lines, a series of lines or arcs. Points are represented by coordinates.

Lines are represented by a series of points having coordinates related by an equation. Polylines are a series of interconnected lines which can be used to create an object. Arcs are similar to lines and can be created by a set of points with coordinate pairs. Each one of these basic entities can be given a name or label.

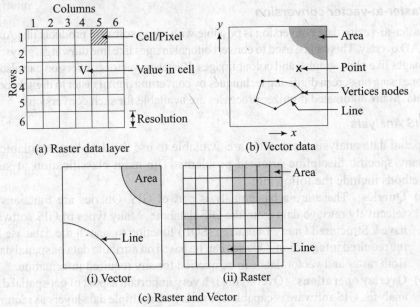

Fig. 24.7 Data models

These basic entities can be used to create an area or a volume or any other object by describing the interrelationships.

In representing real-world objects using vector data, the scale of the map is very important. A road may be represented as a polyline or an arc in a certain scale map. For a different scale and purpose, the same road may have to be represented as a set of parallel lines. Depending upon the scale, many objects like buildings may be represented as points.

Object-oriented data Another way to represent data is as objects and classes which have certain relationships. An object is an entity having certain properties and some operations can be performed on them. Many such geographic objects can be put in a class. Road can be a class where objects like state highways, national highways, express ways, and village roads can be included.

Advantages and disadvantages of the data models

The following are the advantages and disadvantages of the data models:

Raster data   Raster data has the advantage that it is a simple data structure comparable with various forms of images created, has efficient overlaying procedures, can represent continuous data efficiently, and spatial variability can be easily accounted for. The disadvantage of raster data is the large memory requirement, difficulty in deriving topological relationships, and difficulty in editing and transformation of data.

Vector data   This requires less memory space, easy to edit, discrete objects can be easily represented, more accurate for mapping and transformations and topological relationships. The disadvantage of vector data are that it is more complex, cannot represent continuum easily, overlay operations are difficult, less amenable to represent many forms of images, and less efficient in describing spatial variations.

Raster-to-vector conversion

Raster-to-vector data conversion is possible with many software products like Auto-CAD overlay. This can be used to convert bitonal images like line drawings, grayscale images like photographs, and colour images. Raster-to-vector conversion can also be done using on-screen digitizing techniques by converting digital raster to digital vector data. Many automated digitizer processes are available for such conversions.

GIS Analysis

Spatial data analysis techniques are available to use the data-base available in many specific discipline-oriented problems. The main classification of such methods include the following:

(a) **Queries**   These are a basic analysis part of GIS. Queries are functions to selectively retrieve data from the GIS database. Many types of GIS software have a Structured Query Language (SQL) function to search the database for the required information. Queries can be based on attribute data or spatial data. Both raster and vector data can be queried for any required information.

(b) **Overlay operations**   Overlaying is a very important aspect of geospatial data analysis. GIS software is capable of overlaying multiple data layers as required in topological structures. Overlaying of polygon data over polygon data is the most popular and useful overlaying operation. The overlaying function is to overlay any of the data models, raster or vector, to create new thematic structures. Points, lines and polylines, polygons can be overlaid on polygons to get new topological themes.

(c) **Spatial measurement**   This can also be of many types. Measurement of distance between two points is the most common function. You can also measure feature densities based on any requirement.

(d) **Surface analysis**   Given a surface in any of the data models, the surface characteristics can be analysed to find many different features of the surface. As an example one can find a contour line on a given surface by querying for points having the same elevation.

(e) **Network analysis**   This is a useful function to study networks like roads, railways, communication, waste disposal systems, water supply systems, and so on. Network analysis provides a range of tools and techniques to analyse networks for capacity, connectivity, flow patterns, and efficiency.

(f) **Statistical analysis**   With data availability for discrete points, the analysis is basically an interpolation technique for spatial distribution of the data. Many algorithms of interpolation for deterministic and probabilistic analysis are available for various types of analysis.

(g) Visualization   This provides tools for representation of data in 2-D and 3-D presentations through maps, images, and even virtual reality models.

These analysis techniques enable the user to find a large amount of information and interpolated results, maps, etc. from the database.

Integration of remote sensing and GIS

The application of remote sensing techniques has been described earlier. The data from remote sensing forms a very important part of GIS applications. Remote sensing is important to study spatial and temporal phenomena. The data from remote sensing can thus be ideally combined with the spatial data in GIS, many important information can be generated which are vital for studies in many areas. Remote sensing, combined with GIS, gives a very powerful tool to study various phenomena and take important decisions that support developmental activities and for creating a sustainable development model. GIS and remote sensing thus complement each other in bringing the real world as close as possible. This supplemented with real-world field studies will make things much easier for us in decision-making.

Applications of GIS   GIS finds wide applications in many areas. Wherever locational studies are required, GIS provides the best tools for study and research. GIS has revolutionized the use of geographic (spatial) data in a wide variety of formats for a wide spectrum of applications. The applications include business and industry, state agencies, military, and academic research. GIS has become an essential tool for understanding the geography of the world.

Surveying and mapping are the primary areas of GIS application. Resource and environment management and design of public utility services are areas where GIS can play a significant role. Forest resource management, crop management, mining and mineral exploration, and transportation-system design and development are areas where GIS can play an important part.

GIS can also be used by the military for effective management of defence services. Research areas in engineering science and humanities are other areas where GIS can play a crucial role. Land records and land management is a critical area for planners. Weather analysis and prediction of weather patterns can also be done using GIS software.

The application areas are spreading over a wide range of fields such as topographic mapping, cartography, water and water resources management, and transportation engineering. Analysis of many phenomena on a time scale for various purposes in vegetation and crop studies, crime mapping, population patterns, and various environmental studies are possible with GIS.

Some of the major application areas of GIS can be listed as follows:

Natural resources management

- Forest inventory and modelling, forest cover analysis
- Time-based studies on natural resources
- Wildlife management
- Land and water body management
- Locational study and mapping of endangered flora and fauna
- Monitoring environmental hazards
- Impact analysis of natural and man-made processes with respect to the resources

Mineral resources management
- Exploration of oil, coal and other mineral resources
- Assessment of status of mineral deposits
- Continuous monitoring of the impact of resources extraction
- Rehabilitation of explored sites and their reuse
- Impact of exploration on the environment

Transportation planning
- Routing of transportation network, optimization and demand modelling
- Transport corridor analysis, drainage systems
- Fleet monitoring and navigation
- Highway design, route analysis, and customization
- Analysis of accident prone areas and corrective measures
- Landslides and flood analysis for transport design

Environmental planning and management
- Modelling tool and data and procedures for environmental impact assessment
- Ground water status and movement monitoring
- Air and water quality monitoring and management
- Coastal land analysis and coastal area management
- Site locations for waste disposal
- Flood zone mapping and flood mitigation measures
- Ecosystem monitoring and trend studies
- Environmental hazard status studies, predictive modelling, and evaluation

Socio-economic development
- Demographic studies and analysis
- Population studies and migration patterns
- Forecasting and analysis of socio-economic changes
- Impact studies of population expansion and urban and rural clusters on environment

Local area management
- Urban planning, zoning, and area management
- Land use, inventory, and utilization
- Growth monitoring and resource allocation
- Cadastral mapping
- Tax records management
- Engineering analysis for local area development
- Crime mapping

Public utilities management
- Modelling and optimization of communication and distribution network for power, water supply, gas, and waste disposal lines
- Public asset inventory and management
- Design of recreation facilities and other public services
- Designing health care facilities and their location

Disaster management
- Preparation of incidence records and maps
- Vulnerability studies and design of mitigation measures
- Design of disaster preparedness measures

- Disaster response and recovery planning
- Damage assessment and mitigation monitoring
- Planning of temporary shelters and relocation programmes

Business applications
- Market studies for location of plant, retailing outlets, and service networks
- Demand forecast and monitoring
- Development of tourism potential areas
- Route analysis for postal services and other delivery systems
- Real estate investment and management.

Summary

Remote sensing is a common phenomenon. Human vision is an example of remote sensing. Remote sensing through aerial photography was popular in earlier times. Satellite remote sensing has become more viable and common today. Remote sensing uses the visible, near infrared, and radio frequencies to collect data. It has the advantage of being able to cover large and difficult areas in the sky. Very high resolution sensors developed over the years have now made it possible to use this technology for a wide variety of applications. Remote sensing is used in planning and taking logical decisions for a variety of developmental activities. Land use and land cover analysis, water resources management, weather analysis, and environmental management are some of the areas where remote sensing finds wide applications. India has a number of satellites dedicated to remote sensing and managed by the National Remote Sensing Agency (NRSA).

Satellite-based positioning systems are used in aviation and general navigation control on land, water, and the sky. The GPS is the most popular positioning system. This system is managed by the US Department of Defence and is available to civilian users as well. The satellite configuration includes 24 satellites in nearly circular orbits at a height of about 26,500 km. The satellites are in six orbits and from any point on earth five to eight satellites are in the range of a receiver. Using the satellite signals, one can locate the position on the earth in terms of latitude, longitude, and the height above a selected datum. Developments in electronics have made it possible to use hand-held receivers for satellite signals. In surveying, the positioning systems can be used to find the coordinates of points, measure base lines, etc. Repeated measurements can improve accuracy. The limitation of the method is the necessity of having a terrain where satellite signals can be received. Instruments have been developed which can use the GPS for surveying.

The GIS (geographic information system) finds wide applications in many areas. The essential factor in the system is the spatial attributes of data, which are manipulated, analysed, and visually displayed. The GIS can be applied for topographic mapping, cartography, and many other applications in business, industry, environmental studies, and research. GIS software is supplied by many vendors for use in stand-alone or networked systems.

Exercises

Multiple-Choice Questions

1. Remote sensing is
   (a) collecting information without being in contact with the objects
   (b) measuring angles
   (c) measuring heights
   (d) using a total station to collect data about the terrain

2. Passive remote sensing is when
   (a) the remote sensing is done from the ground
   (b) the remote sensing is done from an aircraft
   (c) satellites are used for remote sensing
   (d) external energy source is used for remote sensing
3. In active remote sensing,
   (a) an internal energy source is used for remote sensing
   (b) an external energy source is used for remote sensing
   (c) continuous emission of energy is used
   (d) continuous receiving of radiation is done
4. Scattering is
   (a) when the source emits energy in all directions
   (b) when the emitted and received radiation are different
   (c) diffusion of radiation due to atmosphere
   (d) a defect in the scanning system
5. In the interaction of EM energy with a surface, the term volume phenomenon is used for
   (a) scattering   (b) reflection   (c) absorption   (d) emission
6. Spectral resolution means
   (a) the frequency of receiving radiations
   (b) the wavelengths to which the remote sensing system is sensitive
   (c) sensitivity of the system to small changes in radiation
   (d) the ability of the system to distinguish details in the images
7. Temporal resolution in remote sensing refers to
   (a) the frequency of receiving radiations
   (b) the wavelengths to which the remote sensing system is sensitive
   (c) sensitivity of the system to small changes in radiation
   (d) the ability of the system to distinguish details in the images
8. The global positioning system operated by the US department of Defence uses
   (a) 6 satellites   (b) 12 satellites   (c) 18 satellites   (d) 24 satellites
9. The position of a point can be located in GPS on receiving signals from at least
   (a) 1 satellite   (b) 2 satellites   (c) 3 satellites   (d) 4 satellites
10. A major requirement of GIS application is
    (a) the need to manage spatially referenced data
    (b) images from satellites
    (c) vector data structure
    (d) rastor data structure

Review Questions

1. Define remote sensing and illustrate with some examples.
2. Describe briefly the aerial photographic method and the satellite-based method of remote sensing.
3. Write a brief note on remote sensing in India.
4. Describe briefly the application areas of remote sensing. Give one or two case studies illustrating the applications.
5. Write a brief note on satellite-based positioning systems.
6. Write a brief note on GPS.
7. Explain briefly how GPS works to determine the position coordinates of a point.
8. Write a brief note on the applications of GPS in surveying.
9. Write a brief note on GIS.
10. Write briefly about the applications of GIS.

SI Units and Conversion Factors

1.1 Definition of SI and Supplementary Units

(a) Metre (m) It is the length equal to 1,650,763.73 wavelengths in vacuum of the radiation corresponding to the transition between the levels p10 and 5d5 of the krypton-86 atom.

(b) Kilogram (kg) It is equal to the mass of the international prototype of the kilogram kept at the International Bureau of Weights and Measures, Sèvres, France.

(c) Second (s) It is the duration of 9,192,631,770 periods of radiation corresponding to the transition between the levels of the ground state of the calcium-133 atom, unperturbed by external fields.

(d) Ampere (A) It is the constant current which, if maintained in two straight parallel conductors of infinite length and negligible circular cross section placed 1 m apart in vacuum, will produce between them a force equal to 2×10^{-7} N/m.

(e) Kelvin (K) It is the fraction 1/273.16 of the thermodynamic temperature of the triple point of water.

(f) Candela (cd) It is the luminous intensity, in the perpendicular direction, of the surface of 1/600,000 m^2 of a black body at the temperature of freezing platinum under a pressure of 101,325 N/m^2.

(g) Mole (mol) It is the quantity of matter in a system consisting of as many elementary particles as there are atoms in 0.012 kg of carbon-12.

(h) Radian (rad) One radian is the angle that, having its vertex at the centre of a circle, cuts off an arc on the circumference of the circle equal in length to that of the radius.

(i) Steradian (sr) One steradian is the angle that, having its vertex at the centre of a sphere, cuts off an area of the surface of the sphere equal to that of a square with sides of length equal to the radius of the sphere.

Basic units

Quantity	Symbol	SI unit	Symbol for unit
Length	l	metre	m
Mass	m	kilogram	kg
Time	t	second	s
Electric current	I	ampere	A
Thermodynamic temperature	T	kelvin	K
Luminous intensity	I_v	candela	Cd
Quantity of substance	N	mole	mol

Supplementary units

Quantity	Symbol	SI unit	Symbol for unit
Plane angle	θ	radian	rad
Solid angle	ω	steradian	sr

Prefixes for units

Prefix	Symbol	Multiplying factor
tera	T	10^{12}
giga	G	10^{9}
mega	M	10^{6}
kilo	k	10^{3}
hecto	H	10^{2}
deca	Da	10^{1}
deci	D	10^{-1}
centi	C	10^{-2}
milli	M	10^{-3}
micro	μ	10^{-6}
nano	N	10^{-9}
pico	P	10^{-12}
femto	F	10^{-15}
atto	A	10^{-18}

Other permitted units and their multiples and submultiples

Quantity	Name of unit	Symbol	Value
Time	minute	min	1 min = 60 s
	hour	h	1 h = 60 min
	day	d	1 d = 24 h
Plane angle	degree	°	$1° = (2\pi/360)$ rad
	minute	′	$1' = (1°/60)$
	second	″	$1'' = (1'/60)$
Volume	litre	L	$1\ L = 1\ dm^3$
Mass	tonne	ton	$1\ ton = 10^3\ kg$

Some derived units with special names

Quantity	Symbol	SI unit	Symbol for unit	Relation to other units
Force	F	newton	N	$kg\ m\ s^{-2}$
Energy	E, W	joule	J	N m

Quantity	Symbol	SI unit	Symbol for unit	Relation to other units
Power	P	watt	W	$J\,s^{1}$
Electric charge	Q	coulomb	C	$A\,s$
Electric potential	V	volt	V	$J\,Q^{-1}$
Electric capacitance	C	farad	F	$Q\,V^{-1} = s\,\Omega^{-1}$
Electric resistance	R	ohm	Ω	$V\,A^{-1}$
Magnetic flux	φ	weber	Wb	$V\,s$
Magnetic flux density	B	tesla	T	$Wb\,m^{-1}$
Inductance	M, L	henry	H	$V\,s\,A^{-1}$
Frequency	F	hertz	Hz	s^{-1}
Temperature	θ, t	degree Celsius	°C	$t\,°C$
Luminous flux	Φ	lumen	lm	$cd\,sr$
Illumination	E	lux	Lx	$lm\,m^{-2}$

Some derived units and their symbols/some permitted units

Physical quantity	SI unit	Symbol
Area	square metre	m^{2}
Volume	cubic metre	m^{3}
Mass density	kilogram per cubic metre	kg/m^{3} or $kg\,m^{-3}$
Speed, velocity	metre per second	M/s or $m\,s^{-1}$
Angular velocity	radian per second	rad/s or $rad\,s^{-1}$
Acceleration	metre per second squared	M/s^{2} or $m\,s^{-2}$
Angular acceleration	radian per second squared	rad/s^{2} or $rad\,s^{-2}$
Pressure, stress	newton per square metre bar normal atmosphere	N/m^{2} or $N\,m^{-2}$ bar ($10^{4}\,Pa$) atm (1 atm = 101,325 Pa)
Energy	electron volt	E V (1.6021×10^{-12} J)
Mass of an atom	atomic mass unit	μ (1.6604×10^{-27} kg)
Length	astronomical unit parsec angstrom	AU (1.496×10^{8} km) pc (3.093×10^{12} km) Å (= 0.1 nm)
Area	hectare are	ha ($10^{4}\,m^{2}$) are ($10^{2}\,m^{2}$)
Velocity	knot	1 nautical mile/hour

1.2 Conversion Factors

Length

1 in. = 25.4 mm
1 ft = 12 in. = 0.3048 m
1 yd = 3 ft = 0.9144 m
1 mile = 1609.3 m
1 mile (nautical) = 1853 m

Area

1 sq. in. = 645.16 sq. mm
1 sq. ft = 0.0929 sq. m
1 sq. yd = 0.8361 sq. m
1 acre = 4046.4 sq. m
1 sq. mile = 2.59 sq. km

Volume

1 cu. in. = 16387.1 cu. mm
1 cu. ft = 0.02832 cu. m
1 cu. yd = 0.7646 cu. m
1 gallon (imperial) = 4.546 L
1 gallon (US) = 3.785 L

Mass

1 oz = 28.35 g
1 lb = 0.4536 kg
1 ton (English) = 1016 kg
1 ton (US) = 907.2 kg

Pressure/stress

1 lb/sq. in. = 6896.43 N/sq. m
1 lb/sq. ft = 47.8728 N/sq. m

Energy and power

1 ft lb = 0.1383 kg fm = 1.3567 N m
1 BTU = 0.252 kcal
1 HPh = 0.7463 kWh
1 BTU/s = 0.252 kcal/s
1 h.p. = 0.7463 kW

Thermal

1 BTU/cuft = 9.547 kcal/cu. m
1 BTU/lb = 0.556 kcal/kg

APPENDIX 2

Conventional Symbols

	Feature	Symbol
1.	Chain line	
2.	Boundary line	
3.	Fence	or
4.	Barbed wire	—x—x—x—
5.	Pipe railing	o—o—o—o—o
6.	Store fence	
7.	Hedge fence	△△△△△△△
8.	Power line	
9.	Telephone line	T T T T T
10.	Wall	
11.	Gate	
12.	Bridge	Rail / Road
13.	Stream	
14.	Pond	
15.	House	
16.	Triangulation station	
17.	Traverse station	
18.	Dam	
19.	Church	
20.	North	

	Feature	Symbol
21.	Footpath	
22.	Metalled road	
23.	Unmetalled road	
24.	Single rail line	++++++++++
25.	Double rail line	
26.	Cultivated land	
27.	Trees	
28.	Tunnel	
29.	Road in embankment	
30.	Road in cutting	
31.	Temple	
32.	Mosque	
33.	Benchmark	B+M
34.	Open well	

APPENDIX

3

Areas and Volumes

Areas

1. Square

   Area = a^2

   $d = a\sqrt{2}$

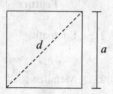

2. Rectangle

   Area = ab

   $d = \sqrt{a^2 + b^2}$

3. Parallelogram

   Area = $ah = ab \sin \alpha$

   $d_1 = \sqrt{(a + h\cot\alpha)^2 + h^2}$

   $d_2 = \sqrt{(a - h\cot\alpha)^2 + h^2}$

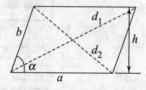

4. Trapezium

   Area = $\dfrac{a+b}{2}\,h$

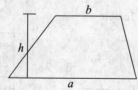

5. Triangle

   Area = $\dfrac{ah}{2}$

   $= \sqrt{s(s-a)(s-b)(s-c)}$

$s = \dfrac{a+b+c}{2}$

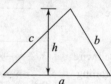

6. Equilateral triangle

   Area = $\dfrac{\sqrt{3}}{4}\,a^2$

   $h = \dfrac{\sqrt{3}}{2}\,a$

7. Regular pentagon

   Area = $\dfrac{5}{8}r^2\sqrt{10 + 2\sqrt{5}}$

   $a = \dfrac{r}{2}\sqrt{10 - 2\sqrt{5}}$

   (r = radius of circumscribing circle)

8. Regular hexagon

   Area = $\dfrac{3\sqrt{3}}{2}\,a^2$

   $d = 2a = 1.155s$

   $s = 0.866d$

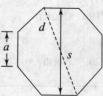

9. Regular octagon

   Area = $2aS = 0.83a^2$

   $a = 0.415s$

   $s = 0.924d$

   $d = 1.083s$

10. Circle

Area $= \dfrac{\pi d^2}{4} = \pi r^2$

Perimeter $= 2\pi r = \pi d$

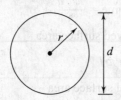

11. Hollow circular section

Area $= \dfrac{\pi \left(D^2 - d^2\right)}{4}$

$ = \pi t \,(d + t)$

$t = \dfrac{D - d}{2}$

12. Sector of circle

Area $= \dfrac{\pi r^2 \alpha^\circ}{360} = \dfrac{r^2}{2}\,\alpha$ (rad)

$ = \dfrac{br}{2}$

$b = \dfrac{\pi r \alpha^\circ}{180}$

13. Segment of circle

$s = 2r \sin\left(\dfrac{\alpha}{2}\right)$

Area $= \dfrac{h}{6s}\,(3h^2 + 4s^2)$

$ = \dfrac{r^2}{2}\,(\alpha - \sin \alpha)$

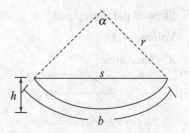

$r = \dfrac{h}{2} + \dfrac{s^2}{8h}$

$h = r\left(1 - \cos\dfrac{\alpha}{2}\right) = \dfrac{s}{2}\,\tan\dfrac{\alpha}{4}$

(α in radians)

14. Ellipse

Area $= \dfrac{\pi dD}{4} = \pi ab$

Perimeter $= \dfrac{\pi(D + d)}{2}$

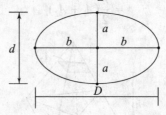

Volumes

15. Cube

Volume $= a\sqrt{3}$

Surface area $= 6a^2$

diagonal $= d\sqrt{3}$

16. Rectangular parallelepiped

Volume $= abc$

Surface area $= 2(ab + ac + bc)$

diagonal $= \sqrt{a^2 + b^2 + c^2}$

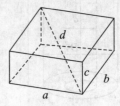

17. Skewed parallelepiped

 Volume = Ah

 A = base area

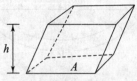

18. Pyramid

 Volume = $A\dfrac{h}{3}$

 (A = base area)

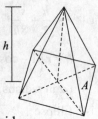

19. Frustum of pyramid

 Volume = $\dfrac{h}{3}(A_1 + A_2 + \sqrt{A_1 A_2})$

 $\approx \dfrac{h}{2}(A_1 + A_2)$

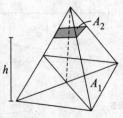

20. Cylinder

 Volume = $\dfrac{\pi}{4}d^2 h$

 Curved surface area

 $A_1 = 2\pi r h$

 Total surface area

 $A = 2\pi r (r + h)$

21. Hollow cylinder

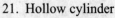

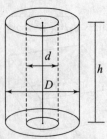

 Volume = $\dfrac{\pi h}{4}(D^2 - d^2)$

22. Cone

 Volume = $\dfrac{\pi r^2 h}{3}$

 Curved surface area

 $A_1 = \pi r l$

 Total surface area

 $A = \pi r(r + l)$

 $l = \sqrt{r^2 + h^2}$

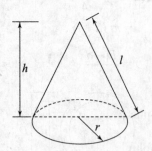

23. Frustum of cone

 Volume = $\dfrac{\pi h}{12}(D^2 + Dd + d^2)$

 Curved surface area

 $A_1 = \dfrac{\pi l}{2}(D + d)$

 $l = \sqrt{\left(\dfrac{D - d}{2}\right)^2 + h^2}$

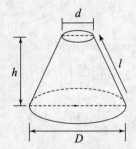

24. Sphere

 Volume = $\dfrac{4}{3}\pi r^3 = \dfrac{1}{6}\pi d^3$

$= 4.189r^3$

Surface area $= 4\pi r^2 = \pi d^2$

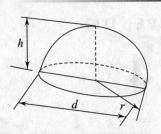

25. Spherical frustum

Volume $= \dfrac{\pi h}{6}(3a^2 + 3b^2 + h^2)$

Surface area $= \pi(2rh + a^2 + b^2)$

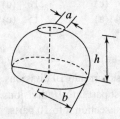

27. Sector of sphere

Volume $= \dfrac{2}{3}\pi r^2 h$

Surface area $= \dfrac{\pi r}{2}(4h + d)$

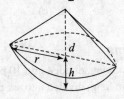

26. Spherical segment

Volume $= \dfrac{\pi h}{6} = \left(\dfrac{3}{4}d^2 + h^2\right)$

Curved surface area

$A \quad = 2\pi rh$

$\quad = \dfrac{\pi\left(d^2 + 4h^2\right)}{4}$

28. Torus

Volume $= \dfrac{\pi^2 D d^2}{4}$

Surface area $= \pi^2 D d$

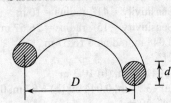

APPENDIX

Answers

Chapter 1

Multiple-Choice Questions

1. (c)   2. (d)   3. (b)   4. (a)   5. (d)
6. (a)   7. (d)   8. (a)   9. (c)   10. (c)
11. (c)   12. (c)   13. (d)   14 (a)   15. (c)
16. (d)

Problems

1. 20 cm long, divided into 10 parts. Divide the leftmost part into 10 divisions.
2. 30 cm long, divided into six parts. Divide the leftmost part into 10 divisions.
3. 25 cm long, divided into 20 parts. Divide the leftmost section into 10 parts.
4. Length of nine divisions of the main scale divided into 10 parts in the vernier
5. Length of nine divisions of the main scale divided into 10 parts in the vernier
6. (a) 39 divisions of main scale divided into 40 parts in the vernier; (b) 59 divisions of the main scale divided into 60 parts in the vernier
7. 59 main scale divisions divided into 60 parts in the vernier
8. Sensitivity = 41″, radius = 10 m
9. Sensitivity = 15″, radius = 26.67 m
10. 0.048 m, radius = 16.67 m
11. 21″
12. (a) 41.25 m, (b) 10.3 m
13. (a) 13.75 m, 0.044 m; (b) 6.87 m, 0.088 m
14. 14″, 30 m

Chapter 2

Multiple-Choice Questions

1. (c)   2. (d)   3. (a)   4. (c)   5. (b)
6. (c)   7. (c)   8. (d)   9. (b)   10. (d)
11. (c)   12. (c)   13. (d)   14. (a)   15. (c)
16. (d)   17. (c)   18. (b)   19. (a)   20. (a)
21. (c)   22. (b)   23. (a)   24. (c)   25. (b)
26. (b)   27. (a)   28. (b)   29. (c)   30. (a)
31. (a)   32. (a)   33. (b)   34. (b)   35. (b)
36. (b)   37. (b)   38. (d)   39. (c)   40. (c)

Problems

1. 128.54 m   2. 1244.74 m   3. 1630.206 m   4. 382.13 m
5. 126.426 m × 189.67 m   6. 199,866.7 sq. m
7. (a) 347.267 m, (b) 133,539 sq. m   8. 12,598 cu.m

9. (a) 0.115 m, (b) 0.103 m, (c) 0.054 m   10. 6.6°
11. Slope corrections: 0.066 m, 0.15 m; 267.832 m   12. 15° 28′ 59″
13. 310.7 m   14. 261.317 m   15. 86.82 m
16. 0.1 m, 2772.07 sq. m   17. 0.009 m   18. 0.516 m
19. 0.003 m   20. 0.003   21. 758.546 m   22. 235 N
23. 0.0264 m   24. Parallel: 0.125 cm on paper, 1.25 m on ground
     Perpendicular: 0.004 cm on paper, 0.04 m on ground
25. 5.74 m
26. 1 in 11.5   27. 13.25   28. 2.5°   29. 1.8°
30. 250.32 m   31. 142.23 m   32. 124.45 m, 309.1 m, 95.34 m
33. 242.4 m   34. 98.45 m   35. 4975.5 sq. m
36. 6450.6 sq. m

Chapter 3

Multiple-Choice Questions

1. (d)   2. (c)   3. (a)   4. (c)   5. (b)
6. (a)   7. (c)   8. (b)   9. (a)   10. (c)
11. (c)   12. (d)   13. (d)   14. (a)   15. (a)
16. (b)   17. (a)   18. (a)   19. (c)   20. (c)
21. (b)   22. (a)   23. (b)   24. (d)   25. (d)
26. (c)   27. (a)   28. (a)   29. (a)   30. (a)

Problems

1. N67° 30′E, N81° 15′W, S56° 05′E, N90° 00′E, N33° 30′W, S00° 00′E
2. 314° 30′, 247° 30′, 151° 15′, 266° 32′, 48° 34′
3. 116° 10′, 23° 48′, 70° 10′, 123° 36′, 71° 50′, 57° 40′
4. 14° 35′, 65° 09′, 98° 48′, 92° 30′, 70° 08′, 61° 42′
5. BC: 68° 05′, CD: 116° 35′, DE: 206° 35′, EA: 275° 20′
6. BC: S80° 30′W, CD: N51° 05′E, DE: S60° 10′E, EA: N73° 14′E
7. BC: N90° 00′E, CD: S00° 00′E, DE: N90° 00′W, EF: – N00° 00′E, FA: N52° 31′W
8. 204° 55′, 324° 55′
9. S70° 15′E, S1° 45′W, S73° 45′W, N34° 15′W
10. N79° 55′E, S55° 05′E, S10° 05′W, S34° 55′W, S79° 55′W, N55° 05′W, N10° 05′W
11. A (0, 405.7), B (330, 405.7), C (506, 250), D (506, 0), E (158, 0)
12. Bearings: BC = N50° 30′E, CD = N54° 15′W, DA = S33° 12′W; deflection angles: A
     = 78° 12′, B = 84° 30′, C = 104° 45′, D = 92° 33′
13. Corrected bearings: QR = 117° 00′, RQ = 297° 00′, RS = 167° 50′, SR = 347° 50′
     SP = 277° 00′
14. Corrected bearings: TS = 43° 40′, TP = 304° 30′, PT = 124° 30′, PQ = 74° 25′, QP =
     254° 25′, QR = 115° 45′
15. C: 40′, D: 1°; bearings: CD: 160° 10′, DA: 269° 20′
16. Magnitude of local attraction at R: 1° 15′, S: 1° 30′, T: 1°
17. Magnitude of local attraction at A: 00° 45′, B: 1° 15′
18. Magnitude of local attraction at C: 00° 40′, D: 1° 30′
19. N54° 45′E, N25° 30′W, S81° 15′E, S17° 30′W
20. N52° 15′W
21. 53° 00′   22. S50° 50′W

Chapter 4

Multiple-Choice Questions

1. (b) 2. (b) 3. (b) 4. (a) 5. (b)
6. (a) 7. (b) 8. (b) 9. (c) 10. (d)
11. (c) 12. (c) 13. (b) 14. (a) 15. (c)

Problems

1. 146° 25′ 10″, 233° 40′ 30″, 301° 04′ 20″, 349° 23′ 00″
2. 225° 35′ 00″, 78° 14′ 20″, 81° 58′ 50″, 139° 41′ 00″, 125° 25′ 10″, 69° 05′ 40″
3. 77° 29′ 20″, 55° 49′ 20″, 83° 51′ 40″, 48° 59′ 20″, 93° 50′ 20″
4. 157° 23′ 40″, 109° 41′ 20″, 91° 39′ 40″, 83° 46′ 50″, 97° 28′ 30″
5. 132° 15′ 40″, 213° 34′ 10″, 297° 34′ 20″, 31° 34′ 40″
6. 115° 51′ 20″, 204° 11′ 40″, 300° 24′ 50″, 22° 56′ 20″

Chapter 5

Multiple-Choice Questions

1. (b) 2. (c) 3. (c) 4. (a) 5. (c)
6. (a) 7. (b) 8. (a) 9. (c) 10. (c)

Problems

1. (a) 1.5 mm, (b) 0.015 mm   2. 1/1920

Chapter 6

Multiple-Choice Questions

1. (b) 2. (c) 3. (b) 4. (a) 5. (b)
6. (b) 7. (d) 8. (c) 9. (b) 10. (b)
11. (b) 12. (d) 13. (d) 14. (c) 15. (b)
16. (b) 17. (b) 18. (c) 19. (d) 20. (d)
21. (d) 22. (d) 23. (b) 24. (c) 25. (b)

Problems

1. 283.16 m, N49° 13′ 12″W   2. 58° 52′ 48″   3. 200.1 m, 76°
4. 61° 10′, 153° 20′, 200° 56′, 280° 06′, and 20° 20′   5. Correct
6. Included angles: $\angle A = 43°\ 20′$, $\angle B = 75°\ 45′$, $\angle C = 43°\ 30′$, $\angle D = 197°\ 25′$; deflection angles: 136° 40′ (R) at A, 104° 15′ (R) at B, 136° 30′ (R) at C, 17° 25′ (L) at D; tangent lengths: – 9.4 cm at A, – 39.4 cm at B, 9.5 cm at C, 3.1 cm at D. (Negative sign indicates that 10 cm is to be measured not on extended line but by measuring it backwards.)
7. 0.58 m, S44° 43′ 12″W
8. Traverse does not close; error in latitude = 0.59 m; error in departure = 0.39 m; error of closure = 0.707 m in first quadrant at bearing N33° 28′ E
10. 178.2 m, 152° 32′   11. 231.94 m, 272.55 m
12. 280° 27′, 222° 52′   13. 128.42 m, N78° 16′ W
14. 200.2 m, 130.6 m   15. 144° 24′, 11° 30′
16. 265.5 m, N20° 30′ E   17. N75° 15′ E, S65° 09′ ′E
18. Both surveys give the correct result.
19. 428.54 m, N27° 52′ 30″W; 1372.1 m, N53° 08′ 20″E
20. 321.13 m, (3546.54, 3587.67)

Chapter 7

Multiple-Choice Questions

1. (c)	2. (c)	3. (b)	4. (c)	5. (d)
6. (d)	7. (c)	8. (d)	9. (b)	10. (a)
11. (b)	12. (c)	13. (c,d)	14. (d)	15. (b)
16. (b)	17. (d)	18. (c)	19. (d)	20. (a)
21. (b)	22. (c)	23. (a)	24. (d)	25. (d)
26. (d)	27. (c)	28. (c)	29. (c)	30. (b)

Problems

1. RL of last point = 99.745
2. RL of last point = 186.465
3. RL of first point = (–) 3.97
4. RL of first point = 203.735
5. RL of last point = 301.305
6. RL of first point = 328.295
7. RL of first point = (–)1.58
8. RL of last point = 98.95
9. Missing readings = 1.215 (BS), 0.865 (FS), 0.985 (BS), 0.625 (FS)
10. Missing readings = 1.315 (IS), 1.215 (BS), 1.105 (FS), 1.015 (FS)
11. (a) (i) 0.1130, (ii) 0.2543; (b) (i) 0.0161, (ii) 0.0363
12. 1.009 m   13. 0.199 m   14. 33.38 m, 18′   15. 242.3 m
16. 39.3 km   17. 172 m   18. 10.001 km   19. 1′ 15″ (upwards)
20. 27″, 1.98 km   21. 1.175 m   22. 192.201, 13″ (downwards)
23. 1.438, 16″   24. 202.532, 0.014 m
25. 187.55 m   26. 695 m

Chapter 8

Multiple-Choice Questions

1. (c)	2. (a)	3. (d)	4. (b)	5. (a)
6. (d)	7. (c)	8. (b)	9. (d)	10.
11. (b)	12. (d)	13. (c)	14. (d)	15. (a)
16. (c)	17. (d)	18. (d)	19. (a)	20. (c)
21. (b)	22. (c)	23. (b)	24. (a)	25. (c)
26. (a)	27. (b)			

Problems

1. 260,280 sq. m   2. 159,532 sq. m   3. 109,320 sq. m   4. 161,648 sq. m
5. 611.52 sq. m   6. 483.61, 501.95, 506.56 sq. m
7. 436, 446.7, 447.25 sq. m   8. 224.8, 2230.4 sq. m
9. 366, 367.2 sq. m 10. 683.5, 686.33 sq. m   11. 174.95 sq. m
12. 396.8 sq. m   13. 1796.5 sq. cm 14. 426.74 sq. cm   15. 3048 sq. cm
16. 2020 sq. cm   17. 120, 30, 3600 sq. cm   18. 8410.5 hectares

Chapter 9

Multiple-Choice Questions

1. (c)	2. (a)	3. (d)	4. (c)	5. (b)
6. (c)	7. (a)	8. (a)	9. (a)	10. (b)
11. (c)	12. (a)	13. (b)	14 (a)	15. (a)
16. (c)	17. (a)			

Chapter 10

Multiple-Choice Questions

1. (d)	2. (c)	3. (c)	4. (c)	5. (a)
6. (b)	7. (b)	8. (b)	9. (c)	10. (d)
11. (d)	12. (b)	13. (a)	14 (d)	15. (c)
16. (a)	17. (c)	18. (d)	19 (a)	20. (b)
21. (c)				

Problems

1. 32.5 m^2 2. 48.97 m^2 3. $1.9 \text{ m}^2, 8.2 \text{ m}^2$
4. $1.125 \text{ m}^2, 10.125 \text{ m}^2$ 5. 60.89 m^2 6. 32.85 m^2
7. 31.38 m^2 8. 24.5 m^2 9. 36.4 m^2 10. 34.1 m^2
11. 1100 m^3 12. 3342 m^3 13. $6070.8 \text{ m}^3, 6083 \text{ m}^3$
14. $2.55 \text{ m}, 3.45 \text{ m}, 1482.5 \text{ m}^3, 1464.3 \text{ m}^3, 18.22 \text{ m}^3$ 15. 2945.8 m^3
16. $2506.3 \text{ m}^3, 2504.3 \text{ m}^3$ 17. $18, 843 \text{ m}^3$
18. $4046.8 \text{ m}^3, 4016 \text{ m}^3, 4.3 \text{ m}^3$ 19. 3677 m^3 20. 82.8 m^3
21. 1.04 m^3 22. $10.2 \times 10^6 \text{ m}^3$ 23. $10, 800 \text{ m}^3$
24. $13, 610 \text{ m}^3$ 25. 2100 m^3
26. Volume of borrow material $= 7200 \text{ m}^3$, max. haul distance $= 389.63 \text{ m}$

Chapter 11

Multiple-Choice Questions

1. (c)	2. (d)	3. (d)	4. (a)	5. (b)
6. (c)	7. (d)	8. (a)	9. (c)	10. (c)

Chapter 12

Multiple-Choice Questions

1. (b)	2. (c)	3. (b)	4. (b)	5. (c)
6. (b)	7. (d)	8. (a)		

Problems

1. No, 7′ 23″ (downwards), 2.175 m
2. No, 4′ 18″ (downwards), 2.815 m, 1.960 m
3. No, 21″ (upwards), 2.245 m, 1.27 m
4. No, 1′ 19″ (downwards), 1.361 m, 2.391 m
5. No, 1′ 43″ (upwards), 1.335 m, 2.180 m

Chapter 13

Multiple-Choice Questions

1. (d)	2. (c)	3. (c)

Chapter 14

Multiple-Choice Questions

1. (b)	2. (d.)	3. (c)	4. (c)	5. (a)
6. (a)	7. (c)	8. (c)	9. (c)	10. (d)
11. (d)	12. (a)	13. (b)	14. (c)	

Problems

1. 183.3 m, 111.035 m
2. $K = 100, C = 0.4$ 3. $i = 1.998$ mm
4. 124.79 m, 892.655 m
5. 190.08 m, 940.299 m
6. 18.75 m
7. 168.86 m, 1 in 20 (approx.)
8. Yes, the plot is a regular hexagon.
9. $K = 100, C = 0$
10. $K = 100.07, C = 0.22$
11. 160.04 m   12.13.6
13. 840.614. 97.575 m, 1055.395 m
15. 54.516 m
16. 36.24 m, 1 in 40 (approx.)
17. 2° 34′   18.0.011 m

Chapter 15

Multiple-Choice Questions

1. (a)	2. (d)	3. (d)	4. (a)	5. (b)
6. (c)	7. (d)	8. (b)	9. (b)	10. (b)
11. (b)	12. (c)	13. (c)	14. (c)	15. (b)
16. (d)	17. (c)	18. (d)	19. (c)	20. (d)
21. (d)	22. (d)	23. (d)	24. (d)	25. (d)
26. (a)	27. (b)	28. (a)	29. (b)	30. (b)
31. (b)	32. (d)	33. (b)	34. (b)	35. (c)
36. (d)	37. (d)	38. (d)	39. (d)	40. (b)
41. (d)				

Problems

1. 381.97 m, 382.01 m
2. 208.14 m, 336.78 m, 67.63 m, 54.71 m
3. 69° 19′ 39″, 264.13 m, 434.5 m
4. 498.03 m, 674.52 m, 218.49 m, 147.96 m
5. 375.19 m, 606.23 m, 102.23 m
6. 570.31 m, 463.93 m, 53.97 m, 49.3 m
7. Mid-ordinate = 56.27 m, (20, 9.52), (40, 18.09), (60, 25.7), (80, 32.38) (100, 38.1)

8. Mid-ordinate = 37.5 m

X	20	40	60	80	100	120	140
Ordinate	36.78	34.62	31	25.82	19.03	10.48	0
Chainage	1325.5	1305.8	1285.8	1265.32	1244.3	1222.57	1200

9. $R = 286.48$ m, 1300.5 m, 1080.68 m, $L = 375$ m

X	20	40	60	80	100	120	140
Radial offset	0.698	2.792	6.283	11.17	17.45	25.13	32.37

10.

X	20	40	60	80	100	120	140	160	180
Offset	0.25	1.00	2.25	4.0	6.25	9.0	12.25	16	20.25

11. $R = 202.45$ m

X	20	40	60	80	100
Offset	0.93	3.99	9.09	16.47	26.42

12. (140, 16.52), from T_1 at 70.78 m   13. 0.39, 0.94, 1.0, 1.0, 1.0

14.

Chord	11.95	20	20	20	20
Angle	1° 05′	2° 59′ 36″	4° 54′ 15″	6° 48′ 51″	8° 43′ 27″

15.

Chord	11.75	20	20	20	20
Angle	50′ 30″	2° 16′ 27″	3° 42′ 23″	5° 8′ 21″	6° 34′ 18″

16. 1341.79 m, 1535.46 m   17. 1214.33 m, 1578.67 m

18. 08.53 m, 758.53 m, 975.19 m   19. 818.39 m, 1177.5 m, 1343.26 m

20. 382 m   21. 139.12 m   22. 687.98 m, 986.23 m, 1403.36 m

23. 1012.02 m, 1519.48 m

24. 700.53 m, 957.17 m, 1189.67 m, 1° 27′ 42″, 2° 57′ 42″, 4° 27′ 42″, 5° 57′ 42″, 7° 27′ 42″

25. 219.78 m, 418.6 m, 635 m, 854.78 m, 1273.38 m

26. 380.23 m, 445.5 m

27. 240 m, (20, 0.83), (40, 3.33), (60, 7.5), (80, 13.33), (100, 20.83)   28. 359.84 m

29. 152.33 m   30. 453 m, 272.45 m, 6.83 m, 1852.345 m   31. 78 m

32. 272.45 m, 400 m, 6.83 m, 693.8 m   33. 367 m, 256.21 m

34.

α	2°	4°	8°	10°	12°
β	61.8	87.3	122.86	138.86	152.14

35. $241.2\sqrt{\sin 2\alpha}$, 162.8 m   36. 720 m

37.

X	0	20	40	60	80	100	120	140
RL	383.74	383.905	384.04	384.145	384.22	384.265	384.28	384.265

38.

X	20	40	60	80	100
RL	1233.625	1233.5	1233.425	1233.4	1233.425

39. 714.5 m; chainages: $T_1 = 515$ m, $T_2 = 1785$ m; RLs: $T_1 = 87.3$ m, $T_2 = 93.85$ m

40. 398.56 m

Chapter 16

Multiple-Choice Questions

1. (d)	2. (d)	3. (b)	4. (d)	5. (b)
6. (b)	7. (c)	8. (d)	9. (c)	10. (a)
11. (d)	12. (b)	13. (d)	14. (c)	

Problems

1. 524.39 m	2. 1091.82 m	3. 1181.48 m	4. 1364.9 m
5. 2385.746 m	6. 1° 22′ 03″, 2° 14′ 09″		7. 1° 30′ 42″
8. 1296.7 m	9. 49.71 m	10. 1120.61 m, 1088.14 m	

Chapter 17

Multiple-Choice Questions

1. (b)	2. (d)	3. (b)	4. (d)	5. (d)
6. (d)	7. (b)	8. (b)	9. (c)	10. (b)
11. (b)				

Problems

1. 4.95   2. 616.3   3. (a) 0.75, (b) 0.643
4. $R = 7.1$ from triangles ABC and ACD   5. 5.582 m
6. 8.684 m   7. 8.95 m   8. 35.13° C   9. 185.194 m
10. 0.00694 m   11. 29.961 m   14. 112.43 m   15. 69° 34′ 35″

Chapter 18

Multiple-Choice Questions

1. (b)   2. (c)   3. (c)   4. (d)   5. (d)
6. (c)   7. (c)   8. (d)   9. (a)   10. (c)
11. (b)   12. (b)   13. (d)   14 (d)   15. (b)
16. (b)   17. (d)   18. (d)   19 (c)   20. (d)
21. (c)   22. (d)   23. (a)   24. (c)

Problems

1. 6/5   2. 1, 12   3. 42° 32′ 16.472″   4. 42° 32′ 18″, weight = 10
5. 24° 42′ 32.57″   6. $A = 46° 20′ 10.465″, B = 25° 32′ 18.67″$
7. $A = 60° 12′ 35.82″, B = 56° 32′ 42.54″, C = 63° 14′ 41.64″$
8. $\angle 1 = 92° 30′ 12.14″, \angle 2 = 87° 25′ 42.72″, \angle 3 = 96° 37′ 54.42″, \angle 4 = 83° 26′$ 10.72″
12. 48° 24′ 34.71″
13. $A = 62° 30′ 36.67″, B = 58° 40′ 51.33″, C = 71° 42′ 40.67″$
14. $A = 90° 32′ 37.26″, B = 89° 28′ 42.96″, C = 98° 44′ 53.89″, D = 81° 13′ 45.89″$
15. $A = 57° 12′ 32.99″, B = 64° 42′ 11.71″, C = 54° 05′ 15.3″$
16. $b = 7302.58$ m, $c = 7197.398$ m   21. 217.815 ± 0.042 m
22. Q = 139.937 m, R = 138.799 m, S = 141.171 m, T = 139.334 m
23. A to B = 1.472, B to D = 2.019, D to A = 3.491, D to C = 1.298, C to B = 0.722

Chapter 19

Multiple-Choice Questions

1. (b)   2. (b)   3. (c)   4. (b)   5. (a)
6. (b)   7. (b)   8. (c)

Problems

1. (342.2 m, 333.55 m)   2. (773.42 m, 866.54 m)
3. − 0.2 m, 1.10 m, 2.5 m, 5.19 m
4. PS = 1529.6 m, QS = 1195.59 m, RS = 417.35 m

Chapter 20

Multiple-Choice Questions

1. (b)   2. (c)   3. (b)   4. (c)

Chapter 21

Multiple-Choice Questions

1. (a)   2. (b)   3. (a)   4. (d   5. (c)
6. (b)   7. (d)   8. (c)   9. (a   10. (c)
11. (a)   12. (a)   13. (c)   14. (c   15. (d)

16. (a) 17. (c) 18. (d) 19. (a 20. (c)
21. (d) 22. (b) 23. (b)

Problems

1. (a) (i) 34° 12′, (ii) 55° 20′; (b) (i) 17° 46′, (ii) 55° 24′ 2. 3411.84 km
3. 2979.8 km 4. (a) 19°, (b) 34° 45′ 5. (a) 77° 15′, (b) 89° 15′
6. 3° 7. Delination = 67° 05′, latitude = 45° 35′N
8. 47° 44′ 24″, 96° 31′ 36″ 9. 9° 26′ 42″, 135° 9′ 42″
10. 62° 16′ 34″, 38° 1′ 18″ 11. 8° 11′ 12″, 51° 0′ 40″
12. 39° 11′ 31″W, 134° 1′ 38″ 13. 108° 51′ 24″E, 19h 11 m
14. (a) 8 h 03 m, (b)11h 41 m 30s
15. (a) 6 h 37 m 5.02s; (b) 10 h 10 m 49.66s; (c) 153°, 343°; (d) 5 h 14 m, 11 h 0 m 6.6 s
16. (a) 13 h 02 m 24 s, (b) 13 h 0 m 40s 17. 10 h 33 m 6.54 s
18. 16 h 34 m 4.08 s 19. 9 h 58 m 55.28s 20. 12° 00′ 29.12″
21. 152° 19′ 40.9″ 22. 64° 00′ 5.72″ 23. 65° 24′ 33″

Chapter 22

Multiple-Choice Questions

1. (b) 2. (c) 3. (a) 4. (c) 5. (a)
6. (a) 7. (a) 8. (a)

Problems

1. 136.94 m, 147.75 m; RL = 117.6 m 2. 594 m, 96.53 m
3. (a) 1760 m; (b) 1/11,700 4. 55 5. 738
6. 1/120 7. 77.2 m 8. 80.39 m

Chapter 23

Multiple-Choice Questions

1. (a) 2. (b) 3. (a) 4. (a) 5. (a)
6. (d) 7. (d) 8. (b)

Chapter 24

Multiple-Choice Questions

1. (a) 2. (d) 3. (a) 4. (c) 5. (c)
6. (b) 7. (a) 8. (d) 9. (d) 10. (a)

Index